Autodata

SERIES

Diagnostic Trouble Codes

Engine Management Systems

Asian and European Vehicles 1992-2004

2005 EDITION

Caution

The information used to compile this manual has been sourced from the vehicle manufacturers and was believed to be the latest available in February 2005 Manufacturers' information is liable to change at any time and there may have been updates or revisions which applied to the content of this manual but were not available at the time of compilation. Later revisions published by the manufacturers' may contain revisions which would be applied retrospectively to the information contained in this manual.

The information contained in this manual applies only to standard models and does not apply to vehicles fitted with equipment other than the standard production options.

Autodata Publications Inc.

Published in the U.S.A. by
Autodata Publications Inc. under license.
Manufactured in the U.S.A.

Autodata Publications Inc.
6301 Bandel Road NW, #403
Rochester, MN 55901

Phone: 800 305-0338
email: autodata@autodatapubs.com
www.autodatapubs.com

Product No.: 05-350	ISBN: 1-893026-27-2	March 2005

Contents

This manual is a comprehensive single source of information on diagnostic trouble codes for engine management systems for Asian and European vehicles introduced or revised during the period 1992 to 2004.

The manual is part of a series from Autodata and deals specifically with trouble codes related to the engine management system. It has been written and presented in a way which enables any professional automotive technician, with appropriate skills and competence, to make accurate diagnoses of fuel and ignition system related components and circuits.

Detailed knowledge of self-diagnosis or fuel injection and ignition systems is not required in order to make full use of this manual. With a basic understanding of fuel and electrical systems, tests and diagnoses can be made using the minimum of specialised test equipment.

Where possible, procedures for accessing and erasing trouble codes without special diagnostic equipment are given, but increasingly some form of 'scan tool' is required to successfully read the fault memories of control modules. This particularly applies to OBD-II systems where 'freeze frame' data and manufacturer 'PID' (Parameter Identification) information is accessed.

Each chapter covers a range of models sharing the same trouble code table and lists the codes in numerical or alphabetical order along with their fault locations and probable causes.

The probable causes column identifies the various different areas of the system that should be investigated in addition to the primary component listed in the fault location column.

If a trouble code has been logged, the fault location can be looked up in the trouble code table. This will give suggested probable causes, which should be the primary areas for checking. Most of these probable causes will suggest checking the wiring as well as the primary component in the circuit. The multi-plugs, wiring continuity and insulation should all be verified before replacing components.

Index

Model	Year	Engine identification	System	Page
ACURA				
Integra 1.7L	1992-93	B17A1	Honda PGM-FI	108
Integra 1.8L	1992-93	B18A1	Honda PGM-FI	108
Integra 1.8L	1994-95	B18B1	Honda PGM-FI	111
Integra 1.8L VTEC	1994-95	B18C1	Honda PGM-FI	111
Integra 1.8L	1996-01	B18B1	Honda PGM-FI	114
Integra 1.8L VTEC	1996-01	B18C1/C5	Honda PGM-FI	114
Legend 3.2L	1991-95	C32A1	Honda PGM-FI	108
MDX 3.5L	2000-04	J35A3/5	Honda PGM-FI	119
NSX/NSX-T 3.0L	1991-04	C30A1	Honda PGM-FI	125
NSX/NSX-T 3.2L	1997-04	C32B1	Honda PGM-FI	125
RSX 2.0L	2002-04	K20A2/3	Honda PGM-FI	119
SLX 3.2L	1996-97	6VD1	Isuzu MFI	131
SLX 3.5L	1998-99	6VE1	Isuzu MFI	131
TSX 2.4L	2004	K24A2	PGM-FI	134
Vigor 2.5L	1992-94	G26A1	Honda PGM-FI	108
2.2CL	1997	F22B1	Honda PGM-FI	136
2.3CL	1998-99	F23A1	Honda PGM-FI	136
2.5TL	1995-98	G25A4	Honda PGM-FI	141
3.0CL	1997-99	J30A1	Honda PGM-FI	146
3.2CL	2000-03	J32A1/2	Honda PGM-FI	119
3.2TL	1996-98	C32A6	Honda PGM-FI	151
3.2TL	1999-03	J32A1	PGM-FI	157
3.2TL	2004	J32A3	PGM-FI	134
3.5RL	1996-04	C35A1	Honda PGM-FI	151
AUDI				
A4 1.8L Turbo	1997-04	4 cylinder	Bosch Motronic	161
A4 1.8L Turbo	1997-04	4 cylinder	Siemens Simos	161
A4 2.8L	1995-02	6 cylinder	Bosch Motronic	161
A4 2.8L	1995-02	6 cylinder	Siemens Simos	161
A4 2.8L	1995-02	6 cylinder	VAG MPI/Digifant	161
A4 3.0L	2002-04	6 cylinder	Bosch Motronic	161
A4 Cabriolet 1.8L Turbo	2003-04	4 cylinder	Bosch Motronic	161
A4 Cabriolet 3.0L	2003-04	6 cylinder	Bosch Motronic	161
A6 2.8L	1995-01	6 cylinder	Bosch Motronic	161
A6 2.8L	1995-01	6 cylinder	Siemens Simos	161

Model	Year	Engine identification	System	Page
A6 2.8L	1995-01	6 cylinder	VAG MPI/Digifant	161
A6 3.0L	2002-04	6 cylinder	Bosch Motronic	161
A6 3.0L	2002-04	6 cylinder	Siemens Simos	161
A6 4.2L	2000-02	8 cylinder	Bosch Motronic	161
A6 4.2L	2000-04	8 cylinder	Siemens Simos	161
A6 4.2L	2000-04	8 cylinder	VAG MPI/Digifant	161
A6/Allroad 2.7L Turbo	2000-04	6 cylinder	Bosch Motronic	161
A6/Allroad 2.7L Turbo	2000-04	6 cylinder	Siemens Simos	161
A8 3.7L/4.2L	1997-04	8 cylinder	Bosch Motronic	161
A8 3.7L/4.2L	1997-04	8 cylinder	Siemens Simos	161
A8 3.7L/4.2L	1997-04	8 cylinder	VAG MPI/Digifant	161
Cabriolet 2.8L	1994-98	6 cylinder	Bosch Motronic	161
Cabriolet 2.8L	1994-98	6 cylinder	Siemens Simos	161
Cabriolet 2.8L	1994-98	6 cylinder	VAG MPI/Digifant	161
RS6 4.2L	2002-03	8 cylinder	Bosch Motronic	161
S4 2.2L	1992-95	AAN	Bosch Motronic	201
S4 2.7L Turbo	2000-02	6 cylinder	Bosch Motronic	161
S4 2.7L Turbo	2000-02	6 cylinder	Siemens Simos	161
S4 4.2L	2004	8 cylinder	Bosch Motronic	161
S4 Cabriolet 4.2L	2004	8 cylinder	Bosch Motronic	161
S6 2.2L	1995	AAN	Bosch Motronic	201
S6 4.2L	2002-03	8 cylinder	Bosch Motronic	161
S8 4.2L	2001-03	8 cylinder	Bosch Motronic	161
S8 4.2L	2001-03	8 cylinder	Siemens Simos	161
TT 1.8L Turbo	2000-04	4 cylinder	Bosch Motronic	161
TT 1.8L Turbo	2000-04	4 cylinder	Siemens Simos	161
TT 3.2L	2004	6 cylinder	Bosch Motronic	161
TT Roadster 1.8L Turbo	2004	4 cylinder	Bosch Motronic	161
TT Roadster 3.2L	2004	6 cylinder	Bosch Motronic	161
100 2.8L	1992-94	6 cylinder	Bosch Motronic	161
100 2.8L	1992-94	6 cylinder	Siemens Simos	161
100 2.8L	1992-94	6 cylinder	VAG MPI/Digifant	161

BMW

Model	Year	Engine identification	System	Page
X3 2.5L (E83)	2004	25 6S 5	Siemens MS43/MS S52/ Motronic ME 9.2.1/2	205
X3 3.0L (E83)	2004	30 6S 3	Siemens MS43/MS S52/ Motronic ME 9.2.1/2	205
X5 (E53)	1999-02	8 cylinder	Siemens MS40/41/42	234

Index

Index

Model	Year	Engine identification	System	Page
DAEWOO				
Lanos 1.6L	**2000-02**	A16DM	IEFI-6/ITMS-6F	**252**
Nubira 2.0L	**2000-02**	T20XED	IEFI-6/ITMS-6F	**252**
HONDA				
Accord 2.2L	**1990-93**	F22A1/4/6	Honda PGM-FI	**256**
Accord 2.2L	**1994-95**	F22B1/2	Honda PGM-FI	**258**
Accord 2.2L	**1996-97**	F22B1/2	Honda PGM-FI	**261**
Accord 2.3L	**1998-02**	F23A1/4/5	Honda PGM-FI	**266**
Accord 2.4L	**2003-04**	F24A4	Honda PGM-FI	**266**
Accord 2.7L	**1995-97**	C27A4	Honda PGM-FI	**261**
Accord 3.0L	**1998-04**	J30A1	Honda PGM-FI	**266**
Civic 1.5L	**1992-95**	D15Z1	Honda PGM-FI	**281**
Civic 1.5L 8V/16V	**1992-95**	D15B7/B8	Honda PGM-FI	**281**
Civic 1.6L	**1992-95**	D16Z6	Honda PGM-FI	**281**
Civic 1.6L GX	**1998-00**	D16B5	Honda PGM-FI	**284**
Civic 1.6L SOHC	**1996-00**	D16Y5	Honda PGM-FI	**284**
Civic 1.7L	**2001-04**	D17A1/A2/A6/A7	Honda PGM-FI	**266**
Civic 2.0L	**2002-04**	K20A3	Honda PGM-FI	**266**
Civic del Sol 1.5L/1.6L	**1993**	D15B7/D16Z6	Honda PGM-FI	**281**
Civic del Sol 1.5L SOHC	**1994-95**	D15B7	Honda PGM-FI	**258**
Civic del Sol 1.6L	**1994-95**	D16Z6	Honda PGM-FI	**258**
Civic del Sol 1.6L DOHC	**1994-95**	B16A3	Honda PGM-FI	**258**
Civic del Sol 1.6L DOHC	**1996-98**	B16A2	Honda PGM-FI	**284**
Civic/del Sol 1.6L	**1996-00**	D16Y8	Honda PGM-FI	**284**
Civic/del Sol 1.6L SOHC	**1996-00**	D16Y7	Honda PGM-FI	**284**
CR-V 2.0L	**1997-01**	B20B4	Honda PGM-FI	**290**
CR-V 2.4L	**2002-04**	K24A1	Honda PGM-FI	**266**
Element 2.4L	**2003-04**	K24A4	PGM-FI	**294**
Odyssey 2.2L	**1995**	F22B6	Honda PGM-FI	**258**
Odyssey 2.2L	**1996-97**	F22B6	Honda PGM-FI	**261**
Odyssey 2.3L	**1998**	F23A7	Honda PGM-FI	**261**
Odyssey 3.5L	**1999-04**	J35A1/A4	Honda PGM-FI	**266**
Passport 2.6L	**1994-95**	4ZE1	Isuzu MFI	**296**
Passport 2.6L	**1996-97**	4ZE1	Isuzu MFI	**299**
Passport 3.2L	**1994-95**	6VD1	Isuzu MFI	**301**
Passport 3.2L	**1996-02**	6VD1	Isuzu MFI	**303**

Index

HYUNDAI

Index

JAGUAR

KIA

Model	Year	Engine identification	System	Page
Rio 1.5L	2001-02	A5D	Mazda EGI	408
Sedona 3.5L	2002-04	VIN code digit 8 = 1	Mazda EGI	408
Sephia 1.6L	1993-96	B6 SOHC/DOHC	Mazda EGI	414
Sephia 1.6L	1996-97	B6 DOHC	Mazda EGI	408
Sephia 1.8L	1995-97	BP DOHC	Mazda EGI	408
Sephia 1.8L	1998-01	T8 DOHC	Mazda EGI	408
Sorento 3.5L	2003-04	VIN code digit 8 = 3	Mazda EGI	408
Spectra 1.8L	2001-04	T8 DOHC	Mazda EGI	408
Spectra 2.0L	2004	G4GC	Mazda EGI	408
Sportage 2.0L	1995-96	FE	Mazda EGI	414
Sportage 2.0L	1996-02	FE DOHC	Mazda EGI	408

LAND ROVER

Model	Year	Engine identification	System	Page
Discovery 4.0L	1998-02	36D/42D/46D	Bosch M5.2.1/MEMS3	417
Freelander 2.5L	2001-02	KV6	Bosch M5.2.1/MEMS3	417
Range Rover 4.0L	1998-00	36D/42D/46D	Bosch M5.2.1/MEMS3	417
Range Rover 4.6L	1998-02	36D/42D/46D	Bosch M5.2.1/MEMS3	417

LEXUS

Model	Year	Engine identification	System	Page
ES300 3.0L	1992-93	3VZ-FE	Toyota TCCS	420
ES300 3.0L	1994-03	1MZ-FE	Toyota SFI	423
ES330 3.3L	2004	3MZ-FE	SFI	423
GS300 3.0L	1993-95	2JZ-GE	Toyota TCCS	427
GS300 3.0L	1996-04	2JZ-GE	Toyota SFI	430
GS400 4.0L	1998-00	1UZ-FE	Toyota SFI	423
GS430 4.3L	2001-04	3UZ-FE	Toyota SFI	423
GX470 4.7L	2003-04	2UZ-FE	SFI	432
IS300 3.0L	2001-04	2JZ-GE	Toyota SFI	430
LS400 4.0L	1993-94	1UZ-FE	Toyota TCCS	436
LS400 4.0L	1995-00	1UZ-FE	Toyota SFI	423
LS430 4.3L	2001-04	3UZ-FE	Toyota SFI	423
LX470 4.7L	1998-04	2UZ-FE	SFI	432
RX300 3.0L	1999-02	1MZ-FE	Toyota SFI	423
RX300 3.0L	1999-03	1MZ-FE	SFI	432
RX330 3.3L	2004	3MZ-FE	SFI	432
SC300 3.0L	1992-95	2JZ-GE	Toyota TCCS	427
SC300 3.0L	1996-00	2JZ-GE	Toyota SFI	430

Index

MAZDA

Model	Year	Engine identification	System	Page
626 2.5L	1993-95	KL V6	Mazda EGI	463
626 2.5L	1996-02	FS/KL	Mazda EGI/Ford EEC	441
929 3.0L	1993-95	JE	Mazda EGI	463

MERCEDES-BENZ

Model	Year	Engine identification	System	Page
C-Class (202)	2000	4 cylinder	Bosch ME 2.1	466
C-Class (203)	2000-04	4 cylinder	Siemens ME SIM4	466
CL (215)	2000-04	8 cylinder	Bosch ME 2.0/2.8	466
CLK (208)	2000-02	6/8 cylinder	Bosch ME 2.0/2.8	466
CLK (209)	2003-04	6/8 cylinder	Bosch ME 2.8	466
CLK320 (208)	1998-02	6 cylinder	Bosch ME 1.0/2.0/ME-SFI 2.1	471
CL500/600 (140)	1993-99	8/12 cylinder	Bosch ME 1.0/2.0/ME-SFI 2.1	471
C220/280 (202)	1994-00	6 cylinder	Bosch/Siemens	474
C230K/240/280 (202)	1993-00	4/6 cylinder	Bosch ME 1.0/2.0/ME-SFI 2.1	471
E-Class (210)	2000-03	6/8 cylinder	Bosch ME 2.0/2.8	466
E-Class (211)	2002-04	6/8 cylinder	Bosch ME 2.8	466
E280/320 (124)	1993-02/95	6 cylinder	Bosch/Siemens	474
E320 (210)	1997-02	6 cylinder	Bosch ME 1.0/2.0/ME-SFI 2.1	471
M-Class (163)	2000-04	6/8 cylinder	Bosch ME 2.0/2.8	466
S-Class (220)	2000-04	8 cylinder	Bosch ME 2.0/2.8	466
SL (129)	2002-03	6/8 cylinder	Bosch ME 2.0/2.8	466
SL (230)	2004	8 cylinder	Bosch ME 2.8	466
SLK (170)	2000-04	4/6 cylinder	Siemens ME SIM4/Bosch ME 2.8	466
SLK 230 (170)	1999-02	4 cylinder	Bosch ME 1.0/2.0/ME-SFI 2.1	471
SL320 (129)	1993-97	6 cylinder	Bosch/Siemens	474
SL500/600 (129)	1995-02	8/12 cylinder	Bosch ME 1.0/2.0/ME-SFI 2.1	471
S320 (140)	1994-99	6 cylinder	Bosch/Siemens	474
S320/420/500/600	1993-99	6/8 cylinder	Bosch ME 1.0/2.0/ME-SFI 2.1	471

MINI

Model	Year	Engine identification	System	Page
Cooper Convertible 1.6L	2004	W10B	Siemens EMS 2000	489
Cooper S 1.6L	2002-04	W11B	Siemens EMS 2000	489
Cooper S Convertible 1.6L	2004	W11B	Siemens EMS 2000	489
Cooper 1.6L	2002-04	W10B	Siemens EMS 2000	489

Index

MITSUBISHI

Model	Year	Engine identification	System	Page
Diamante 3.0L	1992-93	6G72	Mitsubishi MPI	492
Diamante 3.0L	1994-95	6G72	Mitsubishi MPI	495
Diamante 3.0L	1996	6G72	Mitsubishi MPI	498
Diamante 3.5L	1997-04	6G74	Mitsubishi MPI	498
Eclipse 1.8L	1990-94	4G37	Mitsubishi MPI	501
Eclipse 2.0L/Turbo	1992-94	4G63	Mitsubishi MPI	501
Eclipse 2.0L	1995-99	420A	Mitsubishi MPI	504
Eclipse 2.0L Turbo	1995-99	4G63	Mitsubishi MPI	498
Eclipse 2.4L	1996-04	4G64	Mitsubishi MPI	498
Eclipse 3.0L	2000-04	6G72	Mitsubishi MPI	498
Endeavor 3.8L	2004	6G75	Mitsubishi MPI	498
Expo LRV 1.8L – California	1994	4G93	Mitsubishi MPI	508
Expo LRV 1.8L – California	1995-96	4G93	Mitsubishi MPI	511
Expo LRV 1.8L – Federal	1994-95	4G93	Mitsubishi MPI	508
Expo LRV 1.8L – Federal	1996	4G93	Mitsubishi MPI	511
Expo/LRV 1.8L	1992-93	4G93	Mitsubishi MPI	492
Expo LRV 2.4L	1994-95	4G64	Mitsubishi MPI	508
Expo LRV 2.4L FWD – California	1994	4G64	Mitsubishi MPI	508
Expo/LRV 2.4L	1992-93	4G64	Mitsubishi MPI	492
Expo/LRV 2.4L FWD – California	1995-96	4G64	Mitsubishi MPI	511
Expo/LRV 2.4L – Federal	1996	4G64	Mitsubishi MPI	511
Expo 2.4L	1994-95	4G64	Mitsubishi MPI	508
Expo 2.4L FWD – California	1994	4G64	Mitsubishi MPI	508
Expo 2.4L 4X4 – California	1996	4G64	Mitsubishi MPI	511
Galant 2.0L 16V DOHC	1989-93	4G63	Mitsubishi MPI	501
Galant 2.0L 8V SOHC	1989-92	4G63	Mitsubishi MPI	501
Galant 2.4L	1994-95	4G64	Mitsubishi MPI	495
Galant 2.4L	1996-04	4G64	Mitsubishi MPI	498
Galant 3.0L	1999-04	6G72	Mitsubishi MPI	498
Galant 3.8L	2004	6G75	Mitsubishi MPI	498
Lancer Evolution 2.0L	2003-04	4G63	Mitsubishi MPI	498
Lancer Sportback 2.4L	2002-04	4G69	Mitsubishi MPI	498
Lancer 2.0L	2002-04	4G94	Mitsubishi MPI	498
Mighty Max Pickup 2.4L	1990-95	4G64	Mitsubishi MPI	513
Mighty Max Pickup 2.4L	1996	4G64	Mitsubishi MPI	498
Mighty Max Pickup 3.0L	1990-96	6G72	Mitsubishi MPI	513
Mirage 1.5L	1989-94	4G15	Mitsubishi MPI	501

Model	Year	Engine identification	System	Page
Mirage 1.5/1.8L	1995-02	4G15/4G93	Mitsubishi MPI	511
Mirage 1.8L – California	1994	4G93	Mitsubishi MPI	495
Mirage 1.8L – Federal	1994-95	4G93	Mitsubishi MPI	495
Montero Sport 2.4L	1997-99	4G64	Mitsubishi MPI	498
Montero Sport 3.0L	1996-02	6G72	Mitsubishi MPI	498
Montero Sport 3.5L	1996-04	6G74	Mitsubishi MPI	498
Montero 3.0L	1989-93	6G72	Mitsubishi MPI	513
Montero 3.0L	1994-95	6G72	Mitsubishi MPI	516
Montero 3.0L	1996-04	6G72	Mitsubishi MPI	498
Montero 3.5L	1996-04	6G74	Mitsubishi MPI	498
Montero 3.8L	2004	6G75	Mitsubishi MPI	498
Outlander 2.4L	2002-04	4G69	Mitsubishi MPI	498
Precis 1.5L	1990-93	G4AJ	Mitsubishi MPI	501
3000GT	1991-93	6G72 DOHC	Mitsubishi MPI	492
3000GT	1994-95	6G72 DOHC	Mitsubishi MPI	495
3000GT	1996-99	6G72	Mitsubishi MPI	498
3000GT Turbo	1991-93	6G72 DOHC	Mitsubishi MPI	492
3000GT Turbo	1994-95	6G72 DOHC	Mitsubishi MPI	495
3000GT Turbo	1996-99	6G72	Mitsubishi MPI	498

NISSAN

Model	Year	Engine identification	System	Page
Altima 2.4L	1993-94	KA24DE	Nissan ECCS	518
Altima 2.4L	1995	KA24DE	Nissan ECCS	521
Altima 2.4L	1996-01	KA24DE	Nissan ECCS	526
Altima 2.5L	2002	QR25DE	Nissan ECCS	536
Altima 3.5L	2002	VQ35DE	Nissan ECCS	536
Armada 5.6L	2004	VK56DE	ECCS	545
Axxess 2.4L	1990	KA24E	Nissan ECCS	550
Frontier 2.4L	1998-02	KA24DE	Nissan ECCS	526
Frontier 3.3L	1999-02	VG33E/VG33ER	Nissan ECCS	553
Maxima 3.0L	1987-94	VG30E	Nissan ECCS	563
Maxima 3.0L	1992-94	VE30DE	Nissan ECCS	566
Maxima 3.0L	1995	VQ30DE	Nissan ECCS	521
Maxima 3.0L	1995-02	VQ30DE	Nissan ECCS	553
Murano 3.5L	2003-04	VQ35DE	ECCS	545
NX 1.6L	1991-94	GA16DE	Nissan ECCS	518
NX 2.0L	1991-94	SR20DE	Nissan ECCS	518
Pathfinder 3.0L	1990-95	VG30E	Nissan ECCS	550

Index

Model	Year	Engine identification	System	Page
300ZX 3.0L	1990-95	VG30DE	Nissan ECCS	566
300ZX 3.0L	1996	VG30DE	Nissan ECCS	553
300ZX 3.0L Turbo	1987-89	VG30E/VG30ET	Nissan ECCS	563
300ZX 3.0L Turbo	1990-95	VG30DETT	Nissan ECCS	566
300ZX 3.0L Turbo	1996	VG30DETT	Nissan ECCS	553
350Z 3.5L	2003-04	VQ35DE	ECCS	545

SAAB

Model	Year	Engine identification	System	Page
9-3 2.0L Turbo	1999-02	4 cylinder	Trionic T5/T7	569
9-3 2.3L Turbo	1999-02	4 cylinder	Trionic T5/T7	569
9-5 2.3L Turbo	1999-02	4 cylinder	Trionic T5/T7	569
900 2.0L Turbo	1992-98	4 cylinder	Trionic	575
900 2.3L/2.5L	1993-98	4 cylinder	Bosch Motronic 2.8.1 & 2.10.2/3	581
9000 2.3L/Turbo	1992-94	4 cylinder	Bosch LH-Jetronic 2.4/2.4.2	578
9000 2.3L/Turbo	1993-98	4 cylinder	Trionic	575
9000 3.0L V6	1995-97	6 cylinder	Motronic 2.8.1	585

SCION

Model	Year	Engine identification	System	Page
xA 1.5L	2004	1NZ-FE	SFI	589
xB 1.5L	2004	1NZ-FE	SFI	589

SUBARU

Model	Year	Engine identification	System	Page
Baja 2.5L	2003-04	VIN code digit 6 = 6	Subaru MFI/MPI	591
Baja 2.5L Turbo	2004	VIN code digit 6 = 6	Subaru MFI/MPI	591
Forester 2.5L	1998-04	VIN code digit 6 = 6	Subaru MFI	591
Forester 2.5L Turbo	2004	VIN code digit 6 = 6	Subaru MFI/MPI	591
Impreza 1.8L	1993-94	VIN digit 6 = 2	Subaru MPFI	598
Impreza 1.8L	1995-97	VIN code digit 6 = 2	Subaru MFI	591
Impreza 2.0L Turbo	2002-04	VIN code digit 6 = 2	Subaru MFI	591
Impreza 2.2L	1996-01	VIN code digit 6 = 4	Subaru MFI	591
Impreza 2.5L	1998-04	VIN code digit 6 = 6	Subaru MFI	591
Impreza 2.5L Turbo	2004	VIN code digit 6 = 7	Subaru MFI/MPI	591
Legacy 2.2L	1990-94	VIN digit 6 = 6	Subaru MPFI	601
Legacy 2.2L	1995-99	VIN code digit 6 = 4	Subaru MFI	591
Legacy 2.2L Turbo	1991-94	VIN digit 6 = 6	Subaru MPFI	601
Legacy 2.5L	1996-04	VIN code digit 6 = 6	Subaru MFI	591

Index

Model	Year	Engine identification	System	Page
Loyale 1.8L	1989-94	VIN digit 6 = 4 & 5	Subaru SPFI/MPFI	604
Loyale 1.8L Turbo	1989-94	VIN digit 6 = 4 & 5	Subaru SPFI/MPFI	604
Outback 2.5L	1996-04	VIN code digit 6 = 6	Subaru MFI	591
Outback 3.0L	2002-04	VIN code digit 6 = 8	Subaru MFI	591

SUZUKI

Model	Year	Engine identification	System	Page
Aerio SX 2.0L	2003	J20	Suzuki EPI	607
Aerio 2.0L	2003	J20	Suzuki EPI	607
Esteem 1.6L	1995	G16	Suzuki EPI	610
Esteem 1.6L	1996-98	G16	Suzuki EPI	607
Grand Vitara 2.5L	1999-03	H25	Suzuki EPI	607
Samurai 1.3L	1990-94	G13	Suzuki EPI	613
Sidekick 1.6L	1991-95	G16	Suzuki EPI	610
Sidekick 1.6L 16V	1992-95	G16	Suzuki EPI	610
Sidekick 1.6L 16V	1996-98	G16	Suzuki EPI	607
Sidekick 1.8L 16V	1996-98	J18	Suzuki EPI	607
Swift 1.3L	1989-95	G13	Suzuki EPI	613
Swift 1.3L	1996-98	G13	Suzuki EPI	607
Vitara 1.6L	1999-02	G16	Suzuki EPI	607
Vitara 2.0L	1999-03	J20	Suzuki EPI	607
XL-7 2.7L	2001-03	H27	Suzuki EPI	607
X-90 1.6L 16V	1996-98	G16	Suzuki EPI	607

TOYOTA

Model	Year	Engine identification	System	Page
Avalon 3.0L	1995-04	1MZ-FE	Toyota SFI	616
Camry 2.2L	1992-95	5S-FE	Toyota TCCS	620
Camry 2.2L	1996-01	5S-FE	Toyota SFI	624
Camry 2.4L	2002-04	2AZ-FE	Toyota SFI	616
Camry 3.0L	1992-93	3VZ-FE	Toyota TCCS	628
Camry 3.0L	1994-04	1MZ-FE	Toyota SFI	616
Camry 3.3L	2004	3MZ-FE	Toyota SFI	616
Camry Solara 2.2L	1999-01	5S-FE	Toyota SFI	624
Camry Solara 2.4L	2002-04	2AZ-FE	Toyota SFI	616
Camry Solara 3.0L	1999-04	1MZ-FE	Toyota SFI	616
Camry Solara 3.3L	2004	3MZ-FE	Toyota SFI	616
Camry Wagon 3.0L	1992-93	3VZ-FE	Toyota TCCS	628
Celica 1.6L	1990-93	4A-FE	Toyota TCCS	631

Model	Year	Engine identification	System	Page
Celica 1.8L	1994-95	7A-FE	Toyota TCCS	634
Celica 1.8L	1996-97	7A-FE	Toyota SFI	624
Celica 1.8L	2001-04	1ZZ-FE/2ZZ-GE	Toyota SFI	624
Celica 2.2L	1990-93	5S-FE	Toyota TCCS	631
Celica 2.2L	1994-95	5S-FE	Toyota TCCS	634
Celica 2.2L	1996-01	5S-FE	Toyota SFI	624
Corolla 1.6L	1989-92	4A-FE	Toyota TCCS	631
Corolla 1.6L	1993-95	4A-FE	Toyota TCCS	634
Corolla 1.6L	1996-97	4A-FE	Toyota SFI	624
Corolla 1.8L	1993-95	7A-FE	Toyota TCCS	634
Corolla 1.8L	1996-97	7A-FE	Toyota SFI	624
Corolla 1.8L	1998-04	1ZZ-FE	Toyota SFI	624
Cressida 3.0L	1990-92	7M-GE	Toyota TCCS	620
Echo 1.5L	2000-04	1NZ-FE	Toyota SFI	616
Highlander 2.4L	2001-04	2AZ-FE	Toyota SFI	637
Highlander 3.0L	2001-03	1MZ-FE	Toyota SFI	637
Highlander 3.3L	2004	3MZ-FE	Toyota SFI	637
Land Cruiser 4.0L	1988-90	3F-E	Toyota TCCS	641
Land Cruiser 4.0L	1991-94	3F-E	Toyota TCCS	641
Land Cruiser 4.5L	1993-94	1FZ-FE	Toyota TCCS	644
Land Cruiser 4.5L	1995-97	1FZ-FE	Toyota TCCS	647
Land Cruiser 4.7L	1998-04	2UZ-FE	Toyota SFI	649
Matrix 1.8L	2003-04	1ZZ-FE/2ZZ-GE	Toyota SFI	624
MR2 1.8L	2000-04	1ZZ-FE	Toyota SFI	624
MR2 2.2L	1991-95	5S-FE	Toyota TCCS	652
Paseo 1.5L	1992-95	5E-FE	Toyota TCCS	652
Paseo 1.5L	1995-98	5E-FE	Toyota SFI	624
Pickup 3.0L	1992-95	3VZ-E	Toyota TCCS	655
Previa 2.4L	1991-95	2TZ-FE	Toyota TCCS	658
Previa 2.4L	1995-97	2TZ-FZE	Toyota SFI	660
RAV4 2.0L	1996-02	3S-FE	Toyota SFI	624
RAV4 2.0L	2002-03	1AZ-FE	Toyota SFI	637
RAV4 2.4L	2004	2AZ-FE	Toyota SFI	637
Sequoia 4.7L	2001-04	2UZ-FE	Toyota SFI	649
Sienna 3.0L	1998-03	1MZ-FE	Toyota SFI	616
Sienna 3.3L	2004	3MZ-FE	Toyota SFI	616
Supra 3.0L	1993-95	2JZ-GE	Toyota TCCS	662
Supra 3.0L	1996-98	2JZ-GE	Toyota TCCS	647
Supra 3.0L Turbo	1993-95	2JZ-GTE	Toyota TCCS	662

Index

Model	Year	Engine identification	System	Page
Supra 3.0L Turbo	1996-98	2JZ-GTE	Toyota TCCS	647
Tacoma 2.4L	1995-04	2RZ-FE	Toyota SFI	660
Tacoma 2.7L	1995-04	3RZ-FE	Toyota SFI	660
Tacoma 3.4L	1995-04	5VZ-FE	Toyota SFI	637
Tercel 1.5L	1990-94	3E-E	Toyota TCCS	665
Tercel 1.5L	1995-98	5E-FE	Toyota SFI	624
Tundra 3.4L	2000-04	5VZ-FE	Toyota SFI	637
Tundra 4.7L	2000-04	2UZ-FE	Toyota SFI	649
T100 Pickup 2.7L	1994-98	3RZ-FE	Toyota SFI	660
T100 Pickup 3.0L	1993-94	3VZ-E	Toyota TCCS	655
T100 Pickup 3.4L	1995-98	5VZ-FE	Toyota SFI	637
4Runner 2.7L	1995-00	3RZ-FE	Toyota SFI	660
4Runner 3.0L	1992-95	3VZ-E	Toyota TCCS	655
4Runner 3.4L	1996-02	5VZ-FE	Toyota SFI	637
4Runner 4.0L	2003-04	1GR-FE	Toyota SFI	637
4Runner 4.7L	2003-04	2UZ-FE	Toyota SFI	649

VOLKSWAGEN

Model	Year	Engine identification	System	Page
Cabrio	1991-02	4/6 cylinder	Bosch Motronic	667
Cabrio	1991-02	4/6 cylinder	Siemens Simos	667
Cabrio	1991-02	4/6 cylinder	VAG Digifant	667
Corrado	1992-95	4/6 cylinder	Bosch Motronic	667
Corrado	1992-95	4/6 cylinder	Siemens Simos	667
Corrado	1992-95	4/6 cylinder	VAG Digifant	667
Euro Van	2004	6 cylinder	Bosch Motronic	667
Golf	1991-04	4/6 cylinder	Bosch Motronic	667
Golf	1991-04	4/6 cylinder	Siemens Simos	667
Golf	1991-04	4/6 cylinder	VAG Digifant	667
Jetta	1991-04	4/6 cylinder	Bosch Motronic	667
Jetta	1991-04	4/6 cylinder	Siemens Simos	667
Jetta	1991-04	4/6 cylinder	VAG Digifant	667
Jetta 2.0L	1992	4 cylinder	Bosch KE-Motronic	707
New Beetle	1998-04	4 cylinder	Bosch Motronic	667
New Beetle	1998-04	4 cylinder	Siemens Simos	667
New Beetle Cabriolet	2003-04	4 cylinder	Bosch Motronic	667
New Beetle Cabriolet	2003-04	4 cylinder	Siemens Simos	667
Passat	1991-04	4/6 cylinder	Bosch Motronic	667
Passat	1991-04	4/6 cylinder	Siemens Simos	667

Model	Year	Engine identification	System	Page
Passat	**1991-04**	4/6 cylinder	VAG Digifant	**667**
Passat 2.0L	**1992-93**	4 cylinder	Bosch KE-Motronic	**707**
Phaeton	**2004**	BAP, BGH, BGJ	Bosch Motronic ME7.1.1 SFI	**711**
Touareg	**2004**	6 cylinder	Bosch Motronic	**667**
Touareg	**2004**	8 cylinder	Bosch Motronic	**667**
Touareg	**2004**	10 cylinder	Bosch EDC 16	**667**

VOLVO

Model	Year	Engine identification	System	Page
C70 2.3L Turbo	**1998**	B5234T3	Bosch Motronic 4.4	**716**
C70 2.3L/2.4L Turbo	**1999-04**	B5234T3/T9/B5244T/T7	Denso/Motronic ME 7.0	**722**
S40/V40 2.0L Turbo	**2000-04**	B4204T2/T3/T4	Siemens EMS 2000	**731**
S40/V40 2.4L	**2004**	B5244S4	Denso/Motronic ME 7.0	**722**
S40/V40 2.5L Turbo	**2004**	B5254T3	Denso/Motronic ME 7.0	**722**
S60 2.3L/2.4L/2.5L Turbo	**2001-04**	B5234T3/B5244T3/ B5254T2/T4	Denso/Motronic ME 7.0	**722**
S60 2.4L	**2001-04**	B5244S	Denso/Motronic ME 7.0	**722**
S80 2.5L/2.8L/2.9L Turbo	**1999-04**	B5254T2/B6284T/B6294T	Denso/Motronic ME 7.0	**722**
S80 2.9L	**1999-04**	B6294S/S2	Denso/Motronic ME 7.0	**722**
S90/V90 2.9L Turbo	**1997-98**	B6304FS2	Bosch Motronic 4.4	**716**
V70 2.3L Turbo	**1998**	B5234T3/T6	Bosch Motronic 4.4	**716**
V70 2.3L/2.4L/2.5L Turbo	**1998-04**	B5234T3/T8/B5244T/T2/ T3/B5254T2/T4	Denso/Motronic ME 7.0	**722**
V70 2.4L	**1998-04**	B5244S/S6	Denso/Motronic ME 7.0	**722**
XC70 2.5L Turbo	**2003-04**	B5254T2	Denso/Motronic ME 7.0	**722**
XC90 2.5L/2.9L Turbo	**2003-04**	B5254T2/B6294T	Denso/Motronic ME 7.0	**722**
240 2.3L	**1992-93**	4 cylinder	Bosch LH-Jetronic 2.4/3.2	**736**
740 2.3L	**1992**	4 cylinder	Bosch LH-Jetronic 2.4/3.2	**736**
740 2.3L Turbo	**1992**	4 cylinder	Bosch LH-Jetronic 2.4/3.2	**736**
850 2.3L Turbo	**1994-97**	4 cylinder	Bosch Motronic 4.4/ Siemens Fenix 5.1	**741**
940 2.3L/Turbo	**1992-95**	4 cylinder	Bosch LH-Jetronic 2.4/3.2	**736**
960 3.0L	**1992**	6 cylinder	Bosch Motronic 1.8/4.3	**747**
960 3.0L	**1992**	6 cylinder	Siemens Fenix 5.2	**747**

About this manual

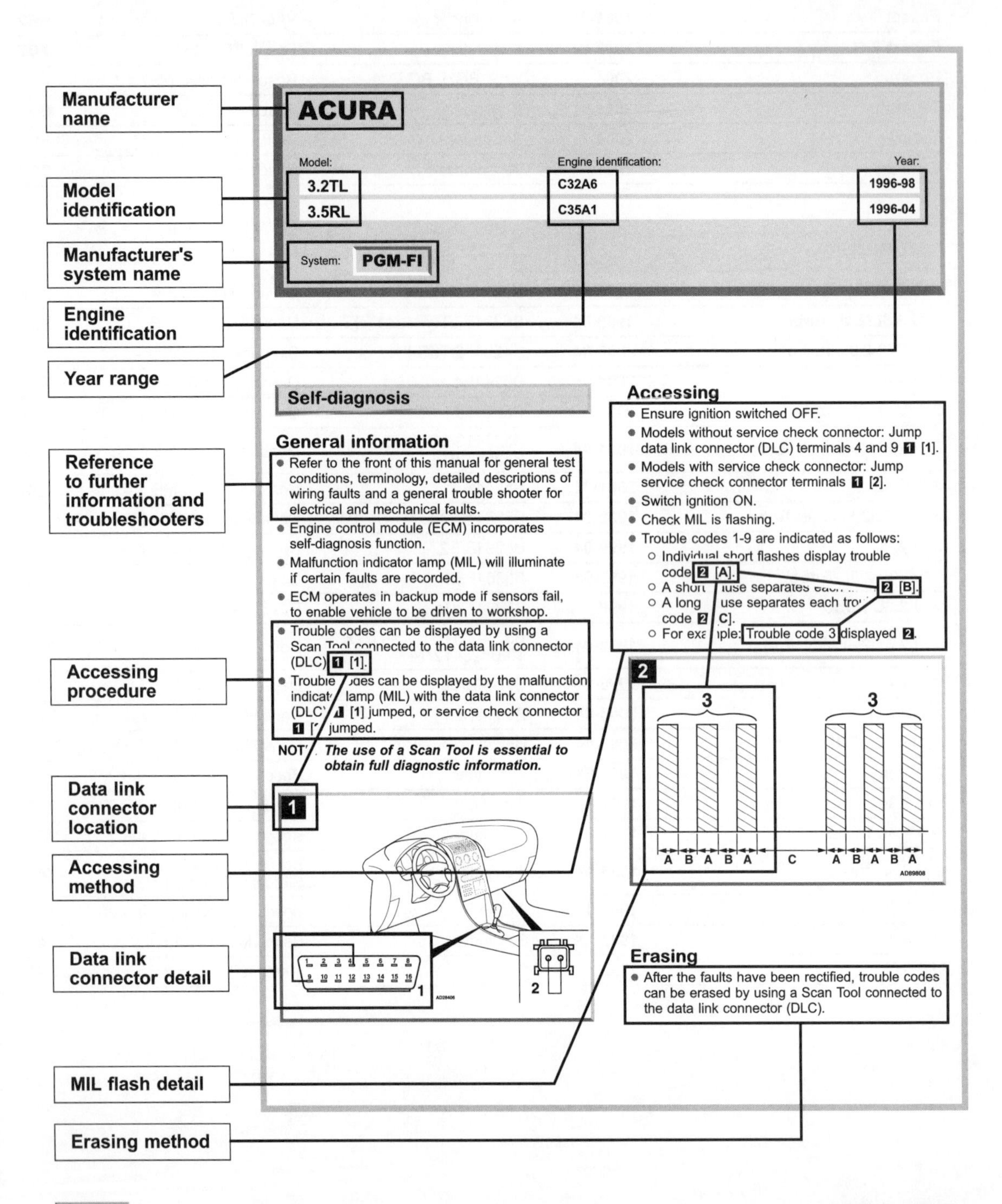

Manufacturer name
Model identification
Manufacturer's system name
Engine identification
Year range
Reference to further information and troubleshooters
Accessing procedure
Data link connector location
Accessing method
Data link connector detail
MIL flash detail
Erasing method
ACURA
Model:
3.2TL
3.5RL
Engine identification:
C32A6
C35A1
Year:
1996-98
1996-04
System: PGM-FI
Self-diagnosis
General information
Refer to the front of this manual for general test conditions, terminology, detailed descriptions of wiring faults and a general trouble shooter for electrical and mechanical faults.
Engine control module (ECM) incorporates self-diagnosis function.
Malfunction indicator lamp (MIL) will illuminate if certain faults are recorded.
ECM operates in backup mode if sensors fail, to enable vehicle to be driven to workshop.
Trouble codes can be displayed by using a Scan Tool connected to the data link connector (DLC) 1 [1].
Trouble codes can be displayed by the malfunction indicator lamp (MIL) with the data link connector (DLC) 1 [1] jumped, or service check connector 1 [2] jumped.
NOTE: The use of a Scan Tool is essential to obtain full diagnostic information.
Accessing
Ensure ignition switched OFF.
Models without service check connector: Jump data link connector (DLC) terminals 4 and 9 1 [1].
Models with service check connector: Jump service check connector terminals 1 [2].
Switch ignition ON.
Check MIL is flashing.
Trouble codes 1-9 are indicated as follows:
Individual short flashes display trouble code 2 [A].
A short pause separates each flash 2 [B].
A long pause separates each trouble code 2 [C].
For example: Trouble code 3 displayed 2.
Erasing
After the faults have been rectified, trouble codes can be erased by using a Scan Tool connected to the data link connector (DLC).
AD28406
AD89808

Trouble code type

Reference to standard OBD-II trouble code tables

OBD or MIL flash code

OBD-II trouble code

Primary fault location

Probable cause

3.2TL • 3.5RL

ACURA

Trouble code identification

Flash code	OBD-II code	Fault location	Probable cause
–	All P0, P2 and U0 codes	Refer to OBD-II trouble code tables at the front of this manual	–
1	–	Heated oxygen sensor (HO2S) – LH front – circuit/voltage low/high	Wiring, fuel system, HO2S, ECM
		oxygen sensor (HO2S) – RH front – circuit/voltage low/high	Wiring, HO2S, fuel system, ECM
3	–	Manifold absolute pressure (MAP) sensor – circuit/voltage low	Wiring, MAP sensor, ECM
3	–	Manifold absolute pressure (MAP) sensor – circuit/voltage high	Wiring open circuit, MAP sensor, ECM
4	–	Crankshaft position (CKP) sensor 1 – range/performance problem/circuit malfunction	Wiring, CKP sensor, valve timing, ECM
5	–	Manifold absolute pressure (MAP) sensor – range/performance problem	Hose leak/blockage, MAP sensor
5	P1128	Manifold absolute pressure (MAP) sensor – pressure lower than expected	MAP sensor
5	P1129	Manifold absolute pressure (MAP) sensor – pressure higher than expected	MAP sensor
6	–	Engine coolant temperature (ECT) sensor – circuit/voltage low/high	Wiring, ECT sensor, ECM
7	–	Throttle position (TP) sensor – circuit/voltage low/high	Wiring, TP sensor, ECM
7	P1121	Throttle position (TP) sensor – position lower than expected	TP sensor
7	P1122	Throttle position (TP) sensor – position higher than expected	TP sensor
8	P1361	Camshaft position (CMP) sensor 1 – intermittent	Wiring, CMP sensor
8	P1362	Camshaft position (CMP) sensor 1 – no signal	Wiring, CMP sensor
9	P1381	Camshaft position (CMP) sensor 1 – range/performance problem	CMP sensor
9	P1382	Camshaft position (CMP) sensor 1 – malfunction	Wiring, CMP sensor, ECM
10	–	Intake air temperature (IAT) sensor – range/performance problem/circuit/voltage low/high	Wiring short circuit, IAT sensor, ECM
12	P1491	Exhaust gas recirculation (EGR) system – valve lift insufficient	Wiring, EGR valve/position sensor, EGR solenoid, hose leak/blockage, ECM
12	P1498	Exhaust gas recirculation (EGR) valve position sensor – voltage high	Wiring, EGR valve/position sensor, ECM

Safety precautions

Airbags (Supplementary Restraint System – SRS)

Many of the models covered by this manual are fitted with airbags as standard equipment. When working on a vehicle fitted with such a system, extreme caution must be taken to avoid accidental firing of the airbag, which could result in personal injury.

Unauthorised repairs to the system could render it inoperative, or cause it to inflate accidentally.

NOTE: ***All related wiring is encased in a yellow outer covering.***

When the engine is started the AIRBAG warning lamp should go out after approximately 5-10 seconds, if not this indicates a fault in the system. The system should be checked and the fault corrected by a competent technician before any other work is undertaken.

- NEVER attempt to test the system using a multi-meter.
- NEVER tamper with or disconnect the airbag wiring harness.
- NEVER make extra connections to any part of the system wiring harness or terminals.
- ALWAYS ensure that the airbag wiring harness has not been trapped or damaged in any way when working on adjacent components or systems.

Electrical

CAUTION: ***To prevent the engine starting and to avoid damaging the catalytic converter(s), disconnect the fuel injector valve multi-plug(s) before cranking tests.***

- ALWAYS ensure that the battery is properly connected before attempting to start the engine.
- DO NOT attempt to start the engine using a source in excess of 12 volts, such as a fast charger (16 volts) or by connecting two batteries in series (24 volts). ALWAYS disconnect the battery before charging it.
- DO NOT disconnect the battery while the engine is running.
- DO NOT connect the battery with reverse polarity.
- DO NOT disconnect or touch the HT leads when the engine is being cranked or when it is running.
- DO NOT connect or disconnect the engine control module (ECM), or any other component of the fuel injection system while the ignition is switched ON.
- DO NOT disconnect ECM Multi-plug within 30 seconds of switching ignition OFF.
- DO NOT connect or disconnect multi-meters, voltmeters, ammeters or ohmmeters with the ignition switched ON.
- DO NOT reverse the polarity of the fuel pump.
- ALWAYS ensure that all electrical connections are in good condition and making good contact, PARTICULARLY the ECM connector.
- ALWAYS disconnect the ignition coil, ECM, fuel pump relay/fuse before carrying out a compression test.
- DO NOT flash a wire or circuit to ground to check that continuity exists.
- Modern ignition systems operate at very high voltages and these high voltages can severely damage transistorised components such as a wrist-watch if electrical contact is made. Wearers of heart pacemaker devices, therefore, should not at any time carry out work involving ignition systems. In addition to the danger from electric shock, further hazards can arise through sudden uncontrolled body movement causing involuntary contact with moving parts of the engine, i.e. fan blades, pulleys and drive belts.
- ALWAYS ensure that any replacement fuel or ignition system parts are correct for the application in question. Many units share common external features, but differ internally.

Mechanical

CAUTION: ***To minimise fire risk, fuel system must be depressurised before disconnecting any fuel lines or fuel system components.***

- ALWAYS disconnect the distributor before carrying out a fuel pump pressure or delivery check.
- AVOID the risk of fire – ALWAYS disconnect the ignition coil supply and ground the coil HT lead, so that NO HT spark can be emitted, before checking the fuel injector valves, or any other component of the fuel injection system likely to result in the presence of fuel in or around the engine bay.
- AVOID the risk of fire – NEVER work on the fuel injection system when SMOKING or close to a NAKED FLAME.
- ALWAYS keep a fire extinguisher close at hand when working on the fuel injection system.
- ALWAYS ensure that test equipment, leads, tools and especially items of clothing, are clear of moving parts and are not liable to fall into the engine bay, due to vibration, when the engine is running.

A/C	Air conditioning
AIR	Secondary air
APP	Accelerator pedal position
ASM	Auto shift manual
ASR	Acceleration skid control
AT	Automatic transmission
ATF	Automatic transmission fluid
AWD	All wheel drive
BARO	Barometric pressure
Batt	Battery
BBDC	Before bottom dead centre
BCM	Body control module
BPP	Brake pedal position
Cat	Catalytic converter
CID	Cylinder identification
CKP	Crankshaft position
CMP	Camshaft position
CPP	Clutch pedal position
CPU	Central processing unit
CTP	Closed throttle position
DERM	Diagnostic energy reserve module
DI	Direct ignition
DLC	Data link connector
DOHC	Double overhead camshaft
DPI	Dual plug inhibit
DRB	Data read-out box (Chrysler)
DVM	Digital voltmeter
ECCS	Electronic concentrated control system
ECI	Electronically controlled injection
ECM	Engine control module
ECT	Engine coolant temperature (sensor)
EDIS	Electronic distributorless ignition system
EEC	Electronic engine control
EFI	Electronic fuel injection
EFP	Electronic accelerator pedal
EGI	Electronic gasoline injection
EGR	Exhaust gas recirculation
EGRT	Exhaust gas recirculation temperature
EGRTVV	Exhaust gas recirculation thermal vacuum valve
Eng.	Engine
EPI	Electronic petrol injection
EPROM	Electronically programmable read only memory
EOT	Engine oil temperature
EPT	Exhaust pressure transducer
EST	Electronic spark timing
ETC	Electronic traction control
ETS	Electronic throttle system
ETV	Electronic throttle valve
EVAP	Evaporative emission
FICM	Fuel injector control module
FL	Fuse link
FP	Fuel pump
FPCM	Fuel pump control module
FWD	Front wheel drive
GVW	Gross vehicle weight
HEI	High energy ignition
HEX	Hexadecimal
HO	High output
HO2S	Heated oxygen sensor
HT	High tension
I/O	Input-output unit
IAC	Idle air control
IAT	Intake air temperature
IC	Ignition control
IC	Integrated circuit
ICM	Ignition control module
IDM	Ignition diagnostic module
IFZ	Infra-red remote control for central locking
Ign	Ignition
IMA	Idling mixture adjustment sensor
IMRC	Intake manifold runner control
ISC	Idle speed control
JTEC	Jeep/Truck engine controller
KAM	Keep alive memory
KOEO	Key on engine off
KOER	Key on engine running
LED	Light emitting diode
LH	Left hand
MAF	Mass air flow
MAP	Manifold absolute pressure

Abbreviations

MECS	Mazda engine control system
MFI	Multi-port fuel injection
MIL	Malfunction indicator lamp
MPFI	Multi-port fuel injection
MPI	Multi-port injection
MT	Manual transmission
N	Neutral position, automatic transmission
NP	Neutral position
NTC	Negative temperature coefficient
O2S	Oxygen sensor
OHC	Overhead camshaft
OSS	Output shaft speed
P	Park position, automatic transmission
PAIR	Pulsed secondary air injection
PCV	Positive crankcase ventilation
PGM-FI	Programmed fuel injection
PID	Parameter identification
PIP	Profile ignition pickup
PNP	Park/neutral position
PROM	Programmable read only memory
PS	Power steering
PSI	Pounds per square inch
PSP	Power steering pressure
PTC	Positive temperature coefficient
PWM	Pulse width modulated
RAM	Random access memory
RH	Right hand
RHD	Right-hand drive
ROM	Read only memory
RPM	Revolutions per minute
SAE	Society of Automotive Engineers
SBEC	Single board engine controller
SDM	Sensing & diagnostic module
SEFI	Sequential electronic fuel injection
SFI	Sequential fuel injection
SHO	Super high output
SOHC	Single overhead camshaft
SPFI	Single-point fuel injection
SPI	Single port injection
SPOUT	Spark output
SS	Shift solenoid
ST	Start
STI	Self test input
STO	Self test output
TAC	Throttle actuator control
TBI	Throttle body injection
TC	Turbocharger
TCC	Torque converter clutch
TCM	Transmission control module
TCS	Traction control system
TFP	Transmission fluid pressure
TFT	Transmission fluid temperature
TP	Throttle position
TPM	Tempomat cruise control
TPS	Throttle position switch/sensor
TR	Transmission range
TSS	Turbine shaft speed
TWC	Three-way catalytic converter
VAF	Volume air flow sensor
VICS	Variable intake control system
VIN	Vehicle identification number
VIS	Variable intake system
VRIS	Variable resonance intake system
VSS	Vehicle speed sensor
VSV	Vacuum switching valve
WOT	Wide open throttle
8V	Eight valve
10V	Ten valve
12V	Twelve valve
16V	Sixteen valve
20V	Twenty valve
24V	Twenty four valve
1X	1 pulse per crankshaft revolution
2X	2 pulses per crankshaft revolution
3X	3 pulses per crankshaft revolution
4X	4 pulses per crankshaft revolution
6X	6 pulses per crankshaft revolution
7X	7 pulses per crankshaft revolution
18X	18 pulses per crankshaft revolution
24X	24 pulses per crankshaft revolution
58X	58 pulses per crankshaft revolution

General trouble shooter

Despite the sophistication of modern engine management systems, basic mechanical faults can still cause unsatisfactory starting, running and driveability problems as well as the possibility of recording numerous trouble codes.

Before assuming that a problem is electronic, it should be established that the engine is in good mechanical condition and that the basic fuel and electrical circuits are satisfactory.

Check the following:

Mechanical

- Compression pressure
- Valve clearances
- Oil filler cap sealing
- No excessive crankcase fumes
- Manifold vacuum
- Valve timing
- Not burning excess oil

Electrical

- Battery fully charged
- Ignition coil(s)
- HT leads (where applicable)
- Engine ground connections
- Plug leads connected correctly
- Ignition timing
- Adequate spark at the spark plugs
- Engine control module (ECM) ground connections

Fuel system

- Air filter for blockage
- Fuel delivery rate
- Fuel regulated pressure
- Vacuum hoses not split or collapsed
- Fuel filter for blockage
- Fuel system pressure
- Fuel injector spray pattern
- Catalytic converter and exhaust system for blockage

Terminology and locations

Throughout this manual the following standard descriptions and terminology have been used for identification purposes, together with SAE J1930 component descriptions.

- ECM – The acronym used to describe either an engine control module (ECM) or a powertrain control module (PCM).
- Bank 1 – Cylinder bank or group including No.1 cylinder (e.g. cylinders 1, 2 & 3 of a six cylinder engine)
- Bank 2 – Cylinder bank or group not including No.1 cylinder (e.g. cylinders 4, 5 & 6 of a six cylinder engine)
- HO2S 1 – Heated oxygen sensor (HO2S) single or nearest to engine (in front of catalytic converter)
- HO2S 2 – Heated oxygen sensor (HO2S) after catalytic converter
- KS 1 – Knock sensor (KS) single or nearest No.1 cylinder
- Left-hand (LH) and right-hand (RH) – As seen from the driver's seat facing forward

Typical sensor locations and descriptions:

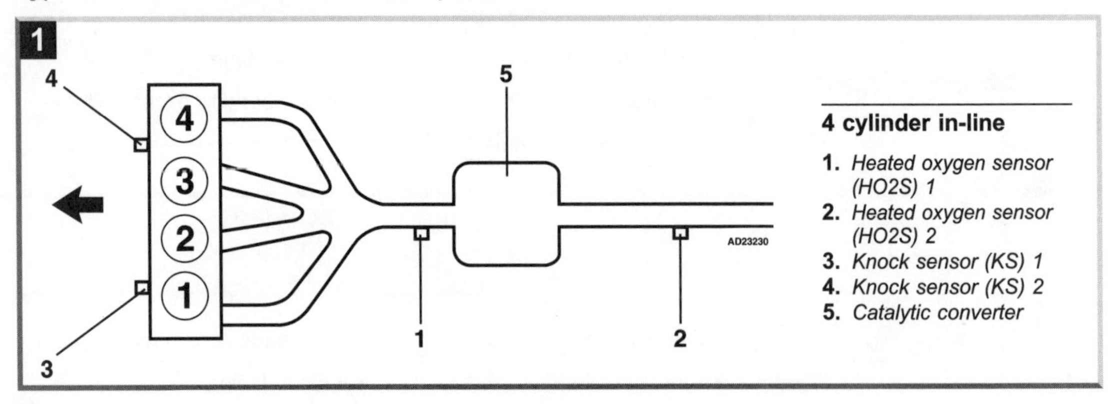

4 cylinder in-line

1. *Heated oxygen sensor (HO2S) 1*
2. *Heated oxygen sensor (HO2S) 2*
3. *Knock sensor (KS) 1*
4. *Knock sensor (KS) 2*
5. *Catalytic converter*

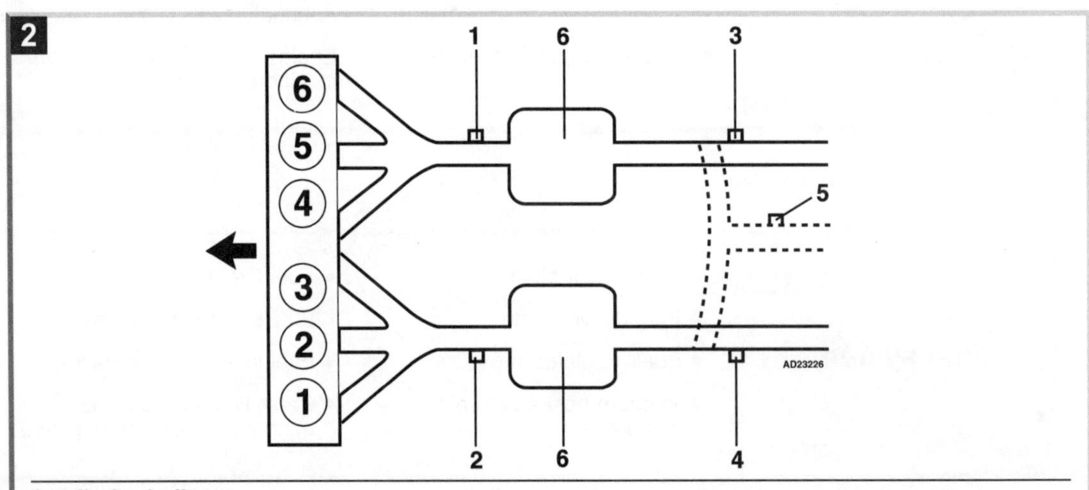

6 cylinder in-line

1. *Heated oxygen sensor (HO2S) 1, bank 2*
2. *Heated oxygen sensor (HO2S) 1, bank 1*
3. *Heated oxygen sensor (HO2S) 2, bank 2*
4. *Heated oxygen sensor (HO2S) 2, bank 1*
5. *Heated oxygen sensor (HO2S) 2 (if only single heated oxygen sensor (HO2S) after cat)*
6. *Catalytic converter*

3

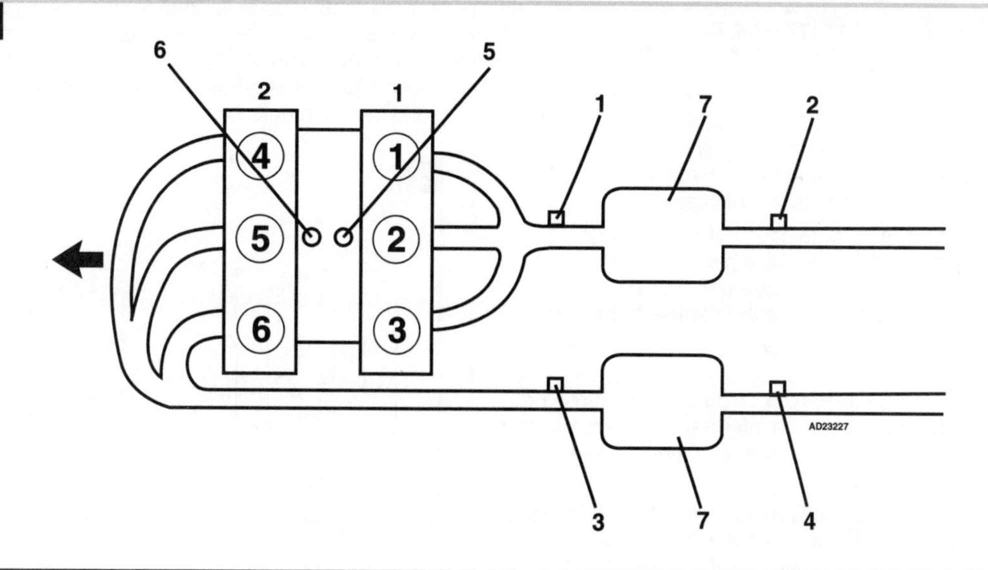

V6

1. *Heated oxygen sensor (HO2S) 1, bank 1*
2. *Heated oxygen sensor (HO2S) 2, bank 1*
3. *Heated oxygen sensor (HO2S) 1, bank 2*
4. *Heated oxygen sensor (HO2S) 2, bank 2*
5. *Knock sensor (KS) 1*
6. *Knock sensor (KS) 2*
7. *Catalytic converter*

4

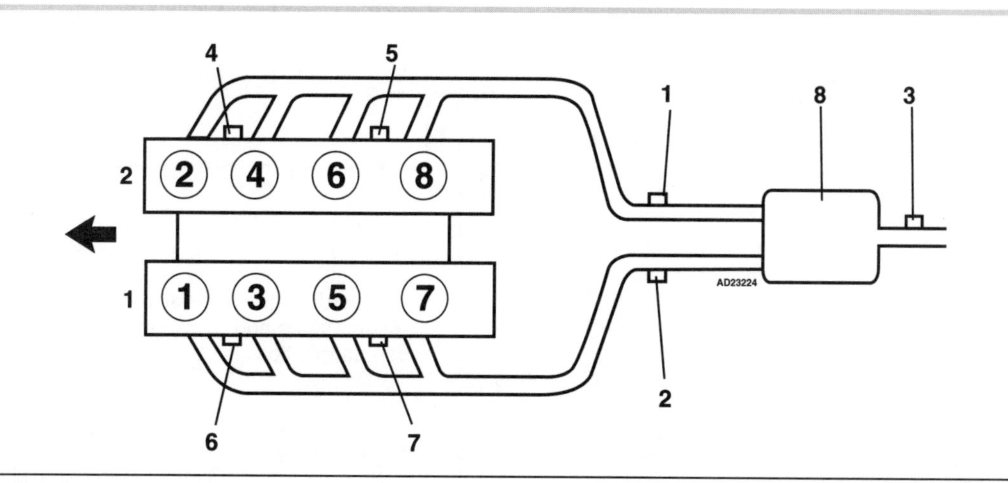

V8

1. *Heated oxygen sensor (HO2S) 1, bank 2*
2. *Heated oxygen sensor (HO2S) 1, bank 1*
3. *Heated oxygen sensor (HO2S) 2*
4. *Knock sensor (KS) 1, bank 2*
5. *Knock sensor (KS) 2, bank 2*
6. *Knock sensor (KS) 1, bank 1*
7. *Knock sensor (KS) 2, bank 1*
8. *Catalytic converter*

How to use this manual

General information

All the fuel and ignition systems covered by this manual are controlled by an engine control module (ECM) which incorporates a self-diagnosis function. These ECMs are capable of detecting certain faulty signals and storing them for access later.

In many cases the detection of a faulty signal will cause the ECM to invoke a backup mode, which adopts a mean signal value (programmed into the ECM memory), to enable the vehicle to be driven until the fault can be diagnosed and rectified.

In many early models faults are signalled by numerical trouble codes comprising two, three or four digits, which can be interpreted by reference to the code tables in this manual. Later models have OBD-II 'P-codes'.

Trouble codes on early models may be displayed by an LED on the ECM, or on the fascia mounted malfunction indicator lamp (MIL). If this is not possible then an LED tester can be used at the data link connector (DLC). If access to trouble codes is possible in this manner it is described along with the relevant code table.

Most later models that comply with OBD-II regulations require a scan tool to enable the trouble code information to be displayed.

OBD-I systems

- Prior to 1988 there was no uniformity in the display, format or method of retrieval of diagnostic trouble codes, each vehicle manufacturer developing a dedicated system with special diagnostic equipment and varying amounts of data logging. Some systems having just a few one or two digit codes while others had more comprehensive three and four digit coding systems.
- In many cases the codes for these early systems could be displayed by connecting an LED tester between the appropriate terminals of the diagnostic socket and counting the flashes or groups of flashes.
- For example **1** represents three short flashes, with a short pause separating each flash, displaying trouble code 3. If more than one code has been logged, a long pause would be used to separate each code.
- If the trouble code is greater than 9, long flashes are used for the tens and short flashes for the units. For example **2** one long flash, and two short flashes separated by a short pause, indicating trouble code 12.

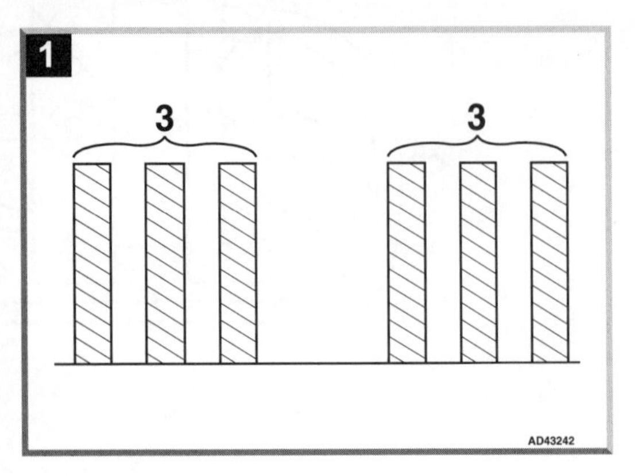

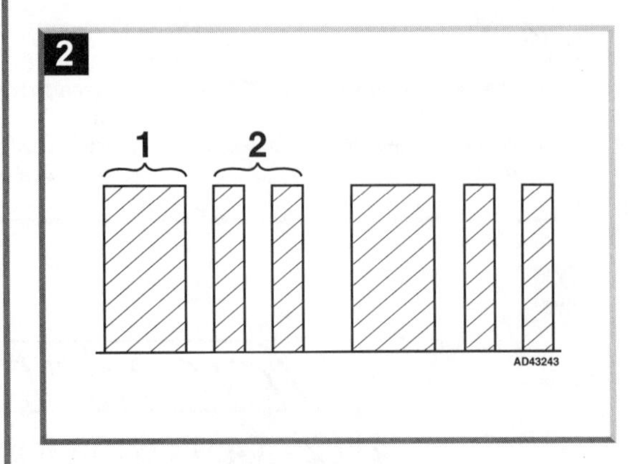

- In 1988 the California Air Resources Board (CARB) introduced legislation covering all new cars sold in California. These new laws governed on-board diagnostic requirements for emissions related faults and began the process of standardization.
- These regulations were known as onboard diagnostic systems - phase I (OBD-I) and were subsequently adopted by the vehicle manufacturers for States other than California.
- The basic requirements of OBD-I were as follows:
- A malfunction indicator lamp (MIL), to provide a visible warning that an emissions related fault had been logged by the engine control module (ECM).

- Provision for the logging and storage, for subsequent retrieval, of diagnostic trouble codes (DTC)/flash codes for emissions related faults.
- The location of the diagnostic socket (data link connector), the method of access to the trouble codes and the configuration and scope of the self-diagnosis was left to individual vehicle manufacturers.
- This resulted in an almost complete lack of standardization, requiring the technician to adopt different procedures and to use different equipment for each vehicle.

OBD-II systems

- In 1995 the California Air Resources Board (CARB), after consultation with the Society of Automotive Engineers (SAE), introduced further legislation covering all new cars sold in California.
- These regulations are known as OBD-II and provide standard diagnostic procedures and protocols which have subsequently been adopted for all States. They closely define trouble code logging and retrieval. More significantly they require monitoring of the operation of the system and components, in addition to detecting component failures.
- OBD-II also covers systems other than engine management.

Basic requirements of OBD-II

Configuration of 16 pin data link connector (DLC)

- The 16 pins of the DLC are arranged in two parallel rows, with certain pin connections defined by the OBD-II regulations and the rest allocated by the vehicle manufacturer – 3
- The pins are designated as follows:

TERMINAL	DESIGNATION
Pin 1	Manufacturer specific
Pin 2	SAE J1850 bus +ve
Pin 3	Manufacturer specific
Pin 4	Chassis ground
Pin 5	Signal ground
Pin 6	CAN data bus, high – ISO 15765-4
Pin 7	K-line – ISO 9141-2/ISO 14230-4
Pin 8	Manufacturer specific
Pin 9	Manufacturer specific
Pin 10	SAE J1850 bus -ve
Pin 11	Manufacturer specific
Pin 12	Manufacturer specific
Pin 13	Manufacturer specific
Pin 14	CAN data bus, low – ISO 15765-4
Pin 15	L-line – ISO 9141-2/ISO 14230-4
Pin 16	Battery positive

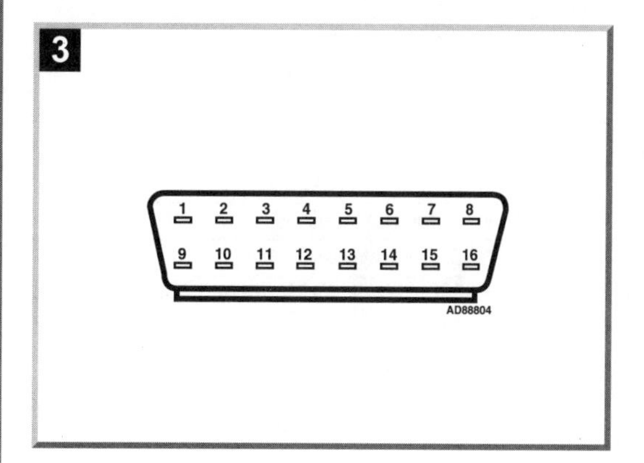

Location for the data link connector (DLC)

- The standard location for the DLC is defined as being between the driver's end of the instrument panel and 300 mm beyond the vehicle center line, accessible from the driver's seat.
- The recommended location is between the steering column and the vehicle center line.
- Access to the connector should not require special tools and should permit one-handed connection and disconnection.

NOTE: ***Some exemptions were granted for existing vehicle designs.***

How to use this manual

OBD-II trouble code format

- Standard, 5 element alpha-numeric format, trouble codes, made up as follows:

Letter prefix defines system group

B	Body
C	Chassis
P	Powertrain
U	Network communications

First digit defines code type

Body/chassis

0	SAE defined (OBD-II)
1	Manufacturer defined
2	Manufacturer defined
3	For future allocation

Powertrain

0	SAE defined (OBD-II)
1	Manufacturer defined
2	SAE defined (OBD-II)
3	P3000-P3399 - SAE defined (OBD-II), P3400-P3999 - Manufacturer defined

Network communications

0	SAE defined (OBD-II)
1	Manufacturer defined
2	Manufacturer defined
3	For future allocation

Second digit/letter defines system area

Powertrain – P0/P1 code

0	Fuel, air or emission control
1	Fuel or air
2	Fuel or air
3	Ignition system or misfire
4	Emission control
5	Vehicle speed, idle speed control or auxiliary inputs
6	Computer or auxiliary outputs
7	Transmission
8	Transmission
9	Transmission
A	Hybrid propulsion
B	For future allocation
C	For future allocation
D	For future allocation
E	For future allocation
F	For future allocation

Powertrain – P2 code

0	Fuel, air or emission control
1	Fuel, air or emission control
2	Fuel, air or emission control
3	Ignition system or misfire
4	Emission control
5	Auxiliary inputs
6	Computer or auxiliary outputs
7	Transmission
8	For future allocation
A	Fuel, air or emission control

Powertrain – P3 code

0	Fuel, air or emission control
1	Fuel, air or emission control
2	Fuel, air or emission control
3	Ignition system or misfire
4	Cylinder deactivation
5	For future allocation
6	For future allocation
7	For future allocation
8	For future allocation
9	For future allocation

Network communications

0	Network electrical
1	Network communications
2	Network communications
3	Network software
4	Network data

Third and fourth digits define specific fault

- These digits define general malfunction, range or performance of component or system, low or high input.

Example code P0108

- This code is logged when a high input is detected in the circuit of the manifold absolute pressure (MAP) sensor or barometric pressure (BARO) sensor.
- P - System – Powertrain
- 0 - Code type – OBD-II
- 1 - System identification (fuel and air)
- 08 - MAP/BARO sensor - circuit, high input

Standard areas monitored:

The following areas are monitored continuously:

- Misfire detection
- Fuel system performance
- Component performance

The following areas are monitored once per 'trip'

NOTE: ***The definition of a 'trip' varies, but is basically – key ON, vehicle driven, key OFF.***

- Catalytic converter
- Evaporative emission system
- Secondary air (AIR) system
- Air conditioning system
- Heated oxygen sensor (HO2S)
- Oxygen sensor (O2S) heater
- Exhaust gas recirculation (EGR) system
- In cases where flash codes and OBD-II P codes are listed for the same model range, only one flash code will be generated for a particular component, but there may be a list of many P codes related to this component. Alternatively there may be some P codes with no equivalent flash code(s).
- These OBD-II P codes will provide more specific details about the fault location, wiring, voltage signals and other information relative to the circuit or system.

Fault logging and drive cycles with OBD-II

- Fault logging is based on inputs received during each 'trip' which may comprise a number of 'drive cycles'.
- Each 'drive cycle' is initiated when the engine is started and is terminated when the engine is switched off.
- For a complete 'trip' to be completed the following, complex, conditions need to be met (and may take several 'drive cycles' to complete):
  - □ Before starting the engine, coolant temperature should be below 50°C and +/-6°C of ambient temperature.
    - ■ This ensures that the ECM logs a cold start.
  - □ Allow engine to idle for 2-3 minutes with an electrical load switched on (such as headlamps or heated rear window).
    - ■ This will ensure that the misfire monitoring and fuel trim monitoring programmes run.
  - □ Drive vehicle up to at least 50 mph and maintain this speed for at least 3 minutes.
    - ■ Misfire and fuel trim monitoring will run.
  - □ Allow the speed to reduce to around 20 mph without changing gear, using the brakes or the clutch.
    - ■ Fuel trim monitoring will run.
  - □ Accelerate to at least 50 mph and maintain for at least 5 minutes.
    - ■ Misfire and fuel trim monitoring will run.
  - □ Allow the speed to reduce to around 20 mph without changing gear, using the brakes or the clutch.
    - ■ Fuel trim monitoring will run.
- Each 'trip' starts when the engine is started and continues through a number of 'drive cycles' until all the OBD-II monitors have completed a self-test.
- The OBD-II monitoring programme provides operating tolerance checks on all emissions related sensors and actuators. Some components/circuits are continuously monitored and some are only activated under predetermined operating conditions.
- Misfire monitoring detects irregularities in the crankshaft position (CKP) sensor signal pattern and identifies which cylinder misfired and if the misfire is frequent enough to cause catalytic converter damage due to excess internal temperatures. If this is the case the malfunction indicator lamp (MIL) will flash.

- If the misfire is likely to increase emissions above the OBD-II limits and occurs on each of two consecutive 'trips' the MIL will flash. If the misfire is absent during the next three 'trips' the MIL will be extinguished.
- When a trouble code is logged a number of data parameters are stored with it, to assist in accurate fault diagnosis. These are:
  - □ Vehicle speed
  - □ Engine coolant temperature
  - □ Engine rpm
  - □ Engine load
  - □ Oxygen sensor system status – open/closed loop
  - □ Distance since fault first logged
  - □ Long term fuel trim (LTFT) level
- OBD-II scan tools are capable of capturing this 'snapshot' of data and displaying (or printing) it for analysis to aid the fault diagnosis process.

Emissions problems and fuel trim

- Short and long term fuel trim refers to the strategy used to reduce exhaust emissions after the basic computation of injection period, using engine load as the major parameter.
- Both front and rear oxygen sensor signals are used to fine tune the fuel/air mixture by increasing or decreasing the injection period +/-25% above or below the basic level.
- Any fault requiring a correction beyond this level will result in a trouble code being logged.
- When the engine is new and running satisfactorily a level of fuel trim will be established – represented by 100%.
- The fuel trim will oscillate the injection period +/-5% above and below the mean level [**A**].
- Manufacturing and in-service tolerances of load sensors (MAP, MAF or VAF) and injectors in particular and faults such as intake air leaks will affect the fuel/air mixture and cause the fuel trim to quickly compensate.
- An intake air leak for instance would result in the injection period increasing, for example to 115-125%. This level will also oscillate +/-5% as before [**B**].
- This new short term fuel trim (STFT) level will be stored in the ECM if it is established as a new basic mixture level. It will then become a long term fuel trim (LTFT) correction and results in the correct mixture level immediately after starting, even when the HO2S has had insufficient time to heat up.
- Long term fuel trim (LTFT) values can be erased by disconnecting the ECM power supply for a suitable length of time.
- If the memory is not erased after repairs the ECM will eventually learn the new LTFT values, but this will take some time and probably cause high emissions and some driveability problems.
- For example if an intake air leak causes the engine to run lean this will be compensated by a change in the LTFT value, which will be stored in the ECM memory.
- After the leak has been repaired this memorised LTFT value will still be used to compute the injection period, resulting in excessively rich running, until new LTFT values have been learned.

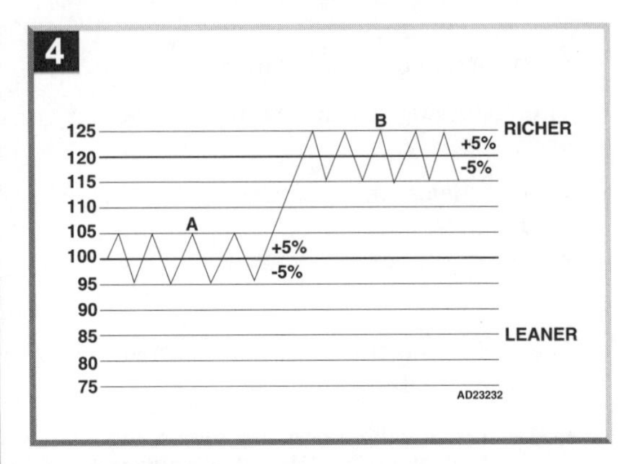

Probable causes

Wiring

- Wiring – refers not only to the wiring harness, but also to any associated component multi-plugs, relay plates, welded, crimped or soldered joints and bulkhead connectors.
- Short to ground – refers to wiring that has a 'leak' to ground somewhere along its length e.g. if the insulation has chaffed through and the wiring is in contact with the body or engine.
- Short to positive – refers to wiring that would normally not be carrying 12 volts, but is shorted to the 12 volt supply e.g. a sensor wire that would normally be fed with approx. 5 volts from the ECM, shorted to a 12 volt battery positive supply.

- Open circuit – refers to wiring that is either disconnected or broken, but is not shorted to either ground or a positive supply.
- Poor connection – refers to a loose, intermittent or high resistance connection.

Mechanical

- Mechanical fault - refers to items such as: fuel leaks, air leaks, crankcase pressure and cylinder compression. Refer to the general trouble shooting section at the front of this manual for further information.

After rectification

- OBD-II monitoring is very sensitive and replacement components may still be identified as faulty if their specifications do not exactly match the originals.
- The trouble code memory must be completely erased and a 'trip' completed.
- The trouble code memory should be accessed to establish that no further codes have been logged.

General test procedures

- Find the model related chapter from vehicle manufacturer, model name, engine code, year of manufacture and system.

Observe the following relevant test conditions:

- All auxiliary equipment, including air conditioning, switched OFF.
- Ensure battery voltage fully charged.
- Ensure battery ground wires in good condition.
- If vehicle fitted with engine malfunction indicator lamp (MIL):
- Check MIL operation – if MIL does not flash or illuminate, replace bulb or repair circuit.

NOTE: ***If MIL remains illuminated or flashes, this indicates a fault, but if it goes out there could still be trouble codes logged in the ECM. Scan tool may not respond if MIL is not working.***

- Engine at normal operating temperature.
- Carry out road test.
- Operate accelerator pedal several times over complete travel range.
- Allow engine to idle.
- If engine does not start: Briefly crank engine and leave ignition switched ON.
- Observe any model specific information.
- Find the location of the data link connector (DLC).
- Jump DLC terminals, connect an LED tester or scan tool, as appropriate and observe any special conditions. If a scan tool is used, follow the manufacturer's operating instructions.
- Follow procedure for accessing trouble codes and note any codes displayed.
- Refer to the appropriate trouble code table. If the vehicle has an OBD-II system the standard SAE defined trouble codes are listed in a table at the front of this manual - refer to relevant section.

NOTE: ***The trouble code may not identify the exact cause of any problem, but serves as a guide to the component or system to be investigated.***

- Refer to probable causes and carry out systematic check of components and circuits until cause of problem is identified and rectified. Where the probable cause is a non-specific engine fault, fuel system or ignition system fault, refer to the General Trouble Shooter section at the front of this manual.
- Check for further trouble codes.
- Follow procedure for erasing trouble codes.

NOTE: ***If the battery is disconnected to erase trouble codes, data stored, such as radio security codes and ECM adaptive memory may be lost.***

- Road test vehicle and recheck for any logged trouble codes.

NOTE: ***If a trouble code is output, but not listed in the appropriate code table, suspect a control module fault. In the case of signal or component malfunction, the ECM may substitute a fixed value from its internal limited operating strategy (LOS) software. This enables the vehicle to be driven in limp-home mode until the fault can be rectified. Depending on the component or circuit in question, it may not be obvious to the driver that the engine management system is operating in limp-home mode.***

OBD-II P0 trouble code table

- The following list covers all SAE defined standard P zero codes allocated at the time of publication.
- All OBD-II codes starting with P zero have standard meanings irrespective of vehicle make or model.
- For OBD-II P1 codes, refer to model specific chapters.
- For OBD-II P2 and U0 codes, refer to relevant trouble code table at the front of this manual.

Trouble code	Fault location	Probable cause
P0000	No fault found	–
P0001	Fuel volume regulator control – circuit open	**Wiring, regulator control solenoid**
P0002	Fuel volume regulator control – circuit range/ performance	**Wiring, regulator control solenoid**
P0003	Fuel volume regulator control – circuit low	**Wiring short to ground, regulator control solenoid**
P0004	Fuel volume regulator control – circuit high	**Wiring open circuit/short to positive, regulator control solenoid**
P0005	Fuel shut-off valve control – circuit open	**Wiring open circuit, fuel shut-off valve**
P0006	Fuel shut-off valve control – circuit low	**Wiring short to ground, fuel shut-off valve**
P0007	Fuel shut-off valve control – circuit high	**Wiring short to positive, fuel shut-off valve**
P0008	Engine position system, bank 1 – performance	**Mechanical fault**
P0009	Engine position system, bank 2 – performance	**Mechanical fault**
P0010	Camshaft position (CMP) actuator, intake/left/front, bank 1 – circuit malfunction	**Wiring, CMP actuator, ECM**
P0011	Camshaft position (CMP), intake/left/front, bank 1 – timing over-advanced/system performance	**Valve timing, engine mechanical fault, CMP actuator**
P0012	Camshaft position (CMP), intake/left/front, bank 1 – timing over-retarded	**Valve timing, engine mechanical fault, CMP actuator**
P0013	Camshaft position (CMP) actuator, intake/left/front, bank 1 – circuit malfunction	**Wiring, CMP actuator, ECM**
P0014	Camshaft position (CMP) actuator, exhaust/right/ rear, bank 1 – timing over-advanced/system performance	**Valve timing, engine mechanical fault, CMP actuator**
P0015	Camshaft position (CMP) actuator, exhaust/right/ rear, bank 1 – timing over-retarded	**Valve timing, engine mechanical fault, CMP actuator**
P0016	Crankshaft position/camshaft position, bank 1 sensor A – correlation	**Wiring, CKP sensor, CMP sensor, mechanical fault**
P0017	Crankshaft position/camshaft position, bank 1 sensor B – correlation	**Wiring, CKP sensor, CMP sensor, mechanical fault**
P0018	Crankshaft position/camshaft position, bank 2 sensor A – correlation	**Wiring, CKP sensor, CMP sensor, mechanical fault**
P0019	Crankshaft position/camshaft position, bank 2 sensor B – correlation	**Wiring, CKP sensor, CMP sensor, mechanical fault**
P0020	Camshaft position (CMP) actuator, intake/left/front, bank 2 – circuit malfunction	**Wiring, CMP actuator, ECM**
P0021	Camshaft position (CMP), intake/left/front, bank 2 – timing over-advanced/system performance	**Valve timing, engine mechanical fault, CMP actuator**
P0022	Camshaft position (CMP), intake/left/front, bank 2 – timing over-retarded	**Valve timing, engine mechanical fault, CMP actuator**
P0023	Camshaft position (CMP) actuator, exhaust/right/ rear, bank 2 – circuit malfunction	**Wiring, CMP actuator, ECM**

OBD-II P0 trouble code table

Trouble code	Fault location	Probable cause
P0024	Camshaft position (CMP), exhaust/right/rear, bank 2 – timing over-advanced/system performance	**Valve timing, engine mechanical fault, CMP actuator**
P0025	Camshaft position (CMP), exhaust/right/rear, bank 2 – timing over-retarded	**Valve timing, engine mechanical fault, CMP actuator**
P0026	Intake valve control solenoid circuit, bank 1 – range/performance	**Wiring, intake valve control solenoid**
P0027	Exhaust valve control solenoid circuit, bank 1 – range/performance	**Wiring, exhaust valve control solenoid**
P0028	Intake valve control solenoid circuit, bank 2 – range/performance	**Wiring, intake valve control solenoid**
P0029	Exhaust valve control solenoid circuit, bank 2 – range/performance	**Wiring, exhaust valve control solenoid**
P0030	Heated oxygen sensor (HO2S) 1, bank 1, heater control – circuit malfunction	**Wiring, HO2S, ECM**
P0031	Heated oxygen sensor (HO2S) 1, bank 1, heater control – circuit low	**Wiring short to ground, HO2S, ECM**
P0032	Heated oxygen sensor (HO2S) 1, bank 1, heater control – circuit high	**Wiring short to positive, HO2S, ECM**
P0033	Turbocharger (TC) wastegate regulating valve – circuit malfunction	**Wiring, TC wastegate regulating valve, ECM**
P0034	Turbocharger (TC) wastegate regulating valve – circuit low	**Wiring short to ground, TC wastegate regulating valve, ECM**
P0035	Turbocharger (TC) wastegate regulating valve – circuit high	**Wiring short to positive, TC wastegate regulating valve, ECM**
P0036	Heated oxygen sensor (HO2S) 2, bank 1, heater control – circuit malfunction	**Wiring, HO2S, ECM**
P0037	Heated oxygen sensor (HO2S) 2, bank 1, heater control – circuit low	**Wiring short to ground, HO2S, ECM**
P0038	Heated oxygen sensor (HO2S) 2, bank 1, heater control – circuit high	**Wiring short to positive, HO2S, ECM**
P0039	Turbo/super charger bypass valve, control circuit – range/performance	**Wiring, bypass valve**
P0040	Oxygen sensor signals swapped, bank 1 sensor 1/ bank 2 sensor 1	**Wiring**
P0041	Oxygen sensor signals swapped, bank 1 sensor 2/ bank 2 sensor 2	**Wiring**
P0042	Heated oxygen sensor (HO2S) 3, bank 1, heater control – circuit malfunction	**Wiring, HO2S, ECM**
P0043	Heated oxygen sensor (HO2S) 3, bank 1, heater control – circuit low	**Wiring short to ground, HO2S, ECM**
P0044	Heated oxygen sensor (HO2S) 3, bank 1, heater control – circuit high	**Wiring short to positive, HO2S, ECM**
P0045	Turbo/super charger boost control solenoid – circuit open	**Wiring, boost control solenoid**
P0046	Turbo/super charger boost control solenoid – circuit range/performance	**Wiring, boost control solenoid, mechanical fault**
P0047	Turbo/super charger boost control solenoid – circuit low	**Wiring short to ground, boost control solenoid**

OBD-II P0 trouble code table

Trouble code	Fault location	Probable cause
P0048	Turbo/super charger boost control solenoid – circuit high	**Wiring short to positive, boost control solenoid**
P0049	Turbo/super charger turbine – over-speed	**Mechanical fault**
P0050	Heated oxygen sensor (HO2S) 1, bank 2, heater control – circuit malfunction	**Wiring, HO2S, ECM**
P0051	Heated oxygen sensor (HO2S) 1, bank 2, heater control – circuit low	**Wiring short to ground, HO2S, ECM**
P0052	Heated oxygen sensor (HO2S) 1, bank 2, heater control – circuit high	**Wiring short to positive, HO2S, ECM**
P0053	Heated oxygen sensor (HO2S), bank 1, sensor 1 – heater resistance	**Wiring, HO2S**
P0054	Heated oxygen sensor (HO2S), bank 1, sensor 2 – heater resistance	**Wiring, HO2S**
P0055	Heated oxygen sensor (HO2S), bank 1, sensor 3 – heater resistance	**Wiring, HO2S**
P0056	Heated oxygen sensor (HO2S) 2, bank 2, heater control – circuit malfunction	**Wiring, HO2S, ECM**
P0057	Heated oxygen sensor (HO2S) 2, bank 2, heater control – heater circuit low	**Wiring short to ground, HO2S, ECM**
P0058	Heated oxygen sensor (HO2S) 2, bank 2, heater control – circuit high	**Wiring short to positive, HO2S, ECM**
P0059	Heated oxygen sensor (HO2S), bank 2, sensor 1 – heater resistance	**Wiring, HO2S**
P0060	Heated oxygen sensor (HO2S), bank 2, sensor 2 – heater resistance	**Wiring, HO2S**
P0061	Heated oxygen sensor (HO2S), bank 2, sensor 3 – heater resistance	**Wiring, HO2S**
P0062	Heated oxygen sensor (HO2S) 3, bank 2, heater control – circuit malfunction	**Wiring, HO2S, ECM**
P0063	Heated oxygen sensor (HO2S) 3, bank 2, heater control – circuit low	**Wiring short to ground, HO2S, ECM**
P0064	Heated oxygen sensor (HO2S) 3, bank 2, heater control – circuit high	**Wiring short to positive, HO2S, ECM**
P0065	Air assisted injector – range/performance problem	**Air assisted injector**
P0066	Air assisted injector – circuit malfunction/circuit low	**Wiring short to ground, air assisted injector, ECM**
P0067	Air assisted injector – circuit high	**Wiring short to positive, air assisted injector, ECM**
P0068	Manifold pressure (MAP) sensor/mass air flow (MAF) sensor – throttle position correlation	**Wiring, MAP sensor, MAF sensor, mechanical fault**
P0069	Manifold pressure (MAP) sensor – barometric pressure correlation	**MAF sensor, mechanical fault**
P0070	Outside air temperature sensor – circuit malfunction	**Wiring, outside air temperature sensor, ECM**
P0071	Outside air temperature sensor – range/ performance problem	**Outside air temperature sensor**
P0072	Outside air temperature sensor – low input	**Wiring short to ground, outside air temperature sensor, ECM**

OBD-II P0 trouble code table

Trouble code	Fault location	Probable cause
P0073	Outside air temperature sensor – high input	Wiring short to positive, outside air temperature sensor, ECM
P0074	Outside air temperature sensor – circuit intermittent	Wiring, poor connection, outside air temperature sensor, ECM
P0075	Intake valve control solenoid, bank 1 – circuit malfunction	Wiring, intake valve control solenoid, ECM
P0076	Intake valve control solenoid, bank 1 – circuit low	Wiring short to ground, intake valve control solenoid, ECM
P0077	Intake valve control solenoid, bank 1 – circuit high	Wiring short to positive, intake valve control solenoid, ECM
P0078	Exhaust valve control solenoid, bank 1 – circuit malfunction	Wiring, exhaust valve control solenoid, ECM
P0079	Exhaust valve control solenoid, bank 1 – circuit low	Wiring short to ground, exhaust valve control solenoid, ECM
P0080	Exhaust valve control solenoid, bank 1 – circuit high	Wiring short to positive, exhaust valve control solenoid, ECM
P0081	Intake valve control solenoid, bank 2 – circuit malfunction	Wiring, intake valve control solenoid, ECM
P0082	Intake valve control solenoid, bank 2 – circuit low	Wiring short to ground, intake valve control solenoid, ECM
P0083	Intake valve control solenoid, bank 2 – circuit high	Wiring short to positive, intake valve control solenoid, ECM
P0084	Exhaust valve control solenoid, bank 2 – circuit malfunction	Wiring, exhaust valve control solenoid, ECM
P0085	Exhaust valve control solenoid, bank 2 – circuit low	Wiring short to ground, exhaust valve control solenoid, ECM
P0086	Exhaust valve control solenoid, bank 2 – circuit high	Wiring short to positive, exhaust valve control solenoid, ECM
P0087	Fuel rail/system pressure – too low	Fuel pump, fuel pressure regulator, fuel supply pipe blockage, mechanical fault
P0088	Fuel rail/system pressure – too high	Fuel pump, fuel pressure regulator, fuel return pipe blockage, mechanical fault
P0089	Fuel pressure regulator – performance problem	Fuel pressure regulator, mechanical fault
P0090	Fuel metering solenoid – open circuit	Wiring open circuit, fuel metering solenoid, ECM
P0091	Fuel metering solenoid – short to ground	Wiring short to ground, fuel metering solenoid, ECM
P0092	Fuel metering solenoid – short to positive	Wiring short to positive, fuel metering solenoid, ECM
P0093	Fuel system leak detected – large leak	Wiring, fuel pressure sensor, mechanical fault
P0094	Fuel system leak detected – small leak	Wiring, fuel pressure sensor, mechanical fault
P0095	Intake air temperature (IAT) sensor 2 – circuit malfunction	Wiring, poor connection, IAT sensor, ECM
P0096	Intake air temperature (IAT) sensor 2 – circuit range/performance	Wiring, poor connection, IAT sensor, ECM
P0097	Intake air temperature (IAT) sensor 2 – circuit low input	Wiring short to ground, IAT sensor, ECM

OBD-II P0 trouble code table

Trouble code	Fault location	Probable cause
P0098	Intake air temperature (IAT) sensor 2 – circuit high input	**Wiring short to positive, IAT sensor, ECM**
P0099	Intake air temperature (IAT) sensor 2 – circuit intermittent/erratic	**Wiring, poor connection, IAT sensor, ECM**
P0100	Mass air flow (MAF) sensor/volume air flow (VAF) sensor – circuit malfunction	**Wiring, MAF/VAF sensor, ECM**
P0101	Mass air flow (MAF) sensor/volume air flow (VAF) sensor – range/performance problem	**Intake leak/blockage, MAF/VAF sensor**
P0102	Mass air flow (MAF) sensor/volume air flow (VAF) sensor – low input	**Wiring short to ground, MAF/VAF sensor, ECM**
P0103	Mass air flow (MAF) sensor/volume air flow (VAF) sensor – high input	**Wiring short to positive, MAF/VAF sensor, ECM**
P0104	Mass air flow (MAF) sensor/volume air flow (VAF) sensor – circuit intermittent	**Wiring, poor connection, MAF/VAF sensor, ECM**
P0105	Manifold absolute pressure (MAP) sensor/barometric pressure (BARO) sensor – circuit malfunction	**Wiring, MAP sensor, BARO sensor, ECM**
P0106	Manifold absolute pressure (MAP) sensor/ barometric pressure (BARO) sensor – range/ performance problem	**Intake/exhaust leak, wiring, MAP sensor, BARO sensor**
P0107	Manifold absolute pressure (MAP) sensor/ barometric pressure (BARO) sensor – low input	**Wiring short to ground, MAP sensor, BARO sensor, ECM**
P0108	Manifold absolute pressure (MAP) sensor/ barometric pressure (BARO) sensor – high input	**Wiring short to positive, MAP sensor, BARO sensor, ECM**
P0109	Manifold absolute pressure (MAP) sensor/ barometric pressure (BARO) sensor – circuit intermittent	**Wiring, poor connection, MAP sensor, BARO sensor, ECM**
P0110	Intake air temperature (IAT) sensor – circuit malfunction	**Wiring, IAT sensor, ECM**
P0111	Intake air temperature (IAT) sensor – range/ performance problem	**IAT sensor**
P0112	Intake air temperature (IAT) sensor – low input	**Wiring short to ground, IAT sensor, ECM**
P0113	Intake air temperature (IAT) sensor – high input	**Wiring open circuit/short to positive, ground wire defective, IAT sensor, ECM**
P0114	Intake air temperature (IAT) sensor – circuit intermittent	**Wiring, poor connection, IAT sensor, ECM**
P0115	Engine coolant temperature (ECT) sensor – circuit malfunction	**Wiring, ECT sensor, ECM**
P0116	Engine coolant temperature (ECT) sensor – range/ performance problem	**Coolant thermostat, poor connection, wiring, ECT sensor**
P0117	Engine coolant temperature (ECT) sensor – low input	**Coolant thermostat, wiring short to ground, ECT sensor**
P0118	Engine coolant temperature (ECT) sensor – high input	**Coolant thermostat, wiring open circuit/short to positive, ground wire defective, ECT sensor**
P0119	Engine coolant temperature (ECT) sensor – circuit intermittent	**Wiring, poor connection, ECT sensor, ECM**
P0120	Throttle position (TP) sensor A/accelerator pedal position (APP) sensor A – circuit malfunction	**Wiring, TP/APP sensor, ECM**
	Throttle position (TP) switch A/accelerator pedal position (APP) switch A – circuit malfunction	**Wiring, TP/APP switch, ECM**

OBD-II P0 trouble code table

Trouble code	Fault location	Probable cause
P0121	Throttle position (TP) sensor A/accelerator pedal position (APP) sensor A – range/performance problem	**Accelerator cable adjustment, TP/APP sensor**
	Throttle position (TP) switch A/accelerator pedal position (APP) switch A – range/performance problem	**Accelerator cable adjustment, TP/APP switch**
P0122	Throttle position (TP) sensor A/accelerator pedal position (APP) sensor A – low input	**Wiring short to ground, TP/APP sensor, ECM**
	Throttle position (TP) switch A/accelerator pedal position (APP) switch A – low input	**Wiring short to ground, TP/APP switch, ECM**
P0123	Throttle position (TP) sensor A/accelerator pedal position (APP) sensor A – high input	**Wiring short to positive, TP/APP sensor, ECM**
	Throttle position (TP) switch A/accelerator pedal position (APP) switch A – high input	**Wiring short to positive, TP/APP switch, ECM**
P0124	Throttle position (TP) sensor A/accelerator pedal position (APP) sensor A – circuit intermittent	**Wiring, poor connection, TP/APP sensor, ECM**
	Throttle position (TP) switch A/accelerator pedal position (APP) switch A – circuit intermittent	**Wiring, poor connection, TP/APP switch, ECM**
P0125	Insufficient coolant temperature for closed loop fuel control	**Wiring, cooling system, coolant thermostat, ECT sensor**
P0126	Insufficient coolant temperature for stable operation	**Wiring, cooling system, coolant thermostat, ECT sensor**
P0127	Intake air temperature – too high	**Wiring short to ground, IAT sensor 2, mechanical fault, ECM**
P0128	Coolant thermostat – coolant temp below thermostat regulating temperature	**Mechanical fault**
P0129	Barometric pressure – too low	**Wiring, BARO sensor, mechanical fault**
P0130	Heated oxygen sensor (HO2S) 1, bank 1 – circuit malfunction	**Heating inoperative, poor connection, wiring, HO2S**
	Oxygen sensor (O2S) 1, bank 1 – circuit malfunction	**Wiring, O2S, ECM**
P0131	Heated oxygen sensor (HO2S) 1, bank 1 – low voltage	**Exhaust leak, wiring short to ground, HO2S, ECM**
	Oxygen sensor (O2S) 1, bank 1 – low voltage	**Exhaust leak, wiring short to ground, O2S, ECM**
P0132	Heated oxygen sensor (HO2S) 1, bank 1 – high voltage	**Wiring short to positive, HO2S, ECM**
	Oxygen sensor (O2S) 1, bank 1 – high voltage	**Wiring short to positive, O2S, ECM**
P0133	Heated oxygen sensor (HO2S) 1, bank 1 – slow response	**Heating inoperative, wiring, HO2S**
	Oxygen sensor (O2S) 1, bank 1 – slow response	**Wiring, O2S**
P0134	Heated oxygen sensor (HO2S) 1, bank 1 – no activity detected	**Wiring open circuit, heating inoperative, HO2S**
	Oxygen sensor (O2S) 1, bank 1 – no activity detected	**Wiring, O2S**
P0135	Heated oxygen sensor (HO2S) 1, bank 1, heater control – circuit malfunction	**Fuse, wiring, HO2S, ECM**
P0136	Heated oxygen sensor (HO2S) 2, bank 1 – circuit malfunction	**Heating inoperative, wiring, HO2S, ECM**
	Oxygen sensor (O2S) 2, bank 1 – circuit malfunction	**Wiring, O2S, ECM**

OBD-II P0 trouble code table

Trouble code	Fault location	Probable cause
P0137	Heated oxygen sensor (HO2S) 2, bank 1 – low voltage	Exhaust leak, wiring short to ground, HO2S, ECM
	Oxygen sensor (O2S) 2, bank 1 – low voltage	Exhaust leak, wiring short to ground, O2S, ECM
P0138	Heated oxygen sensor (HO2S) 2, bank 1 – high voltage	Wiring short to positive, HO2S, ECM
	Oxygen sensor (O2S) 2, bank 1 – high voltage	Wiring short to positive, O2S, ECM
P0139	Heated oxygen sensor (HO2S) 2, bank 1 – slow response	Heating inoperative, wiring, HO2S
	Oxygen sensor (O2S) 2, bank 1 – slow response	Wiring, O2S
P0140	Heated oxygen sensor (HO2S) 2, bank 1 – no activity detected	Wiring, heating inoperative, HO2S, ECM
	Oxygen sensor (O2S) 2, bank 1 – no activity detected	Wiring, O2S, ECM
P0141	Heated oxygen sensor (HO2S) 2, bank 1, heater control – circuit malfunction	Wiring, HO2S, ECM
P0142	Heated oxygen sensor (HO2S) 3, bank 1 – circuit malfunction	Wiring, HO2S, ECM
P0143	Heated oxygen sensor (HO2S) 3, bank 1 – low voltage	Exhaust leak, wiring short to ground, HO2S, ECM
	Oxygen sensor (O2S) 3, bank 1 – low voltage	Exhaust leak, wiring short to ground, O2S, ECM
P0144	Heated oxygen sensor (HO2S) 3, bank 1 – high voltage	Wiring short to positive, HO2S, ECM
	Oxygen sensor (O2S) 3, bank 1 – high voltage	Wiring short to positive, O2S, ECM
P0145	Heated oxygen sensor (HO2S) 3, bank 1 – slow response	Heating inoperative, wiring, HO2S
	Oxygen sensor (O2S) 3, bank 1 – slow response	Wiring, O2S
P0146	Heated oxygen sensor (HO2S) 3, bank 1 – no activity detected	Wiring, HO2S, ECM
	Oxygen sensor (O2S) 3, bank 1 – no activity detected	Wiring, O2S, ECM
P0147	Heated oxygen sensor (HO2S) 3, bank 1, heater control – circuit malfunction	Wiring, HO2S, ECM
P0148	Fuel delivery error	Fuel pump/fuel injection pump
P0149	Fuel timing error	Fuel pump/fuel injection pump
P0150	Heated oxygen sensor (HO2S) 1, bank 2 – circuit malfunction	Wiring, HO2S, ECM
	Oxygen sensor (O2S) 1, bank 2 – circuit malfunction	Wiring, O2S, ECM
P0151	Heated oxygen sensor (HO2S) 1, bank 2 – low voltage	Exhaust leak, wiring short to ground, HO2S, ECM
	Oxygen sensor (O2S) 1, bank 2 – low voltage	Exhaust leak, wiring short to ground, O2S, ECM
P0152	Heated oxygen sensor (HO2S) 1, bank 2 – high voltage	Wiring short to positive, HO2S, ECM
	Oxygen sensor (O2S) 1, bank 2 – high voltage	Wiring short to positive, O2S, ECM
P0153	Heated oxygen sensor (HO2S) 1, bank 2 – slow response	Heating inoperative, wiring, HO2S
	Oxygen sensor (O2S) 1, bank 2 – slow response	Wiring, O2S

Trouble code	Fault location	Probable cause
P0154	Heated oxygen sensor (HO2S) 1, bank 2 – no activity detected	Wiring, HO2S, ECM
	Oxygen sensor (O2S) 1, bank 2 – no activity detected	Wiring, O2S, ECM
P0155	Heated oxygen sensor (HO2S) 1, bank 2, heater control – circuit malfunction	Wiring, HO2S, ECM
P0156	Heated oxygen sensor (HO2S) 2, bank 2 – circuit malfunction	Heating inoperative, wiring, HO2S, ECM
	Oxygen sensor (O2S) 2, bank 2 – circuit malfunction	Wiring, O2S, ECM
P0157	Heated oxygen sensor (HO2S) 2, bank 2 – low voltage	Exhaust leak, wiring short to ground, HO2S, ECM
	Oxygen sensor (O2S) 2, bank 2 – low voltage	Exhaust leak, wiring short to ground, O2S, ECM
P0158	Heated oxygen sensor (HO2S) 2, bank 2 – high voltage	Wiring short to positive, HO2S, ECM
	Oxygen sensor (O2S) 2, bank 2 – high voltage	Wiring short to positive, O2S, ECM
P0159	Heated oxygen sensor (HO2S) 2, bank 2 – slow response	Heating inoperative, wiring, HO2S
	Oxygen sensor (O2S) 2, bank 2 – slow response	Wiring, O2S
P0160	Heated oxygen sensor (HO2S) 2, bank 2 – no activity detected	Wiring, HO2S, ECM
	Oxygen sensor (O2S) 2, bank 2 – no activity detected	Wiring, O2S, ECM
P0161	Heated oxygen sensor (HO2S) 2, bank 2, heater control – circuit malfunction	Wiring, HO2S, ECM
P0162	Heated oxygen sensor (HO2S) 3, bank 2 – circuit malfunction	Wiring, HO2S, ECM
	Oxygen sensor (O2S) 3, bank 2 – circuit malfunction	Wiring, O2S, ECM
P0163	Heated oxygen sensor (HO2S) 3, bank 2 – low voltage	Exhaust leak, wiring short to ground, HO2S, ECM
	Oxygen sensor (O2S) 3, bank 2 – low voltage	Exhaust leak, wiring short to ground, O2S, ECM
P0164	Heated oxygen sensor (HO2S) 3, bank 2 – high voltage	Wiring short to positive, HO2S, ECM
	Oxygen sensor (O2S) 3, bank 2 – high voltage	Wiring short to positive, O2S, ECM
P0165	Heated oxygen sensor (HO2S) 3, bank 2 – slow response	Heating inoperative, wiring, HO2S
	Oxygen sensor (O2S) 3, bank 2 – slow response	Wiring, O2S
P0166	Heated oxygen sensor (HO2S) 3, bank 2 – no activity detected	Wiring, HO2S, ECM
	Oxygen sensor (O2S) 3, bank 2 – no activity detected	Wiring, O2S, ECM
P0167	Heated oxygen sensor (HO2S) 3, bank 2, heater control – circuit malfunction	Wiring, HO2S, ECM
P0168	Fuel temperature – too high	Wiring, fuel temperature sensor, mechanical fault
P0169	Incorrect fuel composition	Wiring, fuel composition sensor, mechanical fault

OBD-II P0 trouble code table

Trouble code	Fault location	Probable cause
P0170	Fuel trim (FT), bank 1 – malfunction	**Intake leak, AIR system, fuel pressure/pump, injector(s), EVAP canister purge valve, HO2S**
P0171	System too lean, bank 1	**Intake/exhaust leak, AIR system, MAF/VAF sensor, fuel pressure/pump, injector(s), HO2S**
P0172	System too rich, bank 1	**Intake blocked, EVAP canister purge valve, fuel pressure, EGR system, injector(s), HO2S**
P0173	Fuel trim (FT), bank 2 – malfunction	**Intake leak, AIR system, fuel pressure/pump, injector(s), EVAP canister purge valve, HO2S**
P0174	System too lean, bank 2	**Intake/exhaust leak, fuel pressure/pump, injector(s), AIR system, hose connection(s)**
P0175	System too rich, bank 2	**Intake blocked, EVAP canister purge valve, fuel pressure, EGR system, injector(s), HO2S**
P0176	Fuel composition sensor – circuit malfunction	**Wiring, fuel composition sensor, ECM**
P0177	Fuel composition sensor – range/performance problem	**Fuel composition sensor**
P0178	Fuel composition sensor – low input	**Wiring short to ground, fuel composition sensor, ECM**
P0179	Fuel composition sensor – high input	**Wiring short to positive, fuel composition sensor, ECM**
P0180	Fuel temperature sensor A – circuit malfunction	**Wiring, fuel temperature sensor, ECM**
P0181	Fuel temperature sensor A – range/performance problem	**Fuel temperature sensor**
P0182	Fuel temperature sensor A – low input	**Wiring short to ground, fuel temperature sensor, ECM**
P0183	Fuel temperature sensor A – high input	**Wiring short to positive, fuel temperature sensor, ECM**
P0184	Fuel temperature sensor A – circuit intermittent	**Wiring, poor connection, fuel temperature sensor, ECM**
P0185	Fuel temperature sensor B – circuit malfunction	**Wiring, fuel temperature sensor, ECM**
P0186	Fuel temperature sensor B – range/performance problem	**Fuel temperature sensor**
P0187	Fuel temperature sensor B – low input	**Wiring short to ground, fuel temperature sensor, ECM**
P0188	Fuel temperature sensor B – high input	**Wiring short to positive, fuel temperature sensor, ECM**
P0189	Fuel temperature sensor B – circuit intermittent	**Wiring, poor connection, fuel temperature sensor, ECM**
P0190	Fuel rail pressure (FRP) sensor – circuit malfunction	**Wiring, fuel rail pressure sensor, ECM**
P0191	Fuel rail pressure (FRP) sensor – range/ performance problem	**Wiring, FRP sensor**
P0192	Fuel rail pressure (FRP) sensor – low input	**Wiring short to ground, FRP sensor**
P0193	Fuel rail pressure (FRP) sensor – high input	**Wiring short to positive, FRP sensor**
P0194	Fuel rail pressure (FRP) sensor – circuit intermittent	**Wiring, poor connection, FRP sensor**
P0195	Engine oil temperature (EOT) sensor – circuit malfunction	**Wiring, EOT sensor, ECM**

Trouble code	Fault location	Probable cause
P0196	Engine oil temperature (EOT) sensor – range/ performance problem	EOT sensor
P0197	Engine oil temperature (EOT) sensor – low input	Wiring short to ground, EOT sensor
P0198	Engine oil temperature (EOT) sensor – high input	Wiring short to positive, EOT sensor
P0199	Engine oil temperature (EOT) sensor – circuit intermittent	Wiring, poor connection, EOT sensor, ECM
P0200	Injector – circuit malfunction	Wiring, injector, ECM
P0201	Injector 1 – circuit malfunction	Wiring, injector, ECM
P0202	Injector 2 – circuit malfunction	Wiring, injector, ECM
P0203	Injector 3 – circuit malfunction	Wiring, injector, ECM
P0204	Injector 4 – circuit malfunction	Wiring, injector, ECM
P0205	Injector 5 – circuit malfunction	Wiring, injector, ECM
P0206	Injector 6 – circuit malfunction	Wiring, injector, ECM
P0207	Injector 7 – circuit malfunction	Wiring, injector, ECM
P0208	Injector 8 – circuit malfunction	Wiring, injector, ECM
P0209	Injector 9 – circuit malfunction	Wiring, injector, ECM
P0210	Injector 10 – circuit malfunction	Wiring, injector, ECM
P0211	Injector 11 – circuit malfunction	Wiring, injector, ECM
P0212	Injector 12 – circuit malfunction	Wiring, injector, ECM
P0213	Cold start injector 1 – circuit malfunction	Wiring, cold start injector, ECM
P0214	Cold start injector 2 – circuit malfunction	Wiring, cold start injector, ECM
P0215	Fuel shut-off solenoid – circuit malfunction	Wiring, fuel shut-off solenoid, ECM
P0216	Fuel injection timing control – circuit malfunction	Wiring, fuel injection timing control solenoid, ECM
P0217	Engine over temperature condition	Wiring, cooling system, coolant thermostat, ECT sensor
P0218	Transmission over temperature condition	Wiring, TFT sensor, ECM
P0219	Engine over speed condition	Incorrect gear change
P0220	Throttle position (TP) sensor B/accelerator pedal position (APP) sensor B – circuit malfunction	Wiring, TP/APP sensor, ECM
	Throttle position (TP) switch B/accelerator pedal position (APP) switch B – circuit malfunction	Wiring, TP/APP switch, ECM
P0221	Throttle position (TP) sensor B/accelerator pedal position (APP) sensor B – range/performance problem	Accelerator cable adjustment, TP/APP sensor
	Throttle position (TP) switch B/accelerator pedal position (APP) switch B – range/performance problem	Accelerator cable adjustment, TP/APP switch
P0222	Throttle position (TP) sensor B/accelerator pedal position (APP) sensor B – low input	Wiring short to ground, TP/APP sensor, ECM
	Throttle position (TP) switch B/accelerator pedal position (APP) switch B – low input	Wiring short to ground, TP/APP switch, ECM
P0223	Throttle position (TP) sensor B/accelerator pedal position (APP) sensor B – high input	Wiring short to positive, TP/APP sensor, ECM
	Throttle position (TP) switch B/accelerator pedal position (APP) switch B – high input	Wiring short to positive, TP/APP switch, ECM

OBD-II P0 trouble code table

Trouble code	Fault location	Probable cause
P0224	Throttle position (TP) sensor B/accelerator pedal position (APP) sensor B – circuit intermittent	Wiring, poor connection, TP/APP sensor, ECM
	Throttle position (TP) switch B/accelerator pedal position (APP) switch B – circuit intermittent	Wiring, poor connection, TP/APP switch, ECM
P0225	Throttle position (TP) sensor C/accelerator pedal position (APP) sensor C – circuit malfunction	Wiring, TP/APP sensor, ECM
	Throttle position (TP) switch C/accelerator pedal position (APP) switch C – circuit malfunction	Wiring, TP/APP switch, ECM
P0226	Throttle position (TP) sensor C/accelerator pedal position (APP) sensor C – range/performance problem	Accelerator cable adjustment, TP/APP sensor
	Throttle position (TP) switch C/accelerator pedal position (APP) switch C – range/performance problem	Accelerator cable adjustment, TP/APP switch
P0227	Throttle position (TP) sensor C/accelerator pedal position (APP) sensor C – low input	Wiring short to ground, TP/APP sensor, ECM
	Throttle position (TP) switch C/accelerator pedal position (APP) switch C – low input	Wiring short to ground, TP/APP switch, ECM
P0228	Throttle position (TP) sensor C/accelerator pedal position (APP) sensor C – high input	Wiring short to positive, TP/APP sensor, ECM
	Throttle position (TP) switch C/accelerator pedal position (APP) switch C – high input	Wiring short to positive, TP/APP switch, ECM
P0229	Throttle position (TP) sensor C/accelerator pedal position (APP) sensor C – circuit intermittent	Wiring, poor connection, TP/APP sensor, ECM
	Throttle position (TP) switch C/accelerator pedal position (APP) switch C – circuit intermittent	Wiring, poor connection, TP/APP switch, ECM
P0230	Fuel pump relay – circuit malfunction	Wiring, fuel pump relay, ECM
P0231	Fuel pump relay – circuit low	Wiring short to ground, fuel pump relay, ECM
P0232	Fuel pump relay – circuit high	Wiring short to positive, fuel pump relay, ECM
P0233	Fuel pump relay – circuit intermittent	Wiring, poor connection, fuel pump relay, ECM
P0234	Engine boost condition – limit exceeded	Hose connection(s), wiring, TC wastegate regulating valve, TC wastegate
P0235	Engine boost condition – limit not reached	Hose connection(s), wiring, TC wastegate regulating valve, TC wastegate, TC
P0236	Manifold absolute pressure (MAP) sensor A, TC system – range/performance problem	Intake/exhaust leak, hose connection(s), MAP sensor
P0237	Manifold absolute pressure (MAP) sensor A, TC system – low input	Wiring short to ground, MAP sensor, ECM
P0238	Manifold absolute pressure (MAP) sensor A, TC system – high input	Wiring short to positive, MAP sensor, ECM
P0239	Manifold absolute pressure (MAP) sensor B, TC system – circuit malfunction	Wiring, MAP sensor, ECM
P0240	Manifold absolute pressure (MAP) sensor B, TC system – range/performance problem	Intake/exhaust leak, hose connection(s), MAP sensor
P0241	Manifold absolute pressure (MAP) sensor B, TC system – low input	Wiring short to ground, MAP sensor, ECM
P0242	Manifold absolute pressure (MAP) sensor B, TC system – high input	Wiring short to positive, MAP sensor, ECM

Trouble code	Fault location	Probable cause
P0243	Turbocharger (TC) wastegate regulating valve A – circuit malfunction	Wiring, TC wastegate regulating valve, ECM
P0244	Turbocharger (TC) wastegate regulating valve A – range/performance problem	TC wastegate regulating valve
P0245	Turbocharger (TC) wastegate regulating valve A – circuit low	Wiring short to ground, TC wastegate regulating valve, ECM
P0246	Turbocharger (TC) wastegate regulating valve A – circuit high	Wiring short to positive, TC wastegate regulating valve, ECM
P0247	Turbocharger (TC) wastegate regulating valve B – circuit malfunction	Wiring, TC wastegate regulating valve, ECM
P0248	Turbocharger (TC) wastegate regulating valve B – range/performance problem	TC wastegate regulating valve
P0249	Turbocharger (TC) wastegate regulating valve B – circuit low	Wiring short to ground, TC wastegate regulating valve, ECM
P0250	Turbocharger (TC) wastegate regulating valve B – circuit high	Wiring short to positive, TC wastegate regulating valve, ECM
P0251	Injection pump A, rotor/cam – circuit malfunction	Wiring, injection pump, ECM
P0252	Injection pump A, rotor/cam – range/performance problem	Injection pump
P0253	Injection pump A, rotor/cam – circuit low	Wiring short to ground, injection pump, ECM
P0254	Injection pump A, rotor/cam – circuit high	Wiring short to positive, injection pump, ECM
P0255	Injection pump A, rotor/cam – circuit intermittent	Wiring, poor connection, injection pump, ECM
P0256	Injection pump B, rotor/cam – circuit malfunction	Wiring, injection pump, ECM
P0257	Injection pump B, rotor/cam – range/performance problem	Injection pump
P0258	Injection pump B, rotor/cam – circuit low	Wiring short to ground, injection pump, ECM
P0259	Injection pump B, rotor/cam – circuit high	Wiring short to positive, injection pump, ECM
P0260	Injection pump B, rotor/cam – circuit intermittent	Wiring, poor connection, injection pump, ECM
P0261	Injector 1 – circuit low	Wiring short to ground, injector, ECM
P0262	Injector 1 – circuit high	Wiring short to positive, injector, ECM
P0263	Cylinder 1 – contribution/balance fault	Wiring, fuel system, ECM
P0264	Injector 2 – circuit low	Wiring short to ground, injector, ECM
P0265	Injector 2 – circuit high	Wiring short to positive, injector, ECM
P0266	Cylinder 2 – contribution/balance fault	Wiring, fuel system, ECM
P0267	Injector 3 – circuit low	Wiring short to ground, injector, ECM
P0268	Injector 3 – circuit high	Wiring short to positive, injector, ECM
P0269	Cylinder 3 – contribution/balance fault	Wiring, fuel system, ECM
P0270	Injector 4 – circuit low	Wiring short to ground, injector, ECM
P0271	Injector 4 – circuit high	Wiring short to positive, injector, ECM
P0272	Cylinder 4 – contribution/balance fault	Wiring, fuel system, ECM
P0273	Injector 5 – circuit low	Wiring short to ground, injector, ECM
P0274	Injector 5 – circuit high	Wiring short to positive, injector, ECM
P0275	Cylinder 5 – contribution/balance fault	Wiring, fuel system, ECM
P0276	Injector 6 – circuit low	Wiring short to ground, injector, ECM
P0277	Injector 6 – circuit high	Wiring short to positive, injector, ECM

OBD-II P0 trouble code table

Trouble code	Fault location	Probable cause
P0278	Cylinder 6 – contribution/balance fault	**Wiring, fuel system, ECM**
P0279	Injector 7 – circuit low	**Wiring short to ground, injector, ECM**
P0280	Injector 7 – circuit high	**Wiring short to positive, injector, ECM**
P0281	Cylinder 7 – contribution/balance fault	**Wiring, fuel system, ECM**
P0282	Injector 8 – circuit low	**Wiring short to ground, injector, ECM**
P0283	Injector 8 – circuit high	**Wiring short to positive, injector, ECM**
P0284	Cylinder 8 – contribution/balance fault	**Wiring, fuel system, ECM**
P0285	Injector 9 – circuit low	**Wiring short to ground, injector, ECM**
P0286	Injector 9 – circuit high	**Wiring short to positive, injector, ECM**
P0287	Cylinder 9 – contribution/balance fault	**Wiring, fuel system, ECM**
P0288	Injector 10 – circuit low	**Wiring short to ground, injector, ECM**
P0289	Injector 10 – circuit high	**Wiring short to positive, injector, ECM**
P0290	Cylinder 10 – contribution/balance fault	**Wiring, fuel system, ECM**
P0291	Injector 11 – circuit low	**Wiring short to ground, injector, ECM**
P0292	Injector 11 – circuit high	**Wiring short to positive, injector, ECM**
P0293	Cylinder 11 – contribution/balance fault	**Wiring, fuel system, ECM**
P0294	Injector 12 – circuit low	**Wiring short to ground, injector, ECM**
P0295	Injector 12 – circuit high	**Wiring short to positive, injector, ECM**
P0296	Cylinder 12 – contribution/balance fault	**Wiring, fuel system, ECM**
P0297	Vehicle over-speed condition	**Wiring, VSS, mechanical fault**
P0298	Engine oil temperature – too high	**Wiring, EOT sensor, mechanical fault**
P0299	Turbo/super charger – low boost	**Mechanical fault**
P0300	Random/multiple cylinder(s) – misfire detected	**Spark plug(s), HT lead(s), injector(s), ignition coil(s), low compression, wiring**
P0301	Cylinder 1 – misfire detected	**Engine mechanical fault, wiring, ignition/fuel system, injector, ECT/MAF sensor, ECM**
P0302	Cylinder 2 – misfire detected	**Engine mechanical fault, wiring, ignition/fuel system, injector, ECT/MAF sensor, ECM**
P0303	Cylinder 3 – misfire detected	**Engine mechanical fault, wiring, ignition/fuel system, injector, ECT/MAF sensor, ECM**
P0304	Cylinder 4 – misfire detected	**Engine mechanical fault, wiring, ignition/fuel system, injector, ECT/MAF sensor, ECM**
P0305	Cylinder 5 – misfire detected	**Engine mechanical fault, wiring, ignition/fuel system, injector, ECT/MAF sensor, ECM**
P0306	Cylinder 6 – misfire detected	**Engine mechanical fault, wiring, ignition/fuel system, injector, ECT/MAF sensor, ECM**
P0307	Cylinder 7 – misfire detected	**Engine mechanical fault, wiring, ignition/fuel system, injector, ECT/MAF sensor, ECM**
P0308	Cylinder 8 – misfire detected	**Engine mechanical fault, wiring, ignition/fuel system, injector, ECT/MAF sensor, ECM**
P0309	Cylinder 9 – misfire detected	**Engine mechanical fault, wiring, ignition/fuel system, injector, ECT/MAF sensor, ECM**
P0310	Cylinder 10 – misfire detected	**Engine mechanical fault, wiring, ignition/fuel system, injector, ECT/MAF sensor, ECM**

OBD-II P0 trouble code table

Trouble code	Fault location	Probable cause
P0311	Cylinder 11 – misfire detected	Engine mechanical fault, wiring, ignition/fuel system, injector, ECT/MAF sensor, ECM
P0312	Cylinder 12 – misfire detected	Engine mechanical fault, wiring, ignition/fuel system, injector, ECT/MAF sensor, ECM
P0313	Misfire detected – low fuel level	Fuel system, mechanical fault
P0314	Single cylinder misfire – cylinder not specified	Engine mechanical fault, wiring, ignition/fuel system, injector
P0315	Crankshaft position system – variation not learned	Engine mechanical fault, wiring
P0316	Misfire detected during startup – first 1000 revolutions	Engine mechanical fault, wiring, ignition/fuel system, injector
P0317	Rough road hardware not present	Wiring, ECM
P0318	Rough road sensor signal A – circuit malfunction	Wiring, rough road sensor A, mechanical fault
P0319	Rough road sensor signal B – circuit malfunction	Wiring, rough road sensor B, mechanical fault
P0320	Crankshaft position (CKP) sensor/engine speed (RPM) sensor – circuit malfunction	Wiring, CKP/RPM sensor, ECM
P0321	Crankshaft position (CKP) sensor/engine speed (RPM) sensor – range/performance problem	Air gap, metal particle contamination, insecure sensor/rotor, wiring, CKP/RPM sensor
P0322	Crankshaft position (CKP) sensor/engine speed (RPM) sensor – no signal	Wiring, CKP/RPM sensor, ECM
P0323	Crankshaft position (CKP) sensor/engine speed (RPM) sensor – circuit intermittent	Wiring, poor connection, CKP/RPM sensor, ECM
P0324	Knock control system error	Wiring, poor connection, KS, ECM
P0325	Knock sensor (KS) 1, bank 1 – circuit malfunction	Wiring, poor connection, KS
P0326	Knock sensor (KS) 1, bank 1 – range/performance problem	Wiring, KS incorrectly tightened, KS
P0327	Knock sensor (KS) 1, bank 1 – low input	Insecure KS, poor connection, wiring short to ground, incorrectly tightened, KS, ECM
P0328	Knock sensor (KS) 1, bank 1 – high input	Wiring short to positive, KS incorrectly tightened, KS, ECM
P0329	Knock sensor (KS) 1, bank 1 – circuit intermittent	Wiring, poor connection, KS, ECM
P0330	Knock sensor (KS) 2, bank 2 – circuit malfunction	Wiring, KS, ECM
P0331	Knock sensor (KS) 2, bank 2 – range/performance problem	Wiring, KS incorrectly tightened, KS
P0332	Knock sensor (KS) 2, bank 2 – low input	Insecure KS, poor connection, wiring short to ground, KS incorrectly tightened, KS, ECM
P0333	Knock sensor (KS) 2, bank 2 – high input	Wiring short to positive, KS incorrectly tightened, KS, ECM
P0334	Knock sensor (KS) 2, bank 2 – circuit intermittent	Wiring, poor connection, KS, ECM
P0335	Crankshaft position (CKP) sensor – circuit malfunction	Wiring, CKP sensor, ECM
P0336	Crankshaft position (CKP) sensor – range/performance problem	Insecure sensor/rotor, air gap, wiring, CKP sensor
P0337	Crankshaft position (CKP) sensor – low input	Wiring short to ground, CKP sensor, ECM
P0338	Crankshaft position (CKP) sensor – high input	Wiring short to positive, CKP sensor, ECM
P0339	Crankshaft position (CKP) sensor – circuit intermittent	Wiring, poor connection, CKP sensor, ECM

OBD-II P0 trouble code table

Trouble code	Fault location	Probable cause
P0340	Camshaft position (CMP) sensor A, bank 1 – circuit malfunction	**Wiring, CMP sensor, ECM**
P0341	Camshaft position (CMP) sensor A, bank 1 – range/performance problem	**Insecure sensor/rotor, air gap, wiring, CMP sensor**
P0342	Camshaft position (CMP) sensor A, bank 1 – low input	**Wiring short to ground, CMP sensor, ECM**
P0343	Camshaft position (CMP) sensor A, bank 1 – high input	**Wiring short to positive, CMP sensor, ECM**
P0344	Camshaft position (CMP) sensor A, bank 1 – circuit intermittent	**Wiring, poor connection, CMP sensor, ECM**
P0345	Camshaft position (CMP) sensor A, bank 2 – circuit malfunction	**Wiring, CMP sensor, ECM**
P0346	Camshaft position (CMP) sensor A, bank 2 – range/performance problem	**Insecure sensor/rotor, air gap, wiring, CMP sensor**
P0347	Camshaft position (CMP) sensor A, bank 2 – low input	**Wiring short to ground, CMP sensor, ECM**
P0348	Camshaft position (CMP) sensor A, bank 2 – high input	**Wiring short to positive, CMP sensor, ECM**
P0349	Camshaft position (CMP) sensor A, bank 2 – circuit intermittent	**Wiring, poor connection, CMP sensor, ECM**
P0350	Ignition coil, primary/secondary – circuit malfunction	**Wiring, ignition coil, ECM**
P0351	Ignition coil A, primary/secondary – circuit malfunction	**Wiring, ignition coil, ECM**
P0352	Ignition coil B, primary/secondary – circuit malfunction	**Wiring, ignition coil, ECM**
P0353	Ignition coil C, primary/secondary – circuit malfunction	**Wiring, ignition coil, ECM**
P0354	Ignition coil D, primary/secondary – circuit malfunction	**Wiring, ignition coil, ECM**
P0355	Ignition coil E, primary/secondary – circuit malfunction	**Wiring, ignition coil, ECM**
P0356	Ignition coil F, primary/secondary – circuit malfunction	**Wiring, ignition coil, ECM**
P0357	Ignition coil G, primary/secondary – circuit malfunction	**Wiring, ignition coil, ECM**
P0358	Ignition coil H, primary/secondary – circuit malfunction	**Wiring, ignition coil, ECM**
P0359	Ignition coil I, primary/secondary – circuit malfunction	**Wiring, ignition coil, ECM**
P0360	Ignition coil J, primary/secondary – circuit malfunction	**Wiring, ignition coil, ECM**
P0361	Ignition coil K, primary/secondary – circuit malfunction	**Wiring, ignition coil, ECM**
P0362	Ignition coil L, primary/secondary – circuit malfunction	**Wiring, ignition coil, ECM**
P0363	Misfire detected – fuelling disabled	**Fuel system, mechanical fault**
P0365	Camshaft position (CMP) sensor B, bank 1 – circuit malfunction	**Wiring, poor connection, CMP sensor, ECM**

Trouble code	Fault location	Probable cause
P0366	Camshaft position (CMP) sensor B, bank 1 – circuit range/performance	Wiring, poor connection, CMP sensor
P0367	Camshaft position (CMP) sensor B, bank 1 – circuit low input	Wiring short to ground, CMP sensor, ECM
P0368	Camshaft position (CMP) sensor B, bank 1 – circuit high input	Wiring short to positive, CMP sensor, ECM
P0369	Camshaft position (CMP) sensor B, bank 1 – circuit intermittent	Wiring, poor connection, ECM
P0370	Timing reference, high resolution signal A – malfunction	Wiring, CKP/RPM/CMP sensor, ECM
P0371	Timing reference, high resolution signal A – too many pulses	Wiring, CKP/RPM/CMP sensor, ECM
P0372	Timing reference, high resolution signal A – too few pulses	Wiring, CKP/RPM/CMP sensor, ECM
P0373	Timing reference, high resolution signal A – intermittent erratic pulses	Wiring, poor connection, CKP/RPM/CMP sensor, ECM
P0374	Timing reference, high resolution signal A – no pulses	Wiring, CKP/RPM/CMP sensor, ECM
P0375	Timing reference, high resolution signal B – malfunction	Wiring, CKP/RPM/CMP sensor, ECM
P0376	Timing reference, high resolution signal B – too many pulses	Wiring, CKP/RPM/CMP sensor, ECM
P0377	Timing reference, high resolution signal B – too few pulses	Wiring, CKP/RPM/CMP sensor, ECM
P0378	Timing reference, high resolution signal B – intermittent erratic pulses	Wiring, poor connection, CKP/RPM/CMP sensor, ECM
P0379	Timing reference, high resolution signal B – no pulses	Wiring, CKP/RPM/CMP sensor, ECM
P0380	Glow plugs, circuit A – malfunction	Wiring, glow plug relay, fuse, glow plugs, ECM
P0381	Glow plug warning lamp – circuit malfunction	Wiring, glow plug warning lamp, ECM
P0382	Glow plugs, circuit B – malfunction	Wiring, glow plug relay, glow plugs, ECM
P0385	Crankshaft position (CKP) sensor B – circuit malfunction	Wiring, CKP sensor, ECM
P0386	Crankshaft position (CKP) sensor B – range/ performance problem	Insecure sensor/rotor, air gap, wiring, CKP sensor
P0387	Crankshaft position (CKP) sensor B – low input	Wiring short to ground, CKP sensor, ECM
P0388	Crankshaft position (CKP) sensor B – high input	Wiring short to positive, CKP sensor, ECM
P0389	Crankshaft position (CKP) sensor B – circuit intermittent	Wiring, poor connection, CKP sensor, ECM
P0390	Camshaft position (CMP) sensor B, bank 2 – circuit malfunction	Wiring, poor connection, CMP sensor, ECM
P0391	Camshaft position (CMP) sensor B, bank 2 – circuit range/performance	Wiring, poor connection, CMP sensor
P0392	Camshaft position (CMP) sensor B, bank 2 – circuit low input	Wiring short to ground, CMP sensor, ECM
P0393	Camshaft position (CMP) sensor B, bank 2 – circuit high input	Wiring short to positive, CMP sensor, ECM

OBD-II P0 trouble code table

Trouble code	Fault location	Probable cause
P0394	Camshaft position (CMP) sensor B, bank 2 – circuit intermittent	**Wiring, poor connection, ECM**
P0400	Exhaust gas recirculation (EGR) system – flow malfunction	**Hose leak/blockage, basic setting not carried out (if applicable), wiring, EGR valve, EGR solenoid, ECM**
P0401	Exhaust gas recirculation (EGR) system – insufficient flow detected	**Hose leak/blockage, basic setting not carried out (if applicable), wiring, EGR valve, EGR solenoid, ECM**
P0402	Exhaust gas recirculation (EGR) system – excessive flow detected	**Hose leak/blockage, basic setting not carried out (if applicable), wiring, EGR valve, EGR solenoid, ECM**
P0403	Exhaust gas recirculation (EGR) – circuit malfunction	**Wiring, EGR solenoid, ECM**
P0404	Exhaust gas recirculation (EGR) system – range/ performance problem	**Hose leak/blockage, wiring, EGR valve/solenoid**
P0405	Exhaust gas recirculation (EGR) valve position sensor A – low input	**Wiring short to ground, EGR valve position sensor, ECM**
P0406	Exhaust gas recirculation (EGR) valve position sensor A – high input	**Wiring short to positive, EGR valve position sensor, ECM**
P0407	Exhaust gas recirculation (EGR) valve position sensor B – low input	**Wiring short to ground, EGR valve position sensor, ECM**
P0408	Exhaust gas recirculation (EGR) valve position sensor B – high input	**Wiring short to positive, EGR valve position sensor, ECM**
P0409	Exhaust gas recirculation (EGR) sensor A – circuit malfunction	**Wiring, poor connection, EGR sensor, ECM**
P0410	Secondary air injection (AIR) system – malfunction	**Wiring, AIR valve, AIR solenoid, ECM**
P0411	Secondary air injection (AIR) system – incorrect flow detected	**AIR pump, AIR valve, AIR hose(s)**
P0412	Secondary air injection (AIR) solenoid A – circuit malfunction	**Wiring, AIR solenoid, ECM**
P0413	Secondary air injection (AIR) solenoid A – open circuit	**Wiring open circuit, AIR solenoid, ECM**
P0414	Secondary air injection (AIR) solenoid A – short circuit	**Wiring short circuit, AIR solenoid, ECM**
P0415	Secondary air injection (AIR) solenoid B – circuit malfunction	**Wiring, AIR solenoid, ECM**
P0416	Secondary air injection (AIR) solenoid B – open circuit	**Wiring open circuit, AIR solenoid, ECM**
P0417	Secondary air injection (AIR) solenoid B – short circuit	**Wiring short circuit, AIR solenoid, ECM**
P0418	Secondary air injection (AIR) pump relay A – circuit malfunction	**Wiring, AIR pump relay, ECM**
P0419	Secondary air injection (AIR) pump relay B – circuit malfunction	**Wiring, AIR pump relay, ECM**
P0420	Catalytic converter system, bank 1 – efficiency below threshold	**Catalytic converter, wiring, HO2S 2**
P0421	Warm up catalytic converter, bank 1 – efficiency below threshold	**Catalytic converter, wiring, HO2S 2**

Trouble code	Fault location	Probable cause
P0422	Main catalytic converter, bank 1 – efficiency below threshold	Catalytic converter, wiring, HO2S 2
P0423	Heated catalytic converter, bank 1 – efficiency below threshold	Catalytic converter, wiring, HO2S 2
P0424	Heated catalytic converter, bank 1 – temperature below threshold	Catalytic converter, wiring, HO2S 2
P0425	Catalytic converter temperature sensor, bank 1	Wiring, poor connection, catalytic converter temperature sensor, ECM
P0426	Catalytic converter temperature sensor, bank 1 – range/performance	Wiring, poor connection, catalytic converter temperature sensor
P0427	Catalytic converter temperature sensor, bank 1 – low input	Wiring short to ground, catalytic converter temperature sensor, ECM
P0428	Catalytic converter temperature sensor, bank 1 – high input	Wiring short to positive, catalytic converter temperature sensor, ECM
P0429	Catalytic converter heater, bank 1 – control circuit malfunction	Wiring, relay, ECM
P0430	Catalytic converter system, bank 2 – efficiency below threshold	Catalytic converter, wiring, HO2S 2
P0431	Warm up catalytic converter, bank 2 – efficiency below threshold	Catalytic converter, wiring, HO2S 2
P0432	Main catalytic converter, bank 2 – efficiency below threshold	Catalytic converter, wiring, HO2S 2
P0433	Heated catalytic converter, bank 2 – efficiency below threshold	Catalytic converter, wiring, HO2S 2
P0434	Heated catalytic converter, bank 2 – temperature below threshold	Catalytic converter, wiring, HO2S 2
P0435	Catalytic converter temperature sensor, bank 2	Wiring, poor connection, catalytic converter temperature sensor, ECM
P0436	Catalytic converter temperature sensor, bank 2 – range/performance	Wiring, poor connection, catalytic converter temperature sensor
P0437	Catalytic converter temperature sensor, bank 2 – low input	Wiring short to ground, catalytic converter temperature sensor, ECM
P0438	Catalytic converter temperature sensor, bank 2 – high input	Wiring short to positive, catalytic converter temperature sensor, ECM
P0439	Catalytic converter heater, bank 2 – control circuit malfunction	Wiring, relay, ECM
P0440	Evaporative emission (EVAP) system – malfunction	Hose connection(s), intake leak, EVAP canister purge valve
P0441	Evaporative emission (EVAP) system – incorrect flow detected	Hose connection(s), intake leak, EVAP canister purge valve
P0442	Evaporative emission (EVAP) system – small leak detected	Hose connection(s), intake leak, EVAP canister, EVAP canister purge valve
P0443	Evaporative emission (EVAP) canister purge valve – circuit malfunction	Wiring, EVAP canister purge valve, ECM
P0444	Evaporative emission (EVAP) canister purge valve – open circuit	Wiring open circuit, EVAP canister purge valve, ECM
P0445	Evaporative emission (EVAP) canister purge valve – short circuit	Wiring short circuit, EVAP canister purge valve, ECM

OBD-II P0 trouble code table

Trouble code	Fault location	Probable cause
P0446	Evaporative emission (EVAP) system, vent control – circuit malfunction	**Wiring, EVAP canister purge valve, ECM**
P0447	Evaporative emission (EVAP) system, vent control – open circuit	**Wiring open circuit, EVAP canister purge valve, ECM**
P0448	Evaporative emission (EVAP) system, vent control – short circuit	**Wiring short circuit, EVAP canister purge valve, ECM**
P0449	Evaporative emission (EVAP) system, vent valve – circuit malfunction	**Wiring, EVAP canister purge valve, ECM**
P0450	Evaporative emission (EVAP) pressure sensor – circuit malfunction	**Wiring, EVAP pressure sensor, ECM**
P0451	Evaporative emission (EVAP) pressure sensor – range/performance problem	**EVAP pressure sensor**
P0452	Evaporative emission (EVAP) pressure sensor – low input	**Wiring short to ground, EVAP pressure sensor, ECM**
P0453	Evaporative emission (EVAP) pressure sensor – high input	**Wiring short to positive, EVAP pressure sensor, ECM**
P0454	Evaporative emission (EVAP) pressure sensor – circuit intermittent	**Wiring, poor connection, EVAP pressure sensor, ECM**
P0455	Evaporative emission (EVAP) system – large leak detected	**Hose connection(s), intake leak, EVAP canister, EVAP canister purge valve**
P0456	Evaporative emission system – very small leak detected	**Mechanical fault, hose connection(s), EVAP pressure sensor**
P0457	Evaporative emission system – leak detected (fuel cap loose/off)	**Mechanical fault, hose connection(s), EVAP pressure sensor**
P0458	Evaporative emission system, purge control valve – circuit low	**Wiring short to ground, EVAP valve**
P0459	Evaporative emission system, purge control valve – circuit high	**Wiring short to positive, EVAP valve**
P0460	Fuel tank level sensor – circuit malfunction	**Wiring, fuel tank level sensor, ECM**
P0461	Fuel tank level sensor – range/performance problem	**Wiring, fuel tank level sensor**
P0462	Fuel tank level sensor – low input	**Wiring short to ground, fuel tank level sensor, ECM**
P0463	Fuel tank level sensor – high input	**Wiring short to positive, fuel tank level sensor, ECM**
P0464	Fuel tank level sensor – circuit intermittent	**Wiring, poor connection, fuel tank level sensor, ECM**
P0465	Evaporative emission (EVAP) canister purge flow sensor – circuit malfunction	**Wiring, EVAP canister purge flow sensor, ECM**
P0466	Evaporative emission (EVAP) canister purge flow sensor – range/performance problem	**EVAP canister purge flow sensor**
P0467	Evaporative emission (EVAP) canister purge flow sensor – low input	**Wiring short to ground, EVAP canister purge flow sensor, ECM**
P0468	Evaporative emission (EVAP) canister purge flow sensor – high input	**Wiring short to positive, EVAP canister purge flow sensor, ECM**
P0469	Evaporative emission (EVAP) canister purge flow sensor – circuit intermittent	**Wiring, poor connection, EVAP canister purge flow sensor, ECM**
P0470	Exhaust gas pressure sensor – circuit malfunction	**Wiring, exhaust gas pressure sensor, ECM**

Trouble code	Fault location	Probable cause
P0471	Exhaust gas pressure sensor – range/performance problem	**Exhaust gas pressure sensor**
P0472	Exhaust gas pressure sensor – low input	**Wiring short to ground, exhaust gas pressure sensor, ECM**
P0473	Exhaust gas pressure sensor – high input	**Wiring short to positive, exhaust gas pressure sensor, ECM**
P0474	Exhaust gas pressure sensor – circuit intermittent	**Wiring, poor connection, exhaust gas pressure sensor, ECM**
P0475	Exhaust gas pressure control valve – circuit malfunction	**Wiring, exhaust gas pressure control valve, ECM**
P0476	Exhaust gas pressure control valve – range/ performance problem	**Exhaust gas pressure control valve**
P0477	Exhaust gas pressure control valve – low input	**Wiring short to ground, exhaust gas pressure control valve, ECM**
P0478	Exhaust gas pressure control valve – high input	**Wiring short to positive, exhaust gas pressure control valve, ECM**
P0479	Exhaust gas pressure control valve – circuit intermittent	**Wiring, poor connection, exhaust gas pressure control valve, ECM**
P0480	Engine coolant blower motor 1 – circuit malfunction	**Wiring, engine coolant blower motor, ECM**
P0481	Engine coolant blower motor 2 – circuit malfunction	**Wiring, engine coolant blower motor, ECM**
P0482	Engine coolant blower motor 3 – circuit malfunction	**Wiring, engine coolant blower motor, ECM**
P0483	Engine coolant blower motor, rationality check – malfunction	**Wiring, engine coolant blower motor, ECM**
P0484	Engine coolant blower motor – circuit over current	**Wiring, engine coolant blower motor, ECM**
P0485	Engine coolant blower motor, power/earth – circuit malfunction	**Wiring, engine coolant blower motor, ECM**
P0486	Exhaust gas recirculation (EGR) valve position sensor B – circuit malfunction	**Wiring, poor connection, EGR valve position sensor, ECM**
P0487	Exhaust gas recirculation (EGR) system, throttle position control – circuit malfunction	**Wiring, poor connection, ECM**
P0488	Exhaust gas recirculation (EGR) system, throttle position control – range/performance	**Wiring, poor connection, ECM**
P0489	Exhaust gas recirculation (EGR) system – circuit low	**Wiring short to ground, EGR valve**
P0490	Exhaust gas recirculation (EGR) system – circuit high	**Wiring short to positive, EGR valve**
P0491	Secondary air injection system, bank 1 – malfunction	**Wiring, AIR solenoid, hose connections, mechanical fault**
P0492	Secondary air injection system, bank 2 – malfunction	**Wiring, AIR solenoid, hose connections, mechanical fault**
P0493	Fan over-speed (clutch locked)	**Fan clutch, mechanical fault**
P0494	Fan speed – low	**Wiring, relay, fan motor, mechanical fault**
P0495	Fan speed – high	**Wiring, relay, fan motor, mechanical fault**
P0496	Evaporative emission system – high purge flow	**Wiring, EVAP valve, mechanical fault**

OBD-II P0 trouble code table

Trouble code	Fault location	Probable cause
P0497	Evaporative emission system – low purge flow	**Wiring, EVAP valve, hoses blocked, mechanical fault**
P0498	Evaporative emission system, vent control – circuit low	**Wiring short to ground, EVAP valve,**
P0499	Evaporative emission system, vent control – circuit high	**Wiring short to positive, EVAP valve**
P0500	Vehicle speed sensor (VSS) – circuit malfunction	**Wiring, VSS, ECM**
P0501	Vehicle speed sensor (VSS) – range/performance problem	**Wiring, speedometer, VSS, CAN data bus**
P0502	Vehicle speed sensor (VSS) – low input	**Wiring short to ground, VSS, ECM**
P0503	Vehicle speed sensor (VSS) – intermittent/erratic/ high input	**Wiring, poor connection, other connected system, instrument panel, VSS**
P0504	Brake switch – A/B correlation	**Wiring, mechanical fault**
P0505	Idle speed control (ISC) system – malfunction	**Wiring, ISC actuator/IAC valve, throttle motor, throttle valve tight/sticking, ECM**
P0506	Idle speed control (ISC) system – rpm lower than expected	**Wiring, ISC actuator/IAC valve, throttle motor, throttle valve tight/sticking, ECM**
P0507	Idle speed control (ISC) system – rpm higher than expected	**Wiring, ISC actuator/IAC valve, throttle motor, throttle valve tight/sticking, ECM**
P0508	Idle air control (IAC) system – circuit low	**Wiring short to ground, IAC valve, ECM**
P0509	Idle air control (IAC) system – circuit high	**Wiring short to positive, IAC valve, ECM**
P0510	Closed throttle position (CTP) switch – circuit malfunction	**Wiring, CTP switch, ECM**
P0511	Idle air control (IAC) system – circuit malfunction	**Wiring, poor connection, IAC valve, ECM**
P0512	Starter request circuit – malfunction	**Wiring, immobilizer system, relay**
P0513	Incorrect immobilizer key	**Immobilizer system**
P0514	Battery temperature sensor – circuit range/ performance	**Wiring, poor connection, battery temperature sensor**
P0515	Battery temperature sensor – circuit malfunction	**Wiring, poor connection, battery temperature sensor**
P0516	Battery temperature sensor – circuit low	**Wiring short to ground, battery temperature sensor, ECM**
P0517	Battery temperature sensor – circuit high	**Wiring short to positive, battery temperature sensor, ECM**
P0518	Idle air control (IAC) system – circuit intermittent	**Wiring, poor connection, IAC valve, ECM**
P0519	Idle air control (IAC) system – circuit performance	**Wiring, poor connection, IAC valve, ECM**
P0520	Engine oil pressure sensor/switch – circuit malfunction	**Wiring, engine oil pressure sensor/switch, ECM**
P0521	Engine oil pressure sensor/switch – range/ performance problem	**Engine oil pressure sensor/switch**
P0522	Engine oil pressure sensor/switch – low voltage	**Wiring short to ground, engine oil pressure sensor/switch, ECM**
P0523	Engine oil pressure sensor/switch – high voltage	**Wiring short to positive, engine oil pressure sensor/switch, ECM**
P0524	Engine oil pressure too low	**Mechanical fault**
P0525	Cruise control, servo control – circuit range/ performance	**Wiring, poor connection, cruise control servo**

Trouble code	Fault location	Probable cause
P0526	Fan speed sensor – circuit malfunction	**Wiring, poor connection, fan speed sensor, ECM**
P0527	Fan speed sensor – circuit range/performance	**Wiring, poor connection, fan speed sensor**
P0528	Fan speed sensor – no signal	**Wiring, poor connection, fan speed sensor, ECM**
P0529	Fan speed sensor – circuit intermittent	**Wiring, poor connection, ECM**
P0530	A/C refrigerant pressure sensor – circuit malfunction	**Wiring, A/C refrigerant pressure sensor, ECM**
P0531	A/C refrigerant pressure sensor – range/ performance problem	**A/C refrigerant pressure sensor**
P0532	A/C refrigerant pressure sensor – low input	**A/C refrigerant pressure too low (incorrectly charged), wiring, A/C refrigerant pressure sensor, ECM**
P0533	A/C refrigerant pressure sensor – high input	**A/C refrigerant pressure too high (cooling fault/ incorrectly charged), wiring, A/C refrigerant pressure sensor, ECM**
P0534	A/C refrigerant charge loss	**A/C leak, wiring, A/C refrigerant pressure sensor**
P0535	A/C evaporator temperature sensor – circuit malfunction	**Wiring, poor connection, A/C evaporator temperature sensor, ECM**
P0536	A/C evaporator temperature sensor – circuit range/ performance	**Wiring, poor connection, A/C evaporator temperature sensor, ECM**
P0537	A/C evaporator temperature sensor – circuit low	**Wiring short to ground, A/C evaporator temperature sensor, ECM**
P0538	A/C evaporator temperature sensor – circuit high	**Wiring short to positive, A/C evaporator temperature sensor, ECM**
P0539	A/C evaporator temperature sensor – circuit intermittent	**Wiring, poor connection, A/C evaporator temperature sensor, ECM**
P0540	Intake air heater A – circuit malfunction	**Wiring, relay, intake air heater**
P0541	Intake air heater A – circuit low	**Wiring short to ground, intake air heater**
P0542	Intake air heater A – circuit high	**Wiring short to positive, intake air heater**
P0543	Intake air heater A – circuit open	**Wiring, intake air heater**
P0544	Exhaust gas recirculation temperature (EGRT) sensor, bank 1 – circuit malfunction	**Wiring, EGRT sensor, ECM**
P0545	Exhaust gas recirculation temperature (EGRT) sensor, bank 1 – low input	**Wiring short to ground, EGRT sensor, ECM**
P0546	Exhaust gas recirculation temperature (EGRT) sensor, bank 1 – high input	**Wiring short to positive, EGRT sensor, ECM**
P0547	Exhaust gas temperature sensor, bank 2 sensor 1 – circuit malfunction	**Wiring, poor connection, exhaust gas temperature sensor, ECM**
P0548	Exhaust gas temperature sensor, bank 2 sensor 1 – circuit low	**Wiring short to ground, exhaust gas temperature sensor, ECM**
P0549	Exhaust gas temperature sensor, bank 2 sensor 1 – circuit high	**Wiring short to positive, exhaust gas temperature sensor, ECM**
P0550	Power steering pressure (PSP) sensor/switch – circuit malfunction	**Wiring, PSP sensor/switch, ECM**
P0551	Power steering pressure (PSP) sensor/switch – range/performance problem	**PAS system, PSP sensor/switch**

Trouble code	Fault location	Probable cause
P0552	Power steering pressure (PSP) sensor/switch – low input	**Wiring short to ground, PSP sensor/switch, ECM**
P0553	Power steering pressure (PSP) sensor/switch – high input	**Wiring short to positive, PSP sensor/switch, ECM**
P0554	Power steering pressure (PSP) sensor/switch – circuit intermittent	**Wiring, poor connection, PSP sensor/switch, ECM**
P0555	Brake booster pressure sensor – circuit malfunction	**Wiring, poor connection, brake booster pressure sensor, ECM**
P0556	Brake booster pressure sensor – circuit range/ performance	**Wiring, poor connection, brake booster pressure sensor, ECM**
P0557	Brake booster pressure sensor – circuit low input	**Wiring short to ground, brake booster pressure sensor, ECM**
P0558	Brake booster pressure sensor – circuit high input	**Wiring short to positive, brake booster pressure sensor, ECM**
P0559	Brake booster pressure sensor – circuit intermittent	**Wiring, poor connection, brake booster pressure sensor, ECM**
P0560	System voltage – malfunction	**Wiring, poor connection, battery, alternator**
P0561	System voltage – unstable	**Wiring, poor connection, battery, alternator**
P0562	System voltage – low	**Wiring, poor connection, battery, alternator**
P0563	System voltage – high	**Alternator**
P0564	Cruise control system, multi-function input A – circuit malfunction	**Wiring, poor connection, multi-function switch, mechanical fault**
P0565	Cruise control master switch, ON signal – malfunction	**Wiring, cruise control master switch, ECM**
P0566	Cruise control master switch, OFF signal – malfunction	**Wiring, cruise control master switch, ECM**
P0567	Cruise control selector switch, RESUME signal – malfunction	**Wiring, cruise control selector switch, ECM**
P0568	Cruise control master switch, SET signal – malfunction	**Wiring, cruise control master switch, ECM**
P0569	Cruise control selector switch, COAST signal – malfunction	**Wiring, cruise control selector switch, ECM**
P0570	Cruise control system, APP sensor signal – malfunction	**Wiring, APP sensor, ECM**
P0571	Cruise/brake switch A – circuit malfunction	**Wiring, cruise/brake switch, ECM**
P0572	Cruise/brake switch A – circuit low	**Wiring short to ground, cruise/brake switch, ECM**
P0573	Cruise/brake switch A – circuit high	**Wiring short to positive, cruise/brake switch, ECM**
P0574	Cruise control system – vehicle speed too high	**Mechanical fault**
P0575	Cruise control system – input circuit malfunction	**Wiring, poor connection, mechanical fault, ECM**
P0576	Cruise control system – input circuit low	**Wiring short to ground**
P0577	Cruise control system – input circuit high	**Wiring short to positive**
P0578	Cruise control system, multi-function input A – circuit stuck	**Wiring, poor connection, multi-function switch, mechanical fault**
P0579	Cruise control system, multi-function input A – circuit range/performance	**Wiring, poor connection, multi-function switch, mechanical fault**

Trouble code	Fault location	Probable cause
P0580	Cruise control system, multi-function input A – circuit low	**Wiring short to ground, multi-function switch, mechanical fault**
P0581	Cruise control system, multi-function input A – circuit high	**Wiring short to positive, multi-function switch, mechanical fault**
P0582	Cruise control system, vacuum control – circuit open	**Wiring, vacuum control solenoid**
P0583	Cruise control system, vacuum control – circuit low	**Wiring short to ground, vacuum control solenoid**
P0584	Cruise control system, vacuum control – circuit high	**Wiring short to positive, vacuum control solenoid**
P0585	Cruise control system, multi-function input A/B – correlation	**Mechanical fault**
P0586	Cruise control system, vent control – circuit open	**Wiring, vent control solenoid**
P0587	Cruise control system, vent control – circuit low	**Wiring short to ground, vent control solenoid**
P0588	Cruise control system, vent control – circuit high	**Wiring short to positive, vent control solenoid**
P0589	Cruise control system, multi-function input B – circuit malfunction	**Wiring, poor connection, multi-function switch, mechanical fault**
P0590	Cruise control system, multi-function input B – circuit stuck	**Wiring, poor connection, multi-function switch, mechanical fault**
P0591	Cruise control system, multi-function input B – circuit range/performance	**Wiring, poor connection, multi-function switch, mechanical fault**
P0592	Cruise control system, multi-function input B – circuit low	**Wiring short to ground, multi-function switch, mechanical fault**
P0593	Cruise control system, multi-function input B – circuit high	**Wiring short to positive, multi-function switch, mechanical fault**
P0594	Cruise control system, servo control – circuit open	**Wiring, servo control solenoid**
P0595	Cruise control system, servo control – circuit low	**Wiring short to ground, servo control solenoid**
P0596	Cruise control system, servo control – circuit high	**Wiring short to positive, servo control solenoid**
P0597	Thermostat heater control system – circuit open	**Wiring, relay, thermostat heater**
P0598	Thermostat heater control system – circuit low	**Wiring short to ground, relay, thermostat heater**
P0599	Thermostat heater control system – circuit high	**Wiring short to positive, relay, thermostat heater**
P0600	CAN data bus – malfunction	**Wiring, connected system, ECM**
P0601	Engine control module (ECM) – memory check sum error	**ECM**
P0602	Engine control module (ECM) – programming error	**ECM**
P0603	Engine control module (ECM) – KAM error	**ECM**
P0604	Engine control module (ECM) – RAM error	**ECM**
P0605	Engine control module (ECM) – ROM error	**ECM**
P0606	Engine control module (ECM) – processor fault	**ECM**
P0607	Control module – performance	**Control module**
P0608	Engine control module (ECM), VSS output A – malfunction	**ECM**
P0609	Engine control module (ECM), VSS output B – malfunction	**ECM**
P0610	Control module – vehicle options error	**Control module**

OBD-II P0 trouble code table

Trouble code	Fault location	Probable cause
P0611	Fuel injector control module – performance	**Fuel injector control module**
P0612	Fuel injector control module – relay control circuit	**Wiring, relay, fuel injector control module**
P0613	Transmission control module (TCM) processor error	**TCM**
P0614	Engine control module (ECM)/transmission control module (TCM) – mismatch	**ECM/TCM**
P0615	Starter relay – circuit malfunction	**Wiring, poor connection, starter relay, ECM**
P0616	Starter relay – circuit low	**Wiring short to ground, starter relay, ECM**
P0617	Starter relay – circuit high	**Wiring short to positive, starter relay, ECM**
P0618	Alternative fuel control module – KAM error	**Alternative fuel control module**
P0619	Alternative fuel control module – RAM/ROM error	**Alternative fuel control module**
P0620	Alternator, control – circuit malfunction	**Wiring, alternator, battery, ECM**
P0621	Alternator warning lamp – circuit malfunction	**Wiring, alternator warning lamp, ECM**
P0622	Alternator, field control – circuit malfunction	**Wiring, alternator, battery, ECM**
P0623	Generator lamp control – circuit malfunction	**Wiring, poor connection, bulb, ECM**
P0624	Fuel cap lamp control – circuit malfunction	**Wiring, poor connection, bulb, ECM**
P0625	Generator field terminal – circuit low	**Wiring short to ground, generator**
P0626	Generator field terminal – circuit high	**Wiring short to positive, generator**
P0627	Fuel pump control – circuit open	**Wiring, relay, fuel pump**
P0628	Fuel pump control – circuit low	**Wiring short to ground, relay, fuel pump**
P0629	Fuel pump control – circuit high	**Wiring short to positive, relay, fuel pump**
P0630	VIN not programmed or mismatch – ECM	**ECM**
P0631	VIN not programmed or mismatch – TCM	**TCM**
P0632	Odometer not programmed – ECM	**ECM**
P0633	Immobilizer key not programmed – ECM	**ECM**
P0634	ECM/TCM – internal temperature too high	**Mechanical fault, ECM/TCM**
P0635	Power steering (PS) control – circuit malfunction	**Wiring, poor connection, PSP switch, ECM**
P0636	Power steering (PS) control – circuit low	**Wiring short to ground, PSP switch, ECM**
P0637	Power Steering (PS) control – circuit high	**Wiring short to positive, PSP switch, ECM**
P0638	Throttle actuator control, bank 1 – range/ performance problem	**Basic setting not carried out (if applicable), ISC actuator/throttle motor, APP sensor**
P0639	Throttle actuator control, bank 2 – range/ performance	**Wiring, throttle control unit**
P0640	Intake air heater control – circuit malfunction	**Wiring, relay, intake air heater**
P0641	Sensor reference voltage A – circuit open	**Wiring short to positive**
P0642	Engine control module (ECM), knock control – defective	**ECM**
P0643	Sensor reference voltage A – circuit high	**Wiring short to positive**
P0644	Driver display, serial communication – circuit malfunction	**Wiring, CAN data bus, ECM**
P0645	Air conditioning (A/C)	**Wiring, A/C system**
P0646	A/C clutch relay control – circuit low	**Wiring short to ground, A/C clutch relay**
P0647	A/C clutch relay control – circuit high	**Wiring short to positive, A/C clutch relay**

Trouble code	Fault location	Probable cause
P0648	Immobilizer lamp control – circuit malfunction	Wiring, poor connection, bulb, ECM
P0649	Cruise control lamp control – circuit	Wiring, poor connection, bulb, ECM
P0650	Malfunction indicator lamp (MIL) – circuit malfunction	Wiring, MIL, ECM
P0651	Sensor reference voltage B – circuit open	Wiring short to positive
P0652	Sensor reference voltage B – circuit low	Wiring short to ground
P0653	Sensor reference voltage B – circuit high	Wiring short to positive
P0654	Engine rpm, output – circuit malfunction	Wiring, ECM
P0655	Engine hot lamp output – circuit malfunction	Wiring, engine hot lamp, ECM
P0656	Fuel level output – circuit malfunction	Wiring, ECM
P0657	Actuator supply voltage – circuit open	Wiring
P0658	Actuator supply voltage – circuit low	Wiring short to ground, actuator
P0659	Actuator supply voltage – circuit high	Wiring short to positive, actuator
P0660	Intake manifold tuning valve, bank 1 – circuit open	Wiring, intake manifold tuning valve
P0661	Intake manifold tuning valve, bank 1 – circuit low	Wiring short to ground, intake manifold tuning valve
P0662	Intake manifold tuning valve, bank 1 – circuit high	Wiring short to positive, intake manifold tuning valve
P0663	Intake manifold tuning valve, bank 2 – circuit open	Wiring, intake manifold tuning valve
P0664	Intake manifold tuning valve, bank 2 – circuit low	Wiring short to ground, intake manifold tuning valve
P0665	Intake manifold tuning valve, bank 2 – circuit high	Wiring short to positive, intake manifold
P0666	ECM/TCM internal temperature sensor – circuit malfunction	Wiring, poor connection, internal temperature sensor, ECM/TCM
P0667	ECM/TCM internal temperature sensor – range/ performance	Wiring, poor connection, internal temperature sensor, ECM/TCM
P0668	ECM/TCM internal temperature sensor – circuit low	Wiring short to ground, internal temperature sensor, ECM/TCM
P0669	ECM/TCM internal temperature sensor – circuit high	Wiring short to positive, internal temperature sensor, ECM/TCM
P0670	Glow plug module control – circuit malfunction	Wiring, poor connection, control module, glow plug, ECM
P0671	Glow plug, cylinder 1 – circuit malfunction	Wiring, poor connection, relay, control module, glow plug, ECM
P0672	Glow plug, cylinder 2 – circuit malfunction	Wiring, poor connection, relay, control module, glow plug, ECM
P0673	Glow plug, cylinder 3 – circuit malfunction	Wiring, poor connection, relay, control module, glow plug, ECM
P0674	Glow plug, cylinder 4 – circuit malfunction	Wiring, poor connection, relay, control module, glow plug, ECM
P0675	Glow plug, cylinder 5 – circuit malfunction	Wiring, poor connection, relay, control module, glow plug, ECM
P0676	Glow plug, cylinder 6 – circuit malfunction	Wiring, poor connection, relay, control module, glow plug, ECM
P0677	Glow plug, cylinder 7 – circuit malfunction	Wiring, poor connection, relay, control module, glow plug, ECM

OBD-II P0 trouble code table

Trouble code	Fault location	Probable cause
P0678	Glow plug, cylinder 8 – circuit malfunction	**Wiring, poor connection, relay, control module, glow plug, ECM**
P0679	Glow plug, cylinder 9 – circuit malfunction	**Wiring, poor connection, relay, control module, glow plug, ECM**
P0680	Glow plug, cylinder 10 – circuit malfunction	**Wiring, poor connection, relay, control module, glow plug, ECM**
P0681	Glow plug, cylinder 11 – circuit malfunction	**Wiring, poor connection, relay, control module, glow plug, ECM**
P0682	Glow plug, cylinder 12 – circuit malfunction	**Wiring, poor connection, relay, control module, glow plug, ECM**
P0683	Glow plug control module communication to ECM – malfunction	**Wiring, poor connection, control module, ECM**
P0684	Glow plug control module communication to ECM – range/performance	**Wiring, poor connection, control module, ECM**
P0685	ECM power relay control – circuit open	**Wiring, ECM relay**
P0686	ECM power relay control – circuit low	**Wiring short to ground, ECM relay, ECM**
P0687	Engine control relay – short to ground	**Wiring short to ground, engine control relay, ECM**
P0688	Engine control relay – short to positive	**Wiring short to positive, engine control relay, ECM**
P0689	ECM power relay sense – circuit low	**Wiring short to ground, ECM relay, ECM**
P0690	ECM power relay sense – circuit high	**Wiring short to positive, ECM relay, ECM**
P0691	Engine coolant blower motor 1 – short to ground	**Wiring short to ground, engine coolant blower motor, ECM**
P0692	Engine coolant blower motor 1 – short to positive	**Wiring short to positive, engine coolant blower motor, ECM**
P0693	Engine coolant blower motor 2 – short to ground	**Wiring short to ground, engine coolant blower motor, ECM**
P0694	Engine coolant blower motor 2 – short to positive	**Wiring short to positive, engine coolant blower motor, ECM**
P0695	Fan 3 control – circuit low	**Wiring short to ground, fan motor**
P0696	Fan 3 control – circuit high	**Wiring short to positive, fan motor**
P0697	Sensor reference voltage C – circuit open	**Wiring short to positive**
P0698	Sensor reference voltage C – circuit low	**Wiring short to ground**
P0699	Sensor reference voltage C – circuit high	**Wiring short to positive**
P0700	Transmission control system – malfunction	**Wiring, TCM**
P0701	Transmission control system – range/performance problem	**Wiring, TCM**
P0702	Transmission control system – electrical	**Wiring, TCM**
P0703	Torque converter/brake switch B – circuit malfunction	**Wiring, torque converter/brake switch, ECM/TCM**
P0704	Clutch pedal position (CPP) switch – circuit malfunction	**Wiring, CPP switch, ECM/TCM**
P0705	Transmission range (TR) sensor/switch, PRNDL input – circuit malfunction	**Wiring, TR sensor/switch, ECM/TCM**
P0706	Transmission range (TR) sensor/switch – range/performance problem	**Wiring, TR sensor/switch**

Trouble code	Fault location	Probable cause
P0707	Transmission range (TR) sensor/switch – low input	**Wiring short to ground, TR sensor/switch, ECM/TCM**
P0708	Transmission range (TR) sensor/switch – high input	**Wiring short to positive, TR sensor/switch, ECM/TCM**
P0709	Transmission range (TR) sensor/switch – circuit intermittent	**Wiring, poor connection, TR sensor/switch, ECM/TCM**
P0710	Transmission fluid temperature (TFT) sensor – circuit malfunction	**Wiring, TFT sensor, ECM, ECM/TCM**
P0711	Transmission fluid temperature (TFT) sensor – range/performance problem	**Wiring, TFT sensor**
P0712	Transmission fluid temperature (TFT) sensor – low input	**Wiring short to ground, TFT sensor, ECM/TCM**
P0713	Transmission fluid temperature (TFT) sensor – high input	**Wiring short to positive, TFT sensor, ECM/TCM**
P0714	Transmission fluid temperature (TFT) sensor – circuit intermittent	**Wiring, poor connection, TFT sensor, ECM/TCM**
P0715	Turbine shaft speed (TSS) sensor – circuit malfunction	**Wiring, TSS sensor, ECM/TCM**
P0716	Turbine shaft speed (TSS) sensor – range/performance problem	**Wiring, TSS sensor**
P0717	Turbine shaft speed (TSS) sensor – no signal	**Wiring, TSS sensor, ECM/TCM**
P0718	Turbine shaft speed (TSS) sensor – circuit intermittent	**Wiring, poor connection, TSS sensor, ECM/TCM**
P0719	Torque converter/brake switch B – circuit low	**Wiring short to ground, torque converter/brake switch, ECM/TCM**
P0720	Output shaft speed (OSS) sensor – circuit malfunction	**Wiring, OSS sensor, ECM/TCM**
P0721	Output shaft speed (OSS) sensor – range/performance problem	**Wiring, OSS sensor**
P0722	Output shaft speed (OSS) sensor – no signal	**Wiring, OSS sensor, ECM/TCM**
P0723	Output shaft speed (OSS) sensor – circuit intermittent	**Wiring, poor connection, OSS sensor, ECM/TCM**
P0724	Torque converter/brake switch B – circuit high	**Wiring short to positive, torque converter/brake switch, ECM/TCM**
P0725	Engine RPM input – circuit malfunction	**Wiring, CKP/RPM sensor, ECM/TCM**
P0726	Engine RPM input – range/performance problem	**Wiring, CKP/RPM sensor**
P0727	Engine RPM input – no signal	**Wiring, CKP/RPM sensor, ECM/TCM**
P0728	Engine RPM input – circuit intermittent	**Wiring, poor connection, CKP/RPM sensor, ECM/TCM**
P0730	Incorrect gear ratio	**Wiring, TR sensor/switch, shift solenoids, transmission mechanical fault**
P0731	Gear 1 – incorrect ratio	**Wiring, TR sensor/switch, shift solenoids, transmission mechanical fault**
P0732	Gear 2 – incorrect ratio	**Wiring, TR sensor/switch, shift solenoids, transmission mechanical fault**
P0733	Gear 3 – incorrect ratio	**Wiring, TR sensor/switch, shift solenoids, transmission mechanical fault**

OBD-II P0 trouble code table

Trouble code	Fault location	Probable cause
P0734	Gear 4 – incorrect ratio	**Wiring, TR sensor/switch, shift solenoids, transmission mechanical fault**
P0735	Gear 5 – incorrect ratio	**Wiring, TR sensor/switch, shift solenoids, transmission mechanical fault**
P0736	Reverse – incorrect ratio	**Wiring, TR sensor/switch, shift solenoids, transmission mechanical fault**
P0737	TCM engine speed output – circuit	**Wiring, TCM**
P0738	TCM engine speed output – circuit low	**Wiring, TCM**
P0739	TCM engine speed output – circuit high	**Wiring, TCM**
P0740	Torque converter clutch (TCC) solenoid – circuit malfunction	**Wiring, TCC solenoid, ECM/TCM**
P0741	Torque converter clutch (TCC) solenoid – performance or stuck off	**Wiring, TCC solenoid**
P0742	Torque converter clutch (TCC) solenoid – stuck on	**Wiring, TCC solenoid**
P0743	Torque converter clutch (TCC) solenoid – electrical	**Wiring, TCC solenoid, ECM/TCM**
P0744	Torque converter clutch (TCC) solenoid – circuit intermittent	**Wiring, poor connection, TCC solenoid, ECM/TCM**
P0745	Transmission fluid pressure (TFP) solenoid – circuit malfunction	**Wiring, TFP solenoid, ECM/TCM**
P0746	Transmission fluid pressure (TFP) solenoid – performance or stuck off	**Wiring, TFP solenoid**
P0747	Transmission fluid pressure (TFP) solenoid – stuck on	**Wiring, TFP solenoid**
P0748	Transmission fluid pressure (TFP) solenoid – electrical	**Wiring, TFP solenoid, ECM/TCM**
P0749	Transmission fluid pressure (TFP) solenoid – circuit intermittent	**Wiring, poor connection, TFP solenoid, ECM/TCM**
P0750	Shift solenoid (SS) A – circuit malfunction	**Wiring, shift solenoid, ECM/TCM**
P0751	Shift solenoid (SS) A – performance or stuck off	**Wiring, shift solenoid**
P0752	Shift solenoid (SS) A – stuck on	**Wiring, shift solenoid**
P0753	Shift solenoid (SS) A – electrical	**Wiring, shift solenoid, ECM/TCM**
P0754	Shift solenoid (SS) A – circuit intermittent	**Wiring, poor connection, shift solenoid, ECM/TCM**
P0755	Shift solenoid (SS) B – circuit malfunction	**Wiring, shift solenoid, ECM/TCM**
P0756	Shift solenoid (SS) B – performance or stuck off	**Wiring, shift solenoid**
P0757	Shift solenoid (SS) B – stuck on	**Wiring, shift solenoid**
P0758	Shift solenoid (SS) B – electrical	**Wiring, shift solenoid, ECM/TCM**
P0759	Shift solenoid (SS) B – circuit intermittent	**Wiring, poor connection, shift solenoid, ECM/TCM**
P0760	Shift solenoid (SS) C – circuit malfunction	**Wiring, shift solenoid, ECM/TCM**
P0761	Shift solenoid (SS) C – performance or stuck off	**Wiring, shift solenoid**
P0762	Shift solenoid (SS) C – stuck on	**Wiring, shift solenoid**
P0763	Shift solenoid (SS) C – electrical	**Wiring, shift solenoid, ECM/TCM**
P0764	Shift solenoid (SS) C – circuit intermittent	**Wiring, poor connection, shift solenoid, ECM/TCM**
P0765	Shift solenoid (SS) D – circuit malfunction	**Wiring, shift solenoid, ECM/TCM**

Trouble code	Fault location	Probable cause
P0766	Shift solenoid (SS) D – performance or stuck off	Wiring, shift solenoid
P0767	Shift solenoid (SS) D – stuck on	Wiring, shift solenoid
P0768	Shift solenoid (SS) D – electrical	Wiring, shift solenoid, ECM/TCM
P0769	Shift solenoid (SS) D – circuit intermittent	Wiring, poor connection, shift solenoid, ECM/TCM
P0770	Shift solenoid (SS) E – circuit malfunction	Wiring, shift solenoid, ECM/TCM
P0771	Shift solenoid (SS) E – performance or stuck off	Wiring, shift solenoid
P0772	Shift solenoid (SS) E – stuck on	Wiring, shift solenoid
P0773	Shift solenoid (SS) E – electrical	Wiring, shift solenoid, ECM/TCM
P0774	Shift solenoid (SS) E – circuit intermittent	Wiring, poor connection, shift solenoid, ECM/TCM
P0775	Pressure control solenoid B – malfunction	Pressure control solenoid
P0776	Pressure control solenoid B – performance or stuck off	Wiring, pressure control solenoid
P0777	Pressure control solenoid B – stuck on	Wiring, pressure control solenoid
P0778	Pressure control solenoid B – electrical malfunction	Wiring, pressure control solenoid
P0779	Pressure control solenoid B – intermittent	Wiring, poor connection, pressure control solenoid
P0780	Gear selection – shift malfunction	Wiring, TR sensor, shift solenoids, transmission mechanical fault
P0781	Gear selection, 1-2 – shift malfunction	Wiring, TR sensor, shift solenoids, transmission mechanical fault
P0782	Gear selection, 2-3 – shift malfunction	Wiring, TR sensor, shift solenoids, transmission mechanical fault
P0783	Gear selection, 3-4 – shift malfunction	Wiring, TR sensor, shift solenoids, transmission mechanical fault
P0784	Gear selection, 4-5 – shift malfunction	Wiring, TR sensor, shift solenoids, transmission mechanical fault
P0785	Shift/timing solenoid – circuit malfunction	Wiring, shift/timing solenoid, ECM/TCM
P0786	Shift/timing solenoid – range/performance problem	Wiring, shift/timing solenoid
P0787	Shift/timing solenoid – low	Wiring short to ground, shift/timing solenoid, ECM/TCM
P0788	Shift/timing solenoid – high	Wiring short to positive, shift/timing solenoid, ECM/TCM
P0789	Shift/timing solenoid – intermittent	Wiring, poor connection, shift/timing solenoid, ECM/TCM
P0790	Transmission mode selection switch – circuit malfunction	Wiring, transmission mode selection switch, ECM/TCM
P0791	Intermediate shaft speed sensor – circuit malfunction	Wiring, poor connection, intermediate shaft speed sensor, ECM/TCM
P0792	Intermediate shaft speed sensor – range/ performance problem	Wiring, poor connection, intermediate shaft speed sensor, ECM/TCM
P0793	Intermediate shaft speed sensor – no signal	Wiring, poor connection, short to ground, intermediate shaft speed sensor, ECM/TCM
P0794	Intermediate shaft speed sensor – intermittent circuit malfunction	Wiring, poor connection, intermediate shaft speed sensor, ECM/TCM

OBD-II P0 trouble code table

Trouble code	Fault location	Probable cause
P0795	Transmission fluid pressure (TFP) solenoid C – circuit malfunction	**Wiring, poor connection, TFP solenoid, ECM/TCM**
P0796	Transmission fluid pressure (TFP) solenoid C – performance or stuck off	**Wiring, poor connection, TFP solenoid, ECM/TCM**
P0797	Transmission fluid pressure (TFP) solenoid C – stuck on	**Wiring, poor connection, TFP solenoid, ECM/TCM**
P0798	Transmission fluid pressure (TFP) solenoid C – electrical malfunction	**Wiring, poor connection, TFP solenoid, ECM/TCM**
P0799	Transmission fluid pressure (TFP) solenoid C – intermittent circuit malfunction	**Wiring, poor connection, ECM/TCM**
P0800	Transfer case control system, MIL request – malfunction	**Wiring, mechanical fault**
P0801	Reverse inhibit circuit – malfunction	**Wiring, poor connection**
P0802	Transmission control system, MIL request – circuit open	**Wiring, mechanical fault**
P0803	1-4 Upshift (Skip shift) solenoid – circuit malfunction	**Wiring, poor connection, upshift solenoid**
P0804	1-4 Upshift (Skip shift) warning lamp – circuit malfunction	**Wiring, poor connection**
P0805	Clutch position sensor – circuit malfunction	**Wiring, poor connection, clutch position sensor, ECM/TCM**
P0806	Clutch position sensor – range/performance problem	**Wiring, poor connection, clutch position sensor, ECM/TCM**
P0807	Clutch position sensor – low input	**Wiring, short to ground, clutch position sensor, ECM/TCM**
P0808	Clutch position sensor – high input	**Wiring, short to positive, clutch position sensor, ECM/TCM**
P0809	Clutch position sensor – intermittent circuit malfunction	**Wiring, poor connection, clutch position sensor, ECM/TCM**
P0810	Clutch position control error	**Wiring, poor connection, ECM/TCM**
P0811	Excessive clutch slip	**Wiring, poor connection, mechanical fault, ECM/TCM**
P0812	Reverse gear – input circuit malfunction	**Wiring, poor connection, ECM/TCM**
P0813	Reverse gear – output circuit malfunction	**Wiring, poor connection, ECM/TCM**
P0814	Transmission range (TR) display – circuit malfunction	**Wiring, poor connection, TR sensor, ECM/TCM**
P0815	Upshift switch – circuit malfunction	**Wiring, poor connection, upshift switch, ECM/TCM**
P0816	Downshift switch – circuit malfunction	**Wiring, poor connection, downshift switch, ECM/TCM**
P0817	Starter disable circuit – malfunction	**Wiring, poor connection, ECM/TCM**
P0818	Driveline disconnect switch – circuit malfunction	**Wiring, poor connection, upshift switch, ECM/TCM**
P0819	Up and down shift switch to transmission range – correlation	**Wiring, poor connection, TR sensor, ECM/TCM**
P0820	Gear lever X-Y position sensor – circuit malfunction	**Wiring, poor connection, gear lever position sensor, ECM/TCM**

Trouble code	Fault location	Probable cause
P0821	Gear lever X position sensor – circuit malfunction	**Wiring, poor connection, gear lever position sensor, ECM/TCM**
P0822	Gear lever Y position sensor – circuit malfunction	**Wiring, poor connection, gear lever position sensor, ECM/TCM**
P0823	Gear lever X position sensor – circuit intermittent	**Wiring, poor connection, gear lever position sensor, ECM/TCM**
P0824	Gear lever Y position sensor – circuit intermittent	**Wiring, poor connection, gear lever position sensor, ECM/TCM**
P0825	Gear lever push-pull switch – circuit malfunction	**Wiring, poor connection, gear lever push-pull switch, ECM/TCM**
P0826	Up and down switch – input circuit	**Wiring, Up/down switch**
P0827	Up and down switch – input circuit low	**Wiring short to ground, Up/down switch**
P0828	Up and down switch – input circuit high	**Wiring short to positive, Up/down switch**
P0829	5-6 Shift	**Mechanical fault**
P0830	Clutch pedal position (CPP) switch A – circuit malfunction	**Wiring, poor connection, CPP switch, ECM/TCM**
P0831	Clutch pedal position (CPP) switch A – low input	**Wiring, short to ground, CPP switch, ECM/TCM**
P0832	Clutch pedal position (CPP) switch A – high input	**Wiring, short to positive, CPP switch, ECM/TCM**
P0833	Clutch pedal position (CPP) switch B – circuit malfunction	**Wiring, poor connection, CPP switch, ECM/TCM**
P0834	Clutch pedal position (CPP) switch B – low input	**Wiring, short to ground, CPP switch, ECM/TCM**
P0835	Clutch pedal position (CPP) switch B – high input	**Wiring, short to positive, CPP switch, ECM/TCM**
P0836	Four wheel drive switch – circuit malfunction	**Wiring, poor connection, four wheel drive switch, ECM/TCM**
P0837	Four wheel drive switch – range/performance problem	**Wiring, poor connection, four wheel drive switch, ECM/TCM**
P0838	Four wheel drive switch – low input	**Wiring, short to ground, four wheel drive switch, ECM/TCM**
P0839	Four wheel drive switch – high input	**Wiring, short to positive, four wheel drive switch, ECM/TCM**
P0840	Transmission fluid pressure (TFP) sensor A – circuit malfunction	**Wiring, poor connection, TFP sensor, ECM/TCM**
P0840	Transmission fluid pressure (TFP) switch A – circuit malfunction	**Wiring, poor connection, TFP switch, ECM/TCM**
P0841	Transmission fluid pressure (TFP) sensor A – range/performance problem	**Wiring, poor connection, TFP sensor, ECM/TCM**
	Transmission fluid pressure (TFP) switch A – range/performance problem	**Wiring, poor connection, TFP switch, ECM/TCM**
P0842	Transmission fluid pressure (TFP) sensor A – low input	**Wiring, short to ground, TFP sensor, ECM/TCM**
	Transmission fluid pressure (TFP) switch A – low input	**Wiring, short to ground, TFP switch, ECM/TCM**
P0843	Transmission fluid pressure (TFP) sensor A – high input	**Wiring, short to positive, TFP sensor, ECM/TCM**
	Transmission fluid pressure (TFP) switch A – high input	**Wiring, short to positive, TFP switch, ECM/TCM**

OBD-II P0 trouble code table

Trouble code	Fault location	Probable cause
P0844	Transmission fluid pressure (TFP) sensor A – intermittent circuit malfunction	**Wiring, poor connection, TFP sensor, ECM/TCM**
	Transmission fluid pressure (TFP) switch A – intermittent circuit malfunction	**Wiring, poor connection, TFP switch, ECM/TCM**
P0845	Transmission fluid pressure (TFP) sensor B – circuit malfunction	**Wiring, poor connection, TFP sensor, ECM/TCM**
	Transmission fluid pressure (TFP) switch B – circuit malfunction	**Wiring, poor connection, TFP switch, ECM/TCM**
P0846	Transmission fluid pressure (TFP) sensor B – range/performance problem	**Wiring, poor connection, TFP sensor, ECM/TCM**
	Transmission fluid pressure (TFP) switch B – range/performance problem	**Wiring, poor connection, TFP switch, ECM/TCM**
P0847	Transmission fluid pressure (TFP) sensor B – low input	**Wiring, short to ground, TFP sensor, ECM/TCM**
	Transmission fluid pressure (TFP) switch B – low input	**Wiring, short to ground, TFP switch, ECM/TCM**
P0848	Transmission fluid pressure (TFP) sensor B – high input	**Wiring, short to positive, TFP sensor, ECM/TCM**
	Transmission fluid pressure (TFP) switch B – high input	**Wiring, short to positive, TFP switch, ECM/TCM**
P0849	Transmission fluid pressure (TFP) sensor B – intermittent circuit malfunction	**Wiring, poor connection, TFP sensor, ECM/TCM**
	Transmission fluid pressure (TFP) switch B – intermittent circuit malfunction	**Wiring, poor connection, TFP switch, ECM/TCM**
P0850	Park/neutral position (PNP) switch – input circuit malfunction	**Wiring, PNP switch, ECM/TCM**
P0851	Park/neutral position (PNP) switch – input circuit low	**Wiring, short to ground, PNP switch, ECM/TCM**
P0852	Park/neutral position (PNP) switch – input circuit high	**Wiring, short to positive, PNP switch, ECM/TCM**
P0853	Drive switch – input circuit malfunction	**Wiring, drive switch, ECM/TCM**
P0854	Drive switch – input circuit low	**Wiring, short to ground, drive switch, ECM/TCM**
P0855	Drive switch – input circuit high	**Wiring, short to positive, drive switch, ECM/TCM**
P0856	Traction control input signal – malfunction	**Wiring, poor connection, ECM/TCM**
P0857	Traction control input signal – range/performance problem	**Wiring, poor connection, ECM/TCM**
P0858	Traction control input signal – low	**Wiring, short to ground, ECM/TCM**
P0859	Traction control input signal – high	**Wiring, short to positive, ECM/TCM**
P0860	Gear shift module communication circuit – malfunction	**Wiring, poor connection, gear shift module, ECM/TCM**
P0861	Gear shift module communication circuit – low input	**Wiring, short to ground, gear shift module, ECM/TCM**
P0862	Gear shift module communication circuit – high input	**Wiring, short to positive, gear shift module, ECM/TCM**
P0863	Transmission control module (TCM) communication circuit – malfunction	**Wiring, poor connection, TCM**
P0864	Transmission control module (TCM) communication circuit – range/performance problem	**Wiring, poor connection, TCM**

Trouble code	Fault location	Probable cause
P0865	Transmission control module (TCM) communication circuit – low input	**Wiring, short to ground, TCM**
P0866	Transmission control module (TCM) communication circuit – high input	**Wiring, short to positive, TCM**
P0867	Transmission fluid pressure (TFP) sensor	**Wiring, poor connection, TFP sensor, ECM/TCM**
P0868	Transmission fluid pressure (TFP) sensor – low	**Wiring, short to ground, TFP sensor, ECM/TCM**
P0869	Transmission fluid pressure (TFP) sensor – high	**Wiring, short to positive, TFP sensor, ECM/TCM**
P0870	Transmission fluid pressure (TFP) sensor C – circuit malfunction	**Wiring, poor connection, TFP sensor, ECM/TCM**
	Transmission fluid pressure (TFP) switch C – circuit malfunction	**Wiring, poor connection, TFP switch, ECM/TCM**
P0871	Transmission fluid pressure (TFP) sensor C – range/performance	**Wiring, poor connection, TFP sensor, ECM/TCM**
	Transmission fluid pressure (TFP) switch C – range/performance	**Wiring, poor connection, TFP switch, ECM/TCM**
P0872	Transmission fluid pressure (TFP) sensor C – circuit low	**Wiring, short to ground, TFP sensor, ECM/TCM**
	Transmission fluid pressure (TFP) switch C – circuit low	**Wiring, short to ground, TFP switch, ECM/TCM**
P0873	Transmission fluid pressure (TFP) sensor C – circuit high	**Wiring, short to positive, TFP sensor, ECM/TCM**
	Transmission fluid pressure (TFP) switch C – circuit high	**Wiring, short to positive, TFP switch, ECM/TCM**
P0874	Transmission fluid pressure (TFP) sensor C – intermittent circuit malfunction	**Wiring, poor connection, TFP sensor, ECM/TCM**
	Transmission fluid pressure (TFP) switch C – intermittent circuit malfunction	**Wiring, poor connection, TFP switch, ECM/TCM**
P0875	Transmission fluid pressure (TFP) sensor D – circuit malfunction	**Wiring, poor connection, TFP sensor, ECM/TCM**
	Transmission fluid pressure (TFP) switch D – circuit malfunction	**Wiring, poor connection, TFP switch, ECM/TCM**
P0876	Transmission fluid pressure (TFP) sensor D – range/performance	**Wiring, poor connection, TFP sensor, ECM/TCM**
	Transmission fluid pressure (TFP) switch D – range/performance	**Wiring, poor connection, TFP switch, ECM/TCM**
P0877	Transmission fluid pressure (TFP) sensor D – circuit low	**Wiring, short to ground, TFP sensor, ECM/TCM**
	Transmission fluid pressure (TFP) switch D – circuit low	**Wiring, short to ground, TFP switch, ECM/TCM**
P0878	Transmission fluid pressure (TFP) sensor D – circuit high	**Wiring, short to positive, TFP sensor, ECM/TCM**
	Transmission fluid pressure (TFP) switch D – circuit high	**Wiring, short to positive, TFP switch, ECM/TCM**
P0879	Transmission fluid pressure (TFP) sensor D – intermittent circuit malfunction	**Wiring, poor connection, TFP sensor, ECM/TCM**
	Transmission fluid pressure (TFP) switch D – intermittent circuit malfunction	**Wiring, poor connection, TFP switch, ECM/TCM**

OBD-II P0 trouble code table

Trouble code	Fault location	Probable cause
P0880	Transmission control module (TCM) – power input signal malfunction	**Wiring, poor connection, TCM**
P0881	Transmission control module (TCM) – power input signal range/performance	**Wiring, poor connection, TCM**
P0882	Transmission control module (TCM) – power input signal low	**Wiring, short to ground, TCM**
P0883	Transmission control module (TCM) – power input signal high	**Wiring, short to positive, TCM**
P0884	Transmission control module (TCM) – power input signal intermittent malfunction	**Wiring, poor connection, TCM**
P0885	Transmission control module (TCM) power relay – control circuit open	**Wiring, poor connection, TCM power relay, TCM**
P0886	Transmission control module (TCM) power relay – control circuit low	**Wiring, short to ground, TCM power relay, TCM**
P0887	Transmission control module (TCM) power relay – control circuit high	**Wiring, short to positive, TCM power relay, TCM**
P0888	Transmission control module (TCM) power relay – sense circuit malfunction	**Wiring, poor connection, TCM power relay, TCM**
P0889	Transmission control module (TCM) power relay – sense circuit range/performance	**Wiring, poor connection, TCM power relay, TCM**
P0890	Transmission control module (TCM) power relay – sense circuit low	**Wiring, short to ground, TCM power relay, TCM**
P0891	Transmission control module (TCM) power relay – sense circuit high	**Wiring, short to positive, TCM power relay, TCM**
P0892	Transmission control module (TCM) power relay – sense circuit intermittent malfunction	**Wiring, poor connection, TCM power relay, TCM**
P0893	Multiple gears engaged	**Mechanical fault**
P0894	Transmission component slipping	**Mechanical fault**
P0895	Shift time too short	**Mechanical fault**
P0896	Shift time too long	**Mechanical fault**
P0897	Transmission fluid deteriorated	**Mechanical fault**
P0898	Transmission control system – MIL request – circuit low	**Wiring, poor connection, short to ground**
P0899	Transmission control system – MIL request – circuit high	**Wiring, poor connection, short to positive**
P0900	Clutch actuator – circuit open	**Wiring, clutch actuator, ECM/TCM**
P0901	Clutch actuator – circuit range/performance	**Wiring, poor connection, clutch actuator, ECM/TCM**
P0902	Clutch actuator – circuit low	**Wiring, short to ground, clutch actuator, ECM/TCM**
P0903	Clutch actuator – circuit high	**Wiring, short to positive, clutch actuator, ECM/TCM**
P0904	Transmission gate select position circuit – malfunction	**Wiring, poor connection, ECM/TCM**
P0905	Transmission gate select position circuit – range/performance	**Wiring, poor connection, ECM/TCM**
P0906	Transmission gate select position circuit – low	**Wiring, short to ground, ECM/TCM**

Trouble code	Fault location	Probable cause
P0907	Transmission gate select position circuit – high	Wiring, short to positive, ECM/TCM
P0908	Transmission gate select position circuit – intermittent circuit malfunction	Wiring, poor connection, ECM/TCM
P0909	Transmission gate select control error	Mechanical fault
P0910	Transmission gate select actuator – circuit open	Wiring, transmission gate select actuator, ECM/TCM
P0911	Transmission gate select actuator – circuit range/ performance	Wiring, poor connection, transmission gate select actuator, ECM/TCM
P0912	Transmission gate select actuator – circuit low	Wiring, short to ground, transmission gate select actuator, ECM/TCM
P0913	Transmission gate select actuator – circuit high	Wiring, short to positive, transmission gate select actuator, ECM/TCM
P0914	Gear shift position circuit – malfunction	Wiring, poor connection, ECM/TCM
P0915	Gear shift position circuit – range/performance	Wiring, poor connection, ECM/TCM
P0916	Gear shift position circuit – low	Wiring, short to ground, ECM/TCM
P0917	Gear shift position circuit – high	Wiring, short to positive, ECM/TCM
P0918	Gear shift position circuit – intermittent malfunction	Wiring, poor connection, ECM/TCM
P0919	Gear shift position control – error	Wiring, poor connection, ECM/TCM
P0920	Gear shift forward actuator – circuit open	Wiring, gear shift forward actuator, ECM/TCM
P0921	Gear shift forward actuator – circuit range/ performance	Wiring, poor connection, gear shift forward actuator, ECM/TCM
P0922	Gear shift forward actuator – circuit low	Wiring, short to ground, gear shift forward actuator, ECM/TCM
P0923	Gear shift forward actuator – circuit high	Wiring, short to positive, gear shift forward actuator, ECM/TCM
P0924	Gear shift reverse actuator – circuit open	Wiring, gear shift reverse actuator, ECM/TCM
P0925	Gear shift reverse actuator – circuit range/ performance	Wiring, poor connection, gear shift reverse actuator, ECM/TCM
P0926	Gear shift reverse actuator – circuit low	Wiring, short to ground, gear shift reverse actuator, ECM/TCM
P0927	Gear shift reverse actuator – circuit high	Wiring, short to positive, gear shift reverse actuator, ECM/TCM
P0928	Gear shift lock solenoid – circuit open	Wiring, gear shift lock solenoid, ECM/TCM
P0929	Gear shift lock solenoid – circuit range/ performance	Wiring, gear shift lock solenoid, ECM/TCM
P0930	Gear shift lock solenoid – circuit low	Wiring, short to ground, gear shift lock solenoid, ECM/TCM
P0931	Gear shift lock solenoid – circuit high	Wiring, short to positive, gear shift lock solenoid, ECM/TCM
P0932	Hydraulic pressure sensor – circuit malfunction	Wiring, poor connection, hydraulic pressure sensor, ECM/TCM
P0933	Hydraulic pressure sensor – range/performance	Wiring, hydraulic pressure sensor, ECM/TCM
P0934	Hydraulic pressure sensor – circuit low input	Wiring, short to ground, hydraulic pressure sensor, ECM/TCM
P0935	Hydraulic pressure sensor – circuit high input	Wiring, short to positive, hydraulic pressure sensor, ECM/TCM

OBD-II P0 trouble code table

Trouble code	Fault location	Probable cause
P0936	Hydraulic pressure sensor – circuit intermittent	**Wiring, poor connection, hydraulic pressure sensor, ECM/TCM**
P0937	Hydraulic oil temperature sensor – circuit malfunction	**Wiring, poor connection, hydraulic oil temperature sensor, ECM/TCM**
P0938	Hydraulic oil temperature sensor – range/ performance	**Wiring, hydraulic oil temperature sensor, ECM/TCM**
P0939	Hydraulic oil temperature sensor – circuit low input	**Wiring, short to ground, hydraulic oil temperature sensor, ECM/TCM**
P0940	Hydraulic oil temperature sensor – circuit high input	**Wiring, short to positive, hydraulic oil temperature sensor, ECM/TCM**
P0941	Hydraulic oil temperature sensor – circuit intermittent	**Wiring, poor connection, hydraulic oil temperature sensor, ECM/TCM**
P0942	Hydraulic pressure unit	**Mechanical fault**
P0943	Hydraulic pressure unit – cycling period too short	**Mechanical fault**
P0944	Hydraulic pressure unit – loss of pressure	**Mechanical fault**
P0945	Hydraulic pump relay – circuit open	**Wiring, hydraulic pump relay, ECM/TCM**
P0946	Hydraulic pump relay – circuit range/performance	**Wiring, hydraulic pump relay, ECM/TCM**
P0947	Hydraulic pump relay – circuit low	**Wiring, short to ground, hydraulic pump relay, ECM/TCM**
P0948	Hydraulic pump relay – circuit high	**Wiring, short to positive, hydraulic pump relay, ECM/TCM**
P0949	ASM – adaptive learning not done	**ECM/TCM**
P0950	ASM control circuit	**Wiring, poor connection, ECM/TCM**
P0951	ASM control circuit – range/performance	**Wiring, poor connection, ECM/TCM**
P0952	ASM control circuit – low	**Wiring, poor connection, short to ground, ECM/TCM**
P0953	ASM control circuit – high	**Wiring, poor connection, short to positive, ECM/TCM**
P0954	ASM – intermittent circuit malfunction	**Wiring, poor connection, ECM/TCM**
P0955	ASM mode circuit – malfunction	**Wiring, poor connection, ECM/TCM**
P0956	ASM mode circuit – range/performance	**Wiring, poor connection, ECM/TCM**
P0957	ASM mode circuit – low	**Wiring, poor connection, short to ground, ECM/TCM**
P0958	ASM mode circuit – high	**Wiring, poor connection, short to positive, ECM/TCM**
P0959	ASM mode circuit – intermittent circuit malfunction	**Wiring, poor connection, ECM/TCM**
P0960	Pressure control (PC) solenoid A – control circuit open	**Wiring, poor connection, pressure control solenoid, ECM/TCM**
P0961	Pressure control (PC) solenoid A – control circuit range/performance	**Wiring, poor connection, pressure control solenoid, ECM/TCM**
P0962	Pressure control (PC) solenoid A – control circuit low	**Wiring, short to ground, pressure control solenoid, ECM/TCM**
P0963	Pressure control (PC) solenoid A – control circuit high	**Wiring, short to positive, pressure control solenoid, ECM/TCM**
P0964	Pressure control (PC) solenoid B – control circuit open	**Wiring, poor connection, pressure control solenoid, ECM/TCM**

Trouble code	Fault location	Probable cause
P0965	Pressure control (PC) solenoid B – control circuit range/performance	**Wiring, poor connection, pressure control solenoid, ECM/TCM**
P0966	Pressure control (PC) solenoid B – control circuit low	**Wiring, short to ground, pressure control solenoid, ECM/TCM**
P0967	Pressure control (PC) solenoid B – control circuit high	**Wiring, short to positive, pressure control solenoid, ECM/TCM**
P0968	Pressure control (PC) solenoid C – control circuit open	**Wiring, poor connection, pressure control solenoid, ECM/TCM**
P0969	Pressure control (PC) solenoid C – control circuit range/performance	**Wiring, poor connection, pressure control solenoid, ECM/TCM**
P0970	Pressure control (PC) solenoid C – control circuit low	**Wiring, short to ground, pressure control solenoid, ECM/TCM**
P0971	Pressure control (PC) solenoid C – control circuit high	**Wiring, short to positive, pressure control solenoid, ECM/TCM**
P0972	Shift solenoid (SS) A – control circuit range/ performance	**Wiring, poor connection, shift solenoid, ECM/TCM**
P0973	Shift solenoid (SS) A – control circuit low	**Wiring, short to ground, shift solenoid, ECM/TCM**
P0974	Shift solenoid (SS) A – control circuit high	**Wiring, short to positive, shift solenoid, ECM/TCM**
P0975	Shift solenoid (SS) B – control circuit range/ performance	**Wiring, poor connection, shift solenoid, ECM/TCM**
P0976	Shift solenoid (SS) B – control circuit low	**Wiring, short to ground, shift solenoid, ECM/TCM**
P0977	Shift solenoid (SS) B – control circuit high	**Wiring, short to positive, shift solenoid, ECM/TCM**
P0978	Shift solenoid (SS) C – control circuit range/ performance	**Wiring, poor connection, shift solenoid, ECM/TCM**
P0979	Shift solenoid (SS) C – control circuit low	**Wiring, short to ground, shift solenoid, ECM/TCM**
P0980	Shift solenoid (SS) C – control circuit high	**Wiring, short to positive, shift solenoid, ECM/TCM**
P0981	Shift solenoid (SS) D – control circuit range/ performance	**Wiring, poor connection, shift solenoid, ECM/TCM**
P0982	Shift solenoid (SS) D – control circuit low	**Wiring, short to ground, shift solenoid, ECM/TCM**
P0983	Shift solenoid (SS) D – control circuit high	**Wiring, short to positive, shift solenoid, ECM/TCM**
P0984	Shift solenoid (SS) E – control circuit range/ performance	**Wiring, poor connection, shift solenoid, ECM/TCM**
P0985	Shift solenoid (SS) E – control circuit low	**Wiring, short to ground, shift solenoid, ECM/TCM**
P0986	Shift solenoid (SS) E – control circuit high	**Wiring, short to positive, shift solenoid, ECM/TCM**

OBD-II P0 trouble code table

Trouble code	Fault location	Probable cause
P0987	Transmission fluid pressure (TFP) sensor E – circuit malfunction	**Wiring, poor connection, TFP sensor, ECM/TCM**
	Transmission fluid pressure (TFP) switch E – circuit malfunction	**Wiring, poor connection, TFP switch, ECM/TCM**
P0988	Transmission fluid pressure (TFP) sensor E – circuit range/performance	**Wiring, poor connection, TFP sensor, ECM/TCM**
	Transmission fluid pressure (TFP) switch E – circuit range/performance	**Wiring, poor connection, TFP switch, ECM/TCM**
P0989	Transmission fluid pressure (TFP) sensor E – circuit low	**Wiring, short to ground, TFP sensor, ECM/TCM**
	Transmission fluid pressure (TFP) switch E – circuit low	**Wiring, short to ground, TFP switch, ECM/TCM**
P0990	Transmission fluid pressure (TFP) sensor E – circuit high	**Wiring, short to positive, TFP sensor, ECM/TCM**
	Transmission fluid pressure (TFP) switch E – circuit high	**Wiring, short to positive, TFP switch, ECM/TCM**
P0991	Transmission fluid pressure (TFP) sensor E – circuit intermittent	**Wiring, poor connection, TFP sensor, ECM/TCM**
	Transmission fluid pressure (TFP) switch E – circuit intermittent	**Wiring, poor connection, TFP switch, ECM/TCM**
P0992	Transmission fluid pressure (TFP) sensor F – circuit malfunction	**Wiring, poor connection, TFP sensor, ECM/TCM**
	Transmission fluid pressure (TFP) switch F – circuit malfunction	**Wiring, poor connection, TFP switch, ECM/TCM**
P0993	Transmission fluid pressure (TFP) sensor F – circuit range/performance	**Wiring, poor connection, TFP sensor, ECM/TCM**
	Transmission fluid pressure (TFP) switch F – circuit range/performance	**Wiring, poor connection, TFP switch, ECM/TCM**
P0994	Transmission fluid pressure (TFP) sensor F – circuit low	**Wiring, short to ground, TFP sensor, ECM/TCM**
	Transmission fluid pressure (TFP) switch F – circuit low	**Wiring, short to ground, TFP switch, ECM/TCM**
P0995	Transmission fluid pressure (TFP) sensor F – circuit high	**Wiring, short to positive, TFP sensor, ECM/TCM**
	Transmission fluid pressure (TFP) switch F – circuit high	**Wiring, short to positive, TFP switch, ECM/TCM**
P0996	Transmission fluid pressure (TFP) sensor F – circuit intermittent	**Wiring, poor connection, TFP sensor, ECM/TCM**
	Transmission fluid pressure (TFP) switch F – circuit intermittent	**Wiring, poor connection, TFP switch, ECM/TCM**
P0997	Shift solenoid (SS) F – control circuit range/ performance	**Wiring, poor connection, shift solenoid, ECM/TCM**
P0998	Shift solenoid (SS) F – control circuit low	**Wiring short to ground, shift solenoid, ECM/TCM**
P0999	Shift solenoid (SS) F – control circuit high	**Wiring short to positive, shift solenoid, ECM/TCM**

OBD-II P2 trouble code table

- The following list covers all P2 codes allocated at the time of publication.
- All OBD-II codes starting with P2 have standard meanings irrespective of vehicle make or model.

Trouble code	Fault location	Probable cause
P2A00	Heated oxygen sensor (HO2S) 1, bank 1 – range/performance	Intake/exhaust leak, fuel pressure, wiring, HO2S
	Oxygen sensor (O2S) 1, bank 1 – range/performance	Intake/exhaust leak, fuel pressure, wiring, O2S
P2A01	Heated oxygen sensor (HO2S) 2, bank 1 – range/performance	Intake/exhaust leak, fuel pressure, wiring, HO2S
	Oxygen sensor (O2S) 2, bank 1 – range/performance	Intake/exhaust leak, fuel pressure, wiring, O2S
P2A02	Heated oxygen sensor (HO2S) 3, bank 1 – range/performance	Intake/exhaust leak, fuel pressure, wiring, HO2S
	Oxygen sensor (O2S) 3, bank 1 – range/performance	Intake/exhaust leak, fuel pressure, wiring, O2S
P2A03	Heated oxygen sensor (HO2S) 1, bank 2 – range/performance	Intake/exhaust leak, fuel pressure, wiring, HO2S
	Oxygen sensor (O2S) 1, bank 2 – range/performance	Intake/exhaust leak, fuel pressure, wiring, O2S
P2A04	Heated oxygen sensor (HO2S) 2, bank 2 – range/performance	Intake/exhaust leak, fuel pressure, wiring, HO2S
	Oxygen sensor (O2S) 2, bank 2 – range/performance	Intake/exhaust leak, fuel pressure, wiring, O2S
P2A05	Heated oxygen sensor (HO2S) 3, bank 2 – range/performance	Intake/exhaust leak, fuel pressure, wiring, HO2S
	Oxygen sensor (O2S) 3, bank 2 – range/performance	Intake/exhaust leak, fuel pressure, wiring, O2S
P2000	Nitrogen oxides (NOx) trap, bank 1 – efficiency below threshold	NOx trap
P2001	Nitrogen oxides (NOx) trap, bank 2 – efficiency below threshold	NOx trap
P2002	Particulate trap, bank 1 – efficiency below threshold	Particulate trap
P2003	Particulate trap, bank 2 – efficiency below threshold	Particulate trap
P2004	Intake manifold air control actuator, bank 1 – actuator stuck open	Wiring, intake manifold air control actuator, mechanical fault
	Intake manifold air control solenoid, bank 1 – solenoid stuck open	Wiring, intake manifold air control solenoid, mechanical fault
P2005	Intake manifold air control actuator, bank 2 – actuator stuck open	Wiring, intake manifold air control actuator, mechanical fault
	Intake manifold air control solenoid, bank 2 – solenoid stuck open	Wiring, intake manifold air control solenoid, mechanical fault
P2006	Intake manifold air control actuator, bank 1 – actuator stuck closed	Wiring, intake manifold air control actuator, mechanical fault
	Intake manifold air control solenoid, bank 1 – solenoid stuck closed	Wiring, intake manifold air control solenoid, mechanical fault
P2007	Intake manifold air control actuator, bank 2 – actuator stuck closed	Wiring, intake manifold air control actuator, mechanical fault
	Intake manifold air control solenoid, bank 2 – solenoid stuck closed	Wiring, intake manifold air control solenoid, mechanical fault

OBD-II P2 trouble code table

Trouble code	Fault location	Probable cause
P2008	Intake manifold air control actuator, bank 1 – open circuit	**Wiring open circuit, intake manifold air control actuator**
	Intake manifold air control solenoid, bank 1 – open circuit	**Wiring open circuit, intake manifold air control solenoid**
P2009	Intake manifold air control actuator, bank 1 – circuit low	**Wiring short to ground, intake manifold air control actuator**
	Intake manifold air control solenoid, bank 1 – circuit low	**Wiring short to ground, intake manifold air control solenoid**
P2010	Intake manifold air control actuator, bank 1 – circuit high	**Wiring short to positive, intake manifold air control actuator**
	Intake manifold air control solenoid, bank 1 – circuit high	**Wiring short to positive, intake manifold air control solenoid**
P2011	Intake manifold air control actuator, bank 2 – open circuit	**Wiring open circuit, intake manifold air control actuator**
	Intake manifold air control solenoid, bank 2 – open circuit	**Wiring open circuit, intake manifold air control solenoid**
P2012	Intake manifold air control actuator, bank 2 – circuit low	**Wiring short to ground, intake manifold air control actuator**
	Intake manifold air control solenoid, bank 2 – circuit low	**Wiring short to ground, intake manifold air control solenoid**
P2013	Intake manifold air control actuator, bank 2 – circuit high	**Wiring short to positive, intake manifold air control actuator**
	Intake manifold air control solenoid, bank 2 – circuit high	**Wiring short to positive, intake manifold air control solenoid**
P2014	Intake manifold air control actuator position sensor/switch, bank 1 – circuit malfunction	**Wiring, intake manifold air control actuator position sensor/switch**
P2015	Intake manifold air control actuator position sensor/switch, bank 1 – range/performance	**Wiring, mechanical fault, intake manifold air control actuator position sensor/switch**
P2016	Intake manifold air control actuator position sensor/switch, bank 1 – circuit low	**Wiring short to ground, intake manifold air control actuator position sensor/switch**
P2017	Intake manifold air control actuator position sensor/switch, bank 1 – circuit high	**Wiring short to positive, intake manifold air control actuator position sensor/switch**
P2018	Intake manifold air control actuator position sensor/switch, bank 1 – circuit intermittent	**Wiring, poor connection, intake manifold air control actuator position sensor/switch**
P2019	Intake manifold air control actuator position sensor/switch, bank 2 – circuit malfunction	**Wiring, intake manifold air control actuator position sensor/switch**
P2020	Intake manifold air control actuator position sensor/switch, bank 2 – range/performance	**Wiring, mechanical fault, intake manifold air control actuator position sensor/switch**
P2021	Intake manifold air control actuator position sensor/switch, bank 2 – circuit low	**Wiring short to ground, intake manifold air control actuator position sensor/switch**
P2022	Intake manifold air control actuator position sensor/switch, bank 2 – circuit high	**Wiring short to positive, intake manifold air control actuator position sensor/switch**
P2023	Intake manifold air control actuator position sensor/switch, bank 2 – circuit intermittent	**Wiring, poor connection, intake manifold air control actuator position sensor/switch**
P2024	Evaporative emission (EVAP) fuel vapour temperature sensor – circuit malfunction	**Wiring, EVAP fuel vapour temperature sensor**

Trouble code	Fault location	Probable cause
P2025	Evaporative emission (EVAP) fuel vapour temperature sensor – range/performance	**Wiring, EVAP fuel vapour temperature sensor**
P2026	Evaporative emission (EVAP) fuel vapour temperature sensor – low voltage	**Wiring short to ground, EVAP fuel vapour temperature sensor**
P2027	Evaporative emission (EVAP) fuel vapour temperature sensor – high voltage	**Wiring short to positive, EVAP fuel vapour temperature sensor**
P2028	Evaporative emission (EVAP) fuel vapour temperature sensor – circuit intermittent	**Wiring, poor connection, EVAP fuel vapour temperature sensor**
P2029	Auxiliary heater (fuel fired) – system disabled	**Auxiliary heater system**
P2030	Auxiliary heater (fuel fired) – performance problem	**Auxiliary heater system**
P2031	Exhaust gas temperature (EGT) sensor 2, bank 1 – circuit malfunction	**Wiring, EGT sensor**
P2032	Exhaust gas temperature (EGT) sensor 2, bank 1 – circuit low	**Wiring short to ground, EGT sensor**
P2033	Exhaust gas temperature (EGT) sensor 2, bank 1 – circuit high	**Wiring short to positive, EGT sensor**
P2034	Exhaust gas temperature (EGT) sensor 2, bank 2 – circuit malfunction	**Wiring, EGT sensor**
P2035	Exhaust gas temperature (EGT) sensor 2, bank 2 – circuit low	**Wiring short to ground, EGT sensor**
P2036	Exhaust gas temperature (EGT) sensor 2, bank 2 – circuit high	**Wiring short to positive, EGT sensor**
P2037	Reductant injection air pressure sensor – circuit malfunction	**Wiring, reductant injection air pressure sensor**
P2038	Reductant injection air pressure sensor – range/performance	**Wiring, reductant injection air pressure sensor**
P2039	Reductant injection air pressure sensor – low input	**Wiring short to ground, reductant injection air pressure sensor**
P2040	Reductant injection air pressure sensor – high input	**Wiring short to positive, reductant injection air pressure sensor**
P2041	Reductant injection air pressure sensor – circuit intermittent	**Wiring, reductant injection air pressure sensor**
P2042	Reductant temperature sensor – circuit malfunction	**Wiring, reductant temperature sensor**
P2043	Reductant temperature sensor – range/performance	**Wiring, reductant temperature sensor**
P2044	Reductant temperature sensor – low input	**Wiring, reductant temperature sensor**
P2045	Reductant temperature sensor – high input	**Wiring, reductant temperature sensor**
P2046	Reductant temperature sensor – circuit intermittent	**Wiring, reductant temperature sensor**
P2047	Reductant injector 1, bank 1 – open circuit	**Wiring, reductant injector**
P2048	Reductant injector 1, bank 1 – circuit low	**Wiring short to ground, reductant injector**
P2049	Reductant injector 1, bank 1 – circuit high	**Wiring short to positive, reductant injector**
P2050	Reductant injector 1, bank 2 – open circuit	**Wiring, reductant injector**
P2051	Reductant injector 1, bank 2 – circuit low	**Wiring short to ground, reductant injector**
P2052	Reductant injector 1, bank 2 – circuit high	**Wiring short to positive, reductant injector**
P2053	Reductant injector 2, bank 1 – open circuit	**Wiring, reductant injector**
P2054	Reductant injector 2, bank 1 – circuit low	**Wiring short to ground, reductant injector**
P2055	Reductant injector 2, bank 1 – circuit high	**Wiring short to positive, reductant injector**

Trouble code	Fault location	Probable cause
P2056	Reductant injector 2, bank 2 – open circuit	Wiring, reductant injector
P2057	Reductant injector 2, bank 2 – circuit low	Wiring short to ground, reductant injector
P2058	Reductant injector 2, bank 2 – circuit high	Wiring short to positive, reductant injector
P2059	Reductant injection air pump – open circuit	Wiring, reductant injection air pump
P2060	Reductant injection air pump – circuit low	Wiring short to ground, reductant injection air pump
P2061	Reductant injection air pump – circuit high	Wiring short to positive, reductant injection air pump
P2062	Reductant supply control – open circuit	Wiring
P2063	Reductant supply control – circuit low	Wiring
P2064	Reductant supply control – circuit high	Wiring
P2065	Fuel gauge tank sensor B – circuit malfunction	Wiring, fuel gauge tank sensor
P2066	Fuel gauge tank sensor B – performance problem	Wiring, fuel gauge tank sensor
P2067	Fuel gauge tank sensor B – circuit low	Wiring short to ground, fuel gauge tank sensor
P2068	Fuel gauge tank sensor B – circuit high	Wiring short to positive, fuel gauge tank sensor
P2069	Fuel gauge tank sensor B – circuit intermittent	Wiring, poor connection, fuel gauge tank sensor
P2070	Intake manifold air control actuator – actuator stuck open	Wiring, intake manifold air control actuator, mechanical fault
	Intake manifold air control solenoid – solenoid stuck open	Wiring, intake manifold air control solenoid, mechanical fault
P2071	Intake manifold air control actuator – actuator stuck closed	Wiring, intake manifold air control actuator, mechanical fault
	Intake manifold air control solenoid – solenoid stuck closed	Wiring, intake manifold air control solenoid, mechanical fault
P2075	Intake manifold air control actuator position sensor/switch – circuit malfunction	Wiring, intake manifold air control actuator position sensor/switch
P2076	Intake manifold air control actuator position sensor/switch – range/performance	Wiring, intake manifold air control actuator position sensor/switch
P2077	Intake manifold air control actuator position sensor/switch – circuit low	Wiring short to ground, intake manifold air control actuator position sensor/switch
P2078	Intake manifold air control actuator position sensor/switch – circuit high	Wiring short to positive, intake manifold air control actuator position sensor/switch
P2079	Intake manifold air control actuator position sensor/switch – circuit intermittent	Wiring, poor connection, intake manifold air control actuator position sensor/switch
P2080	Exhaust gas temperature (EGT) sensor 1, bank 1 – range/performance	Wiring, EGT sensor
P2081	Exhaust gas temperature (EGT) sensor 1, bank 1 – circuit intermittent	Wiring, poor connection, EGT sensor
P2082	Exhaust gas temperature (EGT) sensor 1, bank 2 – range/performance	Wiring, EGT sensor
P2083	Exhaust gas temperature (EGT) sensor 1, bank 2 – circuit intermittent	Wiring, poor connection, EGT sensor
P2084	Exhaust gas temperature (EGT) sensor 2, bank 1 – range/performance	Wiring, EGT sensor
P2085	Exhaust gas temperature (EGT) sensor 2, bank 1 – circuit intermittent	Wiring, poor connection, EGT sensor

Trouble code	Fault location	Probable cause
P2086	Exhaust gas temperature (EGT) sensor 2, bank 2 – range/performance	Wiring, EGT sensor
P2087	Exhaust gas temperature (EGT) sensor 2, bank 2 – circuit intermittent	Wiring, poor connection, EGT sensor
P2088	Camshaft position (CMP) actuator A, bank 1 – circuit low	Wiring short to ground, CMP actuator
P2089	Camshaft position (CMP) actuator A, bank 1 – circuit high	Wiring short to positive, CMP actuator
P2090	Camshaft position (CMP) actuator B, bank 1 – circuit low	Wiring short to ground, CMP actuator
P2091	Camshaft position (CMP) actuator B, bank 1 – circuit high	Wiring short to positive, CMP actuator
P2092	Camshaft position (CMP) actuator A, bank 2 – circuit low	Wiring short to ground, CMP actuator
P2093	Camshaft position (CMP) actuator A, bank 2 – circuit high	Wiring short to positive, CMP actuator
P2094	Camshaft position (CMP) actuator B, bank 2 – circuit low	Wiring short to ground, CMP actuator
P2095	Camshaft position (CMP) actuator B, bank 2 – circuit high	Wiring short to positive, CMP actuator
P2096	Post catalytic converter fuel trim (FT), bank 1 – too lean	Catalytic converter, exhaust leak
P2097	Post catalytic converter fuel trim (FT), bank 1 – too rich	Catalytic converter
P2098	Post catalytic converter fuel trim (FT), bank 2 – too lean	Catalytic converter, exhaust leak
P2099	Post catalytic converter fuel trim (FT), bank 2 – too rich	Catalytic converter
P2100	Throttle actuator control (TAC) motor – open circuit	Wiring, TAC motor
P2101	Throttle actuator control (TAC) motor – range/ performance	Wiring, TAC motor
P2102	Throttle actuator control (TAC) motor – circuit low	Wiring short to ground, TAC motor
P2103	Throttle actuator control (TAC) motor – circuit high	Wiring short to positive, TAC motor
P2104	Throttle actuator control (TAC) system – forced idle mode	Wiring, TAC motor, APP sensor, ECM
P2105	Throttle actuator control (TAC) system – forced engine shut down mode	Wiring, TAC motor, APP sensor, ECM
P2106	Throttle actuator control (TAC) system – forced limited power mode	Wiring, TAC motor, APP sensor, ECM
P2107	Throttle actuator control (TAC) module – processor fault	TAC control module
P2108	Throttle actuator control (TAC) module – performance problem	TAC control module
P2109	Accelerator pedal position (APP) sensor A – minimum stop performance	APP sensor
	Throttle position (TP) sensor A – minimum stop performance	TP sensor, throttle valve tight/sticking

OBD-II P2 trouble code table

Trouble code	Fault location	Probable cause
P2110	Throttle actuator control (TAC) system – forced limited rpm mode	**Wiring, TAC motor, APP sensor, ECM**
P2111	Throttle actuator control (TAC) system – actuator stuck open	**Throttle body, throttle valve tight/sticking**
P2112	Throttle actuator control (TAC) system – actuator stuck closed	**Throttle body, throttle valve tight/sticking**
P2113	Accelerator pedal position (APP) sensor B – minimum stop performance	**APP sensor**
	Throttle position (TP) sensor B – minimum stop performance	**TP sensor, throttle valve tight/sticking**
P2114	Accelerator pedal position (APP) sensor C – minimum stop performance	**APP sensor**
	Throttle position (TP) sensor C – minimum stop performance	**TP sensor, throttle valve tight/sticking**
P2115	Accelerator pedal position (APP) sensor D – minimum stop performance	**APP sensor**
	Throttle position (TP) sensor D – minimum stop performance	**TP sensor, throttle valve tight/sticking**
P2116	Accelerator pedal position (APP) sensor E – minimum stop performance	**Wiring, TP sensor**
	Throttle position (TP) sensor E – minimum stop performance	**Wiring, TP sensor**
P2117	Accelerator pedal position (APP) sensor F – minimum stop performance	**Wiring, APP sensor**
	Throttle position (TP) sensor F – minimum stop performance	**Wiring, TP sensor**
P2118	Throttle actuator control (TAC), throttle motor current – range/performance	**Wiring, throttle motor**
P2119	Throttle actuator control (TAC), throttle valve – range/performance	**Throttle valve tight/sticking, throttle motor**
P2120	Accelerator pedal position (APP) sensor/switch D – circuit malfunction	**Wiring, APP sensor/switch**
	Throttle position (TP) sensor/switch D – circuit malfunction	**Wiring, TP sensor/switch**
P2121	Accelerator pedal position (APP) sensor/switch D – range/performance	**Wiring, APP sensor/switch**
	Throttle position (TP) sensor/switch D – range/performance	**Wiring, TP sensor/switch**
P2122	Accelerator pedal position (APP) sensor/switch D – low input	**Wiring short to ground, APP sensor/switch**
	Throttle position (TP) sensor/switch D – low input	**Wiring short to ground, TP sensor/switch**
P2123	Accelerator pedal position (APP) sensor/switch D – high input	**Wiring short to positive, APP sensor/switch**
	Throttle position (TP) sensor/switch D – high input	**Wiring short to positive, TP sensor/switch**
P2124	Accelerator pedal position (APP) sensor/switch D – circuit intermittent	**Wiring, poor connection, APP sensor/switch**
	Throttle position (TP) sensor/switch D – circuit intermittent	**Wiring, poor connection, TP sensor/switch**

Trouble code	Fault location	Probable cause
P2125	Accelerator pedal position (APP) sensor/switch E – circuit malfunction	**Wiring, APP sensor/switch**
	Throttle position (TP) sensor/switch E – circuit malfunction	**Wiring, TP sensor/switch**
P2126	Accelerator pedal position (APP) sensor/switch E – range/performance	**Wiring, APP sensor/switch**
	Throttle position (TP) sensor/switch E – range/ performance	**Wiring, TP sensor/switch**
P2127	Accelerator pedal position (APP) sensor/switch E – low input	**Wiring short to ground, APP sensor/switch**
	Throttle position (TP) sensor/switch E – low input	**Wiring short to ground, TP sensor/switch**
P2128	Accelerator pedal position (APP) sensor/switch E – high input	**Wiring short to positive, APP sensor/switch**
	Throttle position (TP) sensor/switch E – high input	**Wiring short to positive, TP sensor/switch**
P2129	Accelerator pedal position (APP) sensor/switch E – circuit intermittent	**Wiring, poor connection, APP sensor/switch**
	Throttle position (TP) sensor/switch E – circuit intermittent	**Wiring, poor connection, TP sensor/switch**
P2130	Accelerator pedal position (APP) sensor/switch F – circuit malfunction	**Wiring, APP sensor/switch**
	Throttle position (TP) sensor/switch F – circuit malfunction	**Wiring, TP sensor/switch**
P2131	Accelerator pedal position (APP) sensor/switch F – circuit range/performance	**Wiring, APP sensor/switch**
	Throttle position (TP) sensor/switch F – circuit range/performance	**Wiring, TP sensor/switch**
P2132	Accelerator pedal position (APP) sensor/switch F – low input	**Wiring short to ground, APP sensor/switch**
	Throttle position (TP) sensor/switch F – low input	**Wiring short to ground, TP sensor/switch**
P2133	Accelerator pedal position (APP) sensor/switch F – high input	**Wiring short to positive, APP sensor/switch**
P2133	Throttle position (TP) sensor/switch F – high input	**Wiring short to positive, TP sensor/switch**
P2134	Accelerator pedal position (APP) sensor/switch F – circuit intermittent	**Wiring, poor connection, APP sensor/switch**
	Throttle position (TP) sensor/switch F – circuit intermittent	**Wiring, poor connection, TP sensor/switch**
P2135	Accelerator pedal position (APP) sensor/switch A/B – voltage correlation	**Wiring, APP sensor/switch**
	Throttle position (TP) sensor/switch A/B – voltage correlation	**Wiring, TP sensor/switch**
P2136	Accelerator pedal position (APP) sensor/switch A/C – voltage correlation	**Wiring, APP sensor/switch**
	Throttle position (TP) sensor/switch A/C – voltage correlation	**Wiring, TP sensor/switch**
P2137	Accelerator pedal position (APP) sensor/switch B/C – voltage correlation	**Wiring, APP sensor/switch**
	Throttle position (TP) sensor/switch D/E – voltage correlation	**Wiring, TP sensor/switch**

OBD-II P2 trouble code table

Trouble code	Fault location	Probable cause
P2139	Accelerator pedal position (APP) sensor/switch D/F – voltage correlation	**Wiring, APP sensor/switch**
	Throttle position (TP) sensor/switch D/F – voltage correlation	**Wiring, TP sensor/switch**
P2140	Accelerator pedal position (APP) sensor/switch E/F – voltage correlation	**Wiring, APP sensor/switch**
	Throttle position (TP) sensor/switch E/F – voltage correlation	**Wiring, TP sensor/switch**
P2141	Exhaust gas recirculation (EGR) throttle control valve – circuit low	**Wiring short to ground, EGR throttle control valve**
P2142	Exhaust gas recirculation (EGR) throttle control valve – circuit high	**Wiring short to positive, EGR throttle control valve**
P2143	Exhaust gas recirculation (EGR) vent control – open circuit	**Wiring, EGR vent control**
P2144	Exhaust gas recirculation (EGR) vent control – circuit low	**Wiring short to ground, EGR vent control**
P2145	Exhaust gas recirculation (EGR) vent control – circuit high	**Wiring short to positive, EGR vent control**
P2146	Injector – group A, supply voltage – open circuit	**Wiring, engine control (EC) relay, injector**
P2147	Injector – group A, supply voltage – circuit low	**Wiring short to ground, engine control (EC) relay, injector**
P2148	Injector – group A, supply voltage – circuit high	**Wiring short to positive, engine control (EC) relay, injector**
P2149	Injector – group B, supply voltage – open circuit	**Wiring, engine control (EC) relay, injector**
P2150	Injector – group B, supply voltage – circuit low	**Wiring short to ground, engine control (EC) relay, injector**
P2151	Injector – group B, supply voltage – circuit high	**Wiring short to positive, engine control (EC) relay, injector**
P2152	Injector – group C, supply voltage – open circuit	**Wiring, engine control (EC) relay, injector**
P2153	Injector – group C, supply voltage – circuit low	**Wiring short to ground, engine control (EC) relay, injector**
P2154	Injector – group C, supply voltage – circuit high	**Wiring short to positive, engine control (EC) relay, injector**
P2155	Injector – group D, supply voltage – open circuit	**Wiring, engine control (EC) relay, injector**
P2156	Injector – group D, supply voltage – circuit low	**Wiring short to ground, engine control (EC) relay, injector**
P2157	Injector – group D, supply voltage – circuit high	**Wiring short to positive, engine control (EC) relay, injector**
P2158	Vehicle speed sensor (VSS) B – circuit malfunction	**Wiring, VSS**
P2159	Vehicle speed sensor (VSS) B – range/performance	**Wiring, VSS**
P2160	Vehicle speed sensor (VSS) B – circuit low	**Wiring short to ground, VSS**
P2161	Vehicle speed sensor (VSS) B – circuit intermittent/erratic	**Wiring, poor connection, VSS**
P2162	Vehicle speed sensor (VSS) A/B – correlation	**Wiring, VSS, incorrect tire size**
P2163	Accelerator pedal position (APP) sensor A – maximum stop performance	**Wiring, APP sensor**
	Throttle position (TP) sensor A – maximum stop performance	**Wiring, TP sensor**

Trouble code	Fault location	Probable cause
P2164	Accelerator pedal position (APP) sensor B – maximum stop performance	Wiring, APP sensor
	Throttle position (TP) sensor B – maximum stop performance	Wiring, TP sensor
P2165	Accelerator pedal position (APP) sensor C – maximum stop performance	Wiring, APP sensor
	Throttle position (TP) sensor C – maximum stop performance	Wiring, TP sensor
P2166	Accelerator pedal position (APP) sensor D – maximum stop performance	Wiring, APP sensor
	Throttle position (TP) sensor D – maximum stop performance	Wiring, TP sensor
P2167	Accelerator pedal position (APP) sensor E – maximum stop performance	Wiring, APP sensor
	Throttle position (TP) sensor E – maximum stop performance	Wiring, TP sensor
P2168	Accelerator pedal position (APP) sensor F – maximum stop performance	Wiring, APP sensor
	Throttle position (TP) sensor F – maximum stop performance	Wiring, TP sensor
P2169	Exhaust gas pressure regulator vent solenoid – circuit open	Wiring, exhaust gas pressure regulator vent solenoid
P2170	Exhaust gas pressure regulator vent solenoid – circuit low	Wiring short to ground, exhaust gas pressure regulator vent solenoid
P2171	Exhaust gas pressure regulator vent solenoid – circuit high	Wiring short to positive, exhaust gas pressure regulator vent solenoid
P2172	Throttle actuator control (TAC) system – sudden high airflow detected	Intake system, throttle body
P2173	Throttle actuator control (TAC) system – high airflow detected	Intake system, throttle body
P2174	Throttle actuator control (TAC) system – sudden low airflow detected	Intake system, throttle body
P2175	Throttle actuator control (TAC) system – low airflow detected	Intake system, throttle body
P2176	Throttle actuator control (TAC) system – idle position not learned	Basic setting not carried out
P2177	System too lean off idle, bank 1	Fuel pressure, injectors, intake leak
P2178	System too rich off idle, bank 1	Fuel pressure, injectors, air intake restricted
P2179	System too lean off idle, bank 2	Fuel pressure, injectors, intake leak
P2180	System too rich off idle, bank 2	Fuel pressure, injectors, air intake restricted
P2181	Cooling system performance	Radiator, coolant thermostat, engine coolant blower motor
P2182	Engine coolant temperature (ECT) sensor 2 – circuit malfunction	Wiring, ECT sensor
P2183	Engine coolant temperature (ECT) sensor 2 – range/performance	Wiring, ECT sensor
P2184	Engine coolant temperature (ECT) sensor 2 – circuit low	Wiring short to ground, ECT sensor

OBD-II P2 trouble code table

Trouble code	Fault location	Probable cause
P2185	Engine coolant temperature (ECT) sensor 2 – circuit high	**Wiring short to positive, ECT sensor**
P2186	Engine coolant temperature (ECT) sensor 2 – circuit intermittent/erratic	**Wiring, poor connection, ECT sensor**
P2187	System too lean at idle, bank 1	**Fuel pressure, injectors, intake leak**
P2188	System too rich at idle, bank 1	**Fuel pressure, injectors, air intake restricted**
P2189	System too lean at idle, bank 2	**Fuel pressure, injectors, intake leak**
P2190	System too rich at idle, bank 2	**Fuel pressure, injectors, air intake restricted**
P2191	System too lean at higher load, bank 1	**Fuel pressure, injectors, intake leak**
P2192	System too rich at higher load, bank 1	**Fuel pressure, injectors, air intake restricted**
P2193	System too lean at higher load, bank 2	**Fuel pressure, injectors, intake leak**
P2194	System too rich at higher load, bank 2	**Fuel pressure, injectors, air intake restricted**
P2195	Heated oxygen sensor (HO2S) 1, bank 1 – signal stuck lean	**HO2S, fuel pressure, injectors, intake leak**
	Oxygen sensor (O2S) 1, bank 1 – signal stuck lean	**O2S, fuel pressure, injectors, intake leak**
P2196	Heated oxygen sensor (HO2S) 1, bank 1 – signal stuck rich	**HO2S, fuel pressure, injectors, air intake restricted**
	Oxygen sensor (O2S) 1, bank 1 – signal stuck rich	**O2S, fuel pressure, injectors, air intake restricted**
P2197	Heated oxygen sensor (HO2S) 1, bank 2 – signal stuck lean	**HO2S, fuel pressure, injectors, intake leak**
	Oxygen sensor (O2S) 1, bank 2 – signal stuck lean	**O2S, fuel pressure, injectors, intake leak**
P2198	Heated oxygen sensor (HO2S) 1, bank 2 – signal stuck rich	**HO2S, fuel pressure, injectors, air intake restricted**
	Oxygen sensor (O2S) 1, bank 2 – signal stuck rich	**O2S, fuel pressure, injectors, air intake restricted**
P2199	Intake air temperature (IAT) sensor 1/2 – correlation	**Wiring, IAT sensor**
P2200	Nitrogen oxides (NOx) sensor, bank 1 – circuit malfunction	**Wiring, NOx sensor**
P2201	Nitrogen oxides (NOx) sensor, bank 1 – range/performance	**Wiring, NOx sensor**
P2202	Nitrogen oxides (NOx) sensor, bank 1 – low input	**Wiring short to ground, NOx sensor**
P2203	Nitrogen oxides (NOx) sensor, bank 1 – high input	**Wiring short to positive, NOx sensor**
P2204	Nitrogen oxides (NOx) sensor, bank 1 – intermittent input	**Wiring, poor connection, NOx sensor**
P2205	Nitrogen oxides (NOx) sensor, bank 1, heater control – open circuit	**Wiring, NOx sensor**
P2206	Nitrogen oxides (NOx) sensor, bank 1, heater control – circuit low	**Wiring short to ground, NOx sensor**
P2207	Nitrogen oxides (NOx) sensor, bank 1, heater control – circuit high	**Wiring short to positive, NOx sensor**
P2208	Nitrogen oxides (NOx) sensor, bank 1, heater sense circuit – malfunction	**Wiring, NOx sensor**
P2209	Nitrogen oxides (NOx) sensor, bank 1, heater sense circuit – range/performance	**Wiring, NOx sensor**

Trouble code	Fault location	Probable cause
P2210	Nitrogen oxides (NOx) sensor, bank 1, heater sense circuit – low input	**Wiring short to ground, NOx sensor**
P2211	Nitrogen oxides (NOx) sensor, bank 1, heater sense circuit – high input	**Wiring short to positive, NOx sensor**
P2212	Nitrogen oxides (NOx) sensor, bank 1, heater sense circuit – circuit intermittent	**Wiring, poor connection, NOx sensor**
P2213	Nitrogen oxides (NOx) sensor, bank 2 – circuit malfunction	**Wiring, NOx sensor**
P2214	Nitrogen oxides (NOx) sensor, bank 2 – range/performance	**Wiring, NOx sensor**
P2215	Nitrogen oxides (NOx) sensor, bank 2 – low input	**Wiring short to ground, NOx sensor**
P2216	Nitrogen oxides (NOx) sensor, bank 2 – high input	**Wiring short to positive, NOx sensor**
P2217	Nitrogen oxides (NOx) sensor, bank 2 – intermittent input	**Wiring, poor connection, NOx sensor**
P2218	Nitrogen oxides (NOx) sensor, bank 2, heater control – open circuit	**Wiring, NOx sensor**
P2219	Nitrogen oxides (NOx) sensor, bank 2, heater control – circuit low	**Wiring short to ground, NOx sensor**
P2220	Nitrogen oxides (NOx) sensor, bank 2, heater control – circuit high	**Wiring short to positive, NOx sensor**
P2221	Nitrogen oxides (NOx) sensor, bank 2, heater sense circuit – circuit malfunction	**Wiring, NOx sensor**
P2222	Nitrogen oxides (NOx) sensor, bank 2, heater sense circuit – range/performance	**Wiring, NOx sensor**
P2223	Nitrogen oxides (NOx) sensor, bank 2, heater sense circuit – circuit low	**Wiring short to ground, NOx sensor**
P2224	Nitrogen oxides (NOx) sensor, bank 2, heater sense circuit – circuit high	**Wiring short to positive, NOx sensor**
P2225	Nitrogen oxides (NOx) sensor, bank 2, heater sense circuit – circuit intermittent	**Wiring, poor connection, NOx sensor**
P2226	Barometric pressure (BARO) sensor – circuit malfunction	**Wiring, BARO sensor**
P2227	Barometric pressure (BARO) sensor – range/performance	**Wiring, BARO sensor**
P2228	Barometric pressure (BARO) sensor – circuit low	**Wiring short to ground, BARO sensor**
P2229	Barometric pressure (BARO) sensor – circuit high	**Wiring short to positive, BARO sensor**
P2230	Barometric pressure (BARO) sensor – circuit intermittent	**Wiring, poor connection, BARO sensor**
P2231	Heated oxygen sensor (HO2S) 1, bank 1 – signal circuit shorted to heater circuit	**Wiring, HO2S**
P2232	Heated oxygen sensor (HO2S) 2, bank 1 – signal circuit shorted to heater circuit	**Wiring, HO2S**
P2233	Heated oxygen sensor (HO2S) 3, bank 1 – signal circuit shorted to heater circuit	**Wiring, HO2S**
P2234	Heated oxygen sensor (HO2S) 1, bank 2 – signal circuit shorted to heater circuit	**Wiring, HO2S**
P2235	Heated oxygen sensor (HO2S) 2, bank 2 – signal circuit shorted to heater circuit	**Wiring, HO2S**

OBD-II P2 trouble code table

Trouble code	Fault location	Probable cause
P2236	Heated oxygen sensor (HO2S) 3, bank 2 – signal circuit shorted to heater circuit	**Wiring, HO2S**
P2237	Heated oxygen sensor (HO2S) 1, bank 1, positive current control – open circuit	**Wiring, HO2S**
	Oxygen sensor (O2S) 1, bank 1, positive current control – open circuit	**Wiring, O2S**
P2238	Heated oxygen sensor (HO2S) 1, bank 1, positive current control – circuit low	**Wiring short to ground, HO2S**
	Oxygen sensor (O2S) 1, bank 1, positive current control – circuit low	**Wiring short to ground, O2S**
P2239	Heated oxygen sensor (HO2S) 1, bank 1, positive current control – circuit high	**Wiring short to positive, HO2S**
	Oxygen sensor (O2S) 1, bank 1, positive current control – circuit high	**Wiring short to positive, O2S**
P2240	Heated oxygen sensor (HO2S) 1, bank 2, positive current control – open circuit	**Wiring, HO2S**
	Oxygen sensor (O2S) 1, bank 2, positive current control – open circuit	**Wiring, O2S**
P2241	Heated oxygen sensor (HO2S) 1, bank 2, positive current control – circuit low	**Wiring short to ground, HO2S**
	Oxygen sensor (O2S) 1, bank 2, positive current control – circuit low	**Wiring short to ground, O2S**
P2242	Heated oxygen sensor (HO2S) 1, bank 2, positive current control – circuit high	**Wiring short to positive, HO2S**
	Oxygen sensor (O2S) 1, bank 2, positive current control – circuit high	**Wiring short to positive, O2S**
P2243	Heated oxygen sensor (HO2S) 1, bank 1, reference voltage – open circuit	**Wiring, HO2S**
	Oxygen sensor (O2S) 1, bank 1, reference voltage – open circuit	**Wiring, O2S**
P2244	Heated oxygen sensor (HO2S) 1, bank 1, reference voltage – performance problem	**Wiring, HO2S**
	Oxygen sensor (O2S) 1, bank 1, reference voltage – performance problem	**Wiring, O2S**
P2245	Heated oxygen sensor (HO2S) 1, bank 1, reference voltage – circuit low	**Wiring short to ground, HO2S**
	Oxygen sensor (O2S) 1, bank 1, reference voltage – circuit low	**Wiring short to ground, O2S**
P2246	Heated oxygen sensor (HO2S) 1, bank 1, reference voltage – circuit high	**Wiring short to positive, HO2S**
	Oxygen sensor (O2S) 1, bank 1, reference voltage – circuit high	**Wiring short to positive, O2S**
P2247	Heated oxygen sensor (HO2S) 1, bank 2, reference voltage – open circuit	**Wiring, HO2S**
	Oxygen sensor (O2S) 1, bank 2, reference voltage – open circuit	**Wiring, O2S**

Trouble code	Fault location	Probable cause
P2248	Heated oxygen sensor (HO2S) 1, bank 2, reference voltage – performance problem	Wiring, HO2S
	Oxygen sensor (O2S) 1, bank 2, reference voltage – performance problem	Wiring, O2S
P2249	Heated oxygen sensor (HO2S) 1, bank 2, reference voltage – circuit low	Wiring short to ground, HO2S
	Oxygen sensor (O2S) 1, bank 2, reference voltage – circuit low	Wiring short to ground, O2S
P2250	Heated oxygen sensor (HO2S) 1, bank 2, reference voltage – circuit high	Wiring short to positive, HO2S
	Oxygen sensor (O2S) 1, bank 2, reference voltage – circuit high	Wiring short to positive, O2S
P2251	Heated oxygen sensor (HO2S) 1, bank 1, negative current control – open circuit	Wiring, HO2S
	Oxygen sensor (O2S) 1, bank 1, negative current control – open circuit	Wiring, O2S
P2252	Heated oxygen sensor (HO2S) 1, bank 1, negative current control – circuit low	Wiring short to ground, HO2S
	Oxygen sensor (O2S) 1, bank 1, negative current control – circuit low	Wiring short to ground, O2S
P2253	Heated oxygen sensor (HO2S) 1, bank 1, negative current control – circuit high	Wiring short to positive, HO2S
	Oxygen sensor (O2S) 1, bank 1, negative current control – circuit high	Wiring short to positive, O2S
P2254	Heated oxygen sensor (HO2S) 1, bank 2, negative current control – open circuit	Wiring, HO2S
	Oxygen sensor (O2S) 1, bank 2, negative current control – open circuit	Wiring, O2S
P2255	Heated oxygen sensor (HO2S) 1, bank 2, negative current control – circuit low	Wiring short to ground, HO2S
	Oxygen sensor (O2S) 1, bank 2, negative current control – circuit low	Wiring short to ground, O2S
P2256	Heated oxygen sensor (HO2S) 1, bank 2, negative current control – circuit high	Wiring short to positive, HO2S
	Oxygen sensor (O2S) 1, bank 2, negative current control – circuit high	Wiring short to positive, O2S
P2257	Secondary air injection (AIR) system, control A – circuit low	Wiring short to ground, AIR pump relay, AIR pump, AIR solenoid
P2258	Secondary air injection (AIR) system, control A – circuit high	Wiring short to positive, AIR pump relay, AIR pump, AIR solenoid
P2259	Secondary air injection (AIR) system, control B – circuit low	Wiring short to ground, AIR pump relay, AIR pump, AIR solenoid
P2260	Secondary air injection (AIR) system, control B – circuit high	Wiring short to positive, AIR pump relay, AIR pump, AIR solenoid
P2261	Turbocharger (TC) bypass valve/supercharger (SC) bypass valve	Mechanical fault
P2262	Turbocharger (TC) boost pressure not detected	Mechanical fault
P2263	Turbocharger (TC) boost pressure/supercharger (SC) boost pressure – performance problem	Mechanical fault

OBD-II P2 trouble code table

Trouble code	Fault location	Probable cause
P2264	Fuel/water separator sensor – circuit malfunction	**Wiring, fuel/water separator sensor**
P2265	Fuel/water separator sensor – range/performance	**Wiring, fuel/water separator sensor**
P2266	Fuel/water separator sensor – circuit low	**Wiring short to ground, fuel/water separator sensor**
P2267	Fuel/water separator sensor – circuit high	**Wiring short to positive, fuel/water separator sensor**
P2268	Fuel/water separator sensor – circuit intermittent	**Wiring, poor connection, fuel/water separator sensor**
P2269	Water in fuel	**Water in fuel**
P2270	Heated oxygen sensor (HO2S) 2, bank 1 – signal stuck lean	**Wiring, HO2S, fuel pressure, injectors, intake leak**
	Oxygen sensor (O2S) 2, bank 1 – signal stuck lean	**Wiring, O2S, fuel pressure, injectors, intake leak**
P2271	Heated oxygen sensor (HO2S) 2, bank 1 – signal stuck rich	**Wiring, HO2S, fuel pressure, injectors, air intake restricted**
	Oxygen sensor (O2S) 2, bank 1 – signal stuck rich	**Wiring, O2S, fuel pressure, injectors, air intake restricted**
P2272	Heated oxygen sensor (HO2S) 2, bank 2 – signal stuck lean	**Wiring, HO2S, fuel pressure, injectors, intake leak**
	Oxygen sensor (O2S) 2, bank 2 – signal stuck lean	**Wiring, O2S, fuel pressure, injectors, intake leak**
P2273	Heated oxygen sensor (HO2S) 2, bank 2 – signal stuck rich	**Wiring, HO2S, fuel pressure, injectors, air intake restricted**
	Oxygen sensor (O2S) 2, bank 2 – signal stuck rich	**Wiring, O2S, fuel pressure, injectors, air intake restricted**
P2274	Heated oxygen sensor (HO2S) 3, bank 1 – signal stuck lean	**Wiring, HO2S, fuel pressure, injectors, intake leak**
	Oxygen sensor (O2S) 3, bank 1 – signal stuck lean	**Wiring, O2S, fuel pressure, injectors, intake leak**
P2275	Heated oxygen sensor (HO2S) 3, bank 1 – signal stuck rich	**Wiring, HO2S, fuel pressure, injectors, air intake restricted**
	Oxygen sensor (O2S) 3, bank 1 – signal stuck rich	**Wiring, O2S, fuel pressure, injectors, air intake restricted**
P2276	Heated oxygen sensor (HO2S) 3, bank 2 – signal stuck lean	**Wiring, HO2S, fuel pressure, injectors, intake leak**
	Oxygen sensor (O2S) 3, bank 2 – signal stuck lean	**Wiring, O2S, fuel pressure, injectors, intake leak**
P2277	Heated oxygen sensor (HO2S) 3, bank 2 – signal stuck rich	**Wiring, HO2S, fuel pressure, injectors, air intake restricted**
	Oxygen sensor (O2S) 3, bank 2 – signal stuck rich	**Wiring, O2S, fuel pressure, injectors, air intake restricted**
P2278	Heated oxygen sensor (HO2S) 3, bank 1/heated oxygen sensor (HO2S) 3, bank 2 – signals transposed	**Wiring**
	Oxygen sensor (O2S) 3, bank 1/oxygen sensor (O2S) 3, bank 2 – signals transposed	**Wiring**
P2279	Intake air leak	**Mechanical fault**
P2280	Air leak/blockage between air filter and mass air flow (MAF) sensor	**Mechanical fault**
P2281	Air leak between MAF sensor and throttle body	**Mechanical fault**
P2282	Air leak between throttle body and intake valves	**Mechanical fault**

Trouble code	Fault location	Probable cause
P2283	Injector control pressure sensor – circuit malfunction	Wiring, injector control pressure sensor
P2284	Injector control pressure sensor – range/ performance	Wiring, injector control pressure sensor
P2285	Injector control pressure sensor – circuit low	Wiring short to ground, injector control pressure sensor
P2286	Injector control pressure sensor – circuit high	Wiring short to positive, injector control pressure sensor
P2287	Injector control pressure sensor – circuit intermittent	Wiring, poor connection, injector control pressure sensor
P2288	Injector control pressure – pressure too high	Fuel pressure regulator, injector control pressure sensor
P2289	Injector control pressure, engine off – pressure too high	Fuel pressure regulator
P2290	Injector control pressure – pressure too low	Fuel pressure regulator, injector control pressure sensor
P2291	Injector control pressure, engine cranking – pressure too low	Fuel pressure regulator, injector control pressure sensor
P2292	Injector control pressure – erratic	Fuel pressure regulator, injector control pressure sensor
P2293	Fuel pressure regulator 2 – performance problem	Wiring, fuel pressure regulator
P2294	Fuel pressure regulator 2 – circuit malfunction	Wiring, fuel pressure regulator
P2295	Fuel pressure regulator 2 – circuit low	Wiring short to ground, fuel pressure regulator
P2296	Fuel pressure regulator 2 – circuit high	Wiring short to positive, fuel pressure regulator
P2297	Heated oxygen sensor (HO2S) 1, bank 1 – signal out of range during deceleration	HO2S, intake leak, exhaust leak, injectors
	Oxygen sensor (O2S) 1, bank 1 – signal out of range during deceleration	O2S, intake leak, exhaust leak, injectors
P2298	Heated oxygen sensor (HO2S) 1, bank 2 – signal out of range during deceleration	HO2S, intake leak, exhaust leak, injectors
	Oxygen sensor (O2S) 1, bank 2 – signal out of range during deceleration	O2S, intake leak, exhaust leak, injectors
P2299	Brake pedal position (BPP) switch/accelerator pedal position (APP) sensor – signals incompatible	Wiring, BPP switch, APP sensor
P2300	Ignition coil A, primary circuit – circuit low	Wiring short to ground, ignition coil
P2301	Ignition coil A, primary circuit – circuit high	Wiring short to positive, ignition coil
P2302	Ignition coil A, secondary circuit – malfunction	Wiring, ignition coil
P2303	Ignition coil B, primary circuit – circuit low	Wiring short to ground, ignition coil
P2304	Ignition coil B, primary circuit – circuit high	Wiring short to positive, ignition coil
P2305	Ignition coil B, secondary circuit – malfunction	Wiring, ignition coil
P2306	Ignition coil C, primary circuit – circuit low	Wiring short to ground, ignition coil
P2307	Ignition coil C, primary circuit – circuit high	Wiring short to positive, ignition coil
P2308	Ignition coil C, secondary circuit – malfunction	Wiring, ignition coil
P2309	Ignition coil D, primary circuit – circuit low	Wiring short to ground, ignition coil
P2310	Ignition coil D, primary circuit – circuit high	Wiring short to positive, ignition coil
P2311	Ignition coil D, secondary circuit – malfunction	Wiring, ignition coil

OBD-II P2 trouble code table

Trouble code	Fault location	Probable cause
P2312	Ignition coil E, primary circuit – circuit low	**Wiring short to ground, ignition coil**
P2313	Ignition coil E, primary circuit – circuit high	**Wiring short to positive, ignition coil**
P2314	Ignition coil E, secondary circuit – malfunction	**Wiring, ignition coil**
P2315	Ignition coil F, primary circuit – circuit low	**Wiring short to ground, ignition coil**
P2316	Ignition coil F, primary circuit – circuit high	**Wiring short to positive, ignition coil**
P2317	Ignition coil F, secondary circuit – malfunction	**Wiring, ignition coil**
P2318	Ignition coil G, primary circuit – circuit low	**Wiring short to ground, ignition coil**
P2319	Ignition coil G, primary circuit – circuit high	**Wiring short to positive, ignition coil**
P2320	Ignition coil G, secondary circuit – malfunction	**Wiring, ignition coil**
P2321	Ignition coil H, primary circuit – circuit low	**Wiring short to ground, ignition coil**
P2322	Ignition coil H, primary circuit – circuit high	**Wiring short to positive, ignition coil**
P2323	Ignition coil H, secondary circuit – malfunction	**Wiring, ignition coil**
P2324	Ignition coil I, primary circuit – circuit low	**Wiring short to ground, ignition coil**
P2325	Ignition coil I, primary circuit – circuit high	**Wiring short to positive, ignition coil**
P2326	Ignition coil I, secondary circuit – malfunction	**Wiring, ignition coil**
P2327	Ignition coil J, primary circuit – circuit low	**Wiring short to ground, ignition coil**
P2328	Ignition coil J, primary circuit – circuit high	**Wiring short to positive, ignition coil**
P2329	Ignition coil J, secondary circuit – malfunction	**Wiring, ignition coil**
P2330	Ignition coil K, primary circuit – circuit low	**Wiring short to ground, ignition coil**
P2331	Ignition coil K, primary circuit – circuit high	**Wiring short to positive, ignition coil**
P2332	Ignition coil K, secondary circuit – malfunction	**Wiring, ignition coil**
P2333	Ignition coil L, primary circuit – circuit low	**Wiring short to ground, ignition coil**
P2334	Ignition coil L, primary circuit – circuit high	**Wiring short to positive, ignition coil**
P2335	Ignition coil L, secondary circuit – malfunction	**Wiring, ignition coil**
P2336	Cylinder 1 – above knock threshold	**Ignition timing, knock sensor (KS), fuel quality, mechanical fault**
P2337	Cylinder 2 – above knock threshold	**Ignition timing, knock sensor (KS), fuel quality, mechanical fault**
P2338	Cylinder 3 – above knock threshold	**Ignition timing, knock sensor (KS), fuel quality, mechanical fault**
P2339	Cylinder 4 – above knock threshold	**Ignition timing, knock sensor (KS), fuel quality, mechanical fault**
P2340	Cylinder 5 – above knock threshold	**Ignition timing, knock sensor (KS), fuel quality, mechanical fault**
P2341	Cylinder 6 – above knock threshold	**Ignition timing, knock sensor (KS), fuel quality, mechanical fault**
P2342	Cylinder 7 – above knock threshold	**Ignition timing, knock sensor (KS), fuel quality, mechanical fault**
P2343	Cylinder 8 – above knock threshold	**Ignition timing, knock sensor (KS), fuel quality, mechanical fault**
P2344	Cylinder 9 – above knock threshold	**Ignition timing, knock sensor (KS), fuel quality, mechanical fault**
P2345	Cylinder 10 – above knock threshold	**Ignition timing, knock sensor (KS), fuel quality, mechanical fault**

Trouble code	Fault location	Probable cause
P2346	Cylinder 11 – above knock threshold	**Ignition timing, knock sensor (KS), fuel quality, mechanical fault**
P2347	Cylinder 12 – above knock threshold	**Ignition timing, knock sensor (KS), fuel quality, mechanical fault**
P2400	Evaporative emission (EVAP) leak detection pump, control – open circuit	**Wiring, EVAP leak detection pump**
P2401	Evaporative emission (EVAP) leak detection pump, control – circuit low	**Wiring short to ground, EVAP leak detection pump**
P2402	Evaporative emission (EVAP) leak detection pump, control – circuit high	**Wiring short to positive, EVAP leak detection pump**
P2403	Evaporative emission (EVAP) leak detection pump, sense circuit – open circuit	**Wiring, EVAP leak detection pump**
P2404	Evaporative emission (EVAP) leak detection pump, sense circuit – range/performance	**Wiring, EVAP leak detection pump**
P2405	Evaporative emission (EVAP) leak detection pump, sense circuit – circuit low	**Wiring short to ground, EVAP leak detection pump**
P2406	Evaporative emission (EVAP) leak detection pump, sense circuit – circuit high	**Wiring short to positive, EVAP leak detection pump**
P2407	Evaporative emission (EVAP) leak detection pump, sense circuit – circuit intermittent/erratic	**Wiring, poor connection, EVAP leak detection pump**
P2408	Fuel filler cap warning sensor/switch – circuit malfunction	**Wiring, fuel filler cap warning sensor/switch**
P2409	Fuel filler cap warning sensor/switch – range/ performance	**Wiring, fuel filler cap warning sensor/switch**
P2410	Fuel filler cap warning sensor/switch – circuit low	**Wiring short to ground, fuel filler cap warning sensor/switch**
P2411	Fuel filler cap warning sensor/switch – circuit high	**Wiring short to positive, fuel filler cap warning sensor/switch**
P2412	Fuel filler cap warning sensor/switch – circuit intermittent/erratic	**Wiring, poor connection, fuel filler cap warning sensor/switch**
P2413	Exhaust gas recirculation (EGR) system – performance problem	**Hoses blocked/leaking, EGR solenoid, EGR valve**
P2414	Heated oxygen sensor (HO2S) 1, bank 1 – exhaust sample error	**Exhaust leak, HO2S**
	Oxygen sensor (O2S) 1, bank 1 – exhaust sample error	**Exhaust leak, O2S**
P2415	Heated oxygen sensor (HO2S) 1, bank 2 – exhaust sample error	**Exhaust leak, HO2S**
	Oxygen sensor (O2S) 1, bank 2 – exhaust sample error	**Exhaust leak, O2S**
P2416	Heated oxygen sensor (HO2S) 2, bank 1/heated oxygen sensor (HO2S) 3, bank 1 – signals transposed	**Wiring**
	Oxygen sensor (O2S) 2, bank 1/oxygen sensor (O2S) 3, bank 1 – signals transposed	**Wiring**
P2417	Heated oxygen sensor (HO2S) 2, bank 2/heated oxygen sensor (HO2S) 3, bank 2 – signals transposed	**Wiring**
	Oxygen sensor (O2S) 2, bank 2/oxygen sensor (O2S) 3, bank 2 – signals transposed	**Wiring**

OBD-II P2 trouble code table

Trouble code	Fault location	Probable cause
P2418	Evaporative emission (EVAP) switching valve – open circuit	**Wiring, EVAP switching valve**
P2419	Evaporative emission (EVAP) switching valve – circuit low	**Wiring short to ground, EVAP switching valve**
P2420	Evaporative emission (EVAP) switching valve – circuit high	**Wiring short to positive, EVAP switching valve**
P2421	Evaporative emission (EVAP) vent valve – valve stuck open	**EVAP vent valve**
P2422	Evaporative emission (EVAP) vent valve – valve stuck closed	**EVAP vent valve**
P2423	Hydrocarbon (HC) catalytic converter, bank 1 – efficiency below threshold	**HC catalytic converter**
P2424	Hydrocarbon (HC) catalytic converter, bank 2 – efficiency below threshold	**HC catalytic converter**
P2425	Exhaust gas recirculation (EGR) cooling valve – open circuit	**Wiring, EGR cooling valve**
P2426	Exhaust gas recirculation (EGR) cooling valve – circuit low	**Wiring short to ground, EGR cooling valve**
P2427	Exhaust gas recirculation (EGR) cooling valve – circuit high	**Wiring short to positive, EGR cooling valve**
P2428	Exhaust gas temperature (EGT), bank 1 – temperature too high	–
P2429	Exhaust gas temperature (EGT), bank 2 – temperature too high	–
P2430	Secondary air injection (AIR) system, air flow/ pressure sensor, bank 1 – circuit malfunction	**Wiring, air flow/pressure sensor**
P2431	Secondary air injection (AIR) system, air flow/ pressure sensor, bank 1 – range/performance	**Wiring, air flow/pressure sensor**
P2432	Secondary air injection (AIR) system, air flow/ pressure sensor, bank 1 – circuit low	**Wiring short to ground, air flow/pressure sensor**
P2433	Secondary air injection (AIR) system, air flow/ pressure sensor, bank 1 – circuit high	**Wiring short to positive, air flow/pressure sensor**
P2434	Secondary air injection (AIR) system, air flow/pressure sensor, bank 1 – circuit intermittent/erratic	**Wiring, poor connection, air flow/pressure sensor**
P2435	Secondary air injection (AIR) system, air flow/ pressure sensor, bank 2 – circuit malfunction	**Wiring, air flow/pressure sensor**
P2436	Secondary air injection (AIR) system, air flow/ pressure sensor, bank 2 – range/performance	**Wiring, air flow/pressure sensor**
P2437	Secondary air injection (AIR) system, air flow/ pressure sensor, bank 2 – circuit low	**Wiring short to ground, air flow/pressure sensor**
P2438	Secondary air injection (AIR) system, air flow/ pressure sensor, bank 2 – circuit high	**Wiring short to positive, air flow/pressure sensor**
P2439	Secondary air injection (AIR) system, air flow/pressure sensor, bank 2 – circuit intermittent/erratic	**Wiring, poor connection, air flow/pressure sensor**
P2440	Secondary air injection (AIR) switching valve, bank 1 – valve stuck open	**AIR switching valve**
P2441	Secondary air injection (AIR) switching valve, bank 1 – valve stuck closed	**AIR switching valve**

Trouble code	Fault location	Probable cause
P2442	Secondary air injection (AIR) switching valve, bank 2 – valve stuck open	**AIR switching valve**
P2443	Secondary air injection (AIR) switching valve, bank 2 – valve stuck closed	**AIR switching valve**
P2444	Secondary air injection (AIR) pump, bank 1 – pump stuck on	**AIR pump**
P2445	Secondary air injection (AIR) pump, bank 1 – pump stuck off	**AIR pump**
P2446	Secondary air injection (AIR) pump, bank 2 – pump stuck on	**AIR pump**
P2447	Secondary air injection (AIR) pump, bank 2 – pump stuck off	**AIR pump**
P2500	Alternator warning lamp, L-terminal – circuit low	**Wiring short to ground, alternator, instrument panel**
P2501	Alternator warning lamp, L-terminal – circuit high	**Wiring short to positive, alternator, instrument panel**
P2502	Charging system voltage	**Wiring, alternator, battery**
P2503	Charging system – voltage low	**Wiring, alternator, battery**
P2504	Charging system – voltage high	**Wiring, alternator, battery**
P2505	Engine control module (ECM) – supply voltage	**Wiring, fuses, engine control (EC) relay**
P2506	Engine control module (ECM) – supply voltage, range/performance	**Wiring, fuses, engine control (EC) relay**
P2507	Engine control module (ECM) – supply voltage low	**Wiring short to ground, fuses, engine control (EC) relay**
P2508	Engine control module (ECM) – supply voltage high	**Charging system**
P2509	Engine control module (ECM) – supply voltage, intermittent	**Wiring, fuses, engine control (EC) relay**
P2510	Engine control (EC) relay, sense circuit – range/performance	**Wiring, fuses, EC relay**
P2511	Engine control (EC) relay, sense circuit – circuit intermittent	**Wiring, poor connection, EC relay**
P2512	Event data recorder request – open circuit	**Wiring**
P2513	Event data recorder request – circuit low	**Wiring short to ground**
P2514	Event data recorder request – circuit high	**Wiring short to positive**
P2515	A/C refrigerant pressure sensor B – circuit malfunction	**Wiring, A/C refrigerant pressure sensor**
P2516	A/C refrigerant pressure sensor B – range/performance	**Wiring, A/C refrigerant pressure sensor**
P2517	A/C refrigerant pressure sensor B – circuit low	**Wiring short to ground, A/C refrigerant pressure sensor**
P2518	A/C refrigerant pressure sensor B – circuit high	**Wiring short to positive, A/C refrigerant pressure sensor**
P2519	A/C request A – circuit malfunction	**Wiring, A/C control module, A/C master switch, A/C refrigerant pressure switch/sensor**
P2520	A/C request A – circuit low	**Wiring short to ground, A/C control module, A/C master switch, A/C refrigerant pressure switch/sensor**
P2521	A/C request A – circuit high	**Wiring short to positive, A/C control module**

OBD-II P2 trouble code table

Trouble code	Fault location	Probable cause
P2522	A/C request B – circuit malfunction	**Wiring, A/C control module, A/C master switch, A/C refrigerant pressure switch/sensor**
P2523	A/C request B – circuit low	**Wiring short to ground, A/C control module, A/C master switch, A/C refrigerant pressure switch/sensor**
P2524	A/C request B – circuit high	**Wiring short to positive, A/C control module**
P2525	Vacuum reservoir pressure sensor – circuit malfunction	**Wiring, vacuum reservoir pressure sensor**
P2526	Vacuum reservoir pressure sensor – range/performance	**Wiring, vacuum reservoir pressure sensor, hoses blocked/leaking**
P2527	Vacuum reservoir pressure sensor – circuit low	**Wiring short to ground, vacuum reservoir pressure sensor**
P2528	Vacuum reservoir pressure sensor – circuit high	**Wiring short to positive, vacuum reservoir pressure sensor**
P2529	Vacuum reservoir pressure sensor – circuit intermittent	**Wiring, poor connection, vacuum reservoir pressure sensor**
P2530	Ignition switch, ON position – circuit malfunction	**Wiring, fuse, ignition switch**
P2531	Ignition switch, ON position – circuit low	**Wiring short to ground, fuse, ignition switch**
P2532	Ignition switch, ON position – circuit high	**Wiring short to positive, fuse, ignition switch**
P2533	Ignition switch, ON/start position – circuit malfunction	**Wiring, fuse, ignition switch**
P2534	Ignition switch, ON/start position – circuit low	**Wiring short to ground, fuse, ignition switch**
P2535	Ignition switch, ON/start position – circuit high	**Wiring short to positive, fuse, ignition switch**
P2536	Ignition switch, accessory position – circuit malfunction	**Wiring, fuse, ignition switch**
P2537	Ignition switch, accessory position – circuit low	**Wiring short to ground, fuse, ignition switch**
P2538	Ignition switch, accessory position – circuit high	**Wiring short to positive, fuse, ignition switch**
P2539	Fuel low pressure sensor – circuit malfunction	**Wiring, fuel low pressure sensor**
P2540	Fuel low pressure sensor – range/performance	**Wiring, fuel low pressure sensor**
P2541	Fuel low pressure sensor – circuit low	**Wiring short to ground, fuel low pressure sensor**
P2542	Fuel low pressure sensor – circuit high	**Wiring short to positive, fuel low pressure sensor**
P2543	Fuel low pressure sensor – circuit intermittent	**Wiring, poor connection, fuel low pressure sensor**
P2544	Torque management request, input signal A – malfunction	**Wiring, ECM, TCM**
P2545	Torque management request, input signal A – range/performance	**Wiring, ECM, TCM**
P2546	Torque management request, input signal A – signal low	**Wiring short to ground, ECM, TCM**
P2547	Torque management request, input signal A – signal high	**Wiring short to positive, ECM, TCM**
P2548	Torque management request, input signal B – malfunction	**Wiring, ECM, TCM**
P2549	Torque management request, input signal B – range/performance	**Wiring, ECM, TCM**
P2550	Torque management request, input signal B – signal low	**Wiring short to ground, ECM, TCM**

Trouble code	Fault location	Probable cause
P2551	Torque management request, input signal B – signal high	Wiring short to positive, ECM, TCM
P2552	Throttle/fuel inhibit – circuit malfunction	Wiring
P2553	Throttle/fuel inhibit – range/performance	Wiring
P2554	Throttle/fuel inhibit – circuit low	Wiring short to ground
P2555	Throttle/fuel inhibit – circuit high	Wiring short to positive
P2556	Engine coolant 'low' sensor/switch – circuit malfunction	Wiring, engine coolant 'low' sensor/switch
P2557	Engine coolant 'low' sensor/switch – range/ performance	Wiring, engine coolant 'low' sensor/switch
P2558	Engine coolant 'low' sensor/switch – circuit low	Wiring short to ground, engine coolant 'low' sensor/switch
P2559	Engine coolant 'low' sensor/switch – circuit high	Wiring short to positive, engine coolant 'low' sensor/switch
P2560	Engine coolant level low	Engine coolant level low
P2561	A/C control module – MIL activation requested	A/C control module trouble codes stored
P2562	Turbocharger (TC) boost control position sensor – circuit malfunction	Wiring, TC boost control position sensor
P2563	Turbocharger (TC) boost control position sensor – range/performance	Wiring, TC boost control position sensor
P2564	Turbocharger (TC) boost control position sensor – circuit low	Wiring short to ground, TC boost control position sensor
P2565	Turbocharger (TC) boost control position sensor – circuit high	Wiring short to positive, TC boost control position sensor
P2566	Turbocharger (TC) boost control position sensor – circuit intermittent	Wiring, poor connection, TC boost control position sensor
P2567	Direct ozone reduction catalytic converter temperature sensor – circuit malfunction	Wiring, direct ozone reduction catalytic converter temperature sensor
P2568	Direct ozone reduction catalytic converter temperature sensor – range/performance	Wiring, direct ozone reduction catalytic converter temperature sensor
P2569	Direct ozone reduction catalytic converter temperature sensor – circuit low	Wiring short to ground, direct ozone reduction catalytic converter temperature sensor
P2570	Direct ozone reduction catalytic converter temperature sensor – circuit high	Wiring short to positive, direct ozone reduction catalytic converter temperature sensor
P2571	Direct ozone reduction catalytic converter temperature sensor – circuit intermittent/erratic	Wiring, poor connection, direct ozone reduction catalytic converter temperature sensor
P2572	Direct ozone reduction catalytic converter deterioration sensor	Wiring, direct ozone reduction catalytic converter deterioration sensor
P2573	Direct ozone reduction catalytic converter deterioration sensor – range/performance	Wiring, direct ozone reduction catalytic converter deterioration sensor
P2574	Direct ozone reduction catalytic converter deterioration sensor – circuit low	Wiring short to ground, direct ozone reduction catalytic converter deterioration sensor
P2575	Direct ozone reduction catalytic converter deterioration sensor – circuit high	Wiring short to positive, direct ozone reduction catalytic converter deterioration sensor
P2576	Direct ozone reduction catalytic converter deterioration sensor – circuit intermittent/erratic	Wiring, poor connection, direct ozone reduction catalytic converter deterioration sensor
P2577	Direct ozone reduction catalytic converter – efficiency below threshold	Direct ozone reduction catalytic converter

OBD-II P2 trouble code table

Trouble code	Fault location	Probable cause
P2600	Engine coolant pump motor – open circuit	**Wiring, engine coolant pump relay**
P2601	Engine coolant pump motor – range/performance	**Wiring, engine coolant pump relay**
P2602	Engine coolant pump motor – circuit low	**Wiring short to ground, engine coolant pump relay**
P2603	Engine coolant pump motor – circuit high	**Wiring short to positive, engine coolant pump relay**
P2604	Intake air heater A – range/performance	**Wiring, intake air heater relay, intake air heater**
P2605	Intake air heater A – open circuit	**Wiring, intake air heater relay, intake air heater**
P2606	Intake air heater B – range/performance	**Wiring, intake air heater relay, intake air heater**
P2607	Intake air heater B – circuit low	**Wiring short to ground, intake air heater relay, intake air heater**
P2608	Intake air heater B – circuit high	**Wiring short to positive, intake air heater relay, intake air heater**
P2609	Intake air heater system – performance problem	**Wiring, intake air heater relay, intake air heater**
P2610	Engine control module (ECM) – internal engine off timer performance	**ECM**
P2611	A/C refrigerant distribution valve – open circuit	**Wiring, A/C refrigerant distribution valve**
P2612	A/C refrigerant distribution valve – circuit low	**Wiring short to ground, A/C refrigerant distribution valve**
P2613	A/C refrigerant distribution valve – circuit high	**Wiring short to positive, A/C refrigerant distribution valve**
P2614	Camshaft position (CMP), output signal – open circuit	**Wiring, ECM**
P2615	Camshaft position (CMP), output signal – circuit low	**Wiring short to ground, ECM**
P2616	Camshaft position (CMP), output signal – circuit high	**Wiring short to positive, ECM**
P2617	Crankshaft position (CKP), output signal – open circuit	**Wiring, ECM**
P2618	Crankshaft position (CKP), output signal – circuit low	**Wiring short to ground, ECM**
P2619	Crankshaft position (CKP), output signal – circuit high	**Wiring short to positive, ECM**
P2620	Throttle position (TP), output signal – open circuit	**Wiring, ECM**
P2621	Throttle position (TP), output signal – circuit low	**Wiring short to ground, ECM**
P2622	Throttle position (TP), output signal – circuit high	**Wiring short to positive, ECM**
P2623	Injector control pressure regulator – open circuit	**Wiring, injector control pressure regulator**
P2624	Injector control pressure regulator – circuit low	**Wiring short to ground, injector control pressure regulator**
P2625	Injector control pressure regulator – circuit high	**Wiring short to positive, injector control pressure regulator**
P2626	Heated oxygen sensor (HO2S) 1, bank 1, pumping current trim – open circuit	**Wiring, HO2S, ECM**
	Oxygen sensor (O2S) 1, bank 1, pumping current trim – open circuit	**Wiring, O2S, ECM**
P2627	Heated oxygen sensor (HO2S) 1, bank 1, pumping current trim – circuit low	**Wiring short to ground, HO2S, ECM**
	Oxygen sensor (O2S) 1, bank 1, pumping current trim – circuit low	**Wiring short to ground, O2S, ECM**

Trouble code	Fault location	Probable cause
P2628	Heated oxygen sensor (HO2S) 1, bank 1, pumping current trim – circuit high	Wiring short to positive, HO2S, ECM
	Oxygen sensor (O2S) 1, bank 1, pumping current trim – circuit high	Wiring short to positive, O2S, ECM
P2629	Heated oxygen sensor (HO2S) 1, bank 2, pumping current trim – open circuit	Wiring, HO2S, ECM
	Oxygen sensor (O2S) 1, bank 2, pumping current trim – open circuit	Wiring, O2S, ECM
P2630	Heated oxygen sensor (HO2S) 1, bank 2, pumping current trim – circuit low	Wiring short to ground, HO2S, ECM
	Oxygen sensor (O2S) 1, bank 2, pumping current trim – circuit low	Wiring short to ground, O2S, ECM
P2631	Heated oxygen sensor (HO2S) 1, bank 2, pumping current trim – circuit high	Wiring short to positive, HO2S, ECM
	Oxygen sensor (O2S) 1, bank 2, pumping current trim – circuit high	Wiring short to positive, O2S, ECM
P2632	Fuel pump (FP) B, control – open circuit	Wiring, FP relay, ECM
P2633	Fuel pump (FP) B, control – circuit low	Wiring short to ground, FP relay, ECM
P2634	Fuel pump (FP) B, control – circuit high	Wiring short to positive, FP relay, ECM
P2635	Fuel pump (FP) A – low flow/performance problem	Fuel filter blocked, fuel pump (FP)
P2636	Fuel pump (FP) B – low flow/performance problem	Fuel filter blocked, fuel pump (FP)
P2637	Torque management, feedback signal A – malfunction	Wiring, ECM, TCM
P2638	Torque management, feedback signal A – range/performance	Wiring, ECM, TCM
P2639	Torque management, feedback signal A – signal low	Wiring short to ground, ECM, TCM
P2640	Torque management, feedback signal A – signal high	Wiring short to positive, ECM, TCM
P2641	Torque management, feedback signal B – malfunction	Wiring, ECM, TCM
P2642	Torque management, feedback signal B – range/performance	Wiring, ECM, TCM
P2643	Torque management, feedback signal B – signal low	Wiring short to ground, ECM, TCM
P2644	Torque management, feedback signal B – signal high	Wiring short to positive, ECM, TCM
P2645	Rocker arm actuator A, bank 1 – open circuit	Wiring, rocker arm actuator
P2646	Rocker arm actuator A, bank 1 – performance problem or actuator stuck off	Wiring, rocker arm actuator
P2647	Rocker arm actuator A, bank 1 – actuator stuck on	Rocker arm actuator
P2648	Rocker arm actuator A, bank 1 – circuit low	Wiring short to ground, rocker arm actuator
P2649	Rocker arm actuator A, bank 1 – circuit high	Wiring short to positive, rocker arm actuator
P2650	Rocker arm actuator B, bank 1 – open circuit	Wiring, rocker arm actuator
P2651	Rocker arm actuator B, bank 1 – performance problem or actuator stuck off	Rocker arm actuator
P2652	Rocker arm actuator B, bank 1 – actuator stuck on	Rocker arm actuator
P2653	Rocker arm actuator B, bank 1 – circuit low	Wiring short to ground, rocker arm actuator
P2654	Rocker arm actuator B, bank 1 – circuit high	Wiring short to positive, rocker arm actuator
P2655	Rocker arm actuator A, bank 2 – open circuit	Wiring, rocker arm actuator
P2656	Rocker arm actuator A, bank 2 – performance problem or actuator stuck off	Rocker arm actuator

OBD-II P2 trouble code table

Trouble code	Fault location	Probable cause
P2657	Rocker arm actuator A, bank 2 – actuator stuck on	**Rocker arm actuator**
P2658	Rocker arm actuator A, bank 2 – circuit low	**Wiring short to ground, rocker arm actuator**
P2659	Rocker arm actuator A, bank 2 – circuit high	**Wiring short to positive, rocker arm actuator**
P2660	Rocker arm actuator B, bank 2 – open circuit	**Wiring, rocker arm actuator**
P2661	Rocker arm actuator B, bank 2 – performance problem or actuator stuck off	**Rocker arm actuator**
P2662	Rocker arm actuator B, bank 2 – actuator stuck on	**Rocker arm actuator**
P2663	Rocker arm actuator B, bank 2 – circuit low	**Wiring short to ground, rocker arm actuator**
P2664	Rocker arm actuator B, bank 2 – circuit high	**Wiring short to positive, rocker arm actuator**
P2665	Fuel shut-off solenoid B – open circuit	**Wiring, fuel shut-off solenoid**
P2666	Fuel shut-off solenoid B – circuit low	**Wiring short to ground, fuel shut-off solenoid**
P2667	Fuel shut-off solenoid B – circuit high	**Wiring short to positive, fuel shut-off solenoid**
P2668	Fuel mode indicator lamp – circuit malfunction	**Wiring, fuel mode indicator lamp**
P2669	Actuator supply voltage B – open circuit	**Wiring, ECM**
P2670	Actuator supply voltage B – circuit low	**Wiring short to ground, ECM**
P2671	Actuator supply voltage B – circuit high	**Wiring short to positive, ECM**
P2700	Transmission friction element A, apply time – range/performance	**Transmission mechanical fault, shift solenoid (SS)**
P2701	Transmission friction element B, apply time – range/performance	**Transmission mechanical fault, shift solenoid (SS)**
P27O2	Transmission friction element C, apply time – range/performance	**Transmission mechanical fault, shift solenoid (SS)**
P2703	Transmission friction element D, apply time – range/performance	**Transmission mechanical fault, shift solenoid (SS)**
P2704	Transmission friction element E, apply time – range/performance	**Transmission mechanical fault, shift solenoid (SS)**
P2705	Transmission friction element F, apply time – range/performance	**Transmission mechanical fault, shift solenoid (SS)**
P2706	Shift solenoid (SS) F – circuit malfunction	**Transmission mechanical fault, shift solenoid (SS)**
P2707	Shift solenoid (SS) F – performance problem or solenoid stuck off	**Transmission mechanical fault, shift solenoid (SS)**
P2708	Shift solenoid (SS) F – solenoid stuck on	**Transmission mechanical fault, shift solenoid (SS)**
P2709	Shift solenoid (SS) F – electrical	**Wiring, shift solenoid (SS)**
P2710	Shift solenoid (SS) F – intermittent	**Wiring, poor connection, shift solenoid (SS)**
P2711	Unexpected mechanical gear disengagement	**Operator error, transmission mechanical fault**
P2712	Hydraulic power unit leakage	–
P2713	Transmission fluid pressure (TFP) solenoid D – circuit malfunction	**Wiring, TFP solenoid, TCM**
P2714	Transmission fluid pressure (TFP) solenoid D – performance problem or solenoid stuck off	**Wiring, TFP solenoid, transmission mechanical fault**
P2715	Transmission fluid pressure (TFP) solenoid D – solenoid stuck on	**TFP solenoid, transmission mechanical fault**
P2716	Transmission fluid pressure (TFP) solenoid D – electrical	**Wiring, TFP solenoid**
P2717	Transmission fluid pressure (TFP) solenoid D – circuit intermittent	**Wiring, poor connection, TFP solenoid, TCM**

Trouble code	Fault location	Probable cause
P2718	Transmission fluid pressure (TFP) solenoid D – open circuit	Wiring, TFP solenoid, TCM
P2719	Transmission fluid pressure (TFP) solenoid D – range/performance	TFP solenoid, transmission mechanical fault
P2720	Transmission fluid pressure (TFP) solenoid D – circuit low	Wiring short to ground, TFP solenoid, TCM
P2721	Transmission fluid pressure (TFP) solenoid D – circuit high	Wiring short to positive, TFP solenoid, TCM
P2722	Transmission fluid pressure (TFP) solenoid E – circuit malfunction	Wiring, TFP solenoid, TCM
P2723	Transmission fluid pressure (TFP) solenoid E – performance problem or solenoid stuck off	Wiring, TFP solenoid, transmission mechanical fault
P2724	Transmission fluid pressure (TFP) solenoid E – solenoid stuck on	TFP solenoid, transmission mechanical fault
P2725	Transmission fluid pressure (TFP) solenoid E – electrical	Wiring, TFP solenoid
P2726	Transmission fluid pressure (TFP) solenoid E – circuit intermittent	Wiring, poor connection, TFP solenoid, TCM
P2727	Transmission fluid pressure (TFP) solenoid E – open circuit	Wiring, TFP solenoid, TCM
P2728	Transmission fluid pressure (TFP) solenoid E – range/performance	TFP solenoid, transmission mechanical fault
P2729	Transmission fluid pressure (TFP) solenoid E – circuit low	Wiring short to ground, TFP solenoid, TCM
P2730	Transmission fluid pressure (TFP) solenoid E – circuit high	Wiring short to positive, TFP solenoid, TCM
P2731	Transmission fluid pressure (TFP) solenoid F – circuit malfunction	Wiring, TFP solenoid, TCM
P2732	Transmission fluid pressure (TFP) solenoid F – performance problem or solenoid stuck off	Wiring, TFP solenoid, transmission mechanical fault
P2733	Transmission fluid pressure (TFP) solenoid F – solenoid stuck on	TFP solenoid, transmission mechanical fault
P2734	Transmission fluid pressure (TFP) solenoid F – electrical	Wiring, TFP solenoid
P2735	Transmission fluid pressure (TFP) solenoid F – circuit intermittent	Wiring, poor connection, TFP solenoid, TCM
P2736	Transmission fluid pressure (TFP) solenoid F – open circuit	Wiring, TFP solenoid, TCM
P2737	Transmission fluid pressure (TFP) solenoid F – range/performance	TFP solenoid, transmission mechanical fault
P2738	Transmission fluid pressure (TFP) solenoid F – circuit low	Wiring short to ground, TFP solenoid, TCM
P2739	Transmission fluid pressure (TFP) solenoid F – circuit high	Wiring short to positive, TFP solenoid, TCM
P2740	Transmission fluid temperature (TFT) sensor B – circuit malfunction	Wiring, TFT sensor
P2741	Transmission fluid temperature (TFT) sensor B – circuit range/performance	Wiring, TFT sensor

OBD-II P2 trouble code table

Trouble code	Fault location	Probable cause
P2742	Transmission fluid temperature (TFT) sensor B – circuit low	**Wiring short to ground, TFT sensor**
P2743	Transmission fluid temperature (TFT) sensor B – circuit high	**Wiring short to positive, TFT sensor**
P2744	Transmission fluid temperature (TFT) sensor B – circuit intermittent	**Wiring, poor connection, TFT sensor**
P2745	Transmission intermediate shaft speed sensor B – circuit malfunction	**Wiring, transmission intermediate shaft speed sensor, ECM, TCM**
P2746	Transmission intermediate shaft speed sensor B – range/performance	**Wiring, transmission intermediate shaft speed sensor, ECM, TCM**
P2747	Transmission intermediate shaft speed sensor B – no signal	**Wiring, transmission intermediate shaft speed sensor, ECM, TCM**
P2748	Transmission intermediate shaft speed sensor B – circuit intermittent	**Wiring, poor connection, transmission intermediate shaft speed sensor, ECM, TCM**
P2749	Transmission intermediate shaft speed sensor C – circuit malfunction	**Wiring, transmission intermediate shaft speed sensor, ECM, TCM**
P2750	Transmission intermediate shaft speed sensor C – range/performance	**Wiring, transmission intermediate shaft speed sensor, ECM, TCM**
P2751	Transmission intermediate shaft speed sensor C – no signal	**Wiring, transmission intermediate shaft speed sensor, ECM, TCM**
P2752	Transmission intermediate shaft speed sensor C – circuit intermittent	**Wiring, poor connection, transmission intermediate shaft speed sensor, ECM, TCM**
P2753	Transmission fluid cooler – open circuit	**Wiring, transmission fluid cooler**
P2754	Transmission fluid cooler – circuit low	**Wiring short to ground, transmission fluid cooler**
P2755	Transmission fluid cooler – circuit high	**Wiring short to positive, transmission fluid cooler**
P2756	Torque converter clutch (TCC) pressure control solenoid – circuit malfunction	**Wiring, TCC pressure control solenoid**
P2757	Torque converter clutch (TCC) pressure control solenoid – performance problem or solenoid stuck off	**TCC pressure control solenoid**
P2758	Torque converter clutch (TCC) pressure control solenoid – solenoid stuck on	**TCC pressure control solenoid**
P2759	Torque converter clutch (TCC) pressure control solenoid – electrical fault	**Wiring, TCC pressure control solenoid**
P2760	Torque converter clutch (TCC) pressure control solenoid – circuit intermittent	**Wiring, poor connection, TCC pressure control solenoid**
P2761	Torque converter clutch (TCC) pressure control solenoid – open circuit	**Wiring, TCC pressure control solenoid**
P2762	Torque converter clutch (TCC) pressure control solenoid – range/performance	**Wiring, TCC pressure control solenoid**
P2763	Torque converter clutch (TCC) pressure control solenoid – circuit high	**Wiring short to positive, TCC pressure control solenoid**
P2764	Torque converter clutch (TCC) pressure control solenoid – circuit low	**Wiring short to ground, TCC pressure control solenoid**
P2765	Transmission input shaft speed sensor/turbine shaft speed (TSS) sensor B – circuit malfunction	**Wiring, transmission input shaft speed sensor/ TSS sensor**
P2766	Transmission input shaft speed sensor/turbine shaft speed (TSS) sensor B – range/performance	**Wiring, transmission input shaft speed sensor/ TSS sensor**

Trouble code	Fault location	Probable cause
P2767	Transmission input shaft speed sensor/turbine shaft speed (TSS) sensor B – no signal	Wiring, transmission input shaft speed sensor/ TSS sensor
P2768	Transmission input shaft speed sensor/turbine shaft speed (TSS) sensor B – circuit intermittent	Wiring, poor connection, transmission input shaft speed sensor/TSS sensor
P2769	Torque converter clutch (TCC) – circuit low	Wiring short to ground, TCC
P2770	Torque converter clutch (TCC) – circuit high	Wiring short to positive, TCC
P2771	Four wheel drive, low gear ratio switch – circuit malfunction	Wiring, low gear ratio switch
P2772	Four wheel drive, low gear ratio switch – range/performance	Wiring, low gear ratio switch
P2773	Four wheel drive, low gear ratio switch – circuit low	Wiring short to ground, low gear ratio switch
P2774	Four wheel drive, low gear ratio switch – circuit high	Wiring short to positive, low gear ratio switch
P2775	Transmission gear selection switch, upshift – range/performance	Wiring, transmission gear selection switch
P2776	Transmission gear selection switch, upshift – circuit low	Wiring short to ground, transmission gear selection switch
P2777	Transmission gear selection switch, upshift – circuit high	Wiring short to positive, transmission gear selection switch
P2778	Transmission gear selection switch, upshift – circuit intermittent/erratic	Wiring, poor connection, transmission gear selection switch
P2779	Transmission gear selection switch, downshift – range/performance	Wiring, transmission gear selection switch
P2780	Transmission gear selection switch, downshift – circuit low	Wiring short to ground, transmission gear selection switch
P2781	Transmission gear selection switch, downshift – circuit high	Wiring short to positive, transmission gear selection switch
P2782	Transmission gear selection switch, downshift – circuit intermittent/erratic	Wiring, poor connection, transmission gear selection switch
P2783	Torque converter – temperature too high	Transmission fluid level low, transmission mechanical fault, TCC slipping
P2784	Transmission input shaft speed sensor/turbine shaft speed (TSS) sensor A/B – correlation	Wiring, transmission input shaft speed sensor/ TSS sensor
P2785	Clutch actuator – temperature too high	–
P2786	Gear shift actuator – temperature too high	–
P2787	Clutch – temperature too high	Clutch slipping
P2788	Auto shift manual (ASM) transmission, adaptive learning – at limit	–
P2789	Clutch, adaptive learning – at limit	–
P2790	Gate select direction – circuit malfunction	Wiring
P2791	Gate select direction – circuit low	Wiring short to ground
P2792	Gate select direction – circuit high	Wiring short to positive
P2793	Gear shift direction – circuit malfunction	Wiring
P2794	Gear shift direction – circuit low	Wiring short to ground
P2795	Gear shift direction – circuit high	Wiring short to positive

OBD-II U0 trouble code table

- The following list covers all U0 codes allocated at the time of publication.
- All OBD-II codes starting with U0 have standard meanings irrespective of vehicle make or model.

Trouble code	Fault location
U0001	Controller area network (CAN) data bus, high speed bus
U0002	Controller area network (CAN) data bus, high speed bus – performance problem
U0003	Controller area network (CAN) data bus, high speed bus (+) – open circuit
U0004	Controller area network (CAN) data bus, high speed bus (+) – voltage low
U0005	Controller area network (CAN) data bus, high speed bus (+) – voltage high
U0006	Controller area network (CAN) data bus, high speed bus (-) – open circuit
U0007	Controller area network (CAN) data bus, high speed bus (-) – voltage low
U0008	Controller area network (CAN) data bus, high speed bus (-) – voltage high
U0009	Controller area network (CAN) data bus, high speed bus (-) – shorted to data bus (+)
U0010	Controller area network (CAN) data bus, medium speed bus
U0011	Controller area network (CAN) data bus, medium speed bus – performance problem
U0012	Controller area network (CAN) data bus, medium speed bus (+) – open circuit
U0013	Controller area network (CAN) data bus, medium speed bus (+) – voltage low
U0014	Controller area network (CAN) data bus, medium speed bus (+) – voltage high
U0015	Controller area network (CAN) data bus, medium speed bus (-)
U0016	Controller area network (CAN) data bus, medium speed bus (-) – voltage low
U0017	Controller area network (CAN) data bus, medium speed bus (-) – voltage high
U0018	Controller area network (CAN) data bus, medium speed bus (-) – shorted to data bus (+)
U0019	Controller area network (CAN) data bus, low speed bus
U0020	Controller area network (CAN) data bus, low speed bus – performance problem
U0021	Controller area network (CAN) data bus, low speed bus (+) – open circuit
U0022	Controller area network (CAN) data bus, low speed bus (+) – voltage low
U0023	Controller area network (CAN) data bus, low speed bus (+) – voltage high
U0024	Controller area network (CAN) data bus, low speed bus (-) – open circuit
U0025	Controller area network (CAN) data bus, low speed bus (-) – voltage low
U0026	Controller area network (CAN) data bus, low speed bus (-) – voltage high
U0027	Controller area network (CAN) data bus, low speed bus (-) – shorted to data bus (+)
U0028	Vehicle area network (VAN) data bus A
U0029	Vehicle area network (VAN) data bus A – performance problem
U0030	Vehicle area network (VAN) data bus A (+) – open circuit
U0031	Vehicle area network (VAN) data bus A (+) – voltage low
U0032	Vehicle area network (VAN) data bus A (+) – voltage high
U0033	Vehicle area network (VAN) data bus A (-) – open circuit
U0034	Vehicle area network (VAN) data bus A (-) – voltage low
U0035	Vehicle area network (VAN) data bus A (-) – voltage high
U0036	Vehicle area network (VAN) data bus A (-) – shorted to data bus A (+)
U0037	Vehicle area network (VAN) data bus B
U0038	Vehicle area network (VAN) data bus B – performance problem
U0039	Vehicle area network (VAN) data bus B (+) – open circuit
U0040	Vehicle area network (VAN) data bus B (+) – voltage low
U0041	Vehicle area network (VAN) data bus B (+) – voltage high
U0042	Vehicle area network (VAN) data bus B (-) – open circuit
U0043	Vehicle area network (VAN) data bus B (-) – voltage low

Trouble code	Fault location
U0044	Vehicle area network (VAN) data bus B (-) – voltage high
U0045	Vehicle area network (VAN) data bus B (-) – shorted to data bus B (+)
U0046	Vehicle area network (VAN) data bus C
U0047	Vehicle area network (VAN) data bus C – performance problem
U0048	Vehicle area network (VAN) data bus C (+) – open circuit
U0049	Vehicle area network (VAN) data bus C (+) – voltage low
U0050	Vehicle area network (VAN) data bus C (+) – voltage high
U0051	Vehicle area network (VAN) data bus C (-) – open circuit
U0052	Vehicle area network (VAN) data bus C (-) – voltage low
U0053	Vehicle area network (VAN) data bus C (-) – voltage high
U0054	Vehicle area network (VAN) data bus C (-) – shorted to data bus C (+)
U0055	Vehicle area network (VAN) data bus D
U0056	Vehicle area network (VAN) data bus D – performance problem
U0057	Vehicle area network (VAN) data bus D (+) – open circuit
U0058	Vehicle area network (VAN) data bus D (+) – voltage low
U0059	Vehicle area network (VAN) data bus D (+) – voltage high
U0060	Vehicle area network (VAN) data bus D (-) – open circuit
U0061	Vehicle area network (VAN) data bus D (-) – voltage low
U0062	Vehicle area network (VAN) data bus D (-) – voltage high
U0063	Vehicle area network (VAN) data bus D (-) – shorted to data bus D (+)
U0064	Vehicle area network (VAN) data bus E
U0065	Vehicle area network (VAN) data bus E – performance problem
U0066	Vehicle area network (VAN) data bus E (+) – open circuit
U0067	Vehicle area network (VAN) data bus E (+) – voltage low

Trouble code	Fault location
U0068	Vehicle area network (VAN) data bus E (+) – voltage high
U0069	Vehicle area network (VAN) data bus E (-) – open circuit
U0070	Vehicle area network (VAN) data bus E (-) – voltage low
U0071	Vehicle area network (VAN) data bus E (-) – voltage high
U0072	Vehicle area network (VAN) data bus E (-) – shorted to data bus E (+)
U0073	Control module – data bus Off
U0100	Data bus, engine control module (ECM) A – no communication
U0101	Data bus, transmission control module (TCM) – no communication
U0102	Data bus, transfer box control module – no communication
U0103	Data bus, gear shift module – no communication
U0104	Data bus, cruise control module – no communication
U0105	Data bus, injector control module – no communication
U0106	Data bus, glow plug control module – no communication
U0107	Data bus, throttle actuator control (TAC) module – no communication
U0108	Data bus, alternative fuel control module – no communication
U0109	Data bus, fuel pump (FP) control module – no communication
U0110	Data bus, drive motor control module – no communication
U0111	Data bus, battery energy control module A – no communication
U0112	Data bus, battery energy control module B – no communication
U0113	Data bus, emissions critical control information – no communication
U0114	Data bus, four wheel drive clutch control module – no communication
U0115	Data bus, engine control module (ECM) B – no communication
U0121	Data bus, anti-lock brake system (ABS) control module – no communication

OBD-II U0 trouble code table

Trouble code	Fault location
U0122	Data bus, vehicle dynamics control module – no communication
U0123	Data bus, yaw rate sensor module – no communication
U0124	Data bus, lateral acceleration sensor module – no communication
U0125	Data bus, multi-axis acceleration sensor module – no communication
U0126	Data bus, steering position sensor control module – no communication
U0127	Data bus, tire pressure monitor module – no communication
U0128	Data bus, parking brake control module – no communication
U0129	Data bus, brake system control module – no communication
U0130	Data bus, steering effort control module – no communication
U0131	Data bus, power steering control module – no communication
U0132	Data bus, suspension ride height control module – no communication
U0140	Data bus, body control module (BCM) – no communication
U0141	Data bus, body control module (BCM) A – no communication
U0142	Data bus, body control module (BCM) B – no communication
U0143	Data bus, body control module (BCM) C – no communication
U0144	Data bus, body control module (BCM) D – no communication
U0145	Data bus, body control module (BCM) E – no communication
U0146	Data bus, gateway A – no communication
U0147	Data bus, gateway B – no communication
U0148	Data bus, gateway C – no communication
U0149	Data bus, gateway D – no communication
U0150	Data bus, gateway E – no communication
U0151	Data bus, supplementary restraint system (SRS) control module – no communication
U0152	Data bus, supplementary restraint system (SRS) control module, left – no communication

Trouble code	Fault location
U0153	Data bus, supplementary restraint system (SRS) control module, right – no communication
U0154	Data bus, supplementary restraint system (SRS) occupant sensing control module – no communication
U0155	Data bus, instrumentation control module – no communication
U0156	Data bus, information centre A – no communication
U0157	Data bus, information centre B – no communication
U0158	Data bus, head up display – no communication
U0159	Data bus, parking aid control module A – no communication
U0160	Data bus, audible alert control module – no communication
U0161	Data bus, compass module – no communication
U0162	Data bus, navigation display module – no communication
U0163	Data bus, navigation control module – no communication
U0164	Data bus, A/C control module – no communication
U0165	Data bus, A/C control module, rear
U0166	Data bus, auxiliary heater control module – no communication
U0167	Data bus, immobilizer control module – no communication
U0168	Data bus, alarm system control module – no communication
U0169	Data bus, sunroof control module – no communication
U0170	Data bus, supplementary restraint system (SRS) sensor A – no communication
U0171	Data bus, supplementary restraint system (SRS) sensor B – no communication
U0172	Data bus, supplementary restraint system (SRS) sensor C – no communication
U0173	Data bus, supplementary restraint system (SRS) sensor D – no communication
U0174	Data bus, supplementary restraint system (SRS) sensor E – no communication

Trouble code	Fault location
U0175	Data bus, supplementary restraint system (SRS) sensor F – no communication
U0176	Data bus, supplementary restraint system (SRS) sensor G – no communication
U0177	Data bus, supplementary restraint system (SRS) sensor H – no communication
U0178	Data bus, supplementary restraint system (SRS) sensor I – no communication
U0179	Data bus, supplementary restraint system (SRS) sensor J – no communication
U0180	Data bus, automatic lighting control module – no communication
U0181	Data bus, headlamp level control module – no communication
U0182	Data bus, lamps control module, front – no communication
U0183	Data bus, lamps control module, rear – no communication
U0184	Data bus, radio – no communication
U0185	Data bus, aerial module – no communication
U0186	Data bus, audio unit output amplifier – no communication
U0187	Data bus, digital disc player/changer module A – no communication
U0188	Data bus, digital disc player/changer module B – no communication
U0189	Data bus, digital disc player/changer module C – no communication
U0190	Data bus, digital disc player/changer module D – no communication
U0191	Data bus, television – no communication
U0192	Data bus, personal computer – no communication
U0193	Data bus, digital audio control module A – no communication
U0194	Data bus, digital audio control module B – no communication
U0195	Data bus, subscription entertainment receiver module – no communication
U0196	Data bus, entertainment control module, rear – no communication
U0197	Data bus, telephone control module – no communication
U0198	Data bus, telematics control module – no communication

Trouble code	Fault location
U0199	Data bus, door function control module A – no communication
U0200	Data bus, door function control module B – no communication
U0201	Data bus, door function control module C – no communication
U0202	Data bus, door function control module D – no communication
U0203	Data bus, door function control module E – no communication
U0204	Data bus, door function control module F – no communication
U0205	Data bus, door function control module G – no communication
U0206	Data bus, convertible top control module – no communication
U0207	Data bus, moveable roof control module – no communication
U0208	Data bus, seat adjustment control module A – no communication
U0209	Data bus, seat adjustment control module B – no communication
U0210	Data bus, seat adjustment control module C – no communication
U0211	Data bus, seat adjustment control module D – no communication
U0212	Data bus, steering column control module – no communication
U0213	Data bus, mirror control module A – no communication
U0214	Data bus, remote function actuation – no communication
U0215	Data bus, door contact switch A – no communication
U0216	Data bus, door contact switch B – no communication
U0217	Data bus, door contact switch C – no communication
U0218	Data bus, door contact switch D – no communication
U0219	Data bus, door contact switch E – no communication
U0220	Data bus, door contact switch F – no communication
U0221	Data bus, door contact switch G – no communication

OBD-II U0 trouble code table

Trouble code	Fault location
U0222	Data bus, electric window motor A – no communication
U0223	Data bus, electric window motor B – no communication
U0224	Data bus, electric window motor C – no communication
U0225	Data bus, electric window motor D – no communication
U0226	Data bus, electric window motor E – no communication
U0227	Data bus, electric window motor F – no communication
U0228	Data bus, electric window motor G – no communication
U0229	Data bus, heated steering wheel module – no communication
U0230	Data bus, tailgate control module – no communication
U0231	Data bus, rain sensor control module – no communication
U0232	Data bus, side obstacle detection control module, left – no communication
U0233	Data bus, side obstacle detection control module, right – no communication
U0234	Data bus, convenience recall module – no communication
U0235	Data bus, cruise control front distance range sensor – no communication
U0300	Control module – internal software incompatibility
U0301	Software incompatibility – engine control module (ECM)
U0302	Software incompatibility – transmission control module (TCM)
U0303	Software incompatibility – transfer box control module
U0304	Software incompatibility – gear shift module
U0305	Software incompatibility – cruise control module
U0306	Software incompatibility – injector control module
U0307	Software incompatibility – glow plug control module
U0308	Software incompatibility – throttle actuator control (TAC) module

Trouble code	Fault location
U0309	Software incompatibility – alternative fuel control module
U0310	Software incompatibility – fuel pump (FP) control module
U0311	Software incompatibility – drive motor control module
U0312	Software incompatibility – battery energy control module A
U0313	Software incompatibility – battery energy control module B
U0314	Software incompatibility – four wheel drive clutch control module
U0315	Software incompatibility – anti-lock brake system (ABS) control module
U0316	Software incompatibility – vehicle dynamics control module
U0317	Software incompatibility – parking brake control module
U0318	Software incompatibility – brake system control module
U0319	Software incompatibility – steering effort control module
U0320	Software incompatibility – power steering control module
U0321	Software incompatibility – suspension ride height control module
U0322	Software incompatibility – body control module
U0323	Software incompatibility – instrumentation control module
U0324	Software incompatibility – A/C control module
U0325	Software incompatibility – auxiliary heater control module
U0326	Software incompatibility – immobilizer control module
U0327	Software incompatibility – alarm system control module
U0328	Software incompatibility – steering position sensor control module
U0329	Software incompatibility – steering column control module
U0330	Software incompatibility – tire pressure monitor module
U0331	Software incompatibility – body control module A

Trouble code	Fault location
U0332	Software incompatibility – multi-axis acceleration sensor module
U0400	Invalid data received
U0401	Invalid data received – engine control module (ECM)
U0402	Invalid data received – transmission control module (TCM)
U0403	Invalid data received – transfer box control module
U0404	Invalid data received – gear shift module
U0405	Invalid data received – cruise control module
U0406	Invalid data received – injector control module
U0407	Invalid data received – glow plug control module
U0408	Invalid data received – throttle actuator control (TAC) module
U0409	Invalid data received – alternative fuel control module
U0410	Invalid data received – fuel pump (FP) control module
U0411	Invalid data received – drive motor control module
U0412	Invalid data received – battery energy control module A
U0413	Invalid data received – battery energy control module B
U0414	Invalid data received – four wheel drive clutch control module
U0415	Invalid data received – anti-lock brake system (ABS) control module
U0416	Invalid data received – vehicle dynamics control module
U0417	Invalid data received – parking brake control module
U0418	Invalid data received – brake system control module
U0419	Invalid data received – steering effort control module
U0420	Invalid data received – power steering control module
U0421	Invalid data received – suspension ride height control module
U0422	Invalid data received – body control module

Trouble code	Fault location
U0423	Invalid data received – instrumentation control module
U0424	Invalid data received – A/C control module
U0425	Invalid data received – auxiliary heater control module
U0426	Invalid data received – immobilizer control module
U0427	Invalid data received – alarm system control module
U0428	Invalid data received – steering position sensor control module
U0429	Invalid data received – steering column control module
U0430	Invalid data received – tire pressure monitor module
U0431	Invalid data received – body control module A

ACURA

Model:	Engine identification:	Year:
Integra 1.7L	**B17A1**	**1992-93**
Integra 1.8L	**B18A1**	**1992-93**
Vigor 2.5L	**G26A1**	**1992-94**
Legend 3.2L	**C32A1**	**1991-95**

System: **PGM-FI**

Self-diagnosis

General information

- Refer to the front of this manual for general test conditions, terminology, detailed descriptions of wiring faults and a general trouble shooter for electrical and mechanical faults.
- Engine control module (ECM) incorporates self-diagnosis function.
- Malfunction indicator lamp (MIL) will illuminate if certain faults are recorded.
- ECM operates in backup mode if sensors fail, to enable vehicle to be driven to workshop.
- Trouble codes can be displayed by the malfunction indicator lamp (MIL).

Accessing

- Ensure ignition switched OFF.
- Jump service check connector terminals **1**.
- Switch ignition ON.
- Check MIL is flashing.
- Trouble codes 1-9 are indicated as follows:
  - Individual short flashes display trouble code **2** [**A**].
  - A short pause separates each flash **2** [**B**].
  - A long pause separates each trouble code **2** [**C**].
  - For example: Trouble code 3 displayed **2**.
- Trouble codes greater than 9 are indicated as follows:
  - Long flashes indicate the 'tens' of the trouble code **3** [**A**].
  - Short flashes indicate the 'units' of the trouble code **3** [**C**].
  - A short pause separates each flash **3** [**B**].
  - A long pause separates each trouble code **3** [**D**].
  - For example: Trouble code 12 displayed **3**.
- Count MIL flashes and compare with trouble code table.

NOTE: ***If malfunction indicator lamp (MIL) is ON constantly and no trouble codes are displayed, this indicates an engine control module (ECM) fault.***

- Switch ignition OFF.
- Remove jump lead.

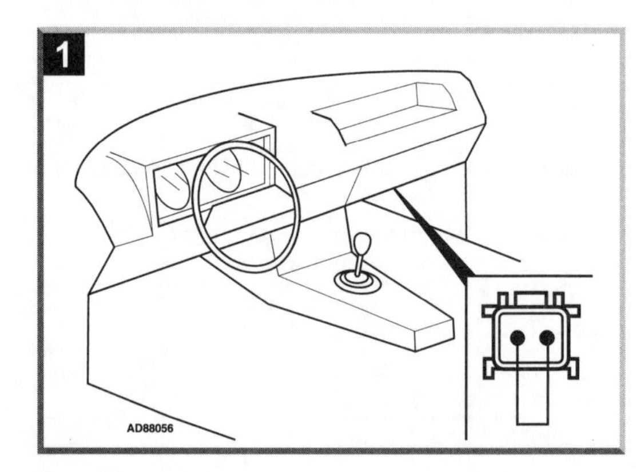

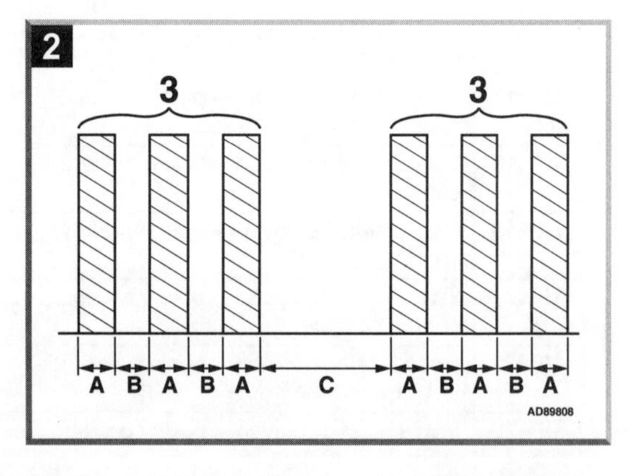

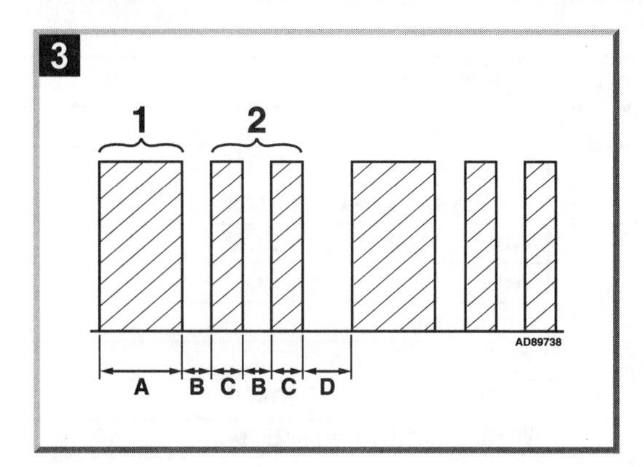

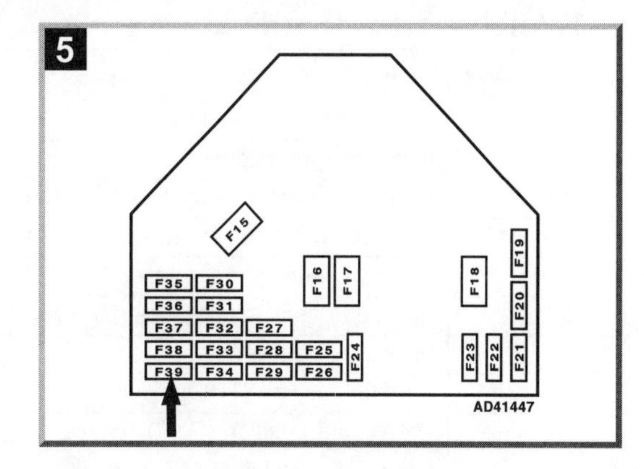

Erasing

- After the faults have been rectified, erase the trouble codes as follows:
- Switch ignition OFF.
- Integra – Remove fuse No.34 (7.5A) from underhood fusebox for 10 seconds minimum **4**.
- Vigor – remove fuse No.39 (7.5A) from underhood fusebox for 10 seconds minimum **5**.
- Legend – remove fuse No.15 (7.5A) from LH fascia fusebox for 10 seconds minimum **6**.
- Reinstall fuse.
- Repeat checking procedure to ensure no data remains in ECM fault memory.

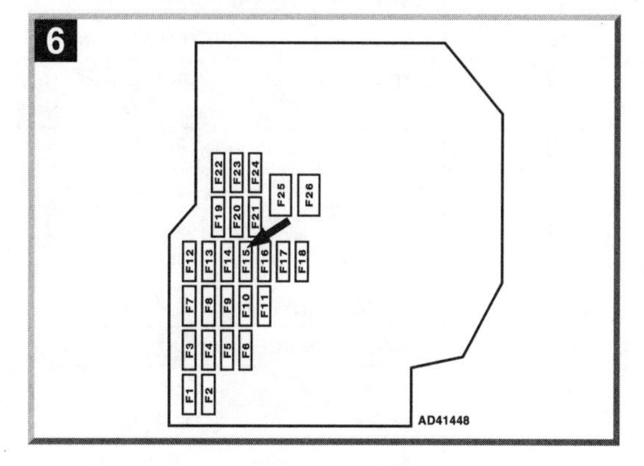

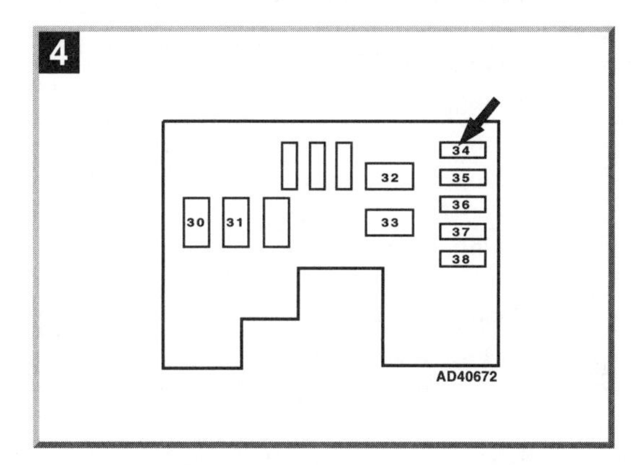

Trouble code identification

Flash code	Fault location	Probable cause
0	Engine control module (ECM)	Wiring, ECM
1	Heated oxygen sensor (HO2S) – LH	Fuel system, wiring, HO2S, ECM
1	Heated oxygen sensor (HO2S) – RH	Fuel system, wiring, HO2S, ECM
3	Manifold absolute pressure (MAP) sensor	Wiring, MAP sensor, ECM
4	Engine speed (RPM) sensor	Wiring, RPM sensor, ECM
5	Manifold absolute pressure (MAP) sensor	Hose leak
6	Engine coolant temperature (ECT) sensor	Wiring, ECT sensor, ECM
7	Throttle position (TP) sensor	Wiring, TP sensor, ECM
8	Crankshaft position (CKP) sensor	Wiring, CKP sensor, ECM
9	Camshaft position (CMP) sensor 1	Wiring, CMP sensor, ECM
10	Intake air temperature (IAT) sensor	Wiring, IAT sensor, ECM
12	Exhaust gas recirculation (EGR) system	Hose leak/blockage, wiring, EGR valve/ solenoid/position sensor, ECM
13	Barometric pressure (BARO) sensor	ECM
14	Idle air control (IAC) valve	Wiring, IAC valve, ECM
15	Ignition output signal	Wiring, ICM, ECM
16	Injectors	Wiring, injector, ECM
17	Vehicle speed sensor (VSS)	Wiring, VSS, ECM
18	Ignition timing adjustment	Wiring, ignition timing adjuster, ECM
20	Electrical load sensor	Wiring, electrical load sensor, ECM
21	Camshaft position (CMP) actuator	Wiring, CMP actuator, ECM
22	Camshaft position (CMP) actuator – oil pressure switch	Wiring, oil pressure switch, ECM
23	Knock sensor (KS) – LH	Wiring, KS, ECM
30	AT/ECM communication signal A	Wiring, ECM
31	AT/ECM communication signal B	Wiring, ECM
35	Traction control/engine control module (ECM) – signal	Wiring, traction control module, ECM
36	Traction control/engine control module (ECM) – signal	Wiring, traction control module, ECM
41	Oxygen sensor heater – LH	Wiring, HO2S, ECM
42	Oxygen sensor heater – RH	Wiring, HO2S, ECM
43	Heated oxygen sensor (HO2S)/fuel supply system – LH	Fuel system, wiring, HO2S, ECM
44	Heated oxygen sensor (HO2S)/fuel supply system – RH	Fuel system, wiring, HO2S, ECM
45	Fuel metering – LH	Fuel system, MAP sensor, injector, ECM
46	Fuel metering – RH	Fuel system, MAP sensor, injector, ECM
53	Knock sensor (KS) – rear/RH	Wiring, KS, ECM
59	Camshaft position (CMP) sensor 2	Wiring, CMP sensor, ECM

ACURA

Model:	Engine identification:	Year:
Integra 1.8L	**B18B1**	**1994-95**
Integra 1.8L VTEC	**B18C1**	**1994-95**

System: **PGM-FI**

Self-diagnosis

General information

- Refer to the front of this manual for general test conditions, terminology, detailed descriptions of wiring faults and a general trouble shooter for electrical and mechanical faults.
- Engine control module (ECM) incorporates self-diagnosis function.
- Malfunction indicator lamp (MIL) will illuminate if certain faults are recorded.
- ECM operates in backup mode if sensors fail, to enable vehicle to be driven to workshop.
- Trouble codes can be displayed by the malfunction indicator lamp (MIL) with the service check connector jumped **1**.

Accessing

- Ensure ignition switched OFF.
- Jump service check connector terminals **1**.
- Switch ignition ON.
- Check MIL is flashing.
- Trouble codes 1-9 are indicated as follows:
  - Individual short flashes display trouble code **2** [**A**].
  - A short pause separates each flash **2** [**B**].
  - A long pause separates each trouble code **2** [**C**].
  - For example: Trouble code 3 displayed **2**.
- Trouble codes greater than 9 are indicated as follows:
  - Long flashes indicate the 'tens' of the trouble code **3** [**A**].
  - Short flashes indicate the 'units' of the trouble code **3** [**C**].
  - A short pause separates each flash **3** [**B**].
  - A long pause separates each trouble code **3** [**D**].
  - For example: Trouble code 12 displayed **3**.
- Count MIL flashes and compare with trouble code table.

NOTE: ***If malfunction indicator lamp (MIL) is ON constantly and no trouble codes are displayed, this indicates an engine control module (ECM) fault.***

- Switch ignition OFF.
- Remove jump lead.

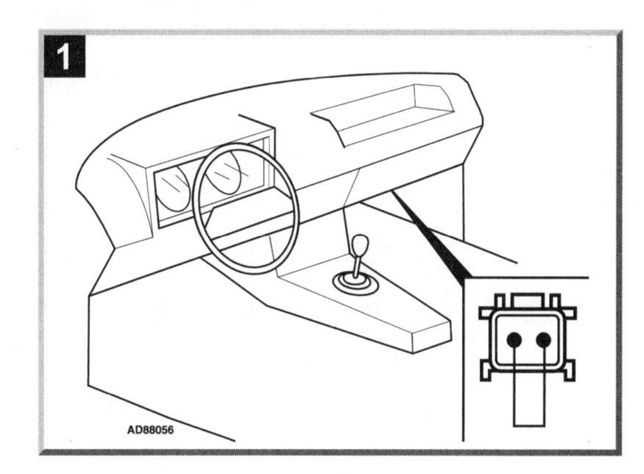

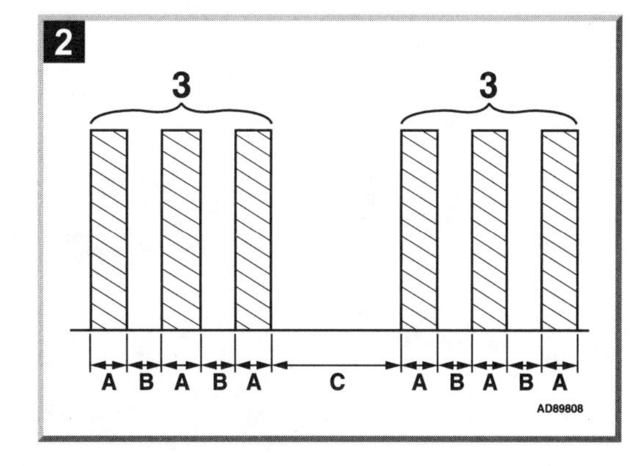

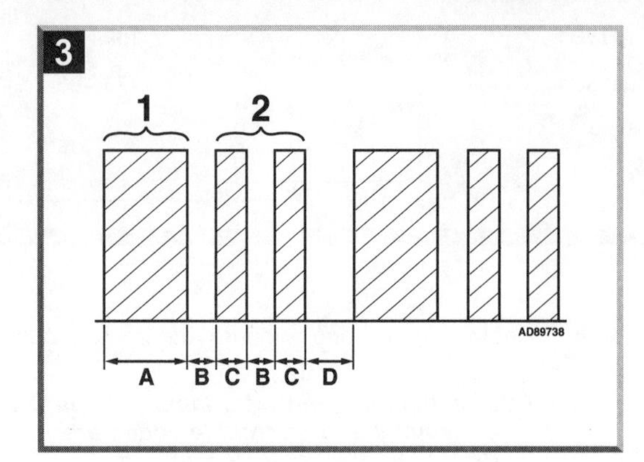

Erasing

- After the faults have been rectified, erase the trouble codes as follows:
- Switch ignition OFF.
- Remove radio fuse (7.5A) from underhood fusebox for 10 seconds minimum **4**.
- Reinstall fuse.
- Repeat checking procedure to ensure no data remains in ECM fault memory.

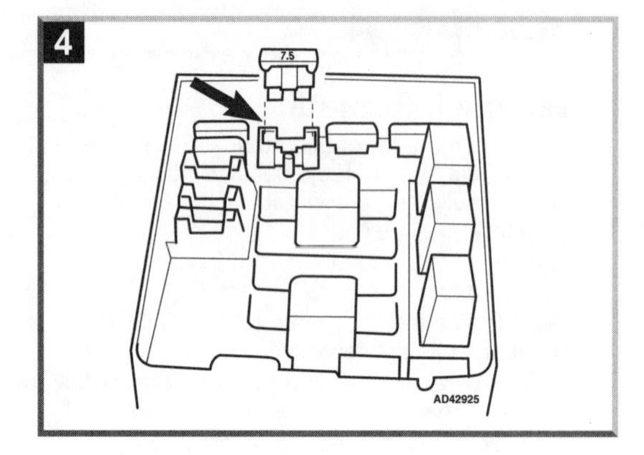

Trouble code identification

Flash code	Fault location	Probable cause
0	Engine control module (ECM)	Wiring, ECM
1	Heated oxygen sensor (HO2S)	Wiring, HO2S, ECM
3	Manifold absolute pressure (MAP) sensor	Wiring, MAP sensor, ECM
4	Crankshaft position (CKP) sensor 1	Wiring, CKP sensor, ECM
6	Engine coolant temperature (ECT) sensor	Wiring, ECT sensor, ECM
7	Throttle position (TP) sensor	Wiring, TP sensor, ECM
8	Crankshaft position (CKP) sensor 2	Wiring, CKP sensor, ECM
9	Camshaft position (CMP) sensor	Wiring, CMP sensor, ECM
10	Intake air temperature (IAT) sensor	Wiring, IAT sensor, ECM
13	Barometric pressure (BARO) sensor	ECM
14	Idle air control (IAC) valve	Wiring, IAC valve, ECM
15	Ignition output signal	Wiring, ICM, ECM

Flash code	Fault location	Probable cause
16	Injectors	Wiring, injector, ECM
17	Vehicle speed sensor (VSS)	Wiring, VSS, ECM
20	Electrical load sensor	Wiring, electrical load sensor, ECM
21	VTEC system – control valve	Wiring, VTEC control valve, ECM
22	VTEC system – pressure switch	Wiring, VTEC pressure switch, ECM
23	Knock sensor (KS)	Wiring, KS, ECM
30	AT/ECM communication signal A	Wiring, ECM
31	AT/ECM communication signal B	Wiring, ECM
41	Heated oxygen sensor (HO2S) – heater circuit	Wiring, HO2S, ECM
43	Fuel supply system	Fuel system, wiring, HO2S, ECM

ACURA

Model:	Engine identification:	Year:
Integra 1.8L	B18B1	1996-01
Integra 1.8L VTEC	B18C1/C5	1996-01

System: **PGM-FI**

Self-diagnosis

General information

- Refer to the front of this manual for general test conditions, terminology, detailed descriptions of wiring faults and a general trouble shooter for electrical and mechanical faults.
- Engine control module (ECM) incorporates self-diagnosis function.
- Malfunction indicator lamp (MIL) will illuminate if certain faults are recorded.
- ECM operates in backup mode if sensors fail, to enable vehicle to be driven to workshop.
- Trouble codes can be displayed by using a Scan Tool connected to the data link connector (DLC) by the malfunction indicator lamp (MIL) with the service check connector jumped **1**.

NOTE: ***The use of a Scan Tool is essential to obtain full diagnostic information.***

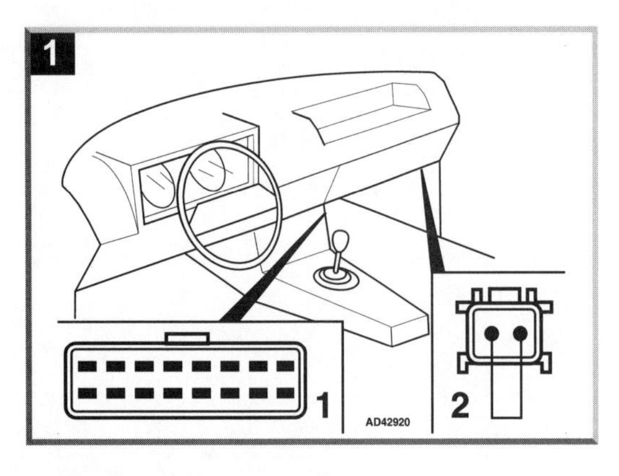

Accessing

- Ensure ignition switched OFF.
- Jump service check connector terminals **1** [2].
- Switch ignition ON.
- Check MIL is flashing.
- Trouble codes 1-9 are indicated as follows:
  - Individual short flashes display trouble code **2** [A].
  - A short pause separates each flash **2** [B].
  - A long pause separates each trouble code **2** [C].
  - For example: Trouble code 3 displayed **2**.
- Trouble codes greater than 9 are indicated as follows:
  - Long flashes indicate the 'tens' of the trouble code **3** [A].
  - Short flashes indicate the 'units' of the trouble code **3** [C].
  - A short pause separates each flash **3** [B].
  - A long pause separates each trouble code **3** [D].
  - For example: Trouble code 12 displayed **3**.
- Count MIL flashes and compare with trouble code table.
- Switch ignition OFF.
- Remove jump lead.

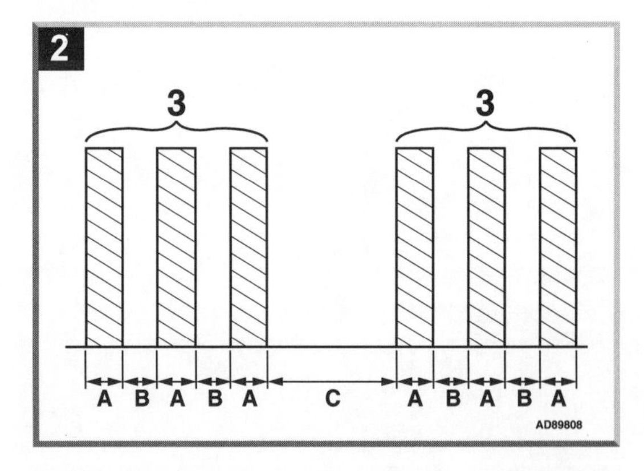

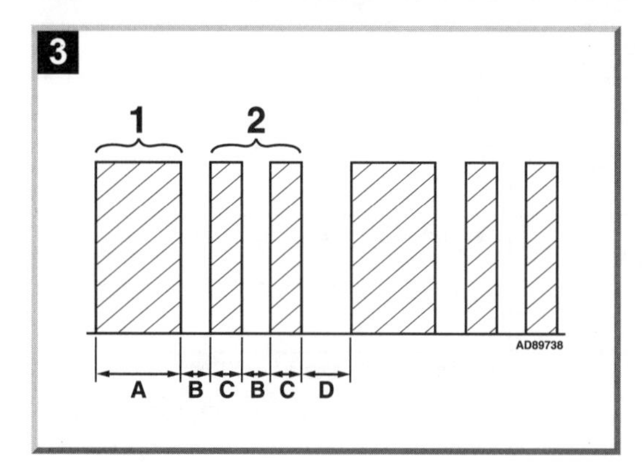

Erasing

- After the faults have been rectified, trouble codes can be erased by using a Scan Tool connected to the data link connector (DLC) or as follows:
- Switch ignition OFF.
- Remove radio fuse (7.5A) from underhood fusebox for 10 seconds minimum **4**.
- Reinstall fuse.
- Repeat checking procedure to ensure no data remains in ECM fault memory.

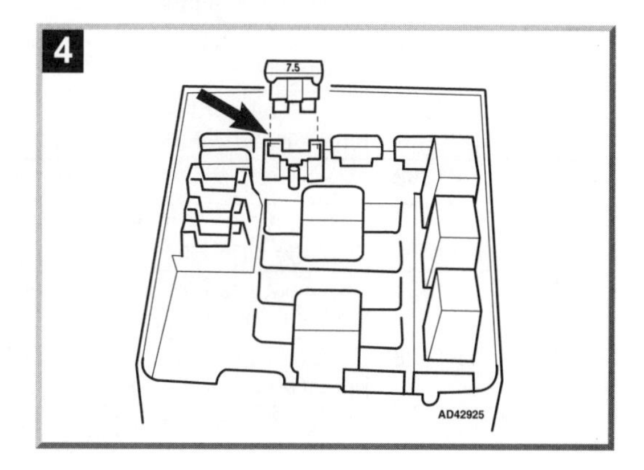

Trouble code identification

Flash code	OBD-II code	Fault location	Probable cause
–	*All P0, P2 and U0 codes*	Refer to OBD-II trouble code tables at the front of this manual	–
1	–	Heated oxygen sensor (HO2S) – front – circuit/voltage high/low	**Wiring short/open circuit, HO2S, fuel system, ECM**
3	–	Manifold absolute pressure (MAP) sensor – circuit/voltage low/high	**Wiring, MAP sensor, ECM**
4	–	Crankshaft position (CKP) sensor – circuit malfunction/range/performance problem	**Wiring, CKP sensor, ECM**
5	–	Manifold absolute pressure (MAP) sensor – range/performance problem	**Hose leak/blockage, MAP sensor**
5	P1128	Manifold absolute pressure (MAP) sensor – pressure lower than expected	**MAP sensor**
5	P1129	Manifold absolute pressure (MAP) sensor – pressure higher than expected	**MAP sensor**
6	–	Engine coolant temperature (ECT) sensor – circuit/voltage low/high	**Wiring short/open circuit, ECT sensor, ECM**
7	–	Throttle position (TP) sensor – circuit/ voltage low/high	**Wiring, TP sensor, ECM**
7	P1121	Throttle position (TP) sensor – position lower than expected	**TP sensor**
7	P1122	Throttle position (TP) sensor – position higher than expected	**TP sensor**

Flash code	OBD-II code	Fault location	Probable cause
8	P1359	Crankshaft position (CKP) sensor – connector disconnection	Wiring
8	P1361	Crankshaft position (CKP) sensor 1 – intermittent signal	CKP sensor
8	P1362	Crankshaft position (CKP) sensor 1 – no signal	Wiring, CKP sensor, ECM
9	P1381	Camshaft position (CMP) sensor – intermittent signal	CMP sensor
9	P1382	Camshaft position (CMP) sensor – no signal	Wiring, CMP sensor, ECM
10	–	Intake air temperature (IAT) sensor – circuit/voltage low/high	Wiring short/open circuit, IAT sensor, ECM
12	P1491	Exhaust gas recirculation (EGR) system – valve lift insufficient	Wiring, EGR valve/position sensor, EGR solenoid, hose leak/blockage, ECM
12	P1498	Exhaust gas recirculation (EGR) valve position sensor – voltage high	Wiring, EGR valve/position sensor, ECM
13	P1106	Barometric pressure (BARO) sensor – range/performance problem	ECM
13	P1107	Barometric pressure (BARO) sensor – circuit/voltage low	Wiring, TCM, ECM
13	P1108	Barometric pressure (BARO) sensor – circuit/voltage high	ECM
14	P1519	Idle control system – malfunction	IAC valve, fast idle thermo valve, throttle body
14	P1508	Idle air control (IAC) valve – circuit failure	Wiring, IAC valve, ECM
17	–	Vehicle speed sensor (VSS) – circuit malfunction	Wiring, VSS, ECM
20	P1297	Electrical load sensor – circuit/voltage low	Wiring, electrical load sensor, ECM
20	P1298	Electrical load sensor – circuit/voltage high	Wiring, electrical load sensor, ECM
22	P1259	VTEC system malfunction	Wiring, VTEC solenoid/pressure switch, ECM
23	–	Knock sensor (KS) – circuit malfunction	Wiring, KS, ECM
30	P1681	AT to ECM – signal A – voltage low	Wiring, TCM, ECM
30	P1682	AT to ECM – signal A – voltage high	Wiring, TCM, ECM
31	P1686	AT to ECM – signal B – voltage low	Wiring, TCM, ECM
31	P1687	AT to ECM – signal B – voltage high	Wiring, TCM, ECM
35	P1676	ABS/TCS control module – communication malfunction	Wiring, ABS/TCS control module, ECM
35	P1678	ABS/TCS control module – communication malfunction	Wiring, ABS/TCS control module, ECM
41	–	Heated oxygen sensor (HO2S) – front – circuit malfunction	Wiring, ECM
41	P1166	Air fuel (A/F) sensor – RSX 1 – heater circuit malfunction	Wiring, A/F sensor, ECM

Flash code	OBD-II code	Fault location	Probable cause
41	P1167	Air fuel (A/F) sensor – RSX 1 – heater system malfunction	**Wiring, A/F sensor, ECM**
45	–	Mixture too lean/rich	**Fuel system, front HO2S, MAP sensor, mechanical fault**
48	P1162	Air fuel (A/F) sensor – RSX 1 – circuit malfunction	**Wiring, A/F sensor, ECM**
54	P1336	Crankshaft position (CKP) sensor 2 – intermittent signal	**CKP sensor**
54	P1337	Crankshaft position (CKP) sensor 2 – no signal	**Wiring, CKP sensor, ECM**
58	P1366	Crankshaft position (CKP) sensor 2 – intermittent signal	**CKP sensor**
58	P1367	Crankshaft position (CKP) sensor 2 – no signal	**Wiring, CKP sensor, ECM**
61	–	Heated oxygen sensor (HO2S) – front – slow response	**HO2S, exhaust system**
61	P1163	Air fuel (A/F) sensor – RSX 1 – slow response	**A/F sensor**
61	P1164	Air fuel (A/F) sensor – RSX 1 – range/ performance problem	**Wiring, A/F sensor, ECM**
63	–	Heated oxygen sensor (HO2S) – rear – slow response/circuit/voltage low/high	**Wiring, HO2S, ECM**
65	–	Heated oxygen sensor (HO2S) – rear – circuit malfunction	**Wiring, ECM**
67	–	Catalytic converter – efficiency below limit	**Catalytic converter, rear HO2S**
70	–	AT – lock-up clutch not engaging/ no gear shift	**Wiring, mainshaft speed sensor, countershaft speed sensor, lock-up control system, shift solenoid (SS) A/B, TCM**
70	P1660	AT to ECM – signal failure	**Wiring, TCM, ECM**
70	P1705	AT – gear shift malfunction	**Wiring, range position switch, TCM**
70	P1705	AT – lock-up clutch not engaging	**Wiring, range position switch, TCM**
70	P1706	AT – gear shift malfunction	**Wiring, range position switch, TCM**
70	P1706	AT – lock-up clutch malfunction	**Wiring, range position switch, TCM**
70	P1753	AT – lock-up clutch not engaging/ disengaging	**Wiring, lock-up control solenoid A, TCM**
70	P1758	AT – lock-up clutch not engaging	**Wiring, lock-up control solenoid B, TCM**
70	P1786	AT – poor gear shift	**Communication wire, ECM, TCM**
70	P1790	AT – lock-up clutch not engaging	**Wiring, TP sensor, TCM**
70	P1791	AT – lock-up clutch not engaging	**Wiring, VSS, TCM**
70	P1792	AT – lock-up clutch not engaging	**Wiring, ECT sensor, TCM**
70	P1794	Automatic transmission – BARO signal	**Wiring, ECM, TCM**

Flash code	OBD-II code	Fault location	Probable cause
71	–	Cylinder No.1 – misfire	**Wiring, injector, ignition system, mechanical fault**
72	–	Cylinder No.2 – misfire	**Wiring, injector, ignition system, mechanical fault**
73	–	Cylinder No.3 – misfire	**Wiring, injector, ignition system, mechanical fault**
74	–	Cylinder No.4 – misfire	**Wiring, injector, ignition system, mechanical fault**
–	P1300	Random misfire	**Fuel/ignition system, MAP sensor, IAC valve**
–	P1399	Random misfire	**Fuel/ignition system, MAP sensor, IAC valve**
86	–	Engine coolant temperature (ECT) sensor – range/performance problem	**ECT sensor, cooling system**
90	P1456	Evaporative emission (EVAP) canister purge system (fuel tank system) – leak detected	**Hose, fuel tank/pressure sensor, fuel filler cap, EVAP valve/bypass solenoid, EVAP two way valve, EVAP canister/vent valve**
90	P1457	Evaporative emission (EVAP) canister purge system (canister system) – leak detected	**Hose, fuel tank/pressure sensor, EVAP valve/bypass solenoid, EVAP two way valve, EVAP canister/vent valve**
91	–	Fuel tank pressure sensor – circuit/voltage low/high	**Wiring, pressure sensor, ECM**
92	–	Evaporative emission (EVAP) canister purge system – incorrect flow	**Wiring, EVAP solenoid, hose leak/ blockage, ECM**
–	P1607	Engine control module (ECM) – internal circuit failure	**ECM**
106	P1077	Intake manifold air control system – low RPM malfunction	**Wiring, intake manifold air control solenoid, ECM**
106	P1078	Intake manifold air control system – high RPM malfunction	**Wiring, intake manifold air control solenoid, ECM**

ACURA

Model:	Engine identification:	Year:
3.2CL	J32A1/2	2000-03
MDX 3.5L	J35A3/5	2000-04
RSX 2.0L	K20A2/3	2002-04

System: **PGM-FI**

Self-diagnosis

General information

- Refer to the front of this manual for general test conditions, terminology, detailed descriptions of wiring faults and a general trouble shooter for electrical and mechanical faults.
- Engine control module (ECM) incorporates self-diagnosis function.
- Malfunction indicator lamp (MIL) will illuminate if certain faults are recorded.
- ECM operates in backup mode if sensors fail, to enable vehicle to be driven to workshop.

Accessing

- Trouble codes can only be displayed by using a Scan Tool connected to the data link connector (DLC) **1** – 3.2CL, **2** – MDX or **3** – RSX.

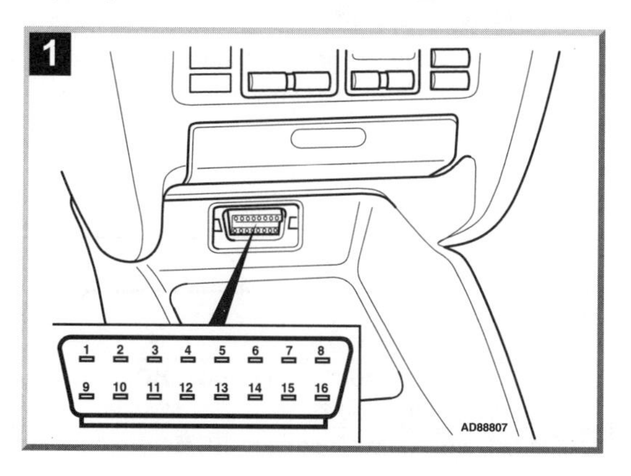

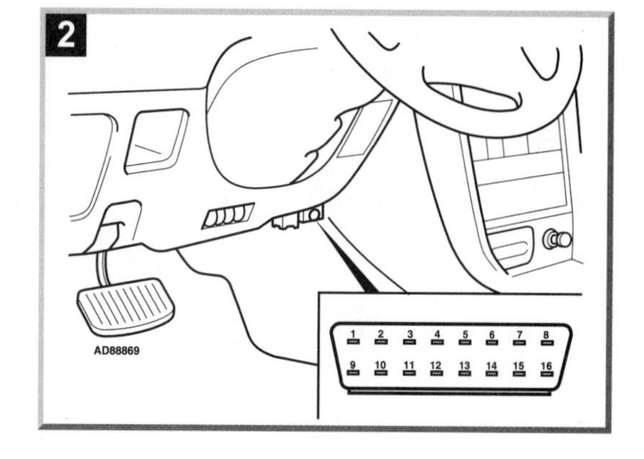

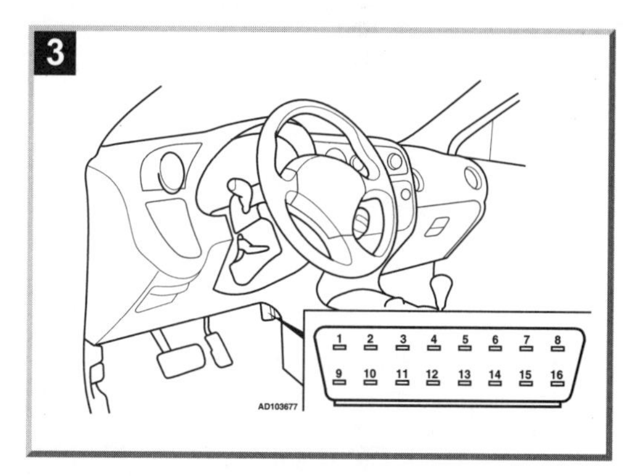

Erasing – →2002

- After the faults have been rectified, trouble codes can be erased by using a Scan Tool connected to the data link connector (DLC) or as follows:
- Switch ignition OFF.
- 3.2CL – Remove clock/backup fuse (15A) from RH fascia fusebox for 10 seconds minimum **4**.
- MDX – Remove clock/backup fuse (7.5A) from RH fascia fusebox for 10 seconds minimum **5**.
- RSX – Remove ECM fuse (15A) from underhood fusebox for 10 seconds minimum **6**.
- Reinstall fuse.
- Repeat checking procedure to ensure no data remains in ECM fault memory.

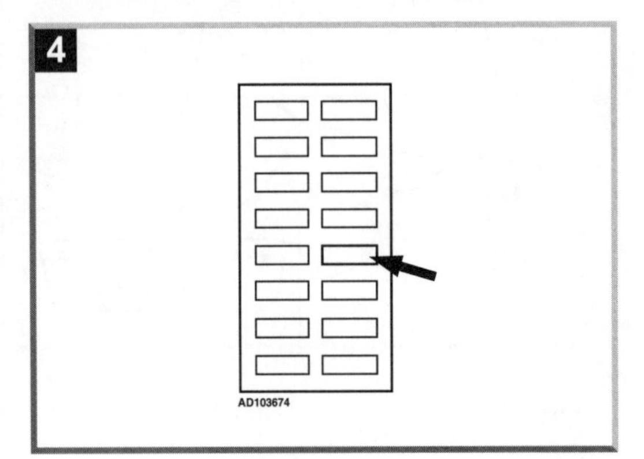

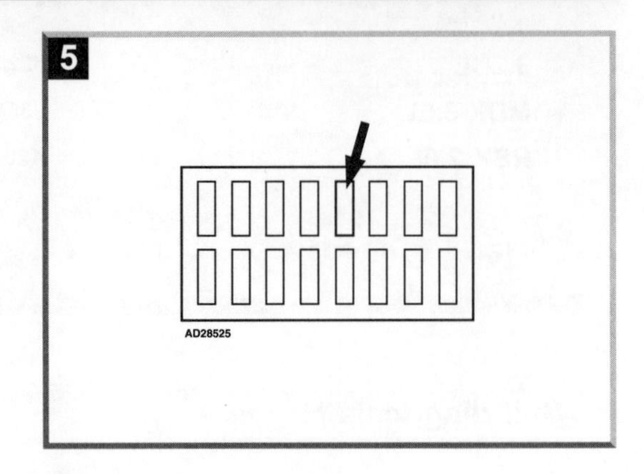

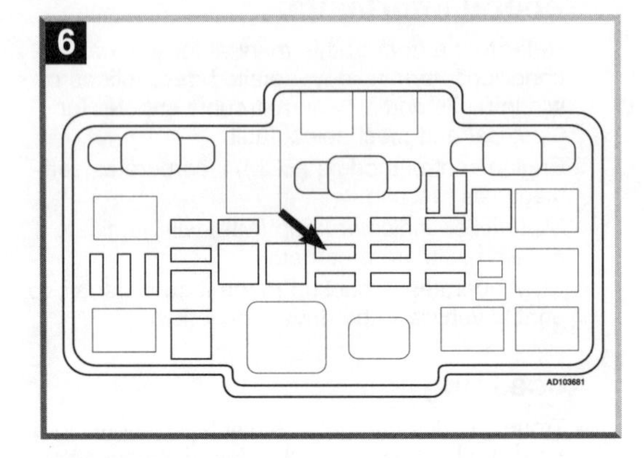

Erasing – 2003→

- Trouble codes can only be erased by using a Scan Tool connected to the data link connector (DLC).

Trouble code identification

Flash code	OBD-II code	Fault location	Probable cause
–	*All P0, P2 and U0 codes*	Refer to OBD-II trouble code tables at the front of this manual	–
1	–	Heated oxygen sensor (HO2S) – front – circuit/voltage high/low	**Wiring short/open circuit, HO2S, fuel system, ECM**
3	–	Manifold absolute pressure (MAP) sensor – circuit/voltage low/high	**Wiring, MAP sensor, ECM**
4	–	Crankshaft position (CKP) sensor – circuit malfunction/range/performance problem	**Wiring, CKP sensor, ECM**
5	–	Manifold absolute pressure (MAP) sensor – range/performance problem	**Hose leak/blockage, MAP sensor**

Flash code	OBD-II code	Fault location	Probable cause
5	P1128	Manifold absolute pressure (MAP) sensor – pressure lower than expected	**MAP sensor**
5	P1129	Manifold absolute pressure (MAP) sensor – pressure higher than expected	**MAP sensor**
6	–	Engine coolant temperature (ECT) sensor – circuit/voltage low/high	**Wiring short/open circuit, ECT sensor, ECM**
7	–	Throttle position (TP) sensor – circuit/ voltage low/high	**Wiring, TP sensor, ECM**
7	P1121	Throttle position (TP) sensor – position lower than expected	**TP sensor**
7	P1122	Throttle position (TP) sensor – position higher than expected	**TP sensor**
8	P1359	Crankshaft position (CKP) sensor – connector disconnection	**Wiring**
8	P1361	Crankshaft position (CKP) sensor 1 – intermittent signal	**CKP sensor**
8	P1362	Crankshaft position (CKP) sensor 1 – no signal	**Wiring, CKP sensor, ECM**
9	P1381	Camshaft position (CMP) sensor – intermittent signal	**CMP sensor**
9	P1382	Camshaft position (CMP) sensor – no signal	**Wiring, CMP sensor, ECM**
10	–	Intake air temperature (IAT) sensor – circuit/ voltage low/high	**Wiring short/open circuit, IAT sensor, ECM**
12	P1491	Exhaust gas recirculation (EGR) system – valve lift insufficient	**Wiring, EGR valve/position sensor, EGR solenoid, hose leak/blockage, ECM**
12	P1498	Exhaust gas recirculation (EGR) valve position sensor – voltage high	**Wiring, EGR valve/position sensor, ECM**
13	P1106	Barometric pressure (BARO) sensor – range/performance problem	**ECM**
13	P1107	Barometric pressure (BARO) sensor – circuit/voltage low	**Wiring, TCM, ECM**
13	P1108	Barometric pressure (BARO) sensor – circuit/voltage high	**ECM**
14	P1519	Idle control system – malfunction	**IAC valve, fast idle thermo valve, throttle body**
14	P1508	Idle air control (IAC) valve – circuit failure	**Wiring, IAC valve, ECM**
17	–	Vehicle speed sensor (VSS) – circuit malfunction	**Wiring, VSS, ECM**
20	P1297	Electrical load sensor – circuit/voltage low	**Wiring, electrical load sensor, ECM**
20	P1298	Electrical load sensor – circuit/voltage high	**Wiring, electrical load sensor, ECM**
22	P1259	VTEC system malfunction	**Wiring, VTEC solenoid/pressure switch, ECM**
23	–	Knock sensor (KS) – circuit malfunction	**Wiring, KS, ECM**
30	P1681	AT to ECM – signal A – voltage low	**Wiring, TCM, ECM**

Flash code	OBD-II code	Fault location	Probable cause
30	P1682	AT to ECM – signal A – voltage high	**Wiring, TCM, ECM**
31	P1686	AT to ECM – signal B – voltage low	**Wiring, TCM, ECM**
31	P1687	AT to ECM – signal B – voltage high	**Wiring, TCM, ECM**
35	P1676	ABS/TCS control module – communication malfunction	**Wiring, ABS/TCS control module, ECM**
35	P1678	ABS/TCS control module – communication malfunction	**Wiring, ABS/TCS control module, ECM**
37	P1656	Electronic stability programme (ESP) control module – communication malfunction	**Wiring, ESP control module**
40	P1683	Throttle motor – default position spring performance problem	**Wiring, throttle sticking/mechanically damaged, throttle motor**
40	P1684	Throttle motor – return spring performance problem	**Wiring, throttle sticking/mechanically damaged, throttle motor**
41	–	Heated oxygen sensor (HO2S) – front – circuit malfunction	**Wiring, ECM**
41	P1166	Air fuel (A/F) sensor – RSX 1 – heater circuit malfunction	**Wiring, A/F sensor, ECM**
41	P1167	Air fuel (A/F) sensor – RSX 1 – heater system malfunction	**Wiring, A/F sensor, ECM**
45	–	Mixture too lean/rich	**Fuel system, front HO2S, MAP sensor, mechanical fault**
48	P1157	Air fuel (A/F) sensor – RSX 1 – circuit malfunction	**Wiring, A/F sensor, ECM**
48	P1162	Air fuel (A/F) sensor – RSX 1 – circuit malfunction	**Wiring, A/F sensor, ECM**
54	P1336	Crankshaft position (CKP) sensor 2 – intermittent signal	**CKP sensor**
54	P1337	Crankshaft position (CKP) sensor 2 – no signal	**Wiring, CKP sensor, ECM**
58	P1366	Crankshaft position (CKP) sensor 2 – intermittent signal	**CKP sensor**
58	P1367	Crankshaft position (CKP) sensor 2 – no signal	**Wiring, CKP sensor, ECM**
61	–	Heated oxygen sensor (HO2S) – front – slow response	**HO2S, exhaust system**
61	P1163	Air fuel (A/F) sensor – RSX 1 – slow response	**A/F sensor**
61	P1164	Air fuel (A/F) sensor – RSX 1 – range/ performance problem	**Wiring, A/F sensor, ECM**
63	–	Heated oxygen sensor (HO2S) – rear – slow response/circuit/voltage low/high	**Wiring, HO2S, ECM**
63	P1710	AT – 2004 MDX – 1st gear hold switch	**Wiring, range position switch, detent bracket/ plunger/spring, ECM**

Flash code	OBD-II code	Fault location	Probable cause
65	–	Heated oxygen sensor (HO2S) – rear – circuit malfunction	**Wiring, ECM**
67	–	Catalytic converter – efficiency below limit	**Catalytic converter, rear HO2S**
70	–	AT – lock-up clutch not engaging/no gear shift	**Wiring, mainshaft speed sensor, countershaft speed sensor, lock-up control system, shift solenoid (SS) A/B, TCM**
70	P1656	Electronic stability programme (ESP) control module – communication malfunction	**Wiring, ESP control module**
70	P1660	AT to ECM – signal failure	**Wiring, TCM, ECM**
70	P1705	AT – gear shift malfunction	**Wiring, range position switch, TCM**
70	P1705	AT – lock-up clutch not engaging	**Wiring, range position switch, TCM**
70	P1706	AT – gear shift malfunction	**Wiring, range position switch, TCM**
70	P1706	AT – lock-up clutch malfunction	**Wiring, range position switch, TCM**
70	P1709	AT – gear selection malfunction	**Wiring, transmission gear selection switch, detent bracket, detent plunger/ spring, ECM**
70	P1710	AT – 1st gear hold switch	**Wiring, range position switch, detent bracket/ plunger/spring, ECM**
70	P1717	AT – gear selection malfunction	**Wiring, range position switch, ECM**
70	P1739	AT – 3rd clutch pressure switch	**Wiring, range position switch, 3rd clutch pressure switch, ECM**
70	P1740	AT – 4th clutch pressure switch	**Wiring, range position switch, 4th clutch pressure switch, ECM**
70	P1750	AT – hydraulic system mechanical malfunction	**Wiring, shift solenoids, ECM, transmission mechanical fault**
70	P1751	AT – hydraulic system mechanical malfunction	**Wiring, shift solenoids, ECM, transmission mechanical fault**
70	P1753	AT – lock-up clutch not engaging/ disengaging	**Wiring, lock-up control solenoid A, TCM**
70	P1758	AT – lock-up clutch not engaging	**Wiring, lock-up control solenoid B, TCM**
70	P1768	Shift solenoid (SS) A	**Wiring, shift solenoid, ECM**
70	P1773	Shift solenoid (SS) B	**Wiring, shift solenoid, ECM**
70	P1778	Shift solenoid (SS) C	**Wiring, shift solenoid, ECM**
70	P1786	AT – poor gear shift	**Communication wire, ECM, TCM**
70	P1790	AT – lock-up clutch not engaging	**Wiring, TP sensor, TCM**
70	P1791	AT – lock-up clutch not engaging	**Wiring, VSS, TCM**
70	P1792	AT – lock-up clutch not engaging	**Wiring, ECT sensor, TCM**
70	P1794	Automatic transmission – BARO signal	**Wiring, ECM, TCM**

Flash code	OBD-II code	Fault location	Probable cause
71	–	Cylinder No.1 – misfire	**Wiring, injector, ignition system, mechanical fault**
72	–	Cylinder No.2 – misfire	**Wiring, injector, ignition system, mechanical fault**
73	–	Cylinder No.3 – misfire	**Wiring, injector, ignition system, mechanical fault**
74	–	Cylinder No.4 – misfire	**Wiring, injector, ignition system, mechanical fault**
–	P1300	Random misfire	**Fuel/ignition system, MAP sensor, IAC valve**
–	P1399	Random misfire	**Fuel/ignition system, MAP sensor, IAC valve**
86	–	Engine coolant temperature (ECT) sensor – range/performance problem	**ECT sensor, cooling system**
90	P1456	Evaporative emission (EVAP) canister purge system (fuel tank system) – leak detected	**Hose, fuel tank/pressure sensor, fuel filler cap, EVAP valve/bypass solenoid, EVAP two way valve, EVAP canister/vent valve**
90	P1457	Evaporative emission (EVAP) canister purge system (canister system) – leak detected	**Hose, fuel tank/pressure sensor, EVAP valve/bypass solenoid, EVAP two way valve, EVAP canister/vent valve**
91	–	Fuel tank pressure sensor – circuit/voltage low/high	**Wiring, pressure sensor, ECM**
91	P1454	Fuel tank pressure sensor – range/performance problem	**Wiring, hose, fuel tank pressure sensor, fuel filler cap, EVAP canister/vent valve, ECM**
92	–	Evaporative emission (EVAP) canister purge system – incorrect flow	**Wiring, EVAP solenoid, hose leak/ blockage, ECM**
–	P1607	Engine control module (ECM) – internal circuit failure	**ECM**
106	P1077	Intake manifold air control system – low RPM malfunction	**Wiring, intake manifold air control solenoid, ECM**
106	P1078	Intake manifold air control system – high RPM malfunction	**Wiring, intake manifold air control solenoid, ECM**
118	P1450	Evaporative emission (EVAP) two way valve, bypass valve – voltage low	**Wiring, EVAP two way valve/bypass solenoid, ECM**
118	P1451	Evaporative emission (EVAP) two way valve, bypass valve – voltage high	**Wiring, EVAP two way valve/bypass solenoid, ECM**
121	P1460	Fuel gauge tank sensor – supply circuit	**Wiring, tank sensor, ECM**

ACURA

Model:	Engine identification:	Year:
NSX/NSX-T 3.0L	C30A1	1991-04
NSX/NSX-T 3.2L	C32B1	1997-04

System: **PGM-FI**

Self-diagnosis

General information

- Refer to the front of this manual for general test conditions, terminology, detailed descriptions of wiring faults and a general trouble shooter for electrical and mechanical faults.
- Engine control module (ECM) incorporates self-diagnosis function.
- Malfunction indicator lamp (MIL) will illuminate if certain faults are recorded.
- ECM operates in backup mode if sensors fail, to enable vehicle to be driven to workshop.
- Trouble codes can be displayed by using a Scan Tool connected to the data link connector (DLC) →2001 **1** [1], 2002→ **1** [2] or by the malfunction indicator lamp (MIL) with the service check connector jumped **1** [3].

NOTE: ***The use of a Scan Tool is essential to obtain full diagnostic information.***

Accessing

- Ensure ignition switched OFF.
- Jump service check connector terminals **1** [3].
- Switch ignition ON.
- Check MIL is flashing.
- Trouble codes 1-9 are indicated as follows:
  - Individual short flashes display trouble code **2** [A].
  - A short pause separates each flash **2** [B].
  - A long pause separates each trouble code **2** [C].
  - For example: Trouble code 3 displayed **2**.
- Trouble codes greater than 9 are indicated as follows:
  - Long flashes indicate the 'tens' of the trouble code **3** [A].
  - Short flashes indicate the 'units' of the trouble code **3** [C].
  - A short pause separates each flash **3** [B].
  - A long pause separates each trouble code **3** [D].
  - For example: Trouble code 12 displayed **3**.
- Count MIL flashes and compare with trouble code table.
- Switch ignition OFF.
- Remove jump lead.

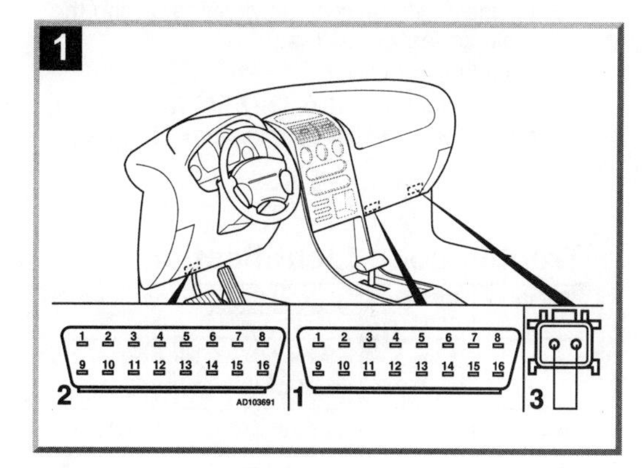

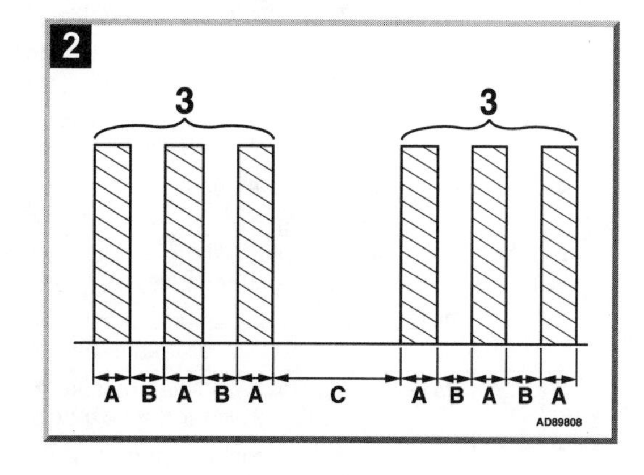

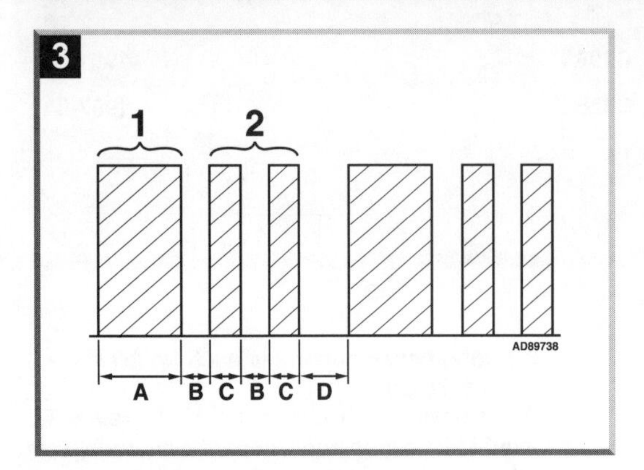

Erasing

- After the faults have been rectified, erase the trouble codes as follows:
- Switch ignition OFF.
- Remove clock fuse No.33 (7.5A) from underhood fusebox for 10 seconds minimum 4.
- Reinstall fuse.
- Repeat checking procedure to ensure no data remains in ECM fault memory.

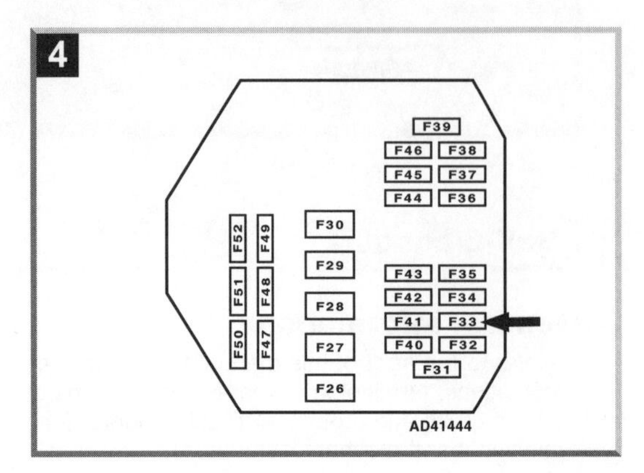

Trouble code identification

Flash code	OBD-II code	Fault location	Probable cause
–	*All P0, P2 and U0 codes*	Refer to OBD-II trouble code tables at the front of this manual	–
0	–	Engine control module –	**Wiring, ECM**
1	–	Heated oxygen sensor (HO2S) – front bank, front sensor – circuit/voltage low/high	**Wiring, HO2S, fuel system, ECM**
2	–	Heated oxygen sensor (HO2S) – rear bank, front sensor – circuit/voltage low/high	**Wiring, HO2S, ECM**
3	–	Manifold absolute pressure (MAP) sensor – circuit/voltage low/high	**Wiring, MAP sensor, ECM**
4	–	Crankshaft position (CKP) sensor 1 – range/performance/circuit malfunction	**Wiring, CKP sensor, valve timing, ECM**
5	–	Manifold absolute pressure (MAP) sensor – range/performance	**Hose leak/blockage, MAP sensor**
5	P1128	Manifold absolute pressure (MAP) sensor – pressure lower than expected	**MAP sensor**
5	P1129	Manifold absolute pressure (MAP) sensor – pressure higher than expected	**MAP sensor**
6	–	Engine coolant temperature (ECT) sensor – circuit/voltage low/high	**Wiring, ECT sensor, ECM**
7	–	Throttle position (TP) sensor – circuit/ voltage low/high	**Wiring, TP sensor, ECM**

Flash code	OBD-II code	Fault location	Probable cause
9	P1382	Camshaft position (CMP) sensor 1 – signal malfunction	**Wiring, CMP sensor, ECM**
10	–	Intake air temperature (IAT) sensor – circuit/ voltage low/high	**Wiring, IAT sensor, ECM**
12	P1491	Exhaust gas recirculation (EGR) system – valve lift insufficient	**Wiring, EGR valve/position sensor, EGR solenoid, hose leak/blockage, ECM**
12	P1498	Exhaust gas recirculation (EGR) valve position sensor – voltage high	**Wiring, EGR valve/position sensor, ECM**
13	P1106	Barometric pressure (BARO) sensor – range/performance	**ECM**
13	P1107	Barometric pressure (BARO) sensor – circuit/voltage low	**ECM**
13	P1108	Barometric pressure (BARO) sensor – circuit/voltage high	**ECM**
14	–	Idle control system – malfunction	**Idle speed, throttle body**
17	–	Vehicle speed sensor (VSS) – voltage low	**Wiring, VSS, ECM**
22	P1279	VTEC control system malfunction – front bank	**Wiring, VTEC solenoid/pressure switch, ECM**
23	–	Knock sensor (KS) – front bank – circuit malfunction	**Wiring, KS, ECM**
31	P1671	AT to ECM signal – no signal	**Wiring, TCM, ECM**
31	P1672	AT to ECM signal – signal failure	**Wiring, TCM, ECM**
34	–	System voltage – malfunction	**Wiring, poor connection, battery**
35	P1676	ECM to traction control system signal – no signal	**Wiring, traction control module, ECM**
35	P1677	ECM to traction control system signal – signal failure	**Wiring, traction control module, ECM**
37	P1246	Accelerator pedal position (APP) sensor 1 – circuit malfunction	**Wiring, APP sensor, ECM**
37	P1247	Accelerator pedal position (APP) sensor 2 – circuit malfunction	**Wiring, APP sensor, ECM**
37	P1248	Accelerator pedal position (APP) sensor 1 & 2 – incorrect signals	**APP sensor**
40	P1241	Throttle actuator control (TAC) motor – circuit 1 – malfunction	**Wiring, TAC motor, ECM**
40	P1242	Throttle actuator control (TAC) motor – circuit 2 – malfunction	**Wiring, TAC motor, ECM**
40	P1243	Throttle position (TP) – insufficient	**Throttle valve, TP sensor, TAC motor**
40	P1244	Closed throttle position (CTP) – insufficient	**Throttle valve, TP sensor**
41	–	Heated oxygen sensor (HO2S) – front bank, front sensor – heater circuit malfunction	**Wiring, ECM**
42	–	Heated oxygen sensor (HO2S) – rear bank, front sensor – heater circuit malfunction	**Wiring, ECM**

Flash code	OBD-II code	Fault location	Probable cause
45	–	Mixture too lean/rich – front bank	**Fuel system, front HO2S, MAP sensor, mechanical fault**
46	–	Mixture too lean/rich – rear bank	**Fuel system, front HO2S, MAP sensor, mechanical fault**
52	P1259	VTEC control system malfunction – rear bank	**Wiring, VTEC solenoid/pressure switch, ECM**
53	–	Knock sensor (KS) – rear bank – circuit malfunction	**Wiring, KS, ECM**
54	P1336	Crankshaft position (CKP) sensor 2 – range/performance	**CKP sensor, valve timing**
54	P1337	Crankshaft position (CKP) sensor 2 – circuit malfunction	**Wiring, CKP sensor, ECM**
59	P1386	Camshaft position (CMP) sensor 2 – intermittent signal	**CMP sensor**
59	P1387	Camshaft position (CMP) sensor 2 – no signal	**Wiring, CMPO sensor, ECM**
60	–	Secondary air injection (AIR) system – malfunction	**Wiring, AIR pump relay, AIR pump, ECM**
60	P1410	Secondary air injection (AIR) system – malfunction	**AIR pump**
60	P1419	Secondary air injection (AIR) system – malfunction	**AIR pipe(s), ECM**
61	–	Heated oxygen sensor (HO2S) – front bank, front sensor – slow response	**HO2S, exhaust system**
62	–	Heated oxygen sensor (HO2S) – rear bank, front sensor – slow response	**HO2S, exhaust system**
63	–	Heated oxygen sensor (HO2S) – front bank, rear sensor – slow response/circuit/voltage low/high	**Wiring, HO2S, ECM**
64	–	Heated oxygen sensor (HO2S) – rear bank, rear sensor – slow response/circuit/voltage low/high	**Wiring, HO2S, ECM**
65	–	Heated oxygen sensor (HO2S) – front bank, rear sensor – heater circuit malfunction	**Wiring, HO2S, ECM**
66	–	Heated oxygen sensor (HO2S) – rear bank, rear sensor – heater circuit malfunction	**Wiring, HO2S, ECM**
67	–	Catalytic converter – front bank – efficiency below limit	**Catalytic converter, rear HO2S**
68	–	Catalytic converter – rear bank – efficiency below limit	**Catalytic converter, rear HO2S**
70	–	AT – lock-up clutch not engaging/no gear shift	**Wiring, mainshaft speed sensor, countershaft speed sensor, lock-up control system, shift solenoid (SS) A/B, TCM**
70	P1705	AT – gear shift malfunction	**Wiring, range position switch, TCM**

Flash code	OBD-II code	Fault location	Probable cause
70	P1706	AT – gear shift malfunction	Wiring, range position switch, TCM
70	P1706	AT – lock-up clutch malfunction	Wiring, range position switch, TCM
70	P1709	AT – gear selection malfunction	Wiring, transmission gear selection switch, ECM
70	P1753	AT – lock-up clutch not engaging/ disengaging	Wiring, lock-up control solenoid A, TCM
70	P1758	AT – lock-up clutch not engaging	Wiring, lock-up control solenoid B, TCM
70	P1768	AT – poor gear shift	Wiring, linear solenoid, TCM
70	P1768	AT – lock-up clutch not engaging	Wiring, linear solenoid, TCM
70	P1788	AT – poor gear shift	Communication wire, ECM, TCM
70	P1790	AT – lock-up clutch not engaging	Wiring, TP sensor, TCM
70	P1791	AT – lock-up clutch not engaging	Wiring, VSS, TCM
70	P1792	AT – lock-up clutch not engaging	Wiring, ECT sensor, TCM
70	P1793	Automatic transmission	Wiring, MAP sensor, TCM
70	P1795	Automatic transmission	Wiring, APP sensor, TCM
71	P1201	Cylinder No.1 – misfire	Wiring, injector, ignition system, mechanical fault
71	P1301	Cylinder No.1 – misfire	Ignition system
72	P1202	Cylinder No.2 – misfire	Wiring, injector, ignition system, mechanical fault
72	P1302	Cylinder No.2 – misfire	Ignition system
73	P1203	Cylinder No.3 – misfire	Wiring, injector, ignition system, mechanical fault
73	P1303	Cylinder No.3 – misfire	Ignition system
74	P1204	Cylinder No.4 – misfire	Wiring, injector, ignition system, mechanical fault
74	P1304	Cylinder No.4 – misfire	Ignition system
75	P1205	Cylinder No.5 – misfire	Wiring, injector, ignition system, mechanical fault
75	P1305	Cylinder No.5 – misfire	Ignition system
76	P1206	Cylinder No.6 – misfire	Wiring, injector, ignition system, mechanical fault
76	P1306	Cylinder No.6 – misfire	Ignition system
–	P1300	Random misfire	EGR/fuel/ignition system, MAP sensor
79	P1316	Spark plug voltage detection – front bank – circuit malfunction	Wiring, voltage detection module, ECM
79	P1317	Spark plug voltage detection – rear bank – circuit malfunction	Wiring, voltage detection module, ECM
79	P1318	Spark plug voltage detection module reset – front bank – reset circuit malfunction	Wiring, voltage detection module, ECM

Flash code	OBD-II code	Fault location	Probable cause
79	P1319	Spark plug voltage detection module reset – rear bank – reset circuit malfunction	**Wiring, voltage detection module, ECM**
80	–	Exhaust gas recirculation (EGR) solenoid – insufficient flow	**EGR solenoid, hose/pipe leak or blockage**
83	P1415	AIR pump electrical current sensor – open/short circuit	**Wiring, AIR pump electrical current sensor, ECM**
83	P1416	AIR pump electrical current sensor – open circuit	**Wiring, AIR pump electrical current sensor, ECM**
86	–	Engine coolant temperature (ECT) sensor – range/performance problem	**ECT sensor, cooling system**
87	P1486	Thermostat – range/performance problem	**Engine coolant blower motor, cooling system**
90	P1456	Evaporative emission (EVAP) canister purge system (fuel tank system) – leak detected	**Hose leak/blockage, fuel tank/pressure sensor, fuel filler cap, EVAP valve/bypass solenoid, EVAP two way valve, EVAP canister/vent valve**
90	P1457	Evaporative emission (EVAP) canister purge system (canister system) – leak detected	**Hose leak/blockage, fuel tank/pressure sensor, EVAP valve/bypass solenoid, EVAP two way valve, EVAP canister/vent valve**
91	–	Fuel tank pressure sensor – circuit/voltage low/high	**Wiring, pressure sensor, ECM**
92	–	Evaporative emission (EVAP) canister purge system – incorrect flow	**EVAP solenoid/flow switch, wiring, throttle body, hose leak/blockage, ECM**
92	P1459	Evaporative emission (EVAP) canister purge system – switch malfunction	**Wiring, flow switch, hose leak/blockage, ECM**
–	P1607	Engine control module (ECM) – internal circuit failure A	**ECM**
–	P1608	Engine control module (ECM) – internal circuit failure B	**ECM**

ACURA

Model:	Engine identification:	Year:
SLX 3.2L	6VD1	1996-97
SLX 3.5L	6VE1	1998-99

System: **Isuzu MFI**

Self-diagnosis

General information

- Refer to the front of this manual for general test conditions, terminology, detailed descriptions of wiring faults and a general trouble shooter for electrical and mechanical faults.
- Engine control module (ECM) incorporates self-diagnosis function.
- Malfunction indicator lamp (MIL) will illuminate if certain faults are recorded.
- ECM operates in backup mode if sensors fail, to enable vehicle to be driven to workshop.

Accessing

- Trouble codes can be displayed by using a Scan Tool connected to the data link connector (DLC) **1** under the LH fascia.

1

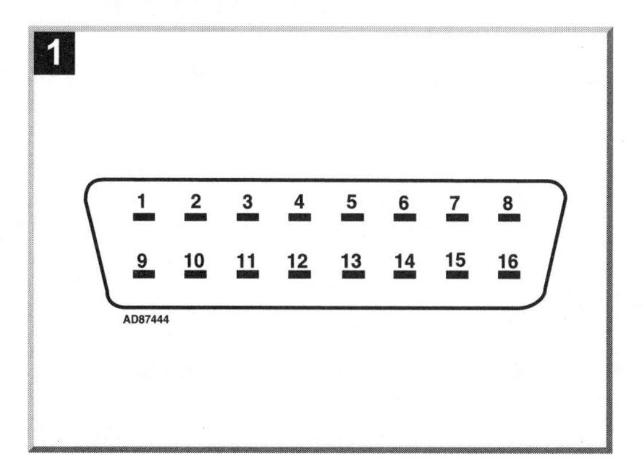

Erasing

- After the faults have been rectified, trouble codes can be erased by using a Scan Tool connected to the data link connector (DLC) or as follows:
- Switch ignition OFF.
- Remove ECM fuse (30A) from underhood fusebox for 30 seconds minimum **2**.
- Reinstall fuse.
- Repeat checking procedure to ensure no data remains in ECM fault memory.

2

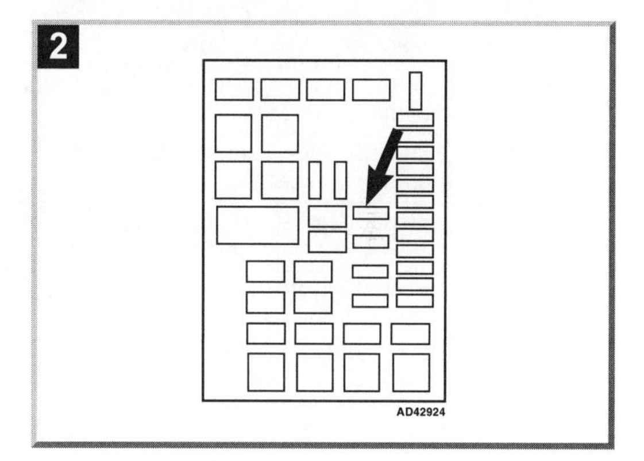

Trouble code identification

OBD-II code	Fault location	Probable cause
All P0, P2 and U0 codes	Refer to OBD-II trouble code tables at the front of this manual	–
P1106	Manifold absolute pressure (MAP) sensor – circuit/intermittent voltage high	**Wiring, MAP sensor**
P1107	Manifold absolute pressure (MAP) sensor – circuit/ intermittent voltage low	**Wiring, MAP sensor**
P1111	Intake air temperature (IAT) sensor – circuit/ intermittent voltage high	**Wiring, IAT sensor**
P1112	Intake air temperature (IAT) sensor – circuit/ intermittent voltage low	**Wiring, IAT sensor**
P1114	Engine coolant temperature (ECT) sensor – circuit/ intermittent voltage low	**Wiring, ECT sensor**
P1115	Engine coolant temperature (ECT) sensor – circuit/ intermittent voltage high	**Wiring, ECT sensor**
P1121	Throttle position (TP) sensor – circuit/ intermittent voltage high	**Wiring, TP sensor**
P1122	Throttle position (TP) sensor – circuit/ intermittent voltage low	**Wiring, TP sensor**
P1133	Heated oxygen sensor (HO2S) – RH front – insufficient switching	**Exhaust system, wiring, HO2S, ECM**
P1134	Heated oxygen sensor (HO2S) – RH front – slow operation	**Exhaust system, wiring, HO2S, ECM**
P1153	Heated oxygen sensor (HO2S) – LH front – insufficient switching	**Exhaust system, wiring, HO2S, ECM**
P1154	Heated oxygen sensor (HO2S) – LH front – slow operation	**Exhaust system, wiring, HO2S, ECM**
P1171	Fuel system – mixture lean under acceleration	**Fuel pump/strainer, incorrect fuel**
P1390	G-sensor – intermittent voltage low	**Wiring, G-sensor**
P1391	G-sensor – performance problem	**Seal damaged/missing, incorrectly installed, wiring, G-sensor, ECM**
P1392	G-sensor – voltage low	**Wiring, G-sensor, ECM**
P1393	G-sensor – voltage high	**Wiring, G-sensor, ECM**
P1394	G-sensor – intermittent voltage high	**Wiring, G-sensor**
P1406	Exhaust gas recirculation (EGR) valve position sensor – circuit problem	**Wiring, EGR valve/position sensor, ECM**
P1441	Evaporative emission (EVAP) canister purge system – flow detected	**Wiring, EVAP solenoid, vacuum switch, ECM**
P1442	Evaporative emission (EVAP) canister purge system – vacuum switch malfunction	**Wiring, vacuum switch, ECM**
P1508	Idle air control (IAC) system – low RPM	**Intake system, throttle body, hose leak, wiring, IAC valve, ECM**
P1509	Idle air control (IAC) system – high RPM	**Hose leak, throttle body, crankcase vent valve, wiring, IAC valve, ECM**

OBD-II code	Fault location	Probable cause
P1618	Engine control module (ECM) – internal fault	**ECM**
P1640	Engine control module (ECM) – internal fault	**Wiring, ECM**
P1790	AT – engine control module (ECM)	**ECM/Programming**
P1792	AT – engine control module (ECM)	**ECM/Programming**
P1835	AT – kick-down switch always ON	**Wiring, kick-down switch, ECM**
P1850	AT – brake band solenoid	**Wiring, brake band solenoid, ECM**
P1860	AT – torque converter clutch (TCC) solenoid	**Wiring, TCC solenoid, ECM**
P1870	Transmission slip	**Range switch, mechanical failure, engine speed signal incorrect**

ACURA

Model:	Engine identification:	Year:
TSX 2.4L	K24A2	2004
TL 3.2L	J32A3	2004

System: **PGM-FI**

Self-diagnosis

General information

- Refer to the front of this manual for general test conditions, terminology, detailed descriptions of wiring faults and a general trouble shooter for electrical and mechanical faults.
- Malfunction indicator lamp (MIL) will illuminate if certain faults are recorded.
- ECM operates in backup mode if sensors fail, to enable vehicle to be driven to repair shop.

Accessing

- Trouble codes can only be displayed by using a Scan Tool connected to the data link connector (DLC) **1**.

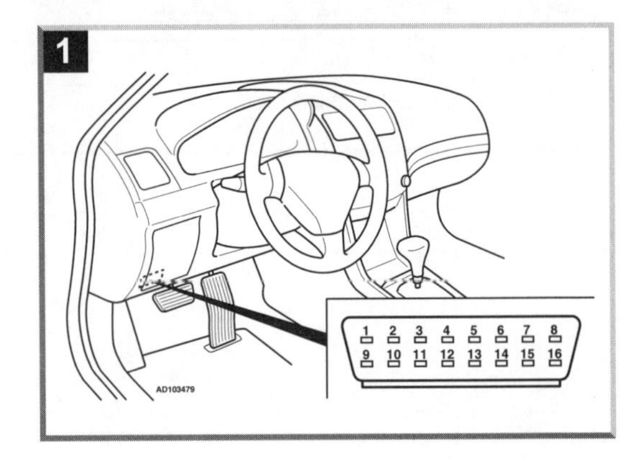

Erasing

- Trouble codes can only be erased by using a Scan Tool connected to the data link connector (DLC) **1**.

Trouble code identification

OBD-II code	Fault location	Probable cause
All P0, P2 and U0 codes	Refer to OBD-II trouble code tables at the front of this manual	–
P1077	Intake manifold air control system – low RPM malfunction	**Wiring, intake manifold air control solenoid, ECM**
P1078	Intake manifold air control system – high RPM malfunction	**Wiring, intake manifold air control solenoid, ECM**
P1128	Manifold absolute pressure (MAP) sensor – pressure lower than expected	**MAP sensor**
P1129	Manifold absolute pressure (MAP) sensor – pressure higher than expected	**MAP sensor**
P1157	Air fuel (A/F) ratio sensor – circuit malfunction	**Wiring, A/F ratio sensor, ECM**
P1297	Electrical load sensor – circuit/voltage low	**Wiring, electrical load sensor, ECM**
P1298	Electrical load sensor – circuit/voltage high	**Wiring, electrical load sensor, ECM**

OBD-II code	Fault location	Probable cause
P1454	Fuel tank pressure sensor – range/ performance problem	**Wiring, hose, fuel tank pressure sensor, fuel filler cap, EVAP canister/vent valve, ECM**
P1683	Throttle motor – default position spring performance problem	**Wiring, throttle sticking/mechanically damaged, throttle motor**
P1684	Throttle motor – return spring performance problem	**Wiring, throttle sticking/mechanically damaged, throttle motor**
P1730	AT – gear shift malfunction	**Transmission fluid, wiring, shift solenoid A/B/D, ECM**
P1731	AT – gear shift malfunction	**Transmission fluid, wiring, shift solenoid E, AT clutch pressure control solenoid A, ECM**
P1732	AT – gear shift malfunction	**Transmission fluid, wiring, shift solenoid B/C, ECM**
P1733	AT – gear shift malfunction	**Transmission fluid, wiring, shift solenoid D, AT clutch pressure control solenoid C, ECM**
P1734	AT – gear shift malfunction	**Transmission fluid, wiring, shift solenoid B/C, ECM**

ACURA

Model:	Engine identification:	Year:
2.2CL	F22B1	1997
2.3CL	F23A1	1998-99

System: **PGM-FI**

Self-diagnosis

General information

- Refer to the front of this manual for general test conditions, terminology, detailed descriptions of wiring faults and a general trouble shooter for electrical and mechanical faults.
- Engine control module (ECM) incorporates self-diagnosis function.
- Malfunction indicator lamp (MIL) will illuminate if certain faults are recorded.
- ECM operates in backup mode if sensors fail, to enable vehicle to be driven to workshop.
- Trouble codes can be displayed by using a Scan Tool connected to the data link connector (DLC) **1** [1] or by the malfunction indicator lamp (MIL) with the service check connector jumped **1** [2].

NOTE: ***The use of a Scan Tool is essential to obtain full diagnostic information.***

Accessing

- Ensure ignition switched OFF.
- Jump service check connector terminals **1** [2].
- Switch ignition ON.
- Check MIL is flashing.
- Trouble codes 1-9 are indicated as follows:
  - Individual short flashes display trouble code **2** [A].
  - A short pause separates each flash **2** [B].
  - A long pause separates each trouble code **2** [C].
  - For example: Trouble code 3 displayed **2**.
- Trouble codes greater than 9 are indicated as follows:
  - Long flashes indicate the 'tens' of the trouble code **3** [A].
  - Short flashes indicate the 'units' of the trouble code **3** [C].
  - A short pause separates each flash **3** [B].
  - A long pause separates each trouble code **3** [D].
  - For example: Trouble code 12 displayed **3**.
- Count MIL flashes and compare with trouble code table.
- Switch ignition OFF.
- Remove jump lead.

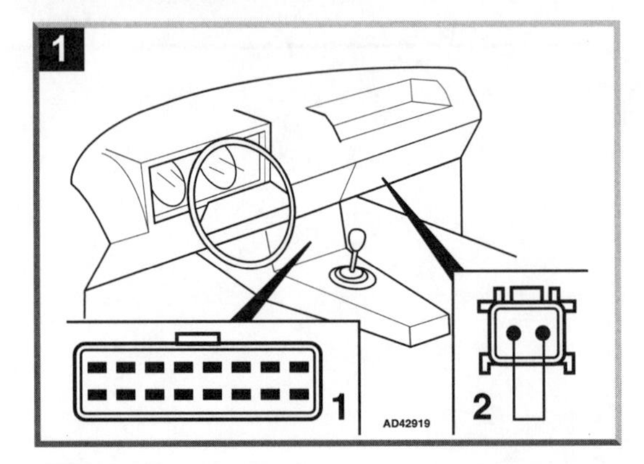

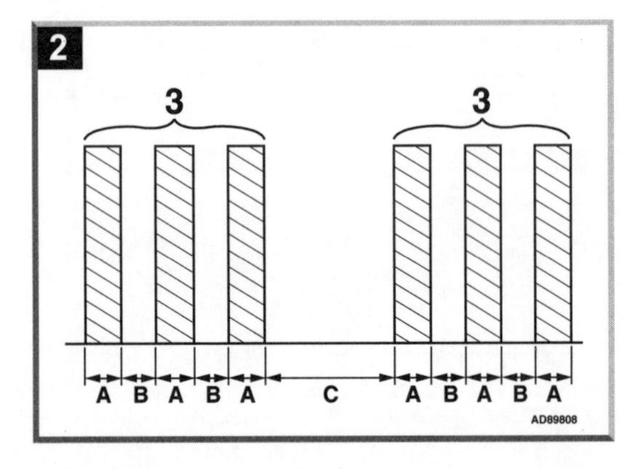

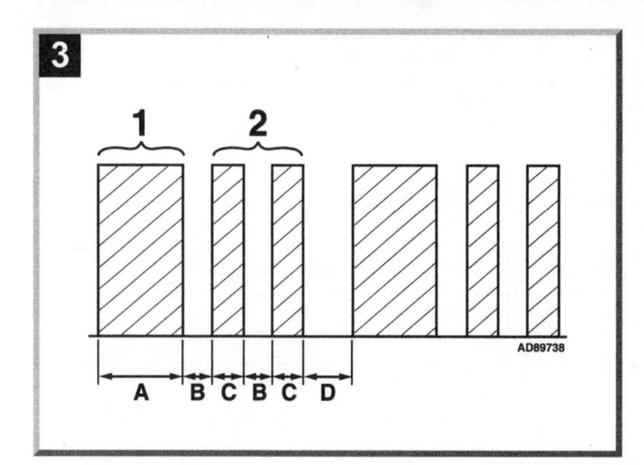

Erasing

- After the faults have been rectified, trouble codes can be erased by using a Scan Tool connected to the data link connector (DLC) or as follows:
- Switch ignition OFF.
- Remove fuse No.39 (7.5A – Radio) from underhood fusebox for 10 seconds minimum 4.
- Reinstall fuse.
- Repeat checking procedure to ensure no data remains in ECM fault memory.

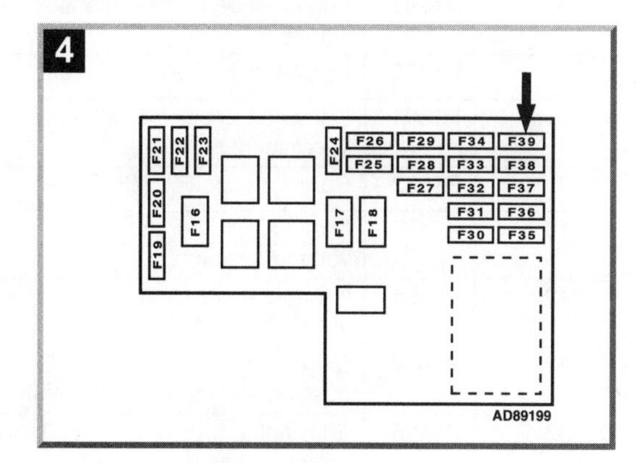

Trouble code identification

Flash code	OBD-II code	Fault location	Probable cause
–	*All P0, P2 and U0 codes*	Refer to OBD-II trouble code tables at the front of this manual	–
1	–	Heated oxygen sensor (HO2S) – front – circuit/voltage low/high	**Wiring, HO2S, ECM**
3	–	Manifold absolute pressure (MAP) sensor – circuit/voltage low/high	**Wiring, MAP sensor, ECM**
4	–	Crankshaft position (CKP) sensor – range/ performance problem/circuit/voltage low	**Wiring, CKP sensor, valve timing, ECM**
5	–	Manifold absolute pressure (MAP) sensor – range/performance problem	**Hose leak/blockage, MAP sensor**
5	P1128	Manifold absolute pressure (MAP) sensor – pressure lower than expected	**MAP sensor**
5	P1129	Manifold absolute pressure (MAP) sensor – pressure higher than expected	**MAP sensor**
6	–	Engine coolant temperature (ECT) sensor – circuit/voltage low/high	**Wiring, ECT sensor, ECM**
7	–	Throttle position (TP) sensor – circuit/ voltage low/high	**Wiring, TP sensor, ECM**
7	P1121	Throttle position (TP) sensor – position lower than expected	**TP sensor**
7	P1122	Throttle position (TP) sensor – position higher than expected	**TP sensor**

Flash code	OBD-II code	Fault location	Probable cause
8	P1359	Crankshaft position (CKP) sensor/engine speed (RPM) sensor – connector disconnection	**Wiring**
8	P1361	Engine speed (RPM) sensor – intermittent signal	**RPM sensor**
8	P1362	Engine speed (RPM) sensor – no signal	**Wiring, RPM sensor, ECM**
9	P1381	Camshaft position (CMP) sensor – intermittent signal	**CMP sensor**
9	P1382	Camshaft position (CMP) sensor – no signal	**Wiring, CMP sensor, ECM**
10	–	Intake air temperature (IAT) sensor – circuit/ voltage low/high	**Wiring, IAT sensor, ECM**
12	P1491	Exhaust gas recirculation (EGR) system – valve lift insufficient	**Wiring, EGR valve/position sensor, EGR solenoid, hose leak/blockage, ECM**
12	P1498	Exhaust gas recirculation (EGR) valve position sensor – voltage high	**Wiring, EGR valve/position sensor, ECM**
13	P1106	Barometric pressure (BARO) sensor – range/performance problem	**ECM**
13	P1107	Barometric pressure (BARO) sensor – circuit/voltage low	**ECM**
13	P1108	Barometric pressure (BARO) sensor – circuit/voltage high	**ECM**
14	–	Idle control system – malfunction	**IAC valve, throttle body**
14	P1508	Idle air control (IAC) valve – circuit failure	**Wiring, IAC valve, ECM**
14	P1519	Idle air control (IAC) valve – circuit failure	**Wiring, IAC valve, ECM**
17	–	Vehicle speed sensor (VSS) – circuit malfunction	**Wiring, VSS, ECM**
20	P1297	Electrical load sensor – circuit/voltage low	**Wiring, electrical load sensor, ECM**
20	P1298	Electrical load sensor – circuit/voltage high	**Wiring, electrical load sensor, ECM**
21	P1253	VTEC system malfunction	**Wiring, VTEC solenoid, ECM**
22	P1259	VTEC system malfunction	**Wiring, VTEC solenoid/pressure switch, ECM**
23	–	Knock sensor (KS) – malfunction	**Wiring, KS, ECM**
41	–	Heated oxygen sensor (HO2S) – front – circuit malfunction	**Wiring, ECM**
45	–	Mixture too lean	**Fuel/exhaust system, front HO2S, MAP sensor, mechanical fault**
45	–	Mixture too rich	**Fuel system, front HO2S, MAP sensor, mechanical fault**
61	–	Heated oxygen sensor (HO2S) – front – slow response	**HO2S, exhaust system**
63	–	Heated oxygen sensor (HO2S) – rear – slow response/circuit/voltage low/high	**Wiring open circuit, HO2S, ECM**

Flash code	OBD-II code	Fault location	Probable cause
65	–	Heated oxygen sensor (HO2S) – rear – heater circuit malfunction	**Wiring, ECM**
67	–	Catalytic converter – efficiency below limit	**Catalytic converter, rear HO2S**
70	–	AT – lock-up clutch not engaging/no gear shift	**Wiring, mainshaft speed sensor, countershaft speed sensor, lock-up control system, shift solenoid (SS) A/B, TCM**
70	P1705	AT – gear shift malfunction	**Wiring, range position switch, ECM**
70	P1705	AT – lock-up clutch not engaging	**Wiring, range position switch, ECM**
70	P1706	AT – gear shift malfunction	**Wiring, range position switch, ECM**
70	P1706	AT – lock-up clutch malfunction	**Wiring, range position switch, ECM**
70	P1738	Automatic transmission	**Wiring, clutch pressure switch 2, ECM**
70	P1739	Automatic transmission	**Wiring, clutch pressure switch 3, ECM**
70	P1753	AT – lock-up clutch not engaging/ disengaging	**Wiring, lock-up control solenoid A, ECM**
70	P1753	AT – no gear shift	**Wiring, lock-up control solenoid A, ECM**
70	P1758	AT – no gear shift	**Wiring, lock-up control solenoid B, ECM**
70	P1768	AT – no gear shift	**Wiring, clutch pressure control solenoid A, ECM**
70	P1773	AT – poor gear shift	**Wiring, clutch pressure control solenoid B, ECM**
70	P1773	AT – lock-up clutch not engaging	**Wiring, clutch pressure control solenoid B, ECM**
70	P1791	AT – lock-up clutch not engaging	**Wiring, VSS, ECM**
71	–	Cylinder No.1 – misfire	**Wiring, injector, ignition system, ignition/fuel/EGR system, MAP sensor, IAC valve, mechanical fault**
72	–	Cylinder No.2 – misfire	**Wiring, injector, ignition system, ignition/fuel/EGR system, MAP sensor, IAC valve, mechanical fault**
73	–	Cylinder No.3 – misfire	**Wiring, injector, ignition system, ignition/fuel/EGR system, MAP sensor, IAC valve, mechanical fault**
74	–	Cylinder No.4 – misfire	**Wiring, injector, ignition system, ignition/fuel/EGR system, MAP sensor, IAC valve, mechanical fault**
80	–	Exhaust gas recirculation (EGR) system – insufficient flow	**Hose leak/blockage, EGR valve**
86	–	Engine coolant temperature (ECT) sensor – range/performance problem	**ECT sensor, cooling system**
90	P1456	Evaporative emission (EVAP) canister purge system (fuel tank system) – leak detected	**Hose, fuel tank/pressure sensor, fuel filler cap, EVAP valve/bypass solenoid, EVAP two way valve, EVAP canister/vent valve**

Flash code	OBD-II code	Fault location	Probable cause
90	P1457	Evaporative emission (EVAP) canister purge system (canister system) – leak detected	**Hose, fuel tank/pressure sensor, EVAP valve/bypass solenoid, EVAP two way valve, EVAP canister/vent valve**
91	–	Fuel tank pressure sensor – circuit/voltage low/high	**Wiring, pressure sensor, ECM**
92	–	Evaporative emission (EVAP) canister purge valve – insufficient flow	**Wiring, EVAP valve, flow switch, throttle body, hose leak/blockage, ECM**
92	P1459	Evaporative emission (EVAP) canister purge system – switch malfunction	**Wiring, flow switch, hose leak/blockage, ECM**
–	P1607	Engine control module (ECM) – internal circuit failure	**ECM**

ACURA

Model: **2.5TL** | Engine identification: **G25A4** | Year: **1995-98**

System: **PGM-FI**

Self-diagnosis

General information

- Refer to the front of this manual for general test conditions, terminology, detailed descriptions of wiring faults and a general trouble shooter for electrical and mechanical faults.
- Engine control module (ECM) incorporates self-diagnosis function.
- Malfunction indicator lamp (MIL) will illuminate if certain faults are recorded.
- ECM operates in backup mode if sensors fail, to enable vehicle to be driven to workshop.
- Trouble codes can be displayed by using a Scan Tool connected to the data link connector (DLC) **1** [1] or by the malfunction indicator lamp (MIL) with the service check connector jumped **1** [2].

NOTE: ***The use of a Scan Tool is essential to obtain full diagnostic information.***

Accessing

- Ensure ignition switched OFF.
- Jump service check connector terminals **1** [2].
- Switch ignition ON.
- Check MIL is flashing.
- Trouble codes 1-9 are indicated as follows:
  - Individual short flashes display trouble code **2** [A].
  - A short pause separates each flash **2** [B].
  - A long pause separates each trouble code **2** [C].
  - For example: Trouble code 3 displayed **2**.
- Trouble codes greater than 9 are indicated as follows:
  - Long flashes indicate the 'tens' of the trouble code **3** [A].
  - Short flashes indicate the 'units' of the trouble code **3** [C].
  - A short pause separates each flash **3** [B].
  - A long pause separates each trouble code **3** [D].
  - For example: Trouble code 12 displayed **3**.
- Count MIL flashes and compare with trouble code table.
- Switch ignition OFF.
- Remove jump lead.

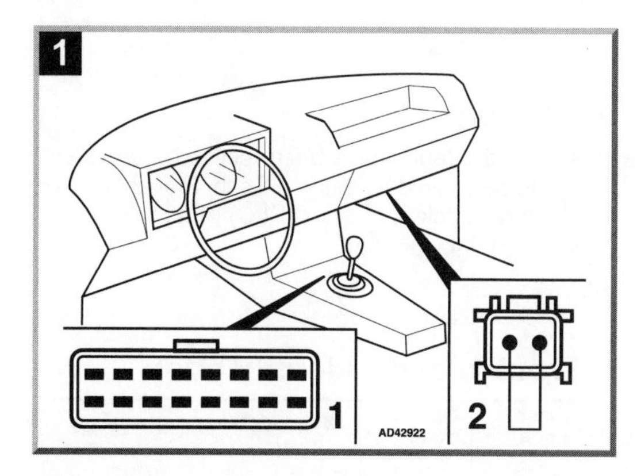

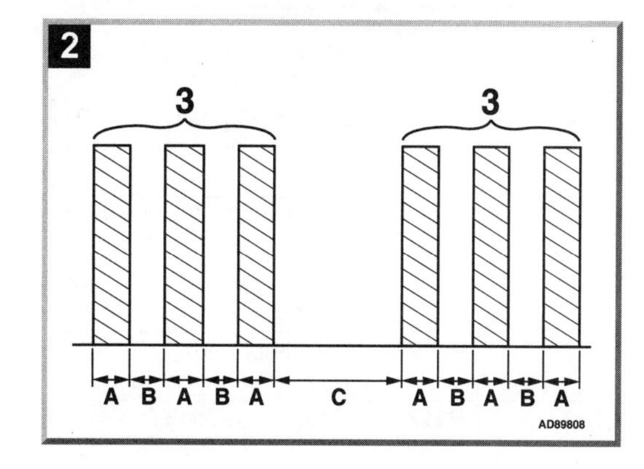

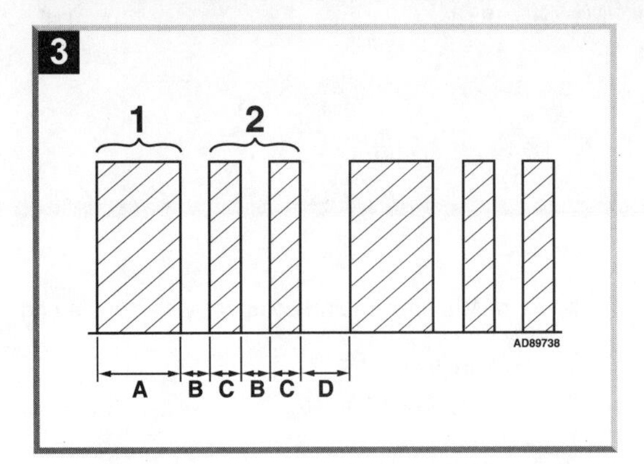

Erasing

- After the faults have been rectified, trouble codes can be erased by using a Scan Tool connected to the data link connector (DLC) or as follows:
- Switch ignition OFF.
- Remove radio fuse No.39 (10A) from underhood fusebox for 10 seconds minimum **4**.
- Reinstall fuse.
- Repeat checking procedure to ensure no data remains in ECM fault memory.

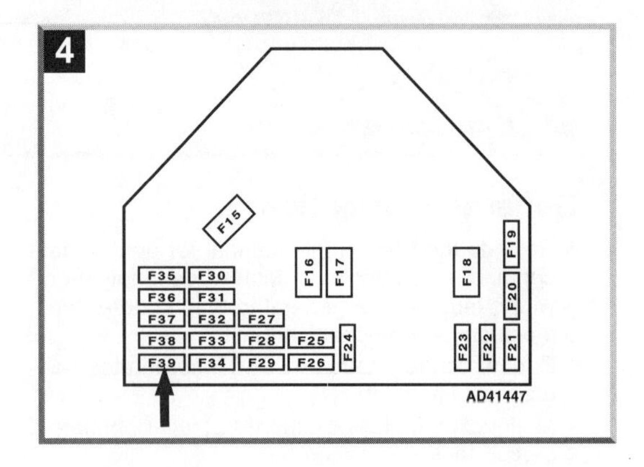

Trouble code identification

Flash code	OBD-II code	Fault location	Probable cause
–	*All P0, P2 and U0 codes*	Refer to OBD-II trouble code tables at the front of this manual	–
1	–	Heated oxygen sensor (HO2S) – front – circuit malfunction	**Wiring, HO2S**
4	–	Crankshaft position (CKP) sensor 1 – range/performance problem/circuit malfunction	**Wiring, CKP sensor, valve timing, ECM**
6	–	Engine coolant temperature (ECT) sensor – circuit/voltage low/high	**Wiring, ECT sensor, ECM**
7	–	Throttle position (TP) sensor – circuit/ voltage low/high	**Wiring, TP sensor, ECM**
7	P1121	Throttle position (TP) sensor – position lower than expected	**TP sensor**
7	P1122	Throttle position (TP) sensor – position higher than expected	**TP sensor**
8	P1359	Crankshaft position (CKP) sensor/engine speed (RPM) sensor – connector disconnection	**Wiring**
8	P1361	Engine speed (RPM) sensor – intermittent signal	**RPM sensor**
8	P1362	Engine speed (RPM) sensor – no signal	**Wiring, RPM sensor, ECM**
9	P1381	Camshaft position (CMP) sensor – intermittent signal	**CMP sensor**

Flash code	OBD-II code	Fault location	Probable cause
9	P1382	Camshaft position (CMP) sensor – no signal	Wiring, CMP sensor, ECM
10	–	Intake air temperature (IAT) sensor – range/ performance problem/circuit/voltage low	Wiring, IAT sensor, ECM
12	P1491	Exhaust gas recirculation (EGR) system – valve lift insufficient	Wiring, EGR valve/position sensor, EGR solenoid, hose leak/blockage, ECM
12	P1498	Exhaust gas recirculation (EGR) valve position sensor – voltage high	Wiring, EGR valve/position sensor, ECM
13	P1106	Barometric pressure (BARO) sensor – range/performance problem	ECM
13	P1107	Barometric pressure (BARO) sensor – circuit/voltage low	ECM
13	P1108	Barometric pressure (BARO) sensor – circuit/voltage high	ECM
14	–	Idle control system – malfunction	IAC valve, fast idle thermo valve, throttle body
14	P1508	Idle air control (IAC) valve – circuit failure	Wiring, IAC valve, ECM
17	–	Vehicle speed sensor (VSS) – voltage low	Wiring, VSS, ECM
20	P1297	Electrical load sensor – circuit/voltage low	Wiring, electrical load sensor, ECM
20	P1298	Electrical load sensor – circuit/voltage high	Wiring, electrical load sensor, ECM
23	–	Knock sensor (KS) – front – voltage low	Wiring, KS, ECM
23	–	Knock sensor (KS) – front – voltage high	Wiring, KS, ECM
30	P1681	AT to ECM – signal A – input low	Wiring, TCM, ECM
30	P1682	AT to ECM – signal A – input high	Wiring, TCM, ECM
31	P1686	AT to ECM – signal B – input low	Wiring, TCM, ECM
31	P1687	AT to ECM – signal B – input high	Wiring, TCM, ECM
41	–	Heated oxygen sensor (HO2S) – heater circuit malfunction	Wiring, ECM
45	–	Mixture too lean/rich	Fuel system, front HO2S, MAF sensor, mechanical fault
50	–	Mass air flow (MAF) sensor – range/ performance problem/voltage low/high	Wiring open circuit, MAF sensor, ECM
50	P1102	Mass air flow (MAF) sensor – signal higher than expected	MAF sensor
50	P1103	Mass air flow (MAF) sensor – signal higher than expected	MAF sensor
53	–	Knock sensor (KS) – rear – voltage low/ high	Wiring, KS, ECM
54	P1336	Crankshaft position (CKP) sensor 2 – intermittent signal interruption	CKP sensor
54	P1337	Crankshaft position (CKP) sensor 2 – no signal	Wiring, CKP sensor, ECM

Flash code	OBD-II code	Fault location	Probable cause
61	–	Heated oxygen sensor (HO2S) – front – slow response	HO2S, exhaust system
63	–	Heated oxygen sensor (HO2S) – rear – slow response	Wiring, HO2S, ECM
65	–	Heated oxygen sensor (HO2S) – rear – heater circuit malfunction	Wiring, ECM
67	–	Catalytic converter – efficiency below limit	Catalytic converter, rear HO2S
70	–	AT – lock-up clutch not engaging/no gear shift	Wiring, mainshaft speed sensor, countershaft speed sensor, lock-up control system, shift solenoid (SS) A/B, TCM
70	P1660	AT to TCM – data line failure	Wiring, ECM, TCM
70	P1705	AT – gear shift malfunction	Wiring, range position switch, TCM
70	P1705	AT – lock-up clutch not engaging	Wiring, range position switch, TCM
70	P1706	AT – gear shift malfunction	Wiring, range position switch, TCM
70	P1706	AT – lock-up clutch malfunction	Wiring, range position switch, TCM
70	P1753	AT – lock-up clutch not engaging/ disengaging	Wiring, lock-up control solenoid A, TCM
70	P1758	AT – no gear shift	Wiring, lock-up control solenoid B, TCM
70	P1768	AT – no gear shift	Wiring, linear solenoid, TCM
70	P1768	AT – lock-up clutch not engaging	Wiring, linear solenoid, TCM
70	P1786	AT – poor gear shift	Communication wire, TCM
70	P1787	AT – lock-up clutch malfunction	Communication wire, TCM
70	P1790	AT – lock-up clutch not engaging	Wiring, TP sensor, TCM
70	P1791	AT – lock-up clutch not engaging	Wiring, VSS, TCM
70	P1792	AT – lock-up clutch not engaging	Wiring, ECT sensor, TCM
70	P1794	Automatic transmission – BARO signal	Wiring, ECM, TCM
71	–	Cylinder No.1 – misfire	Wiring, injector, ignition system, mechanical fault
72	–	Cylinder No.2 – misfire	Wiring, injector, ignition system, mechanical fault
73	–	Cylinder No.3 – misfire	Wiring, injector, ignition system, mechanical fault
74	–	Cylinder No.4 – misfire	Wiring, injector, ignition system, mechanical fault
75	–	Cylinder No.5 – misfire	Wiring, injector, ignition system, mechanical fault
76	–	Random misfire	Ignition/fuel/EGR system, MAF sensor, IAC valve
80	–	Exhaust gas recirculation (EGR) solenoid – insufficient flow	EGR valve, hose/pipe leak or blockage

Flash code	OBD-II code	Fault location	Probable cause
86	–	Engine coolant temperature (ECT) sensor – range/performance problem	**ECT sensor, cooling system**
91	–	Fuel tank pressure sensor – circuit/voltage low/high	**Wiring, pressure sensor, ECM**
92	–	Evaporative emission (EVAP) canister purge system – leak detected/incorrect flow	**Wiring, hose leak/blockage, fuel tank/ pressure sensor, fuel filler cap, EVAP valve/bypass solenoid/flow switch, EVAP two way valve, EVAP canister/vent valve, throttle body, ECM**
92	P1459	Evaporative emission (EVAP) canister purge system – switch malfunction	**Wiring, flow switch, hose leak/blockage, ECM**

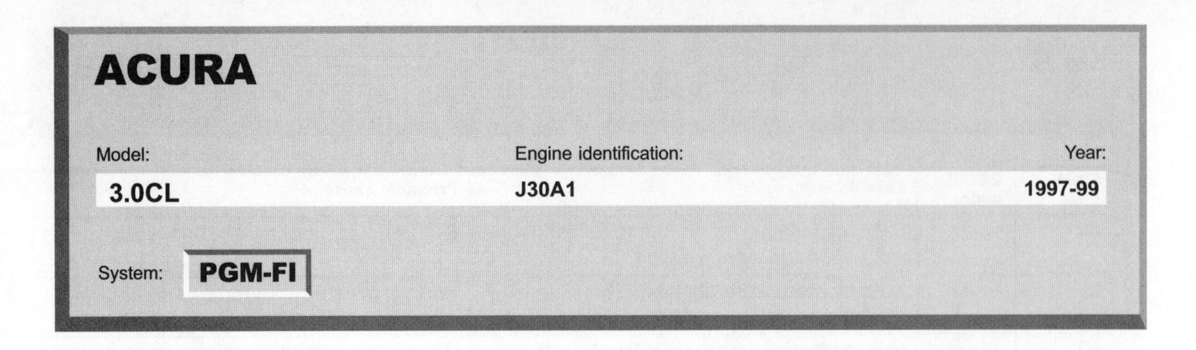

ACURA

Model:	Engine identification:	Year:
3.0CL	J30A1	1997-99

System: **PGM-FI**

Self-diagnosis

General information

- Refer to the front of this manual for general test conditions, terminology, detailed descriptions of wiring faults and a general trouble shooter for electrical and mechanical faults.
- Engine control module (ECM) incorporates self-diagnosis function.
- Malfunction indicator lamp (MIL) will illuminate if certain faults are recorded.
- ECM operates in backup mode if sensors fail, to enable vehicle to be driven to workshop.
- Trouble codes can be displayed by using a Scan Tool connected to the data link connector (DLC) **1** [1] or by the malfunction indicator lamp (MIL) with the service check connector jumped **1** [2].

NOTE: ***The use of a Scan Tool is essential to obtain full diagnostic information.***

Accessing

- Ensure ignition switched OFF.
- Jump service check connector terminals **1** [2].
- Switch ignition ON.
- Check MIL is flashing.
- Trouble codes 1-9 are indicated as follows:
  - Individual short flashes display trouble code **2** [A].
  - A short pause separates each flash **2** [B].
  - A long pause separates each trouble code **2** [C].
  - For example: Trouble code 3 displayed **2**.
- Trouble codes greater than 9 are indicated as follows:
  - Long flashes indicate the 'tens' of the trouble code **3** [A].
  - Short flashes indicate the 'units' of the trouble code **3** [C].
  - A short pause separates each flash **3** [B].
  - A long pause separates each trouble code **3** [D].
  - For example: Trouble code 12 displayed **3**.
- Count MIL flashes and compare with trouble code table.
- Switch ignition OFF.
- Remove jump lead.

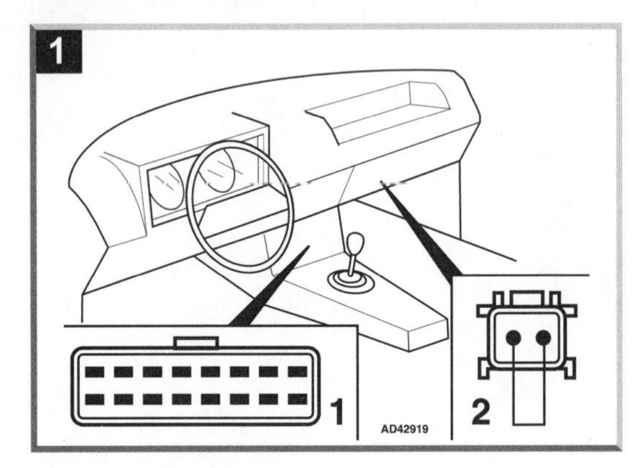

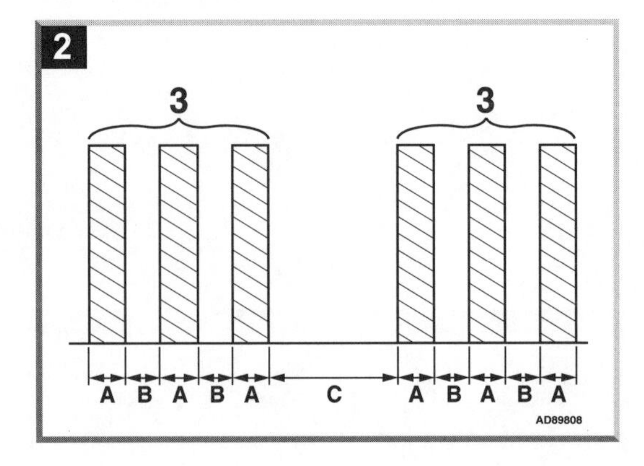

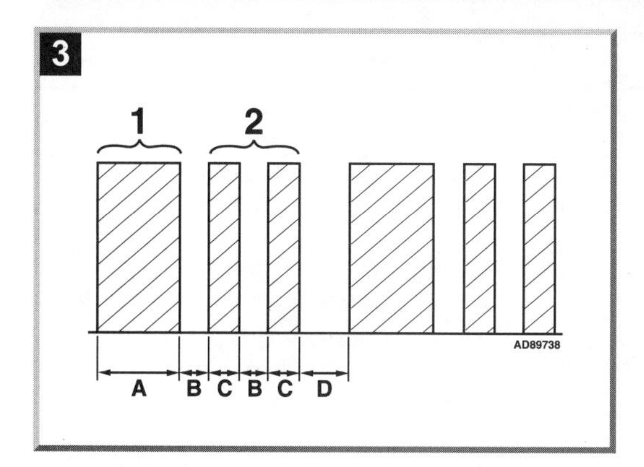

Erasing

- After the faults have been rectified, trouble codes can be erased by using a Scan Tool connected to the data link connector (DLC) or as follows:
- Switch ignition OFF.

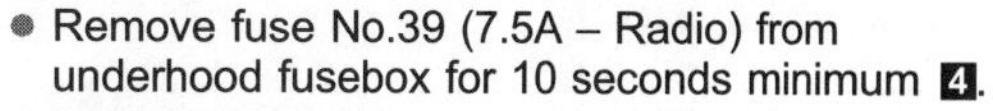

- Remove fuse No.39 (7.5A – Radio) from underhood fusebox for 10 seconds minimum **4**.
- Reinstall fuse.
- Repeat checking procedure to ensure no data remains in ECM fault memory.

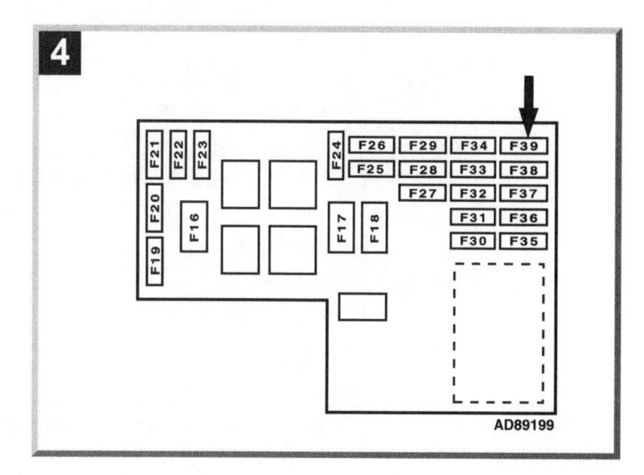

Trouble code identification

Flash code	OBD-II code	Fault location	Probable cause
–	*All P0, P2 and U0 codes*	Refer to OBD-II trouble code tables at the front of this manual	–
1	–	Heated oxygen sensor (HO2S) – front – circuit/voltage low/high	**Wiring, HO2S, fuel system, ECM**
3	–	Manifold absolute pressure (MAP) sensor – circuit/voltage low/high	**Wiring, MAP sensor, ECM**
4	–	Crankshaft position (CKP) sensor – range/performance problem/circuit malfunction	**Wiring, CKP sensor, valve timing, ECM**
5	P1128	Manifold absolute pressure (MAP) sensor – pressure lower than expected	**MAP sensor**
5	P1129	Manifold absolute pressure (MAP) sensor – pressure higher than expected	**MAP sensor**
6	–	Engine coolant temperature (ECT) sensor – circuit/voltage low/high	**Wiring, ECT sensor, ECM**
7	–	Throttle position (TP) sensor – circuit/voltage low/high	**Wiring, TP sensor, ECM**
7	P1121	Throttle position (TP) sensor – position lower than expected	**TP sensor**
7	P1122	Throttle position (TP) sensor – position higher than expected	**TP sensor**
8	P1361	Camshaft position (CMP) sensor 1 – intermittent signal	**CMP sensor**

Flash code	OBD-II code	Fault location	Probable cause
8	P1362	Camshaft position (CMP) sensor 1 – no signal	**Wiring, CMP sensor, ECM**
10	–	Intake air temperature (IAT) sensor – range/ performance problem/circuit/voltage low/ high	**Wiring, IAT sensor, ECM**
12	P1491	Exhaust gas recirculation (EGR) system – valve lift insufficient	**Wiring, EGR valve/lift sensor, EGR solenoid, hose leak/blockage, ECM**
12	P1498	Exhaust gas recirculation (EGR) valve position sensor – voltage high	**Wiring, EGR valve/lift sensor, ECM**
13	P1106	Barometric pressure (BARO) sensor – range/performance problem	**ECM**
13	P1107	Barometric pressure (BARO) sensor – circuit/voltage low	**ECM**
13	P1108	Barometric pressure (BARO) sensor – circuit/voltage high	**ECM**
14	–	Idle control system – malfunction	**IAC valve, fast idle thermo valve, throttle body**
14	P1519	Idle air control (IAC) valve – circuit failure	**Wiring, IAC valve, ECM**
17	–	Vehicle speed sensor (VSS) – circuit malfunction	**Wiring, VSS, ECM**
20	P1297	Electrical load sensor – circuit/voltage low	**Wiring, electrical load sensor, ECM**
20	P1298	Electrical load sensor – circuit/voltage high	**Wiring, electrical load sensor, ECM**
22	P1259	VTEC system malfunction	**Wiring, VTEC solenoid/pressure switch, ECM**
30	P1655	AT – signal failure	**Communication wire, ECM**
41	–	Heated oxygen sensor (HO2S) – front – circuit malfunction	**Wiring, ECM**
45	–	Mixture too lean/rich	**Fuel/exhaust system, front HO2S, MAP sensor, mechanical fault**
58	P1366	Camshaft position (CMP) sensor 2 – intermittent signal	**CMP sensor**
58	P1367	Camshaft position (CMP) sensor 2 – no signal	**Wiring, CMP sensor, ECM**
61	–	Heated oxygen sensor (HO2S) – front – slow response	**HO2S, exhaust system**
63	–	Heated oxygen sensor (HO2S) – rear – slow response circuit/voltage high/low	**Wiring, HO2S, ECM**
65	–	Heated oxygen sensor (HO2S) – rear – circuit malfunction	**Wiring, ECM**
67	–	Catalytic converter – efficiency below limit	**Catalytic converter, rear HO2S**
70	–	AT – lock-up clutch not engaging/no gear shift	**Wiring, mainshaft speed sensor, countershaft speed sensor, lock-up control system, shift solenoid (SS) A/B, TCM**

Flash code	OBD-II code	Fault location	Probable cause
70	P1705	AT – gear shift malfunction	**Wiring, range position switch, TCM**
70	P1705	AT – lock-up clutch not engaging	**Wiring, range position switch, TCM**
70	P1706	AT – gear shift malfunction	**Wiring, range position switch, TCM**
70	P1706	AT – lock-up clutch malfunction	**Wiring, range position switch, TCM**
70	P1738	Automatic transmission	**Wiring, clutch pressure switch 2, TCM**
70	P1739	Automatic transmission	**Wiring, clutch pressure switch 3, TCM**
70	P1753	AT – lock-up clutch not engaging/ disengaging	**Wiring, lock-up control solenoid A, TCM**
70	P1768	AT – no gear shift	**Wiring, clutch pressure control solenoid A, TCM**
70	P1773	AT – poor gear shift	**Wiring, clutch pressure control solenoid B, TCM**
70	P1790	AT – lock-up clutch not engaging	**Wiring, TP sensor, TCM**
70	P1791	AT – lock-up clutch not engaging	**Wiring, VSS, TCM**
71	–	Cylinder No.1 – misfire	**Wiring, injector, ignition/fuel/EGR system, MAP sensor, IAC valve, mechanical fault**
72	–	Cylinder No.2 – misfire	**Wiring, injector, ignition/fuel/EGR system, MAP sensor, IAC valve, mechanical fault**
73	–	Cylinder No.3 – misfire	**Wiring, injector, ignition/fuel/EGR system, MAP sensor, IAC valve, mechanical fault**
74	–	Cylinder No.4 – misfire	**Wiring, injector, ignition/fuel/EGR system, MAP sensor, IAC valve, mechanical fault**
75	–	Cylinder No.5 – misfire	**Wiring, injector, ignition/fuel/EGR system, MAP sensor, IAC valve, mechanical fault**
76	–	Cylinder No.6 – misfire	**Wiring, injector, ignition/fuel/EGR system, MAP sensor, IAC valve, mechanical fault**
–	P1300	Random misfire	**Ignition/fuel/EGR system, MAP sensor, IAC valve**
80	–	Exhaust gas recirculation (EGR) solenoid – insufficient flow	**EGR valve, hose leak/blockage**
86	–	Engine coolant temperature (ECT) sensor – range/performance problem	**ECT sensor, cooling system**
90	P1456	Evaporative emission (EVAP) canister purge system (fuel tank system) – leak detected	**Hose, fuel tank/pressure sensor, fuel filler cap, EVAP valve/bypass solenoid, EVAP two way valve, EVAP canister/vent valve**

Flash code	OBD-II code	Fault location	Probable cause
90	P1457	Evaporative emission (EVAP) canister purge system (canister system) – leak detected	**Hose, fuel tank/pressure sensor, EVAP valve/bypass solenoid, EVAP two way valve, EVAP canister/vent valve**
91	–	Fuel tank pressure sensor – circuit/voltage low/high	**Wiring, pressure sensor, ECM**
–	P1607	Engine control module (ECM) – internal circuit failure	**ECM**

Model:	Engine identification:	Year:
3.2TL	C32A6	1996-98
3.5RL	C35A1	1996-04

System: **PGM-FI**

Self-diagnosis

General information

- Refer to the front of this manual for general test conditions, terminology, detailed descriptions of wiring faults and a general trouble shooter for electrical and mechanical faults.
- Engine control module (ECM) incorporates self-diagnosis function.
- Malfunction indicator lamp (MIL) will illuminate if certain faults are recorded.
- ECM operates in backup mode if sensors fail, to enable vehicle to be driven to workshop.
- Trouble codes can be displayed by using a Scan Tool connected to the data link connector (DLC) **1** [**1**].
- Trouble codes can be displayed by the malfunction indicator lamp (MIL) with the data link connector (DLC) **1** [**1**] jumped, or service check connector **1** [**2**] jumped.

NOTE: ***The use of a Scan Tool is essential to obtain full diagnostic information.***

Accessing

- Ensure ignition switched OFF.
- Models without service check connector: Jump data link connector (DLC) terminals 4 and 9 **1** [**1**].
- Models with service check connector: Jump service check connector terminals **1** [**2**].
- Switch ignition ON.
- Check MIL is flashing.
- Trouble codes 1-9 are indicated as follows:
  - Individual short flashes display trouble code **2** [**A**].
  - A short pause separates each flash **2** [**B**].
  - A long pause separates each trouble code **2** [**C**].
  - For example: Trouble code 3 displayed **2**.
- Trouble codes greater than 9 are indicated as follows:
  - Long flashes indicate the 'tens' of the trouble code **3** [**A**].
  - Short flashes indicate the 'units' of the trouble code **3** [**C**].
  - A short pause separates each flash **3** [**B**].
  - A long pause separates each trouble code **3** [**D**].
  - For example: Trouble code 12 displayed **3**.
- Count MIL flashes and compare with trouble code table.
- Switch ignition OFF.
- Remove jump lead.

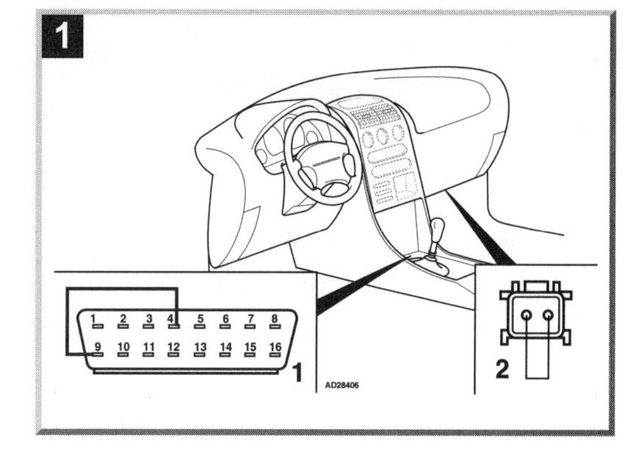

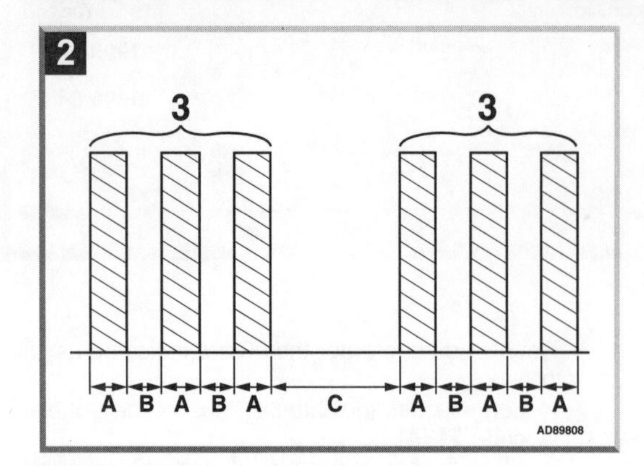

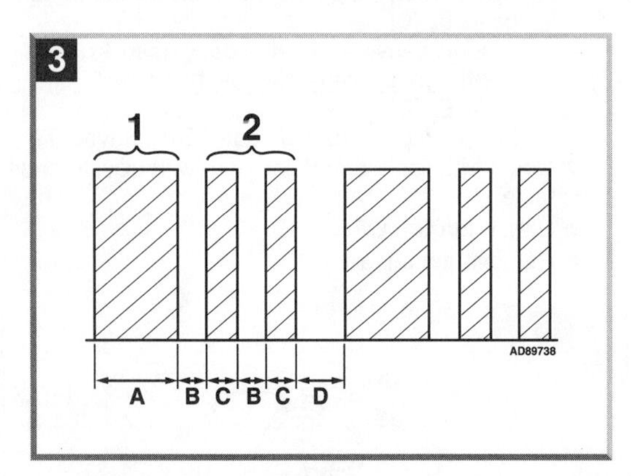

Erasing

- After the faults have been rectified, trouble codes can be erased by using a Scan Tool connected to the data link connector (DLC) or as follows:
- Switch ignition OFF.
- 3.5RL – remove radio fuse (7.5A) from underhood fusebox for 10 seconds minimum 4.
- 3.2TL – remove radio fuse (10A) from underhood fusebox for 10 seconds minimum 4.
- Reinstall fuse.
- Repeat checking procedure to ensure no data remains in ECM fault memory.

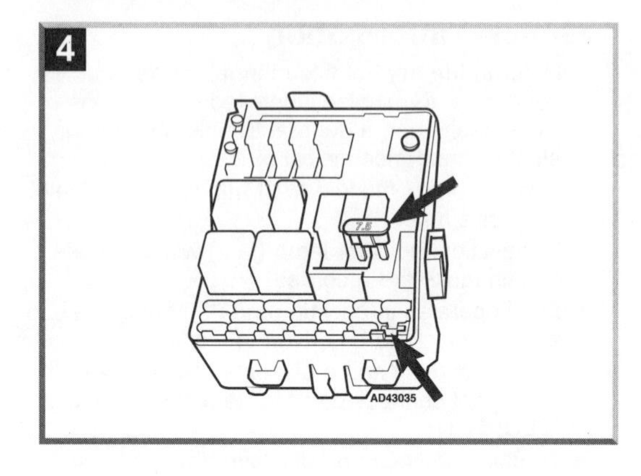

Trouble code identification

Flash code	OBD-II code	Fault location	Probable cause
–	*All P0, P2 and U0 codes*	Refer to OBD-II trouble code tables at the front of this manual	–
1	–	Heated oxygen sensor (HO2S) – LH front – circuit/voltage low/high	**Wiring, fuel system, HO2S, ECM**
2	–	Heated oxygen sensor (HO2S) – RH front – circuit/voltage low/high	**Wiring, HO2S, fuel system, ECM**
3	–	Manifold absolute pressure (MAP) sensor – circuit/voltage low	**Wiring, MAP sensor, ECM**
3	–	Manifold absolute pressure (MAP) sensor – circuit/voltage high	**Wiring open circuit, MAP sensor, ECM**
4	–	Crankshaft position (CKP) sensor 1 – range/performance problem/circuit malfunction	**Wiring, CKP sensor, valve timing, ECM**
5	–	Manifold absolute pressure (MAP) sensor – range/performance problem	**Hose leak/blockage, MAP sensor**

Flash code	OBD-II code	Fault location	Probable cause
5	P1128	Manifold absolute pressure (MAP) sensor – pressure lower than expected	MAP sensor
5	P1129	Manifold absolute pressure (MAP) sensor – pressure higher than expected	MAP sensor
6	–	Engine coolant temperature (ECT) sensor – circuit/voltage low/high	Wiring, ECT sensor, ECM
7	–	Throttle position (TP) sensor – circuit/voltage low/high	Wiring, TP sensor, ECM
7	P1121	Throttle position (TP) sensor – position lower than expected	TP sensor
7	P1122	Throttle position (TP) sensor – position higher than expected	TP sensor
8	P1361	Camshaft position (CMP) sensor 1 – intermittent	Wiring, CMP sensor
8	P1362	Camshaft position (CMP) sensor 1 – no signal	Wiring, CMP sensor
9	P1381	Camshaft position (CMP) sensor 1 – range/performance problem	CMP sensor
9	P1382	Camshaft position (CMP) sensor 1 – malfunction	Wiring, CMP sensor, ECM
10	–	Intake air temperature (IAT) sensor – range/performance problem/circuit/voltage low/high	Wiring short circuit, IAT sensor, ECM
12	P1491	Exhaust gas recirculation (EGR) system – valve lift insufficient	Wiring, EGR valve/position sensor, EGR solenoid, hose leak/blockage, ECM
12	P1498	Exhaust gas recirculation (EGR) valve position sensor – voltage high	Wiring, EGR valve/position sensor, ECM
13	P1106	Barometric pressure (BARO) sensor – range/performance problem	ECM
13	P1107	Barometric pressure (BARO) sensor – circuit/voltage low	ECM
13	P1108	Barometric pressure (BARO) sensor – circuit/voltage high	ECM
14	–	Idle control system – malfunction	IAC valve, fast idle thermo valve, throttle body
14	P1508	Idle air control (IAC) valve – circuit failure	Wiring, IAC valve, ECM
14	P1519	Idle air control (IAC) valve – circuit malfunction	Wiring, IAC valve, ECM
20	P1297	Electrical load control module – voltage low	Wiring, electrical load control module
20	P1298	Electrical load control module – voltage high	Wiring, electrical load control module
22	P1259	Camshaft position (CMP) actuator – circuit malfunction	Wiring, CMP actuator, ECM
23	–	Knock sensor (KS) – single or LH – circuit malfunction	Wiring, KS, ECM

Flash code	OBD-II code	Fault location	Probable cause
34	–	System voltage – malfunction	Wiring, poor connection, battery, alternator
35	P1676	ABS/TCS control module – communication malfunction	Wiring, ABS/TCS control module, ECM
35	P1678	ABS/TCS control module – malfunction	Wiring, ABS/TCS control module, ECM
35	P1690	Traction control system – circuit failure	Wiring, traction control module, ECM
36	P1696	Traction control system – voltage low	Wiring, traction control module, ECM
36	P1697	Traction control system – voltage high	Wiring, traction control module, ECM
41	–	Heated oxygen sensor (HO2S) – LH front – heater circuit malfunction	Wiring, ECM
42	–	Heated oxygen sensor (HO2S) – RH front – heater circuit malfunction	Wiring, ECM
45	–	Mixture too lean/rich – LH bank	Fuel system, LH HO2S, MAP sensor, mechanical fault
46	–	Mixture too lean/rich – RH bank	Fuel system, RH HO2S, MAP sensor, mechanical fault
53	–	Knock sensor (KS) – RH – circuit malfunction	Wiring, KS, ECM
54	P1336	Crankshaft position (CKP) sensor 2 – range/performance problem	CKP sensor, valve timing
54	P1337	Crankshaft position (CKP) sensor 2 – circuit malfunction	Wiring, CKP sensor, ECM
58	P1366	Camshaft position (CMP) sensor 2 – intermittent	Wiring, CMP sensor
58	P1367	Camshaft position (CMP) sensor 2 – no signal	Wiring, CMP sensor
59	P1386	Camshaft position (CMP) sensor 2 – range/performance problem	CMP sensor
59	P1387	Camshaft position (CMP) sensor 2 – malfunction	Wiring, CMP sensor, ECM
61	–	Heated oxygen sensor (HO2S) – LH front – slow response	HO2S, exhaust system
62	–	Heated oxygen sensor (HO2S) – RH front – slow response	HO2S, exhaust system
63	–	Heated oxygen sensor (HO2S) – rear – slow response/circuit/voltage low	Wiring, HO2S, ECM
65	–	Heated oxygen sensor (HO2S) – rear – heater circuit malfunction	Wiring, ECM
67	–	Catalytic converter – efficiency below limit	Catalytic converter, rear HO2S
70	–	AT – lock-up clutch not engaging/no gear shift	Wiring, mainshaft speed sensor, countershaft speed sensor, lock-up control system, shift solenoid (SS) A/B, TCM

Flash code	OBD-II code	Fault location	Probable cause
70	P1656	Electronic stability programme (ESP) control module – communication malfunction	Wiring, ESP control module
70	P1705	AT – gear shift malfunction	Wiring, range position switch, ECM
70	P1705	AT – lock-up clutch not engaging	Wiring, range position switch, ECM
70	P1706	AT – gear shift malfunction	Wiring, range position switch, ECM
70	P1706	AT – lock-up clutch malfunction	Wiring, range position switch, ECM
70	P1709	AT – range position switch	Wiring, range position switch, ECM
70	P1710	AT – 1st gear hold switch	Wiring, range position switch, ECM
70	P1739	AT – 3rd clutch pressure switch	Wiring, range position switch, ECM
70	P1740	AT – 4th clutch pressure switch	Wiring, range position switch, ECM
70	P1750	AT – hydraulic system mechanical malfunction	Wiring, range position switch, ECM
70	P1751	AT – hydraulic system mechanical malfunction	Wiring, range position switch, ECM
70	P1753	AT – lock-up clutch not engaging/ disengaging	Wiring, lock-up control solenoid A, ECM
70	P1758	AT – no gear shift	Wiring, lock-up control solenoid B, ECM
70	P1768	AT – no gear shift	Wiring, linear solenoid, ECM
70	P1773	Shift solenoid (SS) B	Wiring, shift solenoid, ECM
70	P1778	Shift solenoid (SS) C	Wiring, shift solenoid, ECM
70	P1791	AT – lock-up clutch not engaging	Wiring, VSS, ECM
71	P1201	Cylinder No.1 – misfire	Wiring, injector, ignition system, mechanical fault
71	P1301	Cylinder No.1 – misfire	Ignition system
72	P1202	Cylinder No.2 – misfire	Wiring, injector, ignition system, mechanical fault
72	P1302	Cylinder No.2 – misfire	Ignition system
73	P1203	Cylinder No.3 – misfire	Wiring, injector, ignition system, mechanical fault
73	P1303	Cylinder No.3 – misfire	Ignition system
74	P1204	Cylinder No.4 – misfire	Wiring, injector, ignition system, mechanical fault
74	P1304	Cylinder No.4 – misfire	Ignition system
75	P1205	Cylinder No.5 – misfire	Wiring, injector, ignition system, mechanical fault
75	P1305	Cylinder No.5 – misfire	Ignition system
76	P1206	Cylinder No.6 – misfire	Wiring, injector, ignition system, mechanical fault

Flash code	OBD-II code	Fault location	Probable cause
76	P1306	Cylinder No.6 – misfire	Ignition system
–	P1300	Random misfire	EGR/fuel/ignition system, MAP sensor
79	P1316	Spark plug voltage detection – LH bank – circuit malfunction	Wiring, voltage detection module, ECM
79	P1317	Spark plug voltage detection – RH bank – circuit malfunction	Wiring, voltage detection module, ECM
79	P1318	Spark plug voltage detection module reset – LH bank – reset circuit malfunction	Wiring, voltage detection module, ECM
79	P1319	Spark plug voltage detection module reset – RH bank – reset circuit malfunction	Wiring, voltage detection module, ECM
80	–	Exhaust gas recirculation (EGR) solenoid – insufficient flow	EGR solenoid, hose/pipe leak or blockage
86	–	Engine coolant temperature (ECT) sensor – range/performance problem	ECT sensor, cooling system
87	–	Thermostat – range/performance problem	Thermostat, cooling system
87	P1486	Thermostat – range/performance problem	Engine coolant blower motor, cooling system
90	P1456	Evaporative emission (EVAP) canister purge system (fuel tank system) – leak detected	Hose leak/blockage, fuel tank/pressure sensor, EVAP valve/bypass solenoid, EVAP two way valve, EVAP canister/vent valve
90	P1457	Evaporative emission (EVAP) canister purge system (canister system) – leak detected	Hose leak/blockage, fuel tank/ pressure sensor, EVAP valve/bypass solenoid, EVAP two way valve, EVAP canister/vent valve
91	–	Evaporative emission (EVAP) pressure sensor – range/performance problem	Wiring, EVAP pressure sensor
91	–	Fuel tank pressure sensor – circuit/voltage low/high	Wiring open circuit, pressure sensor, ECM
92	–	Evaporative emission (EVAP) canister purge system – incorrect flow	EVAP solenoid/flow switch, wiring, throttle body, hose leak/blockage, ECM
92	P1459	Evaporative emission (EVAP) canister purge system – switch malfunction	Wiring, flow switch, hose leak/blockage, ECM
–	P1399	Random misfire	Wiring, injector, ignition system, mechanical fault
–	P1607	Engine control module (ECM) – internal circuit failure A	ECM

ACURA

Model:	Engine identification:	Year:
3.2TL	J32A1	1999-03

System: **PGM-FI**

Self-diagnosis

General information

- Refer to the front of this manual for general test conditions, terminology, detailed descriptions of wiring faults and a general trouble shooter for electrical and mechanical faults.
- Engine control module (ECM) incorporates self-diagnosis function.
- Malfunction indicator lamp (MIL) will illuminate if certain faults are recorded.
- ECM operates in backup mode if sensors fail, to enable vehicle to be driven to workshop.

Accessing

- Trouble codes can only be displayed by using a Scan Tool connected to the data link connector (DLC) **1** – →2002 or **2** – 2003.

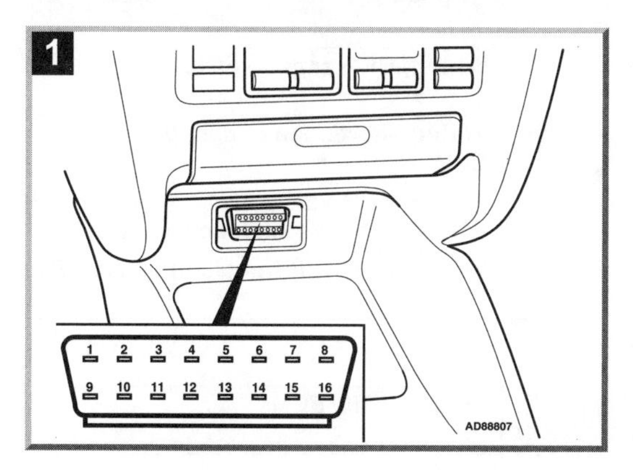

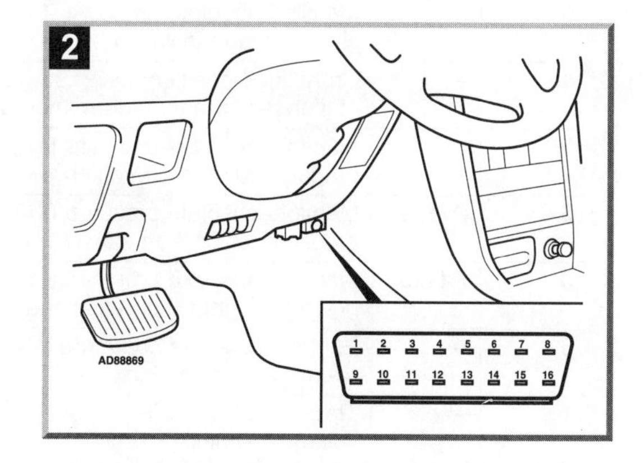

Erasing

- After the faults have been rectified, trouble codes can be erased by using a Scan Tool connected to the data link connector (DLC) or as follows:
- Switch ignition OFF.
- Remove clock/backup fuse (7.5A) from RH fascia fusebox for 10 seconds minimum **3**.
- Reinstall fuse.
- Repeat checking procedure to ensure no data remains in ECM fault memory.

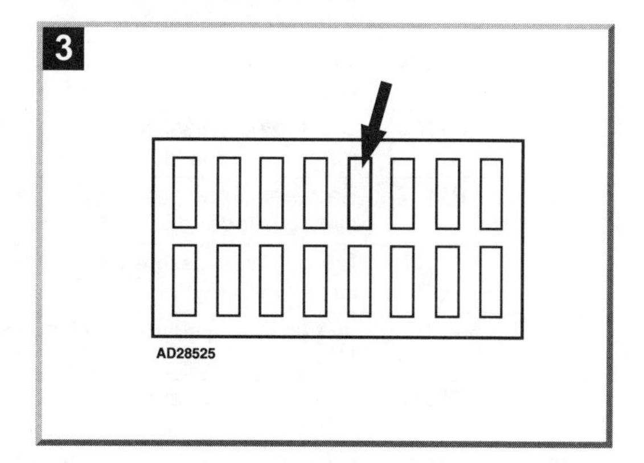

Trouble code identification

Flash code	OBD-II code	Fault location	Probable cause
–	*All P0, P2 and U0 codes*	Refer to OBD-II trouble code tables at the front of this manual	–
1	–	Heated oxygen sensor (HO2S) – front – circuit/voltage high/low	**Wiring short/open circuit, HO2S, fuel system, ECM**
3	–	Manifold absolute pressure (MAP) sensor – circuit/voltage low/high	**Wiring, MAP sensor, ECM**
4	–	Crankshaft position (CKP) sensor – circuit malfunction/range/performance problem	**Wiring, CKP sensor, ECM**
5	–	Manifold absolute pressure (MAP) sensor – range/performance problem	**Hose leak/blockage, MAP sensor**
5	P1128	Manifold absolute pressure (MAP) sensor – pressure lower than expected	**MAP sensor**
5	P1129	Manifold absolute pressure (MAP) sensor – pressure higher than expected	**MAP sensor**
6	–	Engine coolant temperature (ECT) sensor – circuit/voltage low/high	**Wiring short/open circuit, ECT sensor, ECM**
7	–	Throttle position (TP) sensor – circuit/ voltage low/high	**Wiring, TP sensor, ECM**
7	P1121	Throttle position (TP) sensor – position lower than expected	**TP sensor**
7	P1122	Throttle position (TP) sensor – position higher than expected	**TP sensor**
8	P1361	Crankshaft position (CKP) sensor 1 – intermittent signal	**CKP sensor**
8	P1362	Crankshaft position (CKP) sensor 1 – no signal	**Wiring, CKP sensor, ECM**
10	–	Intake air temperature (IAT) sensor – circuit/ voltage low/high	**Wiring short/open circuit, IAT sensor, ECM**
12	P1491	Exhaust gas recirculation (EGR) system – valve lift insufficient	**Wiring, EGR valve/position sensor, EGR solenoid, hose leak/blockage, ECM**
12	P1498	Exhaust gas recirculation (EGR) valve position sensor – voltage high	**Wiring, EGR valve/position sensor, ECM**
13	P1106	Barometric pressure (BARO) sensor – range/performance problem	**ECM**
13	P1107	Barometric pressure (BARO) sensor – circuit/voltage low	**Wiring, TCM, ECM**
13	P1108	Barometric pressure (BARO) sensor – circuit/voltage high	**ECM**
14	–	Idle control system – malfunction	**IAC valve, fast idle thermo valve, throttle body**
14	P1519	Idle control system – malfunction	**IAC valve, fast idle thermo valve, throttle body**
20	P1297	Electrical load sensor – circuit/voltage low	**Wiring, electrical load sensor, ECM**
20	P1298	Electrical load sensor – circuit/voltage high	**Wiring, electrical load sensor, ECM**

Flash code	OBD-II code	Fault location	Probable cause
22	P1259	VTEC system malfunction	**Wiring, VTEC solenoid/pressure switch, ECM**
23	–	Knock sensor (KS) – circuit malfunction	**Wiring, KS, ECM**
35	P1676	ABS/TCS control module – communication malfunction	**Wiring, ABS/TCS control module, ECM**
35	P1678	ABS/TCS control module – communication malfunction	**Wiring, ABS/TCS control module, ECM**
37	P1656	Electronic stability programme (ESP) control module – communication malfunction	**Wiring, ESP control module**
41	–	Heated oxygen sensor (HO2S) – front – circuit malfunction	**Wiring, ECM**
45	–	Mixture too lean/rich	**Fuel system, front HO2S, MAP sensor, mechanical fault**
58	P1366	Crankshaft position (CKP) sensor 2 – intermittent signal	**CKP sensor**
58	P1367	Crankshaft position (CKP) sensor 2 – no signal	**Wiring, CKP sensor, ECM**
61	–	Heated oxygen sensor (HO2S) – front – slow response	**HO2S, exhaust system**
63	–	Heated oxygen sensor (HO2S) – rear – slow response/circuit/voltage low/high	**Wiring, HO2S, ECM**
65	–	Heated oxygen sensor (HO2S) – rear – circuit malfunction	**Wiring, ECM**
67	–	Catalytic converter – efficiency below limit	**Catalytic converter, rear HO2S**
70	–	AT – lock-up clutch not engaging/no gear shift	**Wiring, mainshaft speed sensor, countershaft speed sensor, lock-up control system, shift solenoid (SS) A/B, TCM**
70	P1656	Electronic stability programme (ESP) control module – communication malfunction	**Wiring, ESP control module**
70	P1705	AT – gear shift malfunction	**Wiring, range position switch, TCM**
70	P1706	AT – gear shift malfunction	**Wiring, range position switch, TCM**
70	P1709	AT – gear selection malfunction	**Wiring, transmission gear selection switch, detent bracket, detent plunger/ spring, ECM**
70	P1710	AT – 1st gear hold switch	**Wiring, range position switch, detent bracket/ plunger/spring, ECM**
70	P1738	AT – 2nd clutch pressure switch	**Wiring, range position switch, 2nd clutch pressure switch, ECM**
70	P1739	AT – 3rd clutch pressure switch	**Wiring, range position switch, 3rd clutch pressure switch, ECM**
70	P1740	AT – 4th clutch pressure switch	**Wiring, range position switch, 4th clutch pressure switch, ECM**

Flash code	OBD-II code	Fault location	Probable cause
70	P1750	AT – hydraulic system mechanical malfunction	Wiring, shift solenoids, ECM, transmission mechanical fault
70	P1751	AT – hydraulic system mechanical malfunction	Wiring, shift solenoids, ECM, transmission mechanical fault
70	P1753	AT – lock-up clutch not engaging/ disengaging	Wiring, lock-up control solenoid A, TCM
70	P1768	Shift solenoid (SS) A	Wiring, shift solenoid, ECM
70	P1773	Shift solenoid (SS) B	Wiring, shift solenoid, ECM
70	P1778	Shift solenoid (SS) C	Wiring, shift solenoid, ECM
71	–	Cylinder No.1 – misfire	Wiring, injector, ignition system, mechanical fault
72	–	Cylinder No.2 – misfire	Wiring, injector, ignition system, mechanical fault
73	–	Cylinder No.3 – misfire	Wiring, injector, ignition system, mechanical fault
74	–	Cylinder No.4 – misfire	Wiring, injector, ignition system, mechanical fault
75	–	Cylinder No.5 – misfire	Wiring, injector, ignition system, mechanical fault
76	–	Cylinder No.6 – misfire	Wiring, injector, ignition system, mechanical fault
–	P1300	Random misfire	Fuel/ignition system, MAP sensor, IAC valve
–	P1399	Random misfire	Fuel/ignition system, MAP sensor, IAC valve
80	–	Exhaust gas recirculation (EGR) solenoid – insufficient flow	EGR solenoid, hose/pipe leak or blockage
86	–	Engine coolant temperature (ECT) sensor – range/performance problem	ECT sensor, cooling system
87	P1486	Thermostat – range/performance problem	Engine coolant blower motor, cooling system
90	P1456	Evaporative emission (EVAP) canister purge system (fuel tank system) – leak detected	Hose, fuel tank/pressure sensor, fuel filler cap, EVAP valve/bypass solenoid, EVAP two way valve, EVAP canister/vent valve
90	P1457	Evaporative emission (EVAP) canister purge system (canister system) – leak detected	Hose, fuel tank/pressure sensor, EVAP valve/bypass solenoid, EVAP two way valve, EVAP canister/vent valve
91	–	Fuel tank pressure sensor – circuit/voltage low/high	Wiring, pressure sensor, ECM
–	P1607	Engine control module (ECM) – internal circuit failure	ECM

Model:	**A4 1.8L Turbo • A4 2.8L • A4 3.0L • A4 Cabriolet 1.8L Turbo A4 Cabriolet 3.0L • S4 2.7L Turbo • S4 4.2L • S4 Cabriolet 4.2L 100 2.8L • A6 2.8L • A6/Allroad 2.7L Turbo • A6 3.0L • A6 4.2L • S6/RS6 4.2L A8 3.7L/4.2L • S8 4.2L • Cabriolet 2.8L • TT 1.8L Turbo • TT 3.2L TT Roadster 1.8L Turbo • TT Roadster 3.2L**
Year:	**1992-04**
Engine identification:	**1Z, AAH, AAT, ABC, ABK, ABZ, ACK, ACZ, ADA, ADP, ADR, AEB, AEC, AEH, AEL, AEW, AFB, AFF, AFN, AFY, AGA, AGB, AGN, AGR, AGU, AHA, AHC, AHF, AHH, AHK, AHL, AHU, AJG, AJL, AJM, AJP, AJQ, AKB, AKE, AKG, AKH, AKJ, AKL, AKN, ALF, ALG, ALH, ALW, AMB, AMF, AMU, ANA, ANP, ANY, APB, APF, APG, APP, APR, APS, APT, APU, APX, APY, APZ, AQA, AQD, AQF, AQG, AQH, ARE, ARG, ARJ, ARM, ART, ARU, ARY, ARZ, ASJ, ASV, ATC, ATJ, ATQ, AUA, AUM, AUQ, AUX, AVG, AVK, AVV, AWN, AWP, AYS, AZR, BCY, BEA, BEL, BAM, BAS, BDD, BHE, BHF**
System:	**Bosch EDC/1.3.1/2/3/1.4 • Bosch EDC 15M/P/V • Bosch Motronic • Bosch Motronic M2.4.1 Bosch Motronic M3.2/3.2.1 • Bosch Motronic M3.8.2/3/4 • Bosch Motronic M5.4/.1 Bosch Motronic M5.9.2F • Bosch Motronic ME7.0/7.1/7.1.1/7.5/7.5.1/7.5.10 Bosch MSA 12/15.5 • Siemens Simos • VAG Digifant ML5.7 • VAG MPI**

Data link connector (DLC) locations

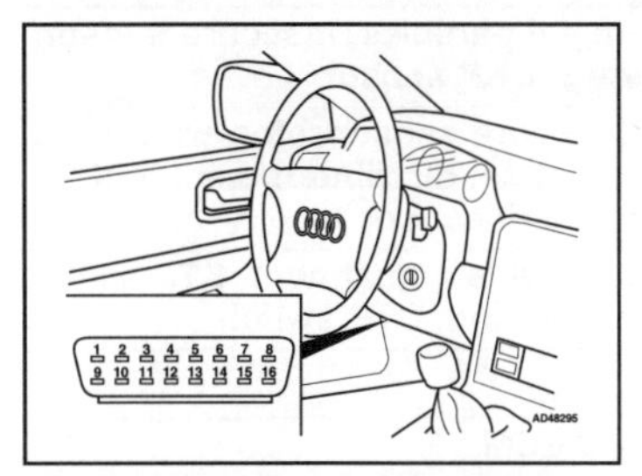

A4/S4 2000→, A6/S6/RS6/Allroad 1997→, A8/S8 2002→ – fascia, driver's side

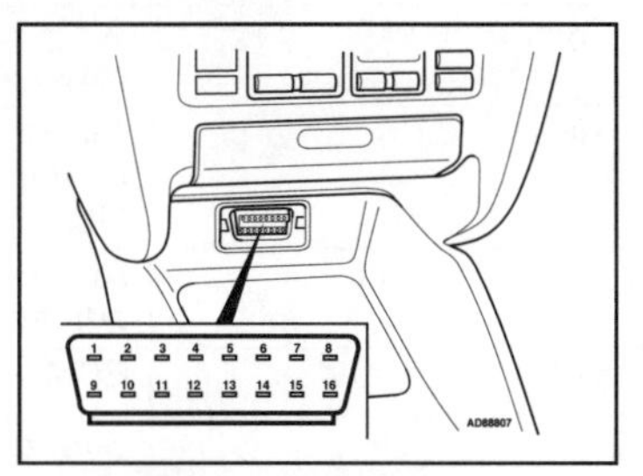

80, Coupé, Cabriolet – in underhood fusebox/relay plate

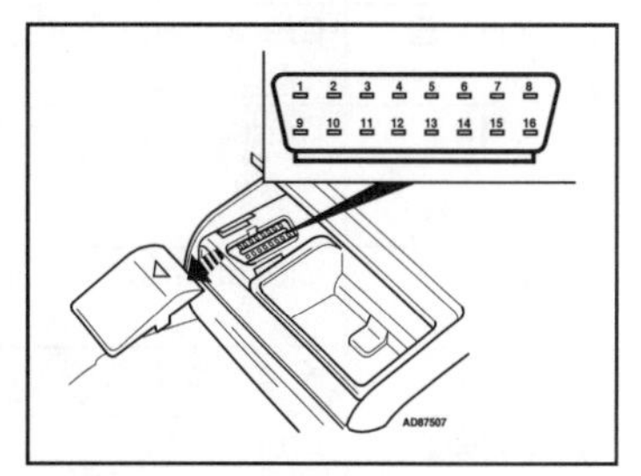

A4/S4 1994-00 – center console, rear

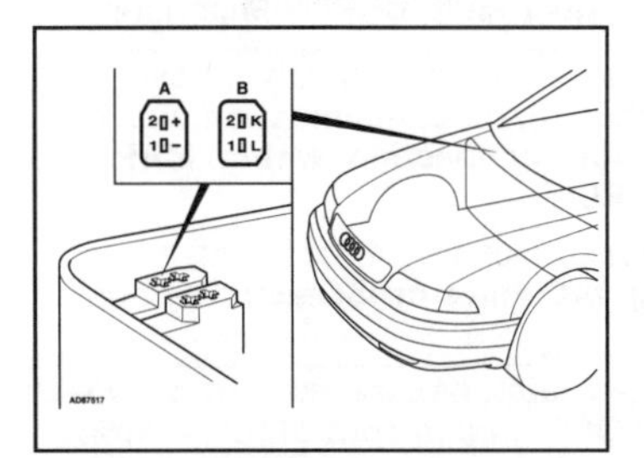

100/A6 →1997 – in underhood relay plate

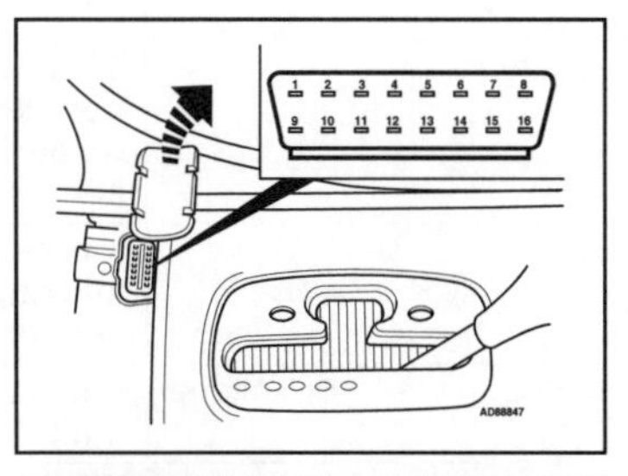

A8/S8 →2002 – under front ashtray

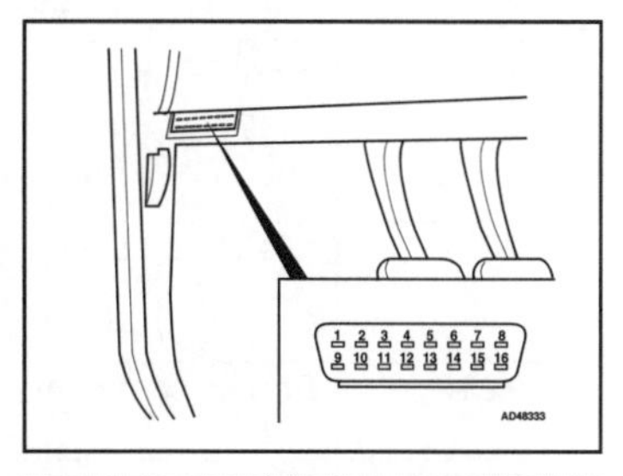

TT – fascia, driver's side

Self-diagnosis

General information

- Refer to the front of this manual for general test conditions, terminology, detailed descriptions of wiring faults and a general trouble shooter for electrical and mechanical faults.
- Carry out road test for at least 10 minutes.
- Automatic transmission in 'P' or 'N'.
- If engine does not start: Crank engine for 6 seconds. Leave ignition switched ON.
- Data link connector (DLC) – 2-pin: (A) black, (B) brown/white.

Accessing and erasing

- The engine control module (ECM) fault memory can only be accessed and erased using diagnostic equipment connected to the data link connector (DLC).

NOTE: ***Self-diagnosis output using 4-digit trouble codes may not display all available diagnostic information (early models).***

Trouble code identification

Scan code 4-digit	Fault location	Probable cause
1111	Engine control module (ECM) – defective	**ECM**
1231	Vehicle speed sensor (VSS)	**Wiring, speedometer, VSS**
1232	Idle speed control (ISC) actuator	**Throttle valve tight/sticking, wiring, multi-plug incorrectly wired, ISC actuator**
2111 1	Crankshaft position (CKP) sensor	**Air gap, metal particles, insecure sensor/rotor, wiring, CKP sensor**
2113	Camshaft position (CMP) sensor	**Air gap, insecure sensor/rotor, wiring, poor connection, fuse, distributor/camshaft alignment, CMP sensor**
2121	Closed throttle position (CTP) switch	**Accelerator cable adjustment, CTP switch adjustment, wiring, CTP switch**
2141	Knock control – control limit exceeded	**ECM**
2142	Knock sensor (KS) 1	**Wiring, KS, ECM**
2144	Knock sensor (KS) 2	**Wiring, KS, ECM**
2212	Throttle position (TP) sensor	**Wiring, poor connection, TP sensor**
2214	Maximum engine RPM exceeded	**Incorrect gear shift, CKP/RPM sensor**
2222	Manifold absolute pressure (MAP) sensor	**TC wastegate regulating valve, hose connection(s), oil contamination, valve timing, poor connection, wiring, MAP sensor, ECM**
2231	Idle speed control (ISC)	**Intake leak/blockage, throttle valve tight/ sticking, IAC valve or ISC actuator/position sensor**
2234	Engine control module (ECM) – supply voltage	**Fuse, alternator, battery, current draw with ignition OFF, engine control relay, wiring**
2242	Mixture adjustment resistor	**Wiring, mixture adjustment resistor**
2243	Instrument panel, fuel consumption signal	**Wiring short to positive, instrument panel**

Scan code 4-digit	Fault location	Probable cause
2312	Engine coolant temperature (ECT) sensor	**Wiring, poor connection, ECT sensor**
2314	Engine/gearbox electrical connection	**Wiring, transmission fault**
2322	Intake air temperature (IAT) sensor	**Wiring, poor connection, IAT sensor**
2323	Volume air flow (VAF) sensor	**Intake leak, wiring, VAF sensor**
2324	Mass air flow (MAF) sensor	**Intake leak, wiring, MAF sensor**
2341	Heated oxygen sensor (HO2S) – lambda control	**Heating inoperative, intake/exhaust leak, misfire, fuel level low, fuel pressure/pump, injector(s), EVAP canister purge valve, MAF sensor filament burn-off, wiring, HO2S**
2342	Heated oxygen sensor (HO2S)	**Wiring, HO2S, heating inoperative, fuel level low**
2411	Exhaust gas recirculation temperature (EGRT) sensor	**Wiring, EGRT sensor**
2412	Intake air temperature (IAT) sensor	**Wiring, poor connection, IAT sensor**
2413	Mixture control (MC)	**Fuel level low, fuel pressure/pump, intake/exhaust leak, misfire, MAF sensor filament burn-off, HO2S, EVAP canister purge valve, injector(s), excessive fuel in engine oil**
3434	Oxygen sensor heater relay	**Wiring, oxygen sensor heater relay**
4312	Exhaust gas recirculation (EGR) solenoid	**Wiring, EGR solenoid**
4332	Engine control module (ECM) – output stages	**Wiring, ECM controlled components**
4343	Evaporative emission (EVAP) canister purge valve	**Wiring, fuse, EVAP canister purge valve**
4411	Injector 1	**Wiring, fuse, injector**
4412	Injector 2	**Wiring, fuse, injector**
4413	Injector 3	**Wiring, fuse, injector**
4414	Injector 4	**Wiring, fuse, injector**
4421	Injector 5	**Wiring, injector**
4422	Injector 6	**Wiring, injector**
4431	Idle air control (IAC) valve	**Wiring, IAC valve**
4433	Fuel pump relay	**Wiring, fuse, fuel pump relay**
4444	No fault found	–

1 Trouble code may be displayed if engine is not idling during self-diagnosis due to missing CKP sensor signal. Ignore trouble code if engine starts.

Scan code 5-digit	OBD-II code	Fault location	Probable cause
–	*All P0, P2 and U0 codes*	Refer to OBD-II trouble code tables at the front of this manual	–
00000	–	No fault found	–
00263	–	Transmission control module (TCM) – incorrect signal	**Wiring short to ground, TCM trouble code(s) stored, ECM**
00268	–	Idle speed control (ISC) actuator	**Wiring, ISC actuator**
00281	–	Vehicle speed sensor (VSS)	**Wiring, speedometer, VSS**
00282	–	Idle speed control (ISC) actuator/throttle motor	**Throttle valve tight/sticking, wiring, multi-plug incorrectly wired, ISC actuator/ throttle motor**
00305	–	Instrument panel, fuel consumption signal	**Wiring, instrument panel**
00513 1	–	Crankshaft position (CKP) sensor	**Air gap, metal particles, insecure sensor/ rotor, wiring, CKP sensor**
00514	–	Crankshaft position (CKP) sensor	**Air gap, insecure sensor/rotor, wiring, poor connection, CKP sensor**
00515	–	Camshaft position (CMP) sensor	**Air gap, insecure sensor/rotor, wiring, poor connection, fuse, distributor/camshaft alignment, CMP sensor**
00516	–	Closed throttle position (CTP) switch	**Accelerator cable adjustment, CTP switch adjustment, throttle valve tight/sticking, wiring, CTP switch**
00518	–	Throttle position (TP) sensor	**Wiring, poor connection, TP sensor**
00519	–	Manifold absolute pressure (MAP) sensor	**TC wastegate regulating valve, hose connection(s), oil contamination, valve timing, poor connection, wiring, MAP sensor, ECM**
00520	–	Mass air flow (MAF) sensor	**Wiring, MAF sensor**
00521	–	Mixture adjustment resistor	**Wiring, mixture adjustment resistor**
00522	–	Engine coolant temperature (ECT) sensor	**Wiring, poor connection, ECT sensor**
00523	–	Intake air temperature (IAT) sensor	**Wiring, poor connection, IAT sensor**
00524	–	Knock sensor (KS) 1	**Wiring, KS, ECM**
00525	–	Heated oxygen sensor (HO2S) 1 – implausible signal	**Wiring, HO2S, heating inoperative, fuel level low**
00526	–	Stop lamp switch	**Wiring, stop lamp switch**
00527	–	Intake air temperature (IAT) sensor	**Wiring, poor connection, IAT sensor**
00528	–	Barometric pressure (BARO) sensor	**Wiring, BARO sensor**
00529	–	Crankshaft position (CKP) sensor	**Wiring, CKP sensor**
00530	–	Idle speed control (ISC) actuator/position sensor	**Wiring, poor connection, multi-plug incorrectly wired, ISC actuator/position sensor**
00532	–	Engine control module (ECM) – supply voltage	**Fuse, alternator, battery, current draw with ignition OFF, engine control relay, wiring**

Scan code 5-digit	OBD-II code	Fault location	Probable cause
00533	–	Idle speed control (ISC)	Intake leak/blockage, throttle valve tight/ sticking, IAC valve or ISC actuator/position sensor
00534	–	Engine oil temperature (EOT) sensor	Wiring, EOT sensor
00535	–	Engine control module (ECM) – knock control 1	Wiring, KS, ECM
00536	–	Engine control module (ECM) – knock control 2	Wiring, KS, ECM
00537	–	Heated oxygen sensor (HO2S) – lambda control	CO adjustment, heating inoperative, intake/exhaust leak, misfire, fuel level low, fuel pressure/pump, injector(s), EVAP canister purge valve, MAP sensor, MAF sensor filament burn-off, wiring, HO2S
00539	–	Fuel temperature sensor	Wiring, fuel temperature sensor
00540	–	Knock sensor (KS) 2	Wiring, KS, ECM
00542	–	Injector needle lift sensor	Air in fuel system, fuel level low, wiring, injector needle lift sensor
00543	–	Maximum engine RPM exceeded	Incorrect gear shift, CKP/RPM sensor, AT fault, ECM
00544	–	Maximum boost pressure exceeded	Hoses interchanged/not connected, hoses blocked/leaking, TC wastegate actuator/ regulating valve, BARO sensor
00545	–	ECM/TCM electrical connection	TCM trouble code(s) stored, wiring, transmission fault
00546	–	Data link connector (DLC) – defective	Wiring
00549	–	Instrument panel, fuel consumption signal	Wiring short to positive, instrument panel
00550	–	Start of injection – control	Wiring, fuel injection timing solenoid, injector needle lift sensor, air in fuel system, fuel level low, pump timing
00552	–	Volume air flow (VAF) sensor	Intake leak, wiring, VAF sensor
00553	–	Mass air flow (MAF) sensor	Intake leak, wiring, MAF sensor
00554	–	Heated oxygen sensor (HO2S) – lambda control	CO adjustment, heating inoperative, intake/exhaust leak, misfire, fuel level low, fuel pressure/pump, injector(s), EVAP canister purge valve, MAP sensor, MAF sensor filament burn-off, wiring, HO2S
00555	–	Heated oxygen sensor (HO2S) 2 – implausible signal	Wiring, HO2S, heating inoperative, fuel level low
00557	–	Power steering pressure (PSP) switch – short to ground	Wiring short to ground, PSP switch
00560	–	Exhaust gas recirculation (EGR) – control difference	Intake leak, EGR valve/solenoid

Scan code 5-digit	OBD-II code	Fault location	Probable cause
00561	–	Mixture control (MC)	**Trouble codes 00525/00533 stored, fuel level low, fuel pressure/pump, intake/exhaust leak, misfire, MAP sensor, MAF sensor filament burn-off, HO2S, EVAP canister purge valve, injector(s), excessive fuel in engine oil**
00575	–	Intake manifold pressure	**Intake leak/blockage, MAP sensor, MAF sensor, TC wastegate actuator/regulating valve, hoses interchanged/not connected, wiring**
00577	–	Knock control, cylinder 1 – control limit exceeded	**Fuel pressure, misfire, intake leak, injector(s)**
00578	–	Knock control, cylinder 2 – control limit exceeded	**Fuel pressure, misfire, intake leak, injector(s)**
00579	–	Knock control, cylinder 3 – control limit exceeded	**Fuel pressure, misfire, intake leak, injector(s)**
00580	–	Knock control, cylinder 4 – control limit exceeded	**Fuel pressure, misfire, intake leak, injector(s)**
00581	–	Knock control, cylinder 5 – control limit exceeded	**Fuel pressure, misfire, intake leak, injector(s)**
00582	–	Knock control, cylinder 6 – control limit exceeded	**Fuel pressure, misfire, intake leak, injector(s)**
00585	–	Exhaust gas recirculation temperature (EGRT) sensor	**Wiring, EGRT sensor**
00586	–	Exhaust gas recirculation (EGR) system – control	**EGR solenoid**
00609	–	Ignition amplifier, primary circuit 1	**Wiring, ignition amplifier**
00610	–	Ignition amplifier, primary circuit 2	**Wiring, ignition amplifier**
00611	–	Ignition amplifier, primary circuit 3	**Wiring, ignition amplifier**
00624	–	AC signal – compressor cut-in	**Wiring, AC system**
00625	–	Vehicle speed signal	**Wiring, speedometer, VSS**
00626	–	Glow plug warning lamp	**Wiring, glow plug warning lamp**
00627	–	Fuel filter water level sensor	**Water in filter, wiring, fuel filter water level sensor**
00628	–	Fuel injection pump control module – engine stop malfunction	**Wiring, fuel injection pump**
00635	–	Heated oxygen sensor (HO2S) 1, bank 1 – heater circuit malfunction	**Wiring, HO2S**
00638	–	ECM/TCM electrical connection	**Wiring, transmission fault**
00640	–	Oxygen sensor heater relay	**Wiring, oxygen sensor heater relay**
00650	–	Clutch pedal position (CPP) switch – short to positive	**Wiring short to positive, CPP switch**
00653	–	Transmission control module (TCM)/TR sensor – implausible TR signal	**Wiring, transmission fault, poor connection**

Scan code 5-digit	OBD-II code	Fault location	Probable cause
00667	–	Outside air temperature signal	Wiring, instrument panel, AC system, outside air temperature sensor
00668	–	Engine control module (ECM) – supply voltage low	Battery, wiring, engine control relay
00670	–	Idle speed control (ISC) actuator position sensor	Wiring, throttle valve, ISC actuator position sensor
00671	–	Cruise control master switch	Wiring, cruise control master switch
00740	–	Camshaft position (CMP) sensor	Wiring, CMP sensor
00741	–	Stop lamp switch/brake pedal position (BPP) switch – implausible ratio	Wiring, both switch positions not synchronised, stop lamp switch, BPP switch
00750	–	Malfunction indicator lamp (MIL) – circuit malfunction	ECM incorrectly coded, wiring, MIL
00758	–	Secondary air injection (AIR) system	AIR solenoid/relay, wiring
00765	–	Fuel quantity adjuster position sensor	Wiring, fuel injection pump
00777	–	Accelerator pedal position (APP) sensor	Incorrectly adjusted, wiring, APP sensor
00792	–	AC pressure switch	Wiring, AC pressure switch
01013	–	AC compressor clutch, load signal	Wiring, AC system
01025	–	Malfunction indicator lamp (MIL)	Wiring, MIL
01028	–	Engine coolant blower motor relay	Wiring, engine coolant blower motor relay
01044	–	Engine control module (ECM) – coding	Incorrectly coded
01050	–	Glow plug monitoring	Fuse, wiring, glow plug relay, glow plugs
01052	–	Fuel lever position sensor	Wiring, fuel lever position sensor
01087	–	Engine control module (ECM) – basic setting	Basic setting not completed, throttle valve tight/sticking
01088	–	Mixture control (MC)	Fuel level low, fuel pressure/pump, MAP sensor, MAF sensor, intake/exhaust leak, EVAP canister purge valve, excessive fuel in engine oil, injector(s)
01117	–	Alternator load signal	Wiring, alternator
01119	–	Gear recognition signal – AT	Wiring
01120	–	Camshaft position (CMP) control, bank 1 – mechanical fault	Wiring, fuse, CMP actuator
01121	–	Camshaft position (CMP) control, bank 2 – mechanical fault	Wiring, fuse, CMP actuator
01126	–	Engine RPM signal	Wiring, CKP/RPM sensor, instrument panel
01163	–	Backfire	Intake leak, wiring, ignition amplifier, injector(s)
01165	–	Idle speed control (ISC) actuator/throttle motor	Accelerator cable adjustment, throttle valve, wiring, ISC actuator/throttle motor, basic setting not carried out
01167	–	Full throttle stop solenoid	Wiring, full throttle stop valve

Scan code 5-digit	OBD-II code	Fault location	Probable cause
01168	–	Idle speed boost solenoid	Wiring, idle speed boost valve
01169	–	Door contact switch, driver's	Wiring, door contact switch
01170	–	Fuel injection timing sensor	Wiring, fuel injection timing sensor
01177	–	Engine control module (ECM) – defective	ECM
01180	–	Engine/AC electrical connection	Wiring
01182	–	Mass air flow (MAF) sensor/throttle position (TP) sensor – signal variation exceeded	Throttle valve sticking, ISC actuator/ throttle motor sticking/mechanically damaged, incorrect throttle control unit, intake leak between MAF sensor and throttle valve, air filter blocked
01183	–	Malfunction indicator lamp (MIL) – circuit malfunction	ECM incorrectly coded, wiring, MIL
01193	–	Engine coolant heater relay 1, low output	Wiring, engine coolant heater relay
01194	–	Engine coolant heater relay 2, high output	Wiring, engine coolant heater relay
01196	–	CAN data bus, TCM – incorrect signal	Wiring
01204	–	Engine coolant temperature (ECT) sensor	Wiring, ECT sensor
01208	–	Engine control module (ECM) – data changed	ECM
01209	–	Alternator speed signal	Wiring, alternator
01235	–	Secondary air injection (AIR) solenoid	Wiring, AIR solenoid
01237	–	Fuel shut-off solenoid	Wiring, fuel shut-off solenoid
01242	–	Engine control module (ECM) – output stages	Wiring, ECM controlled components
01243	–	Intake manifold air control solenoid	Wiring, intake manifold air control solenoid
01247	–	Evaporative emission (EVAP) canister purge valve	Wiring, fuse, EVAP canister purge valve
01249	–	Injector 1	Wiring, fuse, injector
01250	–	Injector 2	Wiring, fuse, injector
01251	–	Injector 3	Wiring, fuse, injector
01252	–	Injector 4	Wiring, fuse, injector
01253	–	Injector 5	Wiring, injector
01254	–	Injector 6	Wiring, injector
01257	–	Idle air control (IAC) valve	Wiring, IAC valve
01259	–	Fuel pump relay	Wiring, fuse, fuel pump relay
01262	–	Turbocharger (TC) wastegate regulating valve	Wiring, fuse, TC wastegate regulating valve
01265	–	Exhaust gas recirculation (EGR) solenoid	Wiring, EGR solenoid
01266	–	Glow plug relay	Wiring, glow plug relay
01268	–	Fuel quantity adjuster	Incorrectly set, wiring, fuel injection pump

Scan code 5-digit	OBD-II code	Fault location	Probable cause
01269	–	Fuel injection timing solenoid	Wiring, fuel injection timing solenoid
01282	–	Intake manifold air control solenoid	Wiring, intake manifold air control solenoid
01283	–	Intake manifold air control actuator	Wiring, intake manifold flap, intake manifold air control actuator
01312	–	CAN data bus – defective	Trouble code(s) stored in other system(s), wiring
01314	–	Engine control module (ECM), cruise control system – defective	ECM
01315	–	CAN data bus, TCM – no signal	TCM trouble code(s) stored, wiring, matching resistor in ECM
01316	–	CAN data bus, ABS – defective	ABS control module incorrectly coded, wiring
01317	–	CAN data bus, instrumentation	Trouble code(s) stored in other system(s), wiring, instrumentation control module
01318	–	Fuel injection pump control module	Data bus wiring, fuel injection pump
01321	–	CAN data bus, SRS	SRS control module trouble code(s) stored, wiring
01375	–	Engine mounting control solenoid, bank 1 & 2	Wiring, engine mounting control solenoid
01376	–	Fuel injection pump position sensor	Wiring, fuel injection pump position sensor
01437	–	Throttle control unit – basic setting	Basic setting not carried out, CTP switch, ISC actuator/position sensor, TP sensor
01440	–	Fuel level signal	Wiring, instrument panel, fuel gauge tank sensor
01441	–	Fuel low level sensor	Wiring, fuel transfer pump, fuel level sensor
01442	–	Engine misfire – fuel pump housing empty	Fuel level low, fuel transfer pump
01575	–	Auxiliary drive – switched OFF	Auxiliary drive overloaded
01613	–	Fuel cooling pump relay – circuit malfunction	Wiring, fuel cooling pump relay
01656	–	SRS control module – crash signal	Wiring
01686	–	Engine coolant blower motor run-on relay	Wiring, engine coolant blower motor run-on relay
01695	–	Fuel temperature sensor – circuit malfunction	Wiring, fuel temperature sensor
16394	–	Camshaft position (CMP) actuator, intake/left/front, bank 1 – circuit malfunction	Wiring, CMP actuator
16395	–	Camshaft position (CMP), intake/left/front, bank 1 – timing over advanced/system performance	Valve timing, camshaft position (CMP) actuator
16396	–	Camshaft position (CMP), intake/left/front, bank 1 – timing over retarded	Valve timing, camshaft position (CMP) actuator
16398	–	Camshaft position (CMP) actuator, exhaust/right/rear, bank 1 – timing over advanced/system performance	Valve timing, camshaft position (CMP) actuator

Scan code 5-digit	OBD-II code	Fault location	Probable cause
16399	–	Camshaft position (CMP) actuator, exhaust/right/rear, bank 1 – timing over retarded	**Valve timing, camshaft position (CMP) actuator**
16414	–	Heated oxygen sensor (HO2S) 1, bank 1, heater control – circuit malfunction	**Wiring open circuit, HO2S**
16415	–	Heated oxygen sensor (HO2S) 1, bank 1, heater control – circuit low	**Wiring short to ground, HO2S**
16416	–	Heated oxygen sensor (HO2S) 1, bank 1, heater control – circuit high	**Wiring short to positive, HO2S**
16474	–	Fuel metering solenoid – open circuit	**Wiring open circuit, fuel metering solenoid**
16475	–	Fuel metering solenoid – short to ground	**Wiring short to ground, fuel metering solenoid**
16476	–	Fuel metering solenoid – short to positive	**Wiring short to positive, fuel metering solenoid**
16485	–	Mass air flow (MAF) sensor, bank 1 – range/performance problem	**Intake leak, wiring, MAF sensor**
16486	–	Mass air flow (MAF) sensor, bank 1 – low input	**Intake leak, air filter blocked, wiring short to ground, fuse, MAF sensor**
16487	–	Mass air flow (MAF) sensor, bank 1 – high input	**Wiring short to positive, ground wire defective, MAF sensor**
16490	–	Manifold absolute pressure (MAP) sensor/ barometric pressure (BARO) sensor – range/performance problem	**Intake/exhaust leak, wiring, MAP sensor, BARO sensor**
16491	–	Manifold absolute pressure (MAP) sensor/ barometric pressure (BARO) sensor – low input	**Wiring short to ground, MAP sensor, BARO sensor**
16492	–	Manifold absolute pressure (MAP) sensor/ barometric pressure (BARO) sensor – high input	**Wiring short to positive, MAP sensor, BARO sensor**
16496	–	Intake air temperature (IAT) sensor – low input	**Wiring short to ground, IAT sensor**
16497	–	Intake air temperature (IAT) sensor – high input	**Wiring open circuit/short to positive, ground wire defective, IAT sensor**
16500	–	Engine coolant temperature (ECT) sensor – range/performance problem	**Coolant thermostat, poor connection, wiring, ECT sensor**
16501	–	Engine coolant temperature (ECT) sensor – low input	**Coolant thermostat, wiring short to ground, ECT sensor**
16502	–	Engine coolant temperature (ECT) sensor – high input	**Coolant thermostat, wiring open circuit/ short to positive, ground wire defective, ECT sensor**
16504	–	Throttle position (TP) sensor – circuit malfunction	**Poor connection, wiring, TP sensor**
16505	–	Throttle position (TP) sensor – range/ performance problem	**Poor connection, TP sensor**
16506	–	Throttle position (TP) sensor – low input	**Signal wire short to ground, supply wire defective, TP sensor**
16507	–	Throttle position (TP) sensor – high input	**Signal wire open circuit/short to positive, ground wire defective, TP sensor**

Scan code 5-digit	OBD-II code	Fault location	Probable cause
16514	–	Heated oxygen sensor (HO2S) 1, bank 1 – circuit malfunction	Heating inoperative, poor connection, wiring, HO2S
16515	–	Heated oxygen sensor (HO2S) 1, bank 1 – voltage low	Wiring short to ground, HO2S
16516	–	Heated oxygen sensor (HO2S) 1, bank 1 – voltage high	Wiring short to positive, HO2S
16517	–	Heated oxygen sensor (HO2S) 1, bank 1 – slow response	Heating inoperative, wiring, HO2S
16518	–	Heated oxygen sensor (HO2S) 1, bank 1 – no activity detected	Wiring open circuit, heating inoperative, HO2S
16519	–	Heated oxygen sensor (HO2S) 1, bank 1 – heater circuit malfunction	Fuse, wiring, HO2S
16520	–	Heated oxygen sensor (HO2S) 2, bank 1 – circuit malfunction	Heating inoperative, wiring, HO2S
16521	–	Heated oxygen sensor (HO2S) 2, bank 1 – low voltage	Wiring short to ground, exhaust leak, HO2S
16522	–	Heated oxygen sensor (HO2S) 2, bank 1 – high voltage	Wiring short to positive, HO2S
16523	–	Heated oxygen sensor (HO2S) 2, bank 1 – slow response	Heating inoperative, wiring, HO2S
16524	–	Heated oxygen sensor (HO2S) 2, bank 1 – no activity detected	Wiring, HO2S
16525	–	Heated oxygen sensor (HO2S) 2, bank 1 – heater circuit malfunction	Wiring, HO2S
16534	–	Heated oxygen sensor (HO2S) 1, bank 2 – circuit malfunction	Wiring, HO2S
16535	–	Heated oxygen sensor (HO2S) 1, bank 2 – low voltage	Wiring short to ground, HO2S
16536	–	Heated oxygen sensor (HO2S) 1, bank 2 – high voltage	Wiring short to positive, HO2S
16537	–	Heated oxygen sensor (HO2S) 1, bank 2 – slow response	Heating inoperative, wiring, HO2S
16538	–	Heated oxygen sensor (HO2S) 1, bank 2 – no activity detected	Wiring, HO2S
16540	–	Heated oxygen sensor (HO2S) 2, bank 2 – circuit malfunction	Heating inoperative, wiring, HO2S
16541	–	Heated oxygen sensor (HO2S) 2, bank 2 – low voltage	Wiring short to ground, exhaust leak, HO2S
16542	–	Heated oxygen sensor (HO2S) 2, bank 2 – high voltage	Wiring short to positive, HO2S
16543	–	Heated oxygen sensor (HO2S) 2, bank 2 – slow response	Heating inoperative, wiring, HO2S
16544	–	Heated oxygen sensor (HO2S) 2, bank 2 – no activity detected	Wiring, HO2S
16545	–	Heated oxygen sensor (HO2S) 2, bank 2 – heater circuit malfunction	Wiring, HO2S

Scan code 5-digit	OBD-II code	Fault location	Probable cause
16554	–	Fuel trim (FT), bank 1 – malfunction	Intake leak, AIR system, fuel pressure/pump, injector(s), EVAP canister purge valve, HO2S
16555	–	System too lean, bank 1	Intake/exhaust leak, AIR system, MAP sensor, MAF sensor, fuel pressure/pump, injector(s), EVAP canister purge valve, HO2S
16556	–	System too rich, bank 1	EVAP canister purge valve, fuel pressure, injector(s), HO2S
16557	–	Fuel trim (FT), bank 2 – malfunction	Fuel pressure/pump, injector(s), AIR system, hose connection(s), intake leak
16558	–	System too lean, bank 2	Fuel pressure/pump, injector(s), intake/exhaust leak, AIR system, hose connection(s)
16559	–	System too rich, bank 2	Fuel pressure, injector(s), EVAP canister purge valve
16575	–	Fuel rail pressure (FRP) sensor – range/performance problem	Wiring, FRP sensor
16576	–	Fuel rail pressure (FRP) sensor – low input	Wiring short to ground, FRP sensor
16577	–	Fuel rail pressure (FRP) sensor – high input	Wiring short to positive, FRP sensor
16578	–	Fuel rail pressure (FRP) sensor – circuit intermittent	Wiring open circuit, FRP sensor
16581	–	Engine oil temperature (EOT) sensor – low input	Wiring short to ground, EOT sensor
16582	–	Engine oil temperature (EOT) sensor – high input	Wiring short to positive, EOT sensor
16585	–	Injector 1 – circuit malfunction	Wiring, injector
16586	–	Injector 2 – circuit malfunction	Wiring, injector
16587	–	Injector 3 – circuit malfunction	Wiring, injector
16588	–	Injector 4 – circuit malfunction	Wiring, injector
16589	–	Injector 5 – circuit malfunction	Wiring, injector
16590	–	Injector 6 – circuit malfunction	Wiring, injector
16603	–	Engine over speed condition	Incorrect gear change
16605	–	Throttle position (TP) sensor B – range/performance problem	Wiring, TP sensor
16606	–	Throttle position (TP) sensor B – low input	Wiring short to ground, TP sensor
16607	–	Throttle position (TP) sensor B – high input	Wiring short to positive, TP sensor
16610	–	Accelerator pedal position (APP) sensor A/B – range/performance problem	Wiring, APP sensor
16611	–	Accelerator pedal position (APP) sensor A – low input	Wiring short to ground, APP sensor

Scan code 5-digit	OBD-II code	Fault location	Probable cause
16612	–	Accelerator pedal position (APP) sensor A – high input	**Wiring short to positive, APP sensor**
16614	–	Fuel pump relay – circuit malfunction	**Wiring, fuel pump relay**
16618	–	Engine boost condition – limit exceeded	**Hose connection(s), wiring, TC wastegate regulating valve, TC wastegate**
16619	–	Engine boost condition – limit not reached	**Hose connection(s), wiring, TC wastegate regulating valve, TC wastegate**
16620	–	Manifold absolute pressure (MAP) sensor A, TC system – range/performance problem	**Intake/exhaust leak, hose connection(s), MAP sensor**
16621	–	Manifold absolute pressure (MAP) sensor A, TC system – low input	**Wiring short to ground, MAP sensor**
16622	–	Manifold absolute pressure (MAP) sensor A, TC system – high input	**Wiring short to positive, MAP sensor**
16627	–	Turbocharger (TC) wastegate regulating valve A – circuit malfunction	**Wiring, TC wastegate regulating valve**
16629	–	Turbocharger (TC) wastegate regulating valve A – circuit low	**Wiring short to ground, TC wastegate regulating valve**
16630	–	Turbocharger (TC) wastegate regulating valve A – circuit high	**Wiring short to positive, TC wastegate regulating valve**
16645	–	Injector 1 – circuit low	**Wiring short to ground, injector**
16646	–	Injector 1 – circuit high	**Wiring short to positive, injector**
16648	–	Injector 2 – circuit low	**Wiring short to ground, injector**
16649	–	Injector 2 – circuit high	**Wiring short to positive, injector**
16651	–	Injector 3 – circuit low	**Wiring short to ground, injector**
16652	–	Injector 3 – circuit high	**Wiring short to positive, injector**
16654	–	Injector 4 – circuit low	**Wiring short to ground, injector**
16655	–	Injector 4 – circuit high	**Wiring short to positive, injector**
16657	–	Injector 5 – circuit low	**Wiring short to ground, injector**
16658	–	Injector 5 – circuit high	**Wiring short to positive, injector**
16660	–	Injector 6 – circuit low	**Wiring short to ground, injector**
16661	–	Injector 6 – circuit high	**Wiring short to positive, injector**
16684	–	Random/multiple cylinder(s) – misfire detected	**Spark plug(s), HT lead(s), injector(s), ignition coil(s), low compression, wiring**
16685	–	Cylinder 1 – misfire detected	**Spark plug, HT lead(s), injector, ignition coil, low compression, wiring**
16686	–	Cylinder 2 – misfire detected	**Spark plug, HT lead(s), injector, ignition coil, low compression, wiring**
16687	–	Cylinder 3 – misfire detected	**Spark plug, HT lead(s), injector, ignition coil, low compression, wiring**
16688	–	Cylinder 4 – misfire detected	**Spark plug, HT lead(s), injector, ignition coil, low compression, wiring**

Scan code 5-digit	OBD-II code	Fault location	Probable cause
16689	–	Cylinder 5 – misfire detected	Spark plug, HT lead(s), injector, ignition coil, low compression, wiring
16690	–	Cylinder 6 – misfire detected	Spark plug, HT lead(s), injector, ignition coil, low compression, wiring
16691	–	Cylinder 7 – misfire detected	Spark plug, HT lead(s), injector, ignition coil, low compression, wiring
16692	–	Cylinder 8 – misfire detected	Spark plug, HT lead(s), injector, ignition coil, low compression, wiring
16705	–	Crankshaft position (CKP) sensor/engine speed (RPM) sensor – range/performance problem	Air gap, metal particles, insecure sensor/ rotor, wiring, CKP/RPM sensor
16706	–	Crankshaft position (CKP) sensor/engine speed (RPM) sensor – no signal	Wiring, CKP/RPM sensor
16709	–	Knock sensor (KS) 1, bank 1 – circuit malfunction	Wiring, poor connection, KS
16710	–	Knock sensor (KS) 1, bank 1 – range/ performance problem	Wiring, KS incorrectly tightened, KS
16711	–	Knock sensor (KS) 1, bank 1 – low input	Insecure KS, poor connection, wiring short to ground, incorrectly tightened, KS
16712	–	Knock sensor (KS) 1, bank 1 – high input	Wiring short to positive, KS incorrectly tightened, KS
16716	–	Knock sensor (KS) 2, bank 2 – low input	Insecure KS, poor connection, wiring short to ground, KS incorrectly tightened, KS
16717	–	Knock sensor (KS) 2, bank 2 – high input	Wiring short to positive, KS incorrectly tightened, KS
16719	–	Crankshaft position (CKP) sensor – circuit malfunction	Wiring, CKP/RPM sensor
16721	–	Crankshaft position (CKP) sensor – low input	Insecure sensor, air gap, wiring short to ground, CKP/RPM sensor
16724	–	Camshaft position (CMP) sensor A, bank 1 – circuit malfunction	Wiring, CMP sensor
16725	–	Camshaft position (CMP) sensor A, bank 1 – range/performance problem	Insecure sensor/rotor, air gap, wiring, CMP sensor
16726	–	Camshaft position (CMP) sensor A, bank 1 – low input	Wiring short to ground, CMP sensor
16727	–	Camshaft position (CMP) sensor A, bank 1 – high input	Wiring short to positive, CMP sensor
16730 2	–	Camshaft position (CMP) sensor A, bank 2 – range/performance problem	Insecure sensor/rotor, air gap, wiring, CMP sensor
16731 2	–	Camshaft position (CMP) sensor A, bank 2 – low input	Wiring short to ground, CMP sensor
16732 2	–	Camshaft position (CMP) sensor A, bank 2 – high input	Wiring short to positive, CMP sensor
16735	–	Ignition coil, cylinder 1, primary/ secondary – circuit malfunction	Wiring, ignition amplifier, ignition coil
16736	–	Ignition coil, cylinder 2, primary/ secondary – circuit malfunction	Wiring, ignition amplifier, ignition coil

Scan code 5-digit	OBD-II code	Fault location	Probable cause
16737	–	Ignition coil, cylinder 3, primary/ secondary – circuit malfunction	Wiring, ignition amplifier, ignition coil
16738	–	Ignition coil, cylinder 4, primary/ secondary – circuit malfunction	Wiring, ignition amplifier, ignition coil
16764	–	Glow plugs – circuit A malfunction	Wiring, glow plug relay, fuse, glow plugs
16784	–	Exhaust gas recirculation (EGR) system – flow malfunction	Basic setting not carried out, EGR valve/ solenoid
16785	–	Exhaust gas recirculation (EGR) system – insufficient flow detected	Hose leak/blockage, basic setting not carried out, EGR valve/solenoid
16786	–	Exhaust gas recirculation (EGR) system – excessive flow detected	EGR valve/solenoid, basic setting not carried out
16787	–	Exhaust gas recirculation (EGR) system – circuit malfunction	Wiring, EGR valve/solenoid
16788	–	Exhaust gas recirculation (EGR) system – range/performance problem	Hose connection(s), wiring, EGR valve/ solenoid
16791	–	Exhaust gas recirculation (EGR) valve position sensor – low input	Wiring short to ground, EGR valve position sensor
16792	–	Exhaust gas recirculation (EGR) valve position sensor – high input	Wiring short to positive, EGR valve position sensor
16795	–	Secondary air injection (AIR) system – incorrect flow detected	AIR pump, AIR valve, AIR hose(s)
16796	–	Secondary air injection (AIR) solenoid A – circuit malfunction	Wiring, AIR solenoid
16802	–	Secondary air injection (AIR) pump relay A – circuit malfunction	Wiring, AIR pump relay
16804	–	Catalytic converter system, bank 1 – efficiency below threshold	Catalytic converter
16806	–	Main catalytic converter, bank 1 – efficiency below threshold	Catalytic converter
16814	–	Catalytic converter system, bank 2 – efficiency below threshold	Catalytic converter
16824	–	Evaporative emission (EVAP) system – malfunction	Hose connection(s), intake leak, EVAP canister purge valve
16825	–	Evaporative emission (EVAP) system – incorrect flow detected	Hose connection(s), intake leak, EVAP canister purge valve
16826	–	Evaporative emission (EVAP) system – small leak detected	Hose connection(s), intake leak, EVAP canister, EVAP canister purge valve
16827	–	Evaporative emission (EVAP) canister purge valve – circuit malfunction	Wiring, EVAP canister purge valve
16828	–	Evaporative emission (EVAP) canister purge valve – open circuit	Wiring open circuit, EVAP canister purge valve
16829	–	Evaporative emission (EVAP) canister purge valve – short circuit	Wiring short circuit, EVAP canister purge valve
16839	–	Evaporative emission (EVAP) system – large leak detected	Hose connection(s), intake leak, EVAP canister, EVAP canister purge valve
16845	–	Fuel tank level sensor – range/ performance problem	Wiring, fuel tank level sensor

Scan code 5-digit	OBD-II code	Fault location	Probable cause
16864	–	Engine coolant blower motor 1 – circuit malfunction	**Wiring, engine coolant blower motor**
16865	–	Engine coolant blower motor 2 – circuit malfunction	**Wiring, engine coolant blower motor**
16885	–	Vehicle speed sensor (VSS) – range/ performance problem	**Wiring, speedometer, VSS, CAN data bus**
16887	–	Vehicle speed sensor (VSS) – intermittent/ erratic/high input	**Wiring, other connected system, instrument panel, VSS**
16890	–	Idle speed control (ISC) system – rpm lower than expected	**Throttle control unit**
16891	–	Idle speed control (ISC) system – rpm higher than expected	**Throttle control unit**
16894	–	Closed throttle position (CTP) switch – circuit malfunction	**Wiring, CTP switch**
16916	–	AC refrigerant pressure sensor – low input	**AC refrigerant pressure too low (incorrectly charged), wiring, AC refrigerant pressure sensor**
16917	–	AC refrigerant pressure sensor – high input	**AC refrigerant pressure too high (cooling fault/incorrectly charged), wiring, AC refrigerant pressure sensor**
16928	–	Exhaust gas recirculation temperature (EGRT) sensor, bank 1 – circuit malfunction	**Wiring, EGRT sensor**
16929	–	Exhaust gas recirculation temperature (EGRT) sensor, bank 1 – low input	**Wiring short to ground, EGRT sensor**
16930	–	Exhaust gas recirculation temperature (EGRT) sensor, bank 1 – high input	**Wiring short to positive, EGRT sensor**
16935	–	Power steering pressure (PSP) sensor/ switch – range/performance problem	**Wiring, PSP switch**
16944	–	System voltage – malfunction	**Fuse(s), battery, wiring, engine control relay**
16946	–	System voltage – low	**Fuse(s), battery, wiring, engine control relay**
16947	–	System voltage – high	**Alternator, wiring**
16952	–	Cruise control master/selector switch, SET signal – malfunction	**Wiring, cruise control master/selector switch**
16955	–	Stop lamp switch/brake pedal position (BPP) switch – circuit malfunction	**Wiring, stop lamp switch, BPP switch**
16983	–	CAN data bus – malfunction	**Trouble code(s) stored in other system, wiring**
16984	–	CAN data bus – malfunction	**Trouble code(s) stored in other system, wiring**
16985	–	Engine control module (ECM) – memory check sum error	**ECM**
16987	–	Engine control module (ECM) – KAM error	**ECM**
16988	–	Engine control module (ECM) – RAM error	**ECM**

Scan code 5-digit	OBD-II code	Fault location	Probable cause
16989	–	Engine control module (ECM) – ROM error	ECM
16990	–	Engine control module (ECM) – PCM processor fault	ECM
17022	–	Throttle actuator control, bank 1 – range/ performance problem	Basic setting not carried out, throttle control unit, APP sensor
17026	–	Engine control module (ECM), knock control – defective	ECM
17029	–	Air conditioning	Wiring, AC system
17034	–	Malfunction indicator lamp (MIL) – circuit malfunction	Wiring, MIL
17040	–	Instrument panel, fuel consumption signal – circuit malfunction	Wiring
17071	–	Engine control relay – short to ground	Wiring short to ground, engine control relay
17072	–	Engine control relay – short to positive	Wiring short to positive, engine control relay
17075	–	Engine coolant blower motor 1 – short to ground	Wiring short to ground, engine coolant blower motor
17076	–	Engine coolant blower motor 1 – short to positive	Wiring short to positive, engine coolant blower motor
17077	–	Engine coolant blower motor 2 – short to ground	Wiring short to ground, engine coolant blower motor
17078	–	Engine coolant blower motor 2 – short to positive	Wiring short to positive, engine coolant blower motor
17091	–	Transmission range (TR) sensor – low input	Wiring short to ground, TR sensor
17092	–	Transmission range (TR) sensor – high input	Wiring short to positive, TR sensor
–	P1009	Mass air flow (MAF) sensor 1/2 – implausible load detection signal	Wiring, MAF sensor(s)
17428	P1020	Fuel pressure – control limit exceeded	Wiring, fuel pressure sensor, fuel pressure control valve, high pressure fuel pump
17431	P1023	Fuel pressure control valve – short to ground	Wiring short to ground, fuel pressure control valve
17432	P1024	Fuel pressure control valve – open circuit	Wiring open circuit, fuel pressure control valve
17433	P1025	Fuel pressure control valve – mechanical fault	Fuel pressure control valve
17434	P1026	Intake manifold air control solenoid – short to positive	Wiring short to positive, intake manifold air control solenoid
17435	P1027	Intake manifold air control solenoid – short to ground	Wiring short to ground, intake manifold air control solenoid
17436	P1028	Intake manifold air control solenoid – open circuit	Wiring open circuit, intake manifold air control solenoid

Scan code 5-digit	OBD-II code	Fault location	Probable cause
17437	P1029	Intake manifold air control valve position sensor – upper limit not reached	**Air control flap tight/sticking, hose connection(s), intake manifold air control actuator**
17438	P1030	Intake manifold air control valve position sensor – lower limit not reached	**Air control flap tight/sticking, hose connection(s), intake manifold air control actuator**
17439	P1031	Intake manifold air control valve position sensor – specification not attained	**Air control flap tight/sticking, hose connection(s), intake manifold air control actuator, intake manifold air control solenoid**
17440 3	P1032	Nitrogen oxides (NOx) sensor – signal too high	**Catalytic converter, wiring short to positive, NOx sensor**
17441 3	P1033	Nitrogen oxides (NOx) sensor – signal too low	**Exhaust leak, wiring short to ground, NOx sensor**
17442 3	P1034	Nitrogen oxides (NOx) sensor – signal outside tolerance	**Catalytic converter, exhaust leak, wiring, NOx sensor**
17443 3	P1035	Nitrogen oxides (NOx) sensor – range/ performance problem	**Wiring, heating inoperative, NOx sensor**
17444 3	P1036	Nitrogen oxides (NOx) heater sensor – short to positive	**Wiring short to positive, NOx sensor**
17445 3	P1037	Nitrogen oxides (NOx) heater sensor – short to ground	**Wiring short to ground, NOx sensor**
17446 3	P1038	Nitrogen oxides (NOx) heater sensor – open circuit	**Wiring open circuit, NOx sensor**
17447	P1039	Injector 1, supply voltage – short circuit	**Wiring, injector**
17448	P1040	Injector 1, supply voltage – circuit malfunction	**Wiring, injector**
17449	P1041	Injector 2, supply voltage – short circuit	**Wiring, injector**
17450	P1042	Injector 2, supply voltage – circuit malfunction	**Wiring, injector**
17451	P1043	Injector 3, supply voltage – short circuit	**Wiring, injector**
17452	P1044	Injector 3, supply voltage – circuit malfunction	**Wiring, injector**
17453	P1045	Injector 4, supply voltage – short circuit	**Wiring, injector**
17454	P1046	Injector 4, supply voltage – circuit malfunction	**Wiring, injector**
17455	P1047	Camshaft position (CMP) actuator, exhaust/right/rear, bank 1 – circuit malfunction	**Wiring, camshaft position (CMP) actuator**
17456	P1048	Camshaft position (CMP) actuator, exhaust/right/rear, bank 1 – short to positive	**Wiring short to positive, camshaft position (CMP) actuator**
17457	P1049	Camshaft position (CMP) actuator, exhaust/right/rear, bank 1 – short to ground	**Wiring short to ground, camshaft position (CMP) actuator**
17458	P1050	Camshaft position (CMP) actuator, exhaust/right/rear, bank 1 – open circuit	**Wiring open circuit, camshaft position (CMP) actuator**

Scan code 5-digit	OBD-II code	Fault location	Probable cause
17471	P1063	Fuel pressure – control limit not reached	**Air in fuel system, fuel leak, fuel lift pump, fuel bypass valve, fuel pressure control valve, high pressure fuel pump**
17472	P1064	Fuel pressure – mechanical fault	**Air in fuel system, fuel leak, fuel lift pump, fuel bypass valve, fuel pressure control valve, high pressure fuel pump**
17473	P1065	Fuel pressure – system deviation	**Air in fuel system, fuel leak, fuel lift pump, fuel bypass valve, fuel pressure control valve, high pressure fuel pump**
17474	P1066	Intake manifold air control solenoid – short to positive	**Wiring short to positive, intake manifold air control solenoid**
17475	P1067	Intake manifold air control solenoid – short to ground	**Wiring short to ground, intake manifold air control solenoid**
17476	P1068	Intake manifold air control solenoid – open circuit	**Wiring open circuit, intake manifold air control solenoid**
17477 [3]	P1069	Nitrogen oxides (NOx) sensor – heater control – short to ground	**Wiring short to ground, NOx sensor**
17478 [3]	P1070	Nitrogen oxides (NOx) sensor – heater control – short to positive	**Wiring short to positive, NOx sensor**
17479 [3]	P1071	Nitrogen oxides (NOx) sensor – heater control – incorrect signal	**Wiring, NOx sensor**
17480 [3]	P1072	Nitrogen oxides (NOx) sensor – heater control – circuit malfunction	**Wiring, NOx sensor**
–	P1073	Mass air flow (MAF) sensor 2 – signal too low	**Wiring, MAF sensor**
–	P1074	Mass air flow (MAF) sensor 2 – signal too high	**Wiring, MAF sensor**
17501	P1093	Mixture control (MC), bank 1 – malfunction	**Fuel pressure/pump, injector(s) intake leak**
17509	P1101	Heated oxygen sensor (HO2S) 1, bank 1 – low voltage/air leak	**Intake/exhaust leak, fuel pressure/pump, wiring short to ground, HO2S**
17510	P1102	Heated oxygen sensor (HO2S) 1, bank 1 – heater short to positive	**Wiring short to positive, HO2S**
17511	P1103	Heated oxygen sensor (HO2S) 1, bank 1 – heater output too low	**Wiring, HO2S**
17513	P1105	Heated oxygen sensor (HO2S) 2, bank 1 – heater short to positive	**Wiring short to positive, HO2S**
17514	P1106	Heated oxygen sensor (HO2S) 1, bank 2 – low voltage/air leak	**Intake/exhaust leak, fuel pressure/pump, wiring short to ground, HO2S**
17515	P1107	Heated oxygen sensor (HO2S) 1, bank 2 – heater short to positive	**Wiring short to positive, HO2S**
17518	P1110	Heated oxygen sensor (HO2S) 2, bank 2 – heater short to positive	**Wiring short to positive, HO2S**
17519	P1111	Heated oxygen sensor (HO2S) control, bank 1 – system too lean	**Intake/exhaust leak, injector blocked, MAP sensor, MAF sensor, fuel pressure/pump, HO2S**

Scan code 5-digit	OBD-II code	Fault location	Probable cause
17520	P1112	Heated oxygen sensor (HO2S) control, bank 1 – system too rich	Excessive fuel in engine oil, injector leaking, fuel pressure, EVAP canister purge valve, MAP sensor, MAF sensor, HO2S
17521	P1113	Heated oxygen sensor (HO2S) 1, bank 1 – heater resistance too high	Wiring, HO2S
17522	P1114	Heated oxygen sensor (HO2S) 2, bank 1 – heater resistance too high	Wiring, HO2S
17523	P1115	Heated oxygen sensor (HO2S) 1, bank 1 – heater short to ground	Wiring short to ground, HO2S
17524	P1116	Heated oxygen sensor (HO2S) 1, bank 1 – heater open circuit	Wiring open circuit, HO2S
17525	P1117	Heated oxygen sensor (HO2S) 2, bank 1 – heater short to ground	Wiring short to ground, HO2S
17526	P1118	Heated oxygen sensor (HO2S) 2, bank 1 – heater open circuit	Wiring open circuit, HO2S
17527	P1119	Heated oxygen sensor (HO2S) 1, bank 2 – heater short to ground	Wiring short to ground, HO2S
17528	P1120	Heated oxygen sensor (HO2S) 1, bank 2 – heater open circuit	Wiring open circuit, HO2S
17529	P1121	Heated oxygen sensor (HO2S) 2, bank 2 – heater short to ground	Wiring short to ground, HO2S
17530	P1122	Heated oxygen sensor (HO2S) 2, bank 2 – heater open circuit	Wiring open circuit, HO2S
17535	P1127	Long term fuel trim, entire speed/load range, bank 1 – system too rich	Excessive fuel in engine oil, fuel pressure, EVAP canister purge valve, injector(s)
17536	P1128	Long term fuel trim, entire speed/load range, bank 1 – system too lean	Fuel pressure/pump, injector(s), intake/ exhaust leak, AIR system, hose leak
17537	P1129	Long term fuel trim, entire speed/load range, bank 2 – system too rich	Excessive fuel in engine oil, fuel pressure, EVAP canister purge valve, injector(s)
17538	P1130	Long term fuel trim, entire speed/load range, bank 2 – system too lean	Fuel pressure/pump, injector(s), intake/ exhaust leak, AIR system, hose leak
17539	P1131	Heated oxygen sensor (HO2S) 1, bank 1 & 2 – heater resistance too high	Wiring, HO2S
17540	P1132	Heated oxygen sensor (HO2S) 1, bank 1 & 2 – heater control – circuit high	Wiring short to positive, HO2S
17541	P1133	Heated oxygen sensor (HO2S) 1, bank 1 & 2 – heater control – circuit low	Wiring open circuit/short to ground, HO2S
17544	P1136	Long term fuel trim, idling, bank 1 – system too lean	Fuel pressure/pump, injector(s), intake/ exhaust leak, AIR system, hose leak
17545	P1137	Long term fuel trim, idling, bank 1 – system too rich	Fuel pressure, injector(s), EVAP canister purge valve
17546	P1138	Long term fuel trim, idling, bank 2 – system too lean	Fuel pressure/pump, injector(s), intake/ exhaust leak, AIR system, hose leak
17547	P1139	Long term fuel trim, idling, bank 2 – system too rich	Fuel pressure, injector(s), EVAP canister purge valve
17548	P1140	Heated oxygen sensor (HO2S) 2, bank 2 – heater resistance too high	Wiring, HO2S

Scan code 5-digit	OBD-II code	Fault location	Probable cause
17549	P1141	Load calculation – implausible value	Wiring, intake leak, MAP sensor, MAF sensor, throttle control unit
17550	P1142	Load calculation – too low	Wiring, MAP sensor, MAF sensor, APP sensor, throttle control unit
17551	P1143	Load calculation – too high	Wiring, intake leak, MAP sensor, MAF sensor, APP sensor, throttle control unit
17552	P1144	Mass air flow (MAF) sensor, bank 1 – open circuit/short to ground	Wiring open circuit/short to ground, MAF sensor
17553	P1145	Mass air flow (MAF) sensor, bank 1 – short to positive	Wiring short to positive, MAF sensor
17554	P1146	Mass air flow (MAF) sensor, bank 1 – supply voltage	Operating voltage too high/low, wiring
17555	P1147	Heated oxygen sensor (HO2S) 1, bank 2 – lambda regulation, system too lean	Intake leak, wiring, HO2S
17556	P1148	Heated oxygen sensor (HO2S) 1, bank 2 – lambda regulation, system too rich	Exhaust leak, wiring, HO2S
17557	P1149	Heated oxygen sensor (HO2S) 1, bank 1 – implausible lambda control value	Exhaust leak, wiring, HO2S
17558	P1150	Heated oxygen sensor (HO2S) 1, bank 2 – implausible lambda control value	Exhaust leak, wiring, HO2S
17559	P1151	Long term fuel trim 1, bank 1 – below lean limit	Fuel pressure/pump, injectors, intake/ exhaust leak, HO2S
17560	P1152	Long term fuel trim 2, bank 1 – below lean limit	Fuel pressure/pump, injectors, intake/ exhaust leak, AIR system, HO2S
17561	P1153	Heated oxygen sensor (HO2S) 2, bank 1 & 2 – interchanged	HO2S 2 on bank 1 & 2 incorrectly installed
17563	P1155	Manifold absolute pressure (MAP) sensor – short to positive	Wiring short to positive, MAP sensor
17564	P1156	Manifold absolute pressure (MAP) sensor – open circuit/short to ground	Wiring open circuit/short to ground, MAP sensor
17565	P1157	Manifold absolute pressure (MAP) sensor – supply voltage	Wiring, MAP sensor
17566	P1158	Manifold absolute pressure (MAP) sensor – range/performance problem	Wiring, hose connection(s), MAP sensor
17567	P1159	Mass air flow (MAF) sensor, bank 1 & 2 – implausible ratio	EGR system, intake leak, wiring, MAF sensor 1/2
17568	P1160	Intake air temperature (IAT) sensor – short to ground	Wiring short to ground, IAT sensor
17569	P1161	Intake air temperature (IAT) sensor – open circuit/short to positive	Wiring open circuit/short to positive, IAT sensor
17570	P1162	Fuel temperature sensor – short to ground	Wiring short to ground, fuel temperature sensor
17571	P1163	Fuel temperature sensor – open circuit/ short to positive	Wiring open circuit/short to positive, fuel temperature sensor
17572	P1164	Fuel temperature sensor – range/ performance problem	Wiring, fuel temperature sensor

Scan code 5-digit	OBD-II code	Fault location	Probable cause
17573	P1165	Long term fuel trim 1, bank 1 – above rich limit	Fuel pressure/pump, injectors, EVAP canister purge valve, EGR system, HO2S, intake/exhaust system
17574	P1166	Long term fuel trim 2, bank 1 – above rich limit	Fuel pressure/pump, injectors, EVAP canister purge valve, EGR system, HO2S, intake/exhaust system
17575	P1167	Mass air flow (MAF) sensor, bank 2 – range/performance problem	Intake leak, wiring, MAF sensor
17576	P1168	Mass air flow (MAF) sensor, bank 2 – low input	Intake leak, air filter blocked, wiring short to ground, fuse, MAF sensor
17577	P1169	Mass air flow (MAF) sensor, bank 2 – high input	Wiring short to positive, ground wire defective, MAF sensor
17578	P1170	Mass air flow (MAF) sensor, bank 2 – supply voltage	Wiring, fuse, engine control relay, injector
17579	P1171	Throttle motor position sensor 2 – range/ performance problem	Wiring, throttle valve tight/sticking, throttle motor position sensor, throttle control unit
17580	P1172	Throttle motor position sensor 2 – low input	Wiring short to ground, throttle motor position sensor, throttle control unit
17581	P1173	Throttle motor position sensor 2 – high input	Wiring short to positive, throttle motor position sensor, throttle control unit
17582	P1174	Fuel measurement system, bank 1 – injection timing incorrect	Fuel pressure/pump, injector(s), intake/ exhaust leak, EGR system, EVAP canister purge valve, HO2S
17584	P1176	Lambda correction after catalyst, bank 1 – control limit reached	Wiring, intake leak, HO2S
17585	P1177	Lambda correction after catalyst, bank 2 – control limit reached	Wiring, intake leak, HO2S
17586	P1178	Heated oxygen sensor (HO2S) 1, bank 1, pump current – open circuit	Wiring open circuit, HO2S
17587	P1179	Heated oxygen sensor (HO2S) 1, bank 1, pump current – short to ground	Wiring short to ground, HO2S
17588	P1180	Heated oxygen sensor (HO2S) 1, bank 1, pump current – short to positive	Wiring short to positive, HO2S
17589	P1181	Heated oxygen sensor (HO2S) 1, bank 1, reference voltage – open circuit	Wiring open circuit, HO2S, HT leads, spark plugs, misfire detection
17590	P1182	Heated oxygen sensor (HO2S) 1, bank 1, reference voltage – short to ground	Wiring short to ground, HO2S, HT leads, spark plugs, misfire detection
17591	P1183	Heated oxygen sensor (HO2S) 1, bank 1, reference voltage – short to positive	Wiring short to positive, HO2S, HT leads, spark plugs, misfire detection
17595	P1187	Heated oxygen sensor (HO2S) 1, bank 1 or 2 – circuit malfunction	Wiring, HO2S
17598	P1190	Heated oxygen sensor (HO2S) 1, bank 1, reference voltage – range/performance problem	Wiring, HO2S, HT leads, spark plugs, misfire detection
17599	P1191	Heated oxygen sensor (HO2S) 1, bank 1 & 2 – interchanged	HO2S 1 on banks 1 & 2 incorrectly installed
17600	P1192	Fuel pressure sensor – supply voltage	Wiring, fuel pressure sensor

Scan code 5-digit	OBD-II code	Fault location	Probable cause
17601	P1193	Fuel pressure sensor – open circuit/short to positive	Wiring open circuit/short to positive
17602	P1194	Fuel pressure control valve – short to positive	Wiring short to positive, fuel pressure control valve
17603	P1195	Fuel pressure control valve – open circuit/short to ground	Wiring open circuit/short to ground, fuel pressure control valve
17604	P1196	Heated oxygen sensor (HO2S) 1, bank 1, heater circuit malfunction	Wiring, HO2S
17605	P1197	Heated oxygen sensor (HO2S) 1, bank 2, heater circuit malfunction	Wiring, HO2S
17606	P1198	Heated oxygen sensor (HO2S) 2, bank 1, heater circuit malfunction	Wiring, HO2S
17607	P1199	Heated oxygen sensor (HO2S) 2, bank 2, heater circuit malfunction	Wiring, HO2S
17609	P1201	Injector 1 – circuit malfunction	Wiring, injector
17610	P1202	Injector 2 – circuit malfunction	Wiring, injector
17611	P1203	Injector 3 – circuit malfunction	Wiring, injector
17612	P1204	Injector 4 – circuit malfunction	Wiring, injector
17613	P1205	Injector 5 – circuit malfunction	Wiring, injector
17614	P1206	Injector 6 – circuit malfunction	Wiring, injector
17615	P1207	Injector 7 – circuit malfunction	Wiring, injector
17616	P1208	Injector 8 – circuit malfunction	Wiring, injector
17621	P1213	Injector 1 – short to positive	Wiring short to positive, injector
17622	P1214	Injector 2 – short to positive	Wiring short to positive, injector
17623	P1215	Injector 3 – short to positive	Wiring short to positive, injector
17624	P1216	Injector 4 – short to positive	Wiring short to positive, injector
17625	P1217	Injector 5 – short to positive	Wiring short to positive, injector
17626	P1218	Injector 6 – short to positive	Wiring short to positive, injector
17627	P1219	Injector 7 – short to positive	Wiring short to positive, injector
17628	P1220	Injector 8 – short to positive	Wiring short to positive, injector
17633	P1225	Injector 1 – short to ground	Wiring short to ground, injector
17634	P1226	Injector 2 – short to ground	Wiring short to ground, injector
17635	P1227	Injector 3 – short to ground	Wiring short to ground, injector
17636	P1228	Injector 4 – short to ground	Wiring short to ground, injector
17637	P1229	Injector 5 – short to ground	Wiring short to ground, injector
17638	P1230	Injector 6 – short to ground	Wiring short to ground, injector
17639	P1231	Injector 7 – short to ground	Wiring short to ground, injector
17640	P1232	Injector 8 – short to ground	Wiring short to ground, injector

Scan code 5-digit	OBD-II code	Fault location	Probable cause
17645	P1237	Injector 1 – open circuit	Wiring open circuit, injector
17646	P1238	Injector 2 – open circuit	Wiring open circuit, injector
17647	P1239	Injector 3 – open circuit	Wiring open circuit, injector
17648	P1240	Injector 4 – open circuit	Wiring open circuit, injector
17649	P1241	Injector 5 – open circuit	Wiring open circuit, injector
17650	P1242	Injector 6 – open circuit	Wiring open circuit, injector
17651	P1243	Injector 7 – open circuit	Wiring open circuit, injector
17652	P1244	Injector 8 – open circuit	Wiring open circuit, injector
17653	P1245	Injector needle lift sensor – short to ground	Wiring short to ground, injector needle lift sensor
17654	P1246	Injector needle lift sensor – range/ performance problem	Injector needle lift sensor, injector pipe defective, fuel level low
17655	P1247	Injector needle lift sensor – open circuit/ short to positive	Wiring open circuit/short to positive, injector needle lift sensor
17656	P1248	Start of injection – control difference	Fuel injection timing solenoid, injector needle lift sensor, fuel level low, pump timing
17658	P1250	Fuel tank level sensor – low input	Fuel level too low
17659	P1251	Fuel injection timing solenoid – short to positive	Wiring short to positive
17660	P1252	Fuel injection timing solenoid – open circuit/short to ground	Wiring open circuit/short to ground, fuel injection timing solenoid
17661	P1253	Instrument panel, fuel consumption signal – short to ground	Wiring short to ground, instrument panel
17662	P1254	Instrument panel, fuel consumption signal – short to positive	Wiring short to positive, instrument panel
17663	P1255	Engine coolant temperature (ECT) sensor – short to ground	Wiring short to ground, ECT sensor
17664	P1256	Engine coolant temperature (ECT) sensor – open circuit/short to positive	Wiring open circuit/short to positive, ECT sensor
17668	P1260	Injector 1 – implausible signal	No control
17669	P1261	Injector 1 – control limit exceeded	Control period too long, wiring, injector
17670	P1262	Injector 1 – control limit not reached	Control period too short, fuel level low, air in fuel system
17671	P1263	Injector 2 – implausible signal	No control
17672	P1264	Injector 2 – control limit exceeded	Control period too long, wiring, injector
17673	P1265	Injector 2 – control limit not reached	Control period too short, fuel level low, air in fuel system
17674	P1266	Injector 3 – implausible signal	No control
17675	P1267	Injector 3 – control limit exceeded	Control period too long, wiring, injector
17676	P1268	Injector 3 – control limit not reached	Control period too short, fuel level low, air in fuel system

Scan code 5-digit	OBD-II code	Fault location	Probable cause
17677	P1269	Injector 4 – implausible signal	No control
17678	P1270	Injector 4 – control limit exceeded	Control period too long, wiring, injector
17679	P1271	Injector 4 – control limit not reached	Control period too short, fuel level low, air in fuel system
17686	P1278	Fuel metering solenoid – short to positive	Wiring short to positive, fuel metering solenoid
17687	P1279	Fuel metering solenoid – open circuit/short to ground	Wiring open circuit/short to ground, fuel metering solenoid
17689	P1281	Fuel metering solenoid – short to ground	Wiring short to ground, fuel metering solenoid
–	P1280	Injector air control valve/solenoid – insufficient flow detected	Wiring, valve/solenoid
–	P1283	Injector air control valve/solenoid – circuit malfunction	Wiring, valve/solenoid
–	P1286	Injector air control system – circuit high	Wiring, valve/solenoid
17690	P1282	Fuel metering solenoid – open circuit	Wiring open circuit, fuel metering solenoid
17695	P1287	Turbocharger (TC) bypass valve – open circuit	Wiring open circuit, TC bypass valve
17696	P1288	Turbocharger (TC) bypass valve – short to positive	Wiring short to positive, TC bypass valve
17697	P1289	Turbocharger (TC) bypass valve – short to ground	Wiring short to ground, TC bypass valve
17698	P1290	Engine coolant temperature (ECT) sensor, ECM controlled cooling system – high input	Wiring, ECT sensor
17699	P1291	Engine coolant temperature (ECT) sensor, ECM controlled cooling system – high input	Wiring, ECT sensor
17700	P1292	Engine coolant thermostat – open circuit	Wiring open circuit, engine coolant thermostat
17701	P1293	Engine coolant thermostat – short to positive	Wiring short to positive, engine coolant thermostat
17702	P1294	Engine coolant thermostat – short to ground	Wiring short to ground, engine coolant thermostat
17703	P1295	Turbocharger (TC), bypass – flow malfunction	TC wastegate regulating valve, hose connection(s), injector
17704	P1296	Engine cooling system – malfunction	ECT sensor, coolant thermostat
17705	P1297	Turbocharger (TC)/throttle valve, hose connection – pressure loss	Hose connection
17707	P1299	Fuel metering solenoid – circuit malfunction	Wiring, fuel metering solenoid
17708	P1300	Random/multiple cylinder(s) – misfire detected	Fuel level low, fuel gauge tank sensor

Scan code 5-digit	OBD-II code	Fault location	Probable cause
17733	P1325	Knock control, cylinder 1 – control limit reached	Poor quality fuel, incorrect fuel, insecure engine component, KS incorrectly tightened/defective, shield wiring open circuit, poor connection
17734	P1326	Knock control, cylinder 2 – control limit reached	Poor quality fuel, incorrect fuel, insecure engine component, KS incorrectly tightened/defective, shield wiring open circuit, poor connection
17735	P1327	Knock control, cylinder 3 – control limit reached	Poor quality fuel, incorrect fuel, insecure engine component, KS incorrectly tightened/defective, shield wiring open circuit, poor connection
17736	P1328	Knock control, cylinder 4 – control limit reached	Poor quality fuel, incorrect fuel, insecure engine component, KS incorrectly tightened/defective, shield wiring open circuit, poor connection
17737	P1329	Knock control, cylinder 5 – control limit reached	Poor quality fuel, incorrect fuel, insecure engine component, KS incorrectly tightened/defective, shield wiring open circuit, poor connection
17738	P1330	Knock control, cylinder 6 – control limit reached	Poor quality fuel, incorrect fuel, insecure engine component, KS incorrectly tightened/defective, shield wiring open circuit, poor connection
17739	P1331	Knock control, cylinder 7 – control limit reached	Poor quality fuel, incorrect fuel, insecure engine component, KS incorrectly tightened/defective, shield wiring open circuit, poor connection
17740	P1332	Knock control, cylinder 8 – control limit reached	Poor quality fuel, incorrect fuel, insecure engine component, KS incorrectly tightened/defective, shield wiring open circuit, poor connection
17743	P1335	Engine torque control 1/2 – limit reached/ exceeded	Throttle control unit, hose(s), TC system, MAP sensor, IAT sensor, MAF sensor, ECT sensor
17744	P1336	Engine torque monitoring – control limit exceeded	Hose(s), TC system, throttle control unit, IAT sensor, MAP sensor, MAF sensor, ECT sensor, APP sensor
17745	P1337	Camshaft position (CMP) sensor, bank 1 – short to ground	Wiring short to ground, CMP sensor
17746	P1338	Camshaft position (CMP) sensor, bank 1 – circuit malfunction	Wiring, CMP sensor
17747	P1339	Crankshaft position (CKP) sensor/engine speed (RPM) sensor – interchanged	Multi-plugs incorrectly connected
17748	P1340	Camshaft position (CMP) sensor 1/bank 1/ crankshaft position (CKP) sensor – out of sequence	Valve timing, CKP/CMP sensor installation, CKP sensor rotor
17749	P1341	Ignition amplifier, primary circuit 1 – short to ground	Wiring short to ground, ignition amplifier, CMP sensor, HT leads, spark plugs

Scan code 5-digit	OBD-II code	Fault location	Probable cause
17750	P1342	Ignition amplifier, primary circuit 1 – short to positive	**Wiring short to positive, ignition amplifier, CMP sensor, HT leads, spark plugs**
17751	P1343	Ignition amplifier, primary circuit 2 – short to ground	**Wiring short to ground, ignition amplifier, CMP sensor, HT leads, spark plugs**
17752	P1344	Ignition amplifier, primary circuit 2 – short to positive	**Wiring short to positive, ignition amplifier, CMP sensor, HT leads, spark plugs**
17753	P1345	Ignition amplifier, primary circuit 3 – short to ground	**Wiring short to ground, ignition amplifier, CMP sensor, HT leads, spark plugs**
17754	P1346	Ignition amplifier, primary circuit 3 – short to positive	**Wiring short to positive, ignition amplifier, CMP sensor, HT leads, spark plugs**
17755	P1347	Camshaft position (CMP) sensor 2/bank 2/ crankshaft position (CKP) sensor – out of sequence	**Valve timing, CKP/CMP sensor installation, CKP sensor rotor**
17756	P1348	Ignition amplifier, primary circuit 1 – open circuit	**Wiring short to ground, ignition amplifier, CMP sensor, HT leads, spark plugs**
17757	P1349	Ignition amplifier, primary circuit 2 – open circuit	**Wiring short to positive, ignition amplifier, CMP sensor, HT leads, spark plugs**
17758	P1350	Ignition amplifier, primary circuit 3 – open circuit	**Wiring short to positive, ignition amplifier, CMP sensor, HT leads, spark plugs**
17759	P1351	Camshaft position (CMP) sensor, bank 1 – range/performance problem	**Ignore trouble code, erase fault memory**
17762	P1354	Fuel quantity adjuster position sensor	**Wiring, fuel injection pump**
17763 4	P1355	Ignition coil/amplifier, cylinder 1 – open circuit	**Wiring open circuit, ignition coil/amplifier**
17764 4	P1356	Ignition coil/amplifier, cylinder 1 – short to positive	**Wiring short to positive, ignition coil/ amplifier**
17765 4	P1357	Ignition coil/amplifier, cylinder 1 – short to ground	**Wiring short to ground, ignition coil/ amplifier**
17766 5	P1358	Ignition coil/amplifier, cylinder 2 – open circuit	**Wiring open circuit, ignition coil/amplifier**
17767 5	P1359	Ignition coil/amplifier, cylinder 2 – short to positive	**Wiring short to positive, ignition coil/ amplifier**
17768 5	P1360	Ignition coil/amplifier, cylinder 2 – short to ground	**Wiring short to ground, ignition coil/ amplifier**
17769	P1361	Ignition coil/amplifier, cylinder 3 – open circuit	**Wiring open circuit, ignition coil/amplifier**
17770	P1362	Ignition coil/amplifier, cylinder 3 – short to positive	**Wiring short to positive, ignition coil/ amplifier**
17771	P1363	Ignition coil/amplifier, cylinder 3 – short to ground	**Wiring short to ground, ignition coil/ amplifier**
17772	P1364	Ignition coil/amplifier, cylinder 4 – open circuit	**Wiring open circuit, ignition coil/amplifier**
17773	P1365	Ignition coil/amplifier, cylinder 4 – short to positive	**Wiring short to positive, ignition coil/ amplifier**
17774	P1366	Ignition coil/amplifier, cylinder 4 – short to ground	**Wiring short to ground, ignition coil/ amplifier**

Scan code 5-digit	OBD-II code	Fault location	Probable cause
17775	P1367	Ignition coil/amplifier, cylinder 5 – open circuit	**Wiring open circuit, ignition coil/amplifier**
17776	P1368	Ignition coil/amplifier, cylinder 5 – short to positive	**Wiring short to positive, ignition coil/ amplifier**
17777	P1369	Ignition coil/amplifier, cylinder 5 – short to ground	**Wiring short to ground, ignition coil/ amplifier**
17778	P1370	Ignition coil/amplifier, cylinder 6 – open circuit	**Wiring open circuit, ignition coil/amplifier**
17779	P1371	Ignition coil/amplifier, cylinder 6 – short to positive	**Wiring short to positive, ignition coil/ amplifier**
17780	P1372	Ignition coil/amplifier, cylinder 6 – short to ground	**Wiring short to ground, ignition coil/ amplifier**
17781	P1373	Ignition coil/amplifier, cylinder 7 – open circuit	**Wiring open circuit, ignition coil/amplifier**
17782	P1374	Ignition coil/amplifier, cylinder 7 – short to positive	**Wiring short to positive, ignition coil/ amplifier**
17783	P1375	Ignition coil/amplifier, cylinder 7 – short to ground	**Wiring short to ground, ignition coil/ amplifier**
17784	P1376	Ignition coil/amplifier, cylinder 8 – open circuit	**Wiring open circuit, ignition coil/amplifier**
17785	P1377	Ignition coil/amplifier, cylinder 8 – short to positive	**Wiring short to positive, ignition coil/ amplifier**
17786	P1378	Ignition coil/amplifier, cylinder 8 – short to ground	**Wiring short to ground, ignition coil/ amplifier**
17793	P1385	Engine control module (ECM) – defective	**ECM**
17794	P1386	Engine control module (ECM), knock control – defective	**ECM**
17795	P1387	Engine control module (ECM), BARO sensor – defective	**ECM**
17796	P1388	Engine control module (ECM), ETS – defective	**ECM**
17796	P1389	Engine control module (ECM) 2 – defective	**ECM**
17799	P1391	Camshaft position (CMP) sensor 2/ bank 2 – short to ground	**Wiring short to ground, CMP sensor**
17800	P1392	Camshaft position (CMP) sensor 2/ bank 2 – open circuit/short to positive	**Wiring open circuit/short to positive, CMP sensor**
17801	P1393	Ignition amplifier, primary circuit 1 – circuit malfunction	**Wiring, ignition amplifier, HT leads, spark plugs**
17802	P1394	Ignition amplifier, primary circuit 2 – circuit malfunction	**Wiring, ignition amplifier, HT leads, spark plugs**
17803	P1395	Ignition amplifier, primary circuit 3 – circuit malfunction	**Wiring, ignition amplifier, HT leads, spark plugs**
17805	P1397	Crankshaft position (CKP) sensor/engine speed (RPM) sensor – control limit reached	**Insecure/damaged rotor, CKP/RPM sensor**

Scan code 5-digit	OBD-II code	Fault location	Probable cause
17806	P1398	Crankshaft position (CKP) sensor/engine speed (RPM) sensor – short to ground	Wiring short to ground, CKP/RPM sensor
17807	P1399	Crankshaft position (CKP) sensor/engine speed (RPM) sensor – short to positive	Wiring short to positive, CKP/RPM sensor
17808	P1400	Exhaust gas recirculation (EGR) valve/ solenoid, bank 1 – circuit malfunction	Wiring, EGR valve
17809	P1401	Exhaust gas recirculation (EGR) valve/ solenoid, bank 1 – short to ground	Wiring short to ground, EGR valve
17810	P1402	Exhaust gas recirculation (EGR) valve/ solenoid, bank 1 – short to positive	Wiring short to positive, EGR valve/ solenoid
17811	P1403	Exhaust gas recirculation (EGR) system – control difference	Basic setting not carried out, EGR system
17812	P1404	Exhaust gas recirculation (EGR) system – basic setting	Basic setting not carried out, EGR system
17815	P1407	Exhaust gas recirculation temperature (EGRT) sensor – low input	Wiring short to ground, EGRT sensor
17816	P1408	Exhaust gas recirculation temperature (EGRT) sensor – high input	Wiring short to positive, ground wire defective, EGRT sensor
17817	P1409	Evaporative emission (EVAP) canister purge valve – circuit malfunction	Wiring, EVAP canister purge valve
17818	P1410	Evaporative emission (EVAP) canister purge valve – short to positive	Wiring short to positive, EVAP canister purge valve
17819	P1411	Secondary air injection (AIR) system, bank 2 – insufficient flow detected	Intake leak, hose(s) blocked/leaking, AIR valve/solenoid
17822	P1414	Secondary air injection (AIR) system, bank 2 – leak detected	Intake leak, hose(s) leaking, AIR valve/ solenoid
17823	P1415	Exhaust gas recirculation (EGR) valve position sensor – lower limit exceeded	Basic setting not carried out
17824	P1416	Exhaust gas recirculation (EGR) valve position sensor – upper limit exceeded	Basic setting not carried out
17828	P1420	Secondary air injection (AIR) valve/ solenoid – circuit malfunction	Wiring, AIR solenoid
17829	P1421	Secondary air injection (AIR) valve/ solenoid – short to ground	Wiring short to ground, AIR valve/solenoid
17830	P1422	Secondary air injection (AIR) valve/ solenoid – short to positive	Wiring short to positive, AIR valve/ solenoid
17831	P1423	Secondary air injection (AIR) system, bank 1 – insufficient flow detected	Hose connection(s), AIR valve/solenoid
17832	P1424	Secondary air injection (AIR) system, bank 1 – leak detected	AIR valve, exhaust leak
17833	P1425	Evaporative emission (EVAP) canister purge valve – short to ground	Wiring short to ground, EVAP canister purge valve
17834	P1426	Evaporative emission (EVAP) canister purge valve – open circuit	Wiring open circuit, EVAP canister purge valve
17835	P1427	Vacuum pump, brakes – short to positive	Wiring short to positive, vacuum pump
17836	P1428	Vacuum pump, brakes – short to ground	Wiring short to ground, vacuum pump

Scan code 5-digit	OBD-II code	Fault location	Probable cause
17837	P1429	Vacuum pump, brakes – open circuit	Wiring open circuit, vacuum pump
17838	P1430	Vacuum pump, brakes – open circuit/short to positive	Wiring open circuit/short to positive, vacuum pump
17839	P1431	Vacuum pump, brakes – open circuit/short to ground	Wiring open circuit/short to ground, vacuum pump
17840	P1432	Secondary air injection (AIR) valve/ solenoid – open circuit	Wiring open circuit, fuse, AIR valve/ solenoid
17841	P1433	Secondary air injection (AIR) pump relay – open circuit	Wiring open circuit, fuse, AIR pump relay
17842	P1434	Secondary air injection (AIR) pump relay – short to positive	Wiring short to positive, AIR pump relay
17843	P1435	Secondary air injection (AIR) pump relay – short to ground	Wiring short to ground, AIR pump relay
17844	P1436	Secondary air injection (AIR) pump relay – circuit malfunction	Wiring, AIR pump relay
17845	P1437	Exhaust gas recirculation (EGR) valve/ solenoid, bank 2 – short to positive	Wiring short to positive, EGR valve/ solenoid
17846	P1438	Exhaust gas recirculation (EGR) valve/ solenoid, bank 2 – open circuit/short to ground	Wiring open circuit/short to ground, EGR valve/solenoid
17847	P1439	Exhaust gas recirculation (EGR) valve position sensor – basic setting	Basic setting not carried out, EGR system
17848	P1440	Exhaust gas recirculation (EGR) valve – open circuit	Wiring open circuit, EGR valve
17849	P1441	Exhaust gas recirculation (EGR) valve/ solenoid, bank 1 – open circuit/short to ground	Wiring open circuit/short to ground, EGR solenoid
17850	P1442	Exhaust gas recirculation (EGR) valve position sensor – high input	Wiring short to positive, EGR valve position sensor
17851	P1443	Exhaust gas recirculation (EGR) valve position sensor – low input	Wiring short to ground, EGR valve position sensor
17852	P1444	Exhaust gas recirculation (EGR) valve position sensor – range/performance problem	Wiring, EGR valve position sensor
17858	P1450	Secondary air injection (AIR) system – short to positive	Wiring short to positive, AIR relay
17859	P1451	Secondary air injection (AIR) system – short to ground	Wiring short to ground, AIR relay
17860	P1452	Secondary air injection (AIR) system – open circuit	Wiring open circuit, AIR relay
17861	P1453	Exhaust gas recirculation temperature (EGRT) sensor 1/bank 1 – open circuit/ short to positive	Wiring open circuit/short to positive, EGRT sensor
17862	P1454	Exhaust gas recirculation temperature (EGRT) sensor 1/bank 1 – short to ground	Wiring short to ground, EGRT sensor
17863	P1455	Exhaust gas recirculation temperature (EGRT) sensor 1/bank 1 – range/ performance problem	Exhaust leak, wiring, EGRT sensor

Scan code 5-digit	OBD-II code	Fault location	Probable cause
17864	P1456	Exhaust gas recirculation temperature (EGRT) control, bank 1 – control limit reached	EGRT sensor
17865	P1457	Exhaust gas recirculation temperature (EGRT) sensor 2/bank 2 – open circuit/ short to positive	Wiring open circuit/short to positive, EGRT sensor
17866	P1458	Exhaust gas recirculation temperature (EGRT) sensor 2/bank 2 – short to ground	Wiring short to ground, EGRT sensor
17867	P1459	Exhaust gas recirculation temperature (EGRT) sensor 2/bank 2 – range/ performance problem	Exhaust leak, wiring, EGRT sensor
17868	P1460	Exhaust gas recirculation temperature (EGRT) control, bank 2 – control limit reached	EGRT sensor
17869	P1461	Exhaust gas recirculation temperature (EGRT) control, bank 1 – range/ performance problem	Exhaust leak/blockage, EGRT sensor
17870	P1462	Exhaust gas recirculation temperature (EGRT) control, bank 2 – range/ performance problem	Exhaust leak/blockage, EGRT sensor
17875	P1467	Evaporative emission (EVAP) canister purge valve – short to positive	Wiring short to positive , EVAP canister purge valve
17876	P1468	Evaporative emission (EVAP) canister purge valve – short to ground	Wiring short to ground, EVAP canister purge valve
17877	P1469	Evaporative emission (EVAP) canister purge valve – open circuit	Wiring open circuit, EVAP canister purge valve
17878	P1470	Evaporative emission (EVAP) leak detection pump/fuel tank vent system – circuit malfunction	Wiring, EVAP leak detection pump
17879	P1471	Evaporative emission (EVAP) leak detection pump – short to positive	Wiring short to positive, EVAP leak detection pump
17880	P1472	Evaporative emission (EVAP) leak detection pump – short to ground	Wiring short to ground, EVAP leak detection pump
17881	P1473	Evaporative emission (EVAP) leak detection pump/fuel tank vent system – Wiring open circuit	Wiring open circuit, EVAP canister purge valve, EVAP leak detection pump
17883	P1475	Evaporative emission (EVAP) leak detection pump/fuel tank vent system – no signal	Wiring, EVAP canister purge valve, EVAP leak detection pump, ECM
17884	P1476	Evaporative emission (EVAP) leak detection pump/fuel tank vent system – vacuum to low	system leak, hose blockage, EVAP canister, EVAP leak detection pump
17885	P1477	Evaporative emission (EVAP) leak detection pump/fuel tank vent system – malfunction	Wiring, hose leak/blockage, EVAP canister purge valve, EVAP canister, EVAP leak detection pump
17886	P1478	Evaporative emission (EVAP) leak detection pump/fuel tank vent system – hose blockage detected	Hose blockage

Scan code 5-digit	OBD-II code	Fault location	Probable cause
17887	P1479	Vacuum system, brakes – mechanical fault	Vacuum pump
17908	P1500	Fuel pump relay – circuit malfunction	Wiring, fuel pump relay
17909	P1501	Fuel pump relay – short to ground	Wiring short to ground, fuel pump relay
17910	P1502	Fuel pump relay – short to positive	Wiring short to positive, fuel pump relay
17911	P1503	Alternator load signal	Wiring, alternator
17912	P1504	Intake system – leak detected	Intake leak, EGR system, EVAP system, hose connection(s), throttle control unit
17913	P1505	Closed throttle position (CTP) switch – does not close	Throttle cable/valve, wiring open circuit/ short to positive, CTP switch adjustment/ defective, ECM
17914	P1506	Closed throttle position (CTP) switch – does not open	Moisture ingress, wiring short to ground, CTP switch adjustment/defective, ECM
17915	P1507	Idle speed control (ISC) – lower limit reached	Throttle control unit/basic setting, intake/ exhaust leak, mechanical fault, AC signals
17916	P1508	Idle speed control (ISC) – upper limit reached	Throttle control unit/basic setting, intake/ exhaust leak, mechanical fault, AC signals
17917	P1509	Idle air control (IAC) valve – circuit malfunction	Wiring, IAC valve
17918	P1510	Idle air control (IAC) valve – short to positive	Wiring open circuit/short to positive, IAC valve
17919	P1511	Intake manifold air control solenoid 1 – current circuit	Wiring, intake manifold air control solenoid
17920	P1512	Intake manifold air control solenoid 1 – short to positive	Wiring short to positive, intake manifold air control solenoid
17921	P1513	Intake manifold air control solenoid 2 – short to positive	Wiring short to positive, intake manifold air control solenoid
17922	P1514	Intake manifold air control solenoid 2 – short to ground	Wiring short to ground, intake manifold air control solenoid
17923	P1515	Intake manifold air control solenoid 1 – short to ground	Wiring short to ground, intake manifold air control solenoid
17924	P1516	Intake manifold air control solenoid 1 – open circuit	Wiring open circuit, intake manifold air control solenoid
17925	P1517	Engine control relay – circuit malfunction	Wiring, engine control relay
17926	P1518	Engine control relay – short to positive	Wiring short to positive, engine control relay
17927	P1519	Camshaft position (CMP) control, bank 1 – malfunction	Cylinder head oil pressure too low, CMP actuator sticking/defective
17928	P1520	Intake manifold air control solenoid 2 – open circuit	Wiring open circuit, intake manifold air control solenoid
17930	P1522	Camshaft position (CMP) control, bank 2 – malfunction	Cylinder head oil pressure too low, CMP actuator sticking/defective
17931	P1523	SRS crash signal received	Airbag triggered
17932	P1524	Fuel pump relay – open circuit/short to ground	Wiring open circuit/short to ground, fuel pump relay

Scan code 5-digit	OBD-II code	Fault location	Probable cause
17933	P1525	Camshaft position (CMP) actuator, bank 1 – circuit malfunction	Wiring, CMP actuator
17934	P1526	Camshaft position (CMP) actuator, bank 1 – short to positive	Wiring short to positive, CMP actuator
17935	P1527	Camshaft position (CMP) actuator, bank 1 – short to ground	Wiring short to ground, CMP actuator
17936	P1528	Camshaft position (CMP) actuator, bank 1 – open circuit	Wiring open circuit, CMP actuator
17937	P1529	Camshaft position (CMP) actuator – short to positive	Wiring short to positive, CMP actuator
17938	P1530	Camshaft position (CMP) actuator – short to ground	Wiring short to ground, CMP actuator
17939	P1531	Camshaft position (CMP) actuator – open circuit	Wiring open circuit, CMP actuator
17940	P1532	Idle control – lean running speed below specification	Throttle control unit
17941	P1533	Camshaft position (CMP) actuator, bank 2 – circuit malfunction	Wiring, CMP actuator
17942	P1534	Camshaft position (CMP) actuator, bank 2 – short to positive	Wiring short to positive, CMP actuator
17943	P1535	Camshaft position (CMP) actuator, bank 2 – short to ground	Wiring short to ground, CMP actuator
17944	P1536	Camshaft position (CMP) actuator, bank 2 – open circuit	Wiring open circuit, CMP actuator
17945	P1537	Fuel shut-off solenoid – malfunction	Fuel shut-off solenoid (leaking/sticking)
17946	P1538	Fuel shut-off solenoid – open circuit/short to ground	Wiring open circuit/short to ground, fuel shut-off solenoid
17947	P1539	Clutch pedal position (CPP) switch – range/performance problem	Wiring, CPP switch
17948	P1540	Vehicle speed signal – high input	Excessive vehicle speed, instrument panel defective
17949	P1541	Fuel pump relay – open circuit	Wiring open circuit, fuel pump relay
17950	P1542	Throttle motor position sensor 1 – range/ performance problem	Throttle valve requires cleaning, wiring, throttle motor position sensor
17951	P1543	Throttle motor position sensor 1 – low input	Wiring short to ground, throttle motor position sensor
17952	P1544	Throttle motor position sensor 1 – high input	Wiring short to positive, throttle motor position sensor
17953	P1545	Throttle valve control – malfunction	Throttle valve tight/sticking, wiring, throttle control unit
17954	P1546	Turbocharger (TC) wastegate regulating valve – short to positive	Wiring short to positive, TC wastegate regulating valve
17955	P1547	Turbocharger (TC) wastegate regulating valve – short to ground	Wiring short to ground, TC wastegate regulating valve
17956	P1548	Turbocharger (TC) wastegate regulating valve – open circuit	Wiring open circuit, TC wastegate regulating valve

Scan code 5-digit	OBD-II code	Fault location	Probable cause
17957	P1549	Turbocharger (TC) wastegate regulating valve – open circuit/short to ground	Wiring open circuit/short to ground, TC wastegate regulating valve
17958	P1550	Turbocharger (TC) pressure – control difference	Intake/exhaust leak, hoses interchanged/ not connected, MAP sensor, TC wastegate regulating valve, turbocharger (TC) wastegate actuator, TC
17961	P1553	Manifold absolute pressure (MAP) sensor/ barometric pressure (BARO) sensor – range/performance problem	Intake/exhaust leak, EGR system, EVAP canister purge valve, throttle control unit, wiring, MAP sensor, BARO sensor
17962	P1554	Throttle control unit – basic setting conditions	Basic setting conditions not met
17963	P1555	Turbocharger (TC) pressure – upper limit exceeded	Hoses interchanged/not connected, TC wastegate regulating valve, turbocharger (TC) wastegate actuator, TC
17964	P1556	Turbocharger (TC) pressure – control limit not reached	TC wastegate regulating valve, intake leak, TC defective
17965	P1557	Turbocharger (TC) pressure – control limit exceeded	Hose connection interchanged/not connected
17966	P1558	Idle speed control (ISC) actuator/throttle motor – circuit malfunction	Wiring, ISC actuator/throttle motor
17967	P1559	Throttle control unit – basic setting malfunction	Accelerator pedal or starter motor operated during basic setting
17968	P1560	Maximum engine RPM exceeded	Incorrect gear shift, wiring open circuit, CKP/RPM sensor
17969	P1561	Fuel quantity adjuster – control difference	Wiring, fuel injection pump
17970	P1562	Fuel quantity adjuster – upper stop value	Fuel quantity adjuster blocked/defective, stop value reached
17971	P1563	Fuel quantity adjuster – lower stop value	Fuel quantity adjuster blocked/defective, lower stop value reached
17972	P1564	Throttle control unit – voltage low during basic setting	Battery, wiring
17973	P1565	Throttle control unit – lower stop not reached	Throttle valve tight/sticking, ISC actuator
17974	P1566	AC compressor, load signal – implausible signal	Wiring, AC system
17976	P1568	Throttle control unit – mechanical fault	Throttle valve tight/sticking
17977	P1569	Cruise control master switch	Wiring, cruise control master switch
17978	P1570	Engine control module (ECM) – immobilizer active	Incorrect/damaged key, incorrectly coded, ECM/immobilizer replacement without coding, wiring, immobilizer defective
17979	P1571	Engine mounting control solenoid, bank 2 – short to positive	Wiring short to positive, engine mounting control solenoid
17980	P1572	Engine mounting control solenoid, bank 2 – short to ground	Wiring short to ground, engine mounting control solenoid
17981	P1573	Engine mounting control solenoid, bank 2 – open circuit	Wiring open circuit, engine mounting control solenoid

Scan code 5-digit	OBD-II code	Fault location	Probable cause
17983	P1575	Engine mounting control solenoid, bank 1 – short to positive	**Wiring short to positive, engine mounting control solenoid**
17984	P1576	Engine mounting control solenoid, bank 1 – short to ground	**Wiring short to ground, engine mounting control solenoid**
17985	P1577	Engine mounting control solenoid, bank 1 – open circuit	**Wiring open circuit, engine mounting control solenoid**
17987	P1579	Throttle control unit – basic setting	**Basic setting not carried out**
17988	P1580	Throttle motor, bank 1 – circuit malfunction	**Wiring, throttle motor**
17989	P1581	Throttle control unit – basic setting	**Basic setting not carried out**
17990	P1582	Idle speed adaptation – limit reached	**Intake/exhaust leak, AIR system, fuel pressure/pump, injector(s), EVAP canister purge valve**
17994	P1586	Engine mounting control solenoid, bank 1 & 2 – short to positive	**Wiring short to positive, engine mounting control solenoid**
17997	P1589	AC/heater air temperature control switch – short to ground	**Wiring short to ground, AC/heater air temperature control switch**
17998	P1590	AC/heater air temperature control switch – open circuit	**Wiring open circuit, AC/heater air temperature control switch**
18000	P1592	Barometric pressure (BARO) sensor/ manifold absolute pressure (MAP) sensor – implausible ratio	**TC system, MAP sensor**
18001	P1593	Altitude adaption – signal outside tolerance	**Intake leak, MAF sensor, throttle control unit**
18007	P1599	Idle control – lean running speed above specification	**IAC valve**
18008	P1600	Engine control module (ECM) – supply voltage low from ignition switch	**Battery, alternator, wiring open circuit**
18009	P1601	Engine control module (ECM) – supply voltage	**Wiring, engine control relay**
18010	P1602	Engine control module (ECM) – supply voltage low from battery	**Battery was disconnected, battery discharged, alternator, wiring open circuit, fuse**
18011	P1603	Engine control module (ECM) – defective	**ECM**
18012	P1604	Engine control module (ECM) – defective	**ECM**
18014	P1606	Rough road signal – circuit malfunction	**ABS control module trouble code(s) stored, CAN data bus**
18016	P1608	Power steering pressure (PSP) switch – circuit malfunction	**Wiring, PSP switch**
18017	P1609	Engine control module (ECM) – crash switch-off triggered	**Airbag triggered**
18018	P1610	Engine control module (ECM) – defective	**ECM**
18019	P1611	Malfunction indicator lamp (MIL) – short to ground	**Wiring short to ground**
18020	P1612	Engine control module (ECM) – coding	**Incorrectly coded**

Scan code 5-digit	OBD-II code	Fault location	Probable cause
18021	P1613	Malfunction indicator lamp (MIL) – open circuit/short to positive	Wiring open circuit/short to positive
18023	P1615	Engine oil temperature (EOT) sensor – range/performance problem	Engine oil level, wiring, EOT sensor
18024	P1616	Glow plug warning lamp – short to positive	Wiring short to positive
18025	P1617	Glow plug warning lamp – open circuit/ short to ground	Bulb, wiring open circuit/short to ground
18026	P1618	Glow plug relay – short to positive	Wiring short to positive, glow plug relay
18027	P1619	Glow plug relay – open circuit/short to ground	Wiring open circuit/short to ground, glow plug relay
18028	P1620	Instrument panel, ECT signal – open circuit/short to positive	Wiring open circuit/short to positive, instrument panel
18029	P1621	Instrument panel, ECT signal – short to ground	Wiring short to ground, instrument panel
18030	P1622	Instrument panel, ECT signal – implausible signal	Wiring, instrument panel, ECT sensor
18031	P1623	CAN data bus – no signal	Trouble code(s) stored in other system(s), wiring, matching resistor in ECM
18032	P1624	Malfunction indicator lamp (MIL) – request signal active	Trouble code(s) stored in other system(s)
18033	P1625	CAN data bus, TCM – incorrect signal	TCM trouble code(s) stored, TCM incorrectly coded, wiring, matching resistor in ECM
18034	P1626	CAN data bus, TCM – no signal	TCM trouble code(s) stored, TCM incorrectly coded, wiring, matching resistor in ECM
18037	P1629	CAN data bus, cruise control – no signal	Cruise control trouble code(s) stored, wiring, matching resistor in ECM
18038	P1630	Throttle/accelerator pedal position (APP) sensor 1/2 – low input	Wiring short to ground, APP sensor, TP sensor
18039	P1631	Throttle/accelerator pedal position (APP) sensor 1/2 – high input	Wiring short to positive, APP sensor, TP sensor
18040	P1632	Accelerator pedal position (APP) sensor – supply voltage	Operating voltage too high/low, wiring
18041	P1633	Accelerator pedal position (APP) sensor 2 – low input	Wiring short to ground, APP sensor
18042	P1634	Accelerator pedal position (APP) sensor 2 – high input	Wiring short to positive, APP sensor
18043	P1635	CAN data bus, AC – no signal	AC control module trouble code(s) stored, wiring, matching resistor in ECM
18044	P1636	CAN data bus, SRS – no signal	SRS control module trouble code(s) stored, wiring, matching resistor in ECM
18045	P1637	CAN data bus, electronic CE – no signal	Trouble code(s) stored, wiring, matching resistor in ECM
18047	P1639	Throttle/accelerator pedal position (APP) sensor 1/2 – range/performance problem	Wiring, APP sensor, TP sensor

Scan code 5-digit	OBD-II code	Fault location	Probable cause
18048	P1640	Engine control module (ECM) – defective	ECM
18050	P1642	SRS control module – system malfunction	Trouble code(s) stored
18053	P1645	CAN data bus, 4WD – no signal	4WD trouble code(s) stored, wiring, matching resistor in ECM
–	P1647	CAN data bus, ECM coding	Incorrectly coded ECM
18056	P1648	CAN data bus – defective	Wiring, matching resistor in ECM
18057	P1649	CAN data bus, ABS – no signal	ABS control module trouble code(s) stored, wiring, matching resistor in ECM
18058	P1650	CAN data bus, instrumentation – no signal	Instrumentation control module trouble code(s) stored, wiring, matching resistor in ECM
18060	P1652	Transmission control module (TCM) – system malfunction	Trouble code(s) stored
18061	P1653	ABS control module – system malfunction	Trouble code(s) stored
18062	P1654	Instrumentation control module – system malfunction	Trouble code stored for engine oil level/ temperature sensor
18064	P1656	AC signal – short to ground	Wiring short to ground
18065	P1657	AC signal – short to positive	Wiring
18066	P1658	CAN data bus, cruise control – incorrect signal	Cruise control trouble code(s) stored, wiring, matching resistor in ECM
18067	P1659	Engine coolant blower motor, speed 1 – short to positive	Wiring short to positive, engine coolant blower motor
18068	P1660	Engine coolant blower motor, speed 1 – short to ground	Wiring short to ground, engine coolant blower motor
18069	P1661	Engine coolant blower motor, speed 2 – short to positive	Wiring short to positive, engine coolant blower motor
18070	P1662	Engine coolant blower motor, speed 2 – short to ground	Wiring short to ground, engine coolant blower motor
18071	P1663	Injector, activation – short to positive	Wiring short to positive, ECM
18072	P1664	Injector, activation – current circuit	Wiring open circuit/short to ground
18073	P1665	Injector – mechanical fault	Injector
18074	P1666	Injector 1 – current circuit	Wiring open circuit/short to ground
18075	P1667	Injector 2 – current circuit	Wiring open circuit/short to ground
18076	P1668	Injector 3 – current circuit	Wiring open circuit/short to ground
18077	P1669	Injector 4 – current circuit	Wiring open circuit/short to ground
18080	P1672	Engine coolant blower motor, speed 1 – open circuit/short to ground	Wiring open circuit/short to ground
18082	P1674	CAN data bus, instrumentation – incorrect signal	Wiring, instrumentation control module trouble code(s) stored, matching resistor in ECM
18084	P1676	ETS warning lamp – circuit malfunction	Instrumentation control module trouble code(s) stored, wiring, matching resistor in ECM

Scan code 5-digit	OBD-II code	Fault location	Probable cause
18085	P1677	ETS warning lamp – short to positive	Instrumentation control module trouble code(s) stored, wiring short to positive, matching resistor in ECM
18086	P1678	ETS warning lamp – short to ground	Instrumentation control module trouble code(s) stored, wiring short to ground, matching resistor in ECM
18087	P1679	ETS warning lamp – open circuit	Instrumentation control module trouble code(s) stored, wiring open circuit, matching resistor in ECM
18088	P1680	Limp-home mode – active	Throttle control unit, APP sensor
18089	P1681	Engine control module (ECM) – programming incomplete	ECM
18090	P1682	CAN data bus, ABS – implausible signal	ABS control module trouble code(s) stored, wiring, matching resistor in ECM
18091	P1683	CAN data bus, SRS – implausible signal	SRS control module trouble code(s) stored, wiring, matching resistor in ECM
18098	P1690	Malfunction indicator lamp (MIL) – circuit malfunction	Instrumentation control module trouble code(s) stored, wiring, matching resistor in ECM
18099	P1691	Malfunction indicator lamp (MIL) – open circuit	Instrumentation control module trouble code(s) stored, wiring open circuit, matching resistor in ECM
18100	P1692	Malfunction indicator lamp (MIL) – short to ground	Instrumentation control module trouble code(s) stored, wiring short to ground, matching resistor in ECM
18101	P1693	Malfunction indicator lamp (MIL) – short to positive	Instrumentation control module trouble code(s) stored, wiring short to positive, matching resistor in ECM
18104	P1696	CAN data bus, steering column electronics – incorrect signal	Wiring, matching resistor in ECM
18153	P1745	Solenoid valves, supply voltage – short to positive	Wiring open circuit/short to positive, poor connection, TCM
–	P1746	TCM/solenoid valves, supply voltage – circuit malfunction	Wiring, battery, alternator
18155	P1747	Solenoid valves, supply voltage – open circuit/short to ground	Battery, wiring open circuit/short to ground, poor connection, TCM
18156	P1748	Transmission control module (TCM) – defective	TCM
18158	P1750	Transmission control module (TCM) – supply voltage low	Fuse, wiring, TCM
18159	P1751	Transmission control module (TCM) – supply voltage high	Wiring, alternator, battery(s)
18259	P1851	CAN data bus, ABS – incorrect signal	ABS trouble code(s) stored, wiring, matching resistor in ECM
18261	P1853	CAN data bus, ABS – incorrect signal	ABS trouble code(s) stored, wiring, matching resistor in ECM
18262	P1854	CAN data bus, ABS – defective	ABS trouble code(s) stored, wiring, matching resistor in ECM

Scan code 5-digit	OBD-II code	Fault location	Probable cause
18308	P1900	Engine coolant blower motor, speed 2 – open circuit/short to ground	Wiring open circuit/short to ground, engine coolant blower motor
18309	P1901	Engine coolant blower motor run-on relay – short to positive	Wiring short to positive, engine coolant blower motor run-on relay
18310	P1902	Engine coolant blower motor run-on relay – open circuit/short to ground	Wiring open circuit/short to ground, engine coolant blower motor run-on relay
18311	P1903	Engine coolant hydraulic blower motor solenoid – short to positive	Wiring short to positive, engine coolant hydraulic blower motor solenoid
18312	P1904	Engine coolant hydraulic blower motor solenoid – open circuit/short to ground	Wiring open circuit/short to ground, engine coolant hydraulic blower motor solenoid
18313	P1905	Charge air coolant pump relay – short to positive	Wiring short to positive, charge air coolant pump relay
18314	P1906	Charge air coolant pump relay – open circuit/short to ground	Wiring open circuit/short to ground, charge air coolant pump relay
18315	P1907	Data bus, ECM 1/2 – defective	Wiring
18316	P1908	Data bus, ECM 1/2 – software version monitoring	Data in ECM 1 & ECM 2 does not match
18317	P1909	Data bus, ECM 1/2 – no signal from ECM 1	Wiring, ECM 1
18318	P1910	Data bus, ECM 1/2 – no signal from ECM 2	Wiring, ECM 2
18318	P1911	Data bus, ECM 1/2 – circuit malfunction	Wiring, ECM 1/2
18320	P1912	Brake servo pressure sensor – open circuit/short to positive	Wiring open circuit/short to positive, brake servo pressure sensor
18321	P1913	Brake servo pressure sensor – short to ground	Wiring short to ground, brake servo pressure sensor
18322	P1914	Brake servo pressure sensor – range/ performance problem	Vacuum leak, wiring, brake servo pressure sensor
18328	P1920	Engine mounting control solenoid, bank 1 & 2 – open circuit/short to ground	Wiring open circuit/short to ground, engine mounting control solenoid
18331	P1923	Engine control module (ECM) 2 – malfunction	Trouble code(s) stored
19456	P3000	CAN data bus, instrumentation – glow plug warning lamp	Wiring, matching resistor in ECM
19458	P3002	Accelerator pedal position (APP) sensor – transmission kick-down switch	APP sensor
19459	P3003	Engine coolant heater relay 1, low output	Wiring, engine coolant heater relay
19461	P3005	Engine coolant heater relay 2, high output	Wiring, engine coolant heater relay
19463	P3007	Camshaft position (CMP) sensor – no signal	Air gap, insecure sensor/rotor, wiring, CMP sensor
19464	P3008	Camshaft position (CMP) sensor – signal limit exceeded	Insecure rotor, camshaft alignment
19465	P3009	Fuel cooling pump relay – short to positive	Wiring short to positive, fuel cooling pump relay
19466	P3010	Fuel cooling pump relay – open circuit/ short to ground	Wiring open circuit/short to ground, fuel cooling pump relay

Scan code 5-digit	OBD-II code	Fault location	Probable cause
19467	P3011	Fuel pump relay 1/2 – short to positive	**Wiring short to positive, fuel pump relay**
19468	P3012	Fuel pump relay 1/2 – open circuit/short to ground	**Wiring open circuit/short to ground, fuel pump relay**
19469	P3013	Turbocharger (TC) wastegate regulating valve B – short to positive	**Wiring short to positive, TC wastegate regulating valve**
19470	P3014	Turbocharger (TC) wastegate regulating valve B – open circuit/short to ground	**Wiring open circuit/short to ground, TC wastegate regulating valve**
19471	P3015	Fuel bypass valve – short to positive	**Wiring short to positive, fuel bypass valve**
19472	P3016	Fuel bypass valve – open circuit/short to ground	**Wiring open circuit/short to ground, fuel bypass valve**
19496	P3040	Gear ratio – implausible	**Transmission fault**
19497	P3041	CAN data bus, instrumentation – implausible ECT signal	**Wiring, matching resistor in ECM**
–	P3081	Engine temperature too low	**Allow engine to warm up, cooling system fault**
–	P3092	Engine control module (ECM) – internal fault	**ECM**
–	P3093	Engine control module (ECM) – internal fault	**ECM**
–	P3096	Engine control module (ECM) – internal fault	**ECM**
–	P3097	Engine control module (ECM) – internal fault	**ECM**
–	P3211	Heated oxygen sensor (HO2S) 1, bank 1 – heater circuit malfunction	**Wiring, HO2S, ECM**
–	P3212	Heated oxygen sensor (HO2S) 1, bank 2 – heater circuit malfunction	**Wiring, HO2S, ECM**
19560	P3104	Intake manifold air control solenoid – short to positive	**Wiring short to positive, intake manifold air control solenoid**
19561	P3105	Intake manifold air control solenoid – open circuit/short to ground	**Wiring open circuit/short to ground, intake manifold air control solenoid**
19717	P3262	Heated oxygen sensor (HO2S) 2, bank 1 & 2 – interchanged	**HO2S 2 on bank 1 & 2 incorrectly installed**
65280	–	CAN data bus, ABS – defective	**ABS trouble code(s) stored, wiring, matching resistor in ECM**
65535	–	Engine control module (ECM) – defective	**ECM**

1 Trouble code may be displayed if engine is not idling during self-diagnosis due to missing CKP sensor signal. Ignore trouble code if engine starts.
2 Located at rear of exhaust camshaft.
3 Incorporates heated oxygen sensor (HO2S) 2.
4 May also produce HT voltage for cylinder 4.
5 May also produce HT voltage for cylinder 3.

Model:	**S4 2.2L • S6 2.2L**
Year:	**1992-95**
Engine identification:	**AAN**
System:	**Bosch Motronic M2.3 • Bosch Motronic M2.3.2**

Data link connector (DLC) locations

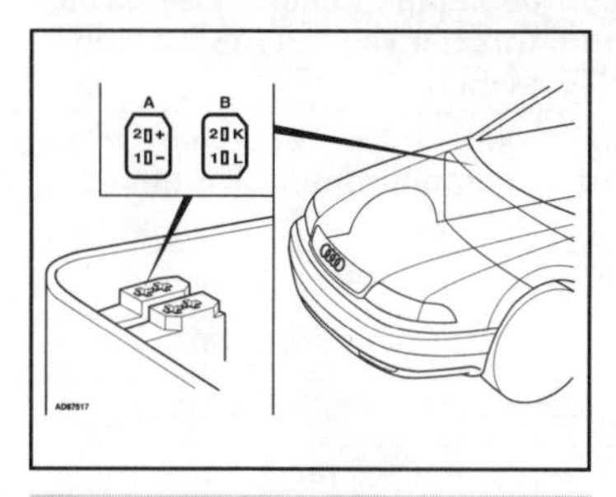

In underhood relay plate

Self-diagnosis

General information

- Refer to the front of this manual for general test conditions, terminology, detailed descriptions of wiring faults and a general trouble shooter for electrical and mechanical faults.

Accessing

- The engine control module (ECM) fault memory can only be accessed using diagnostic equipment connected to the data link connector (DLC).

Erasing

- Ensure ignition switched OFF.
- Disconnect ECM multi-plug or battery ground lead.

NOTE: ***ECM adaptive memory may be erased causing erratic running faults. Carry out road test to allow ECM to re-learn basic values.***

- Diagnostic equipment can also be used to erase data from ECM fault memory.

WARNING: ***Disconnecting the battery may erase memory from electronic units (e.g. radio, clock).***

Trouble code identification

Flash code 4-digit	Fault location	Probable cause
0000	End of test sequence	–
1111	Engine control module (ECM) – defective	ECM
1231	Vehicle speed sensor (VSS)	Wiring, speedometer, VSS

Flash code 4-digit	Fault location	Probable cause
2111	Engine speed (RPM) sensor	Air gap, metal particles, insecure sensor, flywheel ring gear damaged, wiring, CKP/RPM sensor multi-plugs interchanged, RPM sensor
2112	Crankshaft position (CKP) sensor	Air gap, metal particles, insecure sensor, flywheel pin bent/missing, wiring, CKP/RPM sensor multi-plugs interchanged, CKP sensor
2113	Camshaft position (CMP) sensor	Valve timing, distributor alignment, wiring, CMP sensor, CKP sensor flywheel pin bent/ missing, CMP sensor
2121	Closed throttle position (CTP) switch	Accelerator cable adjustment, CTP switch adjustment, throttle valve tight/sticking, wiring, CTP switch
2141	Knock control – control limit exceeded	Poor quality fuel, CKP/RPM sensor wiring, insecure engine component, KS shield wiring
2142	Knock sensor (KS) 1	Wiring, KS loose/defective, ECM
2144	Knock sensor (KS) 2	Wiring, KS loose/defective, ECM
2212	Throttle position (TP) sensor	Wiring, poor connection, TP sensor
2214	Maximum engine RPM exceeded	Incorrect gear shift, AT fault
2222	Manifold absolute pressure (MAP) sensor	Hose leak/blockage, hose fluid collector full, TC wastegate regulating valve, MAP sensor (in ECM)
2223	Barometric pressure (BARO) sensor	Wiring, BARO sensor
2224	Maximum boost pressure exceeded	Hoses interchanged/not connected, hoses blocked/leaking, TC wastegate actuator/ regulating valve, MAP sensor (in ECM)
2231	Idle speed control (ISC)	Intake leak/blockage, wiring, IAC valve, MAF sensor, throttle valve tight/sticking
2234	Engine control module (ECM) – supply voltage	Alternator, battery, current draw with ignition OFF, wiring
2312	Engine coolant temperature (ECT) sensor	Wiring, poor connection, ECT sensor
2314	ECM/TCM electrical connection	TCM trouble code(s) stored, wiring, transmission fault
2322	Intake air temperature (IAT) sensor	Wiring, poor connection, IAT sensor
2324	Mass air flow (MAF) sensor	Intake leak, wiring, MAF sensor
2341	Heated oxygen sensor (HO2S) – lambda control	Fuel level low, fuel pressure/pump, misfire, intake/exhaust leak, wiring, HO2S
2342	Heated oxygen sensor (HO2S)	Intake/exhaust leak, wiring, HO2S, heating inoperative, fuel level low, misfire
2413	Mixture control (MC)	Fuel pressure/pump, misfire, intake/exhaust leak, MAF sensor

Flash code 4-digit	Fault location	Probable cause
4343	Evaporative emission (EVAP) canister purge valve	**Wiring, fuse, EVAP canister purge valve**
4411	Injector 1	**Wiring, fuse, injector**
4412	Injector 2	**Wiring, fuse, injector**
4413	Injector 3	**Wiring, fuse, injector**
4414	Injector 4	**Wiring, fuse, injector**
4421	Injector 5	**Wiring, fuse, injector**
4431	Idle air control (IAC) valve	**Wiring, fuse, IAC valve**
4442	Turbocharger (TC) wastegate regulating valve	**Wiring, fuse, TC wastegate regulating valve**
4444	No fault found	–

VAG code 5-digit	Fault location	Probable cause
00000	No fault found	–
00281	Vehicle speed sensor (VSS)	**Wiring, speedometer, VSS**
00513	Engine speed (RPM) sensor	**Air gap, metal particles, insecure sensor, flywheel ring gear damaged, wiring, CKP/RPM sensor multi-plugs interchanged, RPM sensor**
00514	Crankshaft position (CKP) sensor	**Air gap, metal particles, insecure sensor, flywheel pin bent/missing, wiring, CKP/RPM sensor multi-plugs interchanged, CKP sensor**
00515	Camshaft position (CMP) sensor	**Valve timing, distributor alignment, wiring, CMP sensor, CKP sensor flywheel pin bent/ missing, CMP sensor**
00516	Closed throttle position (CTP) switch	**Accelerator cable adjustment, CTP switch adjustment, throttle valve tight/sticking, wiring, CTP switch**
00518	Throttle position (TP) sensor	**Wiring, poor connection, TP sensor**
00519	Manifold absolute pressure (MAP) sensor	**Hose leak/blockage, hose fluid collector full, TC wastegate regulating valve, MAP sensor (in ECM)**
00522	Engine coolant temperature (ECT) sensor	**Wiring, poor connection, ECT sensor**
00523	Intake air temperature (IAT) sensor	**Wiring, poor connection, IAT sensor**
00524	Knock sensor (KS) 1	**Wiring, KS loose/defective, ECM**
00525	Heated oxygen sensor (HO2S)	**Intake/exhaust leak, wiring, HO2S, heating inoperative, fuel level low, misfire**
00528	Barometric pressure (BARO) sensor	**Wiring, BARO sensor**
00532	Engine control module (ECM) – supply voltage	**Alternator, battery, current draw with ignition OFF, wiring**

VAG code 5-digit	Fault location	Probable cause
00533	Idle speed control (ISC)	Intake leak/blockage, wiring, IAC valve, MAF sensor, throttle valve tight/sticking
00537	Heated oxygen sensor (HO2S) – lambda control	Fuel level low, fuel pressure/pump, misfire, intake/exhaust leak, wiring, HO2S
00540	Knock sensor (KS) 2	Wiring, KS loose/defective, ECM
00543	Maximum engine RPM exceeded	Incorrect gear shift, AT fault
00544	Maximum boost pressure exceeded	Hoses interchanged/not connected, hoses blocked/leaking, TC wastegate actuator/ regulating valve, MAP sensor (in ECM)
00545	ECM/TCM electrical connection	TCM trouble code(s) stored, wiring, transmission fault
00553	Mass air flow (MAF) sensor	Intake leak, wiring, MAF sensor
00561	Mixture control (MC)	Fuel pressure/pump, misfire, intake/exhaust leak, MAF sensor
00577	Knock control, cylinder 1 – control limit exceeded	Poor quality fuel, CKP/RPM sensor wiring, insecure engine component, KS shield wiring
00578	Knock control, cylinder 2 – control limit exceeded	Poor quality fuel, CKP/RPM sensor wiring, insecure engine component, KS shield wiring
00579	Knock control, cylinder 3 – control limit exceeded	Poor quality fuel, CKP/RPM sensor wiring, insecure engine component, KS shield wiring
00580	Knock control, cylinder 4 – control limit exceeded	Poor quality fuel, CKP/RPM sensor wiring, insecure engine component, KS shield wiring
00581	Knock control, cylinder 5 – control limit exceeded	Poor quality fuel, CKP/RPM sensor wiring, insecure engine component, KS shield wiring
01247	Evaporative emission (EVAP) canister purge valve	Wiring, fuse, EVAP canister purge valve
01249	Injector 1	Wiring, fuse, injector
01250	Injector 2	Wiring, fuse, injector
01251	Injector 3	Wiring, fuse, injector
01252	Injector 4	Wiring, fuse, injector
01253	Injector 5	Wiring, fuse, injector
01257	Idle air control (IAC) valve	Wiring, fuse, IAC valve
01262	Turbocharger (TC) wastegate regulating valve	Wiring, fuse, TC wastegate regulating valve
17978	Engine control module (ECM) – immobilizer active	Incorrect/damaged key, incorrectly coded, ECM/immobilizer replacement without coding, wiring, immobilizer defective

Model:	**X3 2.5L (E83) • X3 3.0L (E83) • Z4 2.5L (E85) • Z4 3.0L (E85) Z8 5.0L (E52) • 6 Series 4.4L (E64)**
Year:	**2000-04**
Engine identification:	**25 6S 5, 30 6S 3, 25 6S 5, 30 6S 3, 50 8S 1, N62 B44**
System:	**Siemens MS43 • Siemens MS S52 • Motronic ME 9.2.1 • Motronic ME 9.2.2**

Data link connector (DLC) locations

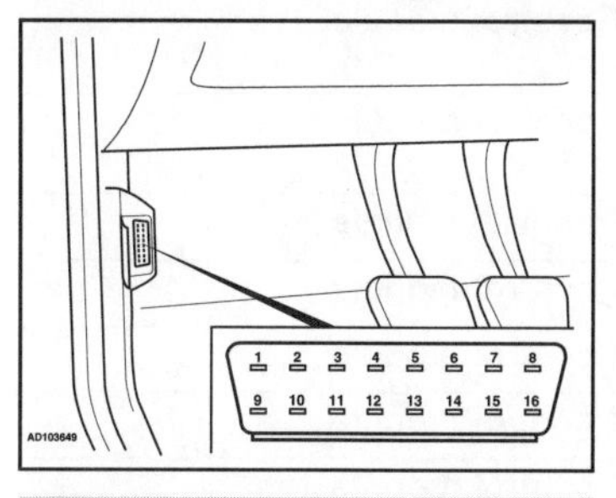

X3 – behind cover, 'A' pillar driver's side

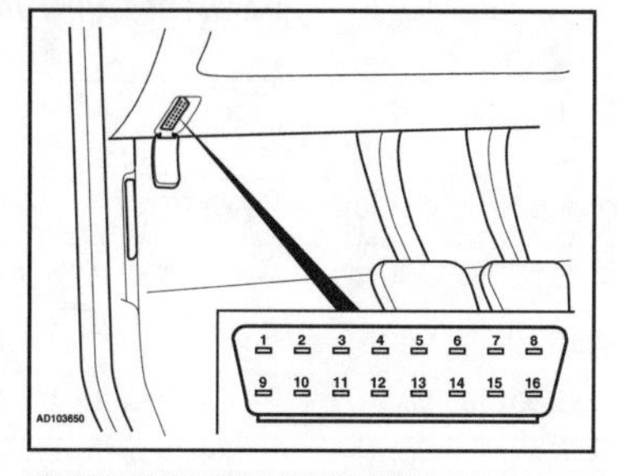

Z4 – driver's footwell

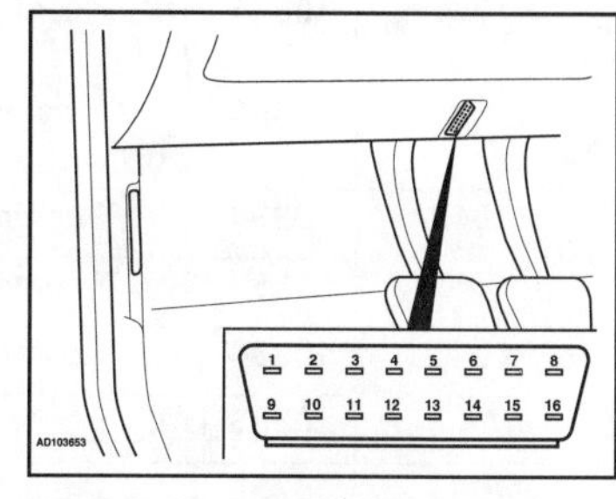

Z8 – driver's footwell

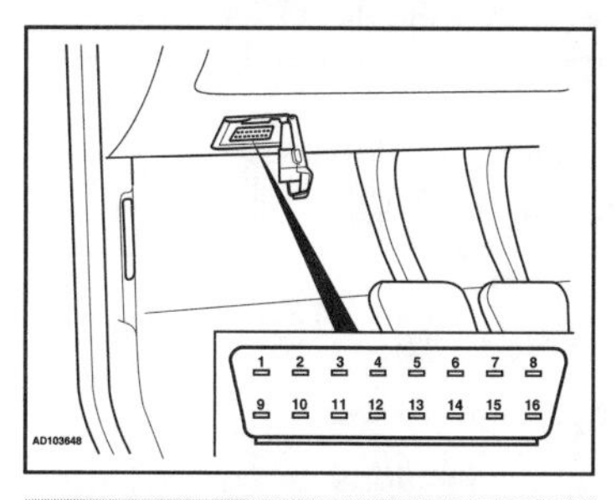

6 Series – driver's footwell

Self-diagnosis

General information

- Refer to the front of this manual for general test conditions, terminology, detailed descriptions of wiring faults and a general trouble shooter for electrical and mechanical faults.
- Engine control module (ECM) incorporates self-diagnosis function.
- Malfunction indicator lamp (MIL) will illuminate if certain faults are recorded.
- ECM operates in backup mode if sensors fail, to enable vehicle to be driven to workshop.

Accessing

- The engine control module (ECM) fault memory can only be accessed and erased using diagnostic equipment connected to the data link connector (DLC).

Erasing

- After the faults have been rectified, trouble codes can only be erased by using a Scan Tool connected to the data link connector (DLC).

Trouble code identification

OBD-II code	BMW type	Fault location	Probable cause
All P0, P2 and U0 codes	–	Refer to OBD-II trouble code tables at the front of this manual	–
P1041	**10334**	VVT control module, bank 1 – EEPROM error	**VVT control module**
	10853	VVT control module, bank 1 – EEPROM error	**VVT control module**
P1042	**10334**	VVT control module, bank 1 – RAM error	**VVT control module**
	10853	VVT control module, bank 1 – RAM error	**VVT control module**
P1043	**10334**	VVT control module, bank 1 – ROM error	**VVT control module**
	10853	VVT control module, bank 1 – ROM error	**VVT control module**
P1044	**10335**	VVT control module, bank 2 – EEPROM error	**VVT control module**
	10854	VVT control module, bank 2 – EEPROM error	**VVT control module**
P1045	**10335**	VVT control module, bank 2 – RAM error	**VVT control module**
	10854	VVT control module, bank 2 – RAM error	**VVT control module**
P1046	**10335**	VVT control module, bank 2 – ROM error	**VVT control module**
	10854	VVT control module, bank 2 – ROM error	**VVT control module**
P1047	**10336**	VVT control circuit, bank 1 – high input	**Wiring, connections**
	10855	VVT control circuit, bank 1 – high input	**Wiring, connections**
P1048	**10336**	VVT control circuit, bank 1 – low input	**Wiring, connections**
	10855	VVT control circuit, bank 1 – low input	**Wiring, connections**
P1049	**10336**	VVT control circuit cables, bank 1 – short circuit	**Wiring, connections**
	10855	VVT control circuit cables, bank 1 – short circuit	**Wiring, connections**
P1050	**10336**	VVT control circuit, bank 1 – circuit malfunction	**Wiring, connections, VVT control module**
	10855	VVT control circuit, bank 1 – circuit malfunction	**Wiring, connections, VVT control module**
P1051	**10337**	VVT control circuit, bank 2 – high input	**Wiring, connections**
	10856	VVT control circuit, bank 2 – high input	**Wiring, connections**

OBD-II code	BMW type	Fault location	Probable cause
P1052	10337	VVT control circuit, bank 2 – low input	Wiring, connections
	10856	VVT control circuit, bank 2 – low input	Wiring, connections
P1053	10337	VVT control circuit cables, bank 2 – short circuit	Wiring, connections
	10856	VVT control circuit cables, bank 2 – short circuit	Wiring, connections
P1054	10337	VVT control circuit, bank 2 – circuit malfunction	Wiring, connections, VVT control module
	10856	VVT control circuit, bank 2 – circuit malfunction	Wiring, connections, VVT control module
P1055	10338	VVT control motor, bank 1 – supply voltage high input	Wiring, connections
	10857	VVT control motor, bank 1 – supply voltage high input	Wiring, connections
P1056	10338	VVT control motor, bank 1 – supply voltage low input	Wiring, connections
	10857	VVT control motor, bank 1 – supply voltage low input	Wiring, connections
P1057	10338	VVT control motor, bank 1 – supply voltage circuit malfunction	Wiring, connections
	10857	VVT control motor, bank 1 – supply voltage circuit malfunction	Wiring, connections
P1058	10339	VVT control motor, bank 2 – supply voltage high input	Wiring, connections
	10858	VVT control motor, bank 2 – supply voltage high input	Wiring, connections
P1059	10339	VVT control motor, bank 2 – supply voltage low input	Wiring, connections
	10858	VVT control motor, bank 2 – supply voltage low input	Wiring, connections
P1060	10339	VVT control motor, bank 2 – supply voltage circuit malfunction	Wiring, connections
	10858	VVT control motor, bank 2 – supply voltage circuit malfunction	Wiring, connections
P1061	10341	VVT limp home request, bank 1 – circuit malfunction	Wiring, connections, ECM
	10859	VVT limp home request, bank 1 – circuit malfunction	Wiring, connections, ECM
P1062	10341	VVT limp home request, bank 1 – full stroke position not reached	Wiring, connections, ECM
	10859	VVT limp home request, bank 1 – full stroke position not reached	Wiring, connections, ECM

OBD-II code	BMW type	Fault location	Probable cause
P1063	10341	VVT limp home request, bank 1 – air mass signal malfunction	**Wiring, connections, mass air sensor, ECM**
	10859	VVT limp home request, bank 1 – air mass signal malfunction	**Wiring, connections, mass air sensor, ECM**
P1064	10342	VVT starting position/parking position comparison – circuit malfunction	**Wiring, connections**
P1064	10342	VVT starting position/parking position comparison – circuit malfunction	**Wiring, connections**
P1065	10227	VVT timeout, no signal – CAN data bus malfunction	**Wiring, connections**
	11741	VVT timeout, no signal – CAN data bus malfunction	**Wiring, connections**
P1066	10227	VVT message monitoring – circuit malfunction	**Wiring, connections**
	11741	VVT message monitoring – circuit malfunction	**Wiring, connections**
P1071	10334	VVT control module, bank 1 – internal malfunction	**VVT control module**
	10853	VVT control module, bank 1 – internal malfunction	**VVT control module**
P1072	10335	VVT control module, bank 2 – internal malfunction	**VVT control module**
	10854	VVT control module, bank 2 – internal malfunction	**VVT control module**
P1087	229	Heated oxygen sensor (HO2S) 1, bank 1 – slow response lean control	**Wiring, intake/exhaust leak, fuel level low, fuel pressure, EVAP canister purge valve, injector(s), HO2S, ECM**
	229	Heated oxygen sensor (HO2S) 1, bank 1 – slow response rich control	**Wiring, intake/exhaust leak, fuel level low, fuel pressure, EVAP canister purge valve, injector(s), HO2S, ECM**
P1089	230	Heated oxygen sensor (HO2S) 1, bank 2 – slow response lean control	**Wiring, intake/exhaust leak, fuel level low, fuel pressure, EVAP canister purge valve, injector(s), HO2S, ECM**
P1094	230	Heated oxygen sensor (HO2S) 1, bank 2 – slow response rich control	**Wiring, intake/exhaust leak, fuel level low, fuel pressure, EVAP canister purge valve, injector(s), HO2S, ECM**
P110F	11201	Ambient air temperature sensor – CAN data bus malfunction	**Wiring, air temperature sensor, ECM**
P1100	57	Mass air flow (MAF) sensor – high input	**Wiring, MAF sensor, ECM**
P111E	12041	Intake air temperature (IAT) sensor 1 – maximum temperature incorrect	**Wiring, IAT sensor, ECM**
P111F	12041	Intake air temperature (IAT) sensor 1 – minimum temperature incorrect	**Wiring, IAT sensor, ECM**
P1115	10122	Ambient air temperature sensor – circuit malfunction	**Wiring, air temperature sensor, ECM**

OBD-II code	BMW type	Fault location	Probable cause
P1117	57	Mass air flow (MAF) sensor/volume air flow (VAF) sensor bank 2 – low input	**Wiring, connector(s), MAF sensor/VAF sensor**
P1118	57	Mass air flow (MAF) sensor/volume air flow (VAF) sensor bank 2 – high input	**Wiring, connector(s), MAF sensor/VAF sensor**
P112A	12001	Engine coolant temperature (ECT) sensor 1 – maximum temperature incorrect	**Wiring, ECT sensor**
P112B	12001	Engine coolant temperature (ECT) sensor 1 – minimum temperature incorrect	**Wiring, ECT sensor**
P1130	10017	Heated oxygen sensor (HO2S) 2, bank 1 – dynamic test malfunction	**Wiring, HO2S, ECM**
	11373	Heated oxygen sensor (HO2S) 2, bank 1 – dynamic test malfunction	**Wiring, HO2S, ECM**
P1131	10023	Heated oxygen sensor (HO2S) 2, bank 2 – dynamic test malfunction	**Wiring, HO2S, ECM**
	11374	Heated oxygen sensor (HO2S) 2, bank 2 – dynamic test malfunction	**Wiring, HO2S, ECM**
P1134	37	Heated oxygen sensor (HO2S) 1, bank 1 – signal malfunction	**Wiring, HO2S, ECM**
P1135	37	Heated oxygen sensor (HO2S) 1, bank 1 – low voltage	**Wiring, HO2S, ECM**
P1136	37	Heated oxygen sensor (HO2S) 1, bank 1 – high voltage	**Wiring, HO2S, ECM**
P1140	73	Mass air flow (MAF) sensor/volume air flow (VAF) sensor – incorrect signal	**Wiring, connector(s), MAF sensor/VAF sensor**
P1151	38	Heated oxygen sensor (HO2S) 1, bank 2 – signal malfunction	**Wiring, HO2S, ECM**
P1152	38	Heated oxygen sensor (HO2S) 1, bank 2 – low voltage	**Wiring, HO2S, ECM**
P1153	38	Heated oxygen sensor (HO2S) 1, bank 2 – high voltage	**Wiring, HO2S, ECM**
P1171	196	Ambient pressure sensor – recognition value incorrect	**Wiring, ambient pressure sensor, ECM**
P1172	196	Ambient pressure sensor – recognition value error	**Wiring, ambient pressure sensor, ECM**
P1173	196	Ambient pressure sensor – recognition learning failed	**Learning procedure not performed**
P1178	216	Heated oxygen sensor (HO2S) 1, bank 1 – slow switching	**Wiring, HO2S, ECM**
P1179	217	Heated oxygen sensor (HO2S) 1, bank 2 – slow switching	**Wiring, HO2S, ECM**
P1184	218	Heated oxygen sensor (HO2S) 1, bank 1 – no activity detected	**Wiring, HO2S**
P1185	219	Heated oxygen sensor (HO2S) 1, bank 1 – no activity detected	**Wiring, HO2S**

OBD-II code	BMW type	Fault location	Probable cause
P1188	174	Fuel trim (FT) sensor 1, bank 1 – range/ performance problem	Wiring, connectors, intake/fuel system, injectors, HO2S, MAF sensor, ECT sensor, EVAP canister purge valve, ECM
P1189	175	Fuel trim (FT) sensor 1, bank 2 – range/ performance problem	Wiring, connectors, intake/fuel system, injectors, HO2S, MAF sensor, ECT sensor, EVAP canister purge valve, ECM
P1204	10277	Heated oxygen sensor (HO2S) 2, bank 1 – circuit malfunction	Wiring, HO2S sensor
P1205	10278	Heated oxygen sensor (HO2S) 2, bank 2 – circuit malfunction	Wiring, HO2S sensor
P1314	227	Fuel mixture – deviation with low fuel	Wiring, fuel level low, fuel pressure, intake/exhaust leak, injector(s), HO2S, EVAP canister purge valve, ECT sensor, ECM
P1314	228	Fuel mixture – deviation with low fuel	Wiring, fuel level low, fuel pressure, intake/exhaust leak, injector(s), HO2S, EVAP canister purge valve, ECT sensor, ECM
P1317	15	'B' Camshaft position (CMP) actuator – range/performance problem	Wiring, CMP actuator
P1327	59	Knock sensor (KS) 2, bank 1 – low input	Wiring, KS incorrectly tightened, KS
P1327	10211	Knock sensor (KS) 2, bank 1 – low input	Wiring, KS incorrectly tightened, KS
	11881	Knock sensor (KS) 2, bank 1 – low input	Wiring, KS incorrectly tightened, KS
P1328	10211	Knock sensor (KS) 2, bank 1 – high input	Wiring, KS incorrectly tightened, KS
	11881	Knock sensor (KS) 2, bank 1 – high input	Wiring, KS incorrectly tightened, KS
P1329	10212	Knock sensor (KS) 3 – low input	Wiring, KS incorrectly tightened, KS
	11882	Knock sensor (KS) 3 – low input	Wiring, KS incorrectly tightened, KS
P1330	10212	Knock sensor (KS) 3 – high input	Wiring, KS incorrectly tightened, KS
	11882	Knock sensor (KS) 3 – high input	Wiring, KS incorrectly tightened, KS
P1332	10213	Knock sensor (KS) 4 – low input	Wiring, KS incorrectly tightened, KS
	11883	Knock sensor (KS) 4 – low input	Wiring, KS incorrectly tightened, KS
P1333	10213	Knock sensor (KS) 4 – high input	Wiring, KS incorrectly tightened, KS
	11883	Knock sensor (KS) 4 – high input	Wiring, KS incorrectly tightened, KS
P1340	213	Multiple cylinder(s) misfire – during start	Wiring, fuel level low, ignition/fuel system, injector(s), spark plug(s), ignition coil(s), mechanical fault
P1341	204	Multiple cylinder(s) misfire – with fuel cut-off	Wiring, fuel level low, ignition/fuel system, intake leak, mechanical fault
P1342	205	Cylinder misfire during start – cylinder 1	Wiring, fuel level low, ignition/fuel system, injector, spark plug, ignition coil, mechanical fault
P1343	196	Cylinder misfire with fuel cut-off – cylinder 1	Wiring, fuel level low, ignition/fuel system, intake leak, mechanical fault

OBD-II code	BMW type	Fault location	Probable cause
P1344	206	Cylinder misfire during start – cylinder 2	Wiring, fuel level low, ignition/fuel system, injector, spark plug, ignition coil, mechanical fault
P1345	197	Cylinder misfire with fuel cut-off – cylinder 2	Wiring, fuel level low, ignition/fuel system, intake leak, mechanical fault
P1346	207	Cylinder misfire during start – cylinder 3	Wiring, fuel level low, ignition/fuel system, injector, spark plug, ignition coil, mechanical fault
P1347	198	Cylinder misfire with fuel cut-off – cylinder 3	Wiring, fuel level low, ignition/fuel system, intake leak, mechanical fault
P1348	208	Cylinder misfire during start – cylinder 4	Wiring, fuel level low, ignition/fuel system, injector, spark plug, ignition coil, mechanical fault
P1349	199	Cylinder misfire with fuel cut-off – cylinder 4	Wiring, fuel level low, ignition/fuel system, intake leak, mechanical fault
P1350	209	Cylinder misfire during start – cylinder 5	Wiring, fuel level low, ignition/fuel system, injector, spark plug, ignition coil, mechanical fault
P1351	200	Cylinder misfire with fuel cut-off – cylinder 5	Wiring, fuel level low, ignition/fuel system, intake leak, mechanical fault
P1352	210	Cylinder misfire during start – cylinder 6	Wiring, fuel level low, ignition/fuel system, injector, spark plug, ignition coil, mechanical fault
P1353	201	Cylinder misfire with fuel cut-off – cylinder 6	Wiring, fuel level low, ignition/fuel system, intake leak, mechanical fault
P1354	211	Cylinder misfire during start – cylinder 7	Wiring, fuel level low, ignition/fuel system, injector, spark plug, ignition coil, mechanical fault
P1355	202	Cylinder misfire with fuel cut-off – cylinder 7	Wiring, fuel level low, ignition/fuel system, intake leak, mechanical fault
P1356	212	Cylinder misfire during start – cylinder 8	Wiring, fuel level low, ignition/fuel system, injector, spark plug, ignition coil, mechanical fault
P1357	203	Cylinder misfire with fuel cut-off – cylinder 8	Wiring, fuel level low, ignition/fuel system, intake leak, mechanical fault
P1378	10238	ECM self-test – Bank 2 knock control malfunction	Wiring, KS, ECM
P1379	10239	ECM self-test – Bank 2 knock control malfunction	Wiring, KS, ECM
P1380	10217	ECM self-test – Bank 2 knock control malfunction	Wiring, KS, ECM
P1381	10215	ECM self-test – Bank 1 knock control malfunction	Wiring, KS, ECM
	11890	ECM self-test – Bank 1 knock control malfunction	Wiring, KS, ECM

OBD-II code	BMW type	Fault location	Probable cause
P1382	10216	ECM self-test – Bank 1 knock control malfunction	Wiring, KS, ECM
	11890	ECM self-test – Bank 1 knock control malfunction	Wiring, KS, ECM
P1386	10214	ECM self-test – Bank 1 knock control malfunction	Wiring, KS, ECM
	11890	ECM self-test – Bank 1 knock control malfunction	Wiring, KS, ECM
P140A	10080	Secondary air injection (AIR) system bank 1, bank 2 – insufficient flow detected	Hose connection(s), AIR valve, AIR pump
	10750	Secondary air injection (AIR) system bank 1, bank 2 – insufficient flow detected	Hose connection(s), AIR valve, AIR pump
P140A	10081	Secondary air injection (AIR) system bank 1, bank 2 – insufficient flow detected	Hose connection(s), AIR valve, AIR pump
	10760	Secondary air injection (AIR) system bank 1, bank 2 – insufficient flow detected	Hose connection(s), AIR valve, AIR pump
P140B	10080	Secondary air injection (AIR) system bank 1, bank 2 – insufficient flow detected	Hose connection(s), AIR valve, AIR pump
	10080	Secondary air injection (AIR) system bank 1, bank 2 – insufficient flow detected	Hose connection(s), AIR valve, AIR pump
P140C	10081	Secondary air injection (AIR) system bank 1, bank 2 – insufficient flow detected	Hose connection(s), AIR valve, AIR pump
	10760	Secondary air injection (AIR) system bank 1, bank 2 – insufficient flow detected	Hose connection(s), AIR valve, AIR pump
P1413	35	Secondary air injection (AIR) pump relay – signal low	Wiring, connector, AIR pump relay
P1413	10084	Secondary air injection (AIR) pump relay – signal low	Wiring, connector, AIR pump relay
	10755	Secondary air injection (AIR) pump relay – signal low	Wiring, connector, AIR pump relay
P1414	35	Secondary air injection (AIR) pump relay – signal high	Wiring, connector, AIR pump relay
P1414	10084	Secondary air injection (AIR) pump relay – signal high	Wiring, connector, AIR pump relay
	10755	Secondary air injection (AIR) pump relay – signal high	Wiring, connector, AIR pump relay
P1423	171	Secondary air injection (AIR) system – insufficient flow detected	Hose connection(s), AIR valve, AIR pump
P1432	171	Secondary air injection (AIR) system – incorrect flow detected	Hose connection(s), AIR valve, AIR pump
P1434	10775	Diagnostic module tank leakage	Wiring, diagnostic module
P1447	10775	Diagnostic module tank leakage – pump current too high during test	Wiring, diagnostic module
P1448	10775	Diagnostic module tank leakage – pump current too low	Wiring, diagnostic module

OBD-II code	BMW type	Fault location	Probable cause
P1449	10775	Diagnostic module tank leakage – pump current too high	**Wiring, diagnostic module**
P1500	211	Idle speed control (ISC) actuator – stuck open	**Idle speed control (ISC) actuator**
P1501	211	Idle speed control (ISC) actuator – stuck closed	**Idle speed control (ISC) actuator**
P1502	2	Idle speed control (ISC) actuator closing solenoid – signal high	**Wiring, connector(s), ISC actuator closing**
P1502	27	Idle speed control (ISC) actuator closing solenoid – signal high	**Wiring, connector(s), ISC actuator closing solenoid, ECM**
P1503	2	Idle speed control (ISC) actuator closing solenoid – signal low	**Wiring, connector(s), ISC actuator closing**
P1503	27	Idle speed control (ISC) actuator closing solenoid – signal low	**Wiring, connector(s), ISC actuator closing solenoid, ECM**
P1504	2	Idle speed control (ISC) actuator closing solenoid – open circuit	**Wiring, connector(s), ISC actuator closing**
P1504	27	Idle speed control (ISC) actuator closing solenoid – open circuit	**Wiring, connector(s), ISC actuator closing solenoid, ECM**
P1506	29	Idle speed control (ISC) actuator opening solenoid – signal high	**Wiring, connector(s), ISC actuator opening**
P1506	53	Idle speed control (ISC) actuator opening solenoid – signal high	**Wiring, connector(s), ISC actuator opening solenoid, ECM**
P1507	29	Idle speed control (ISC) actuator opening solenoid – signal low	**Wiring, connector(s), ISC actuator opening solenoid, ECM**
P1507	53	Idle speed control (ISC) actuator opening solenoid – signal low	**Wiring, connector(s), ISC actuator opening solenoid, ECM**
P1508	29	Idle speed control (ISC) actuator opening solenoid – open circuit	**Wiring, connector(s), ISC actuator opening solenoid, ECM**
P1508	53	Idle speed control (ISC) actuator opening solenoid – open circuit	**Wiring, connector(s), ISC actuator opening solenoid, ECM**
P1509	29	Idle speed control (ISC) actuator – opening malfunction	**Wiring, ISC actuator opening solenoid**
P1512	124	Differential intake manifold circuit – signal low	**Wiring, ECM**
P1513	124	Differential intake manifold circuit – signal high	**Wiring, ECM**
P1515	12160	Engine off timer – incorrect	**Wiring, ECM**
P151F	12110	CAN data bus, vehicle speed sensor – timeout	**Wiring**
P152A	12111	Vehicle speed sensor (VSS) – speed too low compared to reference value	**Wiring, VSS sensor**
P152B	12111	Vehicle speed sensor (VSS) – speed too low compared to reference value	**Wiring, VSS sensor**
P1525	67	'A' Camshaft position (CMP) actuator bank 1 – open circuit	**Wiring, CMP actuator**

OBD-II code	BMW type	Fault location	Probable cause
P1525	72	'A' Camshaft position (CMP) actuator bank 1 – open circuit	Wiring, CMP actuator
P1526	74	'A' Camshaft position (CMP) actuator bank 2 – open circuit	Wiring, CMP actuator
P1526	75	'A' Camshaft position (CMP) actuator bank 2 – open circuit	Wiring, CMP actuator
P1531	21	'B' Camshaft position (CMP) actuator bank 1 – open circuit	Wiring, CMP actuator
P1531	22	'B' Camshaft position (CMP) actuator bank 1 – open circuit	Wiring, CMP actuator
P1532	83	'B' Camshaft position (CMP) actuator bank 2 – open circuit	Wiring, CMP actuator
P1532	84	'B' Camshaft position (CMP) actuator bank 2 – open circuit	Wiring, CMP actuator
P1550	2	Idle speed control (ISC) actuator – closing malfunction	Wiring, ISC actuator
P1552	67	'A' Camshaft position (CMP) actuator bank 1 – open circuit	Wiring, CMP actuator
P1556	72	AC compressor – signal low	Wiring
P1560	22	'B' Camshaft position (CMP) actuator bank 1 – open circuit	Wiring, CMP actuator
P1565	21	Multifunction steering wheel – interface malfunction	Wiring, buttons '+' and '-' pressed together
P1569	74	'A' Camshaft position (CMP) actuator bank 2 – open circuit	Wiring, CMP actuator
P1573	75	'A' Camshaft position (CMP) actuator bank 2 – signal low	Wiring, CMP actuator
P1581	83	'B' Camshaft position (CMP) actuator bank 2 – open circuit	Wiring, CMP actuator
P1585	149	Random/multiple cylinder(s) – misfire detected	Fuel level low, ignition/fuel system, injector(s), intake leak, mechanical fault
P1594	84	'B' Camshaft position (CMP) actuator bank 2 – open circuit	Wiring, CMP actuator
P164A	11891	Knock sensor (KS) – timeout	Wiring, KS sensor
P1602	48	ECM self test – defective	ECM
P1602	122	ECM self test – defective	ECM
P1602	231	ECM self test – defective	ECM
P1603	49	ECM self test – torque monitoring	Wiring, ECM
P1604	50	ECM self test – speed monitoring	Wiring, ECM
P1611	129	Transmission control module – communication malfunction	Wiring, TCM, ECM
P1611	10505	Transmission control module – communication malfunction	Wiring, TCM, ECM

OBD-II code	BMW type	Fault location	Probable cause
P1619	123	Map cooling thermostat control circuit – signal low	Wiring, map cooling thermostat
P1620	123	Map cooling thermostat control circuit – signal high	Wiring, map cooling thermostat
P1624	146	Accelerator pedal position (APP) sensor 1 – circuit malfunction	Wiring, APP sensor, ECM
P1625	147	Accelerator pedal position (APP) sensor 2 – circuit malfunction	Wiring, APP sensor, ECM
P1628	10089	Throttle valve actuator – spring test malfunction bank 1	Wiring, throttle valve actuator, ECM
	11521	Throttle valve actuator – spring test malfunction bank 1	Wiring, throttle valve actuator, ECM
P1629	10089	Throttle valve actuator – spring test malfunction bank 1	Wiring, throttle valve actuator, ECM
	11521	Throttle valve actuator – spring test malfunction bank 1	Wiring, throttle valve actuator, ECM
P1631	10133	Throttle valve actuator – spring test malfunction bank 1	Wiring, throttle valve actuator, ECM
	11520	Throttle valve actuator – spring test malfunction bank 1	Wiring, throttle valve actuator, ECM
P1632	115	Throttle valve – adaptation failure	Wiring, throttle valve adaptation, throttle motor position sensor, ECM
P1633	115	Throttle valve adaptation – limp-home position incorrect	Wiring, throttle valve adaptation, throttle motor position sensor, ECM
P1634	115	Throttle valve adaptation – spring test failed bank 1	Wiring, mechanical fault, ECM
P1634	10133	Throttle valve adaptation – spring test failed bank 1	Wiring, mechanical fault, ECM
	11520	Throttle valve adaptation – spring test failed bank 1	Wiring, mechanical fault, ECM
P1635	115	Throttle valve adaptation – mechanical stop not adapted	Wiring, throttle valve adaptation, throttle motor position sensor, ECM
P1636	109	Throttle valve control circuit bank 1 – circuit malfunction	Wiring, throttle motor position sensor, ECM
P1636	10132	Throttle valve control circuit bank 1 – circuit malfunction	Wiring, throttle motor position sensor, ECM
P1637	162	Throttle position control – control deviation bank 1	Wiring, throttle motor position sensor, ECM
P1637	10130	Throttle position control – control deviation bank 1	Wiring, throttle motor position sensor, ECM
	11505	Throttle position control – control deviation bank 1	Wiring, throttle motor position sensor, ECM
P1638	160	Throttle position control – throttle stuck temporarily bank 1	Wiring, mechanical fault, ECM

OBD-II code	BMW type	Fault location	Probable cause
P1638	10131	Throttle position control – throttle stuck temporarily bank 1	Wiring, mechanical fault, ECM
	11504	Throttle position control – throttle stuck temporarily bank 1	Wiring, mechanical fault, ECM
P1639	161	Throttle position control – throttle stuck permanently bank 1	Wiring, mechanical fault, ECM
P1639	10131	Throttle position control – throttle stuck permanently bank 1	Wiring, mechanical fault, ECM
	11504	Throttle position control – throttle stuck permanently bank 1	Wiring, mechanical fault, ECM
P1640	155	Engine control module (ECM) – RAM/ ROM malfunction	ECM
P1640	156	Engine control module (ECM) – RAM/ ROM malfunction	ECM
P1653	134	Transmission torque – signal incorrect	Wiring, TCM
P1654	134	CAN data bus signal – timeout	Wiring, ECM
P1663	155	Immobilizer control module – incorrect coding	Key learning procedure
P1663	156	Immobilizer control module – incorrect coding	Key learning procedure
P1670	134	Transmission intervention – incorrect signal	Wiring, TCM
P1675	135	Throttle valve actuator, start test – adaptation required	Throttle valve actuator adaptation
P1694	136	Throttle valve actuator, start test – spring test and limp-home position failed	Wiring, mechanical fault, ECM
P3012	10308	Heated oxygen sensor (HO2S) 1, bank 1 – adaptation value too high	Wiring, HO2S, ECM
	11339	Heated oxygen sensor (HO2S) 1, bank 1 – adaptation value too high	Wiring, HO2S, ECM
P3013	10291	Heated oxygen sensor (HO2S) 1, bank 2 – adaptation value too high	Wiring, HO2S, ECM
	11340	Heated oxygen sensor (HO2S) 1, bank 2 – adaptation value too high	Wiring, HO2S, ECM
P3014	10308	Heated oxygen sensor (HO2S) 1, bank 1 – voltage malfunction	Wiring, HO2S, ECM
	11339	Heated oxygen sensor (HO2S) 1, bank 1 – voltage malfunction	Wiring, HO2S, ECM
P3015	10291	Heated oxygen sensor (HO2S) 1, bank 2 – voltage malfunction	Wiring, HO2S, ECM
	11340	Heated oxygen sensor (HO2S) 1, bank 2 – voltage malfunction	Wiring, HO2S, ECM

OBD-II code	BMW type	Fault location	Probable cause
P3016	10013	Heated oxygen sensor (HO2S) 1, bank 1 – calibration malfunction	**Wiring, HO2S, ECM**
	11424	Heated oxygen sensor (HO2S) 1, bank 1 – calibration malfunction	**Wiring, HO2S, ECM**
P3017	10005	Heated oxygen sensor (HO2S) 1, bank 2 – calibration malfunction	**Wiring, HO2S, ECM**
	11425	Heated oxygen sensor (HO2S) 1, bank 2 – calibration malfunction	**Wiring, HO2S, ECM**
P3018	10313	Heated oxygen sensor (HO2S) 1, bank 1 – lambda control malfunction	**Wiring, HO2S**
	11341	Heated oxygen sensor (HO2S) 1, bank 1 – lambda control malfunction	**Wiring, HO2S**
P3019	10292	Heated oxygen sensor (HO2S) 1, bank 2 – lambda control malfunction	**Wiring, HO2S**
	11342	Heated oxygen sensor (HO2S) 1, bank 2 – lambda control malfunction	**Wiring, HO2S**
P3020	10313	Heated oxygen sensor (HO2S) 1, bank 1 – signal voltage malfunction	**Wiring, HO2S**
	11341	Heated oxygen sensor (HO2S) 1, bank 1 – signal voltage malfunction	**Wiring, HO2S**
P3021	10292	Heated oxygen sensor (HO2S) 1, bank 2 – signal voltage malfunction	**Wiring, HO2S**
	11342	Heated oxygen sensor (HO2S) 1, bank 2 – signal voltage malfunction	**Wiring, HO2S**
P3022	10308	Heated oxygen sensor (HO2S) 1, bank 1 – communication malfunction	**Wiring, HO2S**
	11339	Heated oxygen sensor (HO2S) 1, bank 1 – communication malfunction	**Wiring, HO2S**
P3023	10291	Heated oxygen sensor (HO2S) 1, bank 2 – communication malfunction	**Wiring, HO2S**
	11340	Heated oxygen sensor (HO2S) 1, bank 2 – communication malfunction	**Wiring, HO2S**
P3024	10308	Heated oxygen sensor (HO2S) 1, bank 1 – initialization malfunction	**Wiring, HO2S**
	11339	Heated oxygen sensor (HO2S) 1, bank 1 – initialization malfunction	**Wiring, HO2S**
P3025	10291	Heated oxygen sensor (HO2S) 1, bank 2 – initialization malfunction	**Wiring, HO2S**
	11340	Heated oxygen sensor (HO2S) 1, bank 2 – initialization malfunction	**Wiring, HO2S**
P3026	10013	Heated oxygen sensor (HO2S) 1, bank 1 – operating temperature malfunction	**Wiring, HO2S, ECM**
	11424	Heated oxygen sensor (HO2S) 1, bank 1 – operating temperature malfunction	**Wiring, HO2S, ECM**

OBD-II code	BMW type	Fault location	Probable cause
P3027	10005	Heated oxygen sensor (HO2S) 1, bank 2 – operating temperature malfunction	Wiring, HO2S, ECM
	11425	Heated oxygen sensor (HO2S) 1, bank 2 – operating temperature malfunction	Wiring, HO2S, ECM
P3028	10013	Heated oxygen sensor (HO2S) 1, bank 1 – no activity	Wiring, HO2S, ECM
	11424	Heated oxygen sensor (HO2S) 1, bank 1 – no activity	Wiring, HO2S, ECM
P3029	10005	Heated oxygen sensor (HO2S) 1, bank 2 – no activity	Wiring, HO2S, ECM
	11425	Heated oxygen sensor (HO2S) 1, bank 2 – no activity	Wiring, HO2S, ECM
P3200	52615	Transmission control module – CAN chip defective	TCM
P3201	52615	Transmission control module – CAN chip defective	TCM
P3202	52615	Transmission control module – CAN chip defective	TCM
P3203	52619	Local CAN data bus – malfunction	Wiring, connected modules
P3204	52619	Local CAN data bus – malfunction	Wiring, connected modules
P3205	52619	Local CAN data bus – malfunction	Wiring, connected modules
P320D	10621	CAN data bus monitoring – ETC timeout	Wiring, connected modules
P320D	11722	CAN data bus monitoring – ETC timeout	Wiring, connected modules
	11723	CAN data bus monitoring – ETC timeout	Wiring, connected modules
P321E	12151	Ambient pressure sensor – maximum pressure incorrect	Wiring, ambient pressure sensor
P321F	12151	Ambient pressure sensor – minimum pressure incorrect	Wiring, ambient pressure sensor
P3213	10505	CAN data bus monitoring – ETC check malfunction	Wiring, connected modules
	11722	CAN data bus monitoring – ETC check malfunction	Wiring, connected modules
P3213	10621	CAN data bus monitoring – ETC check malfunction	Wiring, connected modules
	11723	CAN data bus monitoring – ETC check malfunction	Wiring, connected modules
P3214	10505	CAN data bus monitoring – ETC check malfunction	Wiring, connected modules
	17722	CAN data bus monitoring – ETC check malfunction	Wiring, connected modules
P322A	12151	Ambient pressure sensor – open circuit	Wiring, ambient pressure sensor
P323C	12151	Ambient pressure sensor – comparison current incorrect	Wiring, ambient pressure sensor

Model:	**X5 3.0L/4.4L/4.6L/4.8L (E53) • 3 Series 2.5L/3.0L (E46)** **5 Series 2.5L/3.0L/4.4L (E39) • 5 Series M5 5.0L (E39)** **5 Series 4.4L (E60) • 7 Series 4.4L (E65/E66)**
Year:	**2000-04**
Engine identification:	**25 6S 5/6, 30 6S 3, 44 8S 1/2, 46 8S 1, 50 8S 1, N62 B44, N62 B48**
System:	**Siemens MS43 • Siemens MS45 • Siemens MS S52 • Motronic ME 7.2** **Motronic ME 9.2.1**

Data link connector (DLC) locations

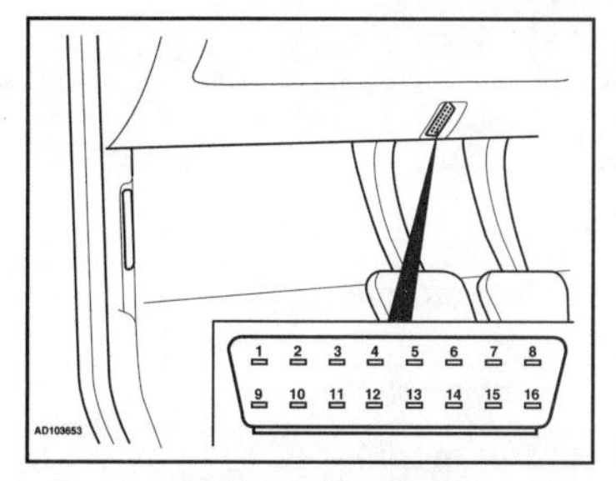

3 Series (E46) – driver's footwell

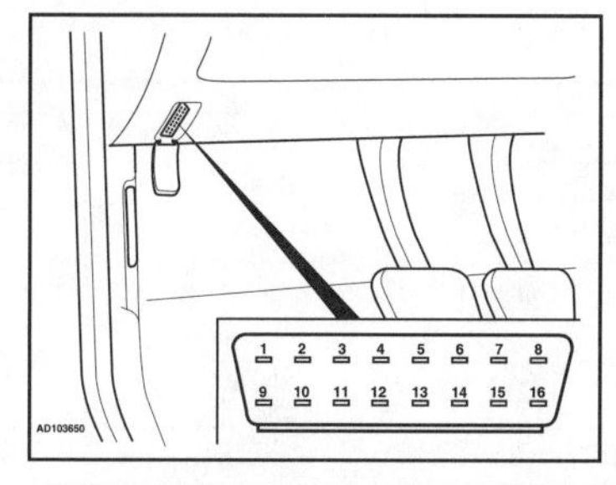

5 Series (E39) – driver's footwell

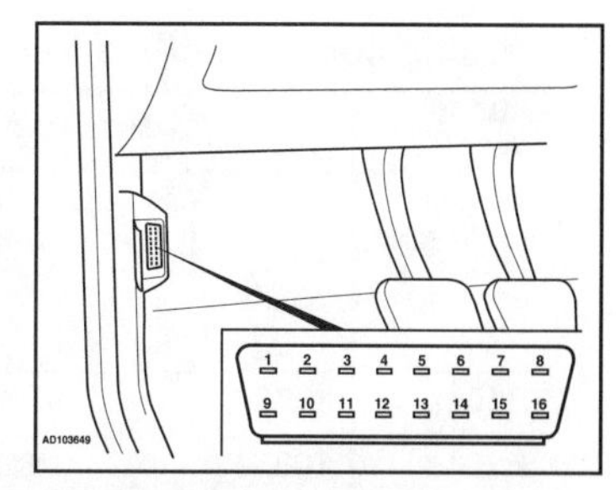

5 Series (E60) – behind cover, 'A' pillar driver's side

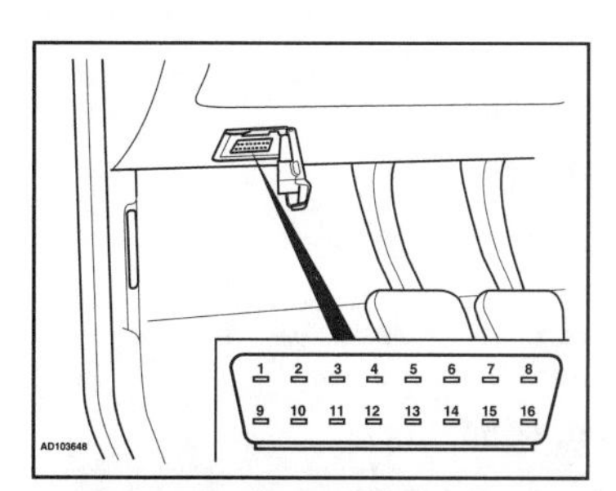

X5 (E53)/7 Series (E65/E66) – driver's footwell

Self-diagnosis

General information

- Refer to the front of this manual for general test conditions, terminology, detailed descriptions of wiring faults and a general trouble shooter for electrical and mechanical faults.
- Engine control module (ECM) incorporates self-diagnosis function.
- Malfunction indicator lamp (MIL) will illuminate if certain faults are recorded.
- ECM operates in backup mode if sensors fail, to enable vehicle to be driven to workshop.

Accessing

- The engine control module (ECM) fault memory can only be accessed and erased using diagnostic equipment connected to the data link connector (DLC).

Erasing

- After the faults have been rectified, trouble codes can only be erased by using a Scan Tool connected to the data link connector (DLC).

Trouble code identification

OBD-II code	BMW type	Fault location	Probable cause
All P0, P2 and U0 codes	–	Refer to OBD-II trouble code tables at the front of this manual	–
P1041	**10334**	VVT control module, bank 1 – EEPROM error	**VVT control module**
P1042	**10334**	VVT control module, bank 1 – RAM error	**VVT control module**
P1043	**10334**	VVT control module, bank 1 – ROM error	**VVT control module**
P1044	**10335**	VVT control module, bank 2 – EEPROM error	**VVT control module**
P1045	**10335**	VVT control module, bank 2 – RAM error	**VVT control module**
P1046	**10335**	VVT control module, bank 2 – ROM error	**VVT control module**
P1047	**10336**	VVT control circuit, bank 1 – high input	**Wiring, connections**
P1048	**10336**	VVT control circuit, bank 1 – low input	**Wiring, connections**
P1049	**10336**	VVT control circuit cables, bank 1 – short circuit	**Wiring, connections**
P1050	**10336**	VVT control circuit, bank 1 – circuit malfunction	**Wiring, connections, VVT control module**
P1051	**10337**	VVT control circuit, bank 2 – high input	**Wiring, connections**
P1052	**10337**	VVT control circuit, bank 2 – low input	**Wiring, connections**
P1053	**10337**	VVT control circuit cables, bank 2 – short circuit	**Wiring, connections**
P1054	**10337**	VVT control circuit, bank 2 – circuit malfunction	**Wiring, connections, VVT control module**
P1055	**10338**	VVT control motor, bank 1 – supply voltage high input	**Wiring, connections**
P1056	**10338**	VVT control motor, bank 1 – supply voltage low input	**Wiring, connections**

OBD-II code	BMW type	Fault location	Probable cause
P1057	10338	VVT control motor, bank 1 – supply voltage circuit malfunction	**Wiring, connections**
P1058	10339	VVT control motor, bank 2 – supply voltage high input	**Wiring, connections**
P1059	10339	VVT control motor, bank 2 – supply voltage low input	**Wiring, connections**
P1060	10339	VVT control motor, bank 2 – supply voltage circuit malfunction	**Wiring, connections**
P1061	10341	VVT limp home request, bank 1 – circuit malfunction	**Wiring, connections, ECM**
P1062	10341	VVT limp home request, bank 1 – full stroke position not reached	**Wiring, connections, ECM**
P1063	10341	VVT limp home request, bank 1 – air mass signal malfunction	**Wiring, connections, mass air sensor, ECM**
P1064	10342	VVT starting position/parking position comparison – circuit malfunction	**Wiring, connections**
P1065	10227	VVT timeout, no signal – CAN data bus malfunction	**Wiring, connections**
P1066	10227	VVT message monitoring – circuit malfunction	**Wiring, connections**
P1071	10334	VVT control module, bank 1 – internal malfunction	**VVT control module**
P1072	10335	VVT control module, bank 2 – internal malfunction	**VVT control module**
P1083	202	Heated oxygen sensor (HO2S) 1, bank 1 – range/performance problem	**Wiring, intake/exhaust leak, fuel level low, fuel pressure, evaporative emission (EVAP) canister purge valve, injector(s), HO2S, ECM**
P1084	202	Heated oxygen sensor (HO2S) 1, bank 1 – range/performance problem	**Wiring, intake/exhaust leak, fuel level low, fuel pressure, evaporative emission (EVAP) canister purge valve, injector(s), HO2S, ECM**
P1085	203	Heated oxygen sensor (HO2S) 1, bank 2 – range/performance problem	**Wiring, intake/exhaust leak, fuel level low, fuel pressure, evaporative emission (EVAP) canister purge valve, injector(s), HO2S, ECM**
P1086	203	Heated oxygen sensor (HO2S) 1, bank 2 – range/performance problem	**Wiring, intake/exhaust leak, fuel level low, fuel pressure, evaporative emission (EVAP) canister purge valve, injector(s), HO2S, ECM**
P1087	229/ 10015	Heated oxygen sensor (HO2S) 1, bank 1 – slow response lean control	**Wiring, intake/exhaust leak, fuel level low, fuel pressure, EVAP canister purge valve, injector(s), HO2S, ECM**
P1088	229/ 10015	Heated oxygen sensor (HO2S) 1, bank 1 – slow response rich control	**Wiring, intake/exhaust leak, fuel level low, fuel pressure, EVAP canister purge valve, injector(s), HO2S, ECM**

OBD-II code	BMW type	Fault location	Probable cause
P1089	230/ 10021	Heated oxygen sensor (HO2S) 1, bank 2 – slow response lean control	**Wiring, intake/exhaust leak, fuel level low, fuel pressure, EVAP canister purge valve, injector(s), HO2S, ECM**
P1094	230/ 10021	Heated oxygen sensor (HO2S) 1, bank 2 – slow response rich control	**Wiring, intake/exhaust leak, fuel level low, fuel pressure, EVAP canister purge valve, injector(s), HO2S, ECM**
P1100	57	Mass air flow (MAF) sensor – high input	**Wiring, MAF sensor, ECM**
P1111	11/ 10125	Engine coolant temperature (ECT) sensor, radiator outlet – malfunction	**Wiring, connector(s), ECT sensor, ECM**
P1112	11/ 10125	Engine coolant temperature (ECT) sensor, radiator outlet – malfunction	**Wiring, connector(s), ECT sensor, ECM**
P1115	10122	Ambient air temperature sensor – circuit malfunction	**Wiring, air temperature sensor, ECM**
P1117	57	Mass air flow (MAF) sensor/volume air flow (VAF) sensor bank2 – low input	**Wiring, connector(s), MAF sensor/VAF sensor**
P1118	57	Mass air flow (MAF) sensor/volume air flow (VAF) sensor bank 2 – high input	**Wiring, connector(s), MAF sensor/VAF sensor**
P1120	114/ 10247	Accelerator pedal position (APP) sensor – circuit malfunction	**Wiring, APP sensor, ECM**
P1121	117/ 10248	Accelerator pedal position (APP) sensor 1 – circuit malfunction	**Wiring, APP sensor, ECM**
P1122	110/ 10231	Accelerator pedal position (APP) sensor 1 – low input	**Wiring short to ground, APP sensor**
P1123	110/ 10231	Accelerator pedal position (APP) sensor 1 – high input	**Wiring short to positive, APP sensor**
P1130	10017	Heated oxygen sensor (HO2S) 2, bank 1 – dynamic test malfunction	**Wiring, HO2S, ECM**
P1131	10023	Heated oxygen sensor (HO2S) 2, bank 2 – dynamic test malfunction	**Wiring, HO2S, ECM**
P1134	13/ 25/37	Heated oxygen sensor (HO2S) 1, bank 1 – signal malfunction	**Wiring, HO2S, ECM**
P1135	13/ 25/37	Heated oxygen sensor (HO2S) 1, bank 1 – low voltage	**Wiring, HO2S, ECM**
P1136	13/ 25/37	Heated oxygen sensor (HO2S) 1, bank 1 – high voltage	**Wiring, HO2S, ECM**
P1137	14/79	Heated oxygen sensor (HO2S) 2, bank 1 – signal malfunction	**Wiring, HO2S, ECM**
P1138	14/79	Heated oxygen sensor (HO2S) 2, bank 1 – low voltage	**Wiring, HO2S, ECM**
P1139	14/79	Heated oxygen sensor (HO2S) 2, bank 1 – high voltage	**Wiring, HO2S, ECM**
P1140	73	Mass air flow (MAF) sensor/volume air flow (VAF) sensor – incorrect signal	**Wiring, connector(s), MAF sensor/VAF sensor**
P1143	215	Heated oxygen sensor (HO2S) 2, bank 1 – signal too high	**Wiring short to positive, catalytic converter, fuel pressure high, HO2S**

OBD-II code	BMW type	Fault location	Probable cause
P1144	215	Heated oxygen sensor (HO2S) 2, bank 1 – signal too low	**Wiring short to ground, poor connection, exhaust leak, fuel pressure low, HO2S**
P1149	216	Heated oxygen sensor (HO2S) 2, bank 2 – signal too high	**Wiring short to positive, catalytic converter, fuel pressure high, HO2S**
P1150	216	Heated oxygen sensor (HO2S) 2, bank 2 – signal too low	**Wiring short to ground, poor connection, exhaust leak, fuel pressure low, HO2S**
P1151	5/38/55	Heated oxygen sensor (HO2S) 1, bank 2 – signal malfunction	**Wiring, HO2S, ECM**
P1152	5/38/55	Heated oxygen sensor (HO2S) 1, bank 2 – low voltage	**Wiring, HO2S, ECM**
P1153	5/38/55	Heated oxygen sensor (HO2S) 1, bank 2 – high voltage	**Wiring, HO2S, ECM**
P1155	4/61	Heated oxygen sensor (HO2S) 2, bank 2 – signal malfunction	**Wiring, HO2S, ECM**
P1156	4/61	Heated oxygen sensor (HO2S) 2, bank 2 – low voltage	**Wiring, HO2S, ECM**
P1157	4/61	Heated oxygen sensor (HO2S) 2, bank 2 – high voltage	**Wiring, HO2S, ECM**
P1158	28	Fuel trim (FT), bank 1 – range/ performance problem	**Wiring, connectors, intake/fuel system, injectors, HO2S, MAF sensor, ECT sensor, EVAP canister purge valve, ECM**
P1159	28	Fuel trim (FT), bank 1 – range/ performance problem	**Wiring, connectors, intake/fuel system, injectors, HO2S, MAF sensor, ECT sensor, EVAP canister purge valve, ECM**
P1160	29	Fuel trim (FT), bank 2 – range/ performance problem	**Wiring, connectors, intake/fuel system, injectors, HO2S, MAF sensor, ECT sensor, EVAP canister purge valve, ECM**
P1161	29	Fuel trim (FT), bank 2 – range/ performance problem	**Wiring, connectors, intake/fuel system, injectors, HO2S, MAF sensor, ECT sensor, EVAP canister purge valve, ECM**
P1168	15	Post catalyst fuel trim (FT), bank 1 – range/performance problem	**Wiring, connectors, intake/fuel system, injectors, HO2S, MAF sensor, ECT sensor, EVAP canister purge valve, ECM**
P1170	21	Post catalyst fuel trim (FT), bank 2 – range/performance problem	**Wiring, connectors, intake/fuel system, injectors, HO2S, MAF sensor, ECT sensor, EVAP canister purge valve, ECM**
P1171	196	Ambient pressure sensor – recognition value incorrect	**Wiring, ambient pressure sensor, ECM**
P1172	196	Ambient pressure sensor – recognition value error	**Wiring, ambient pressure sensor, ECM**
P1173	196	Ambient pressure sensor – recognition learning failed	**Learning procedure not performed**
P1178	216	Heated oxygen sensor (HO2S) 1, bank 1 – slow switching	**Wiring, HO2S, ECM**

OBD-II code	BMW type	Fault location	Probable cause
P1179	217	Heated oxygen sensor (HO2S) 1, bank 2 – slow switching	**Wiring, HO2S, ECM**
P1184	218	Heated oxygen sensor (HO2S) 1, bank 1 – no activity detected	**Wiring, HO2S**
P1185	219	Heated oxygen sensor (HO2S) 1, bank 2 – no activity detected	**Wiring, HO2S**
P1188	174	Fuel trim (FT) sensor 1, bank 1 – range/ performance problem	**Wiring, connectors, intake/fuel system, injectors, HO2S, MAF sensor, ECT sensor, EVAP canister purge valve, ECM**
P1189	175	Fuel trim (FT) sensor 1, bank 2 – range/ performance problem	**Wiring, connectors, intake/fuel system, injectors, HO2S, MAF sensor, ECT sensor, EVAP canister purge valve, ECM**
P1200	24	Fuel trim (FT) upper adaptation, bank 1 – range/performance problem	**Wiring, connectors, intake/fuel system, injectors, HO2S, MAF sensor, ECT sensor, EVAP canister purge valve, ECM**
P1201	24	Fuel trim (FT) upper adaptation, bank 1 – range/performance problem	**Wiring, connectors, intake/fuel system, injectors, HO2S, MAF sensor, ECT sensor, EVAP canister purge valve, ECM**
P1202	25	Fuel trim (FT) upper adaptation, bank 2 – range/performance problem	**Wiring, connectors, intake/fuel system, injectors, HO2S, MAF sensor, ECT sensor, EVAP canister purge valve, ECM**
P1203	25	Fuel trim (FT) upper adaptation, bank 2 – range/performance problem	**Wiring, connectors, intake/fuel system, injectors, HO2S, MAF sensor, ECT sensor, EVAP canister purge valve, ECM**
P1204	10277	Heated oxygen sensor (HO2S) 2, bank 1 – circuit malfunction	**Wiring, HO2S sensor**
P1205	10278	Heated oxygen sensor (HO2S) 2, bank 2 – circuit malfunction	**Wiring, HO2S sensor**
P1222	111/ 10232	Accelerator pedal position (APP) sensor, track 2 – low input	**Wiring short to ground, APP sensor**
P1223	111/ 10232	Accelerator pedal position (APP) sensor, track 2 – high input	**Wiring short to positive, APP sensor**
P1247	10461	Barometric pressure – signal malfunction	**Wiring, ECM**
P1314	202/ 203	Fuel mixture – deviation with low fuel	**Wiring, fuel level low, fuel pressure, intake/exhaust leak, injector(s), HO2S, EVAP canister purge valve, ECT sensor, ECM**
P1314	227/ 228	Fuel mixture – deviation with low fuel	**Wiring, fuel level low, fuel pressure, intake/exhaust leak, injector(s), HO2S, EVAP canister purge valve, ECT sensor, ECM**
P1315	10086	Camshaft position (CMP) sensor, sensor 'A' bank 1 – signal malfunction	**Wiring, CMP sensor, ECM**
P1316	10086	Camshaft position (CMP) sensor, sensor 'A' bank 1 – signal malfunction	**Wiring, CMP sensor, ECM**
P1317	15	'B' Camshaft position (CMP) actuator – range/performance problem	**Wiring, CMP actuator**

OBD-II code	BMW type	Fault location	Probable cause
P1318	10087	Camshaft position (CMP) sensor, sensor 'B' bank 1 – signal malfunction	**Wiring, CMP sensor, ECM**
P1319	10087	Camshaft position (CMP) sensor, sensor 'B' bank 1 – signal malfunction	**Wiring, CMP sensor, ECM**
P1326	10422	Camshaft position (CMP) sensor, sensor 'A' bank 1 – timing reference out of range	**Wiring, CMP sensor, ECM**
P1327	59/211/ 10211	Knock sensor (KS) 2, bank 1 – low input	**Wiring, KS incorrectly tightened, KS**
P1328	211/ 10211	Knock sensor (KS) 2, bank 1 – high input	**Wiring, KS incorrectly tightened, KS**
P1329	10212	Knock sensor (KS) 3 – low input	**Wiring, KS incorrectly tightened, KS**
P1330	10212	Knock sensor (KS) 3 – high input	**Wiring, KS incorrectly tightened, KS**
P1331	10424	Camshaft position (CMP) sensor, sensor 'B' bank 1 – timing reference out of range	**Wiring, CMP sensor, ECM**
P1332	213/ 10213	Knock sensor (KS) 4 – low input	**Wiring, KS incorrectly tightened, KS**
P1333	213/ 10213	Knock sensor (KS) 4 – high input	**Wiring, KS incorrectly tightened, KS**
P1338	10088	Camshaft position (CMP) sensor, sensor 'A' bank 1 – phase position malfunction	**Wiring, CMP sensor, ECM**
P1339	10092	Camshaft position (CMP) sensor, sensor 'B' bank 1 – phase position malfunction	**Wiring, CMP sensor, ECM**
P1340	62/213	Multiple cylinder(s) misfire – during start	**Wiring, fuel level low, ignition/fuel system, injector(s), spark plug(s), ignition coil(s), mechanical fault**
P1341	62/204	Multiple cylinder(s) misfire – with fuel cut-off	**Wiring, fuel level low, ignition/fuel system, intake leak, mechanical fault**
P1342	50/205	Cylinder misfire during start – cylinder 1	**Wiring, fuel level low, ignition/fuel system, injector, spark plug, ignition coil, mechanical fault**
P1342	238/ 10050	Cylinder misfire during start – cylinder 1	**Wiring, fuel level low, ignition/fuel system, injector, spark plug, ignition coil, mechanical fault**
P1343	50/196	Cylinder misfire with fuel cut-off – cylinder 1	**Wiring, fuel level low, ignition/fuel system, intake leak, mechanical fault**
P1343	238/ 10050	Cylinder misfire with fuel cut-off – cylinder 1	**Wiring, fuel level low, ignition/fuel system, intake leak, mechanical fault**
P1344	57/206	Cylinder misfire during start – cylinder 2	**Wiring, fuel level low, ignition/fuel system, injector, spark plug, ignition coil, mechanical fault**
P1344	239/ 10054	Cylinder misfire during start – cylinder 2	**Wiring, fuel level low, ignition/fuel system, injector, spark plug, ignition coil, mechanical fault**

OBD-II code	BMW type	Fault location	Probable cause
P1345	57/197	Cylinder misfire with fuel cut-off – cylinder 2	Wiring, fuel level low, ignition/fuel system, intake leak, mechanical fault
P1345	239/ 10054	Cylinder misfire with fuel cut-off – cylinder 2	Wiring, fuel level low, ignition/fuel system, intake leak, mechanical fault
P1346	55/207	Cylinder misfire during start – cylinder 3	Wiring, fuel level low, ignition/fuel system, injector, spark plug, ignition coil, mechanical fault
P1346	240/ 10052	Cylinder misfire during start – cylinder 3	Wiring, fuel level low, ignition/fuel system, injector, spark plug, ignition coil, mechanical fault
P1347	55/198	Cylinder misfire with fuel cut-off – cylinder 3	Wiring, fuel level low, ignition/fuel system, intake leak, mechanical fault
P1347	240/ 10052	Cylinder misfire with fuel cut-off – cylinder 3	Wiring, fuel level low, ignition/fuel system, intake leak, mechanical fault
P1348	52/208	Cylinder misfire during start – cylinder 4	Wiring, fuel level low, ignition/fuel system, injector, spark plug, ignition coil, mechanical fault
P1348	241/ 10055	Cylinder misfire during start – cylinder 4	Wiring, fuel level low, ignition/fuel system, injector, spark plug, ignition coil, mechanical fault
P1349	52/199	Cylinder misfire with fuel cut-off – cylinder 4	Wiring, fuel level low, ignition/fuel system, intake leak, mechanical fault
P1349	241/ 10055	Cylinder misfire with fuel cut-off – cylinder 4	Wiring, fuel level low, ignition/fuel system, intake leak, mechanical fault
P1350	51/209	Cylinder misfire during start – cylinder 5	Wiring, fuel level low, ignition/fuel system, injector, spark plug, ignition coil, mechanical fault
P1350	242/ 10051	Cylinder misfire during start – cylinder 5	Wiring, fuel level low, ignition/fuel system, injector, spark plug, ignition coil, mechanical fault
P1351	51/200	Cylinder misfire with fuel cut-off – cylinder 5	Wiring, fuel level low, ignition/fuel system, intake leak, mechanical fault
P1351	242/ 10051	Cylinder misfire with fuel cut-off – cylinder 5	Wiring, fuel level low, ignition/fuel system, intake leak, mechanical fault
P1352	54/210	Cylinder misfire during start – cylinder 6	Wiring, fuel level low, ignition/fuel system, injector, spark plug, ignition coil, mechanical fault
P1352	243/ 10053	Cylinder misfire during start – cylinder 6	Wiring, fuel level low, ignition/fuel system, injector, spark plug, ignition coil, mechanical fault
P1353	54/201	Cylinder misfire with fuel cut-off – cylinder 6	Wiring, fuel level low, ignition/fuel system, intake leak, mechanical fault
P1353	243/ 10053	Cylinder misfire with fuel cut-off – cylinder 6	Wiring, fuel level low, ignition/fuel system, intake leak, mechanical fault
P1354	56/211	Cylinder misfire during start – cylinder 7	Wiring, fuel level low, ignition/fuel system, injector, spark plug, ignition coil, mechanical fault

OBD-II code	BMW type	Fault location	Probable cause
P1355	56/202	Cylinder misfire with fuel cut-off – cylinder 7	**Wiring, fuel level low, ignition/fuel system, intake leak, mechanical fault**
P1356	53/212	Cylinder misfire during start – cylinder 8	**Wiring, fuel level low, ignition/fuel system, injector, spark plug, ignition coil, mechanical fault**
P1357	53/203	Cylinder misfire with fuel cut-off – cylinder 8	**Wiring, fuel level low, ignition/fuel system, intake leak, mechanical fault**
P1378	10238	ECM self-test – Bank 2 knock control malfunction	**Wiring, KS, ECM**
P1379	10239	ECM self-test – Bank 2 knock control malfunction	**Wiring, KS, ECM**
P1380	10217	ECM self-test – Bank 2 knock control malfunction	**Wiring, KS, ECM**
P1381	215/ 10215	ECM self-test – Bank 1 knock control malfunction	**Wiring, KS, ECM**
P1382	216/ 10216	ECM self-test – Bank 1 knock control malfunction	**Wiring, KS, ECM**
P1386	10214	ECM self-test – Bank 1 knock control malfunction	**Wiring, KS, ECM**
P140A	10080/ 10081	Secondary air injection (AIR) system bank 1, bank 2 – insufficient flow detected	**Hose connection(s), AIR valve, AIR pump**
P140B	10080	Secondary air injection (AIR) system bank 1, bank 2 – insufficient flow detected	**Hose connection(s), AIR valve, AIR pump**
P140C	10081	Secondary air injection (AIR) system bank 1, bank 2 – insufficient flow detected	**Hose connection(s), AIR valve, AIR pump**
P1413	35/84/ 10084	Secondary air injection (AIR) pump relay – signal low	**Wiring, connector, AIR pump relay**
P1414	35/84/ 10084	Secondary air injection (AIR) pump relay – signal high	**Wiring, connector, AIR pump relay**
P1415	10080	Mass air flow (MAF) sensor/volume air flow (VAF) sensor – circuit malfunction	**Wiring, connector(s), MAF sensor/VAF sensor**
P1429	10201	Diagnostic module tank leakage – heater	**Wiring, diagnostic module**
P1430	10201	Diagnostic module tank leakage – heater low	**Wiring, diagnostic module**
P1431	10201	Diagnostic module tank leakage – heater high	**Wiring, diagnostic module**
P1432	171	Secondary air injection (AIR) system – incorrect flow detected	**Hose connection(s), AIR valve, AIR pump**
P1434	10189	Diagnostic module tank leakage	**Wiring, diagnostic module**
P1444	10186	Diagnostic module tank leakage – pump control open circuit	**Wiring, diagnostic module**
P1445	10186	Diagnostic module tank leakage – pump control signal low	**Wiring, diagnostic module**
P1446	10186	Diagnostic module tank leakage – pump control signal high	**Wiring, diagnostic module**

OBD-II code	BMW type	Fault location	Probable cause
P1447	10189	Diagnostic module tank leakage – pump current too high during solenoid test	Wiring, diagnostic module
P1448	10189	Diagnostic module tank leakage – pump current too low	Wiring, diagnostic module
P1449	10189/ 10775	Diagnostic module tank leakage – pump current too high	Wiring, diagnostic module
P1449	10002	Diagnostic module tank leakage – solenoid control open circuit	Wiring, diagnostic module
P1451	10002	Diagnostic module tank leakage – solenoid control signal low	Wiring, diagnostic module
P1452	10002	Diagnostic module tank leakage – solenoid control signal high	Wiring, diagnostic module
P1453	10084	Secondary air injection (AIR) system – electrical fault	Wiring, connectors, AIR pump relay, AIR pump
P1500	211	Idle speed control (ISC) actuator – stuck open	Idle speed control (ISC) actuator
P1501	211	Idle speed control (ISC) actuator – stuck closed	Idle speed control (ISC) actuator
P1502	2/27/ 10198	Idle speed control (ISC) actuator closing solenoid – signal high	Wiring, connector(s), ISC actuator closing solenoid, ECM
P1503	2/27/ 10198	Idle speed control (ISC) actuator closing solenoid – signal low	Wiring, connector(s), ISC actuator closing solenoid, ECM
P1504	2/27/ 10198	Idle speed control (ISC) actuator closing solenoid – open circuit	Wiring, connector(s), ISC actuator closing solenoid, ECM
P1506	29/53/ 10199	Idle speed control (ISC) actuator opening solenoid – signal high	Wiring, connector(s), ISC actuator opening solenoid, ECM
P1507	29/53/ 10199	Idle speed control (ISC) actuator opening solenoid – signal low	Wiring, connector(s), ISC actuator opening solenoid, ECM
P1508	29/53/ 10199	Idle speed control (ISC) actuator opening solenoid – open circuit	Wiring, connector(s), ISC actuator opening solenoid, ECM
P1509	29	Idle speed control (ISC) actuator – opening malfunction	Wiring, ISC actuator opening solenoid
P1510	10461	Idle speed control valve – stuck	Wiring, mechanical fault
P1511	10270	Differential intake manifold circuit – electrical malfunction	Wiring, ECM
P1512	124/ 10270	Differential intake manifold circuit – signal low	Wiring, ECM
P1513	124/ 10270	Differential intake manifold circuit – signal high	Wiring, ECM
P1523	21/165	'A' Camshaft position (CMP) actuator bank 1 – signal low	Wiring, CMP actuator
P1524	21/165	'A' Camshaft position (CMP) actuator bank 1 – signal high	Wiring, CMP actuator
P1525	21/ 67/72	'A' Camshaft position (CMP) actuator bank 1 – open circuit	Wiring, CMP actuator

OBD-II code	BMW type	Fault location	Probable cause
P1525	165/ 10165	'A' Camshaft position (CMP) actuator bank 1 – open circuit	Wiring, CMP actuator
P1526	74/75/ 166	'A' Camshaft position (CMP) actuator bank 2 – open circuit	Wiring, CMP actuator
P1527	166	'A' Camshaft position (CMP) actuator bank 2 – signal low	Wiring, CMP actuator
P1528	166	'A' Camshaft position (CMP) actuator bank 2 – signal high	Wiring, CMP actuator
P1529	19	'B' Camshaft position (CMP) actuator bank 1 – signal low	Wiring, CMP actuator
P1530	19	'B' Camshaft position (CMP) actuator bank 1 – signal high	Wiring, CMP actuator
P1531	21/22	'B' Camshaft position (CMP) actuator bank 1 – open circuit	Wiring, CMP actuator
P1531	19/ 10173	'B' Camshaft position (CMP) actuator bank 1 – open circuit	Wiring, CMP actuator
P1532	83/84	'B' Camshaft position (CMP) actuator bank 2 – open circuit	Wiring, CMP actuator
P1550	2	Idle speed control (ISC) actuator – closing malfunction	Wiring, ISC actuator
P1552	67	'A' Camshaft position (CMP) actuator bank 1 – open circuit	Wiring, CMP actuator
P1556	72	AC compressor – signal low	Wiring
P1560	22	'B' Camshaft position (CMP) actuator bank 1 – open circuit	Wiring, CMP actuator
P1565	21	Multifunction steering wheel – interface malfunction	Wiring, buttons '+' and '-' pressed together
P1569	74	'A' Camshaft position (CMP) actuator bank 2 – open circuit	Wiring, CMP actuator
P1573	75	'A' Camshaft position (CMP) actuator bank 2 – signal low	Wiring, CMP actuator
P1581	83	'B' Camshaft position (CMP) actuator bank 2 – open circuit	Wiring, CMP actuator
P1585	149	Random/multiple cylinder(s) – misfire detected	Fuel level low, ignition/fuel system, injector(s), intake leak, mechanical fault
P1594	84	'B' Camshaft position (CMP) actuator bank 2 – open circuit	Wiring, CMP actuator
P16A0	10288	Internal control module – checksum error	Internal control module
P16A1	10288	Internal control module – checksum error	Internal control module
P16A2	10288	Internal control module – checksum error	Internal control module
P16A3	10345	Internal control module – memory error	Internal control module
P16A4	10346	Time out control module – knock sensor bus	Wiring, KS sensor, time out control module
P16A5	10347	Time out control module – output stage bus	Wiring, time out control module

OBD-II code	BMW type	Fault location	Probable cause
P16A6	10401	ECM self test – cruise control monitoring	Wiring, ECM
P16A7	10402	ECM self test – mass air flow meter monitoring	Wiring, mass air flow meter, ECM
P16A8	10402	ECM self test – throttle position monitoring	Wiring, throttle position (TP) sensor, ECM
P16A9	10404	ECM self test – speed monitoring reset	Wiring, ECM
P16B0	10405	ECM self test – pedal position monitoring	Wiring, accelerator pedal position (APP) sensor, ECM
P16B1	10410	ECM self test – Idle air control system error	Wiring, idle air control system, ECM
P16B2	10410	ECM self test – Idle air control system error	Wiring, idle air control system, ECM
P16B3	10411	ECM self test – engine/drag/torque control monitoring	Wiring, ECM
P16B4	10411	ECM self test – cruise distance control monitoring	Wiring, ECM
P16B5	10411	ECM self test – automatic manual transmission monitoring	Wiring, ECM
P16B6	10411	ECM self test – ETC monitoring	Wiring, ECM
P16B7	10412	ECM self test – clutch torque monitoring error	Wiring, ECM
P16B8	10412	ECM self test – clutch torque monitoring error	Wiring, ECM
P16B9	10412	ECM self test – torque loss monitoring	Wiring, ECM
P16C0	10412	ECM self test – driving dynamics control monitoring	Wiring, ECM
P16C1	10413	ECM self test – torque monitoring error	Wiring, ECM
P16C2	10417	ECM self test – speed limitation monitoring	Wiring, ECM
P16C3	10418	ECM self test – speed limitation error	Wiring, ECM
P1602	48/122/231	ECM self test – defective	ECM
P1603	49	ECM self test – torque monitoring	Wiring, ECM
P1604	50/51	ECM self test – speed monitoring	Wiring, ECM
P1611	129/10220/10505	Transmission control module – communication malfunction	Wiring, TCM, ECM
P1618	10103	ECM self test – AD-converter monitoring	Wiring, ECM
P1619	123/10140	Map cooling thermostat control circuit – signal low	Wiring, map cooling thermostat
P1620	123/10140	Map cooling thermostat control circuit – signal high	Wiring, map cooling thermostat

OBD-II code	BMW type	Fault location	Probable cause
P1622	10140	Map cooling thermostat control circuit – circuit malfunction	**Wiring, map cooling thermostat**
P1624	146/ 10048	Accelerator pedal position (APP) sensor 1 – circuit malfunction	**Wiring, APP sensor, ECM**
P1625	147/ 10049	Accelerator pedal position (APP) sensor 2 – circuit malfunction	**Wiring, APP sensor, ECM**
P1628	10089	Throttle valve actuator – spring test malfunction bank 1	**Wiring, throttle valve actuator, ECM**
P1629	10089	Throttle valve actuator – spring test malfunction bank 1	**Wiring, throttle valve actuator, ECM**
P1631	10133	Throttle valve actuator – spring test malfunction bank 1	**Wiring, throttle valve actuator, ECM**
P1632	115/ 10134	Throttle valve – adaptation failure	**Wiring, throttle valve adaptation, throttle motor position sensor, ECM**
P1633	115/ 10134	Throttle valve adaptation – limp-home position incorrect	**Wiring, throttle valve adaptation, throttle motor position sensor, ECM**
P1634	115/ 135/ 136	Throttle valve adaptation – spring test failed bank 1	**Wiring, mechanical fault, ECM**
P1634	10133/ 10134/ 10145	Throttle valve adaptation – spring test failed bank 1	**Wiring, mechanical fault, ECM**
P1635	115/ 10134/ 10419	Throttle valve adaptation – mechanical stop not adapted	**Wiring, throttle valve adaptation, throttle motor position sensor, ECM**
P1636	109/ 132/ 10132	Throttle valve control circuit bank 1 – circuit malfunction	**Wiring, throttle motor position sensor, ECM**
P1637	130/ 162	Throttle position control – control deviation bank 1	**Wiring, throttle motor position sensor, ECM**
P1637	10066/ 10130	Throttle position control – control deviation bank 1	**Wiring, throttle motor position sensor, ECM**
P1638	131/ 160	Throttle position control – throttle stuck temporarily bank 1	**Wiring, mechanical fault, ECM**
P1638	10064/ 10131	Throttle position control – throttle stuck temporarily bank 1	**Wiring, mechanical fault, ECM**
P1639	131/ 161	Throttle position control – throttle stuck permanently bank 1	**Wiring, mechanical fault, ECM**
P1639	10065/ 10131	Throttle position control – throttle stuck permanently bank 1	**Wiring, mechanical fault, ECM**
P1640	155/ 156	Engine control module (ECM) – RAM/ ROM malfunction	**ECM**
P1653	134	Transmission torque – signal incorrect	**Wiring, TCM**
P1654	134	CAN data bus signal – timeout	**Wiring, ECM**
P1663	155/ 156	Immobilizer control module – incorrect coding	**Key learning procedure**

OBD-II code	BMW type	Fault location	Probable cause
P1670	134	Transmission intervention – incorrect signal	Wiring, TCM
P1675	135/ 10149	Throttle valve actuator, start test – adaptation required	Throttle valve actuator adaptation
P1694	136/ 10145	Throttle valve actuator, start test – spring test and limp-home position failed	Wiring, mechanical fault, ECM
P3010	10486	Heated oxygen sensor (HO2S) 2, bank 1 – low input after cold start	Wiring, HO2S, ECM
P3011	10487	Heated oxygen sensor (HO2S) 2, bank 2 – low input after cold start	Wiring, HO2S, ECM
P3012	10308	Heated oxygen sensor (HO2S) 1, bank 1 – adaptation value too high	Wiring, HO2S, ECM
P3013	10291	Heated oxygen sensor (HO2S) 1, bank 2 – adaptation value too high	Wiring, HO2S, ECM
P3014	10308	Heated oxygen sensor (HO2S) 1, bank 1 – voltage malfunction	Wiring, HO2S, ECM
P3015	10291	Heated oxygen sensor (HO2S) 1, bank 2 – voltage malfunction	Wiring, HO2S, ECM
P3016	10013	Heated oxygen sensor (HO2S) 1, bank 1 – calibration malfunction	Wiring, HO2S, ECM
P3017	10005	Heated oxygen sensor (HO2S) 1, bank 2 – calibration malfunction	Wiring, HO2S, ECM
P3018	10313	Heated oxygen sensor (HO2S) 1, bank 1 – lambda control malfunction	Wiring, HO2S
P3019	10292	Heated oxygen sensor (HO2S) 1, bank 2 – lambda control malfunction	Wiring, HO2S
P3020	10313	Heated oxygen sensor (HO2S) 1, bank 1 – signal voltage malfunction	Wiring, HO2S
P3021	10292	Heated oxygen sensor (HO2S) 1, bank 2 – signal voltage malfunction	Wiring, HO2S
P3022	10308	Heated oxygen sensor (HO2S) 1, bank 1 – communication malfunction	Wiring, HO2S
P3023	10291	Heated oxygen sensor (HO2S) 1, bank 2 – communication malfunction	Wiring, HO2S
P3024	10308	Heated oxygen sensor (HO2S) 1, bank 1 – initialization malfunction	Wiring, HO2S
P3025	10291	Heated oxygen sensor (HO2S) 1, bank 2 – initialization malfunction	Wiring, HO2S
P3026	10013	Heated oxygen sensor (HO2S) 1, bank 1 – operating temperature malfunction	Wiring, HO2S, ECM
P3027	10005	Heated oxygen sensor (HO2S) 1, bank 2 – operating temperature malfunction	Wiring, HO2S, ECM
P3028	10013	Heated oxygen sensor (HO2S) 1, bank 1 – no activity	Wiring, HO2S, ECM

OBD-II code	BMW type	Fault location	Probable cause
P3029	10005	Heated oxygen sensor (HO2S) 1, bank 2 – no activity	Wiring, HO2S, ECM
P3040	10480	Heated oxygen sensor (HO2S) 2, bank 1 – lean/rich voltage malfunction	Wiring, HO2S, ECM
P3041	10481	Heated oxygen sensor (HO2S) 2, bank 2 – lean/rich voltage malfunction	Wiring, HO2S, ECM
P3198	10205	Engine coolant temperature – malfunction	Wiring, ECT sensor, ECM
P3199	10207	Engine coolant temperature – signal stuck	Wiring, ECT sensor, ECM
P320D	10621	CAN data bus monitoring – ETC timeout	Wiring, connected modules
P3200	52615	Transmission control module – CAN chip defective	TCM
P3201	52615	Transmission control module – CAN chip defective	TCM
P3202	52615	Transmission control module – CAN chip defective	TCM
P3203	52619	Local CAN data bus – malfunction	Wiring, connected modules
P3204	52619	Local CAN data bus – malfunction	Wiring, connected modules
P3205	52619	Local CAN data bus – malfunction	Wiring, connected modules
P3213	10505/ 10621	CAN data bus monitoring – ETC check malfunction	Wiring, connected modules
P3214	10505	CAN data bus monitoring – ETC check malfunction	Wiring, connected modules
P3238	10289	ECM self test – TPU malfunction	Wiring, TPU, ECM

BMW

Model:	**3 Series/Compact (E36) • 3 Series (E46) • 5 Series (E39) 7 Series (E38) • Z3 • 8 Series (E31) • X5 4.4L (E53)**
Year:	**1992-02**
Engine identification:	**19 4E/S 1, 20 6S 2/3/4, 25 6S 3/4, 28 6S 2, 35 8S 1/2, 35 8S 2, 44 8S 1/2, 54 12 1/2**
System:	**Siemens MS40/41/42 • Bosch Motronic M5.2 • Bosch BMS46 • Bosch ME7.2**

Data link connector (DLC) locations

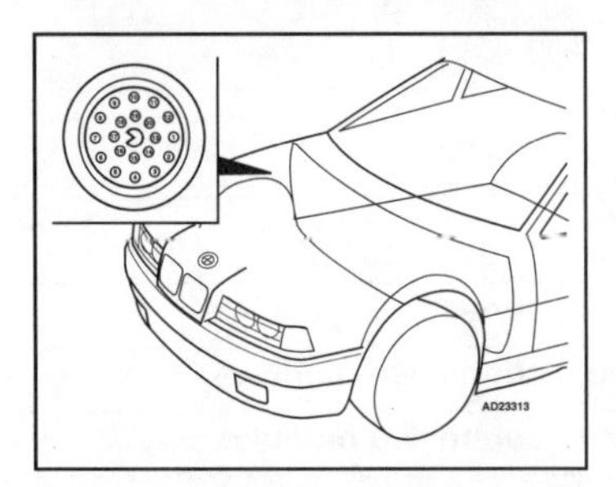

Z3, 3 Series, 5 Series, 7 Series, 8 Series – engine bay RH side

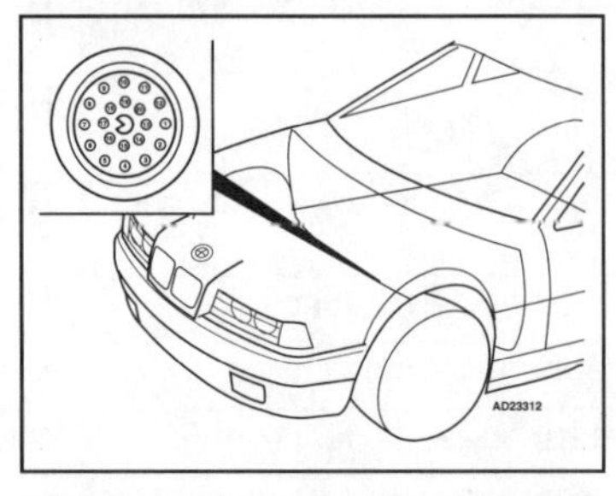

3 Series Compact, X5 – engine bay LH side

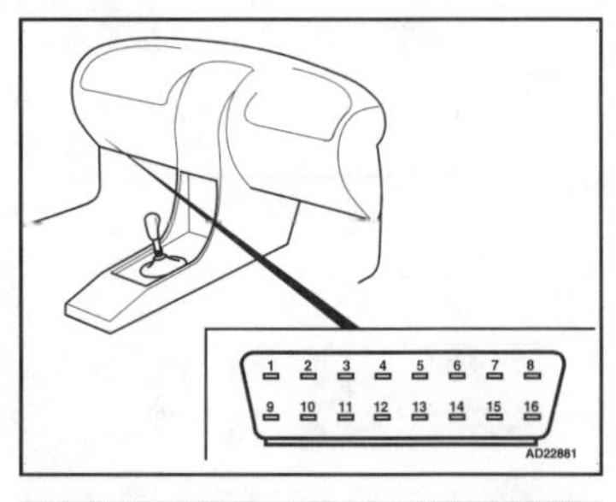

3 Series (E46), Z3, 5 Series (E39), X5 – 2000 →

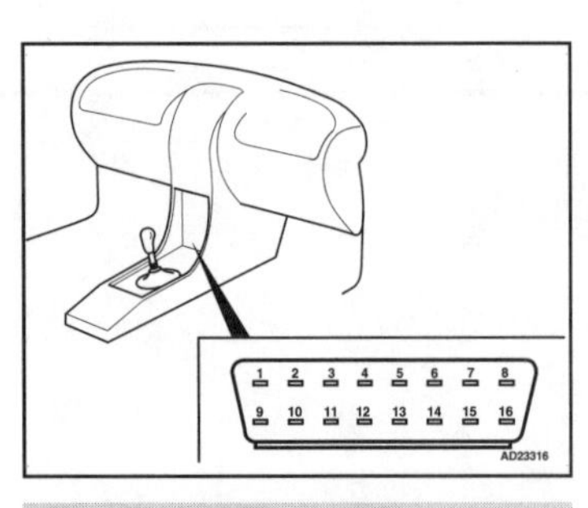

7 Series – 2000 →

Self-diagnosis

General information

- Refer to the front of this manual for general test conditions, terminology, detailed descriptions of wiring faults and a general trouble shooter for electrical and mechanical faults.

Accessing and erasing – Siemens MS40/41

- The engine control module (ECM) fault memory can only be accessed and erased using diagnostic equipment connected to the data link connector (DLC).

NOTE: ***Some models may have both DLCs fitted. DO NOT open 20-pin DLC while using 16-pin DLC.***

Trouble code identification

Scan code	Fault location	Probable cause
–	Engine control module (ECM) – data exchange not possible	**Wiring between DLC and ECM, supply voltage, engine control relay, DLC**
0	Engine control module (ECM) – undefined fault	**Disconnect ECM multi-plug for at least 10 min. – check if trouble code reappears before replacing ECM**
1	Ignition circuit, cyl. 1 – defective	**Wiring, connector, ignition coil, ECM**
2	MS41: Ignition circuit, cyl. 2 – defective	**Wiring, connector, ignition coil, ECM**
	Ignition circuit, cyl. 3 – defective	**Wiring, connector, ignition coil, ECM**
3	MS41: Ignition circuit, cyl. 4 – defective	**Wiring, connector, ignition coil, ECM**
	Ignition circuit, cyl. 5 – defective	**Wiring, connector, ignition coil, ECM**
5	MS41: Ignition circuit, cyl. 6 – defective	**Wiring, connector, ignition coil, ECM**
	Injector, cyl. 6 – circuit malfunction	**Wiring, injector**
6	MS41: Injector, cyl. 2 – circuit malfunction	**Wiring, injector**
	Injector, cyl. 4 – circuit malfunction	**Wiring, injector**
8	MS41: Injector, cyl. 1 – circuit malfunction	**Wiring, injector**
	Mass air flow (MAF) sensor/volume air flow (VAF) sensor – circuit malfunction	**Wiring, connector(s), MAF sensor/VAF sensor**
10	AC control module/ECM communication – malfunction	**Wiring, AC control module**
11	MS41: Engine coolant temperature (ECT) sensor – malfunction	**Wiring, connector(s), ECT sensor**
	Evaporative emission (EVAP) pressure sensor – malfunction	**Wiring, EVAP pressure sensor**
12	Vehicle speed sensor (VSS) – incorrect/no signal	**Wiring, instrument panel, VSS**
14	MS41: Throttle position (TP) sensor – defective	**Wiring, connector(s), TP sensor**
	ECM/transmission control module (TCM) communication – short to ground	**Wiring, ECM, TCM**
16	MS41: Intake air temperature (IAT) sensor – malfunction	**Wiring, connector(s), IAT sensor**
	AC control module – signal	**Wiring, AC condenser blower motor relay, AC refrigerant high pressure switch, AC refrigerant low pressure switch, AC control module**
18	ECM/immobilizer control module communication – signal	**Wiring, immobilizer control module**
20	Malfunction indicator lamp (MIL) – circuit malfunction	**Wiring, bulb, ECM**
21	Camshaft position (CMP) actuator – malfunction	**Wiring, connector, CMP actuator**
22	Injector, cyl. 3 – circuit malfunction	**Wiring, injector**

Scan code	Fault location	Probable cause
23	Injector, cyl. 1 – circuit malfunction	**Wiring, injector**
24	MS41: Injector, cyl. 6 – circuit malfunction	**Wiring, injector**
	AC compressor clutch relay – malfunction	**Wiring, AC compressor clutch relay**
25	MS41: Injector, cyl. 4 – circuit malfunction	**Wiring, injector**
	Oxygen sensor heater 1 – circuit malfunction	**Wiring, connector, engine control relay, HO2S, ECM**
27	Idle speed control (ISC) actuator/idle air control (IAC) valve – malfunction	**Wiring, connector(s), ISC actuator/IAC valve, ECM**
29	Ignition circuit, cyl. 2 – defective	**Wiring, connector, ignition coil, ECM**
30	MS41: Ignition circuit, cyl. 1 – defective	**Wiring, connector, ignition coil, ECM**
	Ignition circuit, cyl. 4 – defective	**Wiring, connector, ignition coil, ECM**
31	MS41: Ignition circuit, cyl. 3 – defective	**Wiring, connector, ignition coil, ECM**
	Ignition circuit, cyl. 6 – defective	**Wiring, connector, ignition coil, ECM**
33	MS41: Ignition circuit, cyl. 5 – defective	**Wiring, connector, ignition coil, ECM**
	Injector, cyl. 5 – circuit malfunction	**Wiring, injector**
35	Secondary air injection (AIR) pump relay – malfunction	**Wiring, connector, AIR pump relay**
46	Fuel level signal	**Fuel level low, wiring, fuel gauge tank sensor**
49	Battery voltage – too low	**Wiring, alternator, battery**
50	Injector, cyl. 2 – circuit malfunction	**Wiring, injector**
51	MS41: Evaporative emission (EVAP) canister purge valve – malfunction	**Wiring, connector(s), EVAP canister purge valve**
	Evaporative emission (EVAP) canister purge valve – malfunction	**Wiring, connector(s), EVAP canister purge valve**
52	Fuel pump relay – malfunction	**Wiring, connector(s), fuel pump relay**
53	MS41: Exhaust gas control solenoid – malfunction	**Wiring, connector, exhaust gas control solenoid**
	Oxygen sensor heater – circuit malfunction	**Wiring, connector, HO2S**
55	MS41: Idle speed control (ISC) actuator/idle air control (IAC) valve – malfunction	**Wiring, connector(s), ISC actuator/IAC valve, ECM**
	Oxygen sensor heater 2 – circuit malfunction	**Wiring, connector, engine control relay, HO2S, ECM**
56	Ignition signal resistor – defective	**Wiring, ignition signal resistor**
57	Knock sensor (KS) 1 – defective	**Wiring, connector, KS incorrectly tightened, KS**
59	Knock sensor (KS) 2 – defective	**Wiring, connector, KS incorrectly tightened, KS**
62	Ignition circuit – secondary circuit	**Wiring, ignition coil(s)**

Scan code	Fault location	Probable cause
63	MS41: Secondary air injection (AIR) solenoid – defective	**Wiring, connector, AIR solenoid**
	Knock sensor (KS) 2 – defective	**Wiring, connector, KS incorrectly tightened, KS**
64	Knock sensor (KS) 1 – defective	**Wiring, connector, KS incorrectly tightened, KS**
65	Camshaft position (CMP) sensor – circuit malfunction	**Wiring, connector, CMP sensor**
68	Mass air flow (MAF) sensor/volume air flow (VAF) sensor – incorrect signal	**Wiring, connector(s), MAF sensor/VAF sensor**
69	MS41: Evaporative emission (EVAP) canister purge valve – electrical fault	**Wiring, connector, EVAP canister purge valve**
	Fuel pump relay – malfunction	**Wiring, connector(s), fuel pump relay**
74	AC compressor clutch relay – malfunction	**Wiring, AC compressor clutch relay**
75	Heated oxygen sensor (HO2S) – circuit malfunction	**Wiring, connector(s), HO2S**
76	MS41: Heated oxygen sensor (HO2S) 1 – circuit malfunction	**Wiring, connector(s), HO2S, ECM**
	Heated oxygen sensor (HO2S) 2 – circuit malfunction	**Wiring, connector(s), HO2S, ECM**
77	Throttle position (TP) sensor – defective	**Wiring, connector(s), TP sensor**
79	Crankshaft position (CKP) sensor/engine speed (RPM) sensor – incorrect/no signal	**Wiring, CKP sensor/RPM sensor**
80	Engine control module (ECM)/ABS control module communication, ASR system	**Wiring, electrical interference, ABS trouble code(s) stored, ECM**
81	Engine coolant temperature (ECT) sensor – malfunction	**Wiring, connector(s), ECT sensor**
82	MS41: Engine control module (ECM)/ABS control module communication, ASR system	**Wiring, electrical interference, ABS trouble code(s) stored, ECM**
	Engine control module (ECM)/ABS control module communication, ASR system	**Wiring, electrical interference, ABS trouble code(s) stored, ECM**
83	Crankshaft position (CKP) sensor/engine speed (RPM) sensor – circuit malfunction	**Wiring, connector, CKP sensor/RPM sensor**
84	Camshaft position (CMP) sensor – range/ performance problem	**Wiring, CMP sensor**
85	Intake air temperature (IAT) sensor – malfunction	**Wiring, connector(s), IAT sensor**
97	Evaporative emission (EVAP) canister purge valve – mechanical fault	**EVAP canister purge valve**
98	Idle speed control (ISC) actuator/idle air control (IAC) valve – mechanical fault	**ISC actuator/IAC valve**
99	Heated oxygen sensor (HO2S) – range/ performance problem	**Wiring, intake/exhaust leak, fuel level low, fuel pressure, evaporative emission (EVAP) canister purge valve, injector(s), HO2S, ECM**

Scan code	Fault location	Probable cause
100	Engine control module (ECM) – defective	**Disconnect ECM multi-plug for at least 10 min. – check if trouble code reappears before replacing ECM**
200	Engine control module (ECM) – defective	**Disconnect ECM multi-plug for at least 5 min. – check if trouble code reappears before replacing ECM**
201	MS41: Heated oxygen sensor (HO2S) 1 – circuit malfunction	**Wiring, connector(s), HO2S**
	Heated oxygen sensor (HO2S) 2 – circuit malfunction	**Wiring, connector(s), HO2S**
202	Heated oxygen sensor (HO2S), cyl. 1-3 – range/performance problem	**Wiring, intake/exhaust leak, fuel level low, fuel pressure, evaporative emission (EVAP) canister purge valve, injector(s), HO2S, ECM**
203	Heated oxygen sensor (HO2S), cyl. 4-6 – range/performance problem	**Wiring, intake/exhaust leak, fuel level low, fuel pressure, evaporative emission (EVAP) canister purge valve, injector(s), HO2S, ECM**
204	Idle speed control (ISC) actuator/idle air control (IAC) valve – idle speed incorrect	**Intake leak, ISC actuator/IAC valve, TP sensor, throttle valve**
209	Immobilizer control module – incorrect signal	**Wiring, incorrectly coded, immobilizer control module**
210	Misfire detected – more than 2 cylinders	**Wiring, ignition coils, spark plugs, ignition signal resistor, ECM**
211	Idle speed control (ISC) actuator/idle air control (IAC) valve – mechanical fault	**ISC actuator/IAC valve**
212	Camshaft position (CMP) actuator – mechanical fault	**Camshaft timing mechanism, CMP actuator**
214	Vehicle speed sensor (VSS) – signal	**Wiring, ECM/ABS control module communication, electrical interference, ABS trouble code(s) stored, ECM**
215	Engine control module (ECM)/ABS control module communication – plausibility	**Wiring, electrical interference, ABS trouble code(s) stored, ECM**
216	CAN data bus, TCM – signal	**Wiring, transmission range (TR) switch, electrical interference, TCM, ECM**
217	CAN data bus, TCM – no data transmission	**Wiring, electrical interference, TCM**
218	CAN data bus, ECM	**Wiring**
219	CAN data bus, ECM	**Wiring**
222	Lambda regulation – defective	**Engine coolant temperature (ECT) sensor, oxygen sensor heater circuit, HO2S, ECM**
227	Mixture control (MC), cyl. 1-3 – range/performance problem	**Wiring, fuel level low, fuel pressure, intake/ exhaust leak, MAF sensor/VAF sensor, injector(s), HO2S, EVAP canister purge valve, ECT sensor, ECM**
228	Mixture control (MC), cyl. 4-6 – range/performance problem	**Wiring, fuel level low, fuel pressure, intake/ exhaust leak, MAF sensor/VAF sensor, injector(s), HO2S, EVAP canister purge valve, ECT sensor, ECM**

Scan code	Fault location	Probable cause
229	Heated oxygen sensor (HO2S) 1 – range/performance problem	**Wiring, intake/exhaust leak, fuel level low, fuel pressure, EVAP canister purge valve, injector(s), HO2S, ECM**
230	Heated oxygen sensor (HO2S) 2 – range/performance problem	**Wiring, intake/exhaust leak, fuel level low, fuel pressure, EVAP canister purge valve, injector(s), HO2S, ECM**
231	Heated oxygen sensor (HO2S) 1 – malfunction	**Wiring, HO2S**
232	Heated oxygen sensor (HO2S) 2 – malfunction	**Wiring, HO2S**
233	Catalytic converter system, bank 1 – efficiency	**Wiring, HO2S, ECM**
234	Catalytic converter system, bank 2 – efficiency	**Wiring, HO2S, ECM**
238	Combustion failure – cyl. 1	**Wiring, ignition/fuel system, ECM**
239	Combustion failure – cyl. 2	**Wiring, ignition/fuel system, ECM**
240	Combustion failure – cyl. 3	**Wiring, ignition/fuel system, ECM**
241	Combustion failure – cyl. 4	**Wiring, ignition/fuel system, ECM**
242	Combustion failure – cyl. 5	**Wiring, ignition/fuel system, ECM**
243	Combustion failure – cyl. 6	**Wiring, ignition/fuel system, ECM**
244	Engine speed (RPM) sensor/crankshaft position (CKP) sensor – circuit malfunction	**Wiring, connector, RPM sensor/CKP sensor**
245	Secondary air injection (AIR) system – bank 1	**Wiring, connector(s), AIR pump relay, AIR solenoid**
246	Secondary air injection (AIR) system – bank 2	**Wiring, connector(s), AIR pump relay, AIR solenoid**
247	Secondary air injection (AIR) solenoid – mechanical fault	**AIR solenoid**
250	Fuel tank vent valve – malfunction	**Wiring, fuel tank vent valve**
251	Fuel tank vent system – malfunction	**Wiring, hoses, fuel tank vent valve**
252	Fuel tank vent system – malfunction	**Wiring, hoses, fuel tank vent valve**
253	Evaporative emission (EVAP) canister purge valve – mechanical fault	**EVAP canister purge valve**
254	Fuel tank vent system – leak detected	**Hose(s), fuel tank vent valve**
255	Evaporative emission (EVAP) canister purge valve – mechanical fault	**EVAP canister purge valve**

Accessing and erasing – except Siemens MS40/41

- The engine control module (ECM) fault memory can only be accessed and erased using diagnostic equipment connected to the data link connector (DLC).

NOTE: ***Some models may have both DLCs fitted. DO NOT open 20-pin DLC while using 16-pin DLC.***

Trouble code identification

OBD-II code	Fault location	Probable cause
All P0, P2 and U0 codes	Refer to OBD-II trouble code tables at the front of this manual	–
P1140	Mass air flow (MAF) sensor/throttle position (TP) sensor communication – plausibility	**Wiring, MAF sensor/TP sensor, ECM**
P1161	Engine oil temperature (EOT) sensor – range/performance problem	**Wiring, connector, EOT sensor, ECM**
P1174	Fuel trim (FT), bank 1 – range/performance problem	**Wiring, connectors, intake/fuel system, injectors, HO2S, MAF sensor, ECT sensor, EVAP canister purge valve, ECM**
P1176	Heated oxygen sensor (HO2S) 1, bank 1 – slow response	**Wiring, HO2S**
P1186	Heated oxygen sensor (HO2S) 2, bank 2 – heater circuit malfunction	**Wiring, connector, fuel pump relay, HO2S**
P1188	Fuel trim (FT), bank 1 – range/performance problem	**Wiring, connectors, intake/fuel system, injectors, HO2S, MAF sensor, ECT sensor, EVAP canister purge valve, ECM**
P1270	Engine torque difference bank 1 to 2 – malfunction	**Trouble code(s) stored, intake leak, mechanical fault, electronic throttle system (ETS)**
P1383	Ignition circuit – malfunction	**Wiring, connectors, spark plug(s), ignition coil(s)**
P1386	Knock sensor(s) (KS) – circuit malfunction	**Wiring, connector, KS incorrectly tightened, KS**
P1396	Crankshaft position (CKP) sensor/engine speed (RPM) sensor 1 & 2 – faulty signal	**CKP/RPM sensor(s), excessive crankshaft clearance, flywheel ring gear damaged**
P1423	Secondary air injection (AIR) system, bank 1 – insufficient flow detected	**Hose connection(s), AIR valve, AIR pump**
P1453	Secondary air injection (AIR) system, bank 1 – electrical fault	**Wiring, connectors, AIR pump relay, AIR pump**
P1470	Evaporative emission (EVAP) canister purge valve – ECM output stage	**Wiring, EVAP canister purge valve, ECM**
P1475	Evaporative emission (EVAP) system – malfunction	**Wiring, connector, EVAP canister purge valve relay, EVAP canister purge valve, ECM**
P1509	Idle speed control (ISC) actuator – malfunction	**Wiring, ISC actuator**

OBD-II code	Fault location	Probable cause
P1511	Intake manifold air control solenoid – circuit malfunction	**Wiring, connector, intake manifold air control solenoid**
P1519	MS40/41/42/46/M5.2: Inlet camshaft control system, end position – plausibility	**Wiring, connector, camshaft position (CMP) sensor, crankshaft position (CKP) sensor, ECM**
	ME7.2: Inlet camshaft control system, bank 1 – plausibility	**Wiring, connector, mechanical fault, camshaft position (CMP) actuator, camshaft position (CMP) sensor(s), crankshaft position (CKP) sensor, ECM**
P1520	Exhaust camshaft control system, end position – plausibility	**Wiring, connector, camshaft position (CMP) sensor, crankshaft position (CKP) sensor, ECM**
P1522	Inlet camshaft control system, camshaft position – plausibility	**Wiring, connector, camshaft position (CMP) actuator, ECM**
	ME7.2: Inlet camshaft control system, bank 2 – plausibility	**Wiring, connector, mechanical fault, camshaft position (CMP) actuator, camshaft position (CMP) sensor(s), crankshaft position (CKP) sensor, ECM**
P1523	Exhaust camshaft control system, camshaft position – plausibility	**Wiring, connector, camshaft position (CMP) actuator, ECM**
P1525	Camshaft position (CMP) actuator, inlet camshaft – ECM output stage	**Wiring, connector, CMP actuator, ECM**
	ME7.2: Inlet camshaft control system, bank 1 – ECM output stage	**Wiring, connectors, CMP actuator, ECM**
P1526	Inlet camshaft control system, bank 2 – ECM output stage	**Wiring, connectors, CMP actuator, ECM**
P1529	Camshaft position (CMP) actuator, exhaust camshaft – ECM output stage	**Wiring, connector, CMP actuator, ECM**
P1550	Idle speed control (ISC) actuator – malfunction	**Wiring, ISC actuator**
P1585	Random/multiple cylinder(s) – misfire detected	**Fuel level low, ignition/fuel system, injector(s), intake leak, mechanical fault**
P1589	Knock sensor(s) (KS) – circuit malfunction	**Wiring, connector, KS incorrectly tightened, KS**
P1593	Intake manifold air control solenoid – ECM output stage	**Wiring, intake manifold air control solenoid, ECM**
P1622	Engine coolant temperature (ECT) sensor, ECM controlled cooling system – ECM output stage	**Wiring, ECT sensor, ECM**
P1690	Malfunction indicator lamp (MIL) – circuit malfunction	**Wiring, MIL**

BMW

Model:	**318i/325i (E36) • 328i (E36) • 525i (E34) • 750i (E32) • 850i/CSI (E31)**
Year:	**1990-96**
Engine identification:	**16 4E 1, 16 4 E2, 18 4E 1, 18 4S 1, 18 4E 2, 20 6S 2, 25 6S 2, 50 12 A, 56 12 1**
System:	**Motronic M1.2/1.7/1.7.2/1.7.3/3.1**

Data link connector (DLC) locations

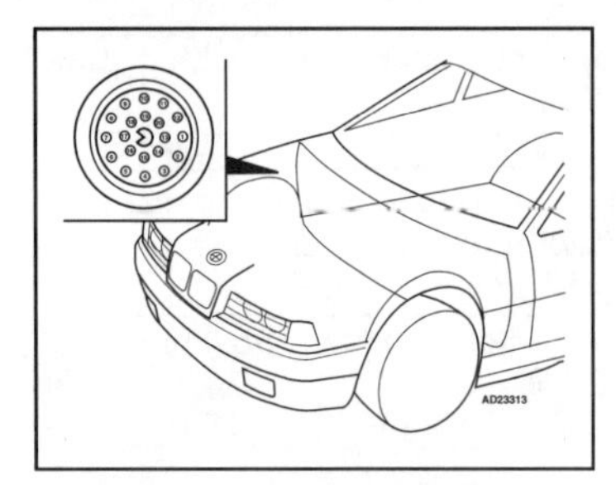

Z3, 3 Series (M1.2/1.7/3.1), 5 Series, 8 Series – engine bay RH side

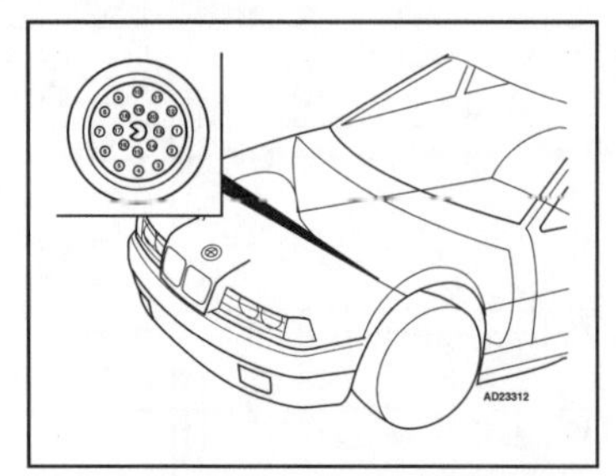

3 Series (M1.7.2/1.7.3), 7 Series – engine bay LH side

Self-diagnosis

General information

- Refer to the front of this manual for general test conditions, terminology, detailed descriptions of wiring faults and a general trouble shooter for electrical and mechanical faults.
- Trouble codes are displayed by using an LED connected to one of the ECM terminals.
- The ECM fault memory can also be checked using diagnostic equipment connected to the data link connector (DLC).

Accessing – flash type

NOTE: ***Flash code cannot be accessed from some engine control modules (ECM).***

- Ensure ignition switched OFF.
- Connect breakout box between ECM and harness multi-plug.
- ECM located in plenum chamber.
- Motronic M1.7/3.1: Connect LED tester between breakout box terminal 8 and battery (12 volt) positive **1**.
- Motronic M1.2: Connect LED tester between breakout box terminal 15 and battery (12 volt) positive **2**.
- Switch ignition ON.
- Depress accelerator pedal fully 5 times within 5 seconds.
- V12: To access trouble codes from ECM LH bank, depress accelerator pedal fully 6 times within 5 seconds.

NOTE: ***V12 engine is controlled by two ECMs, RH bank and LH bank.***

- Count LED flashes. Note trouble codes. Compare with trouble code table.
- Long flash indicates start of trouble code display **3** [**A**].
- Each trouble code consists of four groups of flashes **3** [**B**].

- For example: Trouble code 1211 displayed: Engine control module (ECM) **3** [**B**].
- LED flashes code 0000 **4**, 1000 **5** or 2000 **6** at end of trouble code output.
- Switch ignition OFF.
- Disconnect LED.

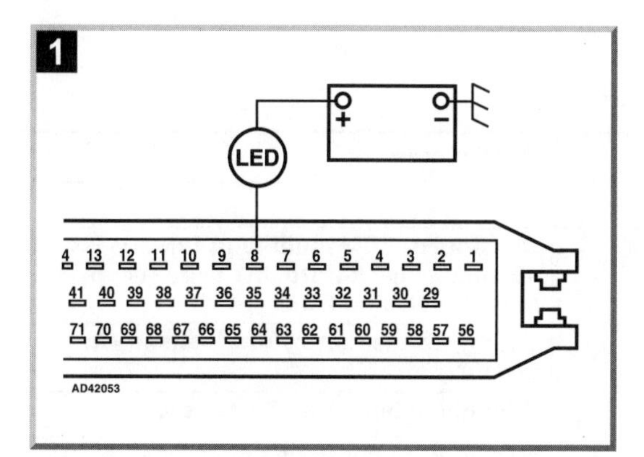

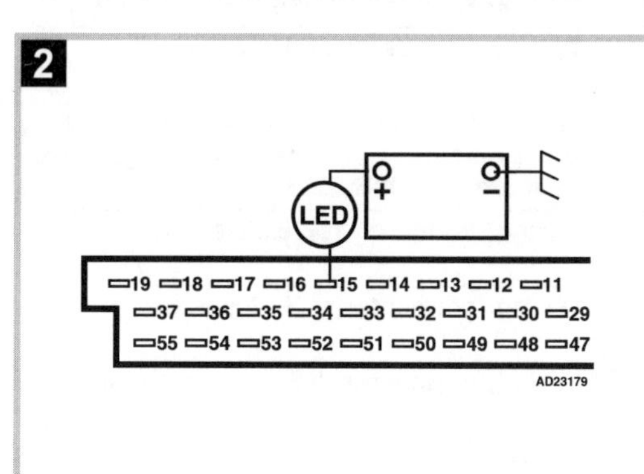

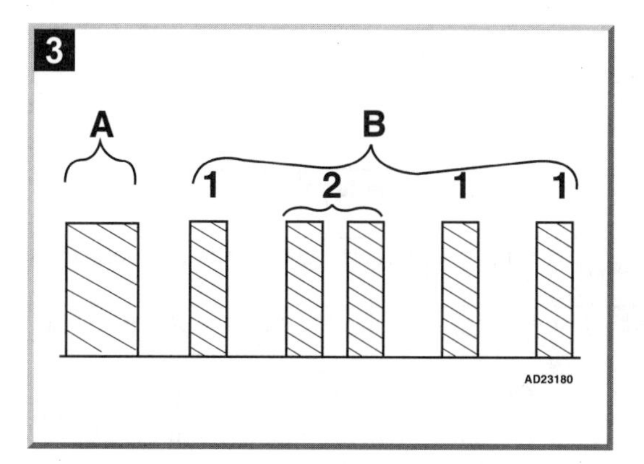

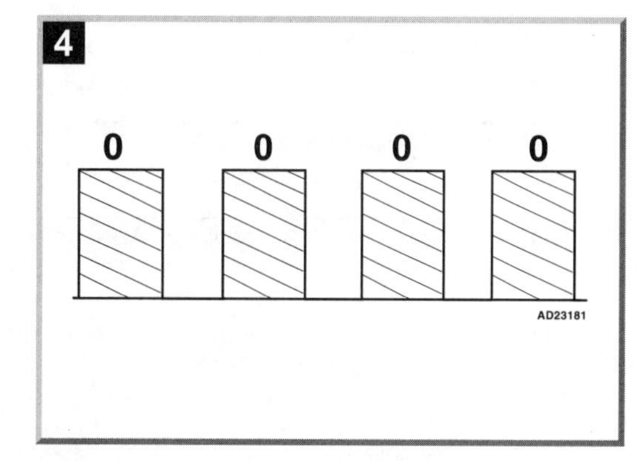

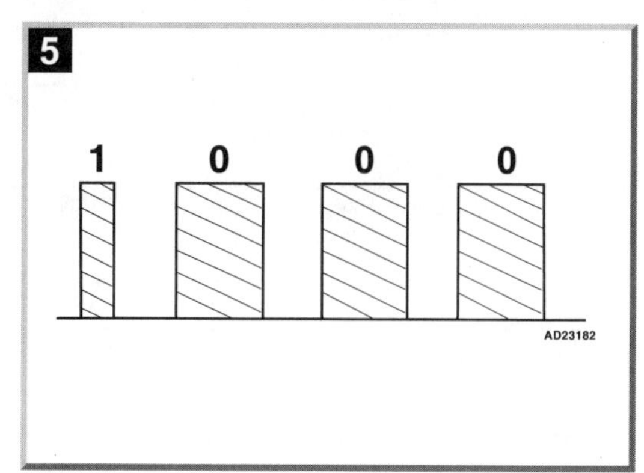

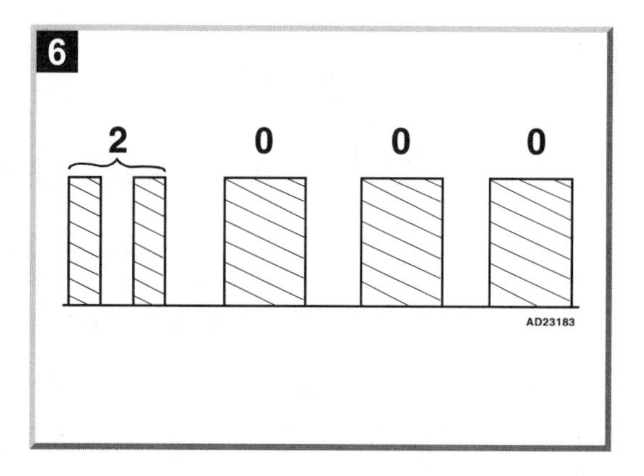

Erasing – flash type

- Depress accelerator pedal fully for 10 seconds minimum when trouble code 0000, 1000 or 2000 (end of output) is flashed.
- Switch ignition OFF.

Trouble code identification

NOTE: V12: Flash codes 1xxx for RH bank, 2xxx LH bank.

Flash code	Fault location	Probable cause
1000	End of trouble code output	–
1211	Engine control module (ECM) – defective	**Disconnect ECM multi-plug for at least 10 min. – check if trouble code reappears before replacing ECM**
1215	Mass air flow (MAF) sensor/volume air flow (VAF) sensor – incorrect signal	**Wiring, connector(s), MAF sensor/VAF sensor**
1216	Throttle position (TP) sensor – defective	**Wiring, connector(s), TP sensor**
1218	Engine control module (ECM) – output stage	**Wiring, ECM controlled components**
1221	Heated oxygen sensor (HO2S) – circuit malfunction	**Wiring, connector(s), HO2S**
1222	Heated oxygen sensor (HO2S) – range/ performance problem	**Wiring, intake/exhaust leak, fuel level low, fuel pressure, EVAP canister purge valve, injector(s), HO2S, ECM**
1223	Engine coolant temperature (ECT) sensor – malfunction	**Wiring, connector(s), ECT sensor**
1224	Intake air temperature (IAT) sensor – malfunction	**Wiring, connector(s), IAT sensor**
1225	Knock sensor (KS) 1 – defective	**Wiring, connector, KS incorrectly tightened**
1226	Knock sensor (KS) 2 – defective	**Wiring, connector, KS incorrectly tightened**
1231	Battery voltage – too low/high	**Wiring, alternator, battery**
1232	Closed throttle position (CTP) switch – malfunction	**Wiring, connector, CTP switch, ECM**
1233	Throttle position (TP) sensor – malfunction	**Wiring, TP sensor, ECM**
1234	Vehicle speed sensor (VSS) – incorrect/no signal	**Wiring, instrument panel, VSS**
1235	Torque converter clutch (TCC) solenoid – defective	**Wiring, TCC solenoid**
1236	Throttle control, traction control system (TCS) – defective	**Wiring, connector(s), throttle control system**
1236 [1]	Engine control module (ECM)/ABS control module communication, ASR system	**Wiring, ABS trouble code(s) stored, ECM**
1237	AC compressor clutch relay – malfunction	**Wiring, AC compressor clutch relay**
1238	Intake manifold air control solenoid – circuit malfunction	**Wiring, connector, intake manifold air control solenoid**

Flash code	Fault location	Probable cause
1238 2	Engine control module (ECM) – unused output stage	**Wiring short to ground/positive, ignore flash code if terminal 18 not connected**
1241	Engine control module (ECM) – no supply voltage	**Wiring, fuse(s), battery**
1241 3	Engine control module (ECM)/ABS control module communication, traction control system (TCS)	**Wiring, ABS trouble code(s) stored, ECM**
1241 2	Engine control module (ECM), electronic throttle system (ETS) – overrun torque control	**Wiring short to positive, ECM**
1242	AC control module/engine control module (ECM) communication – malfunction	**Wiring, AC control module**
1243	Crankshaft position (CKP) sensor/engine speed (RPM) sensor – incorrect/no signal	**Wiring, CKP sensor/RPM sensor**
1244	Camshaft position (CMP) sensor – range/performance problem	**Wiring, CMP sensor**
1245	Engine control module (ECM)/transmission control module (TCM) communication – short to ground	**Wiring, ECM, TCM**
1247	Ignition circuit – primary circuit	**Wiring, ignition coil(s)**
1251	Injector(s), group 1 – circuit malfunction	**Wiring, injector(s)**
1251 2	Injector, cyl. 1 – circuit malfunction	**Wiring, injector**
1252	Injector(s), group 2 – circuit malfunction	**Wiring, injector(s)**
1252 2	Injector, cyl. 5 – circuit malfunction	**Wiring, injector**
1253	Injector, cyl. 3 – circuit malfunction	**Wiring, injector**
1254	Injector, cyl. 6 – circuit malfunction	**Wiring, injector**
1255	Injector, cyl. 2 – circuit malfunction	**Wiring, injector**
1256	Injector, cyl. 4 – circuit malfunction	**Wiring, injector**
1261	Fuel pump relay – malfunction	**Wiring, connector(s), fuel pump relay**
1262	Idle speed control (ISC) actuator/idle air control (IAC) valve – malfunction	**Wiring, connector(s), ISC actuator/IAC valve**
1263	Evaporative emission (EVAP) canister purge valve – malfunction	**Wiring, connector(s), EVAP canister purge valve**
1264	Oxygen sensor heater relay – malfunction	**Wiring, connector, oxygen sensor heater relay**
1265	Malfunction indicator lamp (MIL) – short to ground/positive	**Wiring short to ground/positive, MIL**
1268	Mixture adjustment resistor – signal too low/high	**Wiring, mass air flow (MAF) sensor**
1271	Ignition circuit – defective	**Wiring, connector, ignition coil, ECM**
1271 2	Ignition circuit, cyl. 1 – defective	**Wiring, connector, ignition coil, ECM**
1272	Ignition circuit, cyl. 5 – defective	**Wiring, connector, ignition coil, ECM**

Flash code	Fault location	Probable cause
1273	Ignition circuit, cyl. 3 – defective	**Wiring, connector, ignition coil, ECM**
1274	Ignition circuit, cyl. 6 – defective	**Wiring, connector, ignition coil, ECM**
1275	Ignition circuit, cyl. 2 – defective	**Wiring, connector, ignition coil, ECM**
1276	Ignition circuit, cyl. 4 – defective	**Wiring, connector, ignition coil, ECM**
1278	Transmission control module (TCM) – signal	**Wiring, TCM**
1281	Battery voltage – too low	**Wiring, alternator, battery**
1282	Engine control module (ECM) – internal fault	**Disconnect ECM multi-plug for at least 10 min. – check if trouble code reappears before replacing ECM**
1283	Injector(s) – circuit malfunction	**Wiring, injector(s)**
1285	Engine control module (ECM)/alarm system control module communication – alarm system active	**Wiring short to positive, alarm system control module**
1286	Knock control circuit – defective	**Wiring, connector, KS incorrectly tightened, knock sensor (KS), ECM**
1287	Engine control module (ECM) – electronic throttle system (ETS) signal	**Wiring short to positive, ignore flash code if terminal 62 not connected**
1288	Engine control module (ECM) – internal fault	**ECM**
1288 2	Engine control module (ECM) – automatic stability control (ASR)/overrun torque control defective	**Wiring, ETS, ABS trouble code(s) stored**
1444	No fault found	–
1513	Engine control module (ECM) – unused output stage	**Wiring short to ground/positive, ignore flash code if terminal 18 not connected**
2000	End of trouble code output	–
2211	Engine control module (ECM) – defective	**Disconnect ECM multi-plug for at least 10 min. – check if trouble code reappears before replacing ECM**
2215	Mass air flow (MAF) sensor/volume air flow (VAF) sensor – incorrect signal	**Wiring, connector(s), MAF sensor/ VAF sensor**
2218	Engine control module (ECM) – output stage	**Wiring, ECM controlled components**
2221	Heated oxygen sensor (HO2S) – circuit malfunction	**Wiring, connector(s), HO2S**
2222	Heated oxygen sensor (HO2S) – range/ performance problem	**Wiring, intake/exhaust leak, fuel level low, fuel pressure, evaporative emission (EVAP) canister purge valve, injector(s), HO2S, ECM**
2223	Engine coolant temperature (ECT) sensor – malfunction	**Wiring, connector(s), ECT sensor**
2224	Intake air temperature (IAT) sensor – malfunction	**Wiring, connector(s), IAT sensor**
2231	Battery voltage – too low/high	**Wiring, alternator, battery**
2232	Closed throttle position (CTP) switch – malfunction	**Wiring, connector, CTP switch, ECM**

Flash code	Fault location	Probable cause
2233	Throttle position (TP) sensor – malfunction	**Wiring, TP sensor, ECM**
2234	Vehicle speed sensor (VSS) – incorrect/no signal	**Wiring, instrument panel, VSS**
2235	Torque converter clutch (TCC) solenoid – defective	**Wiring, TCC solenoid**
2236	Engine control module (ECM)/ABS control module communication, automatic stability control (ASR) system	**Wiring, ABS trouble code(s) stored, ECM**
2237	AC compressor clutch relay – malfunction	**Wiring, AC compressor clutch relay**
2241	Engine control module (ECM)/ABS control module communication, traction control system (TCS)	**Wiring, ABS trouble code(s) stored, ECM**
2244	Camshaft position (CMP) sensor – range/ performance problem	**Wiring, CMP sensor**
2245	Engine control module (ECM)/transmission control module (TCM) communication – short to ground	**Wiring, ECM, TCM**
2444	No fault found	–
2251	Injector(s), group 1 – circuit malfunction	**Wiring, injector(s)**
2252	Injector(s), group 2 – circuit malfunction	**Wiring, injector(s)**
2261	Fuel pump relay – malfunction	**Wiring, connector(s), fuel pump relay**
2263	Evaporative emission (EVAP) canister purge valve – malfunction	**Wiring, connector(s), EVAP canister purge valve**
2264	Oxygen sensor heater relay – malfunction	**Wiring, connector, oxygen sensor heater relay**
2265	Malfunction indicator lamp (MIL) – short to ground/positive	**Wiring short to ground/positive, MIL**
2268	Mixture adjustment resistor – signal too low/high	**Wiring, mass air flow (MAF) sensor**
2278	Transmission control module (TCM) – signal	**Wiring, TCM**
2513	Engine control module (ECM) – unused output stage	**Wiring short to ground/positive – ignore flash code if terminal 18 not connected**
4444	No fault found	–

1 Motronic M1.7 (V12), M3.1
2 Motronic M3.1
3 Motronic M1.7 (V12)

Accessing and erasing – scanner type

- The engine control module (ECM) fault memory can also be checked using diagnostic equipment connected to the data link connector (DLC).

Trouble code identification

Scan code	Fault location	Probable cause
–	No fault found	–
0	Engine control module (ECM) – undefined fault	–
1	Engine control module (ECM) – defective	**ECM**
1 [1]	Fuel pump relay – malfunction	**Wiring, connector(s), fuel pump relay**
2	Idle speed control (ISC) actuator/idle air control (IAC) valve – malfunction	**Wiring, connector(s), ISC actuator/IAC valve**
3 [2]	Fuel pump relay – malfunction	**Wiring, connector(s), fuel pump relay**
3	Injector(s), cyl. 1 & 3 or cyl. 2, 4, 6 & 8, 10, 12 – circuit malfunction	**Wiring, injector(s)**
3 [3]	Injector(s), cyl. 2 & 4 – circuit malfunction	**Wiring, injector(s)**
3 [4]	Injector(s), cyl. 1 – circuit malfunction	**Wiring, injector**
4	Injector(s), cyl. 3 – circuit malfunction	**Wiring, injector**
5	Evaporative emission (EVAP) canister purge valve – malfunction	**Wiring, connector(s), EVAP canister purge valve**
5 [4]	Injector(s), cyl. 2 – circuit malfunction	**Wiring, injector**
6	Injector(s) – circuit malfunction	**Wiring, injector(s)**
7	Mass air flow (MAF) sensor/volume air flow (VAF) sensor – incorrect signal	**Wiring, connector(s), MAF sensor/ VAF sensor**
10	Heated oxygen sensor (HO2S) – range/ performance problem	**Wiring, intake/exhaust leak, fuel level low, fuel pressure, evaporative emission (EVAP) canister purge valve, injector(s), ECM**
12	Throttle position (TP) sensor – defective	**Wiring, connector(s), TP sensor**
15	Knock sensor (KS) 1 – defective	**Wiring, connector, KS incorrectly tightened, KS**
16	Injector(s), cyl. 1, 3, 5 & 7, 9, 11 – circuit malfunction	**Wiring, injector(s)**
16 [1]	Camshaft position (CMP) sensor – range/ performance problem	**Wiring, CMP sensor**
17	Injector(s), cyl. 2, 4, 6 & 8, 10, 12 – circuit malfunction	**Wiring, injector(s)**
18	Engine control module (ECM) – unused output stage	**Wiring short to ground/positive, ignore flash code if terminal 18 not connected**
18 [5]	Intake manifold air control solenoid – circuit malfunction	**Wiring, connector, intake manifold air control solenoid**
19	Engine control module (ECM) – unused output stage	**Wiring short to ground/positive, ignore flash code if terminal 18 not connected**

Scan code	Fault location	Probable cause
23	Oxygen sensor heater relay – malfunction	**Wiring, connector, oxygen sensor heater relay**
23 [4]	Ignition circuit, cyl. 2 – defective	**Wiring, connector, ignition coil, ECM**
24	Engine control module (ECM)/transmission control module (TCM) communication	**Wiring short to ground, TCM trouble code(s) stored**
24 [4]	Ignition circuit, cyl. 3 – defective	**Wiring, connector, ignition coil, ECM**
25	Ignition circuit, cyl. 1 – defective	**Wiring, connector, ignition coil, ECM**
26	Battery voltage – too low	**Wiring, alternator, battery**
28	Heated oxygen sensor (HO2S) – circuit malfunction	**Wiring, connector(s), HO2S**
29	Idle speed control (ISC) actuator/idle air control (IAC) valve – malfunction	**Wiring, connector(s), ISC actuator/IAC valve**
31	Injector, cyl. 5 – circuit malfunction	**Wiring, injector**
32	Injector(s), cyl. 2 & 4 – circuit malfunction	**Wiring, injector(s)**
32 [6]	Injector(s), cyl. 1, 3, 5 & 7, 9, 11 – circuit malfunction	**Wiring, injector(s)**
32 [3]	Injector(s), cyl. 1 & 3 – circuit malfunction	**Wiring, injector(s)**
32 [4]	Injector, cyl. 6 – circuit malfunction	**Wiring, injector**
33	Injector, cyl. 4 – circuit malfunction	**Wiring, injector**
36	Evaporative emission (EVAP) canister purge valve – malfunction	**Wiring, connector(s), EVAP canister purge valve**
37	Battery voltage – too low/high	**Wiring, alternator, battery**
37 [1]	Oxygen sensor heater relay – malfunction	**Wiring, connector, oxygen sensor heater relay**
38	Engine control module (ECM)/ABS control module communication, acceleration skid control (ASR) system	**Wiring, ABS trouble code(s) stored, ECM**
41	Mass air flow (MAF) sensor/volume air flow (VAF) sensor – incorrect signal	**Wiring, connector(s), MAF sensor/ VAF sensor**
42	Knock sensor (KS) 2 – defective	**Wiring, connector, KS incorrectly tightened, KS**
44	Intake air temperature (IAT) sensor – malfunction	**Wiring, connector(s), IAT sensor**
45	Engine coolant temperature (ECT) sensor – malfunction	**Wiring, connector(s), ECT sensor**
46	AC condenser blower motor relay – short to positive	**Wiring short to positive, AC condenser blower motor relay, ECM**
46 [4]	Engine control module (ECM) – output stage	**Wiring, ECM controlled components**
48	AC compressor clutch relay – malfunction	**Wiring, AC compressor clutch relay**
50	Ignition circuit, cyl. 4 – defective	**Wiring, connector, ignition coil, ECM**
51	Transmission control module (TCM) – signal	**Wiring, TCM**

Scan code	Fault location	Probable cause
51 [4]	Ignition circuit, cyl. 6 – defective	**Wiring, connector, ignition coil, ECM**
52	Closed throttle position (CTP) switch – malfunction	**Wiring, connector, CTP switch, ECM**
52 [4]	Ignition circuit, cyl. 5 – defective	**Wiring, connector, ignition coil, ECM**
53	Throttle position (TP) sensor/wide open throttle (WOT) switch – malfunction	**Wiring, TP sensor, ECM**
54	Engine control module (ECM)/transmission control module (TCM) communication	**Wiring short to ground, TCM trouble code(s) stored**
54 [1]	Battery voltage – too low/high	**Wiring, alternator, battery**
55	Ignition circuit – defective	**Wiring, connector, ignition coil, ECM**
62	Engine control module (ECM) – electronic throttle system (ETS) signal	**Wiring short to positive, ignore flash code if terminal 62 not connected**
63	Torque converter clutch (TCC) solenoid – defective	**Wiring, TCC solenoid**
64	Engine control module (ECM)/transmission control module (TCM) communication – short to ground	**Wiring, TCM trouble code(s) stored, ECM, TCM**
67	Crankshaft position (CKP) sensor/engine speed (RPM) sensor – incorrect/no signal	**Wiring, CKP sensor/RPM sensor**
70	Heated oxygen sensor (HO2S) – circuit malfunction	**Wiring, connector(s), HO2S**
73	Vehicle speed sensor (VSS) – incorrect/no signal	**Wiring, instrument panel, VSS**
76	Mixture adjustment resistor – signal too low/ high	**Wiring, mass air flow (MAF) sensor**
77	Intake air temperature (IAT) sensor – malfunction	**Wiring, connector(s), IAT sensor**
78	Engine coolant temperature (ECT) sensor – malfunction	**Wiring, connector(s), ECT sensor**
81	Engine control module (ECM)/immobilizer control module communication – malfunction	**Wiring, electrical interference, immobilizer control module, ECM**
81 [4]	Engine control module (ECM)/alarm system control module communication – alarm system active	**Wiring short to positive, alarm system control module**
82	Engine control module (ECM)/ABS control module communication, traction control system (TCS)	**Wiring, ABS trouble code(s) stored, ECM**
82 [4]	Engine control module (ECM), electronic throttle system (ETS) – overrun torque control	**Wiring short to positive, ECM**
83	Throttle control, traction control system (TCS) – defective	**Wiring, connector(s), throttle control system**
83 [4]	Engine control module (ECM) – acceleration skid control (ASR) defective	**Wiring short to positive, ECM**

Scan code	Fault location	Probable cause
85	Air conditioning (AC) control module/engine control module (ECM) communication – malfunction	**Wiring, AC control module**
100	Engine control module (ECM) – output stage	**Wiring, ECM controlled components**
200	Engine control module (ECM) – defective	**Disconnect ECM multi-plug for at least 10 min. – check if trouble code reappears before replacing ECM**
201	Heated oxygen sensor (HO2S) – range/ performance problem	**Wiring, intake/exhaust leak, fuel level low, fuel pressure, evaporative emission (EVAP) canister purge valve, injector(s), HO2S, ECM**
202	Engine control module (ECM) – internal fault	**Disconnect ECM multi-plug for at least 10 min. – check if trouble code reappears before replacing ECM**
203	Ignition circuit – primary circuit	**Wiring, ignition coil(s)**
204	Engine control module (ECM) – acceleration skid control (ASR)/overrun torque control defective	**Wiring, trouble code(s) stored in other system(s)**
206	Knock control circuit – defective	**Wiring, connector, knock sensor (KS), ECM**
207	Knock control circuit – range/performance problem	**Wiring, connector, KS incorrectly tightened, knock sensor (KS)**
220	Engine control module (ECM) – immobilizer active	**Incorrect/damaged key, incorrectly coded, ECM/immobilizer replacement without coding, wiring, immobilizer defective**
300	Engine control module (ECM) – engine cannot be started	**Wiring, engine speed (RPM) sensor, mass air flow (MAF) sensor/volume air flow (VAF) sensor, fuel pump relay, ignition system, injector(s), fuel pressure, throttle control system**

1 Motronic M1.7/1.7.2/1.7.3/3.1
2 Motronic M1.2
3 Motronic M1.7.3
4 Motronic M3.1
5 Motronic M1.7.2/1.7.3
6 Motronic M1.7, V12

DAEWOO

Model:	Engine identification:	Year:
Lanos 1.6L	A16DM	2000-02
Nubira 2.0L	T20XED	2000-02

System: **IEFI-6/ITMS-6F**

Self-diagnosis

General information

- Refer to the front of this manual for general test conditions, terminology, detailed descriptions of wiring faults and a general trouble shooter for electrical and mechanical faults.
- Engine control module (ECM) incorporates self-diagnosis function.
- Malfunction indicator lamp (MIL) will illuminate if certain faults are recorded.
- ECM operates in backup mode if sensors fail, to enable vehicle to be driven to workshop.

Accessing

- Trouble codes can be displayed by using a Scan Tool connected to the data link connector (DLC):
  - Lanos – **1**.
  - Nubira – **2**.

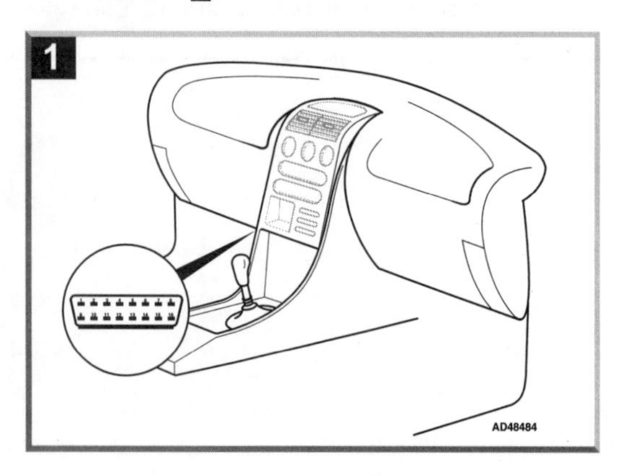

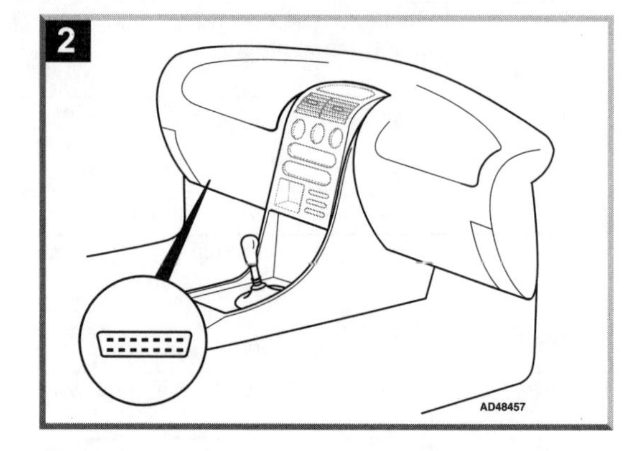

Erasing

- Ensure ignition switched OFF.
- Disconnect battery ground lead.
- Wait 10 seconds minimum.
- Reconnect battery ground lead.

NOTE: ***Disconnecting battery ground lead may also erase memory from electronic units such as radio.***

- The engine control module (ECM) fault memory can also be erased using diagnostic equipment connected to the data link connector (DLC).

Trouble code identification

OBD-II code	Fault location	Probable cause
All P0, P2 and U0 codes	Refer to OBD-II trouble code tables at the front of this manual	–
P1106	Manifold absolute pressure (MAP) sensor – intermittent voltage high	**Wiring short to positive, MAP sensor**
P1107	Manifold absolute pressure (MAP) sensor – intermittent voltage low	**Wiring short to ground, MAP sensor**
P1109	Intake manifold air control valve – circuit malfunction	**Wiring, intake manifold air control valve, ECM**
P1111	Intake air temperature (IAT) sensor – intermittent voltage high	**Wiring short to positive, IAT sensor**
P1112	Intake air temperature (IAT) sensor – intermittent voltage low	**Wiring short to ground, IAT sensor**
P1114	Engine coolant temperature (ECT) sensor – intermittent voltage low	**Wiring short to ground, ECT sensor**
P1115	Engine coolant temperature (ECT) sensor – intermittent voltage high	**Wiring short to positive, ECT sensor**
P1121	Throttle position (TP) sensor – voltage high	**Wiring short to positive, ECT sensor**
P1122	Throttle position (TP) sensor – voltage low	**Wiring short to ground, ECT sensor**
P1130	Oxygen sensor (O2S) 1, bank 1 – too long to respond	**Wiring, O2S**
P1133	Oxygen sensor (O2S) 1, bank 1 – slow response	**Wiring, fuel pressure, leaking injector, MAP sensor, TP sensor, O2S**
P1134	Oxygen sensor (O2S) 1, bank 1 – transition ratio	**Wiring, O2S**
P1167	Oxygen sensor (O2S) 1, bank 1 – rich in deceleration fuel cut-off	**Fuel system, fuel pressure regulator, leaking injector(s)**
P1171	Fuel trim, under load – mixture too lean	**O2S, fuel pressure, fuel system, mechanical fault, ECM**
P1181	Intake manifold air control solenoid – voltage low	**Wiring open circuit/short to ground, intake manifold air control solenoid**
P1182	Intake manifold air control solenoid – voltage high	**Wiring open circuit/short to positive, intake manifold air control solenoid**
P1230	Fuel pump relay – voltage low	**Wiring open circuit/short to ground, fuel pump relay**
P1231	Fuel pump relay – voltage high	**Wiring open circuit/short to positive, fuel pump relay**
P1320	Crankshaft position (CKP) sensor – signal adaptation at limit	**Wiring, CKP sensor**
P1321	Crankshaft position (CKP) sensor – incorrect signal	**Wiring, CKP sensor**
P1336	Crankshaft position (CKP) sensor – variation not learned	**Learning procedure not performed**
P1380	Rough road signal – malfunction	**ABS control module trouble code(s) stored, and or engine misfire present**

OBD-II code	Fault location	Probable cause
P1381	Rough road signal – no communication from ABS CAN data bus	Wiring open circuit, ABS system, and or engine misfire present
P1382	Rough road signal – invalid data	Wiring, ABS system
P1385	Variable reluctance sensor, without ABS – circuit malfunction	Wiring, variable reluctance sensor
P1391	Rough road G-force sensor – implausible signal	Wiring, G-force sensor, ECM
P1392	Rough road G-force sensor – voltage low	Wiring open circuit/short to ground, G-force sensor, ECM
P1393	Rough road G-force sensor – voltage high	Wiring short to positive, G-force sensor, ECM
P1396	WSSD rough road system – malfunction	Wiring, G-force sensor, ECM
P1397	WSSD rough road – detected/malfunction	Wiring, G-force sensor, ABS control module, ECM
P1402	Exhaust gas recirculation (EGR) valve – blocked	Wiring, EGR valve
P1403	Exhaust gas recirculation (EGR) valve – malfunction	Wiring, EGR valve
P1404	Exhaust gas recirculation (EGR) valve – range/ performance problem	Wiring, EGR valve
P1504	Vehicle speed sensor (VSS), AT – no signal	Wiring, transmission control module, VSS
P1511	Idle speed control (ISC) actuator – circuit malfunction	Wiring, ISC actuator
P1512	Idle speed control (ISC) actuator – malfunction	ISC actuator
P1513	Idle speed control (ISC) actuator – functional diagnosis malfunction	Wiring, ISC actuator, ECM
P1537	AC compressor clutch relay – voltage high	Wiring short to positive, AC compressor clutch relay
P1538	AC compressor clutch relay – voltage low	Wiring short to ground, AC compressor clutch relay
P1546	AC compressor clutch, signal – circuit malfunction	Wiring, AC compressor clutch relay
P1601	Engine control module (ECM)/transmission control module – serial peripheral communication malfunction	Wiring, ECM, TCM
P1607	ECM, lower power counter I/C – reset/ malfunction	ECM
P1610	Engine control (EC) relay – voltage high	Wiring open circuit/short to positive, EC relay
P1611	Engine control (EC) relay – voltage low	Wiring short to ground, EC relay
P1618	Serial peripheral interface, ECM – communication diagnostics	ECM/TCM
P1626	Immobilizer system – malfunction	No authorization code

OBD-II code	Fault location	Probable cause
P1627	Transmission control module/engine control module (ECM) – A/D conversion malfunction	TCM/ECM
P1628	Immobilizer system – communication malfunction	Wiring immobilizer control module, engine control module (ECM)
P1629	Immobilizer system – malfunction	Incorrect authorization code
P1631	Immobilizer system – malfunction	Incorrect authorization code
P1635	Engine control module (ECM), sensor reference voltage – low	Wiring, ECM
P1640	Engine control module (ECM), output driver modules – malfunction	Wiring, ECM
P1650	Engine control module (ECM), SPI communication with SIDM chip – malfunction	Wiring, ECM
P1655	Engine control module (ECM), SPI communication with PSVI chip – malfunction	Wiring, ECM
P1660	Malfunction indicator lamp (MIL) – voltage high	Wiring open circuit/short to positive, MIL
P1661	Malfunction indicator lamp (MIL) – voltage low	Wiring short to ground, MIL
P1671	CAN data bus, TCM communication – missing	Wiring, ECM, TCM
P1672	CAN data bus – bus off mode detected	–
P1673	CAN data bus – communication malfunction	Wiring, ECM, TCM

HONDA

Model: **Accord 2.2L**

Engine identification: **F22A1/4/6**

Year: **1990-93**

System: **PGM-FI**

Self-diagnosis

General information

- Refer to the front of this manual for general test conditions, terminology, detailed descriptions of wiring faults and a general trouble shooter for electrical and mechanical faults.
- Engine control module (ECM) incorporates self-diagnosis function.
- Malfunction indicator lamp (MIL) will illuminate if certain faults are recorded.
- ECM operates in backup mode if sensors fail, to enable vehicle to be driven to workshop.
- Trouble codes can be displayed by LED in engine control module (ECM) – 1990 or malfunction indicator lamp (MIL) – 1991-93.

Accessing

- Ensure ignition switched OFF.
- Jump service check connector terminals **1**.
- 1990 – access engine control module (ECM) **2**.
- Switch ignition ON.
- Check ECM LED or MIL is flashing.
- Trouble codes 1-9 are indicated as follows:
  - Individual short flashes display trouble code **3** [**A**].
  - A short pause separates each flash **3** [**B**].
  - A long pause separates each trouble code **3** [**C**].
  - For example: Trouble code 3 displayed **3**.
- Trouble codes greater than 9 are indicated as follows:
  - Long flashes indicate the 'tens' of the trouble code **4** [**A**].
  - Short flashes indicate the 'units' of the trouble code **4** [**C**].
  - A short pause separates each flash **4** [**B**].
  - A long pause separates each trouble code **4** [**D**].
  - For example: Trouble code 12 displayed **4**.
- Count LED or MIL flashes and compare with trouble code table.
- Switch ignition OFF.
- Remove jump lead.

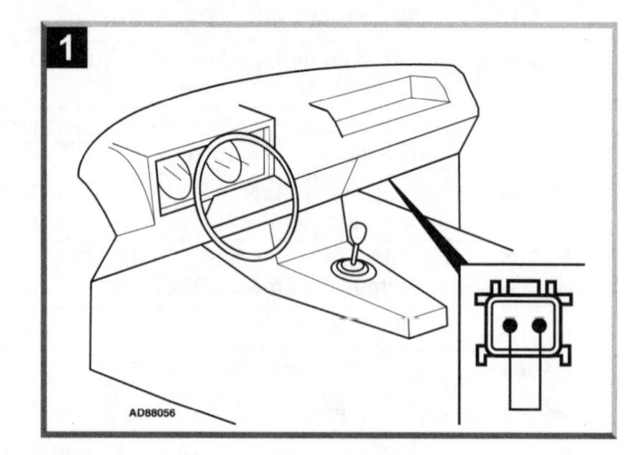

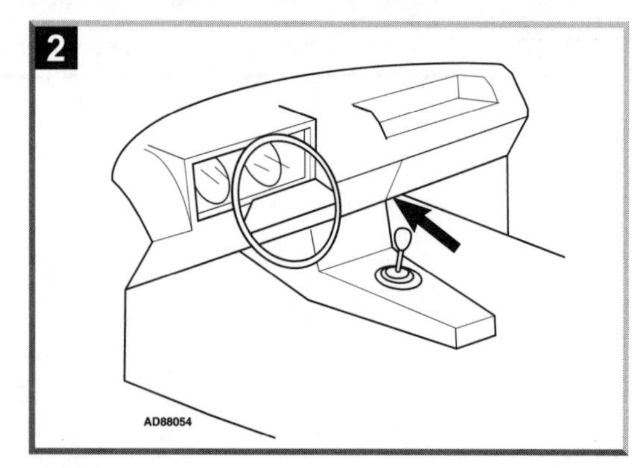

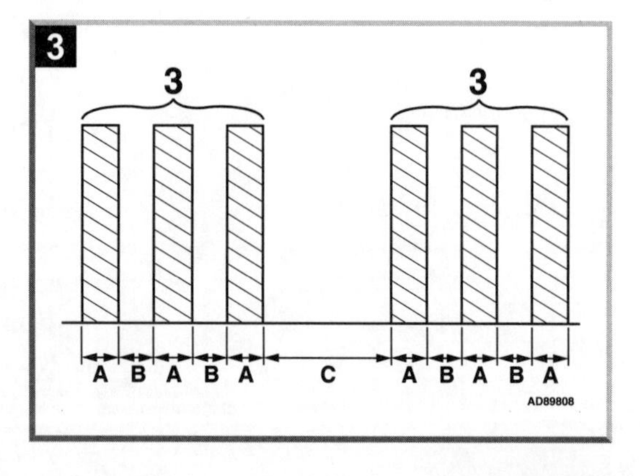

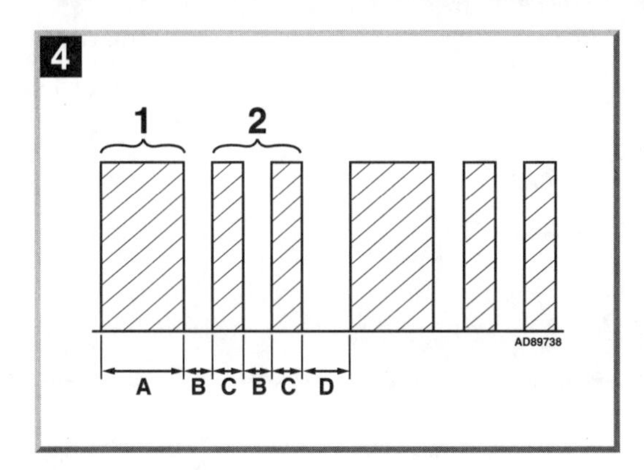

Erasing

- After the faults have been rectified, erase the trouble codes as follows:
- Switch ignition OFF.
- Remove fuse No.24 (7.5A) from underhood fusebox for 10 seconds minimum **5**.
- Reinstall fuse.
- Repeat checking procedure to ensure no data remains in ECM fault memory.

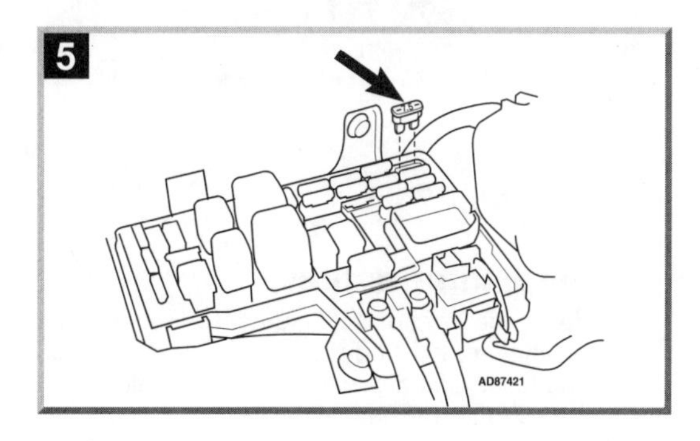

Trouble code identification

Flash code	Fault location	Probable cause
0	Engine control module (ECM)	**Wiring, ECM**
1	Heated oxygen sensor (HO2S)	**Fuel system, wiring, HO2S, ECM**
3	Manifold absolute pressure (MAP) sensor	**Wiring, MAP sensor, ECM**
4	Engine speed (RPM) sensor	**Wiring, RPM sensor, ECM**
5	Manifold absolute pressure (MAP) sensor	**Hose leak**
6	Engine coolant temperature (ECT) sensor	**Wiring, ECT sensor, ECM**
7	Throttle position (TP) sensor	**Wiring, TP sensor, ECM**
8	Crankshaft position (CKP) sensor	**Wiring, CKP sensor, ECM**
9	Camshaft position (CMP) sensor	**Wiring, CMP sensor, ECM**
10	Intake air temperature (IAT) sensor	**Wiring, IAT sensor, ECM**
13	Barometric pressure (BARO) sensor	**ECM**
14	Idle air control (IAC) valve	**Wiring, IAC valve, ECM**
15	Ignition output signal	**Wiring, ignition module, ECM**
16	Injectors	**Wiring, injectors, ECM**
17	Vehicle speed sensor (VSS)	**Wiring, VSS, ECM**
20	Electrical load sensor	**Wiring, electrical load sensor, ECM**
30	AT signal A	**Wiring, ECM**
31	AT signal B	**Wiring, ECM**
41	Oxygen sensor heater	**Wiring, HO2S, ECM**
43	Heated oxygen sensor (HO2S)/fuel supply system	**Fuel system, wiring, HO2S, ECM**

HONDA

Model:	Engine identification:	Year:
Accord 2.2L	**F22B1/2**	**1994-95**
Civic del Sol 1.5L SOHC	**D15B7**	**1994-95**
Civic del Sol 1.6L	**D16Z6**	**1994-95**
Civic del Sol 1.6L DOHC	**B16A3**	**1994-95**
Odyssey 2.2L	**F22B6**	**1995**

System: **PGM-FI**

Self-diagnosis

General information

- Refer to the front of this manual for general test conditions, terminology, detailed descriptions of wiring faults and a general trouble shooter for electrical and mechanical faults.
- Engine control module (ECM) incorporates self-diagnosis function.
- Malfunction indicator lamp (MIL) will illuminate if certain faults are recorded.
- ECM operates in backup mode if sensors fail, to enable vehicle to be driven to workshop.
- Trouble codes can be displayed by the malfunction indicator lamp (MIL) with the service check connector jumped **1**.

Accessing

- Ensure ignition switched OFF.
- Jump service check connector terminals **1**.
- Switch ignition ON.
- Check MIL is flashing.
- Trouble codes 1-9 are indicated as follows:
  - Individual short flashes display trouble code **2** [**A**].
  - A short pause separates each flash **2** [**B**].
  - A long pause separates each trouble code **2** [**C**].
  - For example: Trouble code 3 displayed **2**.
- Trouble codes greater than 9 are indicated as follows:
  - Long flashes indicate the 'tens' of the trouble code **3** [**A**].
  - Short flashes indicate the 'units' of the trouble code **3** [**C**].
  - A short pause separates each flash **3** [**B**].
  - A long pause separates each trouble code **3** [**D**].
  - For example: Trouble code 12 displayed **3**.
- Count MIL flashes and compare with trouble code table.

NOTE: ***If malfunction indicator lamp (MIL) is ON constantly and no trouble codes are displayed, this indicates an engine control module (ECM) fault.***

- Switch ignition OFF.
- Remove jump lead.

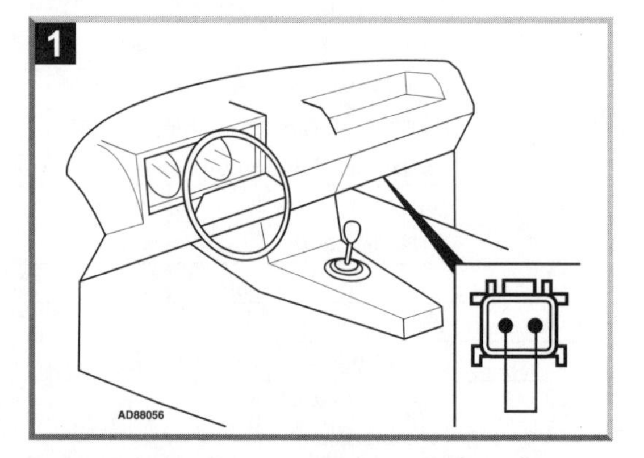

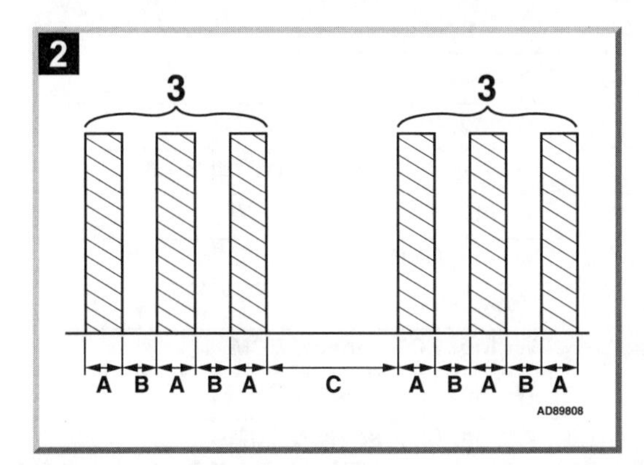

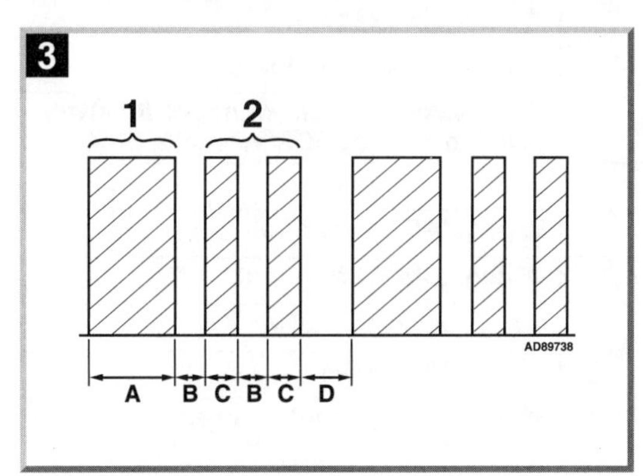

Erasing

- After the faults have been rectified, erase the trouble codes as follows:
- Switch ignition OFF.
- Civic del Sol – remove radio fuse (7.5A) from underhood fusebox for 10 seconds minimum **4**.
- Accord/Odyssey – remove radio fuse (7.5A) from underhood fusebox for 10 seconds minimum **5**.
- Reinstall fuse.
- Repeat checking procedure to ensure no data remains in ECM fault memory.

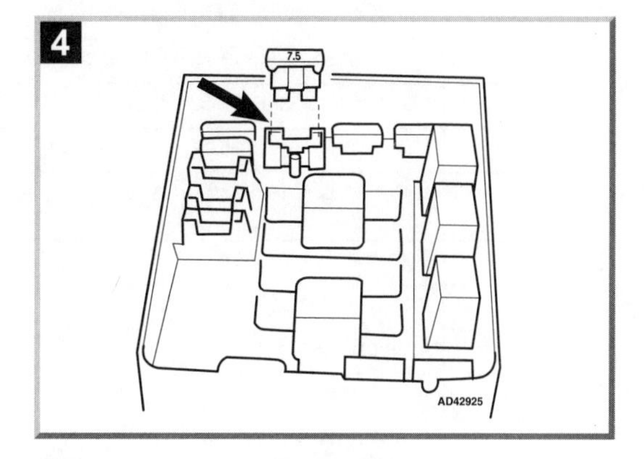

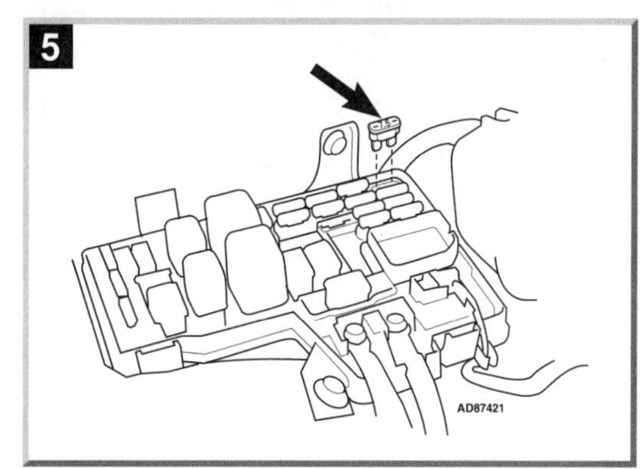

Trouble code identification

Flash code	Fault location	Probable cause
0	Engine control module (ECM)	**Wiring, ECM**
1	Heated oxygen sensor (HO2S)	**Fuel system, wiring, HO2S, ECM**
3	Manifold absolute pressure (MAP) sensor	**Wiring, MAP sensor, ECM**
4	Crankshaft position (CKP) sensor 1	**Wiring, CKP sensor, ECM**
5	Manifold absolute pressure (MAP) sensor	**Hose leak**
6	Engine coolant temperature (ECT) sensor	**Wiring, ECT sensor, ECM**
7	Throttle position (TP) sensor	**Wiring, TP sensor, ECM**
8	Crankshaft position (CKP) sensor 2	**Wiring, CKP sensor, ECM**
9	Camshaft position (CMP) sensor	**Wiring, CMP sensor, ECM**
10	Intake air temperature (IAT) sensor	**Wiring, IAT sensor, ECM**
12	Exhaust gas recirculation (EGR) system	**Hose leak/blockage, wiring, EGR valve/ position sensor, EGR solenoid, ECM**
13	Barometric pressure (BARO) sensor	**ECM**
14	Idle air control (IAC) valve	**Wiring, IAC valve, ECM**
15	Ignition output signal	**Wiring, ICM, ECM**
16	Injectors	**Wiring, injectors, ECM**
17	Vehicle speed sensor (VSS)	**Wiring, VSS, ECM**
19	AT – lock-up control solenoid A/B	**Wiring, lock-up control solenoid A/B, ECM**
20	Electrical load sensor	**Wiring, electrical load sensor, ECM**
21	VTEC system – malfunction	**Wiring, VTEC solenoid, ECM**
22	VTEC system – malfunction	**Wiring, VTEC pressure switch, ECM**
23	Knock sensor (KS)	**Wiring, KS, ECM**
30	AT signal A	**Wiring, ECM**
31	AT signal B	**Wiring, ECM**
41	Heated oxygen sensor (HO2S) – heater circuit	**Wiring, HO2S, ECM**
43	Fuel supply system	**Fuel system, wiring, HO2S, ECM**

HONDA

Model:	Engine identification:	Year:
Accord 2.2L	**F22B1/2**	**1996-97**
Accord 2.7L	**C27A4**	**1995-97**
Odyssey 2.2L	**F22B6**	**1996-97**
Odyssey 2.3L	**F23A7**	**1998**

System: **PGM-FI**

Self-diagnosis

General information

- Refer to the front of this manual for general test conditions, terminology, detailed descriptions of wiring faults and a general trouble shooter for electrical and mechanical faults.
- Engine control module (ECM) incorporates self-diagnosis function.
- Malfunction indicator lamp (MIL) will illuminate if certain faults are recorded.
- ECM operates in backup mode if sensors fail, to enable vehicle to be driven to workshop.
- Trouble codes can be displayed by using a Scan Tool connected to the data link connector (DLC) **1** [**1**] or by the malfunction indicator lamp (MIL) with the service check connector jumped **1** [**2**].

NOTE: ***The use of a Scan Tool is essential to obtain full diagnostic information.***

Accessing

- Ensure ignition switched OFF.
- Jump service check connector terminals **1** [**2**].
- Switch ignition ON.
- Check MIL is flashing.
- Trouble codes 1-9 are indicated as follows:
  - Individual short flashes display trouble code **2** [**A**].
  - A short pause separates each flash **2** [**B**].
  - A long pause separates each trouble code **2** [**C**].
  - For example: Trouble code 3 displayed **2**.
- Trouble codes greater than 9 are indicated as follows:
  - Long flashes indicate the 'tens' of the trouble code **3** [**A**].
  - Short flashes indicate the 'units' of the trouble code **3** [**C**].
  - A short pause separates each flash **3** [**B**].
  - A long pause separates each trouble code **3** [**D**].
  - For example: Trouble code 12 displayed **3**.
- Count MIL flashes and compare with trouble code table.
- Switch ignition OFF.
- Remove jump lead.

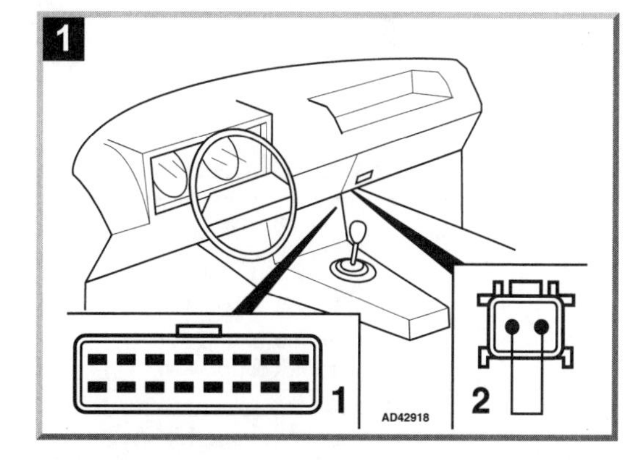

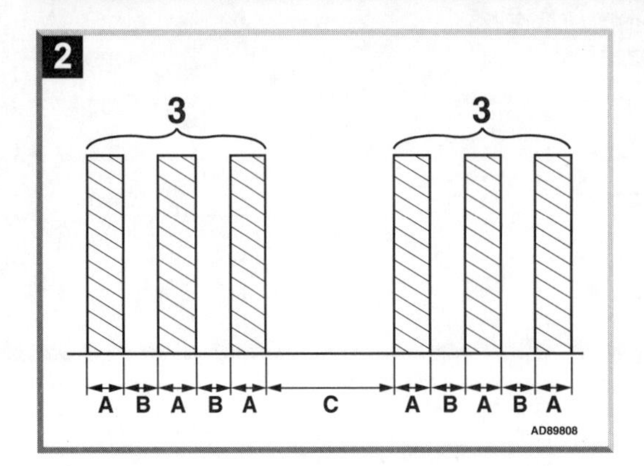

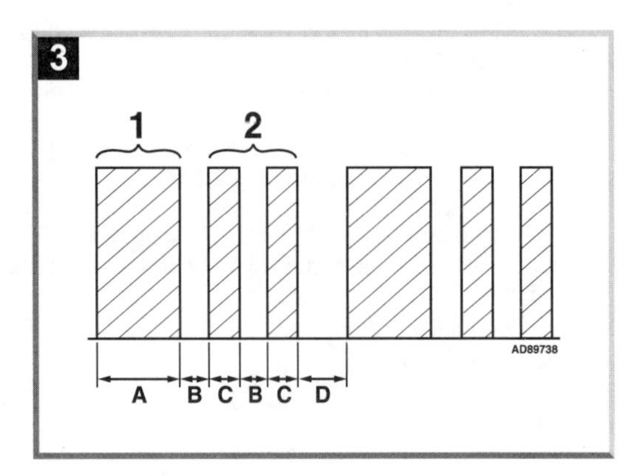

Erasing

- After the faults have been rectified, trouble codes can be erased by using a Scan Tool connected to the data link connector (DLC) or as follows:
- Switch ignition OFF.
- Remove radio fuse (7.5A) from underhood fusebox for 10 seconds minimum **4**.
- Reinstall fuse.
- Repeat checking procedure to ensure no data remains in ECM fault memory.

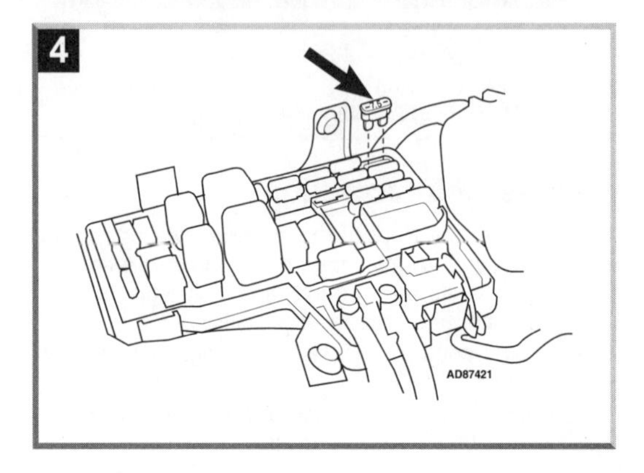

Trouble code identification

Flash code	OBD-II code	Fault location	Probable cause
–	*All P0, P2 and U0 codes*	Refer to OBD-II trouble code tables at the front of this manual	–
1	–	Heated oxygen sensor (HO2S) – front – circuit/voltage low/high	**Wiring open/short circuit, HO2S, fuel system, ECM**
3	–	Manifold absolute pressure (MAP) sensor – circuit/voltage high/low	**Wiring, MAP sensor, ECM**
4	–	Crankshaft position (CKP) sensor – circuit malfunction/range/performance problem	**Wiring, CKP sensor, valve timing, ECM**
5	–	Manifold absolute pressure (MAP) sensor – range/performance problem	**Hose leak, MAP sensor**
5	P1128	Manifold absolute pressure (MAP) sensor – pressure lower than expected	**MAP sensor**
5	P1129	Manifold absolute pressure (MAP) sensor – pressure higher than expected	**MAP sensor**

Flash code	OBD-II code	Fault location	Probable cause
6	–	Engine coolant temperature (ECT) sensor – circuit/voltage low/high	**Wiring short/open circuit, ECT sensor, ECM**
7	–	Throttle position (TP) sensor – circuit/ voltage low/high	**Wiring, TP sensor, ECM**
7	P1121	Throttle position (TP) sensor – position lower than expected	**TP sensor**
7	P1122	Throttle position (TP) sensor – position higher than expected	**TP sensor**
8	P1359	Crankshaft position (CKP) sensor – connector disconnection	**Wiring**
8	P1361	Crankshaft position (CKP) sensor – intermittent signal	**CKP sensor**
8	P1362	Crankshaft position (CKP) sensor – no signal	**Wiring, CKP sensor, ECM**
9	P1381	Camshaft position (CMP) sensor – intermittent signal	**CMP sensor**
9	P1382	Camshaft position (CMP) sensor – no signal	**Wiring, CMP sensor, ECM**
10	–	Intake air temperature (IAT) sensor – circuit/ voltage low/high	**Wiring short/open circuit, IAT sensor, ECM**
12	P1491	Exhaust gas recirculation (EGR) system – valve lift insufficient	**Wiring, EGR valve/position sensor, EGR solenoid, hose leak/blockage, ECM**
12	P1498	Exhaust gas recirculation (EGR) valve position sensor – voltage high	**Wiring, EGR valve/position sensor, ECM**
13	P1106	Barometric pressure (BARO) sensor – range/performance problem	**ECM**
13	P1107	Barometric pressure (BARO) sensor – circuit/voltage low	**ECM**
13	P1108	Barometric pressure (BARO) sensor – circuit/voltage high	**ECM**
14	–	Idle control system – malfunction	**IAC valve, fast idle thermo valve, throttle body**
14	P1508	Idle air control (IAC) valve – circuit failure	**Wiring, IAC valve, ECM**
14	P1519	Idle air control (IAC) valve – circuit failure	**Wiring, IAC valve, ECM**
17	–	Vehicle speed sensor (VSS) – circuit malfunction/range/performance problem	**Wiring, VSS, ECM**
20	P1297	Electrical load sensor – circuit/voltage low	**Wiring, electrical load sensor, ECM**
20	P1298	Electrical load sensor – circuit/voltage high	**Wiring, electrical load sensor, ECM**
21	P1253	VTEC system – malfunction	**Wiring, VTEC solenoid, ECM**
22	P1259	VTEC system – malfunction	**Wiring, VTEC solenoid/pressure switch, ECM**
23	–	Knock sensor (KS) – circuit malfunction	**Wiring, KS, ECM**
30	P1681	AT to ECM – signal A – voltage low	**Wiring, TCM, ECM**
30	P1682	AT to ECM – signal A – voltage high	**Wiring, TCM, ECM**

Flash code	OBD-II code	Fault location	Probable cause
31	P1686	AT to ECM – signal B – voltage low	**Wiring, TCM, ECM**
31	P1687	AT to ECM – signal B – voltage high	**Wiring, TCM, ECM**
41	–	Heated oxygen sensor (HO2S) – front – circuit malfunction	**Wiring, ECM**
45	–	Mixture too lean/rich	**Fuel system, front HO2S, injectors, MAP sensor, wiring, mechanical fault**
61	–	Heated oxygen sensor (HO2S) – front – slow response	**HO2S, exhaust system**
63	–	Heated oxygen sensor (HO2S) – rear – slow response/circuit/voltage low/high	**Wiring, HO2S, ECM**
65	–	Heated oxygen sensor (HO2S) – rear – circuit malfunction	**Wiring, ECM**
67	–	Catalytic converter – efficiency below limit	**Catalytic converter, rear HO2S**
70	–	AT – lock-up clutch not engaging/no gear shift/poor gear shift	**Wiring, mainshaft speed sensor, countershaft speed sensor, lock-up control system, shift solenoid (SS) A/B/C, ECM, TCM**
70	P1660	AT to ECM – data line failure – V6	**Wiring, TCM, ECM**
70	P1705	AT – gear shift malfunction	**Wiring, range position switch, ECM, TCM**
70	P1705	AT – lock-up clutch not engaging	**Wiring, range position switch, ECM, TCM**
70	P1706	AT – gear shift malfunction	**Wiring, range position switch, ECM, TCM**
70	P1706	AT – lock-up clutch malfunction	**Wiring, range position switch, ECM, TCM**
70	P1738	Automatic transmission – except V6	**Wiring, clutch pressure switch 2, ECM**
70	P1739	Automatic transmission – except V6	**Wiring, clutch pressure switch 3, ECM**
70	P1753	AT – lock-up clutch not engaging/ disengaging	**Wiring, lock-up control solenoid A, ECM, TCM**
70	P1758	AT – lock-up clutch not engaging	**Wiring, lock-up control solenoid B, ECM, TCM**
70	P1768	AT – no gear shift – except V6	**Wiring, clutch pressure control solenoid A, ECM**
70	P1768	AT – poor gear shift – V6	**Wiring, linear solenoid, TCM**
70	P1768	AT – lock-up clutch not engaging – V6	**Wiring, linear solenoid, TCM**
70	P1773	AT – no gear shift – except V6	**Clutch pressure control solenoid B, ECM**
70	P1773	AT – lock-up clutch not engaging – except V6	**Clutch pressure control solenoid B, ECM**
70	P1786	AT – poor gear shift – V6	**Communication wire, ECM**
70	P1790	AT – lock-up clutch not engaging – V6	**Wiring, TP sensor, TCM**
70	P1791	AT – lock-up clutch not engaging – V6	**Wiring, VSS, TCM**
70	P1792	AT – lock-up clutch not engaging – V6	**Wiring, ECT sensor, TCM**
70	P1794	Automatic transmission – BARO signal – V6	**Wiring, ECM, TCM**

Flash code	OBD-II code	Fault location	Probable cause
71	–	Cylinder No.1 – misfire	Fuel/ignition/EGR system, injector, wiring, mechanical fault
72	–	Cylinder No.2 – misfire	Fuel/ignition/EGR system, injector, wiring, mechanical fault
73	–	Cylinder No.3 – misfire	Fuel/ignition/EGR system, injector, wiring, mechanical fault
74	–	Cylinder No.4 – misfire	Fuel/ignition/EGR system, injector, wiring, mechanical fault
75	–	Cylinder No.5 – misfire	Fuel/ignition/EGR system, injector, wiring, mechanical fault
76	–	Cylinder No.6 – misfire	Fuel/ignition/EGR system, injector, wiring, mechanical fault
–	P1300	Random misfire	Ignition/fuel/EGR system, MAP sensor, IAC valve
80	–	Exhaust gas recirculation (EGR) solenoid – insufficient flow	EGR valve, hose leak/blockage
86	–	Engine coolant temperature (ECT) sensor – range/performance problem	ECT sensor, cooling system
90	P1456	Evaporative emission (EVAP) canister purge system (fuel tank system) – leak detected	Hose, fuel tank/pressure sensor, fuel filler cap, EVAP valve/bypass solenoid, EVAP two way valve, EVAP canister/vent valve
90	P1457	Evaporative emission (EVAP) canister purge system (canister system) – leak detected	Hose, fuel tank/pressure sensor, EVAP valve/bypass solenoid, EVAP two way valve, EVAP canister/vent valve
91	–	Fuel tank pressure sensor – range/ performance problem/circuit/voltage low/high	Hose leak/blockage, wiring, pressure sensor, ECM
92	–	Evaporative emission (EVAP) canister purge system – incorrect flow	Wiring, EVAP solenoid/flow switch, throttle body, hose leak/blockage, ECM
92	P1459	Evaporative emission (EVAP) canister purge system – switch malfunction	Wiring, EVAP flow switch, hose leak/ blockage, ECM
–	P1607	Engine control module (ECM) – internal circuit failure	ECM

HONDA

Model:	Engine identification:	Year:
Accord 2.3L	F23A1/4/5	1998-02
Accord 2.4L	F24A4	2003-04
Accord 3.0L	J30A1	1998-04
Civic 1.7L	D17A1/A2/A6/A7	2001-04
Civic 2.0L	K20A3	2002-04
CR-V 2.4L	K24A1	2002-04
Odyssey 3.5L	J35A1/A4	1999-04
S2000 2.0L	F20C1	2000-03
S2000 2.2L	F22C1	2004

System: **PGM-FI**

Self-diagnosis

General information

- Refer to the front of this manual for general test conditions, terminology, detailed descriptions of wiring faults and a general trouble shooter for electrical and mechanical faults.
- Engine control module (ECM) incorporates self-diagnosis function.
- Malfunction indicator lamp (MIL) will illuminate if certain faults are recorded.
- ECM operates in backup mode if sensors fail, to enable vehicle to be driven to workshop.
- Trouble codes can be displayed by using a Scan Tool connected to the data link connector (DLC) Accord **1**, CR-V/Odyssey **2**, S2000 2000-01 **3** or Civic/S2000 2002→ **4** or by the malfunction indicator lamp (MIL) with the data link connector (DLC) terminals jumped **5**.

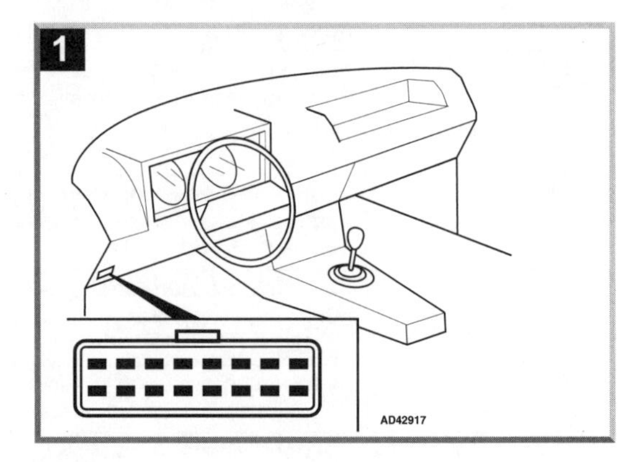

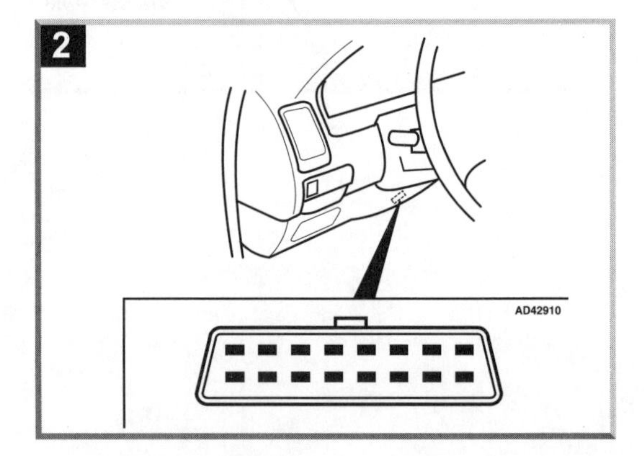

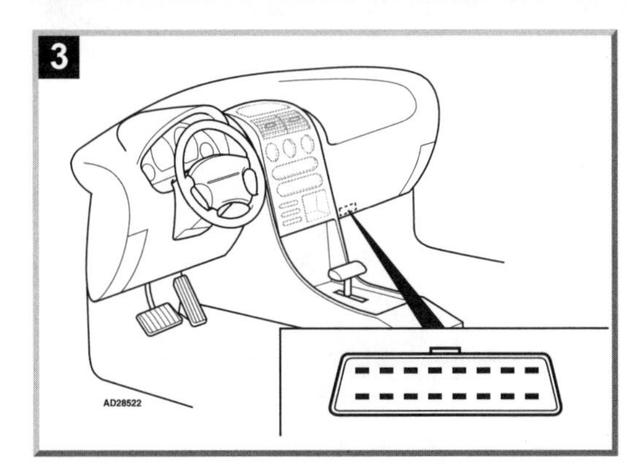

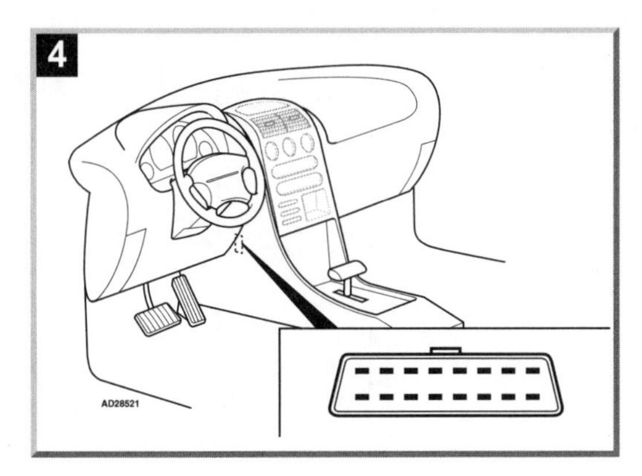

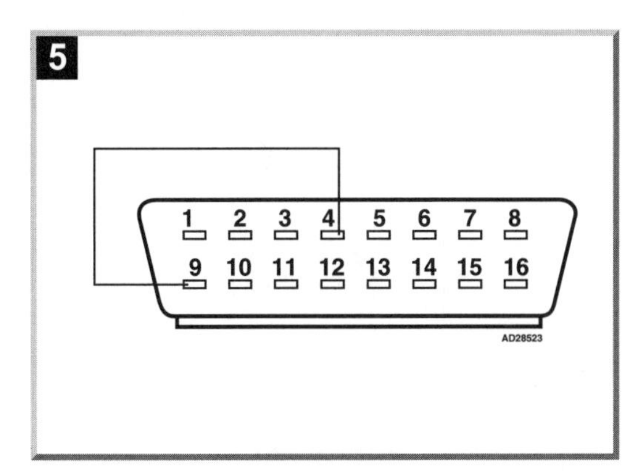

NOTE: ***The use of a Scan Tool is essential to obtain full diagnostic information.***

Accessing

- Ensure ignition switched OFF.
- Jump data link connector (DLC) terminals 4 and 9 **5**.
- Switch ignition ON.
- Check MIL is flashing.
- Trouble codes 1-9 are indicated as follows:
  - Individual short flashes display trouble code **6** [**A**].
  - A short pause separates each flash **6** [**B**].
  - A long pause separates each trouble code **6** [**C**].
  - For example: Trouble code 3 displayed **6**.
- Trouble codes greater than 9 are indicated as follows:
  - Long flashes indicate the 'tens' of the trouble code **7** [**A**].
  - Short flashes indicate the 'units' of the trouble code **7** [**C**].
  - A short pause separates each flash **7** [**B**].
  - A long pause separates each trouble code **7** [**D**].
  - For example: Trouble code 12 displayed **7**.
- Count MIL flashes and compare with trouble code table.

NOTE: ***If a trouble code is displayed but not listed in the trouble code table, suspect engine control module (ECM) fault.***

- Switch ignition OFF.
- Remove jump lead.

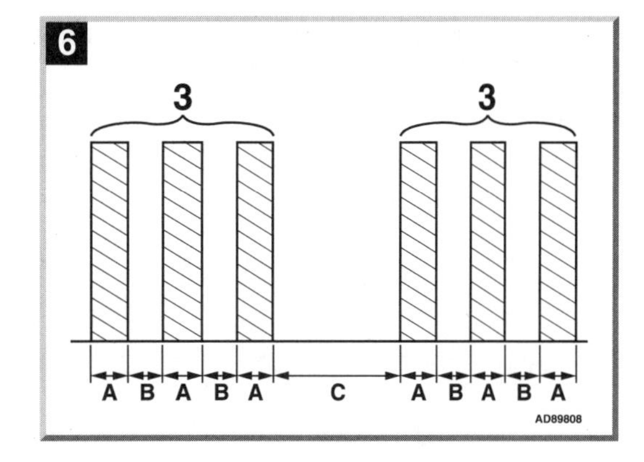

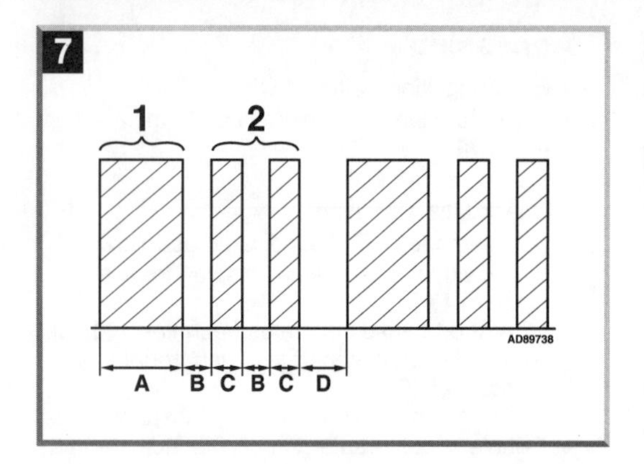

Erasing – except Accord 2003→

- After the faults have been rectified, trouble codes can be erased by using a Scan Tool connected to the data link connector (DLC) or as follows:
- Switch ignition OFF.
- Accord: Remove clock fuse (7.5A) from fascia fusebox for 10 seconds minimum **8**.
- Civic/CR-V: Remove ECM fuse (15A) from underhood fusebox for 10 seconds minimum **9**.
- Odyssey: Remove Clock/Back Up fuse (7.5A) from fascia fusebox for 10 seconds minimum: 1999 – **10** or 2000-04 – **11**.
- S2000: Remove Back Up fuse (7.5A) from fascia fusebox for 60 seconds minimum **12**.
- Reinstall fuse.
- Repeat checking procedure to ensure no data remains in ECM fault memory.

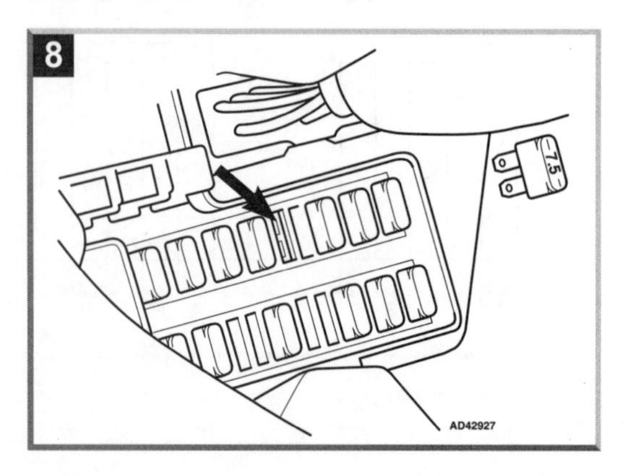

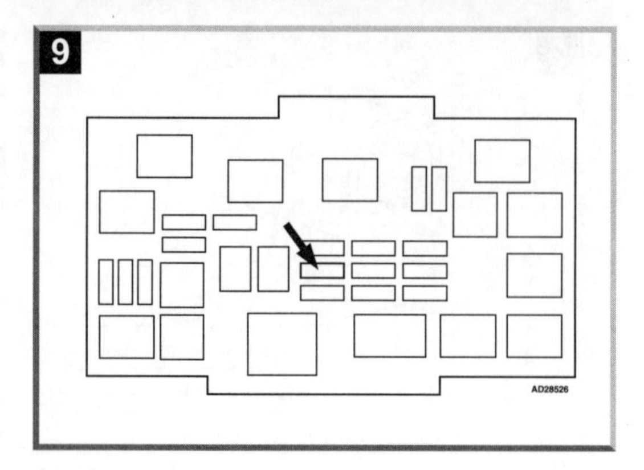

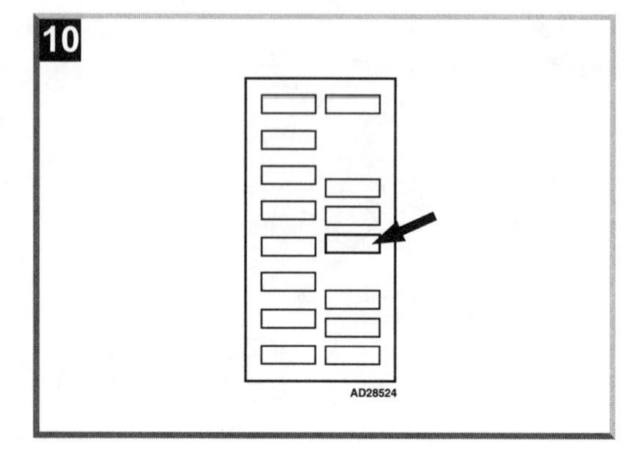

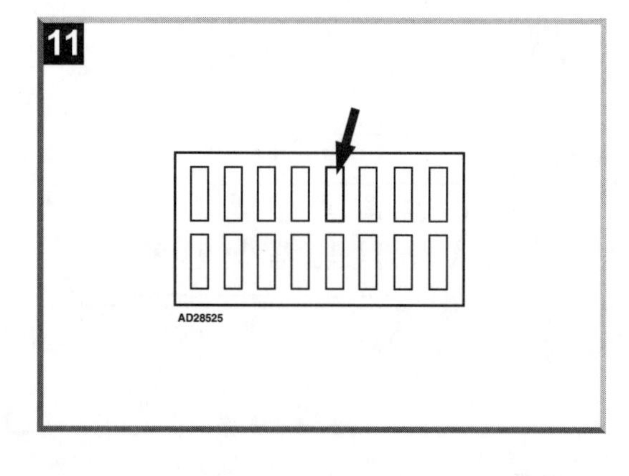

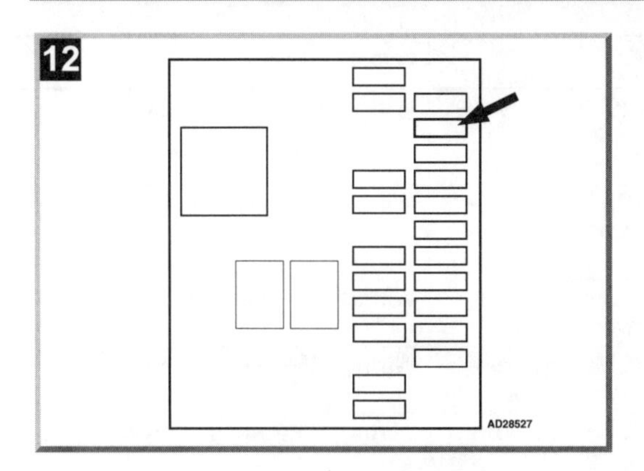

Erasing – Accord 2003 →

- Trouble codes can only be erased by using a Scan Tool connected to the data link connector (DLC) 1.

Trouble code identification – permanent trouble codes

Flash code	Permanent OBD-II code	Fault location	Probable cause
–	*All P0, P2 and U0 codes*	Refer to OBD-II trouble code tables at the front of this manual	–
0	P1630	Transmission control module (TCM) processor error	**TCM**
1	–	Heated oxygen sensor (HO2S) – front – circuit/voltage low/high	**Wiring open/short circuit, HO2S, fuel system, ECM**
3	–	Manifold absolute pressure (MAP) sensor – circuit/voltage high/low	**Wiring, MAP sensor, ECM**
4	–	Crankshaft position (CKP) sensor – circuit malfunction/range/ performance problem	**Wiring, CKP sensor, valve timing, ECM**
5	P1128	Manifold absolute pressure (MAP) sensor – pressure lower than expected	**MAP sensor**
5	P1129	Manifold absolute pressure (MAP) sensor – pressure higher than expected	**MAP sensor**
6	–	Engine coolant temperature (ECT) sensor – circuit/voltage low/high	**Wiring short/open circuit, ECT sensor, ECM**
7	–	Throttle position (TP) sensor – circuit/voltage low/high	**Wiring, TP sensor, ECM**
7	P1121	Throttle position (TP) sensor – position lower than expected	**TP sensor**
7	P1122	Throttle position (TP) sensor – position higher than expected	**TP sensor**
Accord F23A1/4/5: 8	P1359	Crankshaft position (CKP) sensor – top dead center sensor 1 – connector disconnection	**Wiring, ECM**
Except Accord F23A1/4/5: 8	P1359	Camshaft position (CMP) sensor – top dead center sensor 1 – connector disconnection	**Wiring, ECM**

Flash code	Permanent OBD-II code	Fault location	Probable cause
Accord F23A1/4/5: 8	P1361	Crankshaft position (CKP) sensor – top dead center sensor 1 – intermittent signal	**Wiring, CKP sensor, ECM**
Except Accord F23A1/4/5: 8	P1361	Camshaft position (CMP) sensor – top dead center sensor 1 – intermittent signal	**Wiring, CMP sensor, ECM**
Accord F23A1/4/5: 8	P1362	Crankshaft position (CKP) sensor – top dead center sensor 1 – no signal	**Wiring, CKP sensor, ECM**
Except Accord F23A1/4/5: 8	P1362	Camshaft position (CMP) sensor – top dead center sensor 1 – no signal	**Wiring, CMP sensor, ECM**
9	P1381	Camshaft position (CMP) sensor – intermittent signal	**Wiring, CMP sensor, ECM**
9	P1382	Camshaft position (CMP) sensor – no signal	**Wiring, CMP sensor, ECM**
10	–	Intake air temperature (IAT) sensor – circuit/voltage low/high	**Wiring short/open circuit, IAT sensor, ECM**
12	–	Exhaust gas recirculation (EGR) system – range/performance problem	**Hose leak/blockage, wiring, EGR valve/solenoid**
12	–	Exhaust gas recirculation (EGR) valve position sensor A – high input	**Wiring short to positive, EGR valve position sensor, ECM**
12	P1491	Exhaust gas recirculation (EGR) system – valve lift insufficient	**Wiring, EGR valve/position sensor, ECM**
12	P1498	Exhaust gas recirculation (EGR) valve position sensor – voltage high	**Wiring, EGR valve/position sensor, ECM**
13	P1106	Barometric pressure (BARO) sensor – range/performance problem	**ECM**
13	P1107	Barometric pressure (BARO) sensor – circuit/voltage low	**ECM**
13	P1108	Barometric pressure (BARO) sensor – circuit/voltage high	**ECM**
13	P1109	Barometric pressure (BARO) sensor – range/performance problem	**ECM**
14	–	Idle control system – malfunction	**IAC valve, idle speed, hose leak**
14	P1519	Idle air control (IAC) valve – circuit failure	**Wiring, IAC valve, ECM**
17	–	Vehicle speed sensor (VSS) – circuit malfunction/range/ performance problem	**Wiring, speedometer, VSS, ECM**
20	P1297	Electrical load sensor – circuit/ voltage low	**Wiring short circuit, electrical load sensor, ECM**
20	P1298	Electrical load sensor – circuit/ voltage high	**Wiring open circuit, electrical load sensor, ECM**
21	P1253	VTEC system malfunction	**Wiring, VTEC solenoid/pressure switch, ECM**

Flash code	Permanent OBD-II code	Fault location	Probable cause
22	P1257	VTEC system malfunction	Wiring, VTEC solenoid/pressure switch, ECM
22	P1258	VTEC system malfunction	Wiring, VTEC solenoid/pressure switch, ECM
22	P1259	VTEC system malfunction	Wiring, VTEC solenoid/pressure switch, ECM
23	–	Knock sensor (KS) – circuit malfunction	Wiring, KS, ECM
23	P1324	Knock sensor (KS) – circuit/voltage low	Wiring, poor connection, KS, ECM
34	–	System voltage – malfunction	Wiring, poor connection, battery, alternator
35	P1676	ABS/TCS control module – communication malfunction	Wiring, ABS/TCS control module, ECM
35	P1678	ABS/TCS control module – communication malfunction	Wiring, ABS/TCS control module, ECM
39	–	CAN data bus – malfunction	Wiring, connected system, ECM
40	P1683	Throttle motor – default position spring performance problem	Wiring, throttle sticking/ mechanically damaged, throttle motor
40	P1684	Throttle motor – return spring performance problem	Wiring, throttle sticking/ mechanically damaged, throttle motor
Civic D17A6/ K20A3/CR-V/ Accord F23A4: 41	–	Air fuel (A/F) ratio sensor – front – no activity detected	Wiring, A/F ratio sensor, ECM
Except Civic D17A6/K20A3/ CR-V/Accord F23A4: 41	–	Heated oxygen sensor (HO2S) – front – circuit malfunction	Wiring, HO2S, ECM
Civic D17A6/ K20A3/CR-V/ Accord F23A4: 41	P1166	Air fuel (A/F) ratio sensor – front – heater circuit problem	Wiring, A/F ratio sensor, ECM
Except Civic D17A6/K20A3/ CR-V/Accord F23A4: 41	P1166	Heated oxygen sensor (HO2S) – front – heater circuit problem	Wiring, HO2S, ECM
Civic D17A6/ K20A3/CR-V/ Accord F23A4: 41	P1167	Air fuel (A/F) ratio sensor – front – heater circuit malfunction	Wiring, A/F ratio sensor
Except Civic D17A6/K20A3/ CR-V/Accord F23A4: 41	P1167	Heated oxygen sensor (HO2S) – front – heater circuit malfunction	Wiring, HO2S

Flash code	Permanent OBD-II code	Fault location	Probable cause
45	–	Mixture too lean/rich	Fuel system, front HO2S, injectors, MAP sensor, wiring, mechanical fault
Civic D17A6/ K20A3/CR-V/ Accord F24A4: 48	P1157	Air fuel (A/F) ratio sensor – front – circuit malfunction	Wiring, A/F ratio sensor, ECM
Civic D17A6/ K20A3/CR-V: 48	P1158	Air fuel (A/F) ratio sensor – front – low voltage negative terminal	Wiring, A/F ratio sensor, ECM
Civic D17A6/ K20A3/CR-V: 48	P1159	Air fuel (A/F) ratio sensor – front – low voltage positive terminal	Wiring, A/F ratio sensor, ECM
Civic D17A6/ K20A3/Accord F23A4: 48	P1162	Air fuel (A/F) ratio sensor – front – circuit malfunction	Wiring, A/F ratio sensor
Except Civic D17A6/K20A3/ Accord F23A4: 48	P1162	Heated oxygen sensor (HO2S) – front – circuit malfunction	Wiring, HO2S
54	–	Crankshaft position (CKP) sensor B – circuit malfunction	Wiring, poor connection, CKP sensor, ECM
56	P1009	Camshaft position (CMP), intake/left/ front, bank 1 – circuit malfunction/ timing over-advanced/system performance	Wiring, valve timing, engine mechanical fault, CMP actuator, ECM
57	–	Camshaft position (CMP) sensor A, bank 1 – circuit malfunction/range/ performance problem	Insecure sensor/rotor, air gap, wiring, poor connection, CMP sensor, ECM
58	P1366	Camshaft position (CMP) sensor – top dead center sensor 2 – intermittent signal	Wiring, CMP sensor, ECM
58	P1367	Camshaft position (CMP) sensor – top dead center sensor 2 – no signal	Wiring, CMP sensor, ECM
60	–	Secondary air injection (AIR) system – malfunction/incorrect flow detected	Wiring, AIR pump, AIR valve, AIR hose(s), AIR solenoid, ECM
60	P1410	Secondary air injection (AIR) pump – malfunction	Wiring, AIR pump relay, AIR pump, ECM
61	–	Heated oxygen sensor (HO2S) – front – slow response	HO2S, exhaust system
Civic D17A6/ Accord F23A4: 61	P1149	Air fuel (A/F) ratio sensor – front – range/performance problem	A/F ratio sensor
Except Civic D17A6/Accord F23A4: 61	P1149	Heated oxygen sensor (HO2S) – front – range/performance problem	HO2S
Except Civic D17A6/K20A3/ CR-V/Accord F23A4: 61	P1163	Heated oxygen sensor (HO2S) – front – circuit/slow response	HO2S, exhaust system

Flash code	Permanent OBD-II code	Fault location	Probable cause
Civic D17A6/ K20A3/CR-V/ Accord F23A4: 61	P1163	Air fuel (A/F) ratio sensor – front – circuit/slow response	A/F ratio sensor, exhaust system
Except Civic D17A6/K20A3/ CR-V/Accord F23A4: 61	P1164	Heated oxygen sensor (HO2S) – front – range/performance problem	HO2S
Civic D17A6/ K20A3/CR-V/ Accord F23A4: 61	P1164	Air fuel (A/F) ratio sensor – front – range/performance problem	A/F ratio sensor
Except Civic D17A6/K20A3/ Accord F23A4: 61	P1165	Heated oxygen sensor (HO2S) – front – range/performance problem	HO2S
Civic D17A6/ K20A3/Accord F23A4: 61	P1165	Air fuel (A/F) ratio sensor – front – range/performance problem	A/F ratio sensor
63	–	Heated oxygen sensor (HO2S) – rear – slow response/circuit/voltage low/high	Wiring, HO2S, ECM
65	–	Heated oxygen sensor (HO2S) – rear – heater circuit problem	Wiring, HO2S, ECM
67	–	Catalytic converter – efficiency below limit	Catalytic converter, rear HO2S
70	–	AT – lock-up clutch not engaging/no gear shift	Wiring, mainshaft speed sensor, countershaft speed sensor, lock-up control system, shift solenoid (SS) A/B/C, ECM, TCM
70	–	CVT – poor acceleration	Wiring, ignition coil, TCM
70	P1705	AT – gear shift malfunction	Wiring, range position switch, ECM
70	P1705	CVT – gear shift malfunction	Wiring, range position switch, TCM
70	P1706	Automatic transmission	Wiring, range position switch, ECM
70	P1706	CVT – gear shift malfunction	Wiring, range position switch, TCM
70	P1717	AT – gear selection malfunction	Wiring, range position switch, ECM
70	P1730	AT – gear shift malfunction	Transmission fluid, wiring, shift solenoid A/B/D, ECM
70	P1731	AT – gear shift malfunction	Transmission fluid, wiring, shift solenoid E, AT clutch pressure control solenoid A, ECM
70		AT – gear shift malfunction	Transmission fluid, wiring, shift solenoid B/C, ECM

Flash code	Permanent OBD-II code	Fault location	Probable cause
70	P1733	AT – gear shift malfunction	**Transmission fluid, wiring, shift solenoid D, AT clutch pressure control solenoid C, ECM**
70	P1734	AT – gear shift malfunction	**Transmission fluid, wiring, shift solenoid B/C, ECM**
70	P1738	Automatic transmission	**Wiring, clutch pressure switch 2, ECM**
70	P1739	Automatic transmission	**Wiring, clutch pressure switch 3, ECM**
70	P1740	AT – 4th clutch pressure switch	**Wiring, range position switch, 4th clutch pressure switch, ECM**
70	P1750	AT – hydraulic system mechanical malfunction	**Wiring, shift solenoids, ECM, transmission mechanical fault**
70	P1751	AT – hydraulic system mechanical malfunction	**Wiring, shift solenoids, ECM, transmission mechanical fault**
70	P1753	AT – lock-up clutch not engaging/ disengaging	**Wiring, lock-up control solenoid, ECM**
70	P1753	AT – no gear shift	**Wiring, lock-up control solenoid, ECM**
70	P1768	AT – no gear shift	**Wiring, clutch pressure control solenoid A, ECM**
70	P1768	AT – lock-up clutch not engaging	**Wiring, clutch pressure control solenoid A, ECM**
70	P1773	AT – no gear shift	**clutch pressure control solenoid B, ECM**
70	P1773	AT – lock-up clutch not engaging	**clutch pressure control solenoid B, ECM**
70	P1790	AT – lock-up clutch not engaging	**Wiring, TP sensor, TCM**
70	P1792	AT – lock-up clutch not engaging	**Wiring, ECT sensor, TCM**
70	P1793	Automatic transmission	**Wiring, MAP sensor, TCM**
70	P1870	CVT – poor acceleration	**Wiring, shift control linear solenoid, TCM**
70	P1873	CVT – poor acceleration	**Wiring, PH-PL control linear solenoid, TCM**
70	P1876	CVT – poor acceleration	**Wiring, drive pulley pressure control valve**
70	P1877	CVT – poor acceleration	**Wiring, drive pulley pressure control valve**
70	P1878	CVT – poor acceleration	**Wiring, start clutch control linear solenoid, TCM**
70	P1879	CVT – poor acceleration	**Wiring, start clutch control linear solenoid, TCM**
70	P1880	CVT – poor acceleration	**Wiring, start clutch control linear solenoid, TCM**

Flash code	Permanent OBD-II code	Fault location	Probable cause
70	P1881	CVT – poor acceleration	Wiring, start clutch control linear solenoid, TCM
70	P1882	Constantly variable transmission (CVT)	Wiring, inhibitor solenoid, TCM
70	P1885	CVT – poor acceleration	Wiring, drive pulley speed sensor, TCM
70	P1886	CVT – poor acceleration	Wiring, driven pulley speed sensor, TCM
70	P1888	CVT – poor acceleration	Wiring, speed sensor 1, TCM
70	P1889	CVT – poor acceleration	Wiring, speed sensor 2, TCM
70	P1890	CVT – poor acceleration	Shift control system
70	P1891	CVT – poor acceleration	Start clutch control system
70	P1892	CVT – poor acceleration	Wiring, drive pulley pressure control valve, TCM
70	P1893	CVT – poor acceleration	Wiring, drive pulley pressure control valve, TCM
70	P1894	CVT – poor acceleration	Wiring, drive pulley pressure control valve, TCM
70	P1895	CVT – poor acceleration	Wiring, drive pulley pressure control valve, TCM
70	P1896	CVT – poor acceleration	Wiring, drive pulley pressure control valve, TCM
70	P1897	CVT – poor acceleration	Wiring, drive pulley pressure control valve, TCM
70	P1898	CVT – poor acceleration	Wiring, drive pulley pressure control valve, TCM
70	P1899	CVT – poor acceleration	Wiring, drive pulley pressure control valve, TCM
71	–	Cylinder No.1 – misfire	Fuel/ignition/VTEC system, injector, wiring, mechanical fault
72	–	Cylinder No.2 – misfire	Fuel/ignition/VTEC system, injector, wiring, mechanical fault
73	–	Cylinder No.3 – misfire	Fuel/ignition/VTEC system, injector, wiring, mechanical fault
74	–	Cylinder No.4 – misfire	Fuel/ignition/VTEC system, injector, wiring, mechanical fault
75	–	Cylinder No.5 – misfire	Fuel/ignition/VTEC system, injector, wiring, mechanical fault
76	–	Cylinder No.6 – misfire	Fuel/ignition/VTEC system, injector, wiring, mechanical fault
–	P1399	Random misfire	Fuel/ignition system, MAP sensor, IAC valve
80	–	Exhaust gas recirculation (EGR) system – insufficient flow	EGR valve, EGR solenoid, hose leak/blockage

Flash code	Permanent OBD-II code	Fault location	Probable cause
83	–	Secondary air injection (AIR) pump current sensor – circuit/voltage high/ low	**Wiring, current sensor, ECM**
83	P1415	AIR pump electrical current sensor – circuit low	**Wiring, AIR pump electrical current sensor, ECM**
83	P1416	AIR pump electrical current sensor – circuit high	**Wiring, AIR pump electrical current sensor, ECM**
86	–	Engine coolant temperature (ECT) sensor – range/performance problem	**ECT sensor, cooling system**
87	–	Thermostat – range/performance problem	**Thermostat, cooling system**
90	–	Evaporative emission (EVAP) system – small leak detected	**Mechanical fault, hose connection(s), intake leak, EVAP canister, EVAP canister purge valve, EVAP pressure sensor**
90	P1456	Evaporative emission (EVAP) canister purge system (fuel tank system) – leak detected	**Hose, fuel tank/pressure sensor, fuel filler cap, EVAP valve/bypass solenoid, EVAP two way valve, EVAP canister/vent valve, wiring**
90	P1457	Evaporative emission (EVAP) canister purge system (canister system) – leak detected	**Hose, fuel tank/pressure sensor, fuel filler cap, EVAP valve/bypass solenoid, EVAP two way valve, EVAP canister/vent valve, wiring**
91	–	Fuel tank pressure sensor – range/ performance problem/circuit/voltage low/high	**Hose leak/blockage, wiring, pressure sensor, ECM**
91	P1454	Fuel tank pressure sensor – range/ performance problem	**Wiring, hose, fuel tank pressure sensor, fuel filler cap, EVAP canister/vent valve, ECM**
91	P1457	Fuel tank pressure sensor – range/ performance problem	**Hose leak/blockage, wiring, pressure sensor, ECM**
92	–	Evaporative emission (EVAP) canister purge valve – circuit malfunction	**Wiring, EVAP canister purge valve, ECM**
Civic D17A7: 95	–	Fuel rail pressure (FRP) sensor – range/performance problem/ circuit malfunction	**Wiring, FRP sensor**
Civic D17A7: 96	P1182	Fuel temperature sensor – low input	**Wiring short to ground, fuel temperature sensor, ECM**
Civic D17A7: 96	P1183	Fuel temperature sensor – high input	**Wiring short to positive, fuel temperature sensor, ECM**
Civic D17A7: 97	P1192	Fuel tank pressure sensor – low input	**Wiring short to ground, fuel tank sensor**
Civic D17A7: 97	P1193	Fuel tank pressure sensor – high input	**Wiring short to positive, fuel tank sensor**
Civic D17A7: 98	P1187	Fuel tank temperature sensor – low input	**Wiring short to ground, fuel tank temperature sensor, ECM**

Flash code	Permanent OBD-II code	Fault location	Probable cause
Civic D17A7: 98	P1188	Fuel tank temperature sensor – high input	**Wiring short to positive, fuel tank temperature sensor, ECM**
–	P1607	Engine control module (ECM) – internal circuit failure A	**ECM**
101	–	Heated oxygen sensor (HO2S) 3 – circuit/voltage low	**Wiring short circuit, HO2S, ECM**
103	–	Heated oxygen sensor (HO2S) 3 – slow response/circuit/voltage high	**Wiring open circuit, HO2S, ECM**
104	–	Heated oxygen sensor (HO2S) 3 – heater circuit problem	**Wiring, HO2S, ECM**
105	P1420	NOx adsorptive catalyst system – efficiency below limit	**NOx adsorptive three way catalyst**
106	P1077	Intake manifold tuning valve system malfunction – low RPM	**Hose leak/blockage, wiring, intake manifold tuning valve, intake manifold tuning valve actuator, ECM**
106	P1078	Intake manifold tuning valve system malfunction – high RPM	**Hose leak/blockage, wiring, intake manifold tuning valve, intake manifold tuning valve actuator, ECM**
107	–	Intake manifold tuning valve position sensor – circuit/voltage high/low	**Wiring, intake manifold tuning valve position sensor, ECM**
109	P1505	Positive crankcase ventilation (PCV) – air leak	**Hose leak/blockage, PCV system, throttle body**
111	P1130	Heated oxygen sensor (HO2S) 2/3 – faulty components	**HO2S 2, HO2S 3**
117	–	Evaporative emission system, vent control – circuit low/high	**Wiring, EVAP valve**
122	–	Output shaft speed (OSS) sensor – circuit malfunction	**Wiring, OSS sensor, ECM**
131	–	Engine control module (ECM) – KAM error	**ECM**
135	–	ECM power relay control – circuit open	**Wiring, ECM relay**
151	–	Air fuel (A/F) ratio sensor 1, bank 1 – no activity detected	**Wiring, A/F ratio sensor, ECM**
151	–	Air fuel (A/F) ratio sensor 1, bank 1, heater control – circuit malfunction	**Wiring, A/F ratio sensor, ECM**
152	–	Air fuel (A/F) ratio sensor 1, bank 2 – no activity detected	**Wiring, A/F ratio sensor, ECM**
152	–	Air fuel (A/F) ratio sensor 1, bank 2, heater control – circuit malfunction	**Wiring, A/F ratio sensor, ECM**
153	–	System too lean/rich, bank 1	**Intake blocked, intake/exhaust leak, EVAP canister purge valve, fuel pressure, EGR system, injector(s), HO2S/A/F ratio sensor, AIR system, hose connection(s)**

Flash code	Permanent OBD-II code	Fault location	Probable cause
154	–	System too lean/rich, bank 2	**Intake blocked, intake/exhaust leak, EVAP canister purge valve, fuel pressure, EGR system, injector(s), HO2S/A/F ratio sensor, AIR system, hose connection(s)**
157	–	Air fuel (A/F) ratio sensor 1, bank 1 – slow response	**Heating inoperative, wiring, A/F ratio sensor**
158	–	Air fuel (A/F) ratio sensor 1, bank 2 – slow response	**Heating inoperative, wiring, A/F ratio sensor**
161	–	Heated oxygen sensor (HO2S) 2, bank 1 – low/high voltage	**Exhaust leak, wiring, HO2S, ECM**
161	–	Heated oxygen sensor (HO2S) 2, bank 1 – slow response	**Heating inoperative, wiring, HO2S, ECM**
162	–	Heated oxygen sensor (HO2S) 2, bank 2 – low/high voltage	**Exhaust leak, wiring, HO2S, ECM**
162	–	Heated oxygen sensor (HO2S) 2, bank 2 – slow response	**Heating inoperative, wiring, HO2S, ECM**
163	–	Heated oxygen sensor (HO2S) 2, bank 1, heater control – circuit malfunction	**Wiring, HO2S, ECM**
164	–	Heated oxygen sensor (HO2S) 2, bank 2, heater control – circuit malfunction	**Wiring, HO2S, ECM**
165	–	Catalytic converter system, bank 1 – efficiency below threshold	**Catalytic converter, wiring, HO2S 2/A/F ratio sensor 2**
166	–	Catalytic converter system, bank 2 – efficiency below threshold	**Catalytic converter, wiring, HO2S 2/A/F ratio sensor 2**

Trouble code identification – temporary trouble code

Flash code	Temporary OBD-II code	Fault location	Probable cause
–	*All P0, P2 and U0 codes*	Refer to OBD-II trouble code tables at the front of this manual	–
1	–	Heated oxygen sensor (HO2S) – front	**Wiring, HO2S, fuel system, ECM**
5	P1128	Manifold absolute pressure (MAP) sensor – pressure lower than expected	**MAP sensor**
5	P1129	Manifold absolute pressure (MAP) sensor – pressure higher than expected	**MAP sensor**
7	P1121	Throttle position (TP) sensor – position lower than expected	**TP sensor**

Flash code	Temporary OBD-II code	Fault location	Probable cause
7	P1122	Throttle position (TP) sensor – position higher than expected	TP sensor
12	P1491	Exhaust gas recirculation (EGR) system – valve lift insufficient	Wiring, EGR valve/position sensor, ECM
13	P1106	Barometric pressure (BARO) sensor – range/performance problem	ECM
14	–	Idle control system – malfunction	IAC valve, idle speed, hose leak
21	P1257	VTEC system malfunction	Wiring, VTEC solenoid/pressure switch, ECM
22	P1257	VTEC system malfunction	Wiring, VTEC solenoid/pressure switch, ECM
45	–	Mixture too lean/rich	Fuel/EVAP system, front HO2S, injectors, wiring, mechanical fault
56	–	Camshaft position (CMP), intake/left/ front, bank 1 – timing over-advanced/system performance	Valve timing, engine mechanical fault, CMP actuator
60	–	Secondary air injection (AIR) system – malfunction/incorrect flow detected	Wiring, AIR pump, AIR valve, AIR hose(s), AIR solenoid, ECM
60	P01410	Secondary air injection (AIR) pump – malfunction	Wiring, AIR pump relay, AIR pump, ECM
61	–	Heated oxygen sensor (HO2S) – front – slow response	HO2S, exhaust system
Civic D17A6: 61	P1149	Air fuel (A/F) ratio sensor – front – range/performance problem	A/F ratio sensor
Except Civic D17A6: 61	P1149	Heated oxygen sensor (HO2S) – front – range/performance problem	HO2S
Except Civic D17A6/K20A3: 61	P1163	Heated oxygen sensor (HO2S) – front – circuit/slow response	HO2S, exhaust system
Civic D17A6/ K20A3: 61	P1163	Air fuel (A/F) ratio sensor – front – circuit/slow response	A/F ratio sensor, exhaust system
Except Civic D17A6/K20A3: 61	P1164	Heated oxygen sensor (HO2S) – front – range/performance problem	HO2S
Civic D17A6/ K20A3: 61	P1164	Air fuel (A/F) ratio sensor – front – range/performance problem	A/F ratio sensor
Except Civic D17A6/K20A3: 61	P1165	Heated oxygen sensor (HO2S) – front – range/performance problem	HO2S
Civic D17A6/ K20A3: 61	P1165	Air fuel (A/F) ratio sensor – front – range/performance problem	A/F ratio sensor
63	–	Heated oxygen sensor (HO2S) – rear – slow response/circuit/voltage low/high	Wiring, HO2S, ECM
67	–	Catalytic converter – efficiency below limit	Catalytic converter, rear HO2S
71	P1399	Cylinder No.1 – misfire	Fuel/ignition/VTEC system, injector, wiring, mechanical fault

Flash code	Temporary OBD-II code	Fault location	Probable cause
72	P1399	Cylinder No.2 – misfire	**Fuel/ignition/VTEC system, injector, wiring, mechanical fault**
73	P1399	Cylinder No.3 – misfire	**Fuel/ignition/VTEC system, injector, wiring, mechanical fault**
74	P1399	Cylinder No.4 – misfire	**Fuel/ignition/VTEC system, injector, wiring, mechanical fault**
75	P1399	Cylinder No.5 – misfire	**Fuel/ignition/VTEC system, injector, wiring, mechanical fault**
76	P1399	Cylinder No.6 – misfire	**Fuel/ignition/VTEC system, injector, wiring, mechanical fault**
–	P1399	Random misfire	**Fuel/ignition system, MAP sensor, IAC valve**
80	–	Exhaust gas recirculation (EGR) solenoid – insufficient flow	**EGR valve, hose leak/blockage**
86	–	Engine coolant temperature (ECT) sensor – range/performance problem	**ECT sensor, cooling system**
87	P1486	Thermostat – range/performance problem	**Thermostat, cooling system**
90	P1456	Evaporative emission (EVAP) canister purge system (fuel tank system) – leak detected	**Hose, fuel tank/pressure sensor, fuel filler cap, EVAP valve/bypass solenoid, EVAP two way valve, EVAP canister/vent valve, wiring**
90	P1457	Evaporative emission (EVAP) canister purge system (canister system) – leak detected	**Hose, fuel tank/pressure sensor, fuel filler cap, EVAP valve/bypass solenoid, EVAP two way valve, EVAP canister/vent valve, wiring**
91	–	Fuel tank pressure sensor – range/ performance problem/circuit/voltage low/high	**Hose leak/blockage, wiring, pressure sensor, ECM**
101	–	Heated oxygen sensor (HO2S) 3 – circuit/voltage low	**Wiring short circuit, HO2S, ECM**
103	–	Heated oxygen sensor (HO2S) 3 – slow response/circuit/voltage high	**Wiring open circuit, HO2S, ECM**
104	–	Heated oxygen sensor (HO2S) 3 – heater circuit problem	**Wiring, HO2S, ECM**
105	P1420	NOx adsorptive catalyst system – efficiency below limit	**NOx adsorptive three way catalyst**
109	P1505	Positive crankcase ventilation (PCV) – air leak	**Hose leak/blockage, PCV system, throttle body**
111	P1130	Heated oxygen sensor (HO2S) 2/3 – faulty components	**HO2S 2, HO2S 3**

HONDA

Model:	Engine identification:	Year:
Civic 1.5L 8V/16V	**D15B7/B8**	**1992-95**
Civic 1.5/1.6L	**D15Z1/D16Z6**	**1992-95**
Civic del Sol 1.5L/1.6L	**D15B7/D16Z6**	**1993**
Prelude 2.2L	**F22A1/H22A1**	**1992-95**
Prelude 2.3L	**H23A1**	**1992-95**

System: **PGM-FI**

Self-diagnosis

General information

- Refer to the front of this manual for general test conditions, terminology, detailed descriptions of wiring faults and a general trouble shooter for electrical and mechanical faults.
- Engine control module (ECM) incorporates self-diagnosis function.
- Malfunction indicator lamp (MIL) will illuminate if certain faults are recorded.
- ECM operates in backup mode if sensors fail, to enable vehicle to be driven to workshop.
- Trouble codes can be displayed by the malfunction indicator lamp (MIL).

Accessing

- Ensure ignition switched OFF.
- Jump service check connector terminals
  - Civic **1**
  - Prelude **2**
- Switch ignition ON.
- Check MIL is flashing.
- Trouble codes 1-9 are indicated as follows:
  - Individual short flashes display trouble code **3** [**A**].
  - A short pause separates each flash **3** [**B**].
  - A long pause separates each trouble code **3** [**C**].
  - For example: Trouble code 3 displayed **3**.
- Trouble codes greater than 9 are indicated as follows:
  - Long flashes indicate the 'tens' of the trouble code **4** [**A**].
  - Short flashes indicate the 'units' of the trouble code **4** [**C**].
  - A short pause separates each flash **4** [**B**].
  - A long pause separates each trouble code **4** [**D**].
  - For example: Trouble code 12 displayed **4**.
- Count MIL flashes and compare with trouble code table.
- Switch ignition OFF.
- Remove jump lead.

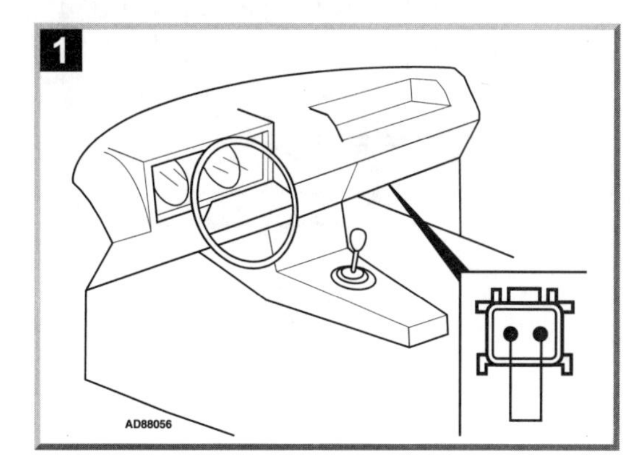

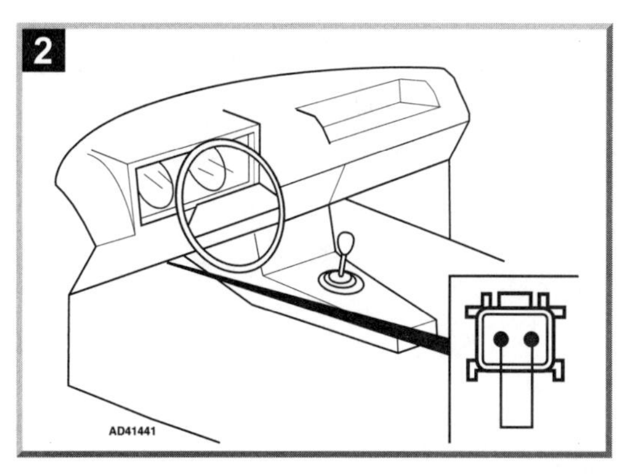

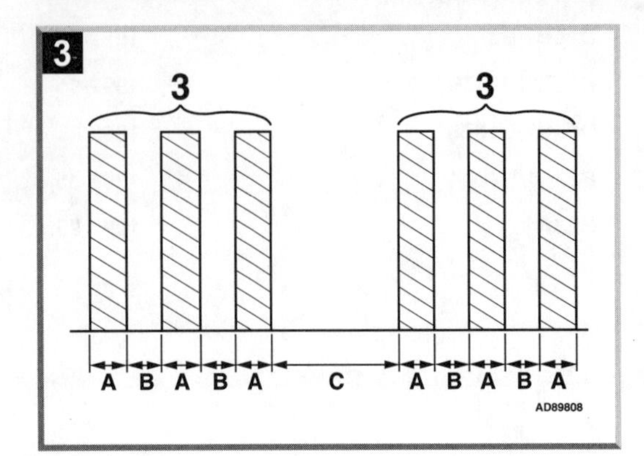

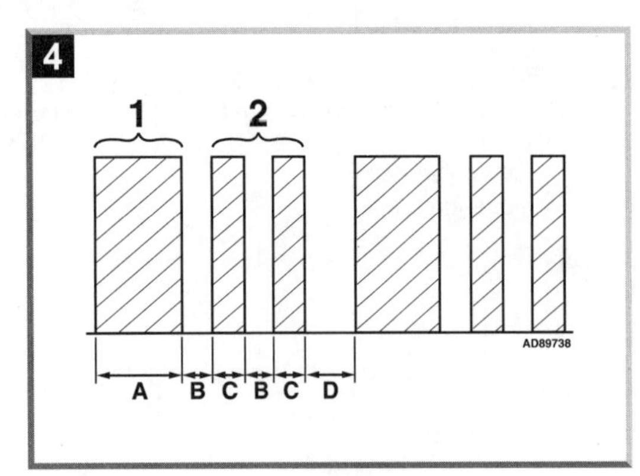

Erasing

- After the faults have been rectified, erase the trouble codes as follows:
- Switch ignition OFF.
- Civic – remove fuse No.32 (7.5A) from underhood fusebox for 10 seconds minimum 5.
- Prelude – remove fuse No.43 (10A) from underhood fusebox for 10 seconds minimum 6.
- Reinstall fuse.
- Repeat checking procedure to ensure no data remains in ECM fault memory.

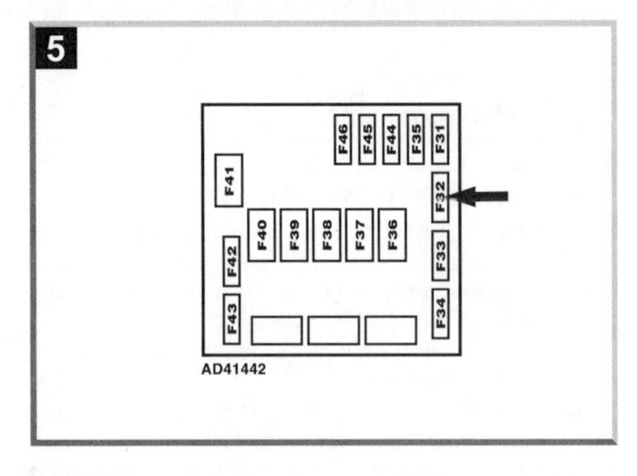

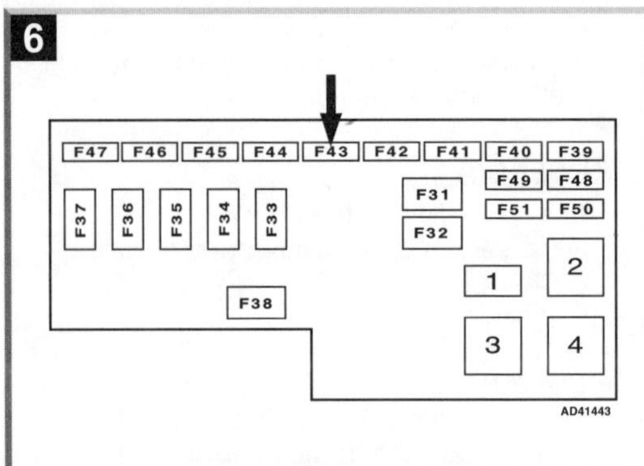

Trouble code identification

Flash code	Fault location	Probable cause
0	Engine control module (ECM)	Wiring, ECM
1	Heated oxygen sensor (HO2S)	Fuel system, wiring, HO2S, ECM
3	Manifold absolute pressure (MAP) sensor	Wiring, MAP sensor, ECM
4	Engine speed (RPM) sensor	Wiring, RPM sensor, ECM
5	Manifold absolute pressure (MAP) sensor	Hose leak
6	Engine coolant temperature (ECT) sensor	Wiring, ECT sensor, ECM
7	Throttle position (TP) sensor	Wiring, TP sensor, ECM
8	Crankshaft position (CKP) sensor	Wiring, CKP sensor, ECM
9	Camshaft position (CMP) sensor	Wiring, CMP sensor, ECM
10	Intake air temperature (IAT) sensor	Wiring, IAT sensor, ECM
12	Exhaust gas recirculation (EGR) system	Hose leak/blockage, wiring, EGR valve/ position sensor, EGR solenoid, ECM
13	Barometric pressure (BARO) sensor	ECM
14	Idle air control (IAC) valve	Wiring, IAC valve, ECM
15	Ignition output signal	Wiring, ICM, ECM
16	Injectors	Wiring, injectors, ECM
17	Vehicle speed sensor (VSS)	Wiring, VSS, ECM
19	Lock-up control solenoid valve (AT)	Wiring, lock-up control solenoid valve, ECM
20	Electrical load sensor	Wiring, electrical load sensor, ECM
21	VTEC system – malfunction	Wiring, VTEC solenoid, ECM
22	VTEC system – malfunction	Wiring, VTEC pressure switch, ECM
23	Knock sensor (KS)	Wiring, KS, ECM
30	AT signal A	Wiring, ECM
31	AT signal B	Wiring, ECM
41	Oxygen sensor heater	Wiring, HO2S, ECM
43	Heated oxygen sensor (HO2S)/fuel supply system	Fuel system, wiring, HO2S, ECM
48	Heated oxygen sensor (HO2S)	Wiring, HO2S, ECM

HONDA

Model:	Engine identification:	Year:
Civic/del Sol 1.6L SOHC	**D16Y7**	**1996-00**
Civic/del Sol 1.6L	**D16Y8**	**1996-00**
Civic 1.6L SOHC	**D16Y5**	**1996-00**
Civic del Sol 1.6L DOHC	**B16A2**	**1996-98**
Civic 1.6 GX	**D16B5**	**1998-00**

System: **PGM-FI**

Self-diagnosis

General information

- Refer to the front of this manual for general test conditions, terminology, detailed descriptions of wiring faults and a general trouble shooter for electrical and mechanical faults.
- Engine control module (ECM) incorporates self-diagnosis function.
- Malfunction indicator lamp (MIL) will illuminate if certain faults are recorded.
- ECM operates in backup mode if sensors fail, to enable vehicle to be driven to workshop.
- Trouble codes can be displayed by using a Scan Tool connected to the data link connector (DLC) **1** [1] or by the malfunction indicator lamp (MIL) with the service check connector jumped **1** [2].

NOTE: ***The use of a Scan Tool is essential to obtain full diagnostic information.***

Accessing

- Ensure ignition switched OFF.
- Jump service check connector terminals **1** [2].
- Switch ignition ON.
- Check MIL is flashing.
- Trouble codes 1-9 are indicated as follows:
  - Individual short flashes display trouble code **2** [A].
  - A short pause separates each flash **2** [B].
  - A long pause separates each trouble code **2** [C].
  - For example: Trouble code 3 displayed **2**.
- Trouble codes greater than 9 are indicated as follows:
  - Long flashes indicate the 'tens' of the trouble code **3** [A].
  - Short flashes indicate the 'units' of the trouble code **3** [C].
  - A short pause separates each flash **3** [B].
  - A long pause separates each trouble code **3** [D].
  - For example: Trouble code 12 displayed **3**.
- Count MIL flashes and compare with trouble code table.

NOTE: ***If a trouble code is displayed but not listed in the trouble code table, suspect engine control module (ECM) fault.***

- Switch ignition OFF.
- Remove jump lead.

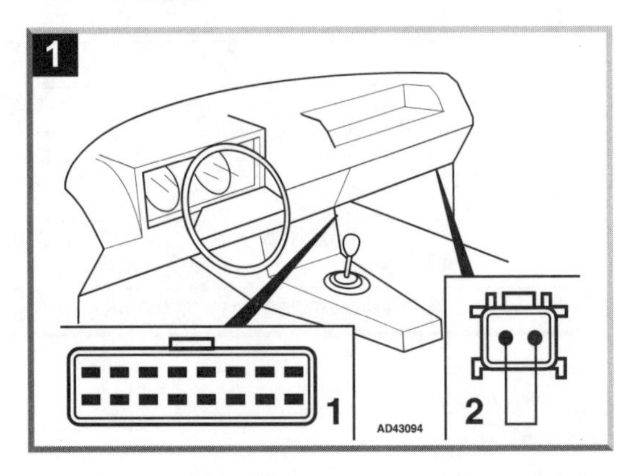

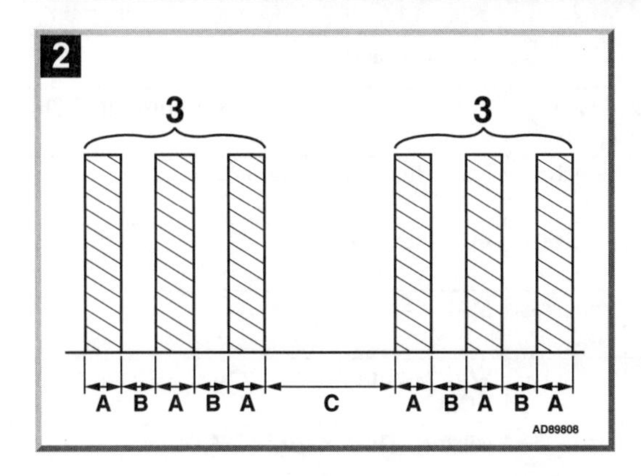

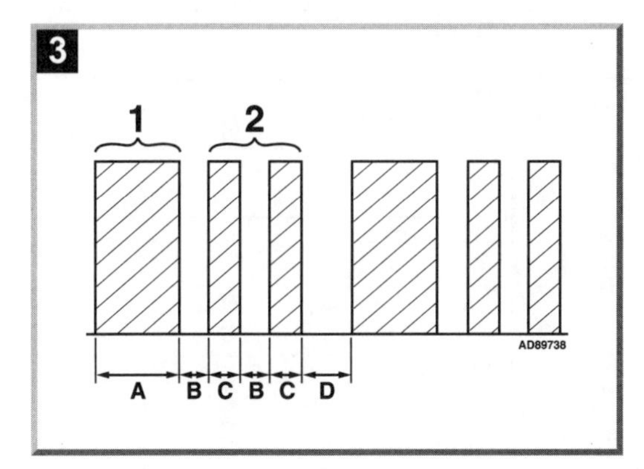

Erasing

- After the faults have been rectified, trouble codes can be erased by using a Scan Tool connected to the data link connector (DLC) 1 [1] or as follows:
- Switch ignition OFF.
- Civic: Remove radio fuse (7.5A) from underhood fusebox for 10 seconds minimum 4.
- Civic del Sol: Remove radio fuse (7.5A) from underhood fusebox for 10 seconds minimum 5.
- Reinstall fuse.
- Repeat checking procedure to ensure no data remains in ECM fault memory.

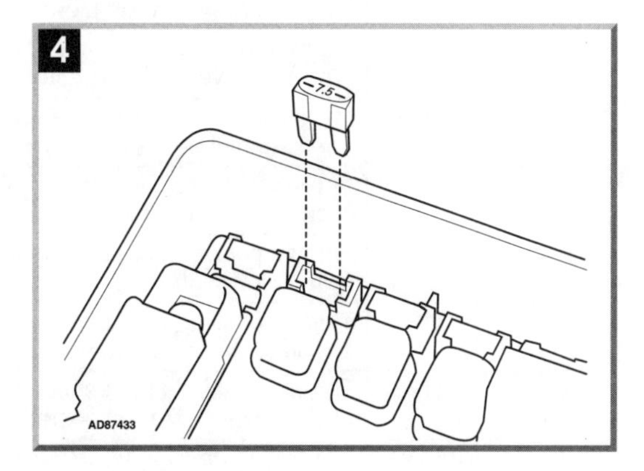

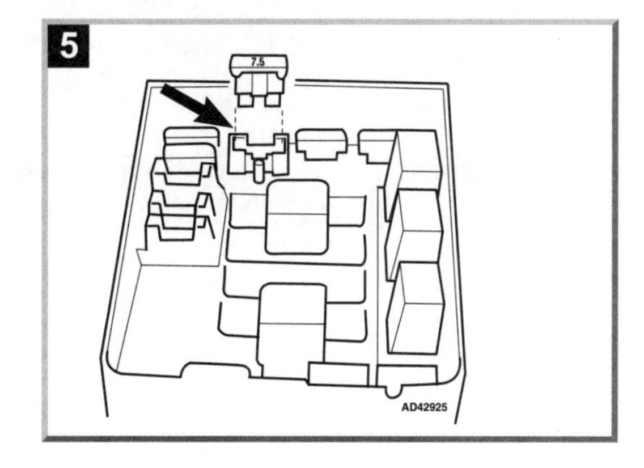

Trouble code identification

Flash code	OBD-II code	Fault location	Probable cause
–	*All P0, P2 and U0 codes*	Refer to OBD-II trouble code tables at the front of this manual	–
1	–	Heated oxygen sensor (HO2S) – front – circuit/voltage low/high	**Wiring open/short circuit, HO2S, fuel system, ECM**
3	–	Manifold absolute pressure (MAP) sensor – circuit/voltage high/low	**Wiring, MAP sensor, ECM**

Flash code	OBD-II code	Fault location	Probable cause
4	–	Crankshaft position (CKP) sensor – circuit malfunction/range/performance problem	**Wiring, CKP sensor, valve timing, ECM**
5	–	Manifold absolute pressure (MAP) sensor – range/performance problem	**Hose leak, MAP sensor**
5	P1128	Manifold absolute pressure (MAP) sensor – pressure lower than expected	**MAP sensor**
5	P1129	Manifold absolute pressure (MAP) sensor – pressure higher than expected	**MAP sensor**
6	–	Engine coolant temperature (ECT) sensor – circuit/voltage low/high	**Wiring short/open circuit, ECT sensor, ECM**
7	–	Throttle position (TP) sensor – circuit/ voltage low/high	**Wiring, TP sensor, ECM**
7	P1121	Throttle position (TP) sensor – position lower than expected	**TP sensor**
7	P1122	Throttle position (TP) sensor – position higher than expected	**TP sensor**
8	P1359	Crankshaft position (CKP) sensor – connector disconnection	**Wiring**
8	P1361	Crankshaft position (CKP) sensor – intermittent signal	**CKP sensor**
8	P1362	Crankshaft position (CKP) sensor – no signal	**Wiring, CKP sensor, ECM**
9	P1381	Camshaft position (CMP) sensor – intermittent signal	**CMP sensor**
9	P1382	Camshaft position (CMP) sensor – no signal	**Wiring, CMP sensor, ECM**
10	–	Intake air temperature (IAT) sensor – circuit/voltage low/high	**Wiring short/open circuit, IAT sensor, ECM**
12	P1491	Exhaust gas recirculation (EGR) system – valve lift insufficient	**Wiring, EGR valve/position sensor, EGR solenoid, hose leak/blockage, ECM**
12	P1498	Exhaust gas recirculation (EGR) valve position sensor – voltage high	**Wiring, EGR valve/position sensor, ECM**
13	P1106	Barometric pressure (BARO) sensor – range/performance problem	**ECM**
13	P1107	Barometric pressure (BARO) sensor – circuit/voltage low	**ECM**
13	P1108	Barometric pressure (BARO) sensor – circuit/voltage high	**ECM**
14	–	Idle control system – malfunction	**IAC valve, throttle body**
14	P1508	Idle air control (IAC) valve – circuit failure	**Wiring, IAC valve, ECM**
14	P1509	Idle air control (IAC) valve – circuit failure	**Wiring, IAC valve, ECM**
17	–	Vehicle speed sensor (VSS) – circuit malfunction/range/performance problem	**Wiring, VSS, ECM**
19	–	AT – lock-up system malfunction	**Lock-up control system**

Flash code	OBD-II code	Fault location	Probable cause
19	P1753	AT – lock-up control solenoid A – circuit failure	Wiring, lock-up control solenoid, ECM
19	P1758	AT – lock-up control solenoid B – circuit failure	Wiring, lock-up control solenoid, ECM
20	P1297	Electrical load sensor – circuit/voltage low	Wiring, electrical load sensor, ECM
20	P1298	Electrical load sensor – circuit/voltage high	Wiring, electrical load sensor, ECM
22	P1259	VTEC system malfunction	Wiring, VTEC solenoid/pressure switch, ECM
23	–	Knock sensor (KS) – circuit malfunction	Wiring, KS, ECM
30	P1655	AT/CVT to ECM signal – failure	Communication wire, ECM
41	–	Heated oxygen sensor (HO2S) – front – circuit malfunction	Wiring, ECM
41	P1166	Heated oxygen sensor (HO2S) – front – heater circuit	Wiring, HO2S, ECM
41	P1167	Heated oxygen sensor (HO2S) – front – heater circuit malfunction	Wiring open circuit, HO2S
45	–	Mixture too lean/rich	Fuel system, front HO2S, injectors, MAP sensor, wiring, mechanical fault
48	P1162	Heated oxygen sensor (HO2S) – front – circuit malfunction	Wiring, HO2S
48	P1168	Heated oxygen sensor (HO2S) – front – voltage low	Wiring short circuit
48	P1169	Heated oxygen sensor (HO2S) – front – voltage high	Wiring
54	P1336	Crankshaft position (CKP) sensor 2 – intermittent signal	CKP sensor
54	P1337	Crankshaft position (CKP) sensor 2 – no signal	Wiring, CKP sensor, ECM
61	–	Heated oxygen sensor (HO2S) – front – slow response	HO2S, exhaust system
61	P1163	Heated oxygen sensor (HO2S) – front – slow response	HO2S
61	P1164	Heated oxygen sensor (HO2S) – front – range/performance problem	HO2S
61	P1165	Heated oxygen sensor (HO2S) – front – range/performance problem	HO2S
63	–	Heated oxygen sensor (HO2S) – rear – slow response/circuit/voltage low/high	Wiring, HO2S, ECM
65	–	Heated oxygen sensor (HO2S) – rear – circuit malfunction	Wiring, ECM
67	–	Catalytic converter – efficiency below limit	Catalytic converter, rear HO2S

Flash code	OBD-II code	Fault location	Probable cause
70	–	AT – lock-up clutch not engaging/no gear shift	**Wiring, mainshaft speed sensor, countershaft speed sensor, lock-up control system, shift solenoid (SS) A/B, ECM, TCM**
70	–	CVT – poor acceleration	**Wiring, ignition coil, TCM**
70	P1705	AT – gear shift malfunction	**Wiring, range position switch, ECM**
70	P1705	AT – lock-up clutch not engaging	**Wiring, range position switch, ECM**
70	P1705	CVT – gear shift malfunction	**Wiring, range position switch, TCM**
70	P1706	AT – gear shift malfunction	**Wiring, range position switch, ECM**
70	P1706	AT – lock-up clutch malfunction	**Wiring, range position switch, ECM**
70	P1706	CVT – gear shift malfunction	**Wiring, range position switch, TCM**
70	P1753	AT – lock-up clutch not engaging/ disengaging	**Wiring, lock-up control solenoid, ECM**
70	P1758	AT – lock-up clutch not engaging	**Wiring, lock-up control solenoid B, ECM**
70	P1768	AT – poor/no gear shift	**Wiring, linear solenoid, ECM**
70	P1768	AT – lock-up clutch not engaging	**Wiring, linear solenoid, ECM**
70	P1790	CVT – kick-down malfunction	**Wiring, TP sensor, TCM**
70	P1790	CVT – gear shift malfunction	**Wiring, TP sensor, TCM**
70	P1791	Constantly variable transmission (CVT)	**Wiring, VSS, TCM**
70	P1793	Constantly variable transmission (CVT)	**Wiring, MAP sensor, TCM**
70	P1870	CVT – poor acceleration	**Wiring, shift control linear solenoid, TCM**
70	P1873	CVT – poor acceleration	**Wiring, PH-PL control linear solenoid, TCM**
70	P1879	CVT – poor acceleration	**Wiring, start clutch control linear solenoid, TCM**
70	P1882	Constantly variable transmission (CVT)	**Wiring, inhibitor solenoid, TCM**
70	P1885	CVT – poor acceleration	**Wiring, drive pulley speed sensor, TCM**
70	P1886	CVT – poor acceleration	**Wiring, driven pulley speed sensor, TCM**
70	P1888	CVT – poor acceleration	**Wiring, secondary gear shaft speed sensor, TCM**
70	P1890	CVT – poor acceleration	**Shift control system**
70	P1891	CVT – poor acceleration	**Start clutch control system**
71	–	Cylinder No.1 – misfire	**Fuel/ignition/EGR system, injector, wiring, mechanical fault**
72	–	Cylinder No.2 – misfire	**Fuel/ignition/EGR system, injector, wiring, mechanical fault**
73	–	Cylinder No.3 – misfire	**Fuel/ignition/EGR system, injector, wiring, mechanical fault**
74	–	Cylinder No.4 – misfire	**Fuel/ignition/EGR system, injector, wiring, mechanical fault**

Flash code	OBD-II code	Fault location	Probable cause
80	–	Exhaust gas recirculation (EGR) solenoid – insufficient flow	EGR valve, hose leak/blockage
86	–	Engine coolant temperature (ECT) sensor – range/performance problem	ECT sensor, cooling system
90	P1456	Evaporative emission (EVAP) canister purge system – leak detected (fuel tank)	Hose, fuel tank/pressure sensor, fuel filler cap, EVAP valve/bypass solenoid, EVAP two way valve, EVAP canister/vent valve
90	P1457	Evaporative emission (EVAP) canister purge system – leak detected (canister)	Hose, fuel tank/pressure sensor, EVAP valve/bypass solenoid, EVAP two way valve, EVAP canister/vent valve
91	–	Fuel tank pressure sensor – circuit/voltage low/high	Hose leak/blockage, wiring, pressure sensor, ECM
92	–	Evaporative emission (EVAP) canister purge system – incorrect flow	Wiring, EVAP solenoid, EVAP canister, hose leak/blockage, ECM
95	–	Fuel rail pressure (FRP) sensor – range/ performance problem/voltage high/low	Wiring open or short to ground, FRP sensor, ECM
96	P1182	Fuel temperature sensor – low input	Wiring short to ground, fuel temperature sensor, ECM
96	P1183	Fuel temperature sensor – high input	Wiring open circuit, fuel temperature sensor, ECM
97	P1192	Fuel tank pressure sensor – low input	Wiring open or short to ground, fuel tank pressure sensor, ECM
97	P1193	Fuel tank pressure sensor – high input	Wiring open circuit, fuel tank pressure sensor, ECM
98	P1187	Fuel tank temperature sensor – low input	Wiring short to ground, fuel tank temperature sensor, ECM
98	P1188	Fuel tank temperature sensor – high input	Wiring open circuit, fuel tank temperature sensor, ECM
–	P1300	Random misfire	Ignition/fuel/EGR system, MAP sensor, IAC valve
–	P1607	Engine control module (ECM) – internal circuit failure	ECM
–	P1705	AT range position switch – circuit failure	Wiring, range position switch, ECM
–	P1706	AT range position switch – no signal	Wiring, range position switch, ECM

HONDA

Model:	Engine identification:	Year:
CR-V	**B20B4**	**1997-01**

System: **PGM-FI**

Self-diagnosis

General information

- Refer to the front of this manual for general test conditions, terminology, detailed descriptions of wiring faults and a general trouble shooter for electrical and mechanical faults.
- Engine control module (ECM) incorporates self-diagnosis function.
- Malfunction indicator lamp (MIL) will illuminate if certain faults are recorded.
- ECM operates in backup mode if sensors fail, to enable vehicle to be driven to workshop.
- Trouble codes can be displayed by using a Scan Tool connected to the data link connector (DLC) **1** [**1**] or by the malfunction indicator lamp (MIL) with the service check connector jumped **1** [**2**].

NOTE: ***The use of a Scan Tool is essential to obtain full diagnostic information.***

Accessing

- Ensure ignition switched OFF.
- Jump service check connector terminals **1** [**2**].
- Switch ignition ON.
- Check MIL is flashing.
- Trouble codes 1-9 are indicated as follows:
  - Individual short flashes display trouble code **2** [**A**].
  - A short pause separates each flash **2** [**B**].
  - A long pause separates each trouble code **2** [**C**].
  - For example: Trouble code 3 displayed **2**.
- Trouble codes greater than 9 are indicated as follows:
  - Long flashes indicate the 'tens' of the trouble code **3** [**A**].
  - Short flashes indicate the 'units' of the trouble code **3** [**C**].
  - A short pause separates each flash **3** [**B**].
  - A long pause separates each trouble code **3** [**D**].
  - For example: Trouble code 12 displayed **3**.
- Count MIL flashes and compare with trouble code table.

NOTE: ***If a trouble code is displayed but not listed in the trouble code table, suspect engine control module (ECM) fault.***

- Switch ignition OFF.
- Remove jump lead.

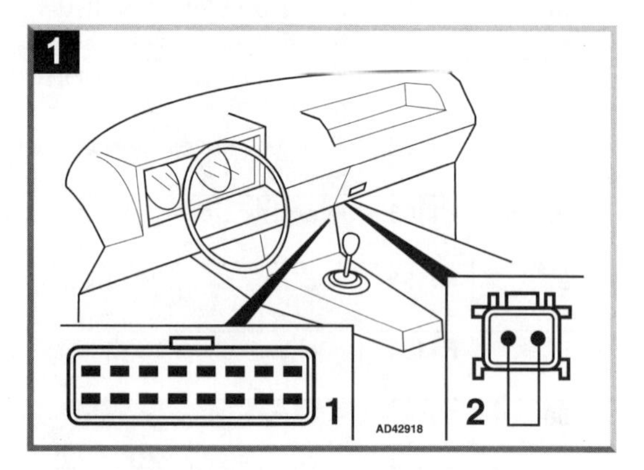

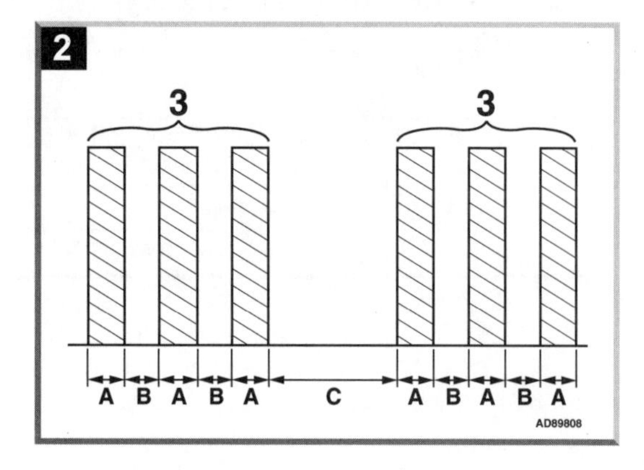

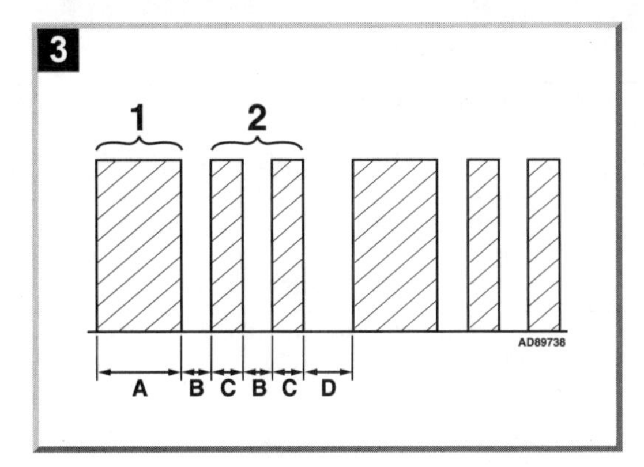

Erasing

- After the faults have been rectified, trouble codes can be erased by using a Scan Tool connected to the data link connector (DLC) or as follows:
- Switch ignition OFF.
- Remove radio fuse (7.5A) from underhood fusebox for 10 seconds minimum 4.
- Reinstall fuse.
- Repeat checking procedure to ensure no data remains in ECM fault memory.

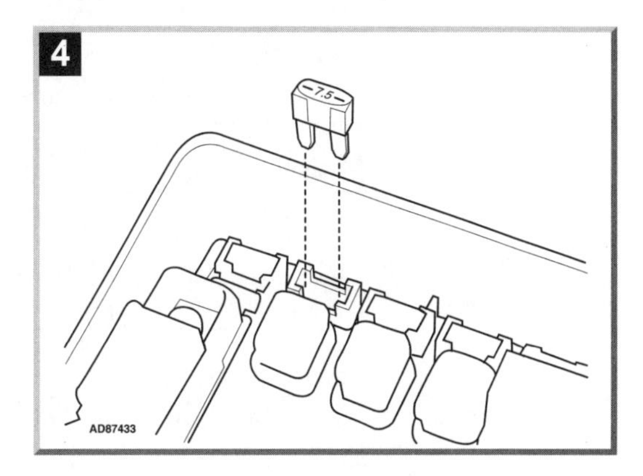

Trouble code identification

Flash code	OBD-II code	Fault location	Probable cause
–	*All P0, P2 and U0 codes*	Refer to OBD-II trouble code tables at the front of this manual	–
1	–	Heated oxygen sensor (HO2S) – front – circuit/voltage low/high	**Wiring open/short circuit, HO2S, fuel system, ECM**
3	–	Manifold absolute pressure (MAP) sensor – circuit/voltage high/low	**Wiring, MAP sensor, ECM**
4	–	Crankshaft position (CKP) sensor – circuit malfunction/range/performance problem	**Wiring, CKP sensor, valve timing, ECM**
5	P1128	Manifold absolute pressure (MAP) sensor – pressure lower than expected	**MAP sensor**
5	P1129	Manifold absolute pressure (MAP) sensor – pressure higher than expected	**MAP sensor**
6	–	Engine coolant temperature (ECT) sensor – circuit/voltage low/high	**Wiring short/open circuit, ECT sensor, ECM**
7	–	Throttle position (TP) sensor – circuit/voltage low/high	**Wiring, TP sensor, ECM**
7	P1121	Throttle position (TP) sensor – position lower than expected	**TP sensor**
7	P1122	Throttle position (TP) sensor – position higher than expected	**TP sensor**
8	P1359	Crankshaft position (CKP) sensor/engine speed (RPM) sensor – connector disconnection	**Wiring**

Flash code	OBD-II code	Fault location	Probable cause
8	P1361	Engine speed (RPM) sensor – intermittent signal	**CKP sensor**
8	P1362	Engine speed (RPM) sensor – no signal	**Wiring, CKP sensor, ECM**
9	P1381	Camshaft position (CMP) sensor – intermittent signal	**CMP sensor**
9	P1382	Camshaft position (CMP) sensor – no signal	**Wiring, CMP sensor, ECM**
10	–	Intake air temperature (IAT) sensor – circuit/ voltage low/high	**Wiring short/open circuit, IAT sensor, ECM**
13	P1106	Barometric pressure (BARO) sensor – range/performance problem	**ECM**
13	P1107	Barometric pressure (BARO) sensor – circuit/voltage low	**ECM**
13	P1108	Barometric pressure (BARO) sensor – circuit/voltage high	**ECM**
14	–	Idle control system – malfunction	**IAC valve, fast idle thermo valve, throttle body**
14	P1508	Idle air control (IAC) valve – circuit failure	**Wiring, IAC valve, ECM**
17	–	Vehicle speed sensor (VSS) – circuit malfunction	**Wiring, VSS, ECM**
20	P1297	Electrical load sensor – circuit/voltage low	**Wiring, electrical load sensor, ECM**
20	P1298	Electrical load sensor – circuit/voltage high	**Wiring, electrical load sensor, ECM**
23	–	Knock sensor (KS) – circuit malfunction	**Wiring, KS, ECM**
41	–	Heated oxygen sensor (HO2S) – front – circuit malfunction	**Wiring, ECM**
45	–	Mixture too lean/rich	**Fuel system, front HO2S, injectors, MAP sensor, wiring, mechanical fault**
54	P1336	Crankshaft position (CKP) sensor 2 – intermittent signal interruption	**CKP sensor**
54	P1337	Crankshaft position (CKP) sensor 2 – no signal	**Wiring, CKP sensor, ECM**
61	–	Heated oxygen sensor (HO2S) – front – slow response	**HO2S, exhaust system**
63	–	Heated oxygen sensor (HO2S) – rear – slow response/circuit/voltage low/high	**Wiring, HO2S, ECM**
65	–	Heated oxygen sensor (HO2S) – rear – circuit malfunction	**Wiring, ECM**
67	–	Catalytic converter – efficiency below limit	**Catalytic converter, rear HO2S**
70	–	AT – lock-up clutch not engaging/no gear shift	**Wiring, mainshaft speed sensor, countershaft speed sensor, lock-up control system, shift solenoid (SS) A/B, ECM**
70	P1705	AT – gear shift malfunction	**Wiring, range position switch, ECM**
70	P1705	AT – lock-up clutch not engaging	**Wiring, range position switch, ECM**

Flash code	OBD-II code	Fault location	Probable cause
70	P1706	AT – gear shift malfunction	**Wiring, range position switch, ECM**
70	P1706	AT – lock-up clutch malfunction	**Wiring, range position switch, ECM**
70	P1753	AT – lock-up clutch not engaging/ disengaging	**Wiring, lock-up control solenoid A, ECM**
70	P1758	AT – lock-up clutch not engaging	**Wiring, lock-up control solenoid B, ECM**
70	P1768	AT – poor gear shift	**Wiring, linear solenoid, ECM**
70	P1768	AT – lock-up clutch not engaging	**Wiring, linear solenoid, ECM**
71	–	Cylinder No.1 – misfire	**Fuel/ignition/EGR system, injector, wiring, mechanical fault**
72	–	Cylinder No.2 – misfire	**Fuel/ignition/EGR system, injector, wiring, mechanical fault**
73	–	Cylinder No.3 – misfire	**Fuel/ignition/EGR system, injector, wiring, mechanical fault**
74	–	Cylinder No.4 – misfire	**Fuel/ignition/EGR system, injector, wiring, mechanical fault**
–	P1300	Random misfire	**Ignition/fuel/EGR system, MAP sensor, IAC valve**
86	–	Engine coolant temperature (ECT) sensor – range/performance problem	**ECT sensor, cooling system**
90	P1456	Evaporative emission (EVAP) canister purge system (fuel tank system) – leak detected	**Hose, fuel tank/pressure sensor, fuel filler cap, EVAP valve/bypass solenoid, EVAP two way valve, EVAP canister/vent valve**
90	P1457	Evaporative emission (EVAP) canister purge system (canister system) – leak detected	**Hose, fuel tank/pressure sensor, EVAP valve/bypass solenoid, EVAP two way valve, EVAP canister/vent valve**
91	–	Fuel tank pressure sensor – circuit/voltage high/low	**Wiring, fuel tank pressure sensor, ECM**
92	–	Evaporative emission (EVAP) canister purge system – incorrect flow	**Wiring, EVAP solenoid, throttle body, hose leak/blockage, ECM**
–	P1607	Engine control module (ECM) – internal circuit failure	**ECM**

HONDA

Model: Element 2.4L

Engine identification: K24A4

Year: 2003-04

System: **PGM-FI**

Self-diagnosis

General information

- Refer to the front of this manual for general test conditions, terminology, detailed descriptions of wiring faults and a general trouble shooter for electrical and mechanical faults.
- Malfunction indicator lamp (MIL) will illuminate if certain faults are recorded.
- ECM operates in backup mode if sensors fail, to enable vehicle to be driven to repair shop.

Accessing

- Trouble codes can only be displayed by using a Scan Tool connected to the data link connector (DLC) **1**.

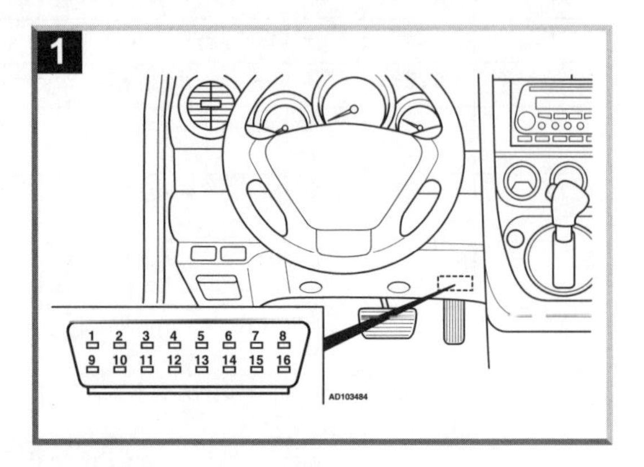

Erasing

- Trouble codes can only be erased by using a Scan Tool connected to the data link connector (DLC) **1**.

Trouble code identification

OBD-II code	Fault location	Probable cause
All P0, P2 and U0 codes	Refer to OBD-II trouble code tables at the front of this manual	–
P1009	VTEC system – malfunction	**Wiring, mechanical fault, VTEC solenoid, engine oil pressure switch, ECM**
P1121	Throttle position (TP) sensor – position lower than expected	**TP sensor**
P1122	Throttle position (TP) sensor – position higher than expected	**TP sensor**
P1128	Manifold absolute pressure (MAP) sensor – pressure lower than expected	**MAP sensor**
P1129	Manifold absolute pressure (MAP) sensor – pressure higher than expected	**MAP sensor**
P1157	Air fuel (A/F) ratio sensor – circuit malfunction	**Wiring, A/F ratio sensor, ECM**
P1297	Electrical load sensor – circuit/voltage low	**Wiring, electrical load sensor, ECM**
P1298	Electrical load sensor – circuit/voltage high	**Wiring, electrical load sensor, ECM**

OBD-II code	Fault location	Probable cause
P1454	Fuel tank pressure sensor – range/performance problem	Wiring, hose, fuel tank pressure sensor, fuel filler cap, EVAP canister/vent valve, ECM
P1731	AT – gear shift malfunction	Transmission fluid, wiring, shift solenoid E, AT clutch pressure control solenoid A, ECM
P1732	AT – gear shift malfunction	Transmission fluid, wiring, shift solenoid B/C, ECM
P1735	AT – gear shift malfunction	Transmission fluid, wiring, shift solenoid B/C/E, AT clutch pressure control solenoid A, ECM
P1736	AT – gear shift malfunction	Transmission fluid, wiring, shift solenoid B/E, AT clutch pressure control solenoid A, ECM

HONDA

Model: **Passport 2.6L**

Engine identification: **4ZE1**

Year: **1994-95**

System: **Isuzu MFI**

Self-diagnosis

General information

- Refer to the front of this manual for general test conditions, terminology, detailed descriptions of wiring faults and a general trouble shooter for electrical and mechanical faults.
- Engine control module (ECM) incorporates self-diagnosis function.
- Malfunction indicator lamp (MIL) will illuminate if certain faults are recorded.
- ECM operates in backup mode if sensors fail, to enable vehicle to be driven to workshop.
- Trouble codes can be displayed by the malfunction indicator lamp (MIL) with the service check connector terminals joined together **1**.

Accessing

- Ensure ignition switched OFF.
- Join together the service check connector terminals **1** [**2**].
- Switch ignition ON. Do NOT start engine.
- MIL should flash.
- Trouble code 12 will be displayed three times.
- Each trouble code will be displayed three times.
- If no trouble codes are stored, code 12 will continue to be displayed.
- Count MIL flashes and compare with trouble code table.
  - Short flashes indicate the 'tens' and 'units' of the trouble code **2**.
  - A 0.4 second pause separates each flash. A 1.2 second pause separates the 'tens' and 'units'. A 3.2 second pause separates each trouble code **2**.
  - For example: Trouble code 23 displayed **2**.
- Count MIL flashes and compare with trouble code table.
- Switch ignition OFF.
- Disconnect the service check connector terminals **1** [**2**].

1

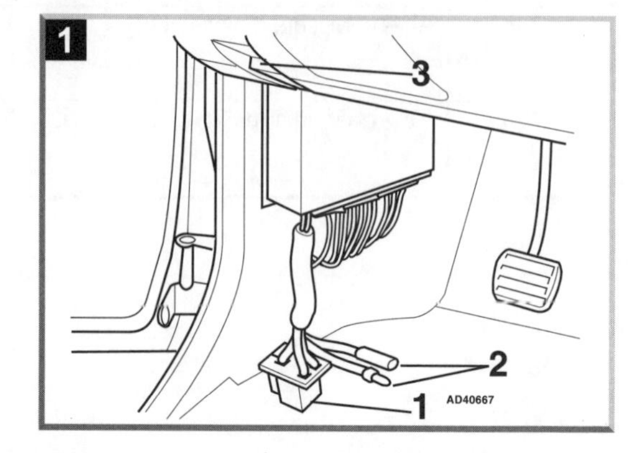

2

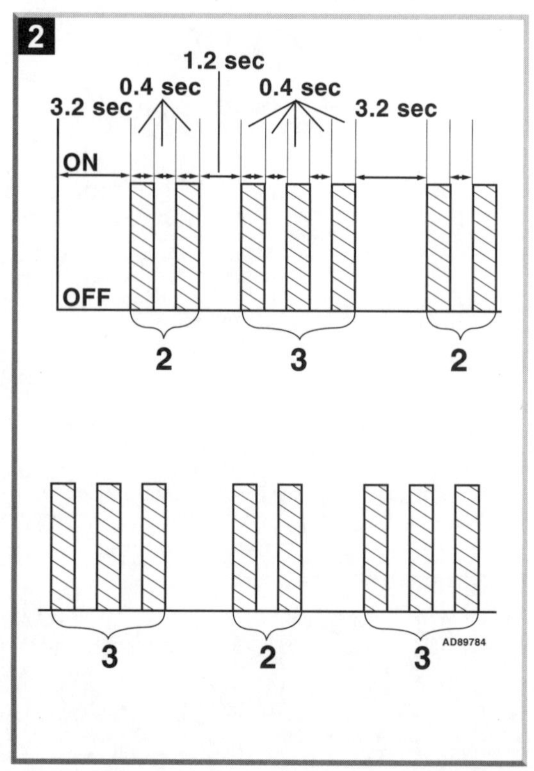

Erasing

- After the faults have been rectified, trouble codes can be erased as follows:
- Ensure ignition switched OFF.
- Remove 60A fuse from underhood fusebox for 30 seconds minimum **3**.
- Reinstall fuse.
- Repeat checking procedure to ensure no data remains in ECM fault memory.

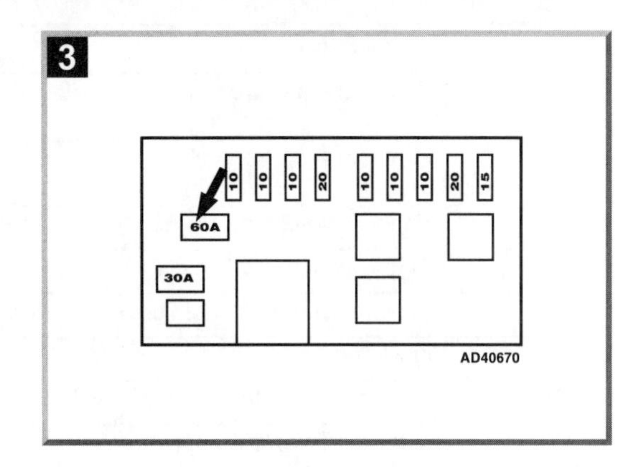

Trouble code identification

Flash code	Fault location	Probable cause
12	Normal condition	–
13	Heated oxygen sensor (HO2S) – open circuit	**Exhaust/secondary air system, wiring, HO2S, ECM**
14	Engine coolant temperature (ECT) sensor – short circuit	**Wiring short circuit, ECT sensor, ECM**
15	Engine coolant temperature (ECT) sensor – open circuit	**Wiring open circuit, ECT sensor, ECM**
21	Closed throttle position (CTP)/wide open throttle (WOT) switch – simultaneous operation	**Wiring, CTP/WOT switch, ECM**
22	Starter signal – no signal	**Vehicle push started, wiring, ECM**
23	Ignition module – short circuit	**Wiring short circuit, ignition module, ECM**
25	Fuel pressure control solenoid – short/open circuit	**Wiring, fuel pressure control solenoid, ECM**
26	Evaporative emission (EVAP) canister purge valve – open/ short circuit	**Wiring, EVAP valve, ECM**
27	Evaporative emission (EVAP) canister purge valve – control circuit	**Wiring, EVAP valve, ECM**
32	Exhaust gas recirculation (EGR) system	**Hose/pipe leak or blockage, EGR valve/ solenoid, throttle body**
33	Fuel injectors – wiring open/short circuit	**Wiring, injector, ECM**
34	Exhaust gas recirculation (EGR) system	**MAP sensor, hose leak/blockage, wiring, ECM**
35	Ignition module – open circuit	**Wiring open circuit, ignition module, ECM**
41	Crankshaft position (CKP) sensor – no signal	**Wiring, CKP sensor, ECM**
43	Closed throttle position (CTP) switch – permanent operation	**Wiring, CTP switch, ECM**

Flash code	Fault location	Probable cause
44	Fuel trim – mixture lean	**Fuel/exhaust system, injectors, wiring, HO2S, MAF sensor, ECM**
45	Fuel trim – mixture rich	**Fuel system, injectors, wiring, HO2S, MAF/ECT sensor, ECM**
51	Engine control module (ECM) – memory failure	**ECM**
52	Engine control module (ECM) – memory failure	**ECM**
53	Fuel pressure control solenoid – control circuit	**Wiring, fuel pressure control solenoid, ECM**
54	Ignition module	**Wiring, ignition module, ECM**
61	Mass air flow (MAF) sensor – voltage low	**Wiring, MAF sensor, ECM**
62	Mass air flow (MAF) sensor – voltage high	**Wiring, MAF sensor, ECM**
63	Vehicle speed sensor (VSS) – no signal	**Wiring, VSS, ECM**
64	Fuel injectors – control circuit	**Wiring, ECM**
65	Wide open throttle (WOT) switch – permanent operation	**Wiring, WOT switch, ECM**

Model: **Passport 2.6L**

Engine identification: **4ZE1**

Year: **1996-97**

System: **Isuzu MFI**

Self-diagnosis

General information

- Refer to the front of this manual for general test conditions, terminology, detailed descriptions of wiring faults and a general trouble shooter for electrical and mechanical faults.
- Engine control module (ECM) incorporates self-diagnosis function.
- Malfunction indicator lamp (MIL) will illuminate if certain faults are recorded.
- ECM operates in backup mode if sensors fail, to enable vehicle to be driven to workshop.

Accessing

- Trouble codes can be displayed by using a Scan Tool connected to the data link connector (DLC) **1** under the LH fascia.

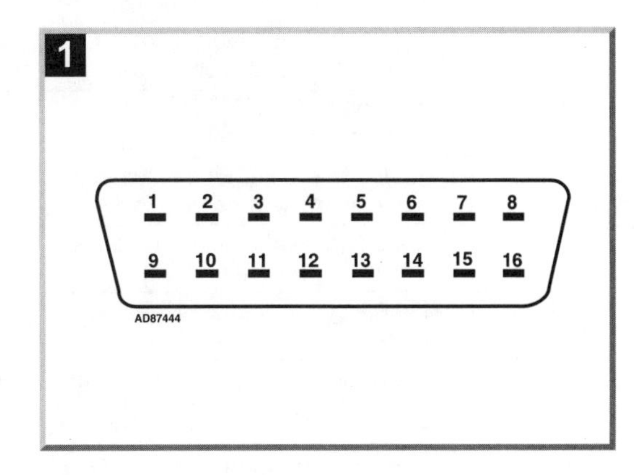

Erasing

- After the faults have been rectified, trouble codes can be erased by using a Scan Tool connected to the data link connector (DLC) **1** or as follows:
- Switch ignition OFF.
- Disconnect battery ground cable for 30 seconds minimum.
- Reconnect battery ground cable.
- Repeat checking procedure to ensure no data remains in ECM fault memory.

NOTE: ***Disconnecting battery ground lead will also erase trouble codes but may erase memory from electronic units such as radio.***

Trouble code identification

OBD-II code	Fault location	Probable cause
All P0, P2 and U0 codes	Refer to OBD-II trouble code tables at the front of this manual	–
P1106	Manifold absolute pressure (MAP) sensor – circuit/intermittent voltage high	**Wiring, MAP sensor, ECM**
P1107	Manifold absolute pressure (MAP) sensor – circuit/intermittent voltage low	**Wiring, MAP sensor, ECM**

OBD-II code	Fault location	Probable cause
P1111	Intake air temperature (IAT) sensor – circuit/ intermittent voltage high	**Wiring, IAT sensor**
P1112	Intake air temperature (IAT) sensor – circuit/ intermittent voltage low	**Wiring, IAT sensor**
P1114	Engine coolant temperature (ECT) sensor – circuit/ intermittent voltage high	**Wiring, ECT sensor**
P1115	Engine coolant temperature (ECT) sensor – circuit/ intermittent voltage low	**Wiring, ECT sensor**
P1121	Throttle position (TP) sensor – circuit/intermittent voltage high	**Wiring, TP sensor**
P1122	Throttle position (TP) sensor – circuit/intermittent voltage low	**Wiring, TP sensor**
P1133	Heated oxygen sensor (HO2S) – insufficient switching	**Exhaust system, wiring, HO2S, ECM**
P1134	Heated oxygen sensor (HO2S) – slow operation	**Exhaust system, wiring, HO2S, ECM**
P1171	Fuel system – mixture lean under acceleration	**Fuel pump/strainer, incorrect fuel**
P1390	G-sensor – intermittent voltage low	**Wiring, G-sensor**
P1391	G-sensor – performance	**Seal damaged/missing, incorrectly installed, wiring, G-sensor, ECM**
P1392	G-sensor – voltage low	**Wiring, G-sensor, ECM**
P1393	G-sensor – voltage high	**Wiring, G-sensor, ECM**
P1394	G-sensor – intermittent voltage high	**Wiring, G-sensor, ECM**
P1406	Exhaust gas recirculation (EGR) valve position sensor – circuit problem	**Wiring, EGR valve/position sensor, ECM**
P1441	Evaporative emission (EVAP) canister purge system – leak detected	**Wiring, EVAP solenoid, vacuum switch, ECM**
P1442	Evaporative emission (EVAP) canister purge system – vacuum switch malfunction	**Wiring, vacuum switch, ECM**
P1640	Engine control module (ECM) – internal fault	**ECM**

HONDA

Model:	Engine identification:	Year:
Passport 3.2L	**6VD1**	**1994-95**

System: **Isuzu MFI**

Self-diagnosis

General information

- Refer to the front of this manual for general test conditions, terminology, detailed descriptions of wiring faults and a general trouble shooter for electrical and mechanical faults.
- Engine control module (ECM) incorporates self-diagnosis function.
- Malfunction indicator lamp (MIL) will illuminate if certain faults are recorded.
- ECM operates in backup mode if sensors fail, to enable vehicle to be driven to workshop.
- Trouble codes can be displayed by the malfunction indicator lamp (MIL) with the data link connector (DLC) terminals jumped **1**.

Accessing

- Ensure ignition switched OFF.
- Jump data link connector (DLC) terminals 1 and 3 **1**.
- Switch ignition ON. Do NOT start engine.
- MIL should flash.
- Trouble code 12 will be displayed three times.
- Each trouble code will be displayed three times.
- If no trouble codes are stored, code 12 will continue to be displayed.
- Count MIL flashes and compare with trouble code table.
  - ○ Short flashes indicate the 'tens' and 'units' of the trouble code **2**.
  - ○ A 0.4 second pause separates each flash. A 1.2 second pause separates the 'tens' and 'units'. A 3.2 second pause separates each trouble code **2**.
  - ○ For example: Trouble code 23 displayed **2**.
- Count MIL flashes and compare with trouble code table.
- Switch ignition OFF.
- Remove jump lead.

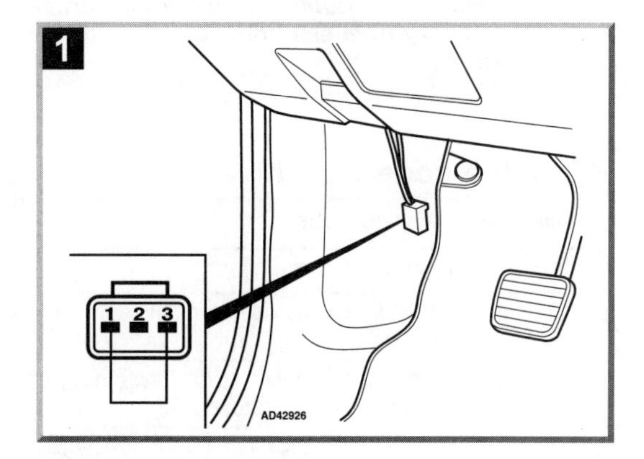

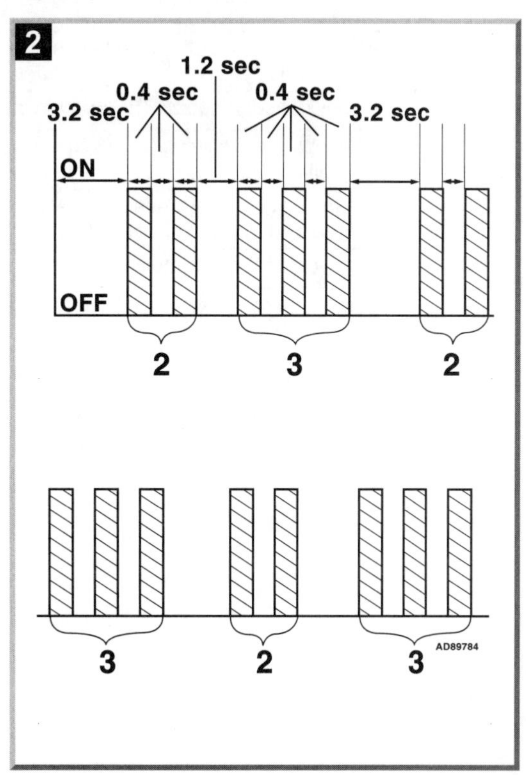

Erasing

- Ensure ignition switched OFF.
- Disconnect battery ground cable for 30 seconds.
- Reconnect battery ground cable.
- Repeat checking procedure to ensure no data remains in ECM fault memory.

NOTE: ***Disconnecting battery ground lead will also erase trouble codes but may erase memory from electronic units such as radio.***

Trouble code identification

Flash code	Fault location	Probable cause
12	Normal condition	–
13	Heated oxygen sensor (HO2S) – short/open circuit	**Wiring, HO2S, ECM**
14	Engine coolant temperature (ECT) sensor – high/low temperature	**Wiring, ECT sensor, ECM**
21	Throttle position (TP) sensor – voltage high/ low	**Wiring, TP sensor, ECM**
23	Intake air temperature (IAT) sensor – high/low temperature	**Wiring, IAT sensor, ECM**
24	Vehicle speed sensor (VSS) – circuit	**Wiring, VSS, ECM**
32	Exhaust gas recirculation (EGR) system – circuit	**Hose/pipe blockage, EGR solenoid/ backpressure transducer, wiring, ECM**
33	Manifold absolute pressure (MAP) sensor – signal voltage or high/low vacuum	**Wiring, MAP sensor, ECM**
42	Ignition control – circuit	**Wiring, ICM, ECM**
44	Heated oxygen sensor (HO2S) – signal – mixture lean	**Fuel/exhaust system, injectors, wiring, ECM**
45	Heated oxygen sensor (HO2S) – signal – mixture rich	**Fuel/EVAP/EGR system, TP/MAP sensor, ECM**
51	Engine control module (ECM) – memory failure	**ECM**

Model: Passport 3.2L

Engine identification: 6VD1

Year: 1996-02

System: **Isuzu MFI**

Self-diagnosis

General information

- Refer to the front of this manual for general test conditions, terminology, detailed descriptions of wiring faults and a general trouble shooter for electrical and mechanical faults.
- Engine control module (ECM) incorporates self-diagnosis function.
- Malfunction indicator lamp (MIL) will illuminate if certain faults are recorded.
- ECM operates in backup mode if sensors fail, to enable vehicle to be driven to workshop.

Accessing

- Trouble codes can be displayed by using a Scan Tool connected to the data link connector (DLC) **1** under the LH fascia.

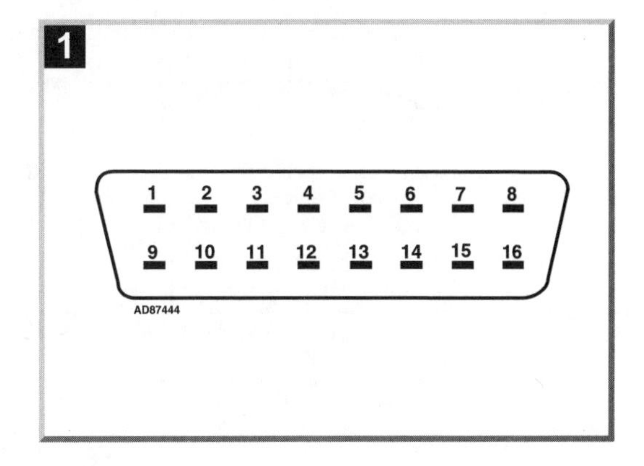

Erasing

- After the faults have been rectified, trouble codes can be erased by using a Scan Tool connected to the data link connector (DLC) **1** or as follows:
- Switch ignition OFF.
- Disconnect battery ground cable for 30 seconds minimum.
- Reconnect battery ground cable.
- Repeat checking procedure to ensure no data remains in ECM fault memory.

NOTE: ***Disconnecting battery ground lead will also erase trouble codes but may erase memory from electronic units such as radio.***

Trouble code identification

OBD-II code	Fault location	Probable cause
All P0, P2 and U0 codes	Refer to OBD-II trouble code tables at the front of this manual	–
P1106	Manifold absolute pressure (MAP) sensor – circuit/intermittent voltage high	**Wiring, MAP sensor**
P1107	Manifold absolute pressure (MAP) sensor – circuit/ intermittent voltage low	**Wiring, MAP sensor**

OBD-II code	Fault location	Probable cause
P1111	Intake air temperature (IAT) sensor – circuit/ intermittent voltage high	**Wiring, IAT sensor**
P1112	Intake air temperature (IAT) sensor – circuit/ intermittent voltage low	**Wiring, IAT sensor**
P1114	Engine coolant temperature (ECT) sensor – circuit/ intermittent voltage low	**Wiring, ECT sensor**
P1115	Engine coolant temperature (ECT) sensor – circuit/ intermittent voltage high	**Wiring, ECT sensor**
P1120	Throttle motor position sensor 1 – circuit malfunction	**Wiring open circuit/short circuit to ground/ short circuit to positive, throttle control unit, ECM**
P1121	Throttle position (TP) sensor – circuit/ intermittent voltage high	**Wiring, TP sensor**
P1122	Throttle position (TP) sensor – circuit/ intermittent voltage low	**Wiring, TP sensor**
P1125	Throttle position (TP) motor – fail safe mode	**Wiring, TP sensor, TP motor, mechanical fault, ECM**
P1133	Heated oxygen sensor (HO2S) – RH front – insufficient switching	**Exhaust system, wiring, HO2S, ECM**
P1134	Heated oxygen sensor (HO2S) – RH front – slow operation	**Exhaust system, wiring, HO2S, ECM**
P1153	Heated oxygen sensor (HO2S) – LH front – insufficient switching	**Exhaust system, wiring, HO2S, ECM**
P1154	Heated oxygen sensor (HO2S) – LH front – slow operation	**Exhaust system, wiring, HO2S, ECM**
P1167	Fuel trim, bank 1 – mixture too rich during deceleration	**Injectors, fuel pump, fuel pressure, ECM**
P1169	Fuel trim, bank 2 – mixture too rich during deceleration	**Injectors, fuel pump, fuel pressure, ECM**
P1171	Fuel system – mixture lean under acceleration	**Fuel pump/strainer, incorrect fuel**
P1220	Throttle motor position sensor 2 – circuit malfunction	**Wiring open circuit/short circuit to ground/ short circuit to positive, throttle control unit, ECM**
P1221	Throttle motor position sensor 1/2 – signal variation	**Wiring, throttle control unit, ECM**
P1271	Accelerator pedal position (APP) sensor 1/2 – signal variation	**Wiring, APP sensor, ECM**
P1272	Accelerator pedal position (APP) sensor 2/3 – signal variation	**Wiring, APP sensor, ECM**
P1273	Accelerator pedal position (APP) sensor 1/3 – signal variation	**Wiring, APP sensor, ECM**
P1275	Accelerator pedal position (APP) sensor 1 – circuit malfunction	**Wiring open circuit/short circuit to ground/ short circuit to positive, APP sensor, ECM**
P1280	Accelerator pedal position (APP) sensor 2 – circuit malfunction	**Wiring open circuit/short circuit to ground/ short circuit to positive, APP sensor, ECM**

OBD-II code	Fault location	Probable cause
P1285	Accelerator pedal position (APP) sensor 3 – circuit malfunction	**Wiring, APP sensor, ECM**
P1290	Throttle position (TP) motor – forced idle mode	**Wiring, APP sensor, TP motor, mechanical fault, ECM**
P1295	Throttle position (TP) motor – power management mode	**Wiring, TP sensor, TP motor, MAP sensor, MAF sensor, ECM**
P1299	Throttle position (TP) motor – forced engine shut down mode	**Wiring, TP sensor, TP motor, MAP sensor, MAF sensor, ECM**
P1310	Ignition control module (ICM) – diagnosis	**Wiring, ignition coil(s), ICM, ECM**
P1311	Ignition control module (ICM) – ignition coil secondary coil signal, circuit 1	**Wiring, ignition coil(s), ICM, ECM**
P1312	Ignition control module (ICM) – ignition coil secondary coil signal, circuit 2	**Wiring, ignition coil(s), ICM, ECM**
P1326	Ignition control module (ICM) – combustion quality input signal	**Wiring, ignition coil(s), ICM, ECM**
P1340	Ignition control module (ICM) – cylinder identification/syncronization	**Wiring, ignition coil(s), ICM, ECM**
P1380	Engine control module (ECM)/ABS control module, rough road signal – communication fault	**Wiring, ABS trouble code(s) stored, ECM**
P1381	Engine control module (ECM)/ABS control module – communication fault	**Wiring, ABS trouble code(s) stored, ECM**
P1390	G-sensor – intermittent voltage low	**Wiring, G-sensor**
P1391	G-sensor – performance	**Seal damaged/missing, incorrectly installed, wiring, G-sensor, ECM**
P1392	G-sensor – voltage low	**Wiring, G-sensor, ECM**
P1393	G-sensor – voltage high	**Wiring, G-sensor, ECM**
P1394	G-sensor – intermittent voltage high	**Wiring, G-sensor**
P1404	Exhaust gas recirculation (EGR) valve – valve closed	**Wiring, EGR valve**
P1406	Exhaust gas recirculation (EGR) valve position sensor – circuit problem	**Wiring, EGR valve/position sensor, ECM**
P1441	Evaporative emission (EVAP) canister purge system – flow detected	**Wiring, EVAP solenoid, ECM**
P1508	Idle air control (IAC) system – low RPM	**Intake system, throttle body, hose leak, wiring, IAC valve, ECM**
P1509	Idle air control (IAC) system – high RPM	**Hose leak, throttle body, crankcase vent valve, wiring, IAC valve, ECM**
P1514	Throttle position (TP) sensor/mass air flow (MAF) sensor – signal variation	**Wiring, TP sensor, MAP sensor, ECM**
P1515	Throttle command/actual throttle position – signal variation	**Wiring, throttle valve sticking, throttle control unit, ECM**
P1516	Throttle position motor – position performance	**Throttle valve tight/sticking, TP sensor, wiring, throttle control unit, ECM**

OBD-II code	Fault location	Probable cause
P1523	Throttle position motor – closed position performance	**Throttle valve tight/sticking, TP sensor, wiring open circuit/short circuit to ground, throttle control unit, ECM**
P1571	Brake pedal position (BPP) switch – no operation	**Wiring, BPP switch**
P1618	Engine control module (ECM) – internal fault	**ECM**
P1625	Engine control module (ECM) – unexpected reset	**Interference from non standard electronics, erase trouble code and re-test**
P1635	Sensor supply voltage – circuit 1 malfunction	**Wiring short circuit to ground/positive, ground wiring open circuit, ECM**
P1639	Sensor supply voltage – circuit 2 malfunction	**Wiring short circuit to ground/positive, ground wiring open circuit, throttle position (TP) sensor, ECM**
P1640	Engine control module (ECM), output driver – internal fault	**Wiring, ECM**
P1646	Sensor supply voltage – circuit C malfunction	**Wiring short circuit to ground/positive, ground wiring open circuit, accelerator pedal position (APP) sensor, ECM**
P1650	Engine control module quad driver/output driver	**Wiring, evaporative emission (EVAP) canister purge valve, intake manifold air control solenoid, ECM**
P1790	Engine control module (ECM) – AT ROM checksum error	**ECM/Programming**
P1792	AT – AT EEPROM checksum error	**ECM/Programming**
P1835	AT – kick-down switch	**Wiring, kick-down switch, ECM**
P1850	AT – brake band solenoid malfunction	**AT trouble code(s) stored**
P1860	AT – torque converter clutch (TCC) solenoid circuit	**Wiring, TCC solenoid, ECM**
P1870	AT – component slipping	**Range switch, mechanical failure, engine speed signal incorrect**

Model: Pilot 3.5L

Engine identification: J35A3

Year: 2003-04

System: PGM-FI

Self-diagnosis

General information

- Refer to the front of this manual for general test conditions, terminology, detailed descriptions of wiring faults and a general trouble shooter for electrical and mechanical faults.
- Malfunction indicator lamp (MIL) will illuminate if certain faults are recorded.
- ECM operates in backup mode if sensors fail, to enable vehicle to be driven to repair shop.

Accessing

- Trouble codes can only be displayed by using a Scan Tool connected to the data link connector (DLC) **1**.

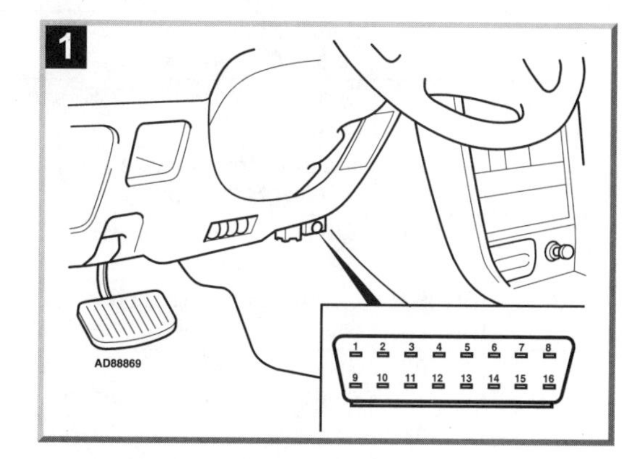

Erasing

- Trouble codes can only be erased by using a Scan Tool connected to the data link connector (DLC) **1**.

Trouble code identification

Flash code	OBD-II code	Fault location	Probable cause
–	*All P0, P2 and U0 codes*	Refer to OBD-II trouble code tables at the front of this manual	–
1	–	Heated oxygen sensor (HO2S) – front – circuit/voltage high/low	**Wiring short/open circuit, HO2S, fuel system, ECM**
3	–	Manifold absolute pressure (MAP) sensor – circuit/voltage low/high	**Wiring, MAP sensor, ECM**
4	–	Crankshaft position (CKP) sensor – circuit malfunction/range/performance problem	**Wiring, CKP sensor, ECM**
5	–	Manifold absolute pressure (MAP) sensor – range/performance problem	**Hose leak/blockage, MAP sensor**
5	P1128	Manifold absolute pressure (MAP) sensor – pressure lower than expected	**MAP sensor**
5	P1129	Manifold absolute pressure (MAP) sensor – pressure higher than expected	**MAP sensor**
6	–	Engine coolant temperature (ECT) sensor – circuit/voltage low/high	**Wiring short/open circuit, ECT sensor, ECM**

Flash code	OBD-II code	Fault location	Probable cause
7	–	Throttle position (TP) sensor – circuit/voltage low/high	**Wiring, TP sensor, ECM**
7	P1121	Throttle position (TP) sensor – position lower than expected	**TP sensor**
7	P1122	Throttle position (TP) sensor – position higher than expected	**TP sensor**
8	–	Camshaft position (CMP) sensor 1 – no signal/intermittent signal	**Wiring, CMP sensor, ECM**
9	–	Output shaft speed (OSS) sensor – circuit malfunction	**Wiring, OSS sensor, ECM**
10	–	Intake air temperature (IAT) sensor – circuit/voltage low/high	**Wiring short/open circuit, IAT sensor, ECM**
12	P1491	Exhaust gas recirculation (EGR) system – valve lift insufficient	**Wiring, EGR valve/position sensor, EGR solenoid, hose leak/blockage, ECM**
12	P1498	Exhaust gas recirculation (EGR) valve position sensor – voltage high	**Wiring, EGR valve/position sensor, ECM**
13	P1106	Barometric pressure (BARO) sensor – range/performance problem	**ECM**
13	P1107	Barometric pressure (BARO) sensor – circuit/voltage low	**Wiring, ECM**
13	P1108	Barometric pressure (BARO) sensor – circuit/voltage high	**ECM**
14	–	Idle control system – malfunction	**IAC valve, fast idle thermo valve, throttle body**
14	P1519	Idle air control (IAC) valve – circuit malfunction	**Wiring, IAC valve, ECM**
15	–	Turbine shaft speed (TSS) sensor – circuit malfunction	**Wiring, TSS sensor, ECM**
20	P1297	Electrical load sensor – circuit/voltage low	**Wiring, electrical load sensor, ECM**
20	P1298	Electrical load sensor – circuit/voltage high	**Wiring, electrical load sensor, ECM**
22	P1259	VTEC system malfunction	**Wiring, VTEC solenoid/pressure switch, ECM**
23	–	Knock sensor (KS) – circuit malfunction	**Wiring, KS, ECM**
26	–	3rd clutch pressure switch – circuit malfunction	**Wiring, poor connection, 3rd clutch pressure switch, ECM**
28	–	Transmission fluid temperature (TFT) sensor – circuit malfunction	**Wiring, TFT sensor, ECM**
40	–	Torque converter clutch (TCC) solenoid – circuit malfunction	**Wiring, TCC solenoid, ECM**
41	–	Heated oxygen sensor (HO2S) – front – heater circuit malfunction	**Wiring, ECM**
45	–	Mixture too lean/rich	**Fuel system, front HO2S, MAP sensor, mechanical fault**
58	–	Camshaft position (CMP) sensor 2 – no signal/intermittent signal	**Wiring, CMP sensor, ECM**

Flash code	OBD-II code	Fault location	Probable cause
61	–	Heated oxygen sensor (HO2S) – front – slow response	**HO2S, exhaust system**
63	–	Heated oxygen sensor (HO2S) – rear – slow response/circuit/voltage low/high	**Wiring, HO2S, ECM**
65	–	Heated oxygen sensor (HO2S) – rear – circuit malfunction	**Wiring, ECM**
67	–	Catalytic converter – efficiency below limit	**Catalytic converter, rear HO2S**
70	–	AT – lock-up clutch not engaging/no gear shift	**Wiring, mainshaft speed sensor, countershaft speed sensor, lock-up control system, shift solenoid (SS) A/B, ECM**
70	P1656	Electronic stability programme (ESP) control module – communication malfunction	**Wiring, ESP control module, ECM**
70	P1705	AT – gear shift malfunction	**Wiring, range position switch, ECM**
70	P1706	AT – gear shift malfunction	**Wiring, range position switch, ECM**
70	P1739	AT – 3rd clutch pressure switch	**Wiring, 3rd clutch pressure switch, ECM**
70	P1740	AT – 4th clutch pressure switch	**Wiring, range position switch, ECM**
70	P1750	AT – hydraulic system mechanical malfunction	**Wiring, range position switch, ECM**
70	P1751	AT – hydraulic system mechanical malfunction	**Wiring, range position switch, ECM**
71	–	Cylinder No.1 – misfire	**Wiring, injector, ignition system, mechanical fault**
72	–	Cylinder No.2 – misfire	**Wiring, injector, ignition system, mechanical fault**
73	–	Cylinder No.3 – misfire	**Wiring, injector, ignition system, mechanical fault**
74	–	Cylinder No.4 – misfire	**Wiring, injector, ignition system, mechanical fault**
75	–	Cylinder No.5 – misfire	**Wiring, injector, ignition system, mechanical fault**
76	–	Cylinder No.6 – misfire	**Wiring, injector, ignition system, mechanical fault**
–	P1399	Random misfire	**Fuel/ignition system, MAP sensor, IAC valve**
80	–	Exhaust gas recirculation (EGR) solenoid – insufficient flow	**EGR solenoid, hose/pipe leak or blockage**
86	–	Engine coolant temperature (ECT) sensor – range/performance problem	**ECT sensor, cooling system**
87	–	Thermostat – range/performance problem	**Thermostat, cooling system**
90	P1456	Evaporative emission (EVAP) canister purge system (fuel tank system) – leak detected	**Hose, fuel tank/pressure sensor, fuel filler cap, EVAP valve/bypass solenoid, EVAP two way valve, EVAP canister/vent valve**

Flash code	OBD-II code	Fault location	Probable cause
90	P1457	Evaporative emission (EVAP) canister purge system (canister system) – leak detected	**Hose, fuel tank/pressure sensor, EVAP valve/bypass solenoid, EVAP two way valve, EVAP canister/vent valve**
91	–	Fuel tank pressure sensor – circuit/voltage low/high	**Wiring, pressure sensor, ECM**
–	P1607	Engine control module (ECM) – internal circuit failure	**ECM**
121	–	Fuel tank level sensor – range/performance problem	**Wiring, fuel tank level sensor**
121	–	Fuel tank level sensor – low input	**Wiring short to ground, fuel tank level sensor, ECM**
121	–	Fuel tank level sensor – high input	**Wiring short to positive, fuel tank level sensor, ECM**

HONDA

Model:	Engine identification:	Year:
Prelude 2.2L	F22A1	1996
Prelude 2.2L	H22A1	1996
Prelude 2.2L	H22A4	1997-02
Prelude 2.3L	H23A1	1996

System: **PGM-FI**

Self-diagnosis

General information

- Refer to the front of this manual for general test conditions, terminology, detailed descriptions of wiring faults and a general trouble shooter for electrical and mechanical faults.
- Engine control module (ECM) incorporates self-diagnosis function.
- Malfunction indicator lamp (MIL) will illuminate if certain faults are recorded.
- ECM operates in backup mode if sensors fail, to enable vehicle to be driven to workshop.
- Trouble codes can be displayed by using a Scan Tool connected to the data link connector (DLC) **1** [**1**] or by the malfunction indicator lamp (MIL) with the service check connector jumped **1** [**2**].

NOTE: ***The use of a Scan Tool is essential to obtain full diagnostic information.***

Accessing

- Ensure ignition switched OFF.
- Jump service check connector terminals **1** [**2**].
- Switch ignition ON.
- Check MIL is flashing.
- Trouble codes 1-9 are indicated as follows:
  - Individual short flashes display trouble code **2** [**A**].
  - A short pause separates each flash **2** [**B**].
  - A long pause separates each trouble code **2** [**C**].
  - For example: Trouble code 3 displayed **2**.
- Trouble codes greater than 9 are indicated as follows:
  - Long flashes indicate the 'tens' of the trouble code **3** [**A**].
  - Short flashes indicate the 'units' of the trouble code **3** [**C**].
  - A short pause separates each flash **3** [**B**].
  - A long pause separates each trouble code **3** [**D**].
  - For example: Trouble code 12 displayed **3**.
- Count MIL flashes and compare with trouble code table.

NOTE: ***If a trouble code is displayed but not listed in the trouble code table, suspect engine control module (ECM) fault.***

- Switch ignition OFF.
- Remove jump lead.

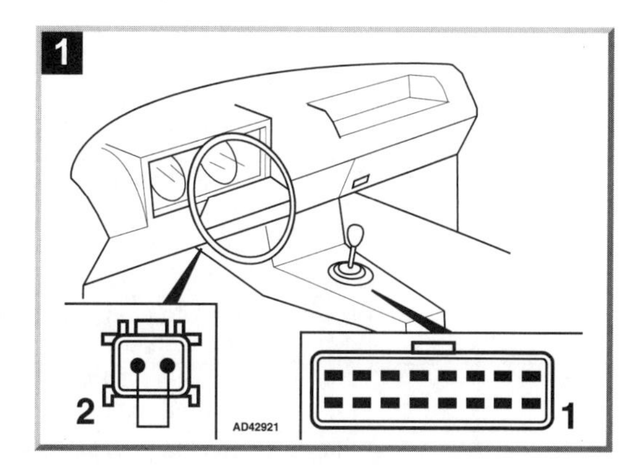

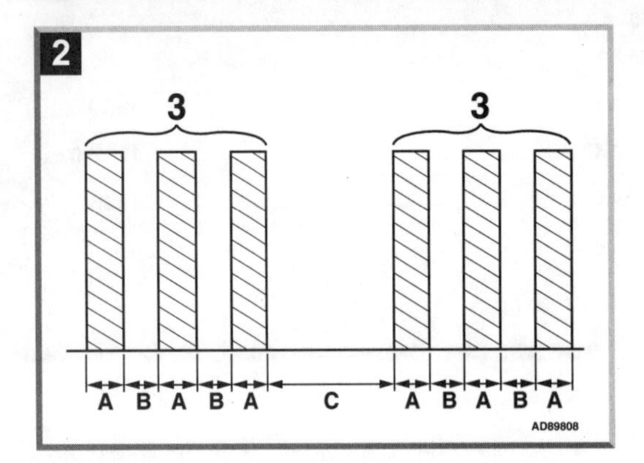

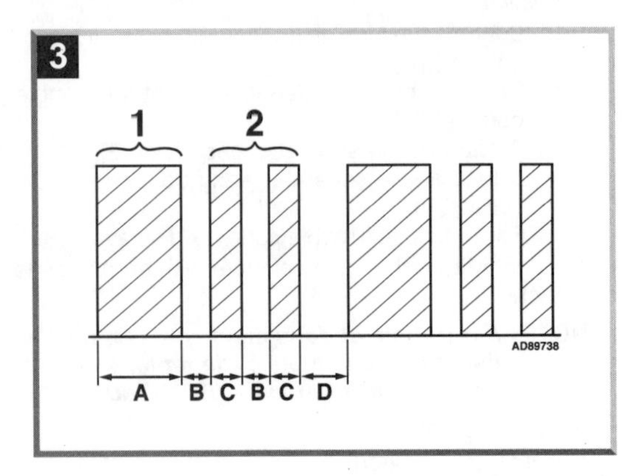

Erasing

- After the faults have been rectified, erase the trouble codes as follows:
- Switch ignition OFF.
- Remove clock/radio fuse No.43 (10A) from underhood fusebox for 10 seconds minimum 4.
- Reinstall fuse.
- Repeat checking procedure to ensure no data remains in ECM fault memory.

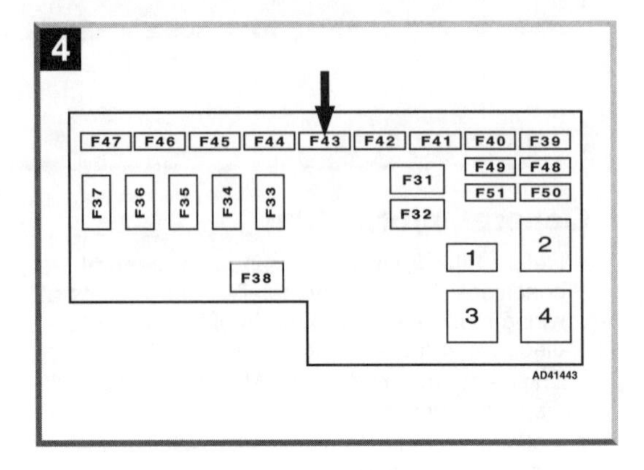

Trouble code identification

Flash code	OBD-II code	Fault location	Probable cause
–	*All P0, P2 and U0 codes*	Refer to OBD-II trouble code tables at the front of this manual	–
1	–	Heated oxygen sensor (HO2S) – front – circuit/voltage low/high	**Wiring open/short circuit, HO2S, fuel system, ECM**
3	–	Manifold absolute pressure (MAP) sensor – circuit/voltage high/low	**Wiring, MAP sensor, ECM**
4	–	Crankshaft position (CKP) sensor – circuit malfunction/range/performance problem	**Wiring, CKP sensor, valve timing, ECM**
5	–	Manifold absolute pressure (MAP) sensor – range/performance problem	**Hose leak, MAP sensor**
5	P1128	Manifold absolute pressure (MAP) sensor – pressure lower than expected	**MAP sensor**
5	P1129	Manifold absolute pressure (MAP) sensor – pressure higher than expected	**MAP sensor**

Flash code	OBD-II code	Fault location	Probable cause
6	–	Engine coolant temperature (ECT) sensor – circuit/voltage low/high	**Wiring short/open circuit, ECT sensor, ECM**
7	–	Throttle position (TP) sensor – circuit/voltage low/high	**Wiring, TP sensor, ECM**
7	P1121	Throttle position (TP) sensor – position lower than expected	**TP sensor**
7	P1122	Throttle position (TP) sensor – position higher than expected	**TP sensor**
8	P1359	Crankshaft position (CKP) sensor – connector disconnection	**Wiring**
8	P1361	Crankshaft position (CKP) sensor – intermittent signal	**CKP sensor**
8	P1362	Crankshaft position (CKP) sensor – no signal	**Wiring, CKP sensor, ECM**
9	P1381	Camshaft position (CMP) sensor – intermittent signal	**CMP sensor**
9	P1382	Camshaft position (CMP) sensor – no signal	**Wiring, CMP sensor, ECM**
10	–	Intake air temperature (IAT) sensor – circuit/voltage low/high	**Wiring short/open circuit, IAT sensor, ECM**
12	P1491	Exhaust gas recirculation (EGR) system – valve lift insufficient	**Wiring, EGR valve/position sensor, EGR solenoid, hose leak/blockage, ECM**
12	P1498	Exhaust gas recirculation (EGR) valve position sensor – voltage high	**Wiring, EGR valve/position sensor, ECM**
13	P1106	Barometric pressure (BARO) sensor – range/performance problem	**ECM**
13	P1107	Barometric pressure (BARO) sensor – circuit/voltage low	**ECM**
13	P1108	Barometric pressure (BARO) sensor – circuit/voltage high	**ECM**
14	–	Idle control system – malfunction	**IAC valve, fast idle thermo valve, throttle body**
14	P1508	Idle air control (IAC) valve – circuit failure	**Wiring, IAC valve, ECM**
17	–	Vehicle speed sensor (VSS) – circuit malfunction/range/performance problem	**Wiring, VSS, ECM**
20	P1297	Electrical load sensor – circuit/voltage low	**Wiring, electrical load sensor, ECM**
20	P1298	Electrical load sensor – circuit/voltage high	**Wiring, electrical load sensor, ECM**
22	P1259	VTEC system malfunction	**Wiring, VTEC solenoid/pressure switch, ECM**
23	–	Knock sensor (KS) – circuit malfunction	**Wiring, KS, ECM**
30	P1655	AT – signal failure	**Communication wire, ECM**
41	–	Heated oxygen sensor (HO2S) – front – circuit malfunction	**Wiring, ECM**
45	–	Mixture too lean/rich	**Fuel system, front HO2S, injectors, MAP sensor, wiring, mechanical fault**

Flash code	OBD-II code	Fault location	Probable cause
61	–	Heated oxygen sensor (HO2S) – front – slow response	HO2S, exhaust system
63	–	Heated oxygen sensor (HO2S) – rear – slow response/circuit/voltage low/high	Wiring, HO2S, ECM
65	–	Heated oxygen sensor (HO2S) – rear – circuit malfunction	Wiring, ECM
67	–	Catalytic converter – efficiency below limit	Catalytic converter, rear HO2S
70	–	AT – lock-up clutch not engaging/no gear shift	Wiring, mainshaft speed sensor, countershaft speed sensor, lock-up control system, shift solenoid (SS) A/B/C, TCM
70	P1705	AT – gear shift malfunction	Wiring, range position switch, TCM
70	P1705	AT – lock-up clutch not engaging	Wiring, range position switch, TCM
70	P1706	AT – gear shift malfunction	Wiring, range position switch, TCM
70	P1706	AT – lock-up clutch malfunction	Wiring, range position switch, TCM
70	P1709	AT – Sport shift mode malfunction	Wiring, mode switch, TCM
70	P1738	Automatic transmission	Wiring, clutch pressure switch 2, TCM
70	P1753	AT – lock-up clutch not engaging	Wiring, lock-up control solenoid, TCM
70	P1753	AT – no gear shift	Wiring, shift control solenoid A, TCM
70	P1768	AT – no gear shift	Wiring, clutch pressure control solenoid A, TCM
70	P1773	AT – no gear shift	Clutch pressure control solenoid B, TCM
70	P1790	AT – lock-up clutch not engaging	Wiring, TP sensor, TCM
70	P1791	AT – lock-up clutch not engaging	Wiring, VSS, TCM
71	–	Cylinder No.1 – misfire	Fuel/ignition/EGR system, injector, wiring, mechanical fault
72	–	Cylinder No.2 – misfire	Fuel/ignition/EGR system, injector, wiring, mechanical fault
73	–	Cylinder No.3 – misfire	Fuel/ignition/EGR system, injector, wiring, mechanical fault
74	–	Cylinder No.4 – misfire	Fuel/ignition/EGR system, injector, wiring, mechanical fault
–	P1300	Random misfire	Ignition/fuel/EGR system, IAC valve
80	–	Exhaust gas recirculation (EGR) solenoid – insufficient flow	EGR valve, hose leak/blockage
86	–	Engine coolant temperature (ECT) sensor – range/performance problem	ECT sensor, cooling system
90	P1456	Evaporative emission (EVAP) canister purge system (fuel tank system) – leak detected	Hose, fuel tank/pressure sensor, fuel filler cap, EVAP valve/bypass solenoid, EVAP two way valve, EVAP canister/vent valve

Flash code	OBD-II code	Fault location	Probable cause
90	P1457	Evaporative emission (EVAP) canister purge system (canister system) – leak detected	**Hose, fuel tank/pressure sensor, EVAP valve/bypass solenoid, EVAP two way valve, EVAP canister/vent valve**
91	–	Fuel tank pressure sensor – circuit/voltage low/high	**Hose leak/blockage, wiring, pressure sensor, ECM**
92	–	Evaporative emission (EVAP) canister purge system – incorrect flow	**Wiring, EVAP solenoid/flow switch, throttle body, hose leak/blockage, ECM**
92	P1459	Evaporative emission (EVAP) canister purge system – switch malfunction	**Wiring, EVAP flow switch, hose leak/ blockage, ECM**
–	P1607	Engine control module (ECM) – internal circuit failure	**ECM**

HYUNDAI

Model: **Accent 1.5L**

Engine identification: **G4EK**

Year: **1995**

System: **ECFI**

Self-diagnosis

General information

- Refer to the front of this manual for general test conditions, terminology, detailed descriptions of wiring faults and a general trouble shooter for electrical and mechanical faults.
- Engine control module (ECM) incorporates self-diagnosis function.
- Malfunction indicator lamp (MIL) will illuminate if certain faults are recorded.
- ECM operates in backup mode if certain sensors fail, to enable car to be driven to workshop.
- Trouble codes can be accessed with suitable code reader connected to the data link connector **1** or displayed with the MIL.

Accessing

- Switch ignition ON. Do NOT start engine.
- Jump data link connector (DLC) terminal 10 to ground for 2.5-7 seconds **1**.
- Malfunction indicator lamp (MIL) should illuminate for approximately 2.5-7 seconds **2** [**A**].
- Malfunction indicator lamp (MIL) should then extinguish **2** [**B**].
- Malfunction indicator lamp (MIL) should then flash.
  - A group of short flashes indicate each of the 4 trouble code digits **2** [**C**].
  - A short pause separates each flash **2** [**D**].
  - A long pause separates each trouble code digit **2** [**E**].
  - For example: Trouble code 1223 is displayed **2**.
- Each trouble code repeats until erased.
- Count MIL flashes and compare with trouble code table.
- To display the next trouble code:
  - Jump data link connector (DLC) terminal 10 to ground for 2.5-7 seconds **1**.
  - When all trouble codes have been displayed (trouble code 3333) jump data link connector (DLC) terminal 10 to ground for 2.5-7 seconds **1**.
- Switch ignition OFF.

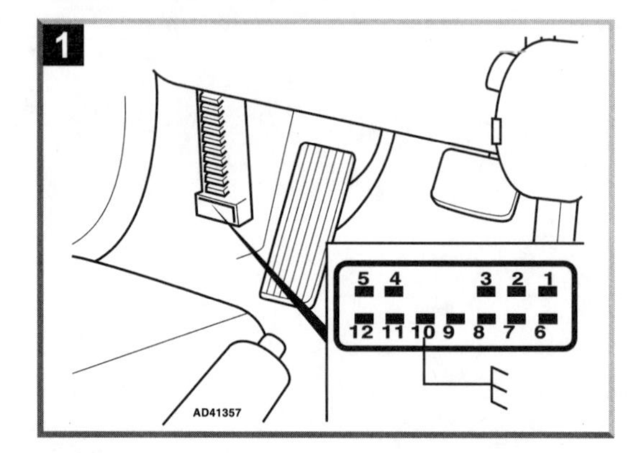

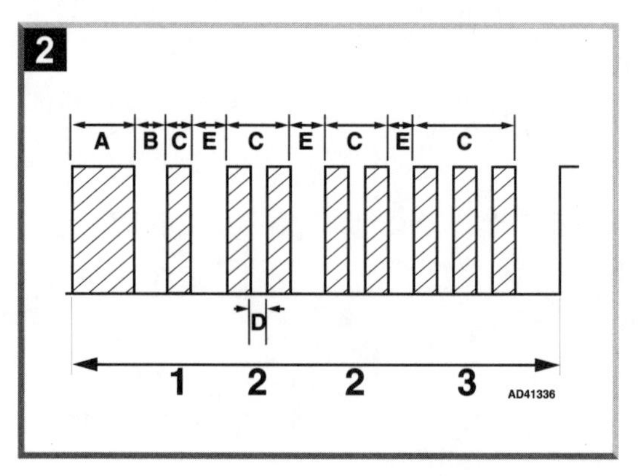

Erasing

- Ensure ignition switched OFF.
- Disconnect battery ground lead for 15 seconds minimum.
- Reconnect battery ground lead.
- Repeat checking procedure to ensure no data remains in ECM fault memory.

NOTE: ***Disconnecting battery ground lead may also erase memory from electronic units such as radio.***

Trouble code identification

Flash code	Scan code	Fault location	Probable cause
1169	13.ECU	Engine control module (ECM) – internal failure	ECM
1233	13.ECU	Engine control module (ECM) – ROM checksum error	ECM
1234	13.ECU	Engine control module (ECM) – internal failure	ECM
3112	41.NO.1 INJECTOR	Injector No.1 failure	Wiring, injector, mechanical fault, ECM
3114	47.ISA	Idle air control (IAC) valve – failure/wiring	Wiring, IAC valve, mechanical fault, ECM
3116	43.NO.3 INJECTOR	Injector No.3 failure	Wiring, injector, mechanical fault, ECM
3117	22.AFS	Mass air flow (MAF) sensor – failure/ wiring	Wiring, MAF sensor, ECM
3122	47.ISA	Idle air control (IAC) valve – failure/wiring	Wiring, IAC valve, mechanical fault, ECM
3128	21.02 SENSOR	Heated oxygen sensor (HO2S) – failure/ wiring	Wiring, HO2 sensor, mechanical fault, fuel system, fuel pressure, ECM
3135	45.PURGE VALVE	Evaporative emission (EVAP) canister purge valve – valve/wiring	Wiring, EVAP valve/solenoid, mechanical fault, ECM
3137	31. BATTERY	Engine control module (ECM) – power supply circuit	Wiring, fuses
3145	23.WTS	Engine coolant temperature (ECT) sensor – failure/wiring	Wiring, ECT sensor, ECM
3146	23.ATS	Intake air temperature (IAT) sensor – short circuit	Wiring, IAT sensor, ECM
3149	33.A/C SWITCH	Air conditioning switch – switch/relay failure	Wiring, A/C compressor, mechanical fault
3153	26.TPS	Throttle position (TP) sensor – failure/ wiring	Wiring, TP sensor, mechanical fault, ECM
3159	29.VEH. SPD.SNSR	Vehicle speed sensor (VSS) – failure/ wiring	Wiring, VSS sensor, mechanical fault, ECM

Flash code	Scan code	Fault location	Probable cause
3211	27.KNOCK SNSR	Knock sensor (KS) – failure/wiring	**Wiring, KS sensor, mechanical fault, ECM**
3222	24.PHASE SENSOR	Camshaft position (CMP) sensor – failure/wiring	**Wiring, CMP sensor, mechanical fault, ECM**
3232	25.CRANK P.SNSR	Crankshaft position (CKP) sensor – failure/wiring	**Wiring, CKP sensor, mechanical fault, ECM**
3234	42.NO.2 INJECTOR	Injector No.2 failure	**Wiring, injector, mechanical fault, ECM**
3235	44.NO.4 INJECTOR	Injector No.4 failure	**Wiring, injector, mechanical fault, ECM**
3241	13.ECU	Engine control module (ECM) – internal failure	**ECM**
3242	13.ECU	Engine control module (ECM) – internal failure	**ECM**
3243	13.ECU	Engine control module (ECM) – internal failure	**ECM**
4133	13.ECU	Engine control module (ECM) – internal failure	**ECM**
4151	81.A/F LEAN	Fuel mixture control – lean mixture	**Wiring, HO2 sensor, fuel system, fuel pressure, mechanical fault, ECM**
4151	82.A/F RICH	Fuel mixture control – rich mixture	**Wiring, HO2 sensor, fuel system, fuel pressure, mechanical fault, ECM**
4152	81.A/F LEAN	Fuel mixture control – lean mixture	**Wiring, HO2 sensor, fuel system, fuel pressure, mechanical fault, ECM**
4152	82.A/F RICH	Fuel mixture control – rich mixture	**Wiring, HO2 sensor, fuel system, fuel pressure, mechanical fault, ECM**
4153	81.A/F LEAN	Fuel mixture control – lean mixture	**Wiring, HO2 sensor, fuel system, fuel pressure, mechanical fault, ECM**
4153	82.A/F RICH	Fuel mixture control – rich mixture	**Wiring, HO2 sensor, fuel system, fuel pressure, mechanical fault, ECM**
3333	–	End of output	–
4444	–	No fault	–

HYUNDAI

Model:	Engine identification:	Year:
Accent 1.5L	G4GK/G4EK/G4EB	1996-02
Accent 1.6L	G4ED	2001-04
Elantra 1.8/2.0L	G4DM/G4GM/G4GF/G4GC	1996-04
Tiburon 1.8/2.0/2.7L	G4GM/G4GF/G4GC/G6BA	1997-04
Sonata 2.0/3.0L	G4AP/G6AT	1996-98
Sonata 2.4/2.5/2.7L	G4JS/G6BV/G6BA	1999-04
Santa Fe 2.4/2.7/3.5L	G4JS/G6BA/G6CU	2001-04
XG300 3.0L	G6CT	2000-01
XG350 3.5L	G6CU	2002-04

System: **MFI (ECFI)**

Self-diagnosis

General information

- Refer to the front of this manual for general test conditions, terminology, detailed descriptions of wiring faults and a general trouble shooter for electrical and mechanical faults.
- Engine control module (ECM) incorporates self-diagnosis function.
- Malfunction indicator lamp (MIL) will illuminate if certain faults are recorded.
- ECM operates in backup mode if sensors fail, to enable vehicle to be driven to workshop.

Accessing

- Trouble codes can be displayed by using a Scan Tool connected to the data link connector (DLC) **1**.

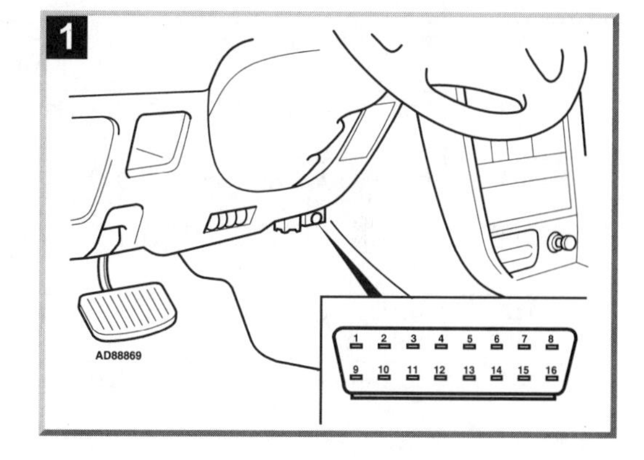

Erasing

- After the faults have been rectified, trouble codes can be erased by using a Scan Tool connected to the data link connector (DLC) or as follows:
- Ensure ignition switched OFF.
- Disconnect battery ground cable for 15 seconds minimum.
- Reconnect battery ground cable.
- Repeat checking procedure to ensure no data remains in ECM fault memory.

NOTE: ***Disconnecting battery ground lead may erase memory from electronic units such as radio.***

Trouble code identification

OBD-II code	Fault location	Probable cause
All P0, P2 and U0 codes	Refer to OBD-II trouble code tables at the front of this manual	–
P1100	Manifold absolute pressure (MAP) sensor EGR – circuit malfunction	**Wiring, MAP sensor**
P1102	Manifold absolute pressure (MAP) sensor EGR mode 3	**Wiring, hoses blocked/leaking, EGR valve, MAP sensor**
P1103	Manifold absolute pressure (MAP) sensor EGR mode 2	**Wiring, hoses blocked/leaking, EGR valve, MAP sensor**
P1110	Electronic throttle system (ETS) – malfunction	**Wiring, ETS**
P1111	Heated oxygen sensor (HO2S) 1, bank 1 – no activity detected	**Heating inoperative, poor connection, wiring, HO2S, ECM**
P1118	Throttle motor – malfunction	**Wiring open/short circuit, throttle motor, ETS**
P1123	Long term fuel trim - additive – mixture too rich	**HO2S, fuel pressure, fuel system, mechanical fault, ECM**
P1124	Long term fuel trim - additive – mixture too lean	**HO2S, fuel pressure, fuel system, mechanical fault, ECM**
P1127	Long term fuel trim – system too rich	**HO2S, fuel pressure, fuel system, mechanical fault, ECM**
P1128	Long term fuel trim – system too lean	**HO2S, fuel pressure, fuel system, mechanical fault, ECM**
P1134	Heated oxygen sensor (HO2S) 1, bank 1 – slow response	**Heating inoperative, HO2S**
P1140	Engine load monitoring (MAF/TP) – signals not plausible	**Wiring, MAF sensor, TP sensor, mechanical fault**
P1147	Accelerator pedal position (APP) sensor 1 – circuit malfunction	**Wiring, APP sensor**
P1151	Accelerator pedal position (APP) sensor 2 – circuit malfunction	**Wiring, APP sensor**
P1152	Accelerator pedal position (APP) sensor 2 – low voltage	**Wiring short to ground, APP sensor**
P1153	Accelerator pedal position (APP) sensor 2 – high voltage	**Wiring short to positive, APP sensor**
P1154	Heated oxygen sensor (HO2S) 1, bank 2 – slow response	**Heating inoperative, HO2S**
P1155	Electronic throttle system – malfunction	**Wiring, ETS limphome valve**
P1159	Intake manifold air control, target position – malfunction	**Unable to get target value**
P1166	Heated oxygen sensor (HO2S) 1, bank 1 – control limit reached	**Wiring open circuit, fuel pressure, exhaust system leak, EVAP system, HO2S**
P1167	Heated oxygen sensor (HO2S) 1, bank 2 – control limit reached	**Wiring open circuit, fuel pressure, exhaust system leak, EVAP system, HO2S**
P1168	Heated oxygen sensor (HO2S) 2, bank 1, heater control – circuit malfunction	**Wiring, HO2S**

OBD-II code	Fault location	Probable cause
P1169	Heated oxygen sensor (HO2S) 2, bank 2, heater control – circuit malfunction	Wiring, HO2S
P1171	Electronic throttle system, WOT – malfunction	ETS stuck open, ECM
P1172	ETS throttle motor – abnormal current	Wiring, throttle motor
P1173	ETS throttle motor – abnormal voltage	Wiring, throttle motor, alternator
P1174	Electronic throttle system, closed throttle 1 – malfunction	ETS stuck, ECM
P1175	Electronic throttle system, closed throttle 2 – malfunction	ETS stuck, ECM
P1176	ETS throttle motor – circuit 1 malfunction	Wiring open/short circuit, throttle motor
P1177	ETS throttle motor – circuit 2 malfunction	Wiring open/short circuit, throttle motor
P1178	Electronic throttle system relay – malfunction	Wiring, ETS relay, ECM
P1184	Heated oxygen sensor (HO2S) 1, bank 2 – no activity detected	Wiring open circuit, heating inoperative, HO2S
P1191	Electronic throttle system, limphome valve – malfunction	–
P1192	Electronic throttle system – limphome valve ON	Target following malfunction
P1193	Electronic throttle system – limphome valve ON	Low RPM
P1194	Electronic throttle system – limphome valve ON	TP sensor 2 malfunction
P1195	Electronic throttle system – limphome valve ON	Target following delay
P1196	Electronic throttle system – limphome valve ON	Close throttle stuck
P1307	Rough road sensor – malfunction	Wiring, Wiring, rough road sensor, ECM
P1308	Rough road sensor – voltage low	Wiring, rough road sensor, ECM
P1309	Rough road sensor – voltage high	Wiring, rough road sensor, ECM
P1330	Ignition spark timing – malfunction	Wiring between ROM change tool and ECM
P1372	Vehicle speed sensor (VSS) – range/ performance problem	Wiring, wheel speed sensor, mechanical fault
P1400	Exhaust gas recirculation (EGR) system – monitor malfunction	Wiring, EGR solenoid, EGR valve, mechanical fault, ECM
P1401	Fuel tank leakage diagnostic module – system malfunction	Fuel tank leakage diagnostic module
P1402	Fuel tank leakage diagnostic module – motor malfunction	Fuel tank leakage diagnostic module motor
P1403	Fuel tank leakage diagnostic module – valve malfunction	Fuel tank leakage diagnostic module valve
P1404	Fuel tank leakage diagnostic module – heater malfunction	Fuel tank leakage diagnostic module
P1440	Evaporative emission (EVAP) system, vent valve – circuit malfunction	Wiring, EVAP canister purge valve, ECM
P1502	Wheel speed sensor – open circuit	Wiring, wheel speed sensor

OBD-II code	Fault location	Probable cause
P1503	Cruise control switch – circuit malfunction	**Wiring, cruise control switch**
P1504	Cruise control switch, CANCEL – circuit malfunction	**Wiring, cruise control switch**
P1505	Idle air control (IAC) valve – opening coil circuit – open circuit	**Wiring, IAC valve**
P1506	Idle air control (IAC) valve – opening coil circuit – short circuit	**Wiring, IAC valve**
P1507	Idle air control (IAC) valve – closing coil circuit – open circuit	**Wiring, IAC valve**
P1508	Idle air control (IAC) valve – closing coil circuit – short circuit	**Wiring, IAC valve**
P1510	Idle air control (IAC) valve – circuit malfunction – open/short circuit	**Wiring, IAC valve, mechanical fault**
P1511	Idle air control (IAC) valve – circuit malfunction – open/short circuit	**Wiring, IAC valve, mechanical fault**
P1513	Idle air control (IAC) valve – opening coil circuit – short circuit	**Wiring, IAC valve, mechanical fault**
P1515	Idle air control (IAC) valve – coil 1 – signal malfunction	**Wiring, IAC valve**
P1516	Idle air control (IAC) valve – coil 2 – signal malfunction	**Wiring, IAC valve**
P1520	Alternator – FR terminal malfunction	**Wiring**
P1521	Power steering pressure (PSP) switch – circuit malfunction	**Wiring, PSP switch**
P1529	Transmission control module (TCM) – communication	**Wiring, TCM, ECM**
P1552	Idle air control (IAC) valve – closing coil circuit – short circuit	**Wiring, IAC valve, mechanical fault**
P1553	Idle air control (IAC) valve – closing coil circuit – open circuit	**Wiring, IAC valve, mechanical fault**
P1586	AT coding signal – not plausible	**Wiring, mechanical fault**
P1602	Transmission control module (TCM) – serial communication problem	**Wiring, TCM, ECM**
P1605	G-sensor – circuit malfunction	**Wiring, G-force sensor, ECM**
P1606	G-sensor – signal not plausible	**Wiring, G-force sensor, ECM**
P1607	Engine control module/Electronic throttle system – communication malfunction	**ECM, ETS**
P1608	Electronic throttle system/Engine control module – communication malfunction	**ETS, ECM**
P1609	Immobilizer control module – malfunction	**Wiring, immobilizer control module, ECM**
P1611	MIL request signal – circuit/voltage low	**Wiring, ECM**
P1613 3	MIL request signal – circuit/voltage high	**Wiring, ECM**

OBD-II code	Fault location	Probable cause
P1613 4	Electronic throttle system, controller – malfunction	**Wiring, ECM to ETS**
P1614 3	MIL request signal – circuit/voltage high	**Wiring, ECM**
P1614 4	Electronic throttle system, EEPROM R/W – malfunction	**ECM to ETS communication error**
P1615	Electronic throttle system, controller – malfunction	**Wiring, ETS to ECM**
P1616	Engine control relay – circuit malfunction	**Wiring, fuse, engine control relay**
P1623	Malfunction indicator lamp (MIL) – circuit malfunction	**Wiring, MIL**
P1624 1	Engine coolant blower motor relay – low speed circuit malfunction	**Wiring, engine coolant blower motor relay**
P1624 2	Transmission control module (TCM) – system malfunction	**Wiring, trouble code(s) stored in other system(s)**
P1625	Engine coolant blower motor relay – high speed circuit malfunction	**Wiring, engine coolant blower motor relay**
P1632 5	CAN data bus – OFF	–
P1632 6	Traction control system – malfunction	–
P1665	Ignition power stage A – circuit malfunction	**Wiring, ignition control module**
P1670	Ignition power stage B – circuit malfunction	**Wiring, ignition control module**
P1690	Immobilizer system – malfunction	**Wiring, immobilizer control module, ECM**
P1691	Immobilizer system – antenna malfunction	**Wiring, antenna, immobilizer control module**
P1693 1	Transmission control module (TCM) MIL request circuit – circuit malfunction	**Wiring, TCM, ECM**
P1693 2	Immobilizer system – transponder malfunction	**Wiring, ECM,**
P1694	Immobilizer system – ECM signal malfunction	**Wiring, ECM,**
P1695	Immobilizer system – ECM EPROM malfunction	**Wiring, ECM,**
P1696	Immobilizer system – ECM EPROM malfunction	**Wiring, ECM,**
P1707	Cruise control, brake pedal position (BPP) switch – circuit malfunction	**Wiring, brake pedal position switch**
P1715	AT pulse generators – open circuit	**Wiring, mechanical fault**
P1750	AT – shift control solenoids	**Wiring, mechanical fault**
P1765	Transmission control module (TCM) – torque reduction malfunction	**Wiring, TCM, ECM**

1 Except Accent
2 Accent
3 Except 3.5 Santa Fe/XG300
4 3.5 Santa Fe/XG300
5 Except XG300
6 XG300

HYUNDAI

Model:	Engine identification:	Year:
Excel 1.5L	G4-J/G4AJ	1990-92
Excel 1.5L	VIN code digit 6 = J	1993-94
Elantra 1.6/1.8L	G4R/G4M	1992-95
Sonata 2.0L	VIN code digit 8 = F	1992-95
Sonata 2.4/3.0L	G4-S/G6-T	1989-95
S Coupe 1.5L	G4AJ/G4DJ	1991-92

System: **ECFI**

Self-diagnosis

General information

- Refer to the front of this manual for general test conditions, terminology, detailed descriptions of wiring faults and a general trouble shooter for electrical and mechanical faults.
- Engine control module (ECM) incorporates self-diagnosis function.
- Malfunction indicator lamp (MIL) will illuminate if certain faults are recorded.
- ECM operates in backup mode if sensors fail, to enable vehicle to be driven to workshop.
- Trouble codes can be accessed with an analogue voltmeter or suitable code reader connected to the data link connector (DLC) **1**.

Accessing

- Ensure ignition switched OFF.
- Connect an analogue voltmeter between data link connector (DLC) 1 and 12 **1**.
- Switch ignition ON. Do NOT start engine.
- Voltmeter needle should be pulsing.
- Trouble codes will be displayed by long or short deflections of the voltmeter needle.
- Trouble codes are indicated as follows:
  - Long deflections indicate the 'tens' of the trouble code **2** [A].
  - Short deflections indicate the 'units' of the trouble code **2** [C].
  - A short pause separates each deflection **2** [B].
  - A long pause separates each trouble code **2** [D].
  - For example: Trouble code 12 displayed **2**.
- Switch ignition OFF.
- Disconnect voltmeter.

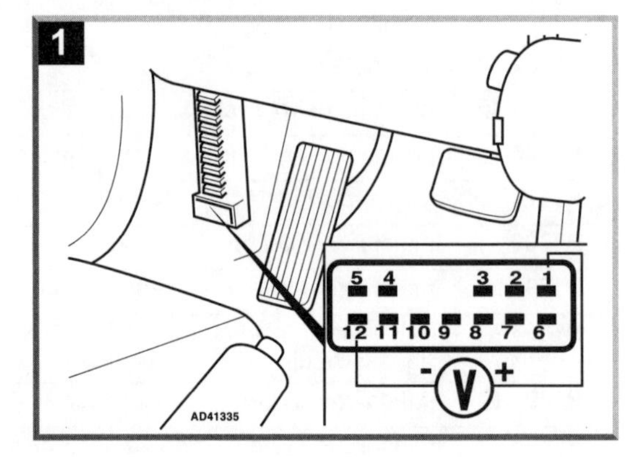

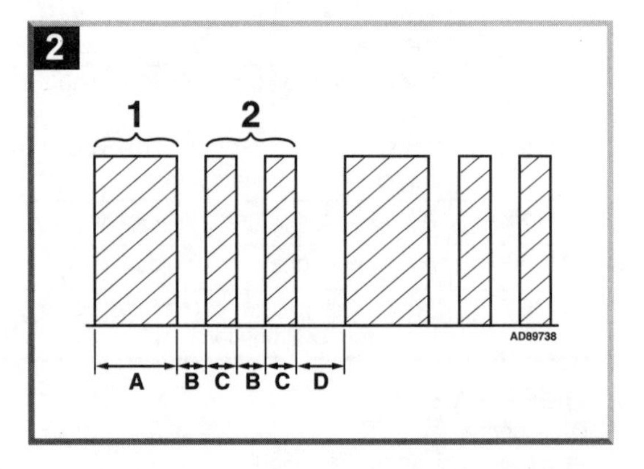

Erasing

- Ensure ignition switched OFF.
- Disconnect battery ground lead.
- Wait 15 seconds minimum.
- Reconnect battery ground lead.
- Repeat checking procedure to ensure no data remains in ECM fault memory.

NOTE: ***Disconnecting battery ground lead may also erase memory from electronic units such as radio.***

Trouble code identification

Flash code	Fault location	Probable cause
11	Oxygen sensor (O2S)/heated oxygen sensor (HO2S) – front	**Wiring, O2S/HO2S, mechanical fault, fuel system, fuel pressure, ECM**
12	Volume air flow (VAF) sensor	**Wiring, VAF sensor, ECM**
13	Intake air temperature (IAT) sensor	**Wiring, IAT sensor, ECM**
14	Throttle position (TP) sensor	**Wiring, TP sensor, mechanical fault, ECM**
15	Idle speed control (ISC) actuator position sensor	**Wiring, ISC sensor, mechanical fault, ECM**
21	Engine coolant temperature (ECT) sensor	**Wiring, ECT sensor, ECM**
22	Crankshaft position (CKP) sensor	**Wiring, CKP sensor, ECM**
23	Camshaft position (CMP) sensor	**Wiring, CMP sensor, ECM**
24	Vehicle speed sensor (VSS)	**Wiring, VSS sensor, ECM**
25	Barometric pressure (BARO) sensor	**Wiring, BARO sensor, ECM**
41	Injectors	**Wiring, injectors, mechanical fault**
42	Fuel pump	**Wiring, fuel pump, fuel system, mechanical fault**
43	Exhaust gas recirculation (EGR) system	**Wiring, EGR valve/solenoid, mechanical fault, ECM**
44	Ignition coil	**Wiring, ignition coil**
59	Heated oxygen sensor (HO2S) – rear	**Wiring, O2S/HO2S, mechanical fault, fuel system, fuel pressure, ECM**
1-1-1-1-1	No faults	–

HYUNDAI

Model:	Engine identification:	Year:
S Coupe 1.5L	**VIN code digit 4 = E**	**1993-95**
S Coupe 1.5L Turbo	**VIN code digit 4 = E**	**1993-95**

System: **ECFI**

Self-diagnosis

General information

- Refer to the front of this manual for general test conditions, terminology, detailed descriptions of wiring faults and a general trouble shooter for electrical and mechanical faults.
- Engine control module (ECM) incorporates self-diagnosis function.
- Malfunction indicator lamp (MIL) will illuminate if certain faults are recorded.
- ECM operates in backup mode if certain sensors fail, to enable car to be driven to workshop.
- Trouble codes can be accessed with suitable code reader connected to the data link connector (DLC) 1 or displayed with the MIL.

Accessing

- Switch ignition ON. Do NOT start engine.
- Jump data link connector (DLC) terminal 10 to ground for 2.5-4 seconds 1.
- Malfunction indicator lamp (MIL) should illuminate for approximately 2.5-7 seconds 2 [A].
- Malfunction indicator lamp (MIL) should then extinguish for approximately 2.5 seconds 2 [B].
- Malfunction indicator lamp (MIL) should then flash.
  - A group of short flashes indicate each of the 4 trouble code digits 2 [C].
  - A short pause separates each flash 2 [D].
  - A long pause separates each trouble code digit 2 [E].
  - For example: Trouble code 1223 is displayed 2.
- Each trouble code repeats until erased.
- Count MIL flashes and compare with trouble code table.
- To display the next trouble code:
- Jump data link connector (DLC) terminal 10 to ground for 2.5-4 seconds 1.
- When all trouble codes have been displayed (trouble code 3333) jump data link connector (DLC) terminal 10 to ground for 2.5-4 seconds 1.
- Switch ignition OFF.

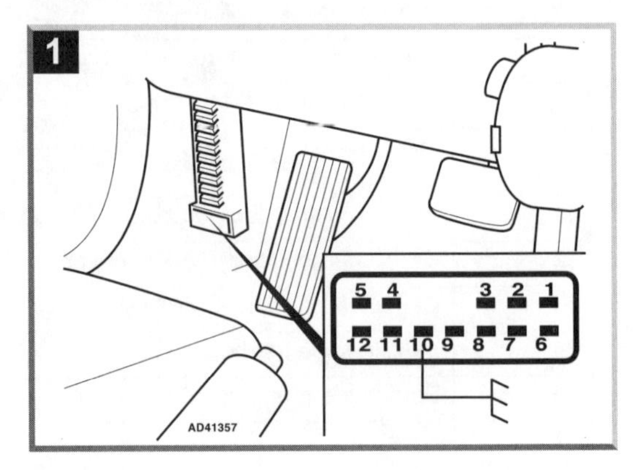

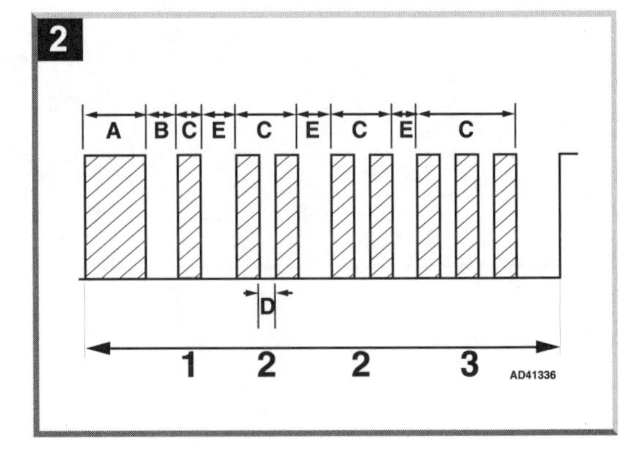

Erasing

- Ensure ignition switched OFF.
- Disconnect battery ground lead for 15 seconds minimum.
- Reconnect battery ground lead.
- Repeat checking procedure to ensure no data remains in ECM fault memory.

NOTE: ***Disconnecting battery ground lead may also erase memory from electronic units such as radio.***

Trouble code identification

Flash code	Fault location	Probable cause
1233	Engine control module (ECM) failure – read only memory	**ECM**
1234	Engine control module (ECM) failure – random access memory circuit	**ECM**
2121	Turbocharger (TC) wastegate regulating valve	**Wiring, wastegate regulating valve, mechanical fault**
3112	Injector No.1	**Wiring, injector, mechanical fault**
3114	Idle air control (IAC) valve – not opening	**Wiring, IAC, mechanical fault, ECM**
3116	Injector No.3	**Wiring, injector, mechanical fault**
3117	Mass air flow (MAF) sensor	**Wiring, MAF sensor, ECM**
3121	Manifold absolute pressure (MAP) sensor	**Wiring, MAP sensor, mechanical fault, ECM**
3122	Idle air control (IAC) valve – not closing	**Wiring, IAC valve, mechanical fault, ECM**
3128	Heated oxygen sensor (HO2S)	**Wiring, HO2S, mechanical fault, fuel system, fuel pressure, ECM**
3135	Evaporative emission (EVAP) canister purge valve	**Wiring, EVAP valve/solenoid, mechanical fault, ECM**
3137	Engine control module (ECM) power supply circuit	**Wiring, fuses**
3145	Engine coolant temperature (ECT) sensor	**Wiring, ECT sensor, ECM**
3149	Air conditioning compressor	**Wiring, AC compressor, mechanical fault**
3152	Turbocharger (TC) pressure too high	**Mechanical fault**
3153	Throttle position (TP) sensor	**Wiring, TP sensor, mechanical fault, ECM**
3211	Knock sensor (KS)	**Wiring, KS sensor, mechanical fault, ECM**
3222	Camshaft position (CMP) sensor	**Wiring, CMP sensor, mechanical fault, ECM**
3224	Engine control module (ECM) failure – knock control circuit	**ECM**
3232	Crankshaft position (CKP) sensor	**Wiring, CKP sensor, mechanical fault, ECM**
3233	Engine control module (ECM) failure – knock control circuit	**ECM**
3234	Injector No.2	**Wiring, injector, mechanical fault**
3235	Injector No.4	**Wiring, injector, mechanical fault**

Flash code	Fault location	Probable cause
3241	Engine control module (ECM) failure – injectors or evaporative emission (EVAP) canister purge valve circuit	ECM
3242	Engine control module (ECM) failure – idle air control (IAC) valve or air conditioning relay circuit	ECM
3333	End of output	–
4151	Fuel mixture control	Wiring, HO2S, fuel system, fuel pressure, mechanical fault, ECM
4152	Fuel mixture control	Wiring, HO2S, fuel system, fuel pressure, mechanical fault, ECM
4153	Fuel mixture control	Wiring, HO2S, fuel system, fuel pressure, mechanical fault, ECM
4154	Fuel mixture control	Wiring, HO2S, fuel system, fuel pressure, mechanical fault, ECM
4155	Engine control module (ECM) failure – injectors/positive crankcase ventilation/ IAC valve/air conditioning relay	ECM
4156	Turbocharger (TC) pressure control	Mechanical fault
4444	No fault	–

Model:	Engine identification:	Year:
G20 2.0L	SR20DE	1991-93
J30 3.0L	VG30DE	1993-95
M30 3.0L	VG30E	1990-92
Q45 4.5L	VH45DE	1990-95

System: **ECCS**

Self-diagnosis

General information

- Refer to the front of this manual for general test conditions, terminology, detailed descriptions of wiring faults and a general trouble shooter for electrical and mechanical faults.
- Engine control module (ECM) incorporates self-diagnosis function.
- Malfunction indicator lamp (MIL) will illuminate if certain faults are recorded.
- ECM operates in backup mode if sensors fail, to enable vehicle to be driven to workshop.
- Trouble codes can be accessed with suitable code reader connected to the data link connector (DLC) 1 – except J30 or 2 – J30.
- Trouble codes can also be displayed by the MIL and the red LED in the ECM.

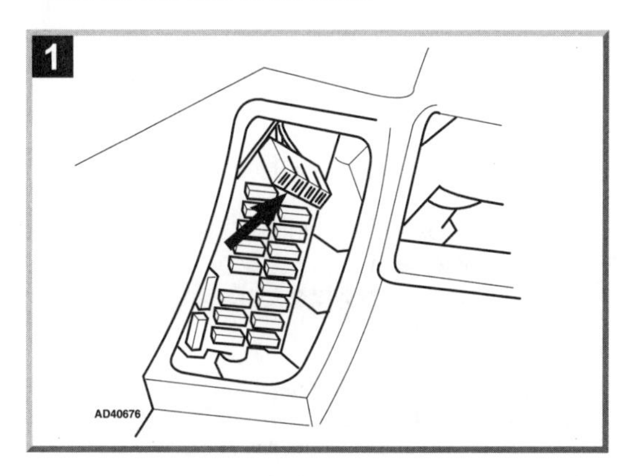

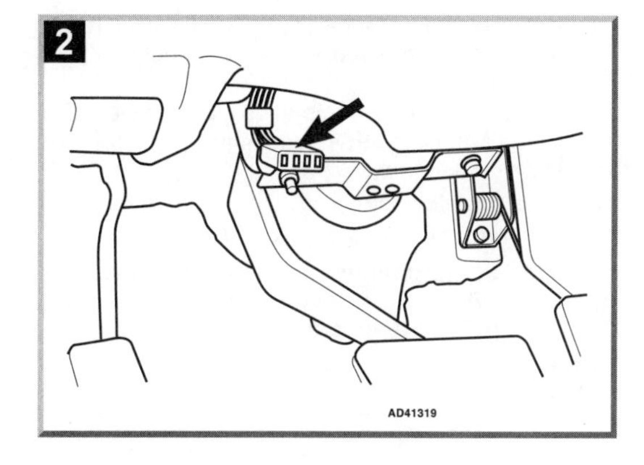

Accessing – Mode 1

- Access engine control module (ECM) 3
- Switch ignition ON.
- Check MIL and red LED in ECM illuminate.
- Start engine. Allow to idle.
- If MIL and LED extinguish, no trouble codes have been recorded.
- If MIL and LED remain illuminated, access trouble codes.

Accessing – Mode 2

- Switch ignition ON. Do NOT start engine.
- Turn diagnostic mode selector on ECM fully clockwise 4.
- Wait at least 2 seconds.
- Turn diagnostic mode selector on ECM fully counterclockwise 4.
- Self-diagnosis now in Mode 2.
- MIL and red LED in ECM should flash.
  - Long flashes indicate the 'tens' of the trouble code 5 [**A**].
  - Short flashes indicate the 'units' of the trouble code 5 [**C**].
  - A short pause separates each flash 5 [**B**].
  - A long pause separates each trouble code 5 [**D**].
  - For example: Trouble code 12 displayed 5.
- Count MIL (and LED) flashes and compare with trouble code table.
- Switch ignition OFF.
- Self-diagnosis will return to Mode 1.

NOTE: ***Starting engine in Self-diagnosis Mode 2 will activate heated oxygen sensor (HO2S) diagnosis.***

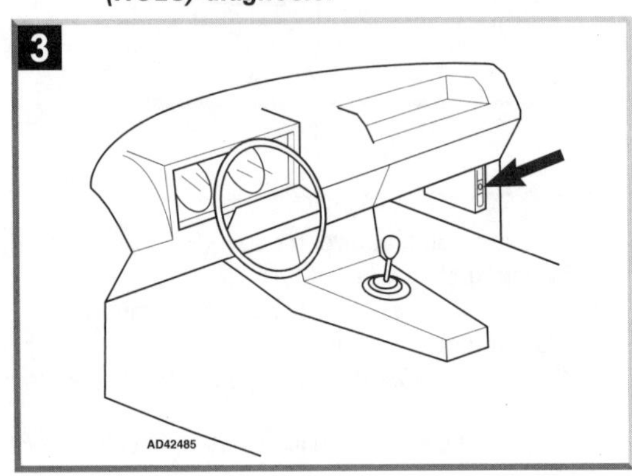

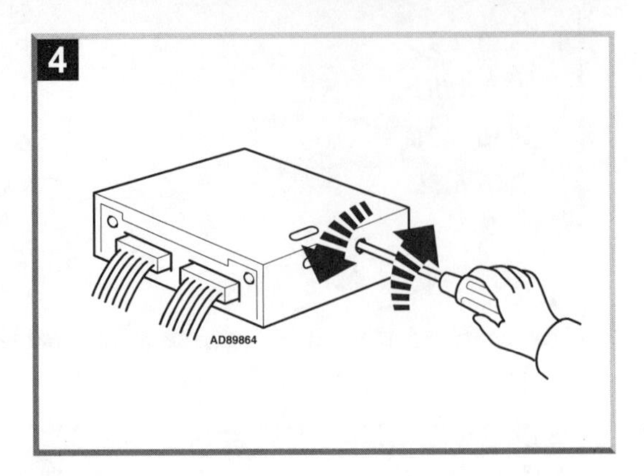

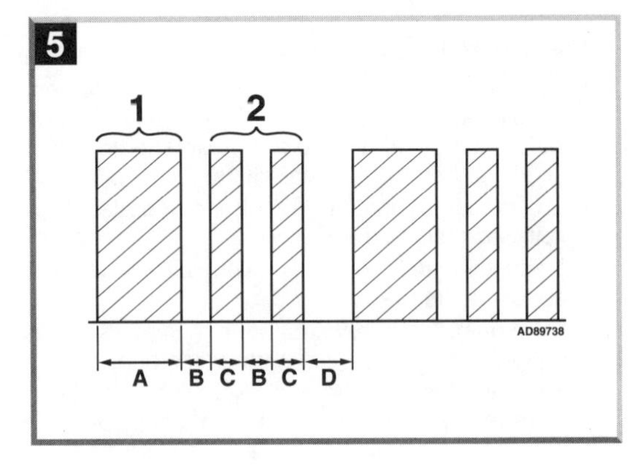

Erasing

- Switch ignition ON. Do NOT start engine.
- Turn diagnostic mode selector on ECM fully clockwise 4.
- Wait at least 2 seconds.
- Turn diagnostic mode selector on ECM fully counterclockwise 4.
- Self-diagnosis now in Mode 2.
- MIL should flash.
- Turn diagnostic mode selector on ECM fully clockwise 4.
- Wait at least 2 seconds.
- Turn diagnostic mode selector on ECM fully counterclockwise 4.
- Self-diagnosis will return to Mode 1.
- Switch ignition OFF.
- Repeat checking procedure to ensure no data remains in ECM fault memory.

Trouble code identification

Flash code	Fault location	Probable cause
11	Crankshaft position (CKP)/camshaft position (CMP) sensor	**Wiring, CKP/CMP sensor, mechanical fault, ECM**
12	Mass air flow (MAF) sensor	**Wiring, MAF sensor, mechanical fault, ECM**
13	Engine coolant temperature (ECT) sensor	**Wiring, ECT sensor, mechanical fault, ECM**
14	Vehicle speed sensor (VSS)	**Wiring, VSS, ECM**
16	Traction control system signal	**Wiring, mechanical fault, ECM**
21	Ignition primary signal	**Wiring, ignition coil, ignition control module**
22	Fuel pump circuit	**Wiring, fuel pump, fuel pump relay**
31	Engine control module (ECM)	**ECM**
32	Exhaust gas recirculation (EGR) system – California	**Wiring, EGR valve, mechanical fault**
33	Heated oxygen sensor (HO2S) – front (4 cyl)	**Wiring, HO2S, mechanical fault, ECM**
33	Heated oxygen sensor (HO2S) – LH (6 cyl)	**Wiring, HO2S, mechanical fault, ECM**
34	Knock sensor (KS)	**Wiring, KS sensor, mechanical fault, ECM**
35	Exhaust gas recirculation temperature (EGRT) sensor	**Wiring, EGRT sensor, mechanical fault**
42	Fuel temperature sensor	**Wiring, sensor, ECM**
43	Throttle position (TP) sensor 1	**Wiring, TP sensor, mechanical fault, ECM**
45	Injector leak – California	**Wiring, injector**
46	Throttle position (TP) sensor 2 – traction control system	**Wiring, TP sensor, mechanical fault, ECM**
51	Injector – California	**Wiring, injector**
53	Heated oxygen sensor (HO2S) – front RH	**Wiring, HO2S, mechanical fault, ECM**
54	AT signal input	**Wiring, mechanical fault**
55	No faults	–

INFINITI

Model:	Engine identification:	Year:
G20 2.0L	SR20DE	1994-96
I30 3.0L	VQ30DE	1996-99
J30 3.0L	VG30DE	1996-98
QX4 3.3L	VG33E	1997-00

System: **ECCS**

Self-diagnosis

General information

- Refer to the front of this manual for general test conditions, terminology, detailed descriptions of wiring faults and a general trouble shooter for electrical and mechanical faults.
- Engine control module (ECM) incorporates self-diagnosis function.
- Malfunction indicator lamp (MIL) will illuminate if certain faults are recorded.
- ECM operates in backup mode if sensors fail, to enable vehicle to be driven to workshop.
- Trouble codes can be displayed by using a Scan Tool connected to the data link connector (DLC) **1** [1], or by the malfunction indicator lamp (MIL).

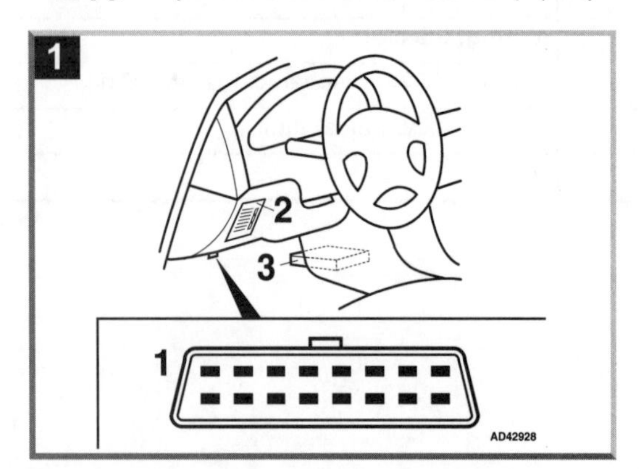

NOTE: ***The use of a Scan Tool is essential to obtain full diagnostic information.***

Accessing – Mode 1

- Switch ignition ON.
- Check MIL illuminates.
- Start engine. Allow to idle.
- If MIL extinguishes, no trouble codes have been recorded.
- If MIL remains illuminated, access trouble codes.

Accessing – Mode 2

- Access engine control module (ECM) **1** [3].
- Switch ignition ON. Do NOT start engine.
- Turn diagnostic mode selector on ECM fully clockwise **2**.
- Wait at least 2 seconds.
- Turn diagnostic mode selector on ECM fully counterclockwise **2**.
- Self-diagnosis now in Mode 2.
- MIL should flash.
- Count flashes and compare with trouble code table.
  - Long flashes indicate the 'hundreds' of the trouble code **3** [**A**].
  - Short flashes indicate the 'units' of the trouble code **3** [**C**].
  - A short pause separates each flash **3** [**B**].
  - A long pause separates each trouble code **3** [**D**].
  - For example: Trouble code 102 displayed **3**.
- Switch ignition OFF.
- Self-diagnosis will return to Mode 1.

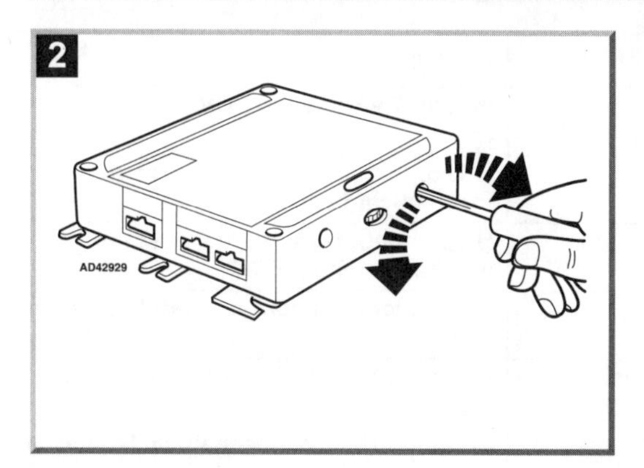

Erasing

- Trouble codes can be erased by using a Scan Tool connected to the data link connector (DLC) or as follows:
- With the ignition switch ON and the engine control module (ECM) in diagnosis test mode 2.
- Turn diagnostic mode selector on ECM fully clockwise **2**.
- Trouble codes are erased.
- Switch ignition OFF.
- Turn diagnostic mode selector on ECM fully counterclockwise **2**.
- Repeat checking procedure to ensure no data remains in ECM fault memory.

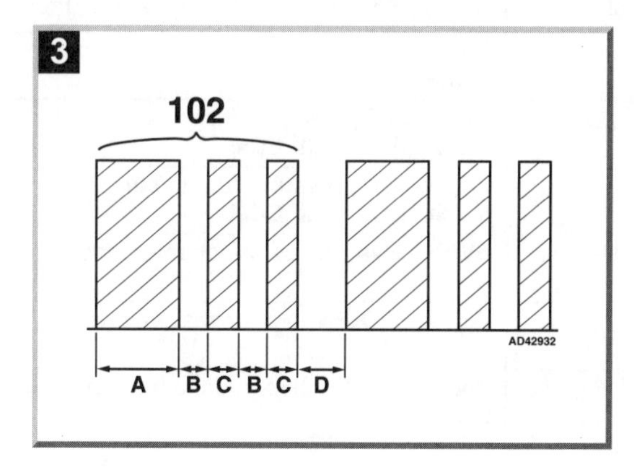

Trouble code identification

Flash code	OBD-II code	Fault location	Probable cause
–	*All P0, P2 and U0 codes*	Refer to OBD-II trouble code tables at the front of this manual	–
0101	–	Camshaft position (CMP) sensor – circuit malfunction	**Wiring, CMP sensor, ECM**
0102	–	Mass air flow (MAF) sensor – circuit malfunction	**Wiring, MAF sensor, ECM**
0103	–	Engine coolant temperature (ECT) sensor – circuit malfunction	**Wiring, ECT sensor, ECM**
0104	–	Vehicle speed sensor (VSS) – circuit malfunction	**Wiring, VSS sensor, ECM**
0107	–	ABS/TCS communication line – malfunction	**Wiring, ABS/TCS ECM, mechanical fault**
0111	P1447	Evaporative emission (EVAP) canister purge control system – malfunction	**Wiring, EVAP valve, mechanical fault, ECM**
0112	P1165	Intake manifold air control system – vacuum check switch circuit malfunction	**Wiring, vacuum check switch, hose leak/ blockage, air control solenoid, ECM**

Flash code	OBD-II code	Fault location	Probable cause
0113	P1443	Evaporative emission (EVAP) canister purge control system – circuit malfunction	**Wiring, EVAP valve, ECM**
0114	–	Fuel trim RH – mixture too rich	**HO2S, fuel pressure, fuel system, mechanical fault, ECM**
0115	–	Fuel trim RH – mixture too lean	**HO2S, fuel pressure, fuel system, mechanical fault, ECM**
0201	P1320	Ignition signal – circuit malfunction	**Wiring, ignition control module, ignition coil**
0203	–	Closed throttle position (CTP) switch – malfunction	**Wiring, CTP switch, mechanical fault, ECM**
0205	–	Idle control system – malfunction	**Wiring, IAC valve, mechanical fault, ECM**
0207	–	ABS/TCS communication line – malfunction	**Wiring, ABS/TCS ECM, mechanical fault**
0208	–	Engine overheating	**Cooling system, wiring, engine coolant blower motor**
0209	–	Fuel trim LH – mixture too rich	**HO2S, fuel pressure, fuel system, mechanical fault, ECM**
0210	–	Fuel trim LH – mixture too lean	**HO2S, fuel pressure, fuel system, mechanical fault, ECM**
0213	P1440	Evaporative emission (EVAP) canister purge control system – small leak	**EVAP valve, hoses, mechanical fault**
0214	P1444	Evaporative emission (EVAP) canister purge control system – malfunction	**Wiring, EVAP valve, hoses, mechanical fault, ECM**
0215	P1446	Evaporative emission (EVAP) canister purge control valve – malfunction	**Wiring, EVAP valve, mechanical fault, ECM**
0301	–	Engine control module (ECM) – internal function failure	**ECM**
0302	–	Exhaust gas recirculation (EGR) system – flow excessively high or low	**Wiring, EGR solenoid valve, mechanical fault**
0303	–	Heated oxygen sensor (HO2S) – LH front – circuit malfunction	**HO2S, ECM**
0304	–	Knock sensor (KS) – circuit	**Wiring, KS sensor, ECM**
0305	P1401	Exhaust gas recirculation (EGR) temperature sensor – circuit malfunction	**Wiring, EGR sensor, ECM**
0306	–	Exhaust gas recirculation (EGR) system – excess flow detected	**EGR valve, mechanical fault**
0307	–	Closed loop control RH – inoperative	**Wiring, HO2S, fuel pressure, fuel system, mechanical fault, ECM**
0307	P1148	Closed loop control LH – inoperative	**Wiring, HO2S, fuel pressure, fuel system, mechanical fault, ECM**
0308	–	Closed loop control LH – inoperative	**Wiring, HO2S, fuel pressure, fuel system, mechanical fault, ECM**
0308	P1168	Closed loop control RH – inoperative	**Wiring, HO2S, fuel pressure, fuel system, mechanical fault, ECM**
0309	P1448	Evaporative emission (EVAP) canister purge control valve – malfunction	**Wiring, EVAP valve, ECM**

Flash code	OBD-II code	Fault location	Probable cause
0311	P1491	Evaporative emission (EVAP) canister purge control system – malfunction	Wiring, EVAP valve, ECM
0312	P1493	Evaporative emission (EVAP) canister purge control valve – malfunction	Wiring, EVAP valve, ECM
0313	–	Heated oxygen sensor (HO2S) – LH rear – voltage high	Wiring, HO2S, fuel pressure, fuel system, mechanical fault, ECM
0314	–	Heated oxygen sensor (HO2S) – LH rear – voltage low	Wiring, HO2S, fuel pressure, fuel system, mechanical fault, ECM
0315	–	Heated oxygen sensor (HO2S) – LH rear – no activity detected	Wiring, HO2S, fuel pressure, fuel system, mechanical fault, ECM
0401	–	Intake air temperature (IAT) sensor – circuit malfunction	Wiring, IAT sensor, ECM
0402	–	Fuel temperature sensor – circuit malfunction	Wiring, temperature sensor, ECM
0403	–	Throttle position (TP) sensor – circuit malfunction	Wiring, TP sensor, ECM
0404	–	ABS/TCS communication line – malfunction	Wiring, ABS/TCS ECM
0407	P1335	Crankshaft position (CKP) sensor – circuit malfunction	Wiring, CKP sensor, ECM
0409	–	Heated oxygen sensor (HO2S) – RH front – slow response	Wiring, HO2S, fuel pressure, fuel system, mechanical fault, ECM
0410	–	Heated oxygen sensor (HO2S) – RH front – voltage high	Wiring, HO2S, fuel pressure, fuel system, mechanical fault, ECM
0411	–	Heated oxygen sensor (HO2S) – RH front – voltage low	Wiring, HO2S, fuel pressure, fuel system, mechanical fault, ECM
0412	–	Heated oxygen sensor (HO2S) – RH front – no activity detected	Wiring, HO2S, fuel pressure, fuel system, mechanical fault, ECM
0413	–	Heated oxygen sensor (HO2S) – LH front – slow response	Wiring, HO2S, fuel pressure, fuel system, mechanical fault, ECM
0414	–	Heated oxygen sensor (HO2S) – LH front – voltage high	Wiring, HO2S, fuel pressure, fuel system, mechanical fault, ECM
0415	–	Heated oxygen sensor (HO2S) – LH front – voltage low	Wiring, HO2S, fuel pressure, fuel system, mechanical fault, ECM
0503	–	Heated oxygen sensor (HO2S) – RH front – circuit malfunction	Wiring, HO2S, ECM
0504	P0600	ECM/TCM – communication malfunction	Wiring, ECM, TCM
0505	–	No self diagnostic failure indicated	–
0509	–	Heated oxygen sensor (HO2S) – LH front – no activity detected	Wiring, HO2S, fuel pressure, fuel system, mechanical fault, ECM
0510	–	Heated oxygen sensor (HO2S) – rear – no activity detected/voltage high	Wiring, HO2S, fuel pressure, fuel system, mechanical fault, ECM
0511	–	Heated oxygen sensor (HO2S) – rear – voltage low	Wiring, HO2S, fuel pressure, fuel system, mechanical fault, ECM
0512	–	Heated oxygen sensor (HO2S) 2, bank 1 – circuit high	Wiring open/short circuit, HO2S, ECM

Flash code	OBD-II code	Fault location	Probable cause
0514	P1402	Exhaust gas recirculation (EGR) temperature system – malfunction	**Wiring, EGR temperature sensor, mechanical fault, ECM**
0515	P0403	EGR volume control valve – malfunction	**Wiring, EGR volume control valve**
0603	–	Cylinder 6 – misfire	**Wiring, ignition system, fuel system, fuel pressure, mechanical fault, ECM**
0604	–	Cylinder 5 – misfire	**Wiring, ignition system, fuel system, fuel pressure, mechanical fault, ECM**
0605	–	Cylinder 4 – misfire	**Wiring, ignition system, fuel system, fuel pressure, mechanical fault, ECM**
0606	–	Cylinder 3 – misfire	**Wiring, ignition system, fuel system, fuel pressure, mechanical fault, ECM**
0607	–	Cylinder 2 – misfire	**Wiring, ignition system, fuel system, fuel pressure, mechanical fault, ECM**
0608	–	Cylinder 1 – misfire	**Wiring, ignition system, fuel system, fuel pressure, mechanical fault, ECM**
0701	–	Random misfire	**Wiring, ignition system, fuel system, EGR valve, fuel pressure, mechanical fault, ECM**
0702	–	Catalytic converter RH – efficiency below limit	**Catalytic converter, rear HO2S**
0703	–	Catalytic converter LH – efficiency below limit	**Catalytic converter, rear HO2S**
0704	–	Evaporative emission (EVAP) canister pressure sensor – malfunction	**Wiring, canister pressure sensor, mechanical fault, ECM**
0705	–	Evaporative emission (EVAP) canister purge control system – small leak	**Wiring, EVAP valve, mechanical fault, ECM**
0707	–	Heated oxygen sensor (HO2S) – rear/ RH rear – slow response/circuit malfunction	**Wiring, HO2S, fuel pressure, fuel system, mechanical fault, ECM**
0708	–	Heated oxygen sensor (HO2S) – LH rear – slow response/circuit malfunction	**Wiring, HO2S, fuel pressure, fuel system, mechanical fault, ECM**
0715	–	Evaporative emission (EVAP) canister purge control system – large leak	**EVAP valve, hoses leaking, fuel filler cap missing, mechanical fault**
0801	P1490	Evaporative emission (EVAP) canister purge control system – malfunction	**Wiring, EVAP solenoid, hoses, mechanical fault, ECM**
0802	–	Crankshaft position (CKP) sensor – circuit malfunction	**Wiring, CKP sensor, ECM**
0803	–	Manifold absolute pressure (MAP) sensor – circuit malfunction	**Wiring, MAP sensor, ECM**
0804	P1605	AT diagnosis communication – signal failure	**Wiring, ECM**
0807	–	Evaporative emission (EVAP) canister purge control system – circuit malfunction	**Wiring, EVAP solenoid, ECM**
0807	P1492	Evaporative emission (EVAP) canister purge control valve – malfunction	**Wiring, EVAP solenoid, hoses, mechanical fault, ECM**
0901	–	Heated oxygen sensor (HO2S) – RH front – heater circuit malfunction	**Wiring, HO2S, ECM**

Flash code	OBD-II code	Fault location	Probable cause
0902	–	Heated oxygen sensor (HO2S) – RH rear – heater circuit malfunction	Wiring, HO2S, ECM
0903	–	Evaporative emission (EVAP) canister purge control system – circuit malfunction	Wiring, EVAP valve, mechanical fault, ECM
0905	P1336	Crankshaft position (CKP) sensor – signal malfunction	Wiring, CKP sensor, ECM
0908	–	Insufficient coolant temperature for closed loop fuel control	Wiring, ECT sensor, mechanical fault, ECM
1001	–	Heated oxygen sensor (HO2S) – LH front – heater circuit malfunction	Wiring, HO2S, ECM
1002	–	Heated oxygen sensor (HO2S) – LH rear – heater circuit malfunction	Wiring, HO2S, ECM
1003	–	Park/neutral position (PNP) switch – circuit malfunction	Wiring, PNP switch, mechanical fault, ECM
1003	P1706	Park neutral position(PNP) switch – malfunction	Wiring, PNP switch, mechanical fault, ECM
1004	P1130	Intake manifold air control solenoid – malfunction	Wiring, control solenoid, intake system, hose leak/blockage, control vacuum check switch, ECM
1005	P1400	Exhaust gas recirculation (EGR) valve – voltage incorrect	Wiring, EGR valve, mechanical fault, ECM
1008	P1445	Evaporative emission (EVAP) canister purge control valve – circuit malfunction	Wiring, EVAP valve, ECM
1101	–	Park/neutral position (PNP) switch – circuit	Wiring, PNP switch, mechanical fault, ECM
1102	–	AT – vehicle speed sensor (VSS)	Wiring, VSS sensor, mechanical fault, ECM
1103	–	AT – first gear selection – malfunction	Wiring, mechanical fault
1104	–	AT – second gear selection – malfunction	Wiring, mechanical fault
1105	–	AT – third gear selection – malfunction	Wiring, mechanical fault
1106	–	AT – fourth gear selection – malfunction	Wiring, mechanical fault
1107	–	AT – signal circuit malfunction	Wiring, mechanical fault
1108	–	AT – shift solenoid A – voltage low	Wiring, shift solenoid, mechanical fault
1201	–	AT – shift solenoid B – voltage low	Wiring, shift solenoid, mechanical fault
1203	P1760	AT – overrun clutch solenoid – voltage low	Wiring, solenoid, mechanical fault
1204	–	Torque converter clutch solenoid – voltage low	Wiring, solenoid, mechanical fault
1205	–	AT – line pressure solenoid – voltage low	Wiring, solenoid, mechanical fault
1206	P1705	AT – throttle position (TP) sensor – voltage high or low	Wiring, TP sensor, mechanical fault, ECM
1207	–	Engine speed (RPM) – incorrect voltage signal ECM to AT control module	Wiring, RPM sensor, mechanical fault, ECM

Flash code	OBD-II code	Fault location	Probable cause
1208	–	Transmission fluid temperature sensor – excessively high or low voltage	Wiring, temperature sensor, mechanical fault, ECM
1302	P1105	MAP/BARO switching valve – circuit malfunction	Wiring, MAP/BARO switching valve
1305	P1220	Fuel pressure (FP) control module – circuit malfunction	Wiring, FP control module
1308	P1900	Engine coolant blower motor – circuit	Wiring, blower motor, ECM

Model:	Engine identification:	Year:
I35 3.5L	VQ35DE	2002-04
QX4 3.5L	VQ35DE	2002-03
Q45 4.5L	VK45DE	2002-04

System: **ECCS**

Self-diagnosis

General information

- Refer to the front of this manual for general test conditions, terminology, detailed descriptions of wiring faults and a general trouble shooter for electrical and mechanical faults.
- Engine control module (ECM) incorporates self-diagnosis function.
- Malfunction indicator lamp (MIL) will illuminate if certain faults are recorded.
- ECM operates in backup mode if sensors fail, to enable vehicle to be driven to workshop.
- Trouble codes can be displayed by using a Scan Tool connected to the data link connector (DLC) **1**, or by the malfunction indicator lamp (MIL).

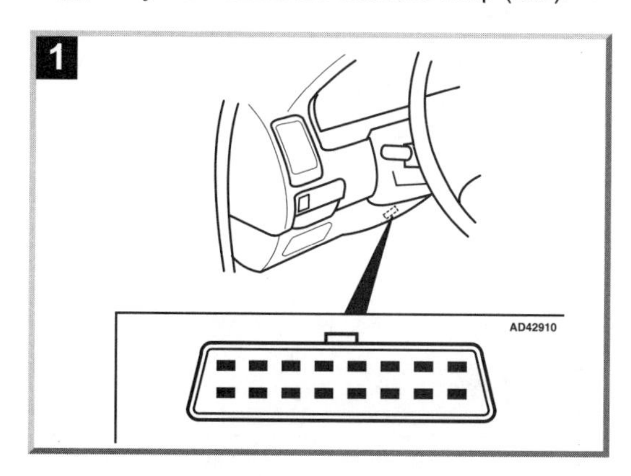

NOTE: ***The use of a Scan Tool is essential to obtain full diagnostic information.***

Accessing – Mode 1

- Switch ignition ON.
- Check MIL illuminates.
- Start engine. Allow to idle.
- If MIL extinguishes, no trouble codes have been recorded.
- If MIL remains illuminated, access trouble codes.

Accessing – Mode 2

- Ensure accelerator pedal is fully released.
- Switch ignition ON. Do NOT start engine.
- Wait 3 seconds.
- Within 5 seconds, repeat the following 5 times:
  - Fully depress accelerator pedal.
  - Fully release accelerator pedal.
- Wait 7 seconds.
- Fully depress accelerator pedal.
- After approximately 10 seconds MIL will start flashing.
- Fully release accelerator pedal.
- Self-diagnosis now in Mode 2.
- Count flashes and compare with trouble code table.
- Each trouble code consists of four groups of one or more flashes.
- The first group of flashes indicate the 'thousands' of the trouble code **2** [**A**].
- The second group of flashes indicate the 'hundreds' of the trouble code **2** [**B**].
- The third group of flashes indicate the 'tens' of the trouble code **2** [**C**].
- The fourth group of flashes indicate the 'units' of the trouble code **2** [**D**].
- Ten flashes in a group indicate '0'.
- A short pause separates each trouble code group **2** [**E**].

- A long pause separates each trouble code 2 [F].
- For example: Trouble code P1130 displayed 2.
- Switch ignition OFF.
- Self-diagnosis will return to Mode 1.

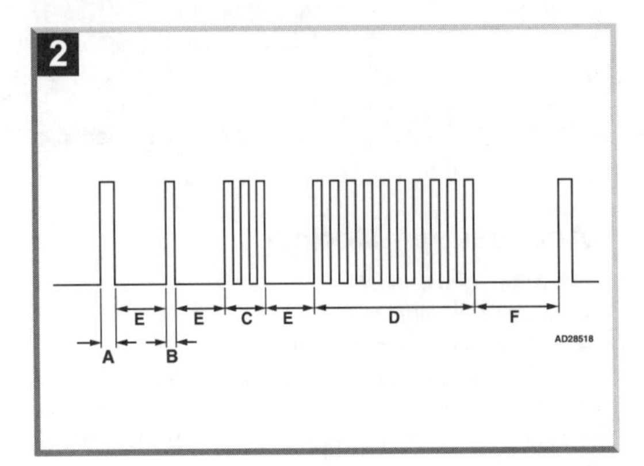

Erasing

- Trouble codes can be erased by using a Scan Tool connected to the data link connector (DLC) or as follows:
- Ensure accelerator pedal is fully released.
- Switch ignition ON. Do NOT start engine.
- Wait 3 seconds.
- Within 5 seconds, repeat the following 5 times:
  - Fully depress accelerator pedal.
  - Fully release accelerator pedal.
- Wait 7 seconds.
- Fully depress accelerator pedal.
- After approximately 10 seconds MIL will start flashing.
- Fully release accelerator pedal.
- Self-diagnosis now in Mode 2.
  - Fully depress accelerator pedal for more than 10 seconds.
  - Fully release accelerator pedal.
  - Trouble code 0000 should be displayed.

Trouble code identification

Flash code	OBD-II code	Fault location	Probable cause
–	*All P0, P2 and U0 codes*	Refer to OBD-II trouble code tables at the front of this manual	–
0000	–	No self diagnostic failure indicated	–
0011	–	Camshaft position (CMP), intake/left/front, bank 1 – timing over-advanced/system performance	**Valve timing, engine mechanical fault, CMP actuator**
0021	–	Camshaft position (CMP), intake/left/front, bank 2 – timing over-advanced/system performance	**Valve timing, engine mechanical fault, CMP actuator**
0031	–	Heated oxygen sensor (HO2S) 1, bank 1, heater control – circuit low	**Wiring short to ground, HO2S, ECM**
0032	–	Heated oxygen sensor (HO2S) 1, bank 1, heater control – circuit high	**Wiring short to positive, HO2S, ECM**
0037	–	Heated oxygen sensor (HO2S) 2, bank 1, heater control – circuit low	**Wiring short to ground, HO2S, ECM**
0038	–	Heated oxygen sensor (HO2S) 2, bank 1, heater control – circuit high	**Wiring short to positive, HO2S, ECM**
0051	–	Heated oxygen sensor (HO2S) 1, bank 2, heater control – circuit low	**Wiring short to ground, HO2S, ECM**
0052	–	Heated oxygen sensor (HO2S) 1, bank 2, heater control – circuit high	**Wiring short to positive, HO2S, ECM**
0057	–	Heated oxygen sensor (HO2S) 2, bank 2, heater control – circuit low	**Wiring short to ground, HO2S, ECM**

Flash code	OBD-II code	Fault location	Probable cause
0058	–	Heated oxygen sensor (HO2S) 2, bank 2, heater control – circuit high	**Wiring short to positive, HO2S, ECM**
0100	–	Mass air flow (MAF) sensor/volume air flow (VAF) sensor – circuit malfunction	**Wiring, MAF/VAF sensor, ECM**
0101	–	Mass air flow (MAF) sensor/volume air flow (VAF) sensor – range/performance problem	**Intake leak/blockage, MAF/VAF sensor**
0102	–	Mass air flow (MAF) sensor/volume air flow (VAF) sensor – low input	**Wiring short to ground, MAF/VAF sensor, ECM**
0103	–	Mass air flow (MAF) sensor/volume air flow (VAF) sensor – high input	**Wiring short to positive, MAF/VAF sensor, ECM**
0105	–	Manifold absolute pressure (MAP) sensor – circuit malfunction	**Wiring, MAP sensor, ECM**
0107	–	Manifold absolute pressure (MAP) sensor – circuit/voltage low	**Wiring, MAP sensor, ECM**
0108	–	Manifold absolute pressure (MAP) sensor – circuit/voltage high	**Wiring open circuit, MAP sensor, ECM**
0110	–	Intake air temperature (IAT) sensor – circuit malfunction	**Wiring, IAT sensor, ECM**
0112	–	Intake air temperature (IAT) sensor – low input	**Wiring short to ground, IAT sensor, ECM**
0113	–	Intake air temperature (IAT) sensor – high input	**Wiring open circuit/short to positive, ground wire defective, IAT sensor, ECM**
0115	–	Engine coolant temperature (ECT) sensor – circuit malfunction	**Wiring, ECT sensor, ECM**
0117	–	Engine coolant temperature (ECT) sensor – circuit/voltage low	**Wiring short circuit, ECT sensor, ECM**
0118	–	Engine coolant temperature (ECT) sensor – circuit/voltage high	**Wiring open circuit, ECT sensor, ECM**
0120	–	Throttle position (TP) sensor A/accelerator pedal position (APP) sensor A – circuit malfunction	**Wiring, TP/APP sensor, ECM**
0121	–	Throttle position (TP) sensor – performance problem	**Wiring, TP sensor, ECM**
0122	–	Throttle position (TP) sensor A/accelerator pedal position (APP) sensor A – low input	**Wiring short to ground, TP/APP sensor, ECM**
0123	–	Throttle position (TP) sensor A/accelerator pedal position (APP) sensor A – high input	**Wiring short to positive, TP/APP sensor, ECM**
0125	–	Insufficient coolant temperature for closed loop fuel control	**Wiring, ECT sensor, mechanical fault, ECM**
0127	–	Intake air temperature (IAT) sensor – temperature too high	**Wiring, IAT sensor, ECM**
0128	–	Thermostat function – performance problem	**Thermostat, ECT, coolant leak**
0130	–	Heated oxygen sensor (HO2S) – LH front – circuit malfunction	**Heating inoperative, poor connection, wiring, HO2S**
0131	–	Heated oxygen sensor (HO2S) – LH front – low voltage	**Exhaust leak, wiring short to ground, HO2S, ECM**

Flash code	OBD-II code	Fault location	Probable cause
0132	–	Heated oxygen sensor (HO2S) – LH front – voltage high	Wiring, HO2S, fuel pressure, fuel system, mechanical fault, ECM
0133	–	Heated oxygen sensor (HO2S) – LH front – slow response	Wiring, HO2S, fuel pressure, fuel system, mechanical fault, ECM
0134	–	Heated oxygen sensor (HO2S) – LH front – no activity detected	Wiring, HO2S, fuel pressure, fuel system, mechanical fault, ECM
0137	–	Heated oxygen sensor (HO2S) – LH rear – low voltage	Exhaust leak, wiring short to ground, HO2S, ECM
0138	–	Heated oxygen sensor (HO2S) – LH rear – voltage high	Wiring, HO2S, fuel pressure, fuel system, mechanical fault, ECM
0139	–	Heated oxygen sensor (HO2S) – LH rear – slow response	Wiring, HO2S, ECM
0140	–	Heated oxygen sensor (HO2S) – LH rear – no activity detected	Wiring, heating inoperative, HO2S, ECM
0141	–	Heated oxygen sensor (HO2S) – LH rear, heater control – circuit malfunction	Wiring, HO2S, ECM
0150	–	Heated oxygen sensor (HO2S) – RH front – circuit malfunction	Wiring, HO2S, ECM
0151	–	Heated oxygen sensor (HO2S) – RH front – low voltage	Exhaust leak, wiring short to ground, HO2S, ECM
0152	–	Heated oxygen sensor (HO2S) – RH front – voltage high	Wiring, HO2S, fuel pressure, fuel system, mechanical fault, ECM
0153	–	Heated oxygen sensor (HO2S) – RH front – slow response	Wiring, HO2S, fuel pressure, fuel system, mechanical fault, ECM
0154	–	Heated oxygen sensor (HO2S) – RH front – no activity detected	Wiring, HO2S, fuel pressure, fuel system, mechanical fault, ECM
0155	–	Heated oxygen sensor (HO2S) – RH front, heater control – circuit malfunction	Wiring, HO2S, ECM
0157	–	Heated oxygen sensor (HO2S) – RH rear – low voltage	Exhaust leak, wiring short to ground, HO2S, ECM
0158	–	Heated oxygen sensor (HO2S) – RH rear – voltage high	Wiring, HO2S, fuel pressure, fuel system, mechanical fault, ECM
0159	–	Heated oxygen sensor (HO2S) – RH rear – slow response	Wiring, HO2S, ECM
0160	–	Heated oxygen sensor (HO2S) – RH rear – no activity detected	Wiring, HO2S, ECM
0161	–	Heated oxygen sensor (HO2S) – RH rear, heater control – circuit malfunction	Wiring, HO2S, ECM
0171	–	Fuel trim LH – mixture too lean	HO2S, fuel pressure, fuel system, mechanical fault, ECM
0172	–	Fuel trim LH – mixture too rich	HO2S, fuel pressure, fuel system, mechanical fault, ECM
0174	–	Fuel trim RH – mixture too lean	HO2S, fuel pressure, fuel system, mechanical fault, ECM
0175	–	Fuel trim RH – mixture too rich	HO2S, fuel pressure, fuel system, mechanical fault, ECM

Flash code	OBD-II code	Fault location	Probable cause
0180	–	Fuel temperature sensor – circuit malfunction	**Wiring, fuel temperature sensor, ECM**
0181	–	Fuel temperature sensor – range/ performance problem	**Wiring**
0182	–	Fuel temperature sensor – voltage low	**Wiring short to ground, fuel temperature sensor, ECM**
0183	–	Fuel temperature sensor – voltage high	**Wiring short to positive, fuel temperature sensor, ECM**
0217	–	Engine over temperature condition	**Wiring, cooling system, coolant thermostat, ECT sensor**
0221	–	Throttle position (TP) sensor – circuit performance problem	**Wiring, throttle motor**
0222	–	Throttle position (TP) sensor 1 – circuit low voltage	**Wiring, throttle motor**
0223	–	Throttle position (TP) sensor 1 – Circuit high voltage	**Wiring, throttle motor**
0226	–	Throttle position (TP) sensor C/accelerator pedal position (APP) sensor C – range/ performance problem	**Accelerator cable adjustment, TP/APP sensor**
0227	–	Throttle position (TP) sensor C/accelerator pedal position (APP) sensor C – low input	**Wiring short to ground, TP/APP sensor, ECM**
0228	–	Throttle position (TP) sensor C/accelerator pedal position (APP) sensor C – high input	**Wiring short to positive, TP/APP sensor, ECM**
0300	–	Random misfire	**Wiring, ignition system, fuel system, EGR valve, fuel pressure, mechanical fault, ECM**
0301	–	Cylinder 1 – misfire	**Wiring, ignition system, fuel system, fuel pressure, mechanical fault, ECM**
0302	–	Cylinder 2 – misfire	**Wiring, ignition system, fuel system, fuel pressure, mechanical fault, ECM**
0303	–	Cylinder 3 – misfire	**Wiring, ignition system, fuel system, fuel pressure, mechanical fault, ECM**
0304	–	Cylinder 4 – misfire	**Wiring, ignition system, fuel system, fuel pressure, mechanical fault, ECM**
0305	–	Cylinder 5 – misfire	**Wiring, ignition system, fuel system, fuel pressure, mechanical fault, ECM**
0306	–	Cylinder 6 – misfire	**Wiring, ignition system, fuel system, fuel pressure, mechanical fault, ECM**
0307	–	Cylinder 7 – misfire	**Wiring, ignition system, fuel system, fuel pressure, mechanical fault, ECM**
0308	–	Cylinder 8 – misfire	**Wiring, ignition system, fuel system, fuel pressure, mechanical fault, ECM**
0325	–	Knock sensor (KS) 1, bank 1 – circuit malfunction	**Wiring, poor connection, KS**

Flash code	OBD-II code	Fault location	Probable cause
0327	–	Knock sensor (KS) 1, bank 1 – low input	**Insecure KS, poor connection, wiring short to ground, incorrectly tightened, KS, ECM**
0328	–	Knock sensor (KS) 1, bank 1 – high input	**Wiring short to positive, KS incorrectly tightened, KS, ECM**
0330	–	Knock sensor (KS) 2, bank 2 – circuit malfunction	**Wiring, KS, ECM**
0332	–	Knock sensor (KS) 2, bank 2 – low input	**Insecure KS, poor connection, wiring short to ground, KS incorrectly tightened, KS, ECM**
0333	–	Knock sensor (KS) 2, bank 2 – high input	**Wiring short to positive, KS incorrectly tightened, KS, ECM**
0335	–	Crankshaft position (CKP) sensor – circuit malfunction	**Wiring, CKP sensor, ECM**
0340	–	Camshaft position (CMP) sensor bank 1 – circuit malfunction	**Wiring, CMP sensor, ECM**
0345	–	Camshaft position (CMP) sensor bank 2 – circuit malfunction	**Wiring, CMP sensor, ECM**
0420	–	Catalytic converter LH – efficiency below limit	**Catalytic converter, rear HO2S**
0430	–	Catalytic converter RH – efficiency below limit	**Catalytic converter, rear HO2S**
0440	–	Evaporative emission (EVAP) system – malfunction	**Hose connection(s), intake leak, EVAP canister purge valve**
0441	–	Evaporative emission (EVAP) canister purge system – incorrect flow	**Wiring, EVAP solenoid, hose leak/ blockage, ECM**
0442	–	Evaporative emission (EVAP) canister purge system – small leak detected	**Wiring, EVAP solenoid, hoses, mechanical fault, ECM**
0443	–	Evaporative emission (EVAP) canister purge valve – circuit malfunction	**Wiring, EVAP canister purge valve, ECM**
0446	–	Evaporative emission (EVAP) system, vent control – circuit malfunction	**Wiring, EVAP canister purge valve, ECM**
0444	–	Evaporative emission (EVAP) canister purge valve – open circuit	**Wiring open circuit, EVAP canister purge valve, ECM**
0445	–	Evaporative emission (EVAP) canister purge valve – short circuit	**Wiring short circuit, EVAP canister purge valve, ECM**
0447	–	Evaporative emission (EVAP) system, vent control – open circuit	**Wiring open circuit, EVAP canister purge valve, ECM**
0450	–	Evaporative emission (EVAP) pressure sensor – circuit malfunction	**Wiring, EVAP pressure sensor, ECM**
0452	–	Fuel tank pressure sensor – circuit/voltage low	**Wiring, pressure sensor, ECM**
0453	–	Fuel tank pressure sensor – circuit/voltage high	**Wiring open circuit, pressure sensor, ECM**
0455	–	Evaporative emission (EVAP) canister purge system – incorrect flow	**Wiring, EVAP solenoid, mechanical fault, ECM**

Flash code	OBD-II code	Fault location	Probable cause
0456	–	Evaporative emission (EVAP) canister purge system (fuel tank system) – leak detected	**Hose, fuel tank/pressure sensor, fuel filler cap, EVAP valve/bypass solenoid, EVAP two way valve, EVAP canister/vent valve**
0460	–	Fuel level sensor – circuit malfunction	**Wiring, sensor, mechanical fault, ECM**
0461	–	Fuel level sensor – circuit performance	**Wiring, sensor, mechanical fault, ECM**
0462	–	Fuel level sensor – voltage low	**Wiring, sensor, mechanical fault, ECM**
0463	–	Fuel level sensor – voltage high	**Wiring, sensor, mechanical fault, ECM**
0464	–	Fuel level sensor – circuit intermittent	**Wiring, poor connection, fuel tank level sensor, ECM**
0500	–	Vehicle speed sensor (VSS) – circuit malfunction	**Wiring, VSS, ECM**
0505	–	Idle control system – malfunction	**Wiring, IAC valve, mechanical fault, ECM**
0506	–	Idle speed control (ISC) system – rpm lower than expected	**Wiring, ISC actuator/IAC valve, throttle motor, throttle valve tight/sticking, ECM**
0507	–	Idle speed control (ISC) system – rpm higher than expected	**Wiring, ISC actuator/IAC valve, throttle motor, throttle valve tight/sticking, ECM**
0510	–	Closed throttle position (CTP) switch – malfunction	**Wiring, CTP switch, mechanical fault, ECM**
0550	–	Power steering pressure (PSP) sensor/ switch – circuit malfunction	**Wiring, PSP sensor/switch, ECM**
0600	–	AT to ECM communication – signal failure	**Wiring, ECM**
0605	–	Engine control module (ECM) – internal function failure	**ECM**
0650	–	Malfunction indicator lamp (MIL) – circuit malfunction	**Wiring, MIL, ECM**
0705	–	Park/neutral position (PNP) switch – circuit malfunction	**Wiring, PNP switch, mechanical fault, ECM**
0710	–	Transmission fluid temperature sensor – excessively high or low voltage	**Wiring, temperature sensor, mechanical fault, ECM**
0720	–	AT – vehicle speed sensor (VSS)	**Wiring, VSS, mechanical fault, ECM**
0725	–	Engine speed (RPM) – incorrect voltage signal ECM to AT control module	**Wiring, RPM sensor, mechanical fault, ECM**
0731	–	AT – first gear selection – malfunction	**Wiring, mechanical fault**
0732	–	AT – second gear selection – malfunction	**Wiring, mechanical fault**
0733	–	AT – third gear selection – malfunction	**Wiring, mechanical fault**
0734	–	AT – fourth gear selection – malfunction	**Wiring, mechanical fault**
0740	–	Torque converter clutch solenoid – voltage low	**Wiring, solenoid, mechanical fault**
0744	–	AT – signal circuit malfunction	**Wiring, mechanical fault**
0745	–	AT – line pressure solenoid – voltage low	**Wiring, solenoid, mechanical fault**

Flash code	OBD-II code	Fault location	Probable cause
0750	–	AT – shift solenoid A – voltage low	Wiring, shift solenoid, mechanical fault
0755	–	AT – shift solenoid B – voltage low	Wiring, shift solenoid, mechanical fault
1065	P1065	Engine control module (ECM) – battery voltage supply	Wiring, fuse, ECM
1102	P1102	MAF sensor – performance problem	Wiring, MAF sensor
1110	P1110	Intake valve timing control solenoid bank 1 – malfunction	Wiring, intake valve control solenoid position sensor, CKP sensor, CMP sensor, ECM
1111	P1111	Intake valve timing control solenoid bank 1 – circuit malfunction	Wiring, intake valve control solenoid, ECM
1119	P1119	Engine coolant temperature (ECT) sensor, radiator – circuit malfunction	Wiring, engine coolant temperature (ECT) sensor
1121	P1121	Throttle motor – malfunction	Throttle motor, ECM
1122	P1122	Throttle motor – circuit malfunction	Wiring, throttle motor, ECM
1123	P1123	Throttle motor relay – circuit malfunction	Wiring, throttle motor relay, ECM
1124	P1124	Throttle motor – short circuit	Wiring, throttle motor, throttle motor relay, ECM
1126	P1126	Throttle motor – open circuit	Wiring, throttle motor, throttle motor relay, ECM
1128	P1128	Throttle motor – short circuit	Wiring, throttle motor, ECM
1130	P1130	Intake manifold air control solenoid – malfunction	Wiring, control solenoid, intake system, hose leak/blockage, control vacuum check switch, ECM
1131	P1131	Intake manifold air control solenoid – malfunction	Wiring, intake manifold air control solenoid
1135	P1135	Intake valve timing control solenoid bank 2 – malfunction	Wiring, intake valve control solenoid position sensor, CKP sensor, CMP sensor, ECM
1136	P1136	Intake valve timing control solenoid bank 2 – circuit malfunction	Wiring, intake valve control solenoid, ECM
1140	P1140	Camshaft position (CMP) actuator LH – circuit malfunction	Wiring, CMP actuator, mechanical fault, ECM
1143	P1143	Heated oxygen sensor (HO2S) 1, bank 1 – voltage low	Wiring, mixture, HO2S, ECM
1144	P1144	Heated oxygen sensor (HO2S) 1, bank 1 – voltage high	HO2S, fuel pressure, fuel system, ECM
1145	P1145	Camshaft position (CMP) actuator RH – circuit malfunction	Wiring, CMP actuator, mechanical fault, ECM
1146	P1146	Heated oxygen sensor (HO2S) 2, bank 1 – voltage low	Wiring, mixture, HO2S, ECM
1147	P1147	Heated oxygen sensor (HO2S) 2, bank 1 – voltage high	Wiring, HO2S, fuel pressure, fuel system, ECM
1148	P1148	Closed loop control bank 1 – inoperative	Wiring, HO2S front, ECM

Flash code	OBD-II code	Fault location	Probable cause
1163	P1163	Heated oxygen sensor (HO2S)	Wiring, HO2S, ECM
1164	P1164	Heated oxygen sensor (HO2S) 1, bank 2 – voltage high	HO2S, fuel pressure, fuel system, ECM
1165	P1165	Intake manifold air control system – vacuum check switch circuit malfunction	Wiring, vacuum check switch, hose leak/ blockage, air control solenoid, ECM
1166	P1166	Heated oxygen sensor (HO2S) 2, bank 2 – voltage high	Wiring, mixture, HO2S, ECM
1167	P1167	Heated oxygen sensor (HO2S) 2, bank 2 – voltage high	Wiring, HO2S, fuel pressure, fuel system, ECM
1168	P1168	Closed loop control LH – inoperative	Wiring, HO2S, fuel pressure, fuel system, mechanical fault, ECM
1211	P1211	ABS/TCS control unit – communication malfunction	Wiring, ABS/TCS control unit, ECM
1212	P1212	ABS/TCS control unit – circuit malfunction	Wiring, ABS/TCS control unit, ECM
1217	P1217	Engine over temperature	Cooling system, engine coolant blower motor
1220	P1220	Fuel pump control module – circuit malfunction	Wiring, fuel pump speed resistor, fuel pump control module
1223	P1223	Throttle position (TP) sensor 2 – voltage low	Wiring, TP sensor
1224	P1224	Throttle position (TP) sensor 2 – voltage high	Wiring, TP sensor
1225	P1225	Throttle motor position sensor – learning problem, voltage low	Wiring, TP sensor 1, TP sensor 2
1226	P1226	Throttle motor position sensor – learning problem	Wiring, TP sensor 1, TP sensor 2
1227	P1227	Accelerator pedal position (APP) sensor 2 – voltage low	Wiring, APP sensor
1228	P1228	Accelerator pedal position (APP) sensor 2 – voltage high	Wiring, APP sensor
1229	P1229	Engine control module (ECM) – sensor supply voltage low/high	Wiring, APP sensor, throttle motor position sensor, MAF sensor, fuel tank pressure sensor, TP sensor(s), refrigerant pressure sensor
1320	P1320	Ignition signal – circuit malfunction	Wiring, ignition coil/module, ignition system, CKP sensor, CMP sensor, ECM
1335	P1335	Crankshaft position (CKP) sensor – circuit malfunction	Wiring, CKP sensor, ECM
1336	P1336	Crankshaft position (CKP) sensor – signal malfunction	Wiring, CKP sensor, ECM
1440	P1440	Evaporative emission (EVAP) canister purge control system – small leak	EVAP valve, hoses, mechanical fault
1442	P1442	Evaporative emission (EVAP) canister purge system – small leak	EVAP canister purge system, hose leak, EVAP pressure sensor/vent valve, fuel filler cap, ECM

Flash code	OBD-II code	Fault location	Probable cause
1444	P1444	Evaporative emission (EVAP) canister purge control system – malfunction	**Wiring, EVAP valve, hoses, mechanical fault, ECM**
1446	P1446	Evaporative emission (EVAP) canister purge control valve – malfunction	**Wiring, EVAP valve, mechanical fault, ECM**
1447	P1447	Evaporative emission (EVAP) canister purge control system – malfunction	**Wiring, EVAP valve, mechanical fault, ECM**
1448	P1448	Evaporative emission (EVAP) canister purge control valve – malfunction	**Wiring, EVAP valve, ECM**
1456	P1456	Evaporative emission (EVAP) canister purge system – very small leak	**EVAP canister purge system, hose leak, EVAP pressure sensor/vent valve, fuel filler cap, ECM**
1464	P1464	Fuel gauge tank sensor – voltage high	**Wiring, fuel gauge tank sensor, ECM**
1480	P1480	Engine coolant blower motor control – circuit malfunction	**Wiring, blower motor speed control solenoid valve, ECM**
1490	P1490	Evaporative emission (EVAP) canister purge control system – malfunction	**Wiring, EVAP valve, mechanical fault, ECM**
1491	P1491	Evaporative emission (EVAP) canister purge control system – malfunction	**Wiring, EVAP valve, ECM**
1564	P1564	Cruise control master switch – circuit malfunction	**Wiring, cruise control master switch, ECM**
1572	P1572	Cruise control brake pedal switch – circuit malfunction	**Wiring, cruise control brake pedal switch, clutch pedal position (CPP) switch, ECM**
1574	P1574	Cruise control vehicle speed sensor – signal malfunction	**Wiring, cruise control vehicle speed sensor, ECM**
1605	P1605	AT diagnosis communication – signal failure	**Wiring, ECM**
1610	P1610	Ignition key/engine control module (ECM) – malfunction	**Incorrect ignition key, ECM**
1615	P1615	Ignition key/immobilizer control module communication – malfunction	**Incorrect ignition key, immobilizer control module**
1705	P1705	AT – throttle position (TP) sensor – voltage high or low	**Wiring, TP sensor, mechanical fault, ECM**
1706	P1706	Park/neutral position (PNP) switch – circuit malfunction	**Wiring, PNP switch, ECM**
1716	P1716	AT – turbine revolution sensor – circuit malfunction	**Wiring, turbine sensor 1, turbine sensor 2, ECM**
1720	P1720	Vehicle speed sensor (VSS) – signal variation	**Wiring, TCM – dtc's stored, ABS/TCS control module – dtc's stored, instrument panel**
1730	P1730	AT – fail safe function – circuit malfunction	**Wiring, transmission clutches, transmission solenoid valves, transmission pressure switches**
1752	P1752	AT – input clutch – solenoid valve malfunction	**Wiring, input clutch solenoid valve**
1754	P1754	AT – input clutch – solenoid valve malfunction	**Wiring, input clutch solenoid valve, transmission pressure switch 3**

Flash code	OBD-II code	Fault location	Probable cause
1757	P1757	AT – front brake – solenoid valve malfunction	**Wiring, front brake solenoid valve**
1759	P1759	AT – front brake – solenoid valve malfunction	**Wiring, front brake solenoid valve, transmission pressure switch 1**
1760	P1760	AT – overrun clutch solenoid – voltage low	**Wiring, solenoid, mechanical fault**
1762	P1762	AT – direct clutch – solenoid valve malfunction	**Wiring, direct clutch solenoid valve**
1764	P1764	AT – direct clutch – solenoid valve malfunction	**Wiring, direct clutch solenoid valve, transmission pressure switch 5**
1767	P1767	AT – high/low reverse – solenoid valve malfunction	**Wiring, high/low reverse solenoid valve**
1769	P1769	AT – high/low reverse – solenoid valve malfunction	**Wiring, high/low reverse solenoid valve, transmission pressure switch 6**
1772	P1772	AT – low coast brake – solenoid valve malfunction	**Wiring, low coast brake solenoid valve**
1774	P1774	AT – low coast brake – solenoid valve malfunction	**Wiring, low coast brake solenoid valve, transmission pressure switch 2**
1780	P1780	AT – shift change – signal malfunction	**Wiring, TCM, ECM**
1800	P1800	Intake manifold air control solenoid – circuit malfunction	**Wiring, intake manifold air control solenoid**
1805	P1805	Brake pedal position (BPP) switch – circuit malfunction	**Wiring, BPP switch**
1000	U1000	CAN communication line – signal malfunction	**Wiring**
1001	U1001	CAN communication line – signal malfunction	**Wiring**

INFINITI

Model:	Engine identification:	Year:
M45 4.5L	VK45DE	2003-04
FX35 3.5L	VQ35DE	2003-04
FX45 4.5L	VK45DE	2003-04
G35 3.5L	VQ35DE	2003-04
QX56 5.6L	VK56DE	2004

System: **ECCS**

Self-diagnosis

General information

- Refer to the front of this manual for general test conditions, terminology, detailed descriptions of wiring faults and a general trouble shooter for electrical and mechanical faults.
- Engine control module (ECM) incorporates self-diagnosis function.
- Malfunction indicator lamp (MIL) will illuminate if certain faults are recorded.
- ECM operates in backup mode if sensors fail, to enable vehicle to be driven to workshop.
- Trouble codes can be displayed by using a Scan Tool connected to the data link connector (DLC) **1**, or by the malfunction indicator lamp (MIL).

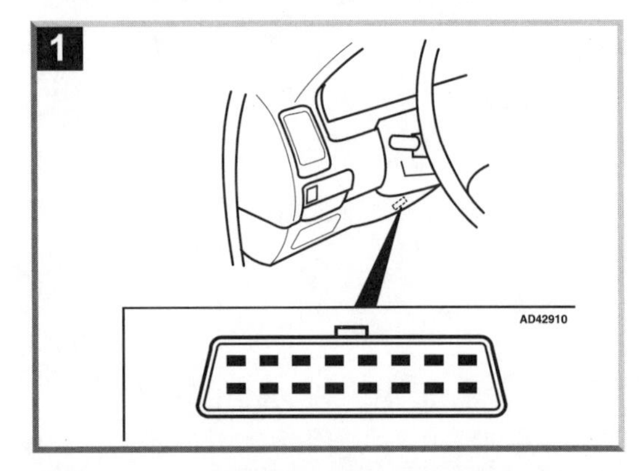

NOTE: ***The use of a Scan Tool is essential to obtain full diagnostic information.***

Accessing – Mode 1

- Switch ignition ON.
- Check MIL illuminates.
- Start engine. Allow to idle.
- If MIL extinguishes, no trouble codes have been recorded.
- If MIL remains illuminated, access trouble codes.

Accessing – Mode 2

- Ensure accelerator pedal is fully released.
- Switch ignition ON. Do NOT start engine.
- Wait 3 seconds.
- Within 5 seconds, repeat the following 5 times:
  - Fully depress accelerator pedal.
  - Fully release accelerator pedal.
- Wait 7 seconds.
- Fully depress accelerator pedal.
- After approximately 10 seconds MIL will start flashing.
- Fully release accelerator pedal.
- Self-diagnosis now in Mode 2.
- Count flashes and compare with trouble code table.
- Each trouble code consists of four groups of one or more flashes.
- The first group of flashes indicate the 'thousands' of the trouble code **2** [**A**].
- The second group of flashes indicate the 'hundreds' of the trouble code **2** [**B**].
- The third group of flashes indicate the 'tens' of the trouble code **2** [**C**].
- The fourth group of flashes indicate the 'units' of the trouble code **2** [**D**].
- Ten flashes in a group indicate '0'.
- A short pause separates each trouble code group **2** [**E**].

- A long pause separates each trouble code **2** [F].
- For example: Trouble code P1130 displayed **2**.
- Switch ignition OFF.
- Self-diagnosis will return to Mode 1.

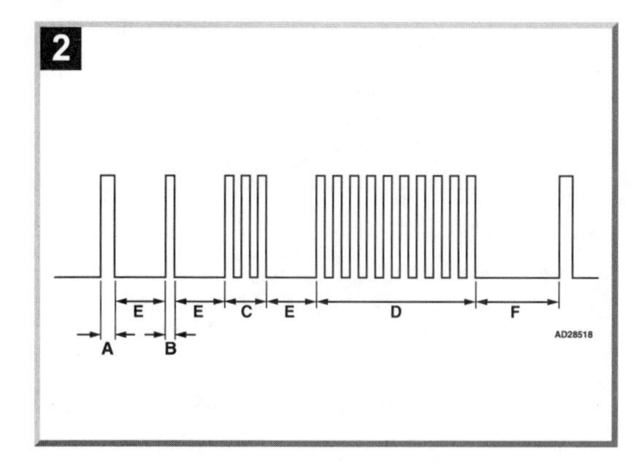

Erasing

- Trouble codes can be erased by using a Scan Tool connected to the data link connector (DLC) or as follows:
- Ensure accelerator pedal is fully released.
- Switch ignition ON. Do NOT start engine.
- Wait 3 seconds.
- Within 5 seconds, repeat the following 5 times:
  - Fully depress accelerator pedal.
  - Fully release accelerator pedal.
- Wait 7 seconds.
- Fully depress accelerator pedal.
- After approximately 10 seconds MIL will start flashing.
- Fully release accelerator pedal.
- Self-diagnosis now in Mode 2.
  - Fully depress accelerator pedal for more than 10 seconds.
  - Fully release accelerator pedal.
  - Trouble code 0000 should be displayed.

Trouble code identification

Flash code	OBD-II code	Fault location	Probable cause
–	*All P0, P2 and U0 codes*	Refer to OBD-II trouble code tables at the front of this manual	–
0000	–	No self diagnostic failure indicated	–
1000	U1000	CAN communication line – signal malfunction	**Wiring**
1001	U1001	CAN communication line – signal malfunction	**Wiring**
1031	P1031	Air fuel (A/F) ratio sensor bank 1, heater – voltage low	**Wiring, A/F sensor**
1032	P1032	Air fuel (A/F) ratio sensor bank 1, heater – voltage high	**Wiring, A/F sensor**
1051	P1051	Air fuel (A/F) ratio sensor bank 2, heater – voltage low	**Wiring, A/F sensor**
1052	P1052	Air fuel (A/F) ratio sensor bank 2, heater – voltage high	**Wiring, A/F sensor**
1065	P1065	Engine control module (ECM) – battery voltage supply	**Wiring, fuse, ECM**
1102	P1102	MAF sensor – performance problem	**Wiring, MAF sensor**
1111	P1111	Intake valve timing control solenoid bank 1 – circuit malfunction	**Wiring, intake valve control solenoid, ECM**
1119	P1119	Engine coolant temperature (ECT) sensor – circuit malfunction	**Wiring, engine coolant temperature (ECT) sensor**

Flash code	OBD-II code	Fault location	Probable cause
1121	P1121	Throttle motor – malfunction	Throttle motor, ECM
1122	P1122	Throttle motor – circuit malfunction	Wiring, throttle motor, ECM
1124	P1124	Throttle motor relay – short circuit	Wiring, throttle motor, throttle motor relay, ECM
1126	P1126	Throttle motor relay – open circuit	Wiring, throttle motor, throttle motor relay, ECM
1128	P1128	Throttle motor – short circuit	Wiring, throttle motor, ECM
1136	P1136	Intake valve timing control solenoid bank 2 – circuit malfunction	Wiring, intake valve control solenoid, ECM
1140	P1140	Intake valve timing control position sensor bank 1 – circuit malfunction	Wiring, intake valve control position sensor, crankshaft (CKP) sensor, camshaft position (CMP) sensor, mechanical fault, ECM
1143	P1143	Heated oxygen sensor (HO2S) 1, bank 1 – voltage low	Wiring, mixture, HO2S, ECM
1144	P1144	Heated oxygen sensor (HO2S) 1, bank 1 – voltage high	HO2S, fuel pressure, fuel system, ECM
1145	P1145	Intake valve timing control position sensor bank 2 – circuit malfunction	Wiring, intake valve control position sensor, crankshaft (CKP) sensor, camshaft position (CMP) sensor, mechanical fault, ECM
1146	P1146	Heated oxygen sensor (HO2S) 2, bank 1 – voltage low	Wiring, mixture, HO2S, ECM
1147	P1147	Heated oxygen sensor (HO2S) 2, bank 1 – voltage low	Wiring, HO2S, fuel pressure, fuel system, ECM
1148	P1148	Closed loop control bank 1 – inoperative	Wiring, HO2S front, ECM
1163	P1163	Heated oxygen sensor (HO2S) 1, bank 2 – voltage low	Wiring, mixture, HO2S, ECM
1164	P1164	Heated oxygen sensor (HO2S) 1, bank 2 – voltage high	HO2S, fuel pressure, fuel system, ECM
1166	P1166	Heated oxygen sensor (HO2S) 2, bank 2 – voltage low	Wiring, mixture, HO2S, ECM
1167	P1167	Heated oxygen sensor (HO2S) 2, bank 2 – voltage low	Wiring, HO2S, fuel pressure, fuel system, ECM
1168	P1168	Closed loop control bank 2 – inoperative	Wiring, HO2S front, ECM
1211	P1211	ABS/TCS control unit – malfunction	Wiring, ABS/TCS control unit, ECM
1212	P1212	ABS/TCS control unit – communication malfunction	Wiring, ABS/TCS control unit, ECM
1217	P1217	Engine over temperature	Cooling system, engine coolant blower motor
1220	P1220	Fuel pump control module – circuit malfunction	Wiring, fuel pump speed resistor, fuel pump control module
1225	P1225	Throttle motor position sensor – learning problem, voltage low	Wiring, TP sensor 1, TP sensor 2

Flash code	OBD-II code	Fault location	Probable cause
1226	P1226	Throttle motor position sensor – learning problem	**Wiring, TP sensor 1, TP sensor 2**
1229	P1229	Engine control module (ECM) – sensor supply voltage low/high	**Wiring, APP sensor, throttle motor position sensor, MAF sensor, fuel tank pressure sensor, TP sensor(s), refrigerant pressure sensor**
1271	P1271	Air fuel (A/F) ratio sensor bank 1 – voltage low	**Wiring, A/F sensor**
1272	P1272	Air fuel (A/F) ratio sensor bank 1 – voltage high	**Wiring, A/F sensor**
1273	P1273	Air fuel (A/F) ratio sensor bank 1 – lean mixture	**A/F sensor, fuel pressure, injectors, intake air leak**
1274	P1274	Air fuel (A/F) ratio sensor bank 1 – rich mixture	**A/F sensor, fuel pressure, injectors**
1276	P1276	Air fuel (A/F) ratio sensor bank 1 – activity	**Wiring, A/F sensor**
1278	P1278	Air fuel (A/F) ratio sensor bank 1 – slow response	**Wiring, fuel pressure, injectors, MAF sensor, PCV, A/F sensor, intake air leaks, exhaust leaks**
1279	P1279	Air fuel (A/F) ratio sensor bank 1 – slow response	**Wiring, fuel pressure, injectors, MAF sensor, PCV, A/F sensor, intake air leaks, exhaust leaks**
1281	P1281	Air fuel (A/F) ratio sensor bank 2 – voltage low	**Wiring, A/F sensor**
1282	P1282	Air fuel (A/F) ratio sensor bank 2 – voltage high	**Wiring, A/F sensor**
1283	P1283	Air fuel (A/F) ratio sensor bank 2 – lean mixture	**A/F sensor, fuel pressure, injectors, intake air leak**
1284	P1284	Air fuel (A/F) ratio sensor bank 2 – rich mixture	**A/F sensor, fuel pressure, injectors**
1286	P1286	Air fuel (A/F) ratio sensor bank 2 – activity	**Wiring, A/F sensor**
1288	P1288	Air fuel (A/F) ratio sensor bank 2 – slow response	**Wiring, fuel pressure, injectors, MAF sensor, PCV, A/F sensor, intake air leaks, exhaust leaks**
1289	P1289	Air fuel (A/F) ratio sensor bank 2 – slow response	**Wiring, fuel pressure, injectors, MAF sensor, PCV, A/F sensor, intake air leaks, exhaust leaks**
1442	P1442	Evaporative emission (EVAP) canister purge system – small leak	**EVAP canister purge system, hose leak, EVAP pressure sensor/vent valve, fuel filler cap, ECM**
1444	P1444	Evaporative emission (EVAP) canister purge control system – malfunction	**Wiring, EVAP valve, hoses, mechanical fault, ECM**
1446	P1446	Evaporative emission (EVAP) canister purge control valve – malfunction	**Wiring, EVAP valve, mechanical fault, ECM**
1448	P1448	Evaporative emission (EVAP) canister purge control valve – malfunction	**Wiring, EVAP valve, ECM**

Flash code	OBD-II code	Fault location	Probable cause
1456	P1456	Evaporative emission (EVAP) canister purge system – very small leak	**EVAP canister purge system, hose leak, EVAP pressure sensor/vent valve, fuel filler cap, ECM**
1464	P1464	Fuel gauge tank sensor – voltage high	**Wiring, fuel gauge tank sensor, ECM**
1480	P1480	Engine coolant blower motor control – circuit malfunction	**Wiring, blower motor speed control solenoid valve, ECM**
1490	P1490	Evaporative emission (EVAP) canister purge control system – malfunction	**Wiring, vacuum cut valve bypass valve, mechanical fault, ECM**
1491	P1491	Evaporative emission (EVAP) canister purge control system – malfunction	**Wiring, vacuum cut valve bypass valve, mechanical fault, ECM**
1564	P1564	Cruise control master switch – circuit malfunction	**Wiring, cruise control master switch, ECM**
1568	P1568	Cruise control master switch – signal out of specification	**Wiring, cruise control master switch, ECM**
1572	P1572	Cruise control brake pedal switch – circuit malfunction	**Wiring, brake pedal position (BPP) switch, cruise control brake pedal switch, ECM**
1574	P1574	Cruise control vehicle speed sensor – signal malfunction	**Wiring, TCM – dtc's stored, ABS/TCS control module – dtc's stored, instrument panel, ECM**
1610	P1610	Ignition key/engine control module (ECM) – malfunction	**Incorrect ignition key, ECM**
1615	P1615	Ignition key/immobilizer control module communication – malfunction	**Incorrect ignition key, immobilizer control module**
1705	P1705	AT – throttle position (TP) sensor – voltage high or low	**Wiring, TP sensor, mechanical fault, ECM**
1706	P1706	Park/neutral position (PNP) switch – circuit malfunction	**Wiring, PNP switch, ECM**
1716	P1716	Automatic transmission turbine revolution sensor – circuit malfunction	**Wiring, turbine sensor 1, turbine sensor 2, ECM**
1720	P1720	Vehicle speed sensor (VSS) – signal variation	**Wiring, TCM – dtc's stored, ABS/TCS control module – dtc's stored, instrument panel**
1730	P1730	Automatic transmission fail safe function – circuit malfunction	**Wiring, transmission clutches, transmission solenoid valves, transmission pressure switches**
1752	P1752	Automatic transmission input clutch – solenoid valve malfunction	**Wiring, input clutch solenoid valve**
1754	P1754	Automatic transmission input clutch – solenoid valve malfunction	**Wiring, input clutch solenoid valve, transmission pressure switch 3**
1757	P1757	Automatic transmission front brake – solenoid valve malfunction	**Wiring, front brake solenoid valve**
1759	P1759	Automatic transmission front brake – solenoid valve malfunction	**Wiring, front brake solenoid valve, transmission pressure switch 1**
1762	P1762	Automatic transmission direct clutch – solenoid valve malfunction	**Wiring, direct clutch solenoid valve**

Flash code	OBD-II code	Fault location	Probable cause
1764	P1764	Automatic transmission direct clutch – solenoid valve malfunction	**Wiring, direct clutch solenoid valve, transmission pressure switch 5**
1767	P1767	Automatic transmission high/low reverse – solenoid valve malfunction	**Wiring, high/low reverse solenoid valve**
1769	P1769	Automatic transmission high/low reverse – solenoid valve malfunction	**Wiring, high/low reverse solenoid valve, transmission pressure switch 6**
1772	P1772	Automatic transmission low coast brake – solenoid valve malfunction	**Wiring, low coast brake solenoid valve**
1774	P1774	Automatic transmission low coast brake – solenoid valve malfunction	**Wiring, low coast brake solenoid valve, transmission pressure switch 2**
1780	P1780	Automatic transmission shift change – signal malfunction	**Wiring, TCM, ECM**
1805	P1805	Brake pedal position (BPP) switch – circuit malfunction	**Wiring, BPP switch**

INFINITI

Model:	Engine identification:	Year:
QX4 3.5L	VQ35DE	2001
I30 3.0L	VQ30DE	2000-01

System: **ECCS**

Self-diagnosis

General information

- Refer to the front of this manual for general test conditions, terminology, detailed descriptions of wiring faults and a general trouble shooter for electrical and mechanical faults.
- Engine control module (ECM) incorporates self-diagnosis function.
- Malfunction indicator lamp (MIL) will illuminate if certain faults are recorded.
- ECM operates in backup mode if sensors fail, to enable vehicle to be driven to workshop.

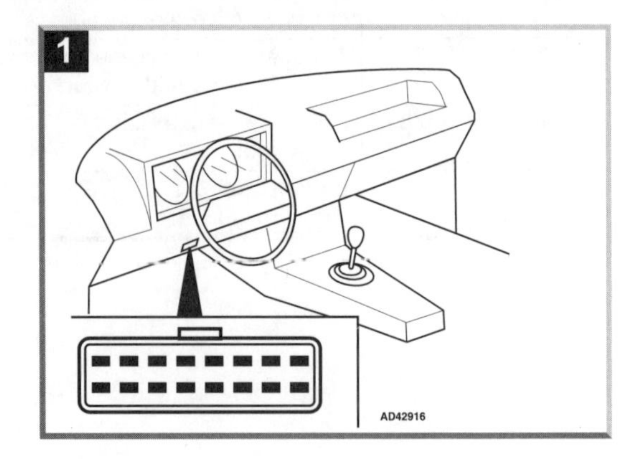

Accessing

- Trouble codes can only be displayed by using a Scan Tool connected to the data link connector (DLC) 1.

Erasing

- After the faults have been rectified, trouble codes can only be erased by using a Scan Tool connected to the data link connector (DLC) 1.

Trouble code identification

OBD-II code	Fault location	Probable cause
All P0, P2 and U0 codes	Refer to OBD-II trouble code tables at the front of this manual	–
P1110	Camshaft position (CMP) actuator, intake, bank 1	**Wiring, CMP actuator, CMP actuator position sensor, CMP sensor, CKP sensor**
P1111	Camshaft position (CMP) actuator, intake, bank 1 – circuit malfunction	**Wiring, CMP actuator**
P1126	Thermostat – stuck open	**Wiring, thermostat, ECT sensor**
P1130	Intake manifold air control solenoid – circuit malfunction	**Wiring, control solenoid, intake system, hose leak/blockage, control vacuum check switch, MAF sensor, TP sensor, CKP sensor**
P1135	Camshaft position (CMP) actuator, intake, bank 2	**Wiring, CMP actuator, CMP actuator position sensor, CMP sensor, CKP sensor**
P1136	Camshaft position (CMP) actuator, intake, bank 2 – circuit malfunction	**Wiring, CMP actuator**

OBD-II code	Fault location	Probable cause
P1140	Camshaft position (CMP) actuator position sensor, intake, bank 2 – circuit malfunction	**Wiring, CMP actuator, CMP actuator position sensor, CMP sensor, CKP sensor**
P1145	Camshaft position (CMP) actuator position sensor, intake, bank 2 – circuit malfunction	**Wiring, CMP actuator, CMP actuator position sensor, CMP sensor, CKP sensor**
P1148	Closed loop control, bank 1	**Wiring, HO2S, HO2S heater**
P1165	Intake manifold air control vacuum check switch	**Wiring, control solenoid, intake system, hose leak/blockage, control vacuum check switch**
P1168	Closed loop control, bank 2	**Wiring, HO2S, HO2S heater**
P1211	TCS function – malfunction	**TCS system, ABS/TCS ECM**
P1212	TCS communication – circuit malfunction	**Wiring, ABS/TCS ECM**
P1217	Engine overheating	**Thermostat, cooling system**
P1320	Ignition signal, primary – circuit malfunction	**Wiring, ignition amplifier, audio unit suppresser CKP sensor**
P1335	Crankshaft position (CKP) sensor – circuit malfunction	**Wiring, CKP sensor**
P1336	Crankshaft position (CKP) sensor – circuit malfunction	**Wiring, CKP sensor, mechanical fault**
P1401	Exhaust gas recirculation temperature (EGRT) sensor – circuit malfunction	**Wiring, EGRT sensor, ECM**
P1402	Exhaust gas recirculation (EGR) system – excessive flow	**Wiring, EGR/EVAP valve, EGR valve, EGRT sensor, EGR back pressure transducer, ECM**
P1440	Evaporative emission (EVAP) canister purge system – leaking/malfunction	**Wiring, fuel tank/cap, hose leak/blockage, EVAP valve/canister, EVAP vent valve, BARO sensor, MAP/BARO switching valve, ECM**
P1441	Evaporative emission (EVAP) canister purge control system – small leak	**EVAP valve, hoses, fuel filler cap, mechanical fault**
P1444	Evaporative emission (EVAP) canister purge valve – circuit malfunction	**Wiring, EVAP canister purge valve, hoses, EVAP pressure sensor, EVAP canister, mechanical fault**
P1446	Evaporative emission (EVAP) canister vent valve – close	**Hoses, EVAP vent valve, EVAP pressure sensor, EVAP canister**
P1447	Evaporative emission (EVAP) canister purge flow monitoring	**EVAP canister purge valve, hoses, EVAP pressure sensor, EVAP canister**
P1448	Evaporative emission (EVAP) canister vent valve – open	**Hoses, EVAP vent valve, EVAP pressure sensor, EVAP canister**
P1464	Fuel gauge tank sensor – circuit malfunction	**Wiring, fuel gauge tank sensor**
P1490	Evaporative emission (EVAP) vacuum cut bypass valve – circuit malfunction	**Wiring, EVAP vacuum cut bypass valve**
P1491	Evaporative emission (EVAP) vacuum cut bypass valve	**Hoses, EVAP vacuum cut bypass valve, EVAP pressure sensor, EVAP canister, EVAP canister vent valve**
P1605	Diagnosis communication, AT – circuit malfunction	**Wiring, TCM**

OBD-II code	Fault location	Probable cause
P1610	Imobilizer	**Wiring, Imobilizer control module, incorrect key**
P1615	Imobilizer	**Wiring, Imobilizer control module, incorrect key**
P1705	Throttle position (TP) sensor, AT – circuit malfunction	**Wiring, TP sensor, mechanical fault, ECM**
P1706	Park neutral position(PNP) switch – circuit malfunction	**Wiring, PNP switch, mechanical fault, ECM**
P1760	Overrun clutch solenoid, AT – circuit malfunction	**Wiring, solenoid, mechanical fault**

INFINITI

Model:	Engine identification:	Year:
Q45 4.1/4.5L	**VH41DE/VH45DE**	**1996-01**

System: **ECCS**

Self-diagnosis

General information

- Refer to the front of this manual for general test conditions, terminology, detailed descriptions of wiring faults and a general trouble shooter for electrical and mechanical faults.
- Engine control module (ECM) incorporates self-diagnosis function.
- Malfunction indicator lamp (MIL) will illuminate if certain faults are recorded.
- ECM operates in backup mode if sensors fail, to enable vehicle to be driven to workshop.
- Trouble codes can be displayed by using a Scan Tool connected to the data link connector (DLC) **1** [1], or by the malfunction indicator lamp (MIL).

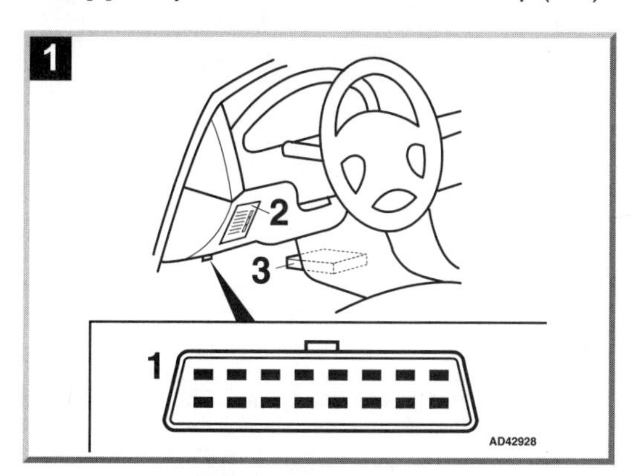

NOTE: ***The use of a Scan Tool is essential to obtain full diagnostic information.***

Accessing – Mode 1

- Switch ignition ON.
- Check MIL illuminates.
- Start engine. Allow to idle.
- If MIL extinguishes, no trouble codes have been recorded.
- If MIL remains illuminated, access trouble codes.

Accessing – Mode 2

- Access engine control module (ECM) **1** [3].
- Switch ignition ON. Do NOT start engine.
- Turn diagnostic mode selector on ECM fully clockwise **2**.
- Wait at least 2 seconds.
- Turn diagnostic mode selector on ECM fully counterclockwise **2**.
- Self-diagnosis now in Mode 2.
- MIL should flash.
- Count flashes and compare with trouble code table.
- Each trouble code consists of four groups of one or more flashes.
- The first group of flashes indicate the 'thousands' of the trouble code **3** [A].
- The second group of flashes indicate the 'hundreds' of the trouble code **3** [B].
- The third group of flashes indicate the 'tens' of the trouble code **3** [C].
- The fourth group of flashes indicate the 'units' of the trouble code **3** [D].
- Ten flashes in a group indicate '0'.
- A short pause separates each trouble code group **3** [E].
- A long pause separates each trouble code **3** [F].
- For example: Trouble code P1130 displayed **3**.
- Switch ignition OFF.
- Self-diagnosis will return to Mode 1.

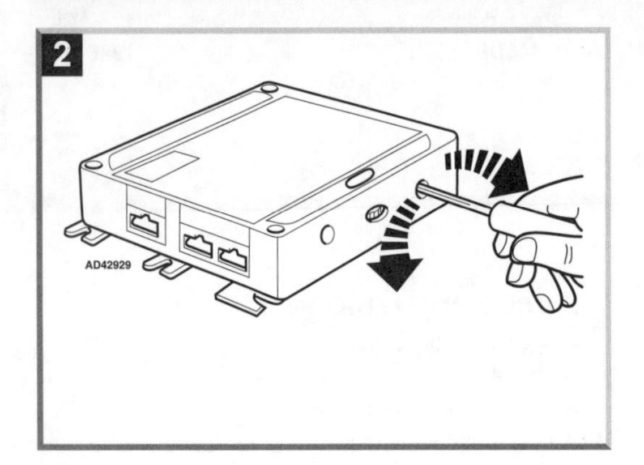

Erasing

- Trouble codes can be erased by using a Scan Tool connected to the data link connector (DLC) or as follows:
- With the ignition switch ON and the engine control module (ECM) in diagnosis test mode 2.
- Turn diagnostic mode selector on ECM fully clockwise **2**.
- Trouble codes are erased.
- Switch ignition OFF.
- Turn diagnostic mode selector on ECM fully counterclockwise **2**.
- Repeat checking procedure to ensure no data remains in ECM fault memory.

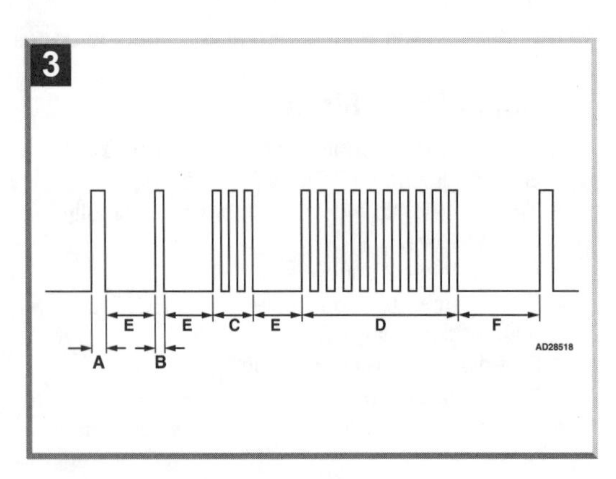

Trouble code identification

Flash code	OBD-II code	Fault location	Probable cause
–	*All P0, P2 and U0 codes*	Refer to OBD-II trouble code tables at the front of this manual	–
0101	–	Camshaft position (CMP) sensor – circuit malfunction	**Wiring, CMP sensor, ECM**
0102	–	Mass air flow (MAF) sensor – circuit malfunction	**Wiring, MAF sensor, ECM**
0103	–	Engine coolant temperature (ECT) sensor – circuit malfunction	**Wiring, ECT sensor, ECM**
0104	–	Vehicle speed sensor (VSS) – circuit malfunction	**Wiring, VSS, ECM**
0106	**P1210**	Traction control system – malfunction	**Wiring, TP sensor, mechanical fault, ECM**
0107	–	ABS/TCS communication – circuit malfunction	**Wiring, ABS/TCS ECM**

Flash code	OBD-II code	Fault location	Probable cause
0110	P1125	Throttle position (TP) sensor 1 & 2 – signal malfunction	**Wiring, TP sensor, mechanical fault, ECM**
0111	P1447	Evaporative emission (EVAP) canister purge control system – malfunction	**Wiring, EVAP valve, mechanical fault, ECM**
0113	P1443	Evaporative emission (EVAP) canister purge control system – circuit malfunction	**Wiring, EVAP valve, ECM**
0114	–	Fuel trim LH – mixture too rich	**HO2S, fuel pressure, fuel system, mechanical fault, ECM**
0115	–	Fuel trim LH – mixture too lean	**HO2S, fuel pressure, fuel system, mechanical fault, ECM**
0201	P1320	Ignition signal – circuit malfunction	**Wiring, ignition control module, ignition coil**
0203	–	Closed throttle position (CTP) switch – malfunction	**Wiring, CTP switch, mechanical fault, ECM**
0205	–	Idle control system – malfunction	**Wiring, IAC valve, mechanical fault, ECM**
0208	–	Overheat – circuit malfunction	**Wiring, cooling fan, mechanical fault**
0209	–	Fuel trim RH – mixture too rich	**HO2S, fuel pressure, fuel system, mechanical fault, ECM**
0210	–	Fuel trim RH – mixture too lean	**HO2S, fuel pressure, fuel system, mechanical fault, ECM**
0212	–	Knock sensor (KS) RH – circuit	**Wiring, KS sensor, ECM**
0213	P1440	Evaporative emission (EVAP) canister purge control system – small leak	**EVAP valve, hoses, mechanical fault**
0214	P1444	Evaporative emission (EVAP) canister purge control system – malfunction	**Wiring, EVAP valve, hoses, mechanical fault, ECM**
0215	P1446	Evaporative emission (EVAP) canister purge control valve – malfunction	**Wiring, EVAP valve, mechanical fault, ECM**
0301	–	Engine control module (ECM) – internal function failure	**ECM**
0302	–	Exhaust gas recirculation (EGR) system – flow excessively high or low	**Wiring, EGR solenoid valve, mechanical fault**
0303	–	Heated oxygen sensor (HO2S) – RH front – circuit malfunction	**HO2S, ECM**
0304	–	Knock sensor (KS) LH – circuit	**Wiring, KS sensor, ECM**
0305	P1401	Exhaust gas recirculation (EGR) temperature sensor – circuit malfunction	**Wiring, EGR sensor, ECM**
0306	–	Exhaust gas recirculation (EGR) system – excess flow detected	**EGR valve, mechanical fault**
0307	–	Closed loop control LH – inoperative	**Wiring, HO2S, fuel pressure, fuel system, mechanical fault, ECM**
0307	P1148	Closed loop control RH – inoperative	**Wiring, HO2S, fuel pressure, fuel system, mechanical fault, ECM**
0308	–	Closed loop control RH – inoperative	**Wiring, HO2S, fuel pressure, fuel system, mechanical fault, ECM**

Flash code	OBD-II code	Fault location	Probable cause
0308	P1168	Closed loop control LH – inoperative	Wiring, HO2S, fuel pressure, fuel system, mechanical fault, ECM
0309	P1448	Evaporative emission (EVAP) canister purge control valve – malfunction	Wiring, EVAP valve, ECM
0311	P1491	Evaporative emission (EVAP) canister purge control system – malfunction	Wiring, EVAP valve, ECM
0312	P1493	Evaporative emission (EVAP) canister purge control valve – malfunction	Wiring, EVAP valve, ECM
0313	–	Heated oxygen sensor (HO2S) – RH rear – voltage high	Wiring, HO2S, fuel pressure, fuel system, mechanical fault, ECM
0314	–	Heated oxygen sensor (HO2S) – RH rear – voltage low	Wiring, HO2S, fuel pressure, fuel system, mechanical fault, ECM
0315	–	Heated oxygen sensor (HO2S) – RH rear – no activity detected	Wiring, HO2S, fuel pressure, fuel system, mechanical fault, ECM
0401	–	Intake air temperature (IAT) sensor – circuit malfunction	Wiring, IAT sensor, ECM
0402	–	Fuel temperature sensor – circuit malfunction	Wiring, temperature sensor, ECM
0403	–	Throttle position (TP) sensor 1 – circuit malfunction	Wiring, TP sensor, ECM
0404	–	ABS/TCS communication – circuit malfunction	Wiring, ABS/TCS ECM
0406	P1120	Throttle position (TP) sensor 2 – circuit malfunction	Wiring, TP sensor, ECM
0407	P1335	Crankshaft position (CKP) sensor – circuit malfunction	Wiring, CKP sensor, ECM
0409	–	Heated oxygen sensor (HO2S) – LH front – slow response	Wiring, HO2S, fuel pressure, fuel system, mechanical fault, ECM
0410	–	Heated oxygen sensor (HO2S) – LH front – voltage high	Wiring, HO2S, fuel pressure, fuel system, mechanical fault, ECM
0411	–	Heated oxygen sensor (HO2S) – LH front – voltage low	Wiring, HO2S, fuel pressure, fuel system, mechanical fault, ECM
0412	–	Heated oxygen sensor (HO2S) – LH front – no activity detected	Wiring, HO2S, fuel pressure, fuel system, mechanical fault, ECM
0413	–	Heated oxygen sensor (HO2S) – RH front – slow response	Wiring, HO2S, fuel pressure, fuel system, mechanical fault, ECM
0414	–	Heated oxygen sensor (HO2S) – RH front – voltage high	Wiring, HO2S, fuel pressure, fuel system, mechanical fault, ECM
0415	–	Heated oxygen sensor (HO2S) – RH front – voltage low	Wiring, HO2S, fuel pressure, fuel system, mechanical fault, ECM
0503	–	Heated oxygen sensor (HO2S) – LH front – circuit malfunction	Wiring, HO2S, ECM
0504	–	AT communication – circuit malfunction	Wiring, TCM
0505	–	No self diagnostic failure indicated	–

Flash code	OBD-II code	Fault location	Probable cause
0509	–	Heated oxygen sensor (HO2S) – RH front – no activity detected	Wiring, HO2S, fuel pressure, fuel system, mechanical fault, ECM
0510	–	Heated oxygen sensor (HO2S) – rear – no activity detected/voltage high	Wiring, HO2S, fuel pressure, fuel system, mechanical fault, ECM
0511	–	Heated oxygen sensor (HO2S) – rear – voltage low	Wiring, HO2S, fuel pressure, fuel system, mechanical fault, ECM
0512	–	Heated oxygen sensor (HO2S) – LH rear – no activity detected	Wiring, HO2S, fuel pressure, fuel system, mechanical fault, ECM
0514	P1402	Exhaust gas recirculation (EGR) temperature system – malfunction	Wiring, EGR temperature sensor, mechanical fault, ECM
0601	–	Cylinder 8 – misfire	Wiring, ignition system, fuel system, fuel pressure, mechanical fault, ECM
0602	–	Cylinder 7 – misfire	Wiring, ignition system, fuel system, fuel pressure, mechanical fault, ECM
0603	–	Cylinder 6 – misfire	Wiring, ignition system, fuel system, fuel pressure, mechanical fault, ECM
0604	–	Cylinder 5 – misfire	Wiring, ignition system, fuel system, fuel pressure, mechanical fault, ECM
0605	–	Cylinder 4 – misfire	Wiring, ignition system, fuel system, fuel pressure, mechanical fault, ECM
0606	–	Cylinder 3 – misfire	Wiring, ignition system, fuel system, fuel pressure, mechanical fault, ECM
0607	–	Cylinder 2 – misfire	Wiring, ignition system, fuel system, fuel pressure, mechanical fault, ECM
0608	–	Cylinder 1 – misfire	Wiring, ignition system, fuel system, fuel pressure, mechanical fault, ECM
0701	–	Random misfire	Wiring, ignition system, fuel system, EGR valve, fuel pressure, mechanical fault, ECM
0702	–	Catalytic converter LH – efficiency below limit	Catalytic converter, rear HO2S
0703	–	Catalytic converter RH – efficiency below limit	Catalytic converter, rear HO2S
0704	–	Evaporative emission (EVAP) canister pressure sensor – malfunction	Wiring, pressure sensor, mechanical fault, ECM
0705	–	Evaporative emission (EVAP) canister purge control system – small leak	Wiring, EVAP valve, mechanical fault, ECM
0707	–	Heated oxygen sensor (HO2S) – rear/LH rear – circuit malfunction/slow response	Wiring, HO2S, ECM
0708	–	Heated oxygen sensor (HO2S) – RH rear – slow response/circuit malfunction	Wiring, HO2S, ECM
0715	P0455	Evaporative emission (EVAP) system – large leak detected	Hose connection(s), intake leak, EVAP canister, EVAP canister purge valve
0801	P1441	Evaporative emission (EVAP) canister purge control system – small leak	Wiring, EVAP valve, mechanical fault, ECM

Flash code	OBD-II code	Fault location	Probable cause
0801	P1490	Evaporative emission (EVAP) canister purge control system – malfunction	**Wiring, EVAP valve, mechanical fault, ECM**
0802	–	Crankshaft position (CKP) sensor – circuit malfunction	**Wiring, CKP sensor, ECM**
0803	–	Manifold absolute pressure (MAP) sensor – circuit malfunction	**Wiring, MAP sensor, ECM**
0804	P1605	AT diagnosis communication – signal failure	**Wiring, TCM**
0805	P1110	Camshaft position (CMP) actuator LH – malfunction	**Wiring, CMP actuator, mechanical fault, ECM**
0807	–	Evaporative emission (EVAP) canister purge control system – circuit malfunction	**Wiring, EVAP solenoid, ECM**
0807	P1492	Evaporative emission (EVAP) canister purge control valve – malfunction	**Wiring, EVAP solenoid, ECM**
0901	–	Heated oxygen sensor (HO2S) – LH front – heater circuit malfunction	**Wiring, HO2S, ECM**
0902	–	Heated oxygen sensor (HO2S) – LH rear – heater circuit malfunction	**Wiring, HO2S, ECM**
0903	–	Evaporative emission (EVAP) canister purge control system – circuit malfunction	**Wiring, EVAP valve, mechanical fault, ECM**
0905	P1336	Crankshaft position (CKP) sensor – signal malfunction	**Wiring, CKP sensor, ECM**
0908	–	Insufficient coolant temperature for closed loop fuel control	**Wiring, ECT sensor, mechanical fault, ECM**
1001	–	Heated oxygen sensor (HO2S) – RH front – heater circuit malfunction	**Wiring, HO2S, ECM**
1002	–	Heated oxygen sensor (HO2S) – RH rear – heater circuit malfunction	**Wiring, HO2S, ECM**
1003	–	Park/neutral position (PNP) switch – circuit malfunction	**Wiring, PNP switch, mechanical fault, ECM**
1003	P1706	Park neutral position (PNP) switch – malfunction	**Wiring, PNP switch, mechanical fault, ECM**
1005	P1400	Exhaust gas recirculation (EGR) valve – voltage incorrect	**Wiring, EGR valve, mechanical fault, ECM**
1008	–	Evaporative emission (EVAP) canister purge control valve – circuit malfunction	**Wiring, EVAP valve**
1008	P1445	Evaporative emission (EVAP) canister purge control valve – circuit malfunction	**Wiring, EVAP valve, ECM**
1101	–	Park/neutral position (PNP) switch – 1998 – circuit	**Wiring, PNP switch, mechanical fault, ECM**
1102	–	AT – vehicle speed sensor (VSS)	**Wiring, VSS, mechanical fault, ECM**
1103	–	AT – first gear selection – malfunction	**Wiring, mechanical fault**
1104	–	AT – second gear selection – malfunction	**Wiring, mechanical fault**
1105	–	AT – third gear selection – malfunction	**Wiring, mechanical fault**
1106	–	AT – fourth gear selection – malfunction	**Wiring, mechanical fault**

Flash code	OBD-II code	Fault location	Probable cause
1107	–	AT – signal circuit malfunction	**Wiring, mechanical fault**
1108	–	AT – shift solenoid A – voltage low	**Wiring, shift solenoid, mechanical fault**
1130	P1130	Intake manifold air control solenoid – malfunction	**Wiring, control solenoid, intake system, hose leak/blockage, control vacuum check switch, ECM**
1201	–	AT – shift solenoid B – voltage low	**Wiring, shift solenoid, mechanical fault**
1203	P1760	AT – overrun clutch solenoid – voltage low	**Wiring, solenoid, mechanical fault**
1204	–	Torque converter clutch solenoid – voltage low	**Wiring, solenoid, mechanical fault**
1205	–	AT – line pressure solenoid – voltage low	**Wiring, solenoid, mechanical fault**
1206	P1705	AT – throttle position (TP) sensor – voltage high or low	**Wiring, TP sensor, mechanical fault, ECM**
1207	–	Engine speed (RPM) – incorrect voltage signal ECM to AT control module	**Wiring, RPM sensor, mechanical fault, ECM**
1208	–	Transmission fluid temperature sensor – excessively high or low voltage	**Wiring, temperature sensor, mechanical fault, ECM**
1301	P1135	Camshaft position (CMP) actuator RH – malfunction	**Wiring, CMP actuator, mechanical fault, ECM**
1302	P1105	MAP – BARO switching valve – circuit malfunction	**Wiring, MAP/BARO switching valve**
1303	P1140	Camshaft position (CMP) actuator LH – circuit malfunction	**Wiring, CMP actuator, mechanical fault, ECM**
1304	P1145	Camshaft position (CMP) actuator RH – circuit malfunction	**Wiring, CMP actuator, mechanical fault, ECM**
1305	P1220	Fuel pressure (FP) control module – circuit malfunction	**Wiring, FP control module**
1308	P1900	Engine coolant blower motor – circuit	**Wiring, blower motor, ECM**

ISUZU

Model:	Engine identification:	Year:
Hombre 2.2L	VIN position 8 = 4	1996-00
Hombre 4.3L	VIN position 8 = X/W	1996-00
Axiom 3.5L	6VE1	2002
Amigo 2.2L	X22SE/Y22SE	1998-00
Amigo 3.2L	6VD1	1998-00
Rodeo 2.2L	X22SE/Y22SE	1998-02
Rodeo 2.6L	4ZE1	1996-97
Rodeo 3.2L	6VD1	1996-02
Trooper 3.2L	6VD1	1996-97
Trooper 3.5L	6VE1	1998-02
VehiCROSS 3.5L	6VE1	1999-01
Ascenda 4.2L	VIN position 8 = S	2003-04
Ascenda 5.3L	VIN position 8 = P	2003-04

System: **MFI**

Self-diagnosis

General information

- Refer to the front of this manual for general test conditions, terminology, detailed descriptions of wiring faults and a general trouble shooter for electrical and mechanical faults.
- Engine control module (ECM) incorporates self-diagnosis function.
- Malfunction indicator lamp (MIL) will illuminate if certain faults are recorded.
- ECM operates in backup mode if sensors fail, to enable vehicle to be driven to workshop.

Accessing

- Trouble codes can be displayed by using a Scan Tool connected to the data link connector (DLC) **1** under the LH fascia.

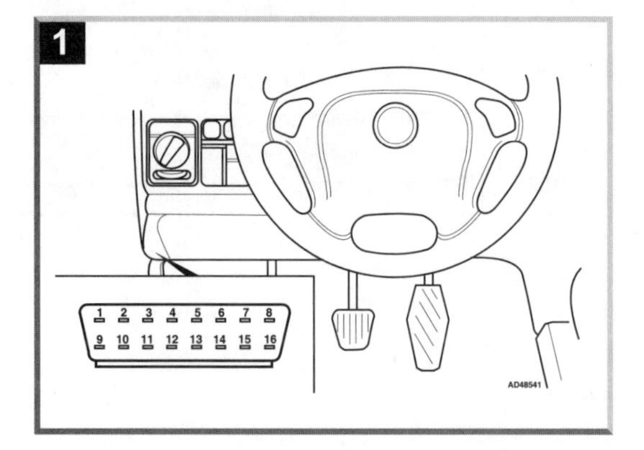

Erasing – except Ascenda 4.2L/5.3L

- After the faults have been rectified, trouble codes can be erased by using a Scan Tool connected to the data link connector (DLC) 1 or as follows:
- Ensure ignition switched OFF.
- Disconnect battery ground cable for 30 seconds minimum.
- Reconnect battery ground cable.
- Repeat checking procedure to ensure no data remains in ECM fault memory.

NOTE: ***Disconnecting battery ground lead will erase trouble codes but may also erase memory from electronic units such as radio.***

Erasing – Ascenda 4.2L/5.3L

- After the faults have been rectified, trouble codes can only be erased by using a Scan Tool connected to the data link connector (DLC) 1.

Trouble code identification

OBD-II code	Fault location	Probable cause
All P0, P2 and U0 codes	Refer to OBD-II trouble code tables at the front of this manual	–
P1106	Manifold absolute pressure (MAP) sensor – circuit/intermittent voltage high	**Wiring, MAP sensor, mechanical fault, ECM**
P1107	Manifold absolute pressure (MAP) sensor – circuit/intermittent voltage low	**Wiring, MAP sensor, mechanical fault, ECM**
P1111	Intake air temperature (IAT) sensor – circuit/intermittent voltage high	**Wiring, IAT sensor, ECM**
P1112	Intake air temperature (IAT) sensor – circuit/intermittent voltage low	**Wiring, IAT sensor, ECM**
P1114	Engine coolant temperature (ECT) sensor – circuit/intermittent voltage low	**Wiring, ECT sensor, ECM**
P1115	Engine coolant temperature (ECT) sensor – circuit/intermittent voltage high	**Wiring, ECT sensor, ECM**
P1120	Throttle motor position sensor 1 – circuit malfunction	**Wiring open circuit/short circuit to ground/ short circuit to positive, throttle control unit, ECM**
P1121	Throttle position (TP) sensor – circuit/intermittent voltage high	**Wiring, TP sensor, mechanical fault, ECM**
P1122	Throttle position (TP) sensor – circuit/intermittent voltage low	**Wiring, TP sensor, mechanical fault, ECM**
P1125 1	Throttle position (TP) motor – fail safe mode	**Wiring, TP sensor, TP motor, mechanical fault, ECM**
P1125 2	Accelerator pedal position (APP) sensor 1/2/3 – signal malfunction	**Wiring, APP sensor, ECM**
P1133 1	Heated oxygen sensor (HO2S) – RH front (6 cyl) – insufficient switching	**Wiring, HO2S, fuel system, fuel pressure, mechanical fault, ECM**
P1133 1	Oxygen sensor (O2S) – front (4 cyl) – insufficient switching	**Wiring, O2S, fuel system, fuel pressure, mechanical fault, ECM**

OBD-II code	Fault location	Probable cause
P1133 2	Oxygen sensor (O2S)/heated oxygen sensor (HO2S) 1, bank 1 – insufficient switching	**Wiring, exhaust system leaking, O2S/HO2S**
P1134 1	Heated oxygen sensor (HO2S) – RH front – slow operation	**Wiring, HO2S, fuel system, fuel pressure, mechanical fault, ECM**
P1134 2	Heated oxygen sensor (HO2S) 1, bank 1 – slow response	**Wiring, oxygen sensor heater, exhaust system leaking, HO2S**
P1137	Heated oxygen sensor (HO2S) 1/2 – voltage incorrect	**Wiring, exhaust system leaking, HO2S**
P1138	Heated oxygen sensor (HO2S) 2 – voltage high	**Wiring, HO2S**
P1153 1	Heated oxygen sensor (HO2S) – LH front – insufficient switching	**Wiring, HO2S, fuel system, fuel pressure, mechanical fault, ECM**
P1153 2	Heated oxygen sensor (HO2S) 1, bank 2 – insufficient switching	**Wiring, oxygen sensor heater, exhaust system leaking, HO2S**
P1154 1	Heated oxygen sensor (HO2S) – LH front – slow operation	**Wiring, HO2S, fuel system, fuel pressure, mechanical fault, ECM**
P1154 2	Heated oxygen sensor (HO2S) 1, bank 2 – slow response	**Wiring, oxygen sensor heater, exhaust system leaking, HO2S**
P1167	Fuel trim, bank 1 – mixture too rich during deceleration	**Injectors, fuel pump, fuel pressure, ECM**
P1169	Fuel trim, bank 2 – mixture too rich during deceleration	**Injectors, fuel pump, fuel pressure, ECM**
P1171	Fuel trim – mixture too lean during acceleration	**Wiring, O2/HO2S, mechanical fault, fuel system, fuel pressure, ECM**
P1220	Throttle motor position sensor 2 – circuit malfunction	**Wiring open circuit/short circuit to ground/ short circuit to positive, throttle control unit, ECM**
P1221	Throttle motor position sensor 1/2 – signal variation	**Wiring, throttle control unit, ECM**
P1258	Engine coolant – excessive temperature	**Wiring, engine coolant blower motor, mechanical fault, ECM**
P1271	Accelerator pedal position (APP) sensor 1/2 – signal variation	**Wiring, APP sensor, ECM**
P1272	Accelerator pedal position (APP) sensor 2/3 – signal variation	**Wiring, APP sensor, ECM**
P1273	Accelerator pedal position (APP) sensor 1/3 – signal variation	**Wiring, APP sensor, ECM**
P1275	Accelerator pedal position (APP) sensor 1 – circuit malfunction	**Wiring open circuit/short circuit to ground/ short circuit to positive, APP sensor, ECM**
P1280	Accelerator pedal position (APP) sensor 2 – circuit malfunction	**Wiring open circuit/short circuit to ground/ short circuit to positive, APP sensor, ECM**
P1285	Accelerator pedal position (APP) sensor 3 – circuit malfunction	**Wiring, APP sensor, ECM**
P1290	Throttle position (TP) motor – forced idle mode	**Wiring, APP sensor, TP motor, mechanical fault, ECM**
P1295	Throttle position (TP) motor – power management mode	**Wiring, TP sensor, TP motor, MAP sensor, MAF sensor, ECM**

OBD-II code	Fault location	Probable cause
P1299	Throttle position (TP) motor – forced engine shut down mode	**Wiring, TP sensor, TP motor, MAP sensor, MAF sensor, ECM**
P1310	Ignition control module (ICM) – diagnosis	**Wiring, ignition coil(s), ICM, ECM**
P1311	Ignition control module (ICM) – ignition coil secondary coil signal, circuit 1	**Wiring, ignition coil(s), ICM, ECM**
P1312	Ignition control module (ICM) – ignition coil secondary coil signal, circuit 2	**Wiring, ignition coil(s), ICM, ECM**
P1326	Ignition control module (ICM) – combustion quality input signal	**Wiring, ignition coil(s), ICM, ECM**
P1336	Crankshaft position (CKP) sensor – signal variation	**Wiring, CKP sensor, mechanical fault, Learning procedure not performed, ECM**
P1340	Ignition control module (ICM) – cylinder identification/syncronization	**Wiring, ignition coil(s), ICM, ECM**
P1345	CKP/CMP signals – signal malfunction	**Wiring, CKP/CMP sensor, mechanical fault**
P1351	Ignition control module (ICM) – signal voltage high	**Wiring, ignition control module, ignition system, mechanical fault, ECM**
P1361	Ignition control module (ICM) – signal not switching	**Wiring, ignition control module, ignition system, mechanical fault, ECM**
P1380	ECM to ABS communication – unusable signal	**Wiring, mechanical fault**
P1381	ECM to ABS communication – misfire detected	**Wiring, mechanical fault**
P1390	G-sensor – intermittent voltage low	**Wiring, G-sensor, mechanical fault, ECM**
P1391	G-sensor – performance	**Wiring, G-sensor, mechanical fault, ECM**
P1392	G-sensor – voltage low	**Wiring, G-sensor, mechanical fault, ECM**
P1393	G-sensor – voltage high	**Wiring, G-sensor, mechanical fault, ECM**
P1394	G-sensor – intermittent voltage high	**Wiring, G-sensor, mechanical fault, ECM**
P1404	Exhaust gas recirculation (EGR) solenoid – valve closed	**Wiring, EGR valve/solenoid/position sensor, mechanical fault, ECM**
P1404	Exhaust gas recirculation (EGR) valve – stuck open	**Wiring open circuit/short circuit to positive, exhaust gas recirculation (EGR) valve position sensor, EGR solenoid**
P1406	Exhaust gas recirculation (EGR) system – circuit problem	**Wiring, EGR valve/solenoid/position sensor, mechanical fault, ECM**
P1415	Secondary air injection (AIR) system, bank 1	**Wiring, hoses blocked/leaking, AIR valve**
P1416	Secondary air injection (AIR) system, bank 2	**Wiring, hoses blocked/leaking, AIR valve**
P1441	Evaporative emission (EVAP) canister purge system – leak detected	**Wiring, EVAP valve/solenoid, hoses, mechanical fault**
P1442	Evaporative emission (EVAP) canister purge switch – voltage high	**Wiring, EVAP valve/solenoid, hoses, mechanical fault**
P1481	Engine coolant blower motor speed – malfunction	**Wiring, engine coolant blower motor, ECM**
P1482	Engine coolant blower motor clutch – incorrect voltage	**Wiring, engine coolant blower motor relay, engine coolant blower motor, ECM**

OBD-II code	Fault location	Probable cause
P1484	Engine coolant blower motor relay – speed signal variation	**Wiring, engine coolant blower motor relay, engine coolant blower motor, ECM**
P1508	Idle air control (IAC) system – low RPM	**Wiring, IAC valve, mechanical fault, ECM**
P1509	Idle air control (IAC) system – high RPM	**Wiring, IAC valve, mechanical fault, ECM**
P1512	Throttle control unit – signal variation	**Wiring, throttle control unit, ECM**
P1514	Throttle position (TP) sensor/mass air flow (MAF) sensor – signal variation	**Wiring, TP sensor, MAP sensor, ECM**
P1515	Throttle command/actual throttle position – signal variation	**Wiring, throttle valve sticking, throttle control unit, ECM**
P1516	Throttle position motor – position performance	**Throttle valve tight/sticking, TP sensor, wiring, throttle control unit, ECM**
P1518	Throttle control module/engine control module (ECM) – communication malfunction	**Wiring, throttle control module, ECM**
P1520	Park/neutral position (PNP) switch – circuit malfunction	**Wiring, PNP switch, mechanical fault**
P1523	Throttle position motor – closed position performance	**Throttle valve tight/sticking, TP sensor, wiring open circuit/short circuit to ground, throttle control unit, ECM**
P1546	A/C compressor clutch relay – circuit malfunction	**Wiring open circuit/short circuit to ground/ short circuit to positive, A/C compressor clutch relay, ECM**
P1571 1	Brake pedal position (BPP) switch – no operation	**Wiring, BPP switch**
P1571 2	Traction control system (TCS), torque request signal – malfunction	**Wiring, poor connections, ABS/TCS**
P1574	Brake pedal position (BPP) switch – circuit malfunction	**Wiring, BPP switch**
P1600 1	Serial communication – malfunction	**ECM**
P1600 2	Engine control module (ECM) – malfunction	**ECM**
P1601	Serial communication – malfunction	**ECM**
P1602	ECM to ABS communication – circuit malfunction	**ECM**
P1618	Engine control module (ECM) – internal fault	**ECM**
P1621	Engine control module (ECM) – internal memory performance	**ECM**
P1625	Engine control module (ECM) – unexpected reset	**Interference from non standard electronics, erase trouble code and re-test**
P1626	Immobilizer system – ECM to body control module communication malfunction	**Wiring, ECM, body control module**
P1627 1	Engine control module (ECM) – analogue/digital converter	**Sensor supply wiring open/short circuit, ECM**
P1627 2	Engine control module (ECM) – malfunction	**ECM**
P1630	Engine control module (ECM), immobilizer code – learning mode	**ECM immobilizer code programming not completed**

OBD-II code	Fault location	Probable cause
P1631	Immobilizer system – password/code incorrect	**Immobilizer module, incorrect key, mechanical fault, ECM**
P1632	Immobilizer system – fueling disabled	**Imobilizer module, incorrect key, mechanical fault, ECM**
P1633	Engine control module (ECM), ignition supply voltage	**Fuse, wiring open circuit, ECM**
P1635 1	Sensor supply voltage – circuit A malfunction	**Wiring short circuit to ground/positive, ground wiring open circuit, ECM**
P1635 2	Engine control module (ECM), sensor supply voltage – circuit malfunction	**Wiring, poor connections, TP sensor 1 , accelerator pedal position (APP) sensor 2, A/C refrigerant pressure sensor, FTP sensor, engine coolant blower motor speed sensor, engine oil pressure sensor, ECM**
P1636	Engine control module (ECM) – RAM stack	**Interference from non standard electronics, erase trouble code and re-test**
P1637	Alternator 'L' terminal – circuit malfunction	**Wiring open circuit/short circuit to ground, alternator, battery temperature sensor, ECM**
P1639 1	Sensor supply voltage – circuit B malfunction	**Wiring short circuit to ground/positive, ground wiring open circuit, throttle position (TP) sensor, ECM**
P1639 2	Engine control module (ECM), sensor supply voltage – circuit malfunction	**Wiring, poor connections, MAP sensor sensor, throttle position (TP) sensor 2, accelerator pedal position (APP) sensor 1, ECM**
P1638	Alternator field circuit– signal out of range	**Wiring, alternator, ECM**
P1640	Engine control module, output driver 1 – supply voltage high	**Wiring, instruments, EVAP canister purge valve, AC compressor relay**
P1646	Sensor supply voltage – circuit C malfunction	**Wiring short circuit to ground/positive, ground wiring open circuit, accelerator pedal position (APP) sensor, ECM**
P1650	Engine control module quad driver/output driver	**Wiring, evaporative emission (EVAP) canister purge valve, intake manifold air control solenoid, ECM**
P1680	Engine control module (ECM) – malfunction	**ECM**
P1681	Engine control module (ECM) – malfunction	**ECM**
P1683	Engine control module (ECM) – malfunction	**ECM**
P1682	Throttle control module/engine control module (ECM) – ignition power supply variation	**Wiring, ECM**
P1689	Traction control system (TCS), delivered torque signal – malfunction	**Wiring, poor connections, ECM**
P1790	Engine control module (ECM) – AT ROM checksum error	**ECM**
P1792	AT – AT EEPROM checksum error	**ECM**
P1810	AT – manual valve position pressure switch malfunction	**Wiring, manual valve position pressure switch**

OBD-II code	Fault location	Probable cause
P1835	AT – kick-down switch	**Wiring, kick-down switch, mechanical fault**
P1860	AT – torque converter clutch (TCC) solenoid circuit	**Wiring, TCC solenoid**
P1870	AT – component slipping	**Mechanical fault**
P1875	AT – component slipping	**Mechanical fault**
U1000	Class 2 communication ID not learned	**ECM**
U1026	4WD AT control module communication – no signal	**Wiring, 4WD AT control module, ECM**
U1041	ECM to ABS communication – no signal	**Wiring, ABS control module, ECM**
U1064	ECM to body control module communication – no signal	**Wiring, body control module, ECM**
U1152	Class 2 communication ID missing	**Wiring, missing control module**
U1300	Class 2 communication – low voltage	**Wiring, ECM**
U1301	Class 2 communication – high voltage	**Wiring, ECM**

1 Except Ascenda
2 Ascenda

Model:	Engine identification:	Year:
Amigo 2.6L	4ZE1	1989-94
Impulse 2.3L	4ZD1	1988-89
Impulse 2.0L Turbo	4ZC1-T	1985-89
Pickup 2.6L	4ZE1	1988-90
Trooper 2.6L	4ZE1	1988-90

System: **I-TEC**

Self-diagnosis

General information

- Refer to the front of this manual for general test conditions, terminology, detailed descriptions of wiring faults and a general trouble shooter for electrical and mechanical faults.
- Engine control module (ECM) incorporates self-diagnosis function.
- Turbocharger (TC) control module incorporates self-diagnosis function.
- Malfunction indicator lamp (MIL) will illuminate if certain faults are recorded.
- ECM operates in backup mode if sensors fail, to enable vehicle to be driven to workshop.
- TC module operates in backup mode if sensors fail, to enable vehicle to be driven to workshop.
- ECM trouble codes can be accessed with suitable code reader connected to the data link connector (DLC) **1** [**1**] or displayed with the malfunction indicator lamp (MIL).
- TC control module trouble codes can be accessed with the TC control module LED.

Accessing – engine control module (ECM)

- Switch ignition ON, but do NOT start engine.
- Check MIL illuminates.
- Connect together the MIL diagnostic connector terminals
  - except Trooper **1** [**2**]
  - Trooper **2** – adjacent to engine control module (ECM), in center console.
- MIL should flash.
- Trouble code 12 will be displayed three times.
- Each trouble code will be displayed three times.
- If no trouble codes are stored, code 12 will continue to be displayed.
- Count MIL flashes and compare with trouble code table.
  - Short flashes indicate the 'tens' and 'units' of the trouble code **3**.
  - A 0.4 second pause separates each flash. A 1.2 second pause separates the 'tens' and 'units'. A 3.2 second pause separates each trouble code **3**.
  - For example: Trouble code 23 displayed **3**.
- Count MIL flashes and compare with trouble code table.
- Switch ignition OFF.
- Disconnect the MIL diagnostic connector terminals.

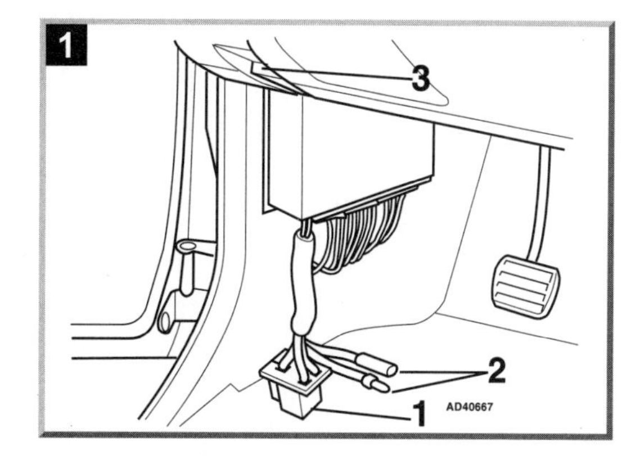

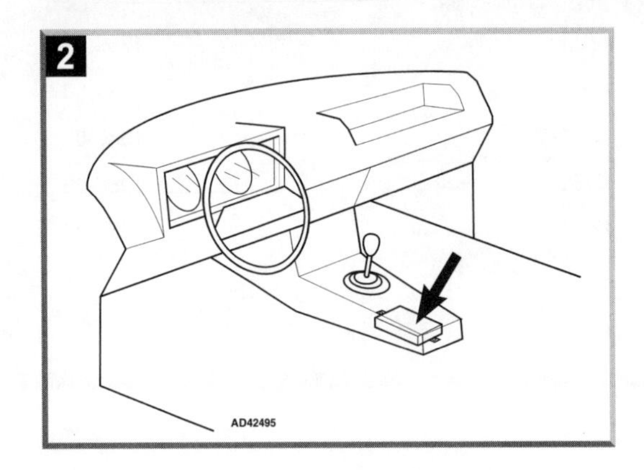

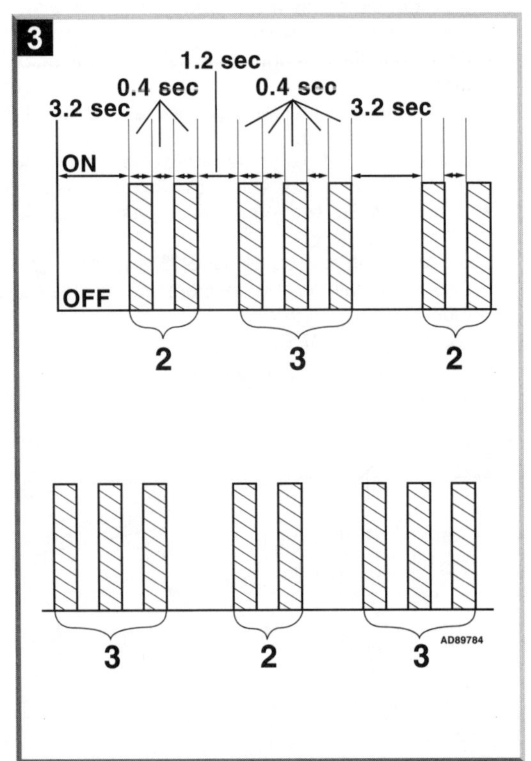

Erasing – engine control module (ECM)

- Ensure ignition switched OFF.
  - Except Amigo – remove 'BACKUP' fuse (10A) **4**.
  - Amigo – remove No.3 fuse (10A) from fascia fusebox **1** [**3**] or main (60A) fuse from underhood fusebox **5**.
- Wait 10 seconds.
- Reinstall fuse.
- Repeat checking procedure to ensure no data remains in ECM fault memory.

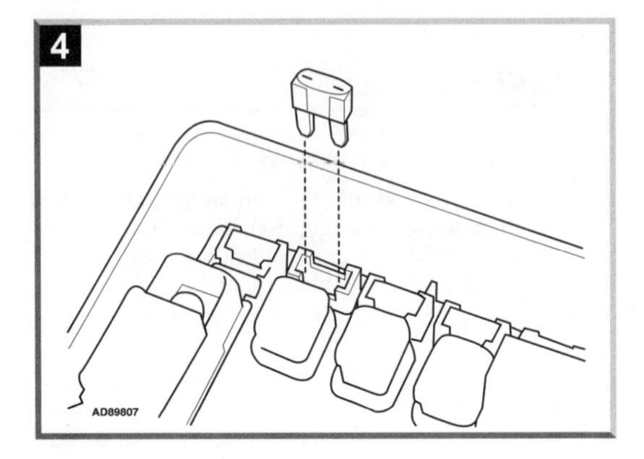

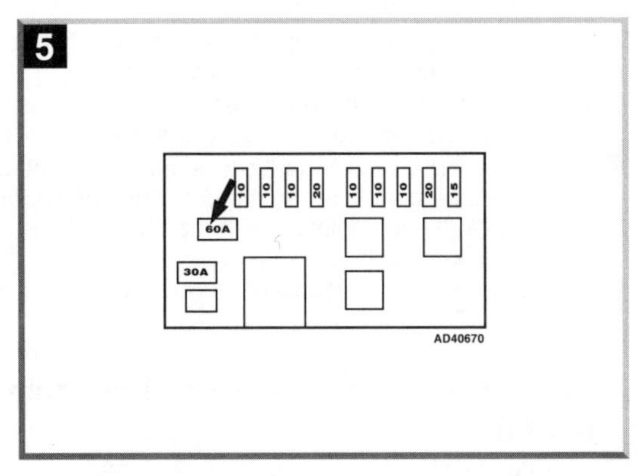

Trouble code identification – engine control module (ECM)

Flash code	Fault location	Probable cause
12	No fault	–
13	Oxygen sensor (O2S)	**Wiring, O2S, mechanical fault, ECM**
14	Engine coolant temperature (ECT) sensor – short circuit	**Wiring, ECT sensor**
15	Engine coolant temperature (ECT) sensor	**Wiring, ECT sensor**
16	Engine coolant temperature (ECT) sensor – open circuit	**Wiring, ECT sensor**
21	Closed throttle position (CTP) switch/wide open throttle (WOT) switch	**Wiring, CTP/WOT switch, mechanical fault, ECM**
22	Starter signal – open circuit	**Wiring, ignition switch, mechanical fault**
23	Ignition module – short circuit to ground	**Wiring, ignition control module, ECM**
25	Fuel pressure regulator control solenoid	**Wiring, solenoid valve, mechanical fault**
26	Evaporative emission (EVAP) canister purge valve	**Wiring, EVAP valve, mechanical fault**
27	Evaporative emission (EVAP) canister purge valve/engine control module (ECM)	**Wiring, EVAP valve, mechanical fault, ECM**
32	Exhaust gas recirculation temperature (EGRT) sensor	**Wiring, EGRT sensor, ECM**
33	Injector	**Wiring, injector, mechanical fault**
34	Exhaust gas recirculation temperature (EGRT) sensor/wiring	**Wiring, EGRT sensor, ECM**
35	Ignition module – open circuit	**Wiring, ignition system, mechanical fault, ECM**
41	Crankshaft position (CKP) sensor	**Wiring, CKP sensor, ECM**
43	Closed throttle position (CTP) switch	**Wiring, CTP switch, mechanical fault, ECM**
44	Fuel trim (mixture) – lean	**Wiring, O2S, mechanical fault, fuel system, fuel pressure, ECM**
45	Fuel trim (mixture) – rich	**Wiring, O2S, mechanical fault, fuel system, fuel pressure, ECM**
51	Engine control module (ECM)	**ECM**
52	Engine control module (ECM)	**ECM**
53	Fuel pressure regulator control solenoid/engine control module (ECM)	**Wiring, solenoid valve, mechanical fault, ECM**
54	Ignition module/wiring	**Wiring, ignition control module, ECM**
55	Engine control module (ECM)	**ECM**
61	Mass air flow (MAF) sensor – signal low	**Wiring, MAF sensor, mechanical fault, ECM**
62	Mass air flow (MAF) sensor – signal high	**Wiring, MAF sensor, mechanical fault, ECM**
63	Vehicle speed sensor (VSS)	**Wiring, VSS sensor, mechanical fault, ECM**
64	Injector/engine control module (ECM)	**Wiring, injector, ECM**

Flash code	Fault location	Probable cause
65	Wide open throttle (WOT) switch	**Wiring, WOT switch, ECM**
66	Knock sensor (KS)	**Wiring, KS sensor, mechanical fault, ECM**
71	Throttle position (TP) sensor – turbo control system	**Wiring, TP sensor, mechanical fault, ECM**
72	Exhaust gas recirculation (EGR) solenoid	**Wiring, EGR valve, mechanical fault, ECM**
73	Exhaust gas recirculation (EGR) solenoid/ engine control module (ECM)	**Wiring, EGR valve, mechanical fault, ECM**

Accessing – turbocharger (TC) control module

- Access turbocharger (TC) control module situated above engine control module (ECM) **1**.
- Switch ignition ON, but do NOT start engine.
- LED should flash **6**.
- LED will illuminate 7 times for half a second (trouble codes blocks 1-7).
- Trouble code will be indicated by rapid flashing of appropriate trouble code block.
- For example **7** indicates trouble code 3, engine coolant temperature (ECT) sensor.
- Cycle will be repeated after approximately 4 seconds.
- LED will now display trouble codes 2 and 6.

NOTE: ***This does not indicate faults at this time.***

- Fully depress, then release accelerator pedal: Stored trouble code 6 will be erased.
- Start engine: Stored trouble code 2 will be erased.
- LED should be flashing.
- LED will illuminate 7 times for half a second (trouble codes blocks 1-7).
- Trouble code will be indicated by rapid flashing of appropriate trouble code block.
- Sequence will be repeated after approximately 4 seconds.
- Compare recorded codes with trouble code table.
- Switch ignition OFF.

6

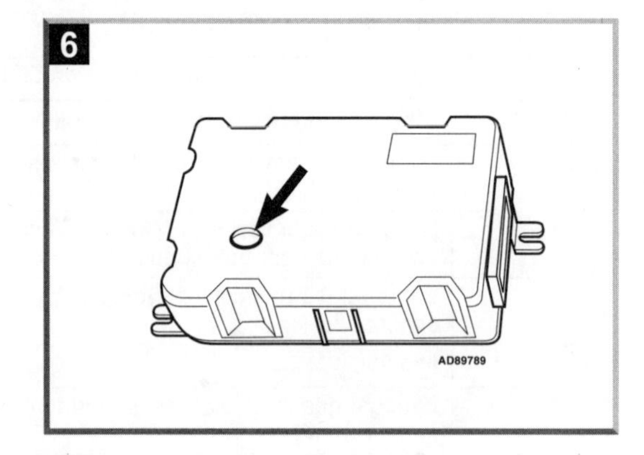

7

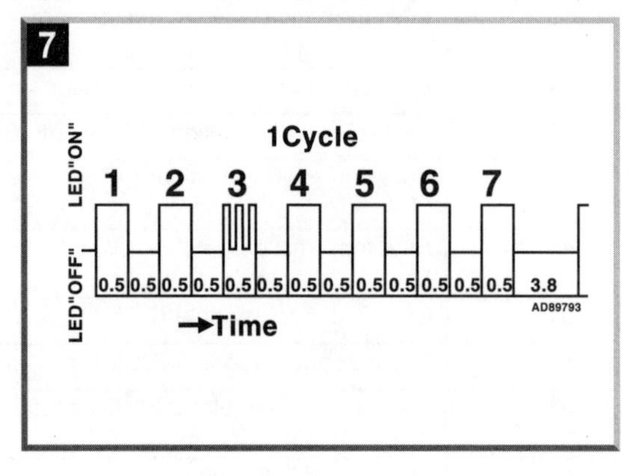

Erasing – turbocharger (TC) control module

- Switch ignition ON, but do NOT start engine.
- LED will display trouble codes 2 and 6 [6].

NOTE: ***This does not indicate faults at this time.***

- Fully depress, then release accelerator pedal: Stored code 6 will be erased.
- Start engine: Stored trouble code 2 will be erased.
- LED should be flashing.
- Ensure that NO trouble codes are displayed.

Trouble code identification – turbocharger (TC) control module

Flash code	Fault location	Probable cause
1	Knock sensor (KS)	**Wiring, KS sensor, mechanical fault**
2	Ignition signal – signal absent	**Wiring, ignition control module, mechanical fault**
3	Engine coolant temperature (ECT) sensor – short or open circuit	**Wiring, ECT sensor, ECM**
4	Turbocharger (TC) wastegate control motor position switch	**Wiring, motor, mechanical fault**
5	Throttle position (TP) sensor – short or open circuit	**Wiring, TP sensor, ECM**
6	Throttle position (TP) sensor – signal absent or abnormal	**Wiring, TP sensor, mechanical fault, ECM**
7	Battery voltage – high or low	**Wiring, fuses, mechanical fault**

ISUZU

Model:	Engine identification:	Year:
Oasis 2.2L	**F22B6**	**1996-97**
Oasis 2.3L	**F23A7**	**1998-99**

System: **PGM-FI**

Self-diagnosis

General information

- Refer to the front of this manual for general test conditions, terminology, detailed descriptions of wiring faults and a general trouble shooter for electrical and mechanical faults.
- Engine control module (ECM) incorporates self-diagnosis function.
- Malfunction indicator lamp (MIL) will illuminate if certain faults are recorded.
- ECM operates in backup mode if sensors fail, to enable vehicle to be driven to workshop.
- Trouble codes can be displayed by using a Scan Tool connected to the data link connector (DLC) **1** [1] or by the malfunction indicator lamp (MIL) with the service check connector jumped **1** [2].

NOTE: ***The use of a Scan Tool is essential to obtain full diagnostic information.***

Accessing

- Ensure ignition switched OFF.
- Jump service check connector terminals **1** [2].
- Switch ignition ON.
- Check MIL is flashing.
- Trouble codes 1-9 are indicated as follows:
  - Individual short flashes display trouble code **2** [A].
  - A short pause separates each flash **2** [B].
  - A long pause separates each trouble code **2** [C].
  - For example: Trouble code 3 displayed **2**.
- Trouble codes greater than 9 are indicated as follows:
  - Long flashes indicate the 'tens' of the trouble code **3** [A].
  - Short flashes indicate the 'units' of the trouble code **3** [C].
  - A short pause separates each flash **3** [B].
  - A long pause separates each trouble code **3** [D].
  - For example: Trouble code 12 displayed **3**.
- Count MIL flashes and compare with trouble code table.
- Switch ignition OFF.
- Remove jump lead.

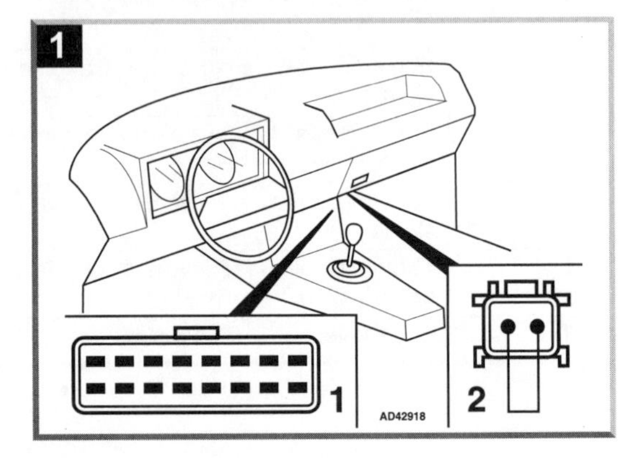

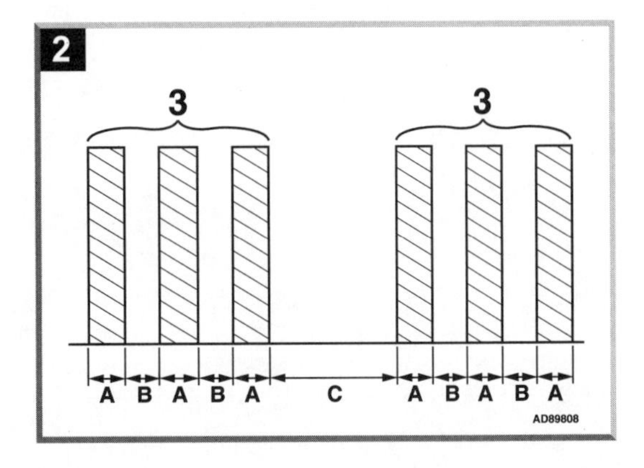

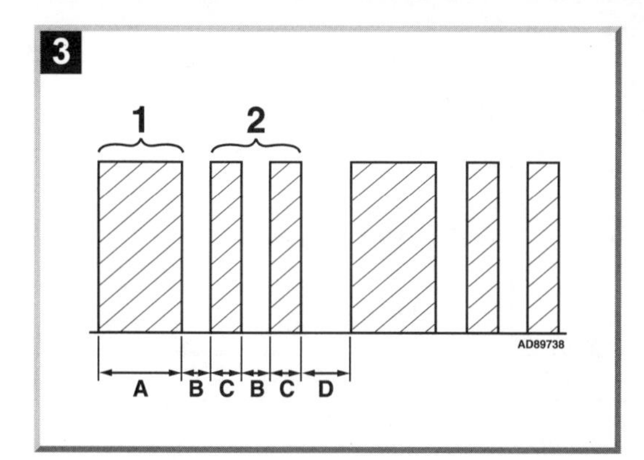

Erasing

- After the faults have been rectified, trouble codes can be erased by using a Scan Tool connected to the data link connector (DLC) or as follows:
- Switch ignition OFF.
- Remove radio fuse (7.5A) from underhood fusebox for 10 seconds minimum 4.
- Reinstall fuse.
- Repeat checking procedure to ensure no data remains in ECM fault memory.

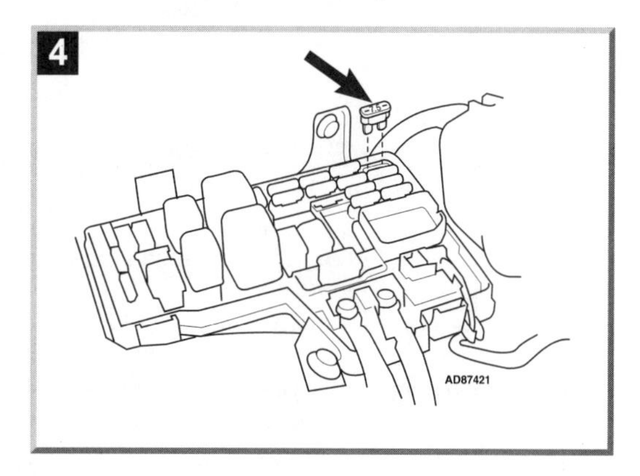

Trouble code identification

Flash code	OBD-II code	Fault location	Probable cause
–	*All P0, P2 and U0 codes*	Refer to OBD-II trouble code tables at the front of this manual	–
1	–	Heated oxygen sensor (HO2S) – front – circuit/voltage low/high	**Wiring, HO2S, fuel system, ECM**
3	–	Manifold absolute pressure (MAP) sensor – circuit/voltage low/high	**Wiring, MAP sensor, ECM**
4	–	Crankshaft position (CKP) sensor – range/ performance problem/circuit malfunction	**Wiring, CKP sensor, valve timing, ECM**
5	–	Manifold absolute pressure (MAP) sensor – range/performance problem	**Hose leak, MAP sensor**
5	P1128	Manifold absolute pressure (MAP) sensor – pressure lower than expected	**MAP sensor**
5	P1129	Manifold absolute pressure (MAP) sensor – pressure higher than expected	**MAP sensor**
6	–	Engine coolant temperature (ECT) sensor – circuit/voltage low/high	**Wiring, ECT sensor, ECM**
7	–	Throttle position (TP) sensor – circuit/voltage low/high	**Wiring, TP sensor, ECM**
7	P1121	Throttle position (TP) sensor – position lower than expected	**TP sensor**
7	P1122	Throttle position (TP) sensor – position higher than expected	**TP sensor**

Flash code	OBD-II code	Fault location	Probable cause
8	P1359	Crankshaft position (CKP) sensor – connector disconnection	**Wiring**
8	P1361	Crankshaft position (CKP) sensor – intermittent signal	**CKP sensor**
8	P1362	Crankshaft position (CKP) sensor – no signal	**Wiring, CKP sensor, ECM**
9	P1381	Camshaft position (CMP) sensor – intermittent signal	**CMP sensor**
9	P1382	Camshaft position (CMP) sensor – no signal	**Wiring, CMP sensor, ECM**
10	–	Intake air temperature (IAT) sensor – range/performance problem/circuit/voltage low/high	**IAT sensor, Wiring, ECM**
12	P1491	Exhaust gas recirculation (EGR) system – valve lift insufficient	**Wiring, EGR valve/position sensor, EGR solenoid, hose leak/blockage, ECM**
12	P1498	Exhaust gas recirculation (EGR) valve position sensor – voltage high	**Wiring, EGR valve/position sensor, ECM**
13	P1106	Barometric pressure (BARO) sensor – range/performance problem	**ECM**
13	P1107	Barometric pressure (BARO) sensor – circuit/voltage low	**ECM**
13	P1108	Barometric pressure (BARO) sensor – circuit/voltage high	**ECM**
14	–	Idle control system – malfunction	**IAC valve, fast idle thermo valve, throttle body**
14	P1508	Idle air control (IAC) valve – circuit failure	**Wiring, IAC valve, ECM**
14	P1519	Idle air control (IAC) valve – circuit failure	**Wiring, IAC valve, ECM**
17	–	Vehicle speed sensor (VSS) – circuit malfunction/ range/performance problem	**Wiring, VSS, ECM**
20	P1297	Electrical load sensor – circuit/voltage low	**Wiring, electrical load sensor, ECM**
20	P1298	Electrical load sensor – circuit/voltage high	**Wiring, electrical load sensor, ECM**
21	P1253	VTEC system – malfunction	**Wiring, VTEC solenoid, ECM**
22	P1259	VTEC system – malfunction	**Wiring, VTEC solenoid/pressure switch, ECM**
23	–	Knock sensor (KS) – circuit malfunction	**Wiring, KS, ECM**
30	P1681	AT to ECM – signal A – voltage low	**Wiring, TCM, ECM**
30	P1682	AT to ECM – signal A – voltage high	**Wiring, TCM, ECM**
31	P1686	AT to ECM – signal B – voltage low	**Wiring, TCM, ECM**
31	P1687	AT to ECM – signal B – voltage high	**Wiring, TCM, ECM**
41	–	Heated oxygen sensor (HO2S) – front – circuit malfunction	**Wiring, ECM**
45	–	Mixture lean/rich	**Fuel/exhaust system, front HO2S, MAP sensor, mechanical fault**

Flash code	OBD-II code	Fault location	Probable cause
61	–	Heated oxygen sensor (HO2S) – front – slow response	**HO2S, exhaust system**
63	–	Heated oxygen sensor (HO2S) – rear – slow response/circuit/voltage low/high	**Wiring, HO2S, ECM**
65	–	Heated oxygen sensor (HO2S) – rear – circuit malfunction	**Wiring, ECM**
67	–	Catalytic converter – efficiency below limit	**Catalytic converter, rear HO2S**
70	–	AT – lock-up clutch not engaging/no gear shift	**Wiring, mainshaft speed sensor, countershaft speed sensor, lock-up control system, shift solenoid (SS) A/B/C, TCM**
70	P1660	AT to ECM – data line failure	**Wiring, TCM, ECM**
70	P1705	AT – gear shift malfunction	**Wiring, range position switch, ECM**
70	P1705	AT – gear shift malfunction	**Wiring, range position switch, TCM**
70	P1705	AT – lock-up clutch not engaging	**Wiring, range position switch, ECM**
70	P1705	AT – lock-up clutch not engaging	**Wiring, range position switch, TCM**
70	P1706	AT – gear shift malfunction	**Wiring, range position switch, ECM**
70	P1706	AT – gear shift malfunction	**Wiring, range position switch, TCM**
70	P1706	AT – lock-up clutch malfunction	**Wiring, range position switch, ECM**
70	P1706	AT – lock-up clutch malfunction	**Wiring, range position switch, TCM**
70	P1738	Automatic transmission	**Wiring, clutch pressure switch 2, ECM**
70	P1739	Automatic transmission	**Wiring, clutch pressure switch 3, ECM**
70	P1753	AT – lock-up clutch not engaging/ disengaging	**Wiring, lock-up control solenoid A, ECM**
70	P1753	AT – lock-up clutch not engaging/ disengaging	**Wiring, lock-up control solenoid A, TCM**
70	P1758	AT – lock-up clutch not engaging	**Wiring, lock-up control solenoid B, ECM**
70	P1758	AT – lock-up clutch not engaging	**Wiring, lock-up control solenoid B, TCM**
70	P1768	AT – no gear shift	**Wiring, clutch pressure control solenoid A, ECM**
70	P1768	AT – poor gear shift	**Wiring, linear solenoid, TCM**
70	P1768	AT – lock-up clutch not engaging	**Wiring, linear solenoid, TCM**
70	P1773	AT – no gear shift	**Wiring, clutch pressure control solenoid B, ECM**
70	P1773	AT – lock-up clutch not engaging	**Wiring, clutch pressure control solenoid B, ECM**
70	P1786	AT – poor gear shift	**Communication wire, ECM**
70	P1790	AT – lock-up clutch not engaging	**Wiring, TP sensor, TCM**
70	P1791	AT – lock-up clutch not engaging	**Wiring, VSS, TCM**

Flash code	OBD-II code	Fault location	Probable cause
70	P1792	AT – lock-up clutch not engaging	**Wiring, ECT sensor, TCM**
70	P1794	Automatic transmission – BARO signal	**Wiring, ECM, TCM**
71	–	Cylinder No.1 – misfire	**Wiring, Ignition/fuel/EGR system, MAP sensor, IAC valve**
72	–	Cylinder No.2 – misfire	**Wiring, Ignition/fuel/EGR system, MAP sensor, IAC valve**
73	–	Cylinder No.3 – misfire	**Wiring, Ignition/fuel/EGR system, MAP sensor, IAC valve**
74	–	Cylinder No.4 – misfire	**Wiring, Ignition/fuel/EGR system, MAP sensor, IAC valve**
–	P1300	Random misfire	**Ignition/fuel/EGR system, MAP sensor, IAC valve**
80	–	Exhaust gas recirculation (EGR) solenoid – insufficient flow	**EGR valve, hose leak/blockage**
86	–	Engine coolant temperature (ECT) sensor – range/performance problem	**ECT sensor, cooling system**
90	P1456	Evaporative emission (EVAP) canister purge system (fuel tank system) – leak detected	**Hose, fuel tank/pressure sensor, fuel filler cap, EVAP valve/bypass solenoid, EVAP two way valve, EVAP canister/vent valve**
90	P1457	Evaporative emission (EVAP) canister purge system (canister system) – leak detected	**Hose, fuel tank/pressure sensor, EVAP valve/bypass solenoid, EVAP two way valve, EVAP canister/vent valve**
91	–	Fuel tank pressure sensor – circuit/voltage low/high	**Wiring, pressure sensor, ECM**
92	–	Evaporative emission (EVAP) canister purge system – incorrect flow	**Wiring, EVAP solenoid/flow switch, throttle body, hose leak/blockage, ECM**
92	P1459	Evaporative emission (EVAP) canister purge system – switch malfunction	**Wiring, EVAP flow switch, hose leak/ blockage, ECM**
–	P1607	Engine control module (ECM) – internal circuit failure	**ECM**

Model:	Engine identification:	Year:
Pickup 2.6L	4ZE1	1993
Rodeo 2.6L	4ZE1	1991-96

System: **I-TEC**

Self-diagnosis

General information

- Refer to the front of this manual for general test conditions, terminology, detailed descriptions of wiring faults and a general trouble shooter for electrical and mechanical faults.
- Engine control module (ECM) incorporates self-diagnosis function.
- Malfunction indicator lamp (MIL) will illuminate if certain faults are recorded.
- ECM operates in backup mode if sensors fail, to enable vehicle to be driven to workshop.
- Trouble codes can be accessed with suitable code reader connected to the data link connector (DLC) **1** [**1**] or displayed with the malfunction indicator lamp (MIL).

Accessing

- Ensure ignition switched OFF.
- Jump diagnostic connector terminals (DLC)
  - except Rodeo 6.95-96 **1** [**2**]
  - Rodeo 6.95-96 **2** [**2**].
- Switch ignition ON. Do NOT start engine.
- MIL should flash.
  - Long flashes indicate the 'tens' of the trouble code **3** [**A**].
  - Short flashes indicate the 'units' of the trouble code **3** [**C**].
  - A short pause separates each flash **3** [**B**].
  - A long pause separates each trouble code **3** [**D**].
  - For example: Trouble code 12 displayed **3**.
- Trouble code 12 will be displayed 3 times.
- Each trouble code will then be displayed 3 times.
- If no trouble codes are stored, code 12 will continue to be displayed.
- Count MIL flashes and compare with trouble code table.
- Switch ignition OFF.
- Remove jump lead.

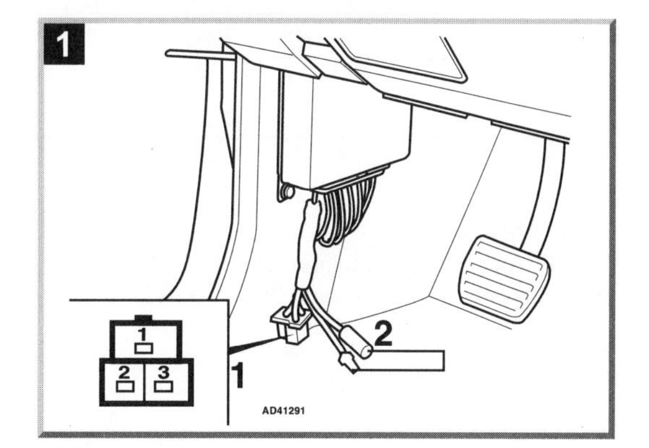

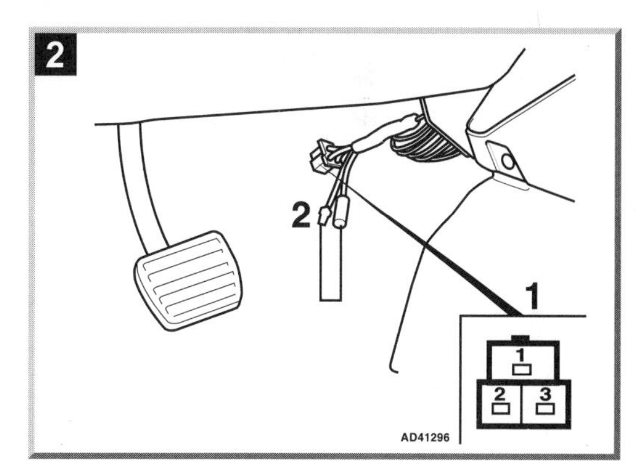

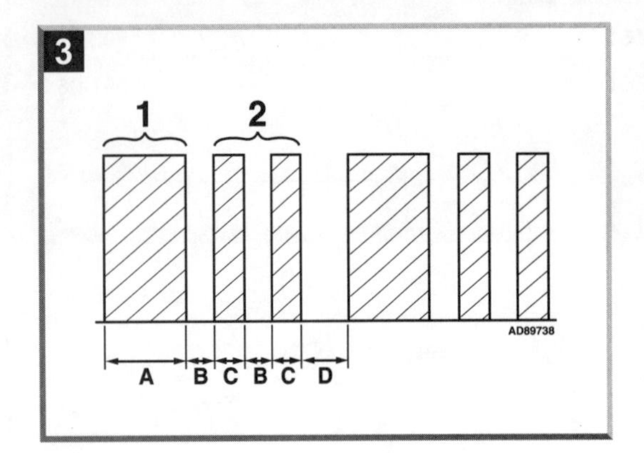

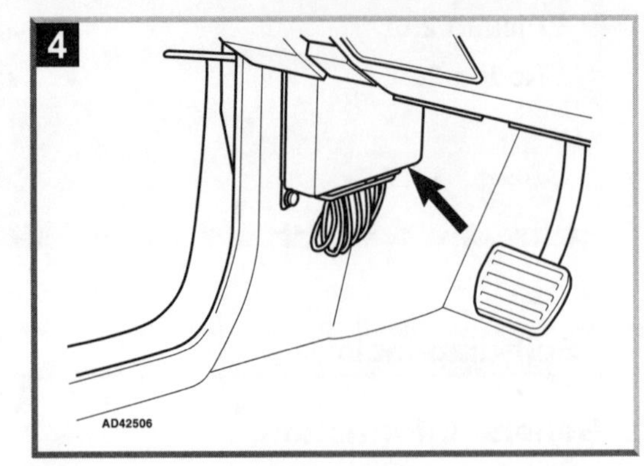

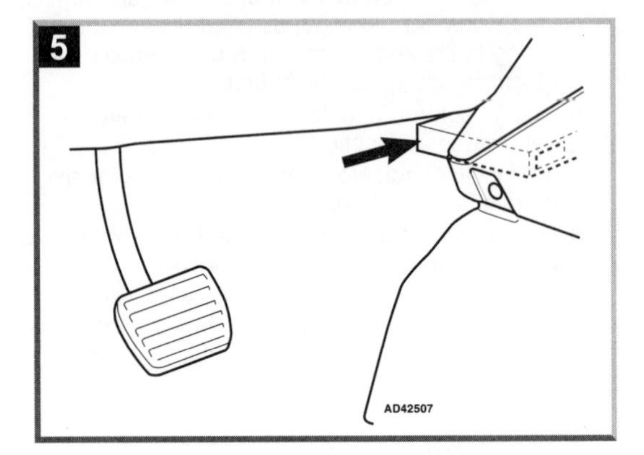

Erasing

- Ensure ignition switched OFF.
- Disconnect engine control module (ECM) harness multi-plugs for 30 seconds
  - except Rodeo 06.95-96 **4**
  - Rodeo 06.95-96 **5**.
- Reconnect multi-plugs.
- Repeat checking procedure to ensure no data remains in ECM fault memory.

Trouble code identification

Flash code	Fault location	Probable cause
12	No fault	–
13	Heated oxygen sensor (HO2S)/oxygen sensor (O2S)	**Wiring, HO2S/O2S, fuel system, fuel pressure, mechanical fault, ECM**
14	Engine coolant temperature (ECT) sensor – short circuit	**Wiring, ECT sensor, ECM**
15	Engine coolant temperature (ECT) sensor – open circuit	**Wiring, ECT sensor, ECM**
21	Throttle position (TP) switch	**Wiring, TP switch, ECM**
22	Starter signal – open circuit	**Wiring, ignition switch, mechanical fault**
23	Ignition module – short circuit to ground	**Wiring, ignition control module**
25	Fuel pressure regulator control solenoid	**Wiring, solenoid valve, mechanical fault, ECM**

Flash code	Fault location	Probable cause
26	Evaporative emission (EVAP) canister purge valve	Wiring, EVAP valve, mechanical fault, ECM
27	Evaporative emission (EVAP) canister purge valve/engine control module (ECM)	Wiring, EVAP valve, mechanical fault, ECM
32	Exhaust gas recirculation (EGR) system	Wiring, EGR valve, mechanical fault, ECM
33	Injector	Wiring, injector, mechanical fault
34	Manifold absolute pressure (MAP) sensor	Wiring, MAP sensor, ECM
35	Ignition module – open circuit	Wiring, ignition control module, ECM
41	Crankshaft position (CKP) sensor	Wiring, CKP sensor, mechanical fault, ECM
43	Throttle position (TP) switch – closed	Wiring, TP switch, mechanical fault
44	Fuel trim – mixture lean	Wiring, O2S, mechanical fault, fuel system, fuel pressure, ECM
45	Fuel trim – mixture rich	Wiring, O2S, mechanical fault, fuel system, fuel pressure, ECM
51	Engine control module (ECM)	ECM
52	Engine control module (ECM)	ECM
53	Fuel pressure regulator control solenoid/engine control module (ECM)	Wiring, solenoid valve, mechanical fault, ECM
54	Ignition module/wiring	Wiring, ignition control module, ECM
61	Mass air flow (MAF) sensor – voltage low	Wiring, MAF sensor, mechanical fault, ECM
62	Mass air flow (MAF) sensor – voltage high	Wiring, MAF sensor, mechanical fault, ECM
63	Vehicle speed sensor (VSS)	Wiring, VSS sensor, mechanical fault, ECM
64	Injector/engine control module (ECM)	Wiring, injector, ECM
65	Throttle position (TP) switch – wide open	Wiring, TP switch, mechanical fault

ISUZU

Model:	Engine identification:	Year:
Pickup 3.1L	**VIN digit 8 = Z**	**1991-94**
Rodeo 3.1L	**VIN digit 8 = Z**	**1991-92**
Trooper 2.8L	**VIN digit 8 = R**	**1991**

System: **MFI**

Self-diagnosis

General information

- Refer to the front of this manual for general test conditions, terminology, detailed descriptions of wiring faults and a general trouble shooter for electrical and mechanical faults.
- Engine control module (ECM) incorporates self-diagnosis function.
- Malfunction indicator lamp (MIL) will illuminate if certain faults are recorded.
- ECM operates in backup mode if sensors fail, to enable vehicle to be driven to workshop.
- Trouble codes can be accessed with suitable code reader or displayed with the malfunction indicator lamp (MIL).

Accessing

- Ensure ignition switched OFF.
- Jump data link connector (DLC) terminals 5 and 6 – **1** Trooper or **2** Rodeo/Pickup.
- Switch ignition ON. Do NOT start engine.
- MIL should flash.
  - Long flashes indicate the 'tens' of the trouble code **3** [**A**].
  - Short flashes indicate the 'units' of the trouble code **3** [**C**].
  - A short pause separates each flash **3** [**B**].
  - A long pause separates each trouble code **3** [**D**].
  - For example: Trouble code 12 displayed **3**.
- Trouble code 12 will be displayed 3 times.
- Each trouble code will then be displayed 3 times.
- If no trouble codes are stored, code 12 will continue to be displayed.
- Count MIL flashes and compare with trouble code table.
- Switch ignition OFF.
- Remove jump lead.

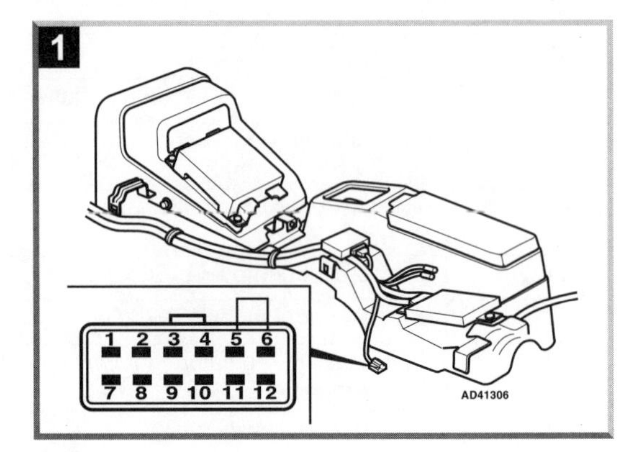

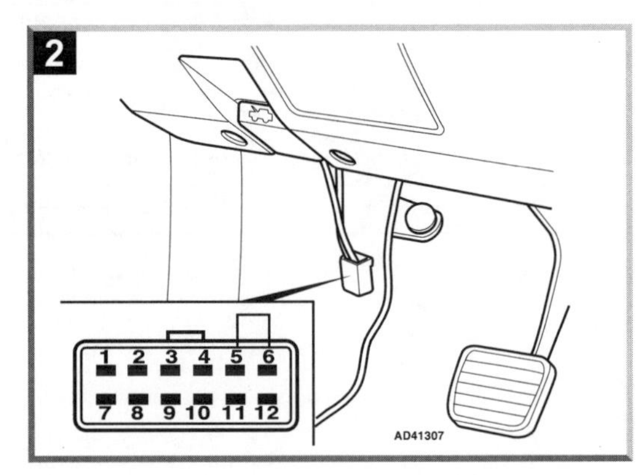

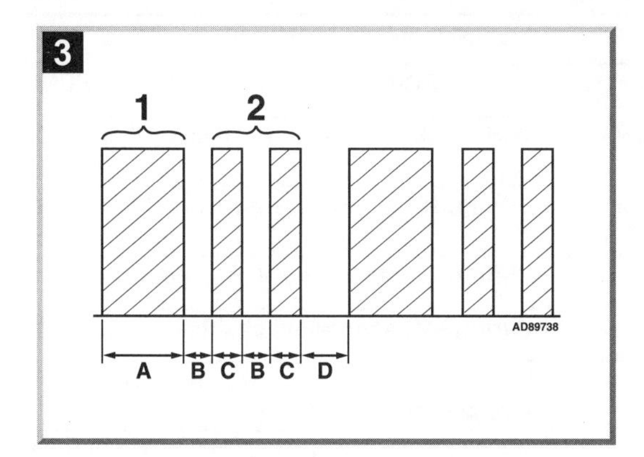

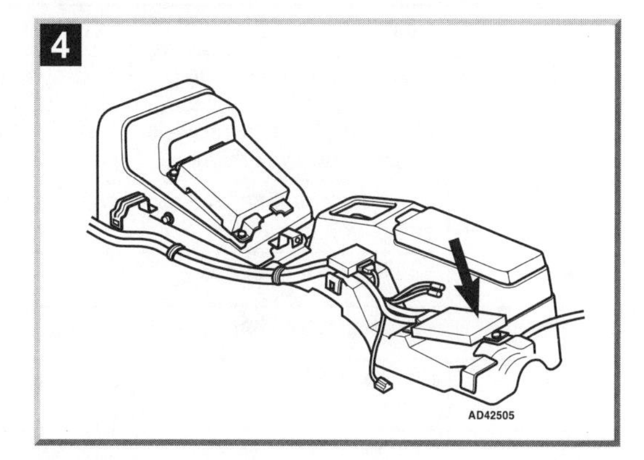

Erasing

- Ensure ignition switched OFF.
- Disconnect engine control module (ECM) harness multi-plugs for 30 seconds.
  - Trooper 4
  - Rodeo/Pickup 5
- Reconnect multi-plugs.
- Repeat checking procedure to ensure no data remains in ECM fault memory.

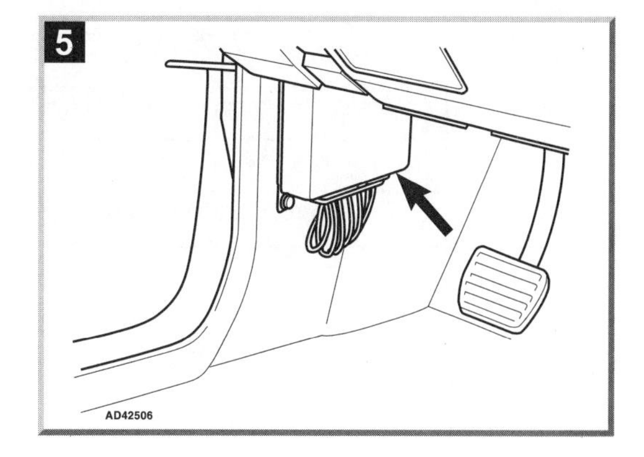

Trouble code identification

Flash code	Fault location	Probable cause
12	Normal condition	–
13	Oxygen sensor (O2S) – open circuit	Wiring, O2S
14	Engine coolant temperature (ECT) sensor – short circuit	Wiring, ECT sensor
15	Engine coolant temperature (ECT) sensor – open circuit	Wiring, ECT sensor
21	Throttle position (TP) sensor – signal voltage high	Wiring, TP sensor, mechanical fault, ECM
22	Throttle position (TP) sensor – signal voltage low	Wiring, TP sensor, mechanical fault, ECM
24	Vehicle speed sensor (VSS)	Wiring, VSS sensor, mechanical fault, ECM

Flash code	Fault location	Probable cause
32	Exhaust gas recirculation (EGR) system	**Wiring, EGR valve, mechanical fault, ECM**
33	Manifold absolute pressure (MAP) sensor – signal voltage high	**Wiring, MAP sensor, mechanical fault, ECM**
34	Manifold absolute pressure (MAP) sensor – signal voltage low	**Wiring, MAP sensor, mechanical fault, ECM**
42	Ignition control circuit	**Wiring, ignition control module, ECM**
43	Knock sensor (KS)	**Wiring, KS sensor, mechanical fault, ECM**
44	Oxygen sensor (O2S) – mixture lean	**Wiring, O2S, mechanical fault, fuel system, fuel pressure, ECM**
45	Oxygen sensor (O2S) – mixture rich	**Wiring, O2S, mechanical fault, fuel system, fuel pressure, ECM**
51	Engine control module (ECM) – memory	**Wiring, ECM**
52	Engine control module (ECM) – memory	**Wiring, ECM**
54	Fuel pump – voltage low	**Wiring, mechanical fault, fuel pump**
55	Engine control module (ECM)	**Wiring, ECM**

Model:	Engine identification:	Year:
Rodeo 3.2L	6VD1	1993-95
Stylus 1.6/1.8L	4XE1-V/W	1991-93
Trooper 3.2L	6VD1-V/W	1992-95

System: **MFI**

Self-diagnosis

General information

- Refer to the front of this manual for general test conditions, terminology, detailed descriptions of wiring faults and a general trouble shooter for electrical and mechanical faults.
- Engine control module (ECM) incorporates self-diagnosis function.
- Malfunction indicator lamp (MIL) will illuminate if certain faults are recorded.
- ECM operates in backup mode if sensors fail, to enable vehicle to be driven to workshop.
- Trouble codes can be accessed with suitable code reader or displayed with the malfunction indicator lamp (MIL).

Accessing

- Ensure ignition switched OFF.
- Jump data link connector (DLC) terminals 1 and 3 as follows:
  - Stylus – **1**.
  - Trooper – **2**.
  - Rodeo – **3**.
- Switch ignition ON. Do NOT start engine.
- MIL should flash.
  - Long flashes indicate the 'tens' of the trouble code **4** [**A**].
  - Short flashes indicate the 'units' of the trouble code **4** [**C**].
  - A short pause separates each flash **4** [**B**].
  - A long pause separates each trouble code **4** [**D**].
  - For example: Trouble code 12 displayed **4**.
- Trouble code 12 will be displayed 3 times.
- Each trouble code will then be displayed 3 times.
- If no trouble codes are stored, code 12 will continue to be displayed.
- Count MIL flashes and compare with trouble code table.
- Switch ignition OFF.
- Remove jump lead.

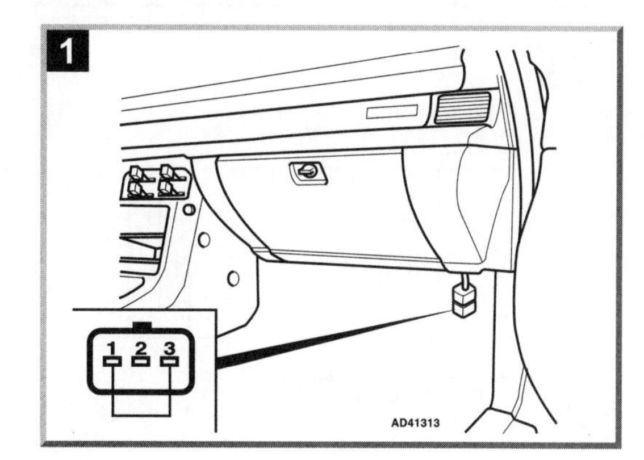

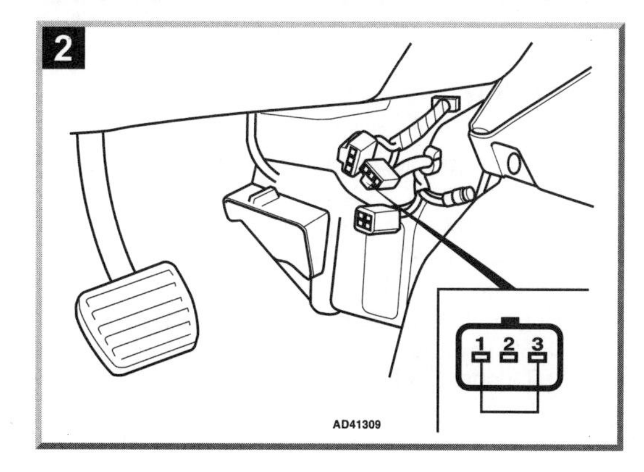

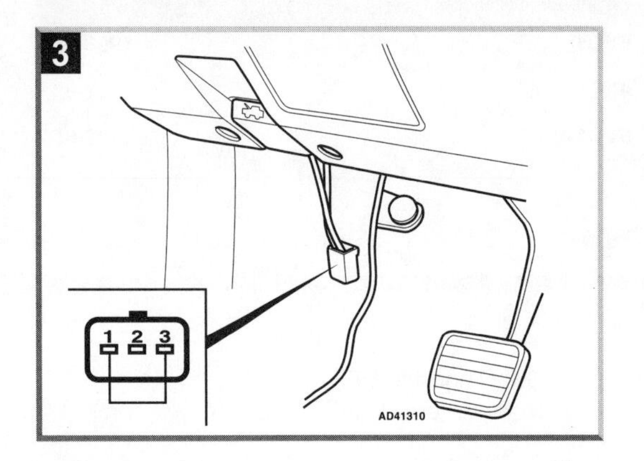

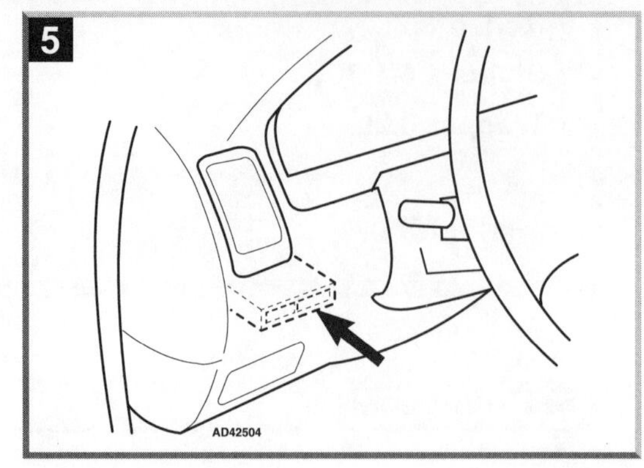

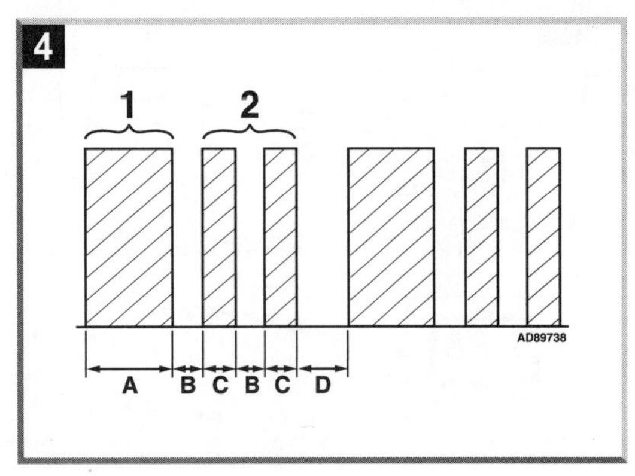

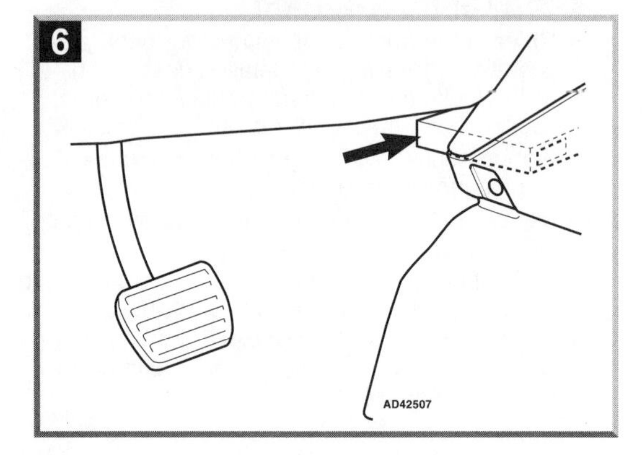

Erasing

- Ensure ignition switched OFF.
- Disconnect engine control module (ECM) harness multi-plugs for 30 seconds.
- Reconnect multi-plugs.
- Repeat checking procedure to ensure no data remains in ECM fault memory.
- Engine control module (ECM) location:
  - Stylus – 5.
  - Trooper – 6.
  - Rodeo – 7.

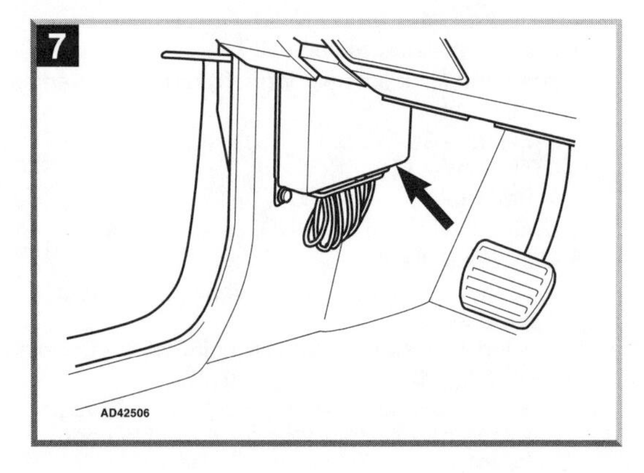

Trouble code identification

Flash code	Fault location	Probable cause
12	Normal condition	–
13	Heated oxygen sensor (HO2S) – short/open circuit	Wiring, HO2S/O2S
14	Engine coolant temperature (ECT) sensor	Wiring, ECT sensor
21	Throttle position (TP) sensor – signal voltage	Wiring, TP sensor, mechanical fault, ECM
23	Intake air temperature (IAT) sensor	Wiring, IAT sensor
24	Vehicle speed sensor (VSS) – circuit	Wiring, VSS sensor, mechanical fault, ECM
32	Exhaust gas recirculation (EGR) system	Wiring, EGR valve, mechanical fault, ECM
33	Manifold absolute pressure (MAP) sensor – signal	Wiring, MAP sensor, mechanical fault, ECM
42	Ignition control signal	Wiring, ignition control module, ECM
43	Knock sensor (KS)	Wiring, KS sensor, mechanical fault, ECM
44	Heated oxygen sensor (HO2S)/oxygen sensor (O2S) – signal – mixture lean	Wiring, HO2S/O2S, mechanical fault, fuel system, fuel pressure, ECM
45	Heated oxygen sensor (HO2S)/oxygen sensor (O2S) – signal – mixture rich	Wiring, HO2S/O2S, mechanical fault, fuel system, fuel pressure, ECM
51	Engine control module (ECM) – memory	Wiring, ECM

JAGUAR

Model:	Engine identification:	Year:
S-Type 3.0L	VIN code digit 11 = F	2000-04
S-Type 4.2L	VIN code digit 11 = H	2000-04
X-Type 2.5L	VIN code digit 11 = X	2002-04
X-Type 3.0L	VIN code digit 11 = W	2002-04

System: **Nippon Denso**

Self-diagnosis

General information

- Refer to the front of this manual for general test conditions, terminology, detailed descriptions of wiring faults and a general trouble shooter for electrical and mechanical faults.
- Malfunction indicator lamp (MIL) will illuminate if certain faults are recorded.
- ECM operates in backup mode if sensors fail, to enable vehicle to be driven to workshop.

Accessing and erasing

- The engine control module (ECM) fault memory can only be accessed and erased using diagnostic equipment connected to the data link connector (DLC):
  - X-Type **1**
  - S-Type **2**

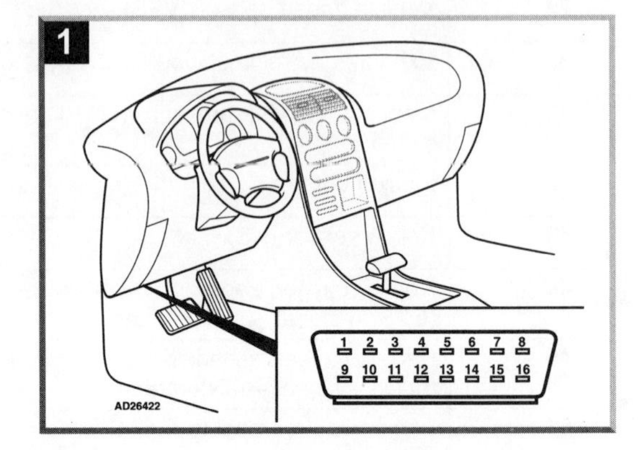

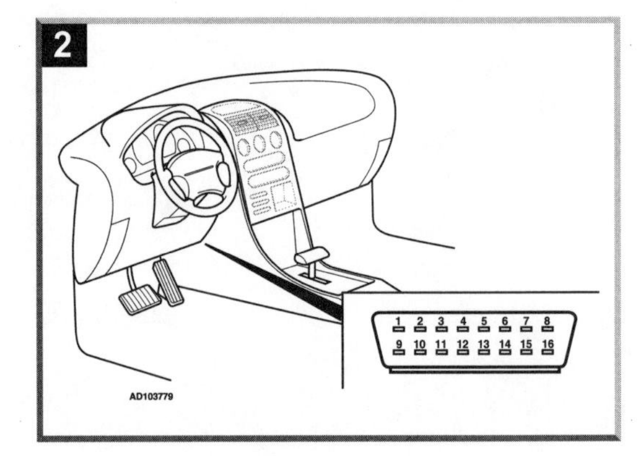

Trouble code identification

OBD-II code	Fault location	Probable cause
All P0, P2 and U0 codes	Refer to OBD-II trouble code tables at the front of this manual	–
P1000	System readiness test not complete	–
P1001	Engine running system self test not complete	–
P1100	Mass air flow (MAF) sensor – intermittent signal	**Wiring, MAF sensor**
P1101	Mass air flow (MAF) sensor – signal out of range	**Wiring, MAF sensor**
P1104	Mass air flow (MAF) sensor – ground circuit malfunction	**Wiring, MAF sensor**
P1107	Manifold absolute pressure (MAP) sensor – signal voltage low	**Wiring, MAP sensor**
P1108	Manifold absolute pressure (MAP) sensor – signal voltage high	**Wiring, MAP sensor**
P1111	OBD checks complete since last memory clear	–
P1112	Intake air temperature (IAT) sensor – intermittent signal	**Wiring, IAT sensor**
P1116	Engine coolant temperature (ECT) sensor – signal out of range	**Wiring, ECT sensor**
P1117	Engine coolant temperature (ECT) sensor – intermittent signal	**Wiring, ECT sensor**
P1121	Throttle position (TP) sensor/mass air flow (MAF) sensor – implausible signal	**Wiring, TP sensor**
P1122	Accelerator pedal position (APP) sensor 1 – signal voltage low	**Wiring, APP sensor**
P1123	Accelerator pedal position (APP) sensor 1 – signal voltage high	**Wiring, APP sensor**
P1124	Throttle position (TP) sensor – signal out of range	**Wiring, TP sensor**
P1127	Heated oxygen sensor (HO2S) 2 (downstream), bank 1 – exhaust not hot enough for self-test	**Engine not at operating temperature, exhaust leak**
P1128	Heated oxygen sensors (HO2S) 1 (upstream), bank 1/2	**Wiring, HO2S 1 (upstream) swapped banks**
P1129	Heated oxygen sensors (HO2S) 2 (downstream), bank 1/2	**Wiring, HO2S 2 (downstream) swapped banks**
P1130	Heated oxygen sensors (HO2S) 1 (upstream), bank 1 – slow response, adaptive fuel at limit	**Wiring, HO2S, intake system leak, exhaust leak, contaminated fuel, ECM**
P1131	Heated oxygen sensors (HO2S) 1 (upstream), bank 1 – slow response, system lean	**Wiring, HO2S, intake system leak, exhaust leak, contaminated fuel, ECM**
P1146	Alternator – charge circuit low	**Wiring, alternator voltage regulator, alternator**
P1215	Accelerator pedal position (APP) sensor 2 – signal voltage low	**Wiring, APP sensor**

OBD-II code	Fault location	Probable cause
P1216	Accelerator pedal position (APP) sensor 2 – signal voltage high	**Wiring, APP sensor**
P1224	Throttle control position – malfunction	**Wiring, TP sensor, throttle motor relay, throttle motor, throttle body**
P1229	Throttle motor – circuit malfunction	**Wiring, throttle motor relay, throttle motor**
P1234	Fuel pump (FP) control module – control module disabled or off line	**Wiring, FP control module , ECM**
P1236	Fuel pump (FP) not responding to ECM commands	**Wiring, FP control module**
P1240	Sensor power supply – circuit malfunction	**Wiring, APP sensor, MAP sensor, TP sensor, fuel tank pressure sensor(s)**
P1241	Sensor power supply – low input	**Wiring open/short to ground, APP sensor, MAP sensor, TP sensor, fuel tank pressure sensor(s)**
P1242	Sensor power supply – high input	**Wiring open/short to positive, APP sensor, MAP sensor, TP sensor, fuel tank pressure sensor(s)**
P1243	Sensor ground circuit – malfunction	**Wiring, APP sensor, MAP sensor, TP sensor, fuel tank pressure sensor(s)**
P1244	Alternator – charge circuit high	**Wiring, alternator voltage regulator, alternator**
P1245	Engine cranking signal – voltage low	**Wiring, starter motor relay, ignition switch**
P1246	Engine cranking signal – voltage high	**Wiring, starter motor relay, ignition switch**
P1250	Throttle valve return spring – malfunction	**Throttle body**
P1251	Throttle motor relay – malfunction	**Wiring, throttle motor relay**
P1254	Throttle limp home spring – malfunction	**Throttle body**
P1260	Vehicle immobilized	**Invalid ignition key, immobilizer system fault**
P1313	Misfire/catalyst damage – bank 1	**Engine mechanical fault, wiring, ignition/fuel system, injector(s), MAP/TP/ECT/IAT sensor, EVAP system, ECM**
P1314	Misfire/catalyst damage – bank 2	**Engine mechanical fault, wiring, ignition/fuel system, injector(s), MAP/TP/ECT/IAT sensor, EVAP system, ECM**
P1316	Misfire	**Wiring, other trouble code(s) stored, ignition/ fuel system, MAP/TP/ECT/IAT sensor, injector(s), ECM**
P1338	Fuel pump (FP) – feedback circuit malfunction	**Wiring, FP control module**
P1340	Camshaft position (CMP) sensor 2, bank 2 – circuit malfunction	**Wiring, air gap, metal particles, insecure sensor, CMP sensor**
P1341	Camshaft position (CMP) sensor 2, bank 2 – circuit range/performance	**Wiring, air gap, metal particles, insecure sensor, CMP sensor**
P1344	Accelerator pedal position (APP) sensor 1/2 – range/performance problem	**Wiring, APP sensor(s)**

OBD-II code	Fault location	Probable cause
P1367	Ignition amplifier, bank 1 – malfunction	**Wiring, ignition amplifier**
P1368	Ignition amplifier, bank 2 – malfunction	**Wiring, ignition amplifier**
P1384	Camshaft position (CMP) actuator, bank 1 – malfunction	**Wiring, oil pressure low/high, mechanical fault, CMP actuator**
P1396	Camshaft position (CMP) actuator, bank 2 – malfunction	**Wiring, oil pressure low/high, mechanical fault, CMP actuator**
P1516	Park/neutral position (PNP) switch, gear change neutral/drive – malfunction	**Wiring, selector cable adjustment, PNP switch, TR sensor**
P1517	Park/neutral position (PNP) switch, cranking – malfunction	**Wiring, selector cable adjustment, PNP switch, TR sensor**
P1532	Intake manifold air control solenoid 2 (lower) – circuit malfunction	**Wiring, intake manifold air control solenoid**
P1549	Intake manifold air control solenoid 1 (upper) – circuit malfunction	**Wiring, intake manifold air control solenoid**
P1571	Brake pedal position (BPP) switch – circuit malfunction	**Wiring, BPP switch**
P1573	CAN data bus, throttle position (TP) – message error	**Wiring, TP sensor**
P1582	Engine control module (ECM) – data stored	**Wiring, inertia switch activated/defective, throttle in limp home mode**
P1601	Incorrect ECM/TCM fitted to vehicle	**ECM/TCM coding, ECM, TCM**
P1603	Transmission control module (TCM) – EEPROM malfunction	**Battery disconnected with ignition ON, wiring, TCM**
P1606	Engine control (EC) relay – malfunction	**Wiring, EC relay**
P1609	Engine control module (ECM) – microprocessor fault	**ECM**
P1611	Engine control module (ECM) – central processor malfunction	**ECM**
P1629	Alternator – field circuit malfunction	**Wiring, alternator voltage regulator, alternator**
P1631	Throttle motor relay, coil circuit OFF – malfunction	**Wiring, throttle motor relay**
P1632	Alternator/charging system – load circuit malfunction	**Wiring, alternator voltage regulator, alternator**
P1633	Engine control module (ECM) – central processor malfunction	**ECM**
P1634	ECM/throttle monitoring – circuit malfunction	**ECM**
P1637	CAN data bus, ECM/ABS/TCS control module – network malfunction	**Wiring, ABS/TCS control module, ECM**
P1638	CAN data bus, ECM/instrumentation control module – network malfunction	**Wiring, instrumentation control module, ECM**
P1642	CAN data bus – circuit malfunction	**Wiring**
P1643	CAN data bus, ECM/TCM – network malfunction	**Wiring, TCM, ECM**

OBD-II code	Fault location	Probable cause
P1646	Heated oxygen sensor (HO2S) 1 (upstream), bank 1 – control malfunction	**Wiring, HO2S, ECM**
P1647	Heated oxygen sensor (HO2S) 1 (upstream), bank 2 – control malfunction	**Wiring, HO2S, ECM**
P1648	Engine control module (ECM) – self test malfunction	**ECM**
P1656	ECM/TP sensor amplifier – circuit malfunction	**ECM**
P1657	Throttle motor relay, coil circuit ON – malfunction	**Wiring, throttle motor relay**
P1658	Throttle motor relay – malfunction	**Wiring, throttle motor relay**
P1699	CAN data bus, ECM/AC control module – network malfunction	**Wiring, AC control module, ECM**
P1710	Transmission control valve solenoids – ground circuit malfunction	**Wiring**
P1745	Transmission low clutch timing solenoid – circuit malfunction	**Wiring, internal wiring, solenoid**
P1746	Transmission timing reduction solenoid – circuit malfunction	**Wiring, internal wiring, solenoid**
P1747	Transmission brake timing solenoid 2/4 – circuit malfunction	**Wiring, internal wiring, solenoid**
P1777	CAN data bus, torque reduction – malfunction	**Wiring, TCM, ECM**
P1780	Transmission gear selection switch D-4 – circuit malfunction	**Wiring, selector cable adjustment, TR sensor, transmission shift control module, TCM**
P1793	TCM, ignition power supply – circuit malfunction	**Wiring, relay**
P1796	CAN data bus, TCM – malfunction	**Wiring, TCM**
P1797	CAN data bus, ECM/TCM – malfunction	**Wiring, TCM, ECM**
P1799	CAN data bus, TCM/ABS/TCS control module – network malfunction	**Wiring, TCM, ABS/TCS control module**

Model:	**XJS 4.0L • XJS 6.0L**
Year:	**1993-94**
Engine identification:	**AJ6, V12**
System:	**Lucas 15/26/36 CU**

Data link connector (DLC) locations

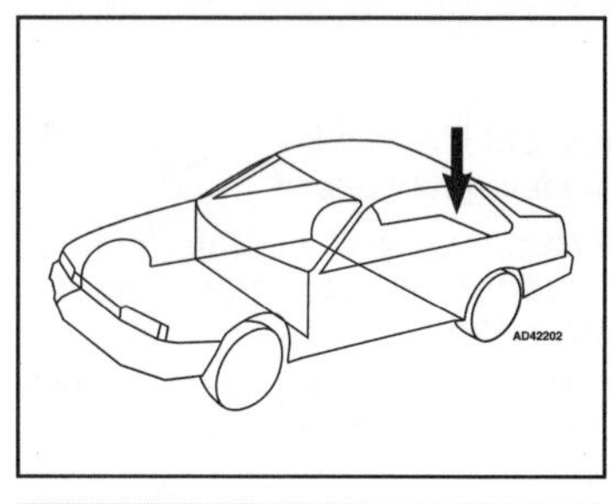

XJS

Self-diagnosis

General information

- Refer to the front of this manual for general test conditions, terminology, detailed descriptions of wiring faults and a general trouble shooter for electrical and mechanical faults.
- Engine control module (ECM) incorporates self-diagnosis function.
- Malfunction indicator lamp (MIL) will illuminate if certain faults are recorded.
- ECM operates in backup mode if sensors fail, to enable vehicle to be driven to workshop.

Accessing

- Switch ignition ON.
- 'CHECK ENGINE' will be displayed if fault has been recorded **1**.
- Switch ignition OFF for 5 seconds minimum.
- Switch ignition ON. DO NOT start engine.
- Trouble code will be displayed on trip computer display **2** or **3**.
- Note trouble codes. Compare with trouble code table.
- If more than one trouble code is stored, trouble codes will be displayed with 2 second pause between codes.
- The ECM fault memory can also be checked using suitable diagnostic equipment connected to the data link connector (DLC).

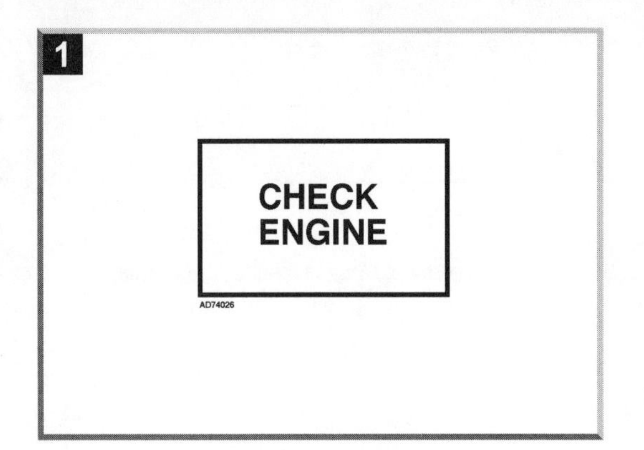

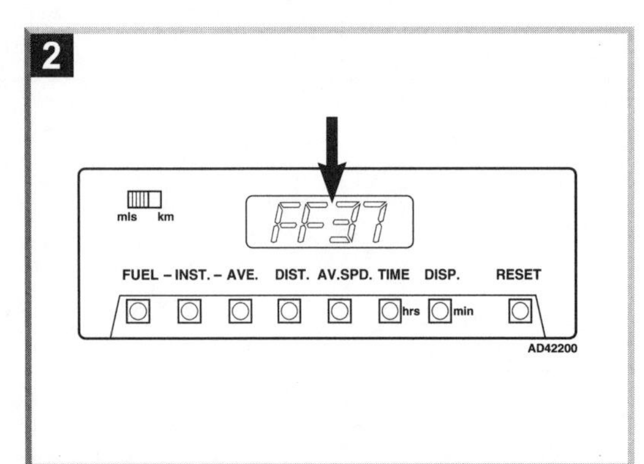

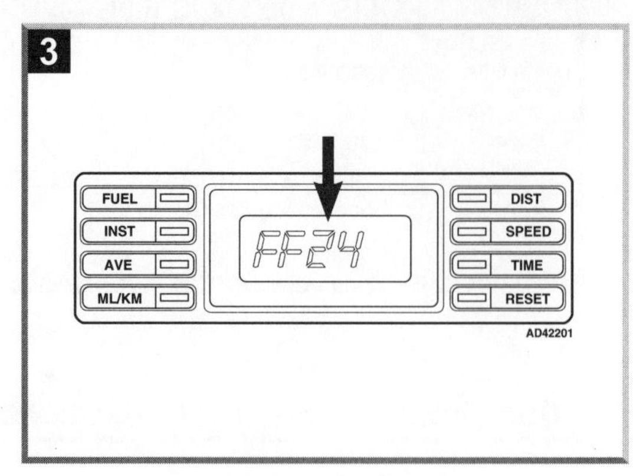

Erasing

- Ensure ignition switched OFF.
- Disconnect battery ground lead.

WARNING: ***Disconnecting the battery may erase memory from electronic units (e.g. radio).***

- Suitable diagnostic equipment can also be used to erase data from ECM fault memory.

Trouble code identification

Trouble code	Fault location	Probable cause
FF11	Throttle position (TP) sensor	Wiring, TP sensor, ECM
FF12	Mass air flow (MAF) sensor	Wiring, MAF sensor, ECM
FF13	Manifold absolute pressure (MAP) sensor	Wiring, MAP sensor, ECM
FF14	Engine coolant temperature (ECT) sensor	Wiring, ECT sensor, ECM
FF16	Intake air temperature (IAT) sensor	Wiring, IAT sensor, ECM
FF17	Throttle position (TP) sensor	Wiring, TP sensor, ECM
FF18	Throttle position (TP) sensor/mass air flow (MAF) sensor/manifold absolute pressure (MAP) sensor	Wiring, TP sensor/MAF sensor/MAP sensor, ECM
FF19	Throttle position (TP) sensor/mass air flow (MAF) sensor/manifold absolute pressure (MAP) sensor	Wiring, TP sensor/MAF sensor/MAP sensor, ECM

Trouble code	Fault location	Probable cause
FF22	Fuel pump – operation	Wiring, fuel pump relay, fuel pump, ECM
FF23	Fuel supply	Wiring, fuel pump relay, fuel pump, fuel line blocked, fuel filter, ECM
FF24	Ignition amplifier – signal	Wiring, ignition amplifier, ECM
FF26	Air leak	Intake manifold
FF29	Engine control module (ECM) – self-check	ECM
FF33	Injector – ECM signal	Wiring, injector, ECM
FF34	Injector/injectors – bank A	Wiring, injector(s), ECM
FF36	Injectors – bank B	Wiring, injectors, ECM
FF37	Exhaust gas recirculation (EGR) solenoid – ECM signal	Wiring, EGR solenoid, ECM
FF39	Exhaust gas recirculation (EGR) valve position sensor	Wiring, EGR valve position sensor, ECM
FF44	Heated oxygen sensor (HO2S)/wiring – bank A	Wiring, HO2S, ECM
FF45	Heated oxygen sensor (HO2S)/ – bank B	Wiring, HO2S, ECM
FF46	Idle air control (IAC) valve – coil 1	Wiring, IAC valve, ECM
FF47	Idle air control (IAC) valve – coil 2	Wiring, IAC valve, ECM
FF48	Idle air control (IAC) valve	Wiring, IAC valve, ECM
FF49	Ballast resistor	Ballast resistor
FF66	Secondary air injection (AIR) system – control	Wiring, AIR pump relay, AIR pump, ECM
FF68	Vehicle speed sensor (VSS)/wiring	Wiring, VSS
FF69	Park/neutral position (PNP) switch/wiring	Wiring, PNP switch
FF77	Engine speed (RPM) signal	Wiring, RPM sensor, ECM
FF89	Evaporative emission (EVAP) canister purge valve – ECM signal	Wiring, EVAP canister purge valve, ECM

JAGUAR

Model:	**XJ6 4.0L • XJ12 6.0L**
Year:	**1991-94**
Engine identification:	**AJ6, V12**
System:	**Lucas 15/36 CU**

Data link connector (DLC) locations

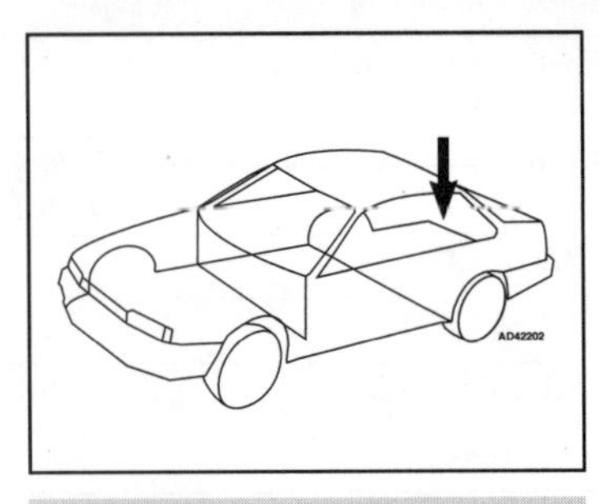

XJ6/XJ12

Self-diagnosis

General information

- Refer to the front of this manual for general test conditions, terminology, detailed descriptions of wiring faults and a general trouble shooter for electrical and mechanical faults.
- Engine control module (ECM) incorporates self-diagnosis function.
- Malfunction indicator lamp (MIL) will illuminate if certain faults are recorded.
- ECM operates in backup mode if sensors fail, to enable vehicle to be driven to workshop.

Accessing

- Switch ignition ON. DO NOT start engine.
- 'CHECK ENGINE' will be displayed if fault has been recorded **1**.
- Switch ignition OFF for 5 seconds minimum.
- Switch ignition ON. DO NOT start engine.
- Depress VCM button **2** [**1**].
- Trouble code will be displayed on VCM display **2** [**2**].
- Note trouble codes. Compare with trouble code table.
- If more than one trouble code is stored, trouble codes will be displayed with 2 second pause between codes.
- VCM display can be cleared by depressing CLEAR button **1** [**3**].
- The ECM fault memory can also be checked using suitable diagnostic equipment connected to the data link connector (DLC).

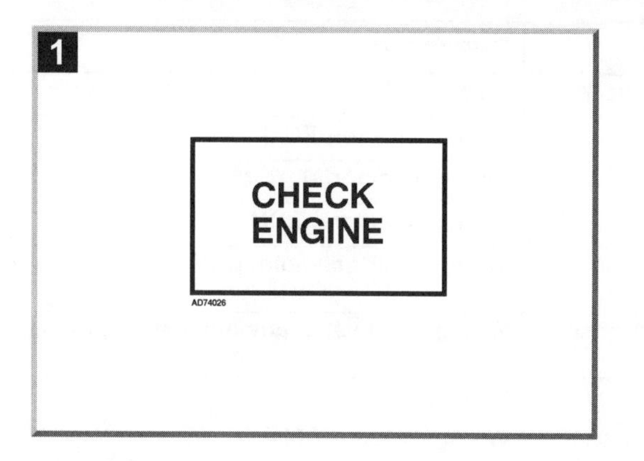

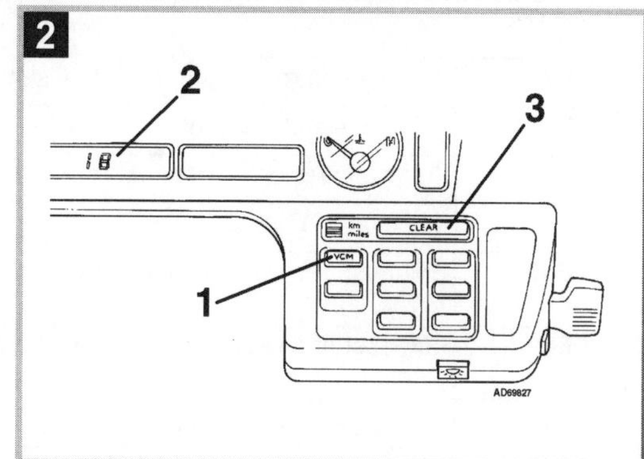

Erasing

- Ensure ignition switched OFF.
- Disconnect battery ground lead.

WARNING: ***Disconnecting the battery may erase memory from electronic units (e.g. radio).***

- Suitable diagnostic equipment can also be used to erase data from ECM fault memory.

Trouble code identification

Trouble code	Fault location	Probable cause
FF11	Throttle position (TP) sensor	**Wiring, TP sensor, ECM**
FF12	Mass air flow (MAF) sensor	**Wiring, MAF sensor, ECM**
FF13	Manifold absolute pressure (MAP) sensor	**Wiring, MAP sensor, ECM**
FF14	Engine coolant temperature (ECT) sensor	**Wiring, ECT sensor, ECM**
FF16	Intake air temperature (IAT) sensor	**Wiring, IAT sensor, ECM**
FF17	Throttle position (TP) sensor	**Wiring, TP sensor, ECM**
FF18	Throttle position (TP) sensor/mass air flow (MAF) sensor/manifold absolute pressure (MAP) sensor	**Wiring, TP sensor/MAF sensor/MAP sensor, ECM**
FF19	Throttle position (TP) sensor/mass air flow (MAF) sensor/manifold absolute pressure (MAP) sensor	**Wiring, TP sensor/MAF sensor/MAP sensor, ECM**
FF22	Fuel pump – operation	**Wiring, fuel pump relay, fuel pump, ECM**
FF23	Fuel supply	**Wiring, fuel pump relay, fuel pump, fuel line blocked, fuel filter, ECM**
FF24	Ignition amplifier – signal	**Wiring, ignition amplifier, ECM**
FF26	Air leak	**Intake manifold**

Trouble code	Fault location	Probable cause
FF29	Engine control module (ECM) – self-check	ECM
FF33	Injector – ECM signal	Wiring, injector, ECM
FF34	Injector/injectors – bank A	Wiring, injector(s), ECM
FF36	Injectors – bank B	Wiring, injectors, ECM
FF37	Exhaust gas recirculation (EGR) solenoid – ECM signal	Wiring, EGR solenoid, ECM
FF39	Exhaust gas recirculation (EGR) valve position sensor	Wiring, EGR valve position sensor, ECM
FF44	Heated oxygen sensor (HO2S)/wiring – bank A	Wiring, HO2S, ECM
FF45	Heated oxygen sensor (HO2S)/wiring – bank B	Wiring, HO2S, ECM
FF46	Idle air control (IAC) valve – coil 1	Wiring, IAC valve, ECM
FF47	Idle air control (IAC) valve – coil 2	Wiring, IAC valve, ECM
FF48	Idle air control (IAC) valve	Wiring, IAC valve, ECM
FF49	Ballast resistor	Ballast resistor
FF66	Secondary air injection (AIR) system – control	Wiring, AIR pump relay, AIR pump, ECM
FF68	Vehicle speed sensor (VSS)/wiring	Wiring, VSS
FF69	Park/neutral position (PNP) switch/wiring	Wiring, PNP switch
FF77	Engine speed (RPM) sensor – signal	Wiring, RPM sensor, ECM
FF89	Evaporative emission (EVAP) canister purge valve – ECM signal	Wiring, EVAP canister purge valve, ECM

JAGUAR

Model:	XJ6 4.0L • XJ12 6.0L • XJS 4.0L • XJS 6.0L
Year:	1995-97
Engine identification:	AJ16, V12
System:	Nippon Denso

Data link connector (DLC) locations

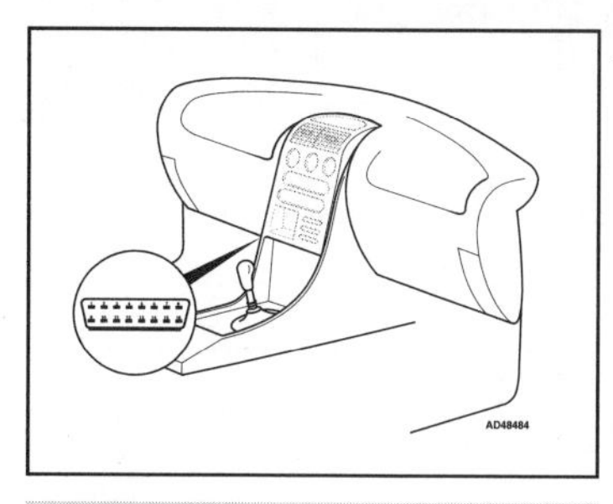

XJ6/XJ12

XJS

Self-diagnosis

General information

- Refer to the front of this manual for general test conditions, terminology, detailed descriptions of wiring faults and a general trouble shooter for electrical and mechanical faults.
- Engine control module (ECM) incorporates self-diagnosis function.
- Malfunction indicator lamp (MIL) will illuminate if certain faults are recorded.
- ECM operates in backup mode if sensors fail, to enable vehicle to be driven to workshop.

Accessing

- The ECM fault memory can only be checked using suitable diagnostic equipment connected to the data link connector (DLC) **1**.

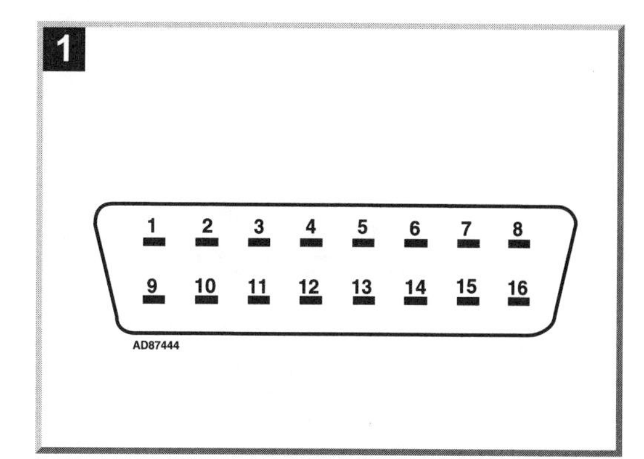

Erasing

- Suitable diagnostic equipment required to erase data from ECM fault memory.

Trouble code identification

OBD-II code	Fault location	Probable cause
All P0, P2 and U0 codes	Refer to OBD-II trouble code tables at the front of this manual	–
P1000	Engine control module (ECM) – internal error	**ECM**
P1111	Engine control module (ECM) – internal error	**ECM**
P1106	Manifold absolute pressure (MAP) sensor – range/performance	**Wiring, MAP sensor**
P1107	Manifold absolute pressure (MAP) sensor – circuit low input	**Wiring short to ground, MAP sensor, ECM**
P1108	Manifold absolute pressure (MAP) sensor – circuit high input	**Wiring short to positive, MAP sensor, ECM**
P1135	Crankshaft position (CKP) sensor	**Wiring, CKP sensor**
P1136	Crankshaft position (CKP) sensor	**Wiring, CKP sensor**
P1137	Heated oxygen sensor (HO2S) 2 – cylinders 1,2,3 – indicates lean	**Wiring, HO2S**
P1138	Heated oxygen sensor (HO2S) 2 – cylinders 1,2,3 – indicates rich	**Wiring, HO2S**
P1157	Heated oxygen sensor (HO2S) 2 – cylinders 4,5,6 – indicates lean	**Wiring, HO2S**
P1158	Heated oxygen sensor (HO2S) 2 – cylinders 4,5,6 – indicates rich	**Wiring, HO2S**
P1171	Weak mixture – cylinders 1,2,3,4,5,6	**Wiring, fuel level, fuel pump, fuel pump relay, injector(s)**
P1172	Rich mixture – cylinders 1,2,3,4,5,6	**Blocked fuel line, injector(s), other trouble code(s) stored**
P1176	Weak mixture – long term fuel trim	**Wiring, fuel pump, blocked fuel filter, injector(s), other trouble code(s) stored**
P1177	Rich mixture – long term fuel trim	**Fuel pressure, blocked fuel line, injector(s), other trouble code(s) stored**
P1178	Weak mixture – long term fuel trim	**Air leak, other trouble code(s) stored**
P1179	Rich mixture – long term fuel trim	**Exhaust leak, other trouble code(s) stored**
P1185	Oxygen sensor heater 1 – cylinders 1,2,3,4,5,6	**Wiring, HO2S**
P1186	Oxygen sensor heater 1 – cylinders 1,2,3,4,5,6	**Wiring, HO2S**
P1187	Oxygen sensor heater 1 – cylinders 1,2,3,4,5,6 – circuit malfunction	**Wiring, HO2S**
P1188	Oxygen sensor heater 1 – cylinders 1,2,3,4,5,6 – resistance	**Wiring, HO2S**
P1189	Oxygen sensor heater 1 – cylinders 1,2,3,4,5,6 – low resistance	**Wiring, HO2S**

OBD-II code	Fault location	Probable cause
P1190	Oxygen sensor heater 1 – cylinders 1,2,3,4,5,6 – low resistance	Wiring, HO2S
P1191	Oxygen sensor heater 2 – cylinders 1,2,3,4,5,6	Wiring, HO2S
P1192	Oxygen sensor heater 2 – cylinders 1,2,3,4,5,6	Wiring, HO2S
P1193	Oxygen sensor heater 2 – cylinders 1,2,3,4,5,6 – circuit malfunction	Wiring, HO2S
P1194	Oxygen sensor heater 2 – cylinders 1,2,3,4,5,6 – resistance	Wiring, HO2S
P1195	Oxygen sensor heater 2 – cylinders 1,2,3,4,5,6 – low resistance	Wiring, HO2S
P1196	Oxygen sensor heater 2 – cylinders 1,2,3,4,5,6 – low resistance	Wiring, HO2S
P1198	Fuel level sensor	Wiring, fuel level sensor
P1199	Fuel level sensor	Wiring, fuel level sensor
P1201	Injector – cylinder 1 – short circuit/open circuit	Wiring, injector
P1202	Injector – cylinder 2 – short circuit/open circuit	Wiring, injector
P1203	Injector – cylinder 3 – short circuit/open circuit	Wiring, injector
P1204	Injector – cylinder 4 – short circuit/open circuit	Wiring, injector
P1205	Injector – cylinder 5 – short circuit/open circuit	Wiring, injector
P1206	Injector – cylinder 6 – short circuit/open circuit	Wiring, injector
P1240	Sensor power supply – malfunction	Wiring, MAP sensor, TP sensor
P1241	Sensor power supply – low input	Wiring, MAP sensor, TP sensor
P1242	Sensor power supply – high input	Wiring, MAP sensor, TP sensor
P1244	Barometric pressure (BARO) sensor, incorporated in ECM – range/performance	Wiring, ECM
P1245	No cranking signal – low signal	Wiring, CKP sensor
P1246	No cranking signal – high signal	Wiring, CKP sensor
P1313	Misfire/catalyst damage – bank 1	Engine mechanical fault, wiring, ignition/fuel system, injector(s), MAP/TP/ECT/IAT sensor, EVAP system, ECM
P1314	Misfire/catalyst damage – bank 2	Engine mechanical fault, wiring, ignition/fuel system, injector(s), MAP/TP/ECT/IAT sensor, EVAP system, ECM
P1315	Persistent misfire	Other trouble code(s) stored (P0301-P0306)
P1316	Misfire	Wiring, other trouble code(s) stored, ignition/ fuel system, MAP/TP/ECT/IAT sensor, injector(s), ECM
P1335	Crankshaft position (CKP) sensor – circuit malfunction	Wiring, CKP sensor, ECM
P1336	Crankshaft position (CKP) sensor – range/ performance	Insecure sensor/rotor, air gap, wiring, CKP sensor

OBD-II code	Fault location	Probable cause
P1361	Ignition coil, cylinder 1 – no activation	**Wiring, ignition coil, engine control relay, ECM**
P1362	Ignition coil, cylinder 2 – no activation	**Wiring, ignition coil, engine control relay, ECM**
P1363	Ignition coil, cylinder 3 – no activation	**Wiring, ignition coil, engine control relay, ECM**
P1364	Ignition coil, cylinder 4 – no activation	**Wiring, ignition coil, engine control relay, ECM**
P1365	Ignition coil, cylinder 5 – no activation	**Wiring, ignition coil, engine control relay, ECM**
P1366	Ignition coil, cylinder 6 – no activation	**Wiring, ignition coil, engine control relay, ECM**
P1367	Ignition amplifier – ECM signal – bank 1	**Wiring, ignition coil relay, ignition amplifier, ECM**
P1368	Ignition amplifier – ECM signal – bank 2	**Wiring, ignition coil relay, ignition amplifier, ECM**
P1371	Ignition coil, cylinder 1 – early activation	**Wiring, ignition coil, engine control relay, ECM**
P1372	Ignition coil, cylinder 2 – early activation	**Wiring, ignition coil, engine control relay, ECM**
P1373	Ignition coil, cylinder 3 – early activation	**Wiring, ignition coil, engine control relay, ECM**
P1374	Ignition coil, cylinder 4 – early activation	**Wiring, ignition coil, engine control relay, ECM**
P1375	Ignition coil, cylinder 5 – early activation	**Wiring, ignition coil, engine control relay, ECM**
P1376	Ignition coil, cylinder 6 – early activation	**Wiring, ignition coil, engine control relay, ECM**
P1400	Exhaust gas recirculation (EGR) solenoid – position control	**Wiring, EGR valve position sensor, EGR solenoid**
P1401	Exhaust gas recirculation (EGR) solenoid – position control	**Wiring, EGR valve position sensor, EGR solenoid**
P1408	Exhaust gas recirculation (EGR) solenoid	**Wiring, EGR solenoid, EGR temperature sensor, EGR solenoid**
P1409	Exhaust gas recirculation (EGR) solenoid	**Wiring, EGR solenoid, EGR temperature sensor, EGR solenoid**
P1440	Evaporative emission (EVAP) canister purge valve – open	**Wiring, EVAP canister purge valve, ECM**
P1441	Evaporative emission (EVAP) canister purge valve	**Wiring, EVAP canister purge valve, hose connection(s)**
P1443	Evaporative emission (EVAP) canister purge valve – circuit malfunction	**Wiring, EVAP canister purge valve, ECM**
P1448	Evaporative emission (EVAP) canister purge valve	**Wiring, EVAP canister purge valve, hose connection(s)**
P1506	Idle air control (IAC) valve – RPM lower than expected	**Wiring, intake system blocked, MAP sensor, IAC valve, fuel pressure, ECM**

OBD-II code	Fault location	Probable cause
P1507	Idle air control (IAC) valve – RPM higher than expected	**Wiring, intake air leak, throttle valve sticking, MAP sensor, ECM**
P1508	Idle air control (IAC) valve – circuit malfunction	**Wiring open circuit, IAC valve, ECM**
P1509	Idle air control (IAC) valve – circuit malfunction	**Wiring short circuit, IAC valve, ECM**
P1512	Closed throttle position (CTP) switch – low input	**Wiring, CTP switch, ECM**
P1513	Closed throttle position (CTP) switch – high input	**Wiring, CTP switch, ECM**
P1514	Park/neutral position (PNP) switch – high load neutral/drive	**Wiring, PNP switch**
P1516	Park/neutral position (PNP) switch – gear chance neutral/drive	**Wiring, PNP switch**
P1517	Park/neutral position (PNP) switch – cranking neutral/drive	**Wiring, PNP switch**
P1607	Engine control module (ECM) – MIL circuit	**Wiring, engine malfunction indicator lamp (MIL), ECM**
P1608	Engine control module (ECM)	**ECM**
P1621	Engine control module (ECM)	**ECM**
P1622	Engine control module (ECM)	**ECM**
P1641	Fuel pump relay 1 – malfunction	**Wiring, fuel pump relay**
P1646	Fuel pump relay 2 – malfunction	**Wiring, fuel pump relay**
P1775	Engine control module (ECM) – TCM signal	**Wiring, ECM**
P1776	Engine control module (ECM) – TCM retard signal	**Wiring, ECM**
P1777	Engine control module (ECM) – TCM retard signal	**Wiring, ECM**

KIA

Model:	Engine identification:	Year:
Amanti 3.5L	G6CU	2004
Optima 2.4/2.5/2.7L	G4JS/G6BV	2001-04
Rio 1.5/1.6L	A5D/A6D	2001-02
Sedona 3.5L	VIN code digit 8 = 1	2002-04
Sephia 1.6/1.8L	B6 DOHC/BP DOHC	1995-97
Sephia 1.8L	T8 DOHC	1998-01
Spectra 1.8L	T8 DOHC	2001-04
Spectra 2.0L	G4GC	2004
Sportage 2.0L	FE DOHC	1996-02
Sorento 3.5L	VIN code digit 8 = 3	2003-04

System: **Mazda EGI**

Self-diagnosis

General information

- Refer to the front of this manual for general test conditions, terminology, detailed descriptions of wiring faults and a general trouble shooter for electrical and mechanical faults.
- Engine control module (ECM) incorporates self-diagnosis function.
- Malfunction indicator lamp (MIL) will illuminate if certain faults are recorded.
- ECM operates in backup mode if sensors fail, to enable vehicle to be driven to workshop.

Accessing

- Trouble codes can be displayed by using a Scan Tool connected to the data link connector (DLC) **1**.

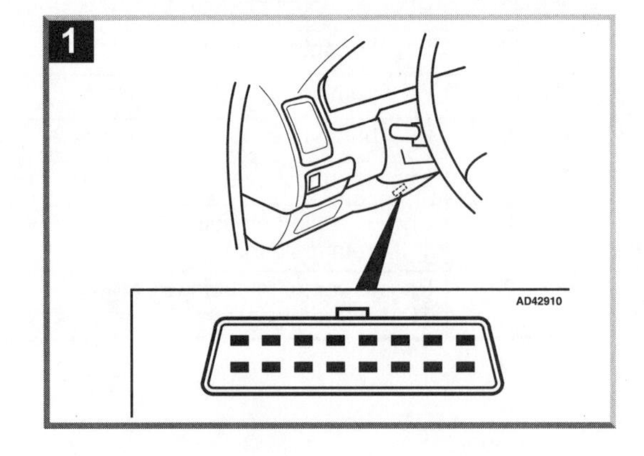

Erasing – except 2004 Spectra

- After the faults have been rectified, trouble codes can be erased by using a Scan Tool connected to the data link connector (DLC) or as follows:
- Ensure ignition switched OFF.
- Disconnect battery ground cable for 20 seconds minimum.
- Reconnect battery ground cable.
- Repeat checking procedure to ensure no data remains in ECM fault memory.

NOTE: ***Disconnecting battery ground lead may also erase memory from electronic units such as radio.***

Erasing – 2004 Spectra

- Trouble codes can only be erased by using a Scan Tool connected to the data link connector (DLC).

Trouble code identification

OBD-II code	Fault location	Probable cause
All P0, P2 and U0 codes	Refer to OBD-II trouble code tables at the front of this manual	–
P1000	System diagnosis incomplete	**Drive cycle incomplete, ECM**
P1100	Manifold absolute pressure (MAP) sensor – for EGR system – circuit malfunction	
P1102	Manifold absolute pressure (MAP) sensor – for EGR system – mode 3	**Wiring, MAP sensor, EGR valve, hoses leaking/blocked, ECM**
P1103	Manifold absolute pressure (MAP) sensor – for EGR system – mode 2	**Wiring, MAP sensor, EGR valve, hoses leaking/blocked, ECM**
P1110	Electronic throttle system (ETS) – malfunction	**Wiring, throttle valve tight/sticking, ETS**
P1115	AT – ECT sensor signal	**Wiring, TCM, ECM**
P1118	Electronic throttle system (ETS), throttle motor – malfunction	**Wiring, throttle motor**
P1121	AT – TP sensor signal	**Wiring, TP sensor, TCM**
P1134	Heated oxygen sensor (HO2S) – front, bank 1 – signal malfunction	**HO2S, ECM**
P1151	Accelerator pedal position (APP) sensor – circuit malfunction	**Wiring, APP sensor, ECM**
P1152	Accelerator pedal position (APP) sensor – low input	**Wiring short to ground, APP sensor, ECM**
P1153	Accelerator pedal position (APP) sensor – high input	**Wiring short to positive, APP sensor, ECM**
P1154	Heated oxygen sensor (HO2S) – front, bank 2 – signal malfunction	**HO2S, ECM**
P1155	Limp-home solenoid – malfunction	**Wiring, limp-home solenoid, ECM**
P1159	Intake manifold air control actuator – circuit malfunction	**Wiring, intake manifold air control actuator, ECM**
P1166	Heated oxygen sensor (HO2S) – front, bank 1 – control limit	**Fuel/intake/ignition system, wiring, HO2S, ECM**
P1167	Heated oxygen sensor (HO2S) – front, bank 2 – control limit	**Fuel/intake/ignition system, wiring, HO2S, ECM**
P1168	Heated oxygen sensor (HO2S), heater – rear, bank 1 – circuit malfunction	**Wiring, HO2S, ECM**
P1169	Heated oxygen sensor (HO2S), heater – rear, bank 2 – circuit malfunction	**Wiring, HO2S, ECM**
P1170	Heated oxygen sensor (HO2S) – front – no signal change	**Fuel/intake/ignition system, wiring, HO2S, ECM**

OBD-II code	Fault location	Probable cause
P1171	Electronic throttle system (ETS), WOT – malfunction	**Wiring, throttle valve tight/sticking, throttle motor, ETS control module**
P1172	Electronic throttle system (ETS), throttle motor – circuit malfunction	**Wiring, throttle motor**
P1173	Electronic throttle system (ETS) – throttle valve command/actual throttle valve position, correlation	**Wiring, throttle valve tight/sticking, ETS**
P1174	Electronic throttle system (ETS), closed throttle – range/performance problem	**Wiring, throttle valve tight/sticking**
P1175	Electronic throttle system (ETS), closed throttle – range/performance problem	**Wiring, throttle valve tight/sticking**
P1176	Electronic throttle system (ETS), throttle motor – circuit 1 malfunction	**Wiring, throttle motor**
P1177	Electronic throttle system (ETS), throttle motor – circuit 2 malfunction	**Wiring, throttle motor**
P1178	Electronic throttle system (ETS) – supply voltage	**Wiring, ETS relay, engine control relay, alternator, battery**
P1192	Limp-home solenoid – signal malfunction	**Wiring, limp-home solenoid, ECM**
P1193	Limp-home solenoid – rpm lower than expected	**Wiring, mechanical fault, limp-home solenoid, throttle motor, throttle valve tight/ sticking, ECM**
P1194	Limp-home solenoid – signal malfunction	**Wiring, TP sensor, limp-home solenoid, ECM**
P1195 1	Manifold absolute pressure (MAP) sensor – for EGR system – circuit malfunction	**Wiring, MAP sensor, ECM**
P1195 2	Limp-home solenoid – signal malfunction	**Wiring, TP sensor, limp-home solenoid, ECM**
P1196	Limp-home solenoid – malfunction	**Wiring, limp-home solenoid, ECM**
P1250	Fuel pressure regulator control solenoid – circuit malfunction	**Wiring, fuel pressure control solenoid, ECM**
P1307	G-sensor – signal malfunction	**Wiring, G-sensor, ECM**
P1308	G-sensor – signal low	**Wiring open/short to ground, G-sensor, ECM**
P1309	G-sensor – signal high	**Wiring open/short to positive, G-sensor, ECM**
P1330	Ignition adjustment connector – short to ground	**Wiring short to ground, ECM**
P1372	Crankshaft position (CKP) sensor – signal malfunction	**Wiring, CKP sensor, mechanical fault, ECM**
P1386	Knock sensor (KS)	**Wiring, KS, ECM**
P1402 3	Exhaust gas recirculation (EGR) valve position sensor – circuit malfunction	**Wiring, EGR valve position sensor, ECM**
P1402 4	Evaporative emission (EVAP) leak detection module pump – circuit malfunction	**Wiring, EVAP leak detection pump**
P1403	Evaporative emission (EVAP) leak detection module solenoid – circuit malfunction	**Wiring, EVAP leak detection module solenoid**
P1404	Evaporative emission (EVAP) leak detection module heater – circuit malfunction	**Wiring, EVAP leak detection heater**

OBD-II code	Fault location	Probable cause
P1446	Evaporative emission (EVAP) leak detection module solenoid – circuit malfunction	**Wiring, EVAP leak detection module solenoid**
P1447	Evaporative emission (EVAP) leak detection module pump – circuit malfunction	**Wiring, EVAP leak detection pump**
P1448	Evaporative emission (EVAP) leak detection module heater – circuit malfunction	**Wiring, EVAP leak detection heater**
P1449	Evaporative emission (EVAP) canister purge valve – circuit malfunction	**Wiring, EVAP valve, ECM**
P1450	Evaporative emission (EVAP) canister purge system – vacuum leak detected	**Hose leak, wiring, EVAP valve, ECM**
P1455	Fuel tank level sensor – circuit malfunction	**Wiring, instrument cluster, fuel tank level sensor, ECM**
P1457	Evaporative emission (EVAP) canister purge valve (low)	**Wiring, EVAP valve, ECM**
P1458	A/C compressor control – signal malfunction	**Wiring, A/C switch, ECM**
P1485	Exhaust gas recirculation (EGR) solenoid (vacuum) – circuit malfunction	**Wiring, EGR solenoid, ECM**
P1486	Exhaust gas recirculation (EGR) solenoid (vent) – circuit malfunction	**Wiring, EGR solenoid, ECM**
P1487	Manifold absolute pressure (MAP) sensor solenoid – EGR system – circuit malfunction	**Wiring, MAP sensor, ECM**
P1496	Exhaust gas recirculation (EGR) valve – circuit 1 malfunction	**Wiring, EGR valve, ECM**
P1497	Exhaust gas recirculation (EGR) valve – circuit 2 malfunction	**Wiring, EGR valve, ECM**
P1498	Exhaust gas recirculation (EGR) valve – circuit 3 malfunction	**Wiring, EGR valve, ECM**
P1499	Exhaust gas recirculation (EGR) valve – circuit 4 malfunction	**Wiring, EGR valve, ECM**
P1500	Vehicle speed signal – malfunction	**Wiring, instrument cluster, TCM**
P1503	Cruise control master switch – malfunction	**Wiring, cruise control master switch, ETS, ECM**
P1504	Cruise control master switch – signal malfunction	**Wiring, cruise control master switch, ETS, ECM**
P1505	Idle air control (IAC) valve – opening coil signal low	**Wiring open/short to ground, IAC valve, ECM**
P1506	Idle air control (IAC) valve – opening coil signal high	**Wiring short to positive, IAC valve, ECM**
P1507	Idle air control (IAC) valve – closing coil signal low	**Wiring open/short to ground, IAC valve, ECM**
P1508	Idle air control (IAC) valve – closing coil signal high	**Wiring short to positive, IAC valve, ECM**
P1510	Idle air control (IAC) valve – coil 1 – circuit malfunction	**Wiring, IAC valve, ECM**

OBD-II code	Fault location	Probable cause
P1511	Idle air control (IAC) valve – coil 2 – circuit malfunction	**Wiring, IAC valve, ECM**
P1521	Power steering pressure (PSP) switch – circuit malfunction	**Wiring, PSP switch, ECM**
P1523	Intake manifold air control solenoid	**Wiring, air control solenoid, ECM**
P1529	MIL request signal – TCM to ECM	**Wiring, ECM, TCM**
P1586	AT coding signal	**Wiring**
P1602	Transmission control module (TCM) – communication malfunction	**Wiring, ECM, TCM**
P1607	Engine control module (ECM)/electronic throttle system (ETS) control module – communication malfunction	**Wiring, ECM, ETS control module**
P1608 1	Engine control module (ECM) – malfunction	**Wiring, ECM**
P1608 2	Electronic throttle system (ETS) control module/ engine control module (ECM) – communication malfunction	**Wiring, ETS control module, ECM**
P1609	Immobilizer – malfunction	**Wiring, poor connection, multifunction control module, ECM**
P1611	MIL request signal – circuit/voltage low	**Wiring open/short circuit, ECM**
P1613 1	Engine control module (ECM) – self test failed	**ECM**
P1613 2	Electronic throttle system (ETS) control module – malfunction	**Wiring, ECM, ETS control module**
P1614 1	MIL request signal – circuit/voltage high	**Wiring short to positive, ECM**
P1614 2	Electronic throttle system (ETS) control module – EEPROM rewrite error	**Wiring, ETS control module**
P1615	Electronic throttle system (ETS) control module – malfunction	**Wiring, ETS control module, ECM**
P1616	Engine control relay – circuit malfunction	**Wiring, engine control relay, ECM**
P1623	Malfunction indicator lamp (MIL) – circuit malfunction	**Wiring, MIL**
P1624 5	MIL request signal – TCM to ECM	**TCM trouble codes stored**
P1624 6	Engine coolant blower motor relay low speed – circuit malfunction	**Wiring, relay, ECM**
P1625	Engine coolant blower motor relay high speed – circuit malfunction	**Wiring, relay, ECM**
P1631	Generator voltage – no charge	**Wiring, generator, ECM**
P1632 1	Battery voltage monitor – circuit malfunction	**Wiring, ECM**
P1632 2	CAN data bus – OFF	**CAN data bus**
P1633	Battery voltage – overcharging	**Wiring, generator, ECM**
P1634	Generator terminal B – open circuit	**Wiring, generator, ECM**
P1640	Engine control relay – voltage low/high	**Wiring, engine control relay, charging system**

OBD-II code	Fault location	Probable cause
P1642	Engine control module – non immobilizer/ incorrect type	**ECM**
P1673	Coolant blower motor – circuit malfunction	**Wiring, blower motor relay, blower motor, ECM**
P1690	Immobilizer control module – malfunction	**Wiring, engine control relay, immobilizer control module**
P1691	Immobilizer read coil	**Wiring, immobilizer read coil, immobilizer control module**
P1693 [7]	MIL – circuit malfunction	**Wiring, MIL, ECM, TCM**
P1693 [8]	Immobilizer key not programmed – ECM	**Wiring, ignition key not matched/damaged, immobilizer read coil, immobilizer control module, ECM**
P1694	CAN data bus, Immobilizer control module – signal malfunction	**Wiring, immobilizer control module, ECM**
P1695	Engine control module (ECM) – EEPROM error	**Wiring, immobilizer control module, ECM**
P1696	Immobilizer control module – invalid data received	**Wiring, immobilizer control module**
P1700	Overdrive lamp – circuit malfunction	**Wiring short to ground, overdrive lamp**
P1707	Cruise/brake switch – circuit malfunction	**Wiring, cruise/brake switch, ECM**
P1765	Transmission control module (TCM) – torque reduction malfunction	**Wiring, TCM, ECM**
P1780	Transmission control module (TCM) – ignition advance circuit malfunction	**Wiring open circuit/short circuit to ground/ positive**
P1794	Supply voltage	**Wiring, fuse, battery**
P1795	4WD low gear position switch – circuit malfunction	**Wiring, low gear position switch, TCM**
P1797	Clutch pedal position (CPP) switch – circuit malfunction	**Wiring, CPP switch, ECM**
P1797	Park/neutral position (PNP) switch – circuit malfunction	**Wiring, PNP switch, ECM**
P1800	Engine torque signal – malfunction	**Wiring, TCM**

[1] Except Amanti
[2] Amanti
[3] Except Spectra 2002-03
[4] Spectra 2002-03
[5] Except Optima
[6] Optima
[7] Except 2004 Spectra
[8] 2004 Spectra

KIA

Model:	Engine identification:	Year:
Sephia 1.6L	**B6 SOHC/DOHC**	**1993-96**
Sportage 2.0L	**FE**	**1995-96**

System: **Mazda EGI**

Self-diagnosis

General information

- Refer to the front of this manual for general test conditions, terminology, detailed descriptions of wiring faults and a general trouble shooter for electrical and mechanical faults.
- Engine control module (ECM) incorporates self-diagnosis function.
- Malfunction indicator lamp (MIL) will illuminate if certain faults are recorded.
- ECM operates in backup mode if sensors fail, to enable car to be driven to workshop.
- Trouble codes can be displayed with the malfunction indicator lamp (MIL) using the data link connector (DLC):
  - Sephia – center of engine firewall.
  - Sportage – near air cleaner housing.

Accessing

- Ensure ignition switched OFF.
- Jump data link connector DLC terminals ENG. TEST and GND **1**.
- Switch ignition ON.
- Sephia – start engine. Allow to idle until at normal operating temperature.
- Sephia – run engine at approximately 2000 rpm for 3 minutes.
- Malfunction indicator lamp (MIL) should flash **2**.
  - Long flashes indicate the 'tens' of the trouble code **3** [**A**].
  - Short flashes indicate the 'units' of the trouble code **3** [**C**].
  - A short pause separates each flash **3** [**B**].
  - A long pause separates each trouble code **3** [**D**].
  - For example: Trouble code 12 displayed **3**.
- Count MIL flashes. Note trouble codes. Compare with trouble code table.
- Switch ignition OFF.
- Remove jump lead.

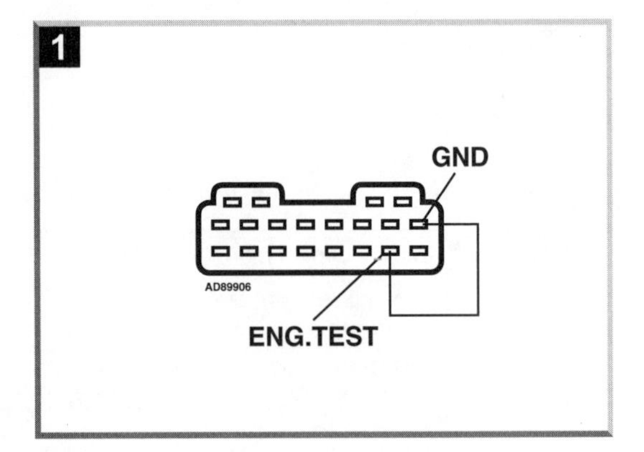

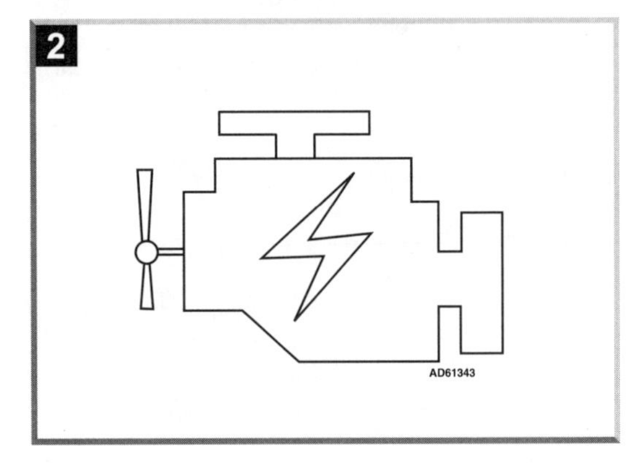

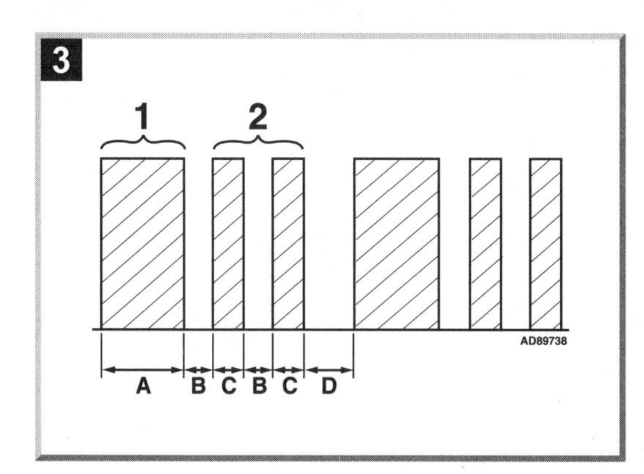

Erasing

- Ensure ignition switched OFF.
- Disconnect battery ground cable for 20 seconds minimum.
- Repeat checking procedure to ensure no trouble codes remain in ECM memory.

NOTE: ***Disconnecting battery ground lead may also erase memory from electronic units such as radio.***

Trouble code identification

Flash code	Fault location	Probable cause
02	Crankshaft position (CKP)/engine speed (RPM) sensor	Wiring, CKP/RPM sensor, ECM
03	Crankshaft position (CKP) sensor/engine speed (RPM) sensor – B6 SOHC	Wiring, CKP/RPM sensor, ECM
03	Camshaft position (CMP) sensor – B6 DOHC	Wiring, CMP sensor, ECM
03	Camshaft position (CMP) sensor – FE	Wiring, CMP sensor, ECM
04	Crankshaft position (CKP) sensor	Wiring, CKP sensor, ECM
07	Crankshaft position (CKP) sensor – reluctor malfunction	Air gap, reluctor
08	Volume air flow (VAF) sensor – B6 SOHC	Wiring, VAF sensor, ECM
08	Mass air flow (MAF) sensor – B6 DOHC	Wiring, MAF sensor, ECM
09	Engine coolant temperature (ECT) sensor	Wiring, ECT sensor, ECM
10	Intake air temperature (IAT) sensor	Wiring, IAT sensor, ECM
12	Throttle position (TP) sensor	Wiring, TP sensor, ECM
14	Barometric pressure (BARO) sensor	ECM
15	Oxygen sensor (O2S)/heated oxygen sensor (HO2S) – no signal	Wiring, O2S/HO2S, ECM
16	Exhaust gas recirculation (EGR) valve position sensor	Wiring, EGR valve position sensor, ECM
17	Oxygen sensor (O2S)/heated oxygen sensor (HO2S) – incorrect signal	Fuel/intake system, wiring, O2S/HO2S, injector, ECM
18	Injector No.1 – circuit malfunction	Wiring, injector, ECM
19	Injector No.2 – circuit malfunction	Wiring, injector, ECM
20	Injector No.3 – circuit malfunction	Wiring, injector, ECM

Flash code	Fault location	Probable cause
21	Injector No.4 – circuit malfunction	**Wiring, injector, ECM**
24	Fuel pump relay – circuit malfunction	**Wiring, fuel pump relay, ECM**
25	Fuel pressure regulator control solenoid 1 or single	**Wiring, fuel pressure regulator control solenoid, ECM**
26	Evaporative emission (EVAP) canister purge valve	**Wiring, EVAP valve, ECM**
28	Exhaust gas recirculation (EGR) solenoid (vacuum)	**Wiring, EGR solenoid, ECM**
29	Exhaust gas recirculation (EGR) solenoid (vent)	**Wiring, EGR solenoid, ECM**
34	Idle air control (IAC) valve	**Wiring, IAC valve, ECM**
35	Fuel pressure regulator control solenoid 2 – B6 DOHC	**Wiring, fuel pressure regulator control solenoid, ECM**
35	Injector malfunction – FE	**Wiring, injector**
36	Mixture control – malfunction	**Fuel/intake system, wiring, HO2S, injector, ECM**
37	Mixture control – malfunction	**Fuel/intake system, wiring, HO2S, injector, ECM**
41	Intake manifold air control solenoid	**Wiring, air control solenoid, ECM**
46	A/C compressor relay – circuit malfunction	**Wiring, A/C compressor relay, ECM**
48	ECM power stage – group 1	**Wiring, injector, EVAP valve, EGR solenoid, power stage in ECM**
49	ECM power stage – group 2	**Wiring, ISC actuator, power stage in ECM**
56	Idle speed control (ISC) actuator – circuit malfunction	**Wiring, ISC actuator, ECM**
57	A/C signal – circuit malfunction	**Wiring, A/C pressure switch, A/C thermo switch, A/C relay, ECM**
73	Vehicle speed sensor (VSS) – circuit malfunction	**Wiring, VSS, ECM**
87	Malfunction indicator lamp (MIL) – circuit malfunction	**Wiring, MIL, ECM**
88	Engine control module (ECM) – internal memory failure	**Wiring, ECM**
99	Supply voltage	**Wiring, battery, ECM**

LAND ROVER

Model:	**Freelander 2.5L • Discovery 4.0L • Range Rover 4.0L Range Rover 4.6L**
Year:	**1998-02**
Engine identification:	**36D, 42D, 46D, K18, KV6**
System:	**Bosch M5.2.1/MEMS3**

Data link connector (DLC) locations

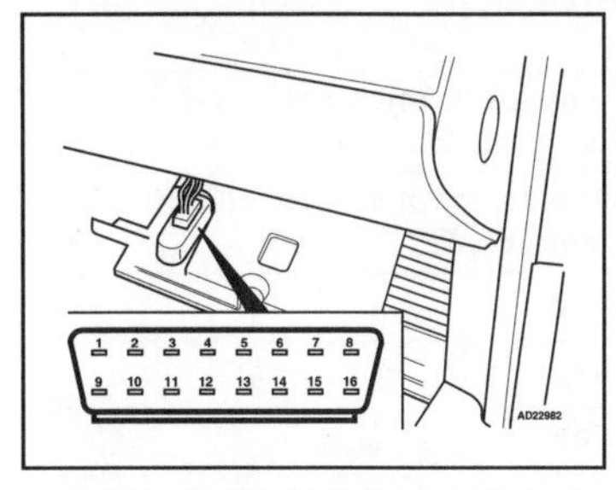

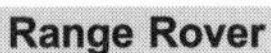

Range Rover

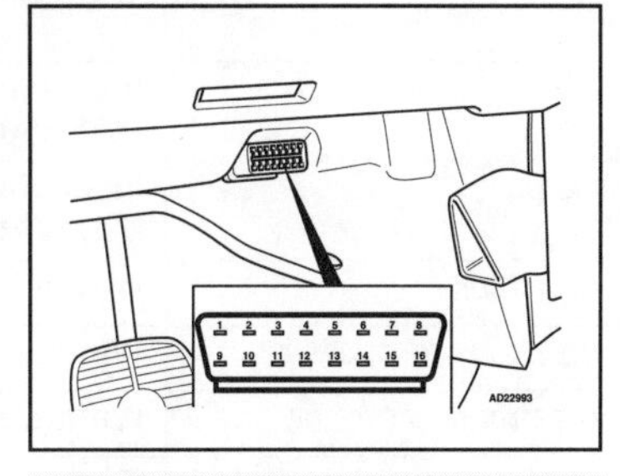

Discovery

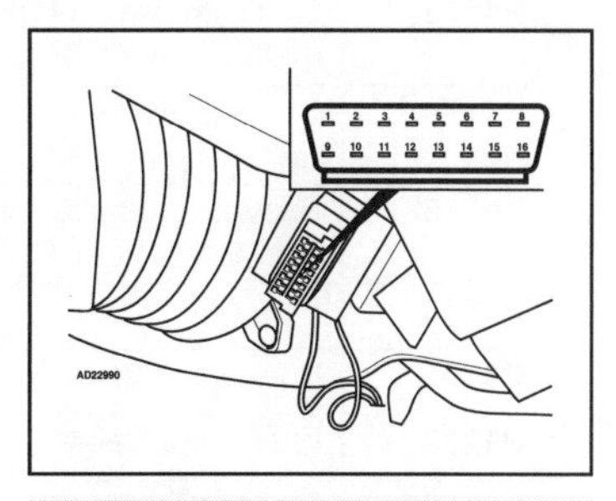

Freelander – under fascia

Self-diagnosis

General information

- Refer to the front of this manual for general test conditions, terminology, detailed descriptions of wiring faults and a general trouble shooter for electrical and mechanical faults.

Accessing and erasing

- The engine control module (ECM) fault memory can only be accessed and erased using diagnostic equipment connected to the data link connector (DLC).

Trouble code identification

OBD-II code	Fault location	Probable cause
All P0, P2 and U0 codes	Refer to OBD-II trouble code tables at the front of this manual	–
P1000	Engine control module (ECM) memory erased – no codes stored	**Memory erased, ECM**
P1117	Radiator outlet engine coolant temperature (ECT) – low input	**Wiring, ECT sensor, ECM**

OBD-II code	Fault location	Probable cause
P1118	Radiator outlet engine coolant temperature (ECT) – high input	**Wiring, ECT sensor, ECM**
P1129	Heated oxygen sensors (HO2S) 1 – sensors transposed	**Wiring**
P1170	Downstream fuel trim (FT), bank 1 – malfunction	**Wiring, HO2S**
P1171	System too lean, bank 1	**Intake/fuel system, injectors, HO2S, MAF/VAF sensor, ECT sensor**
P1172	System too rich, bank 1	**Intake/fuel system, injectors, HO2S, MAF/VAF sensor, ECT sensor**
P1173	Downstream fuel trim (FT), bank 1 – malfunction	**Wiring, HO2S**
P1174	System too lean, bank 2	**Intake/fuel system, injectors, HO2S, MAF/VAF sensor, ECT sensor**
P1175	System too rich, bank 2	**Intake/fuel system, injectors, HO2S, MAF/VAF sensor, ECT sensor**
P1230	Fuel pump relay – open circuit	**Wiring, relay**
P1231	Fuel pump relay – short circuit to positive	**Wiring, relay**
P1232	Fuel pump relay – short circuit to ground	**Wiring, relay**
P1300	Random/multiple cylinder(s) – misfire detected	**Engine mechanical fault, wiring, ignition/fuel system, injector, ECT/MAF sensor, ECM**
P1412	Secondary air injection (AIR) system, bank 1 – malfunction	**Wiring, control valve, air pump, ECM**
P1413	Secondary air injection (AIR) system, bank 1 – control valve open	**Wiring, control valve, ECM**
P1414	Secondary air injection (AIR) system, bank 1 – low flow	**Wiring, control valve, air pump, ECM**
P1415	Secondary air injection (AIR) system, bank 2 – malfunction	**Wiring, control valve, air pump, ECM**
P1416	Secondary air injection (AIR) system, bank 2 – control valve open	**Wiring, control valve, ECM**
P1417	Secondary air injection (AIR) system, bank 2 – low flow	**Wiring, control valve, air pump, ECM**
P1450	Evaporative emission (EVAP) pressure pump – circuit plausibility	**Wiring, pressure pump, ECM**
P1452	Evaporative emission (EVAP) pressure pump – low current	**Wiring, pressure pump, ECM**
P1453	Evaporative emission (EVAP) pressure pump – high current	**Wiring, pressure pump, ECM**
P1451	Leak detection pump	**Wiring, pump, ECM**
P1481	Leak detection pump – heater circuit	**Wiring open circuit, pump heater**
P1482	Leak detection pump – heater circuit	**Wiring short circuit to ground, pump heater**
P1483	Leak detection pump – heater circuit	**Wiring short circuit to positive, pump heater**

OBD-II code	Fault location	Probable cause
P1510	Idle air control (IAC) valve – opening circuit	**IAC valve, wiring short circuit to positive**
P1513	Idle air control (IAC) valve – opening circuit	**IAC valve, wiring, short circuit to ground**
P1514	Idle air control (IAC) valve – opening circuit	**IAC valve, wiring open circuit**
P1551	Idle air control (IAC) valve – closing circuit	**Wiring, IAC valve, wiring open circuit**
P1552	Idle air control (IAC) valve – closing circuit	**Wiring, IAC valve, wiring short circuit to ground**
P1553	Idle air control (IAC) valve – closing circuit	**Wiring, IAC valve, wiring short circuit to positive**
P1590	Antilock brake system (ABS) rough road signal – error message from ABS control unit	**Wiring, ABS system**
P1591	Antilock brake system (ABS) rough road signal – short circuit to ground, open circuit	**Wiring, ABS system**
P1592	Antilock brake system (ABS) rough road signal – short circuit to positive	**Wiring, ABS system**
P1669	Engine control module (ECM) cooling fan – circuit malfunction	**Wiring open circuit, cooling fan motor**
P1670	Engine control module (ECM) cooling fan – circuit low	**Wiring, short circuit to ground, cooling fan motor**
P1671	Engine control module (ECM) cooling fan – circuit high	**Wiring short circuit positive**
P1700	Transfer box – low range signal implausible	**Wiring, transmission range (TR) switch**
P1701	Transfer box – fault signal	**Wiring, transfer box**
P1702	Transfer box signal line – communication error	**Wiring, ECM**
P1703	Transfer box link – signal line	**Wiring, open circuit, short circuit to positive**
P1708	Transfer box link – signal line	**Short circuit to ground**
P1776	Transmission control system – torque interface malfunction	**Wiring, transmission control module (TCM), ECM**

LEXUS

Model:	Engine identification:	Year:
ES300 3.0L	3VZ-FE	1992-93

System: **TCCS**

Self-diagnosis

General information

- Refer to the front of this manual for general test conditions, terminology, detailed descriptions of wiring faults and a general trouble shooter for electrical and mechanical faults.
- Engine control module (ECM) incorporates self diagnosis function.
- Malfunction indicator lamp (MIL) will illuminate if certain faults are recorded.
- ECM operates in backup mode if sensors fail, to enable vehicle to be driven to workshop.
- ECM trouble codes can be accessed using data link connector (DLC) 1 **1**, or data link connector (DLC) 2 **2**.

Preparatory conditions

- Battery voltage 11 V minimum.
- Throttle valve fully closed.
- Transmission in neutral position.
- All accessories including A/C switched OFF.

Accessing

- Switch ignition ON. Do NOT start engine.
- Check MIL illuminates.
- Jump DLC terminals TE1 and E1 **1** or **2**.
- MIL should be flashing.
  - First group of short flashes indicates the 'tens' of the trouble code **3** [**A**].
  - Second group of short flashes indicates the 'units' of the trouble code **3** [**D**].
  - A 1.5 second pause separates 'tens' and 'units' **3** [**C**].
  - A 2.5 second pause separates each trouble code **3** [**E**].
  - For example: Trouble code 32 displayed **3**.
- If no trouble codes are stored, MIL lamp will flash 2 times per second **4**.
- Count MIL flashes and compare with trouble code table.
- Remove jump lead.
- Switch ignition OFF.

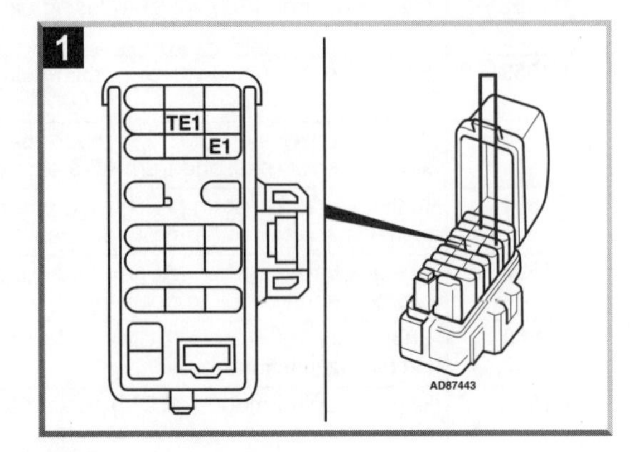

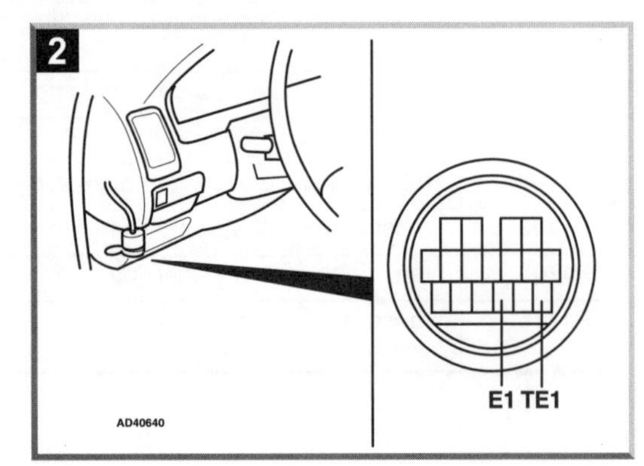

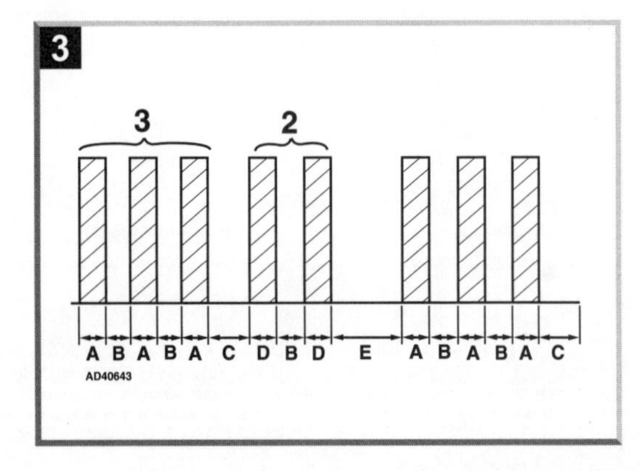

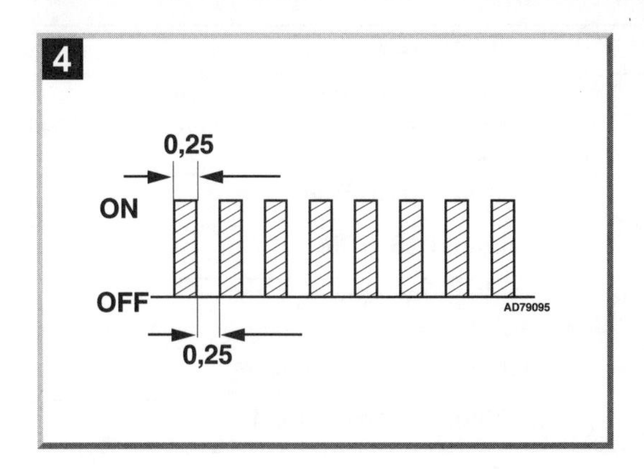

Erasing

- Ensure ignition switched OFF.
- Remove EFi 15A fuse **5**, from underhood fusebox.
- Wait 10 seconds minimum.
- Reinstall fuse.
- Repeat checking procedure to ensure no data remains in ECM fault memory.

NOTE: ***Disconnecting battery ground lead will also erase trouble codes but may erase memory from electronic units such as radio.***

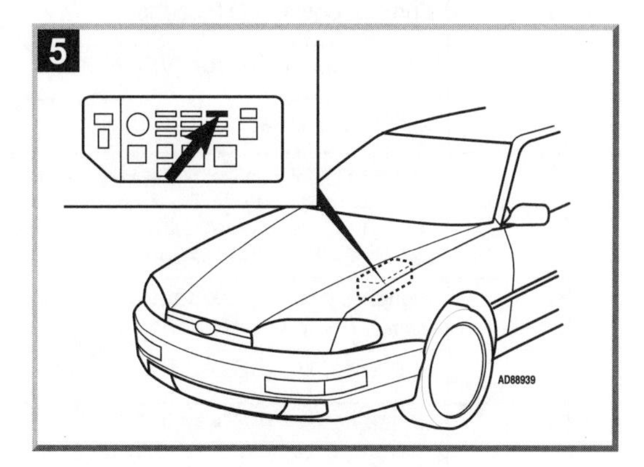

Trouble code identification

Flash code	Fault location	Probable cause
12	Crankshaft position (CKP) sensor	Wiring, CKP sensor, ECM
12	Engine speed (RPM) sensor	Wiring, RPM sensor, ECM
13	Engine speed (RPM) sensor – intermittent signal	Wiring, RPM sensor, ECM
14	Ignition module/Ignition primary confirmation signal	Wiring, ignition module, ECM
16	Automatic transmission signal	ECM
21	Oxygen sensor (O2S) – front LH	Wiring, O2S, ECM
22	Engine coolant temperature (ECT) sensor	Wiring, ECT sensor, ECM
24	Intake air temperature (IAT) sensor	Wiring, IAT sensor, ECM
25	Mixture control – lean	Intake/fuel/ignition system, wiring, front LH/RH O2S, VAF sensor, valve timing, mechanical fault, ECM

Flash code	Fault location	Probable cause
26	Mixture control – rich	**Intake/fuel/ignition system, wiring, VAF sensor, valve timing, mechanical fault, ECM**
27	Heated oxygen sensor (HO2S) – rear	**Wiring, HO2S, ECM**
28	Oxygen sensor (O2S) – front RH	**Wiring, O2S, ECM**
31	Volume air flow (VAF) sensor – incorrect signal – idling	**Wiring, VAF sensor, ECM**
32	Volume air flow (VAF) sensor – incorrect signal	**Wiring, VAF sensor, ECM**
41	Throttle position (TP) sensor	**Wiring, TP sensor, ECM**
42	Vehicle speed sensor (VSS) 1	**Wiring, VSS, ECM**
43	Cranking signal	**Wiring, ignition switch, starter relay, ECM**
51	Switch signal – A/C switch ON during diagnosis	**Wiring, A/C switch, A/C amplifier, ECM**
51	Switch signal – closed throttle position (CTP) switch OFF during diagnosis	**Wiring, CTP switch, ECM**
51	Switch signal – park/neutral position (PNP) switch not in P or N during diagnosis	**Wiring, PNP switch, ECM**
52	Knock sensor (KS) – LH	**Wiring, KS, ECM**
53	Knock sensor (KS) control	**ECM**
55	Knock sensor (KS) – RH	**Wiring, KS, ECM**
71	Exhaust gas recirculation (EGR) system	**Hose/pipe leak or blockage, wiring, EGR valve, EGRT sensor, ECM**

Model:	Engine identification:	Year:
ES300 3.0L	1MZ-FE	1994-03
ES330 3.3L	3MZ-FE	2004
GS400 4.0L	1UZ-FE	1998-00
GS430 4.3L	3UZ-FE	2001-04
LS400 4.0L	1UZ-FE	1995-00
LS430 4.3L	3UZ-FE	2001-04
SC400 4.0L	1UZ-FE	1996-00
SC430 4.3L	3UZ-FE	2002-04

System: **SFI**

Self-diagnosis

General information

- Refer to the front of this manual for general test conditions, terminology, detailed descriptions of wiring faults and a general trouble shooter for electrical and mechanical faults.
- Engine control module (ECM) incorporates self-diagnosis function.
- Malfunction indicator lamp (MIL) will illuminate if certain faults are recorded.
- ECM operates in backup mode if sensors fail, to enable vehicle to be driven to workshop.

Accessing

- Trouble codes can be displayed by using a Scan Tool connected to the data link connector (DLC) **1**.

1

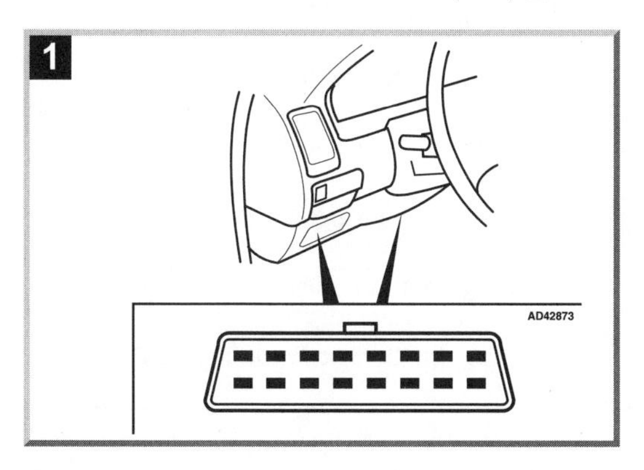

Erasing – except 2004 LS430/SC430

- After the faults have been rectified, trouble codes can be erased by using a Scan Tool connected to the data link connector (DLC) or as follows:
- Ensure ignition switched OFF.
- Disconnect battery ground cable for 60 seconds minimum.
- Reconnect battery ground cable.

NOTE: ***Disconnecting battery ground lead will erase trouble codes but may also erase memory from electronic units such as radio.***

Erasing – 2004 LS430/SC430

- Trouble codes can only be erased by using a Scan Tool connected to the data link connector (DLC).

Trouble code identification

OBD-II code	Fault location	Probable cause
All P0, P2 and U0 codes	Refer to OBD-II trouble code tables at the front of this manual	–
P1100	Barometric pressure (BARO) sensor – circuit malfunction	**ECM**
P1120	Accelerator pedal position (APP) sensor – circuit malfunction	**Wiring, APP sensor, ECM**
P1121	Accelerator pedal position (APP) sensor – range/performance problem	**APP sensor, ECM**
P1125	Throttle actuator control (TAC) motor – circuit malfunction	**Wiring, TAC motor, ECM**
P1126	Throttle actuator control (TAC) clutch – circuit malfunction	**Wiring, magnetic clutch, ECM**
P1127	Throttle actuator control (TAC) power source – circuit malfunction	**Wiring, ECM**
P1128	Throttle actuator control (TAC) lock – malfunction	**TAC motor, throttle body, ECM**
P1129	Throttle actuator control (TAC) system – malfunction	**TAC system, ECM**
P1130	Air fuel (A/F) sensor – LH front – range/performance problem	**Wiring, air fuel (A/F) sensor, ECM**
P1133	Air fuel (A/F) sensor – LH front – response malfunction	**Air fuel (A/F) sensor**
P1135	Air fuel (A/F) sensor – LH front – heater circuit malfunction	**Wiring, air fuel (A/F) sensor, ECM**
P1150	Air fuel (A/F) sensor – RH front – range/performance problem	**Wiring, air fuel (A/F) sensor, ECM**
P1153	Air fuel (A/F) sensor – RH front – response malfunction	**Air fuel (A/F) sensor**
P1155	Air fuel (A/F) sensor – RH front – heater circuit malfunction	**Wiring, air fuel (A/F) sensor, ECM**
P1200	Fuel pump (FP) relay/ECM – circuit malfunction	**Wiring, fuel pump relay, ECM**
P1300	Ignition control – ES300 →98/cylinder Nos.1-6 – circuit malfunction	**Wiring, ignition module, ECM**
P1300	Ignition control – cylinder No.1 – circuit malfunction	**Wiring, ignition coil/module, ECM**
P1305	Ignition control – cylinder No.2 – circuit malfunction	**Wiring, ignition coil/module, ECM**
P1310	Ignition control – cylinder No.3 – circuit malfunction	**Wiring, ignition coil/module, ECM**
P1315	Ignition control – cylinder No.4 – circuit malfunction	**Wiring, ignition coil/module, ECM**
P1320	Ignition control – cylinder No.5 – circuit malfunction	**Wiring, ignition coil/module, ECM**

OBD-II code	Fault location	Probable cause
P1325	Ignition control – cylinder No.6 – circuit malfunction	**Wiring, ignition coil/module, ECM**
P1330	Ignition control – cylinder No.7 – circuit malfunction	**Wiring, ignition coil/module, ECM**
P1335	Crankshaft position (CKP) sensor – no signal	**Wiring, CKP sensor, starter motor, ECM**
P1340	Camshaft position (CMP) sensor 2, bank 1 – GS430/LS430/SC430 03→ – circuit malfunction	**Wiring, CMP, mechanical fault, ECM**
P1340	Ignition control – cylinder No.8 – circuit malfunction	**Wiring, ignition coil/module, ECM**
P1341	Camshaft position (CMP) sensor 2, bank 1 – range/performance problem	**Wiring, CMP, mechanical fault, ECM**
P1345	Variable valve timing sensor – LH bank – circuit malfunction	**Wiring, valve timing sensor, ECM**
P1346	Variable valve timing sensor – LH bank – range/performance problem	**Valve timing, ECM**
P1349	Variable valve timing system – LH bank – malfunction	**Valve timing, oil control valve, valve timing control module, ECM**
P1350	Variable valve timing sensor – RH bank – circuit malfunction	**Wiring, valve timing sensor, ECM**
P1351	Variable valve timing sensor – RH bank – range/performance problem	**Valve timing, ECM**
P1354	Variable valve timing system – RH bank – malfunction	**Valve timing, oil control valve, valve timing control module, ECM**
P1400	Throttle position (TP) sensor 2 – circuit malfunction	**Wiring, TP sensor, ECM**
P1401	Throttle position (TP) sensor 2 – range/performance problem	**TP sensor**
P1410	Exhaust gas recirculation (EGR) valve position sensor – malfunction	**Wiring, EGR valve position sensor, ECM**
P1411	Exhaust gas recirculation (EGR) valve position sensor – range/performance problem	**EGR valve position sensor**
P1500	Starter signal – circuit malfunction	**Wiring, ignition switch, engine control relay, ECM**
P1520	Stop lamp switch – signal malfunction	**Wiring, stop lamp switch, ECM**
P1565	Cruise control switch – circuit malfunction	**Wiring, cruise control switch, ECM**
P1566	Cruise control switch – circuit malfunction	**Wiring, cruise control switch, ECM**
P1600	Engine control module (ECM) – supply voltage	**Wiring, ECM**
P1605	Knock control – malfunction	**ECM**
P1633	Throttle actuator control (TAC) system – malfunction	**ECM**
P1645	Body control module – malfunction	**Body control module, A/C control module, wiring**

OBD-II code	Fault location	Probable cause
P1656	Oil control valve – LH bank – malfunction	**Wiring, oil control valve, ECM**
P1663	Oil control valve – RH bank – malfunction	**Wiring, oil control valve, ECM**
P1705	AT – direct clutch speed sensor	**Wiring, direct clutch speed sensor, ECM**
P1725	Input shaft speed (ISS) sensor/turbine shaft speed (TSS) sensor – circuit malfunction	**Wiring, ISS sensor/TSS sensor, ECM**
P1730	Intermediate shaft speed sensor – AT	**Wiring, intermediate shaft speed sensor, ECM**
P1760	AT – pressure control shift solenoid	**Wiring, shift solenoid, ECM**
P1765	AT – pressure control shift solenoid	**Wiring, shift solenoid, ECM**
P1780	Park/neutral position (PNP) switch – circuit malfunction	**Wiring, PNP switch, ECM**
P1815	Shift control solenoid – malfunction	**Wiring, shift control solenoid, ECM**
P1818	Shift control solenoid – circuit malfunction	**Wiring, shift control solenoid, ECM**
B2785	Ignition switch ON – malfunction	**Ignition switch, engine control relay, wiring**
B2786	Ignition switch OFF – malfunction	**Ignition switch, engine control relay, wiring**
B2791	Key unlock warning switch – malfunction	**Wiring, key unlock warning switch, engine control relay**
B2795	Unrecognized key code	**Key, incorrect key**
B2796	Immobilizer system – no communication	**Key, transponder, wiring, ECM**
B2797	Communication malfunction 1 – interference	**Incorrect key has been used, wiring, ignition module, ECM**
B2798	Communication malfunction 2	**Key, transponder, ignition module, wiring, ECM**
B2799	Immobilizer – malfunction	**Wiring, immobilizer system, ECM**

LEXUS

Model:	Engine identification:	Year:
GS300 3.0L	2JZ-GE	1993-95
SC300 3.0L	2JZ-GE	1992-95

System: **TCCS**

Self-diagnosis

General information

- Refer to the front of this manual for general test conditions, terminology, detailed descriptions of wiring faults and a general trouble shooter for electrical and mechanical faults.
- Engine control module (ECM) incorporates self-diagnosis function.
- Malfunction indicator lamp (MIL) will illuminate if certain faults are recorded.
- ECM operates in backup mode if sensors fail, to enable vehicle to be driven to workshop.
- ECM trouble codes can be accessed using data link connector (DLC) 1 **1**, or data link connector (DLC) 2 **2**.

Preparatory conditions

- Battery voltage 11 V minimum.
- Throttle valve fully closed.
- Transmission in neutral position.
- All accessories including A/C switched OFF.

Accessing

- Switch ignition ON. Do NOT start engine.
- Check MIL illuminates.
- Jump DLC terminals TE1 and E1 **1** or **2**.
- MIL should be flashing.
  - First group of short flashes indicates the 'tens' of the trouble code **3** [**A**].
  - Second group of short flashes indicates the 'units' of the trouble code **3** [**D**].
  - A 1.5 second pause separates 'tens' and 'units' **3** [**C**].
  - A 2.5 second pause separates each trouble code **3** [**E**].
  - For example: Trouble code 32 displayed **3**.
- If no trouble codes are stored, MIL lamp will flash 2 times per second **4**.
- Count MIL flashes and compare with trouble code table.
- Remove jump lead.
- Switch ignition OFF.

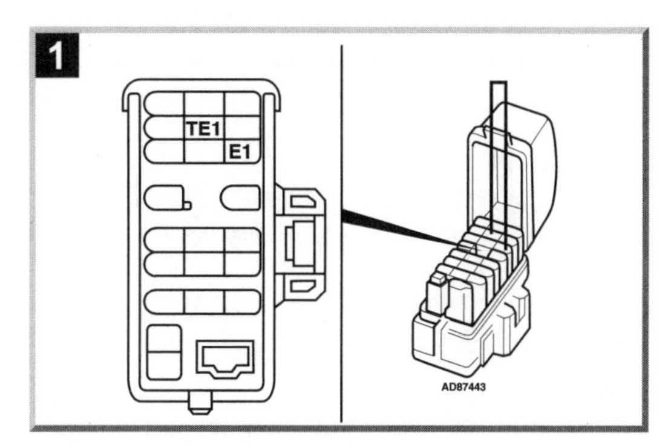

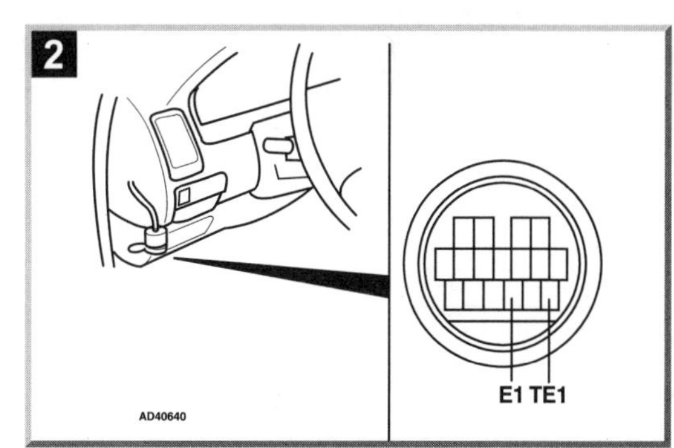

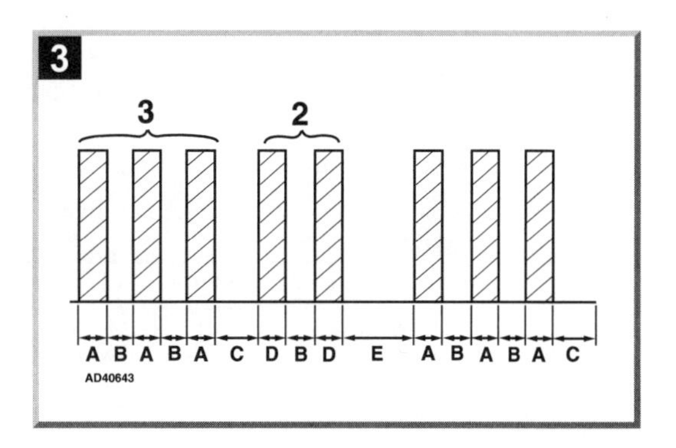

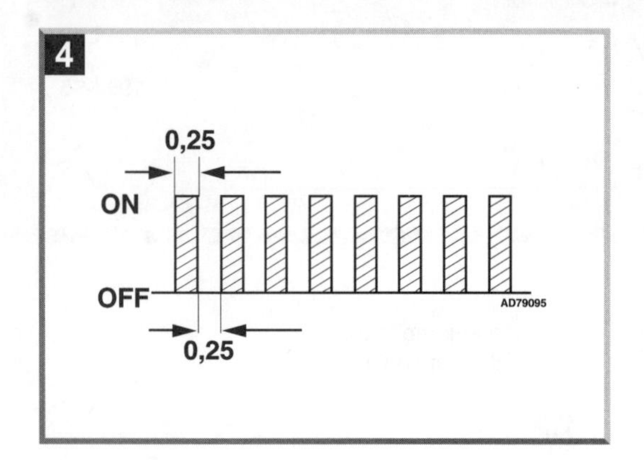

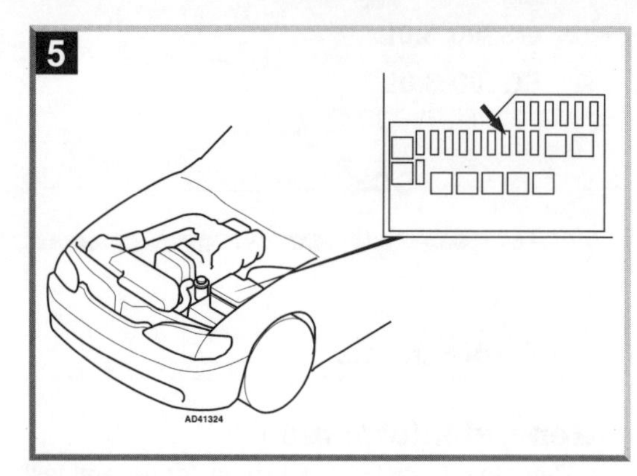

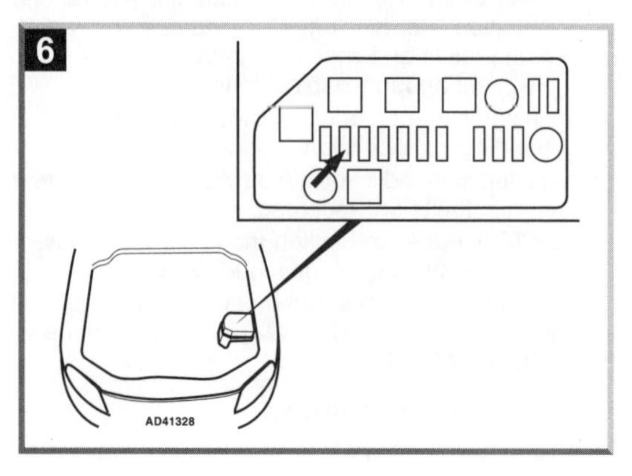

Erasing

- Ensure ignition switched OFF.
- GS300 – remove EFi 20A fuse from underhood fusebox **5**.
- SC300 – remove EFi 30A fuse from underhood fusebox **6**.
- Wait 10 seconds minimum.
- Reinstall fuse.
- Repeat checking procedure to ensure no data remains in ECM fault memory.

NOTE: ***Disconnecting battery ground lead will also erase trouble codes but may erase memory from electronic units such as radio.***

Trouble code identification

Flash code	Fault location	Probable cause
12	Crankshaft position (CKP) sensor	**Wiring, CKP sensor, ECM**
12	Engine speed (RPM) sensor	**Wiring, RPM sensor, ECM**
13	Engine speed (RPM) sensor – intermittent signal	**Wiring, RPM sensor, ECM**
14	Ignition module/Ignition primary confirmation signal	**Wiring, ignition module, ECM**
16	Automatic transmission signal	**ECM**
21	Oxygen sensor (O2S)/heated oxygen sensor (HO2S) – front 1	**Wiring, O2S/HO2S, ECM**
22	Engine coolant temperature (ECT) sensor	**Wiring, ECT sensor, ECM**
24	Intake air temperature (IAT) sensor	**Wiring, IAT sensor, ECM**

Flash code	Fault location	Probable cause
25	Mixture control – lean	**Intake/fuel/ignition system, wiring, front O2S/HO2S, VAF sensor, injector, valve timing, mechanical fault, ECM**
26	Mixture control – rich	**Intake/fuel/ignition system, wiring, VAF sensor, injector, valve timing, mechanical fault, ECM**
27	Heated Oxygen sensor (HO2S) – rear	**Wiring, HO2S, ECM**
28	Oxygen sensor (O2S)/heated oxygen sensor (HO2S) – front 2	**Wiring, O2S/HO2S, ECM**
31	Volume air flow (VAF) sensor	**Wiring, VAF sensor, ECM**
35	Barometric pressure (BARO) sensor	**ECM**
41	Throttle position (TP) sensor 1	**Wiring, TP sensor, ECM**
42	Vehicle speed sensor (VSS) 1	**Wiring, VSS, ECM**
43	Cranking signal	**Wiring, ignition switch, starter relay, ECM**
47	Throttle position (TP) sensor 2 – traction control system	**Wiring, TP sensor, ECM**
51	Switch signal – A/C switch ON during diagnosis	**Wiring, A/C switch, A/C amplifier, ECM**
51	Switch signal – closed throttle position (CTP) switch OFF during diagnosis	**Wiring, CTP switch, ECM**
51	Switch signal – park/neutral position (PNP) switch not in P or N during diagnosis	**Wiring, PNP switch, ECM**
51	Switch signal – transmission range (TR) switch not in P or N during diagnosis	**Wiring, TR switch, ECM**
52	Knock sensor (KS) 1	**Wiring, KS, ECM**
53	Knock sensor (KS) control	**ECM**
55	Knock sensor (KS) 2	**Wiring, KS, ECM**
61	Vehicle speed sensor (VSS) 2 – AT	**Wiring, VSS, ECM**
62	AT shift solenoid 1	**Wiring, shift solenoid, ECM**
63	AT shift solenoid 2	**Wiring, shift solenoid, ECM**
64	AT shift solenoid 3	**Wiring, shift solenoid, ECM**
71	Exhaust gas recirculation (EGR) system	**Hose/pipe leak or blockage, wiring, EGR solenoid, EGR valve, EGRT sensor, ECM**
78	Fuel pump control module (FPCM)	**Wiring, FPCM, fuel pump, ECM**

LEXUS

Model:	Engine identification:	Year:
GS300 3.0L	2JZ-GE	1996-04
IS300 3.0L	2JZ-GE	2001-04
SC300 3.0L	2JZ-GE	1996-00

System: **SFI**

Self-diagnosis

General information

- Refer to the front of this manual for general test conditions, terminology, detailed descriptions of wiring faults and a general trouble shooter for electrical and mechanical faults.
- Engine control module (ECM) incorporates self-diagnosis function.
- Malfunction indicator lamp (MIL) will illuminate if certain faults are recorded.
- ECM operates in backup mode if sensors fail, to enable vehicle to be driven to workshop.

Accessing

- Trouble codes can be displayed by using a Scan Tool connected to the data link connector (DLC) **1**.

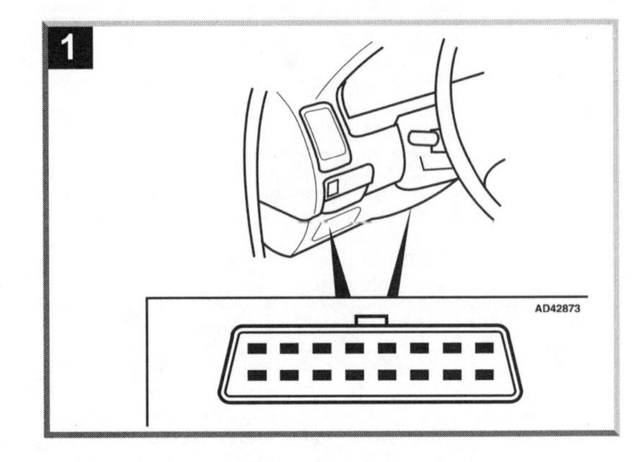

Erasing

- After the faults have been rectified, trouble codes can be erased by using a Scan Tool connected to the data link connector (DLC) or as follows:
- Ensure ignition switched OFF.
- Disconnect battery ground cable for 60 seconds minimum.
- Reconnect battery ground cable.

NOTE: ***Disconnecting battery ground lead will erase trouble codes but may also erase memory from electronic units such as radio.***

Trouble code identification

OBD-II code	Fault location	Probable cause
All P0, P2 and U0 codes	Refer to OBD-II trouble code tables at the front of this manual	–
P1120	Accelerator pedal position (APP) sensor – circuit malfunction	Wiring, APP sensor, ECM
P1121	Accelerator pedal position (APP) sensor – range/performance problem	APP sensor, ECM

OBD-II code	Fault location	Probable cause
P1125	Throttle actuator control (TAC) motor – circuit malfunction	Wiring, TAC motor, ECM
P1126	Throttle actuator control (TAC) clutch – circuit malfunction	Wiring, magnetic clutch, ECM
P1127	Throttle actuator control (TAC) power source – circuit malfunction	Wiring, ECM
P1128	Throttle actuator control (TAC) lock – malfunction	TAC motor, throttle body, ECM
P1129	Throttle actuator control (TAC) system – malfunction	TAC system, ECM
P1200	Fuel pump (FP) relay/ECM – circuit malfunction	Wiring, fuel pump/control module, ECM
P1300	Ignition control – circuit malfunction	Wiring, ignition module, ECM
P1335	Crankshaft position (CKP) sensor – no signal	Wiring, CKP sensor, starter motor, ECM
P1346	Variable valve timing sensor – LH bank – range/performance problem	Valve timing, ECM
P1349	Variable valve timing system – circuit malfunction	Valve timing, oil control valve, valve timing control module, ECM
P1400	Throttle position (TP) sensor 2 – circuit malfunction	Wiring, TP sensor, ECM
P1401	Throttle position (TP) sensor 2 – range/performance problem	TP sensor
P1500	Starter signal – circuit malfunction	Wiring, ignition switch, starter relay, ECM
P1520	Stop lamp switch – signal malfunction	Wiring, stop lamp switch, ECM
P1565	Cruise control switch – circuit malfunction	Wiring, cruise control switch, ECM
P1600	Engine control module (ECM) – supply voltage	Wiring, ECM
P1605	Knock control – malfunction	ECM
P1633	Engine control module (ECM) – throttle actuator control (TAC) system – malfunction	ECM
P1645	Body control module – malfunction	Body control module, A/C control module, wiring
P1656	Oil control valve – malfunction	Wiring, oil control valve, ECM
P1780	Park/neutral position (PNP) switch	Wiring, PNP switch, ECM
B2785	Ignition switch ON – malfunction	Ignition switch, engine control relay, wiring
B2786	Ignition switch OFF – malfunction	Ignition switch, engine control relay, wiring
B2791	Key unlock warning switch – malfunction	Wiring, key unlock warning switch, engine control relay
B2795	Unrecognized key code	Key, incorrect key
B2796	Immobilizer system – no communication	Key, transponder, wiring, ECM
B2797	Communication malfunction 1 – interference	Incorrect key has been used, wiring, ignition module, ECM
B2798	Communication malfunction 2	Key, transponder, ignition module, wiring, ECM

LEXUS

Model:	Engine identification:	Year:
GX470 4.7L	2UZ-FE	2003-04
LX470 4.7L	2UZ-FE	1998-04
RX300 3.0L	1MZ-FE	1999-03
RX330 3.3L	3MZ-FE	2004

System: **SFI**

Self-diagnosis

General information

- Refer to the front of this manual for general test conditions, terminology, detailed descriptions of wiring faults and a general trouble shooter for electrical and mechanical faults.
- Engine control module (ECM) incorporates self-diagnosis function.
- Malfunction indicator lamp (MIL) will illuminate if certain faults are recorded.
- ECM operates in backup mode if sensors fail, to enable vehicle to be driven to workshop.

Accessing

- Trouble codes can be displayed by using a Scan Tool connected to the data link connector (DLC) **1**.

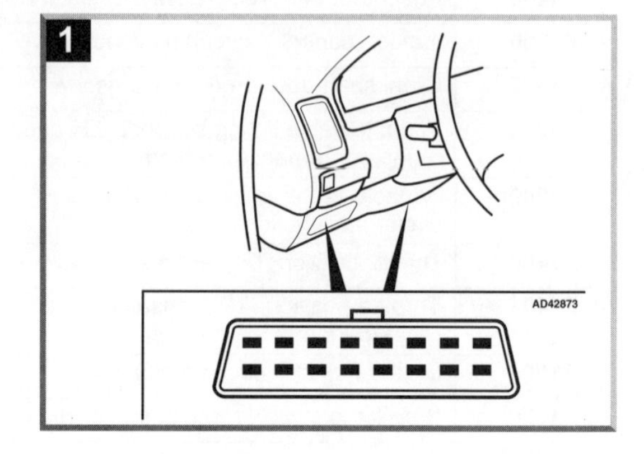

Erasing

- After the faults have been rectified, trouble codes can be erased by using a Scan Tool connected to the data link connector (DLC) or as follows:
- Ensure ignition switched OFF.
- Disconnect battery ground cable for 60 seconds minimum.
- Reconnect battery ground cable.

NOTE: ***Disconnecting battery ground lead will erase trouble codes but may also erase memory from electronic units such as radio.***

Trouble code identification

OBD-II code	Fault location	Probable cause
All P0, P2 and U0 codes	Refer to OBD-II trouble code tables at the front of this manual	–
P1100	Barometric pressure (BARO) sensor – circuit malfunction	**ECM**
P1120	Accelerator pedal position (APP) sensor – circuit malfunction	**Wiring, APP sensor, ECM**
P1121	Accelerator pedal position (APP) sensor – range/performance problem	**APP sensor, ECM**
P1125	Throttle actuator control (TAC) motor – circuit malfunction	**Wiring, TAC motor, ECM**
P1126	Throttle actuator control (TAC) clutch – circuit malfunction	**Wiring, magnetic clutch, ECM**
P1127	Throttle actuator control (TAC) power source – circuit malfunction	**Wiring, ECM**
P1128	Throttle actuator control (TAC) lock – malfunction	**TAC motor, throttle body, ECM**
P1129	Throttle actuator control (TAC) system – malfunction	**TAC system, ECM**
P1130	Air fuel (A/F) sensor – LH front – range/performance problem	**Wiring, air fuel (A/F) sensor, ECM**
P1133	Air fuel (A/F) sensor – LH front – response malfunction	**Air fuel (A/F) sensor**
P1135	Air fuel (A/F) sensor – LH front – heater circuit malfunction	**Wiring, air fuel (A/F) sensor, ECM**
P1150	Air fuel (A/F) sensor – RH front – range/ performance problem	**Wiring, air fuel (A/F) sensor, ECM**
P1153	Air fuel (A/F) sensor – RH front – response malfunction	**Air fuel (A/F) sensor**
P1155	Air fuel (A/F) sensor – RH front – heater circuit malfunction	**Wiring, air fuel (A/F) sensor, ECM**
P1200	Fuel pump (FP) relay/ECM – circuit malfunction	**Wiring, fuel pump relay, ECM**
P1300	Ignition control – cylinder No.1 – circuit malfunction	**Wiring, ignition coil/module, ECM**
P1305	Ignition control – cylinder No.2 – circuit malfunction	**Wiring, ignition coil/module, ECM**
P1310	Ignition control – cylinder No.3 – circuit malfunction	**Wiring, ignition coil/module, ECM**
P1315	Ignition control – cylinder No.4 – circuit malfunction	**Wiring, ignition coil/module, ECM**
P1320	Ignition control – cylinder No.5 – circuit malfunction	**Wiring, ignition coil/module, ECM**
P1325	Ignition control – cylinder No.6 – circuit malfunction	**Wiring, ignition coil/module, ECM**

OBD-II code	Fault location	Probable cause
P1330	Ignition control – cylinder No.7 – circuit malfunction	**Wiring, ignition coil/module, ECM**
P1335	Crankshaft position (CKP) sensor – no signal	**Wiring, CKP sensor, starter motor, ECM**
P1340	Ignition control – cylinder No.8 – circuit malfunction	**Wiring, ignition coil/module, ECM**
P1345	Variable valve timing sensor – LH bank – circuit malfunction	**Wiring, valve timing sensor, ECM**
P1346	Variable valve timing sensor – LH bank – range/performance problem	**Valve timing, ECM**
P1349	Variable valve timing system – LH bank – malfunction	**Valve timing, oil control valve, valve timing control module, ECM**
P1350	Variable valve timing sensor – RH bank – circuit malfunction	**Wiring, valve timing sensor, ECM**
P1351	Variable valve timing sensor – RH bank – range/performance problem	**Valve timing, ECM**
P1354	Variable valve timing system – RH bank – malfunction	**Valve timing, oil control valve, valve timing control module, ECM**
P1400	Throttle position (TP) sensor 2 – circuit malfunction	**Wiring, TP sensor, ECM**
P1401	Throttle position (TP) sensor 2 – range/performance problem	**TP sensor**
P1410	Exhaust gas recirculation (EGR) valve position sensor – malfunction	**Wiring, EGR valve position sensor, ECM**
P1411	Exhaust gas recirculation (EGR) valve position sensor – range/performance problem	**EGR valve position sensor**
P1500	Starter signal – circuit malfunction	**Wiring, ignition switch, engine control relay, ECM**
P1520	Stop lamp switch – signal malfunction	**Wiring, stop lamp switch, ECM**
P1565	Cruise control switch – circuit malfunction	**Wiring, cruise control switch, ECM**
P1566	Cruise control switch – circuit malfunction	**Wiring, cruise control switch, ECM**
P1570	Cruise control laser radar sensor – malfunction	**Cruise control laser radar sensor, cruise control module, ECM**
P1572	Cruise control laser radar sensor – improper aiming of beam axis	**Cruise control laser radar sensor alignment, cruise control laser radar sensor**
P1575	Cruise control/skid control warning buzzer – malfunction	**Wiring, warning buzzer, ABS control module**
P1578	Brake system – malfunction	**ABS system, ABS control module**
P1600	Engine control module (ECM) – supply voltage	**Wiring, ECM**
P1605	Knock control – malfunction	**ECM**
P1615	Data bus, cruise control module – communication error	**CAN data bus, cruise control module, ECM**
P1616	Data bus, cruise control module – communication error	**CAN data bus, ECM, cruise control module**

OBD-II code	Fault location	Probable cause
P1633	Throttle actuator control (TAC) system – malfunction	ECM
P1645	Body control module – malfunction	Body control module, A/C control module, CAN data bus
P1656	Oil control valve – LH bank – malfunction	Wiring, oil control valve, ECM
P1663	Oil control valve – RH bank – malfunction	Wiring, oil control valve, ECM
P1705	AT – direct clutch speed sensor	Wiring, direct clutch speed sensor, ECM
P1765	AT – pressure control shift solenoid	Wiring, shift solenoid, ECM
P1780	Park/neutral position (PNP) switch – circuit malfunction	Wiring, PNP switch, ECM
P1782	Transfer box low ratio switch – circuit malfunction	Wiring, transfer box low ratio switch, ECM
P1783	Transfer box neutral switch – circuit malfunction	Wiring, transfer box neutral switch, ECM
B2785	Ignition switch ON – malfunction	Ignition switch, engine control relay, wiring
B2786	Ignition switch OFF – malfunction	Ignition switch, engine control relay, wiring
B2791	Key unlock warning switch – malfunction	Wiring, key unlock warning switch, engine control relay
B2795	Unrecognized key code	Key, incorrect key
B2796	Immobilizer system – no communication	Key, transponder, wiring, ECM
B2797	Communication malfunction 1 – interference	Incorrect key has been used, wiring, ignition module, ECM
B2798	Communication malfunction 2	Key, transponder, ignition module, wiring, ECM
B2799	Immobilizer – malfunction	Wiring, immobilizer system, ECM

LEXUS

Model:	Engine identification:	Year:
LS400 4.0L	1UZ-FE	1993-94
SC400 4.0L	1UZ-FE	1992-95

System: **TCCS**

Self-diagnosis

General information

- Refer to the front of this manual for general test conditions, terminology, detailed descriptions of wiring faults and a general trouble shooter for electrical and mechanical faults.
- Engine control module (ECM) incorporates self-diagnosis function.
- Malfunction indicator lamp (MIL) will illuminate if certain faults are recorded.
- ECM operates in backup mode if sensors fail, to enable vehicle to be driven to workshop.
- ECM trouble codes can be accessed using data link connector (DLC) 1 [1], or data link connector (DLC) 2 [2].

Preparatory conditions

- Battery voltage 11 V minimum.
- Throttle valve fully closed.
- Transmission in neutral position.
- All accessories including A/C switched OFF.

Accessing

- Switch ignition ON. Do NOT start engine.
- Check MIL illuminates.
- Jump DLC terminals TE1 and E1 [1] or [2].
- MIL should be flashing.
  - First group of short flashes indicates the 'tens' of the trouble code [3] [**A**].
  - Second group of short flashes indicates the 'units' of the trouble code [3] [**D**].
  - A 1.5 second pause separates 'tens' and 'units' [3] [**C**].
  - A 2.5 second pause separates each trouble code [3] [**E**].
  - For example: Trouble code 32 displayed [3].
- If no trouble codes are stored, MIL lamp will flash 2 times per second [4].
- Count MIL flashes and compare with trouble code table.
- Remove jump lead.
- Switch ignition OFF.

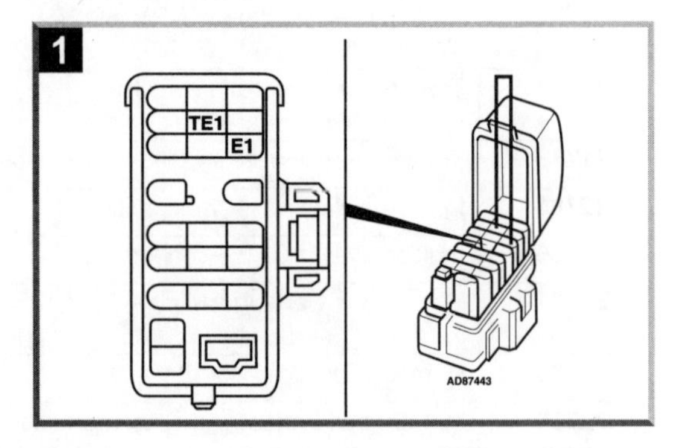

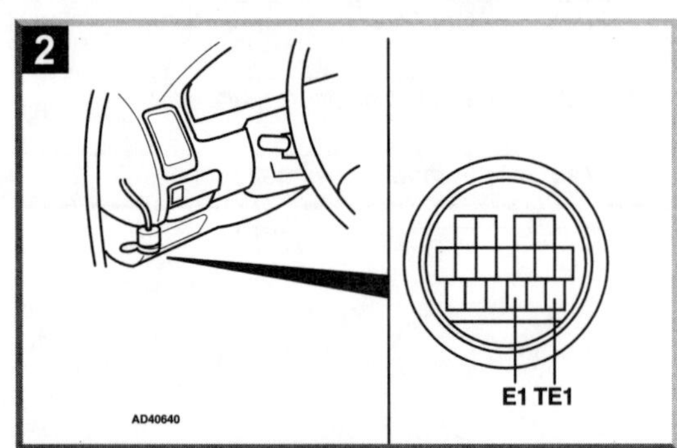

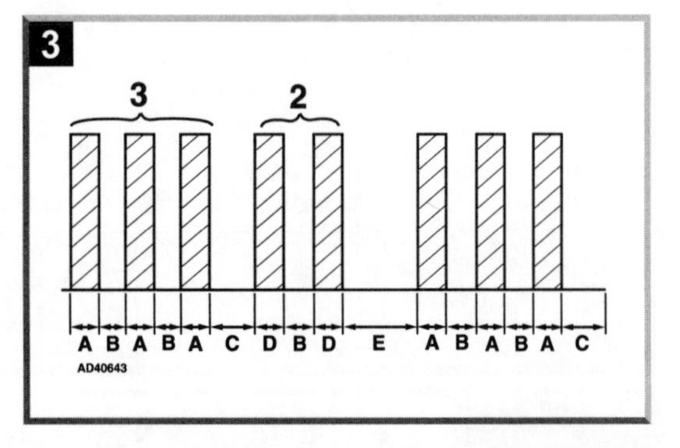

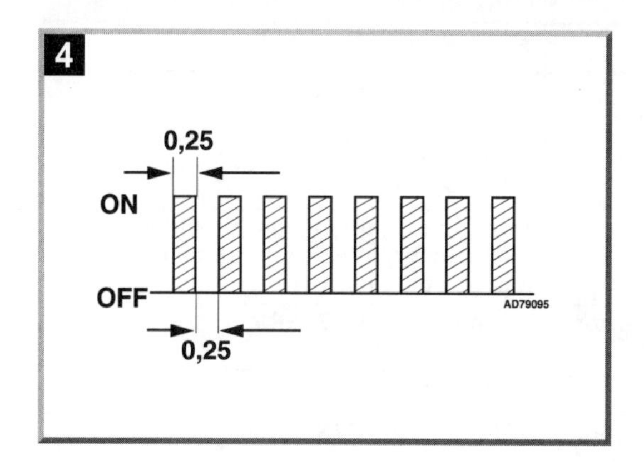

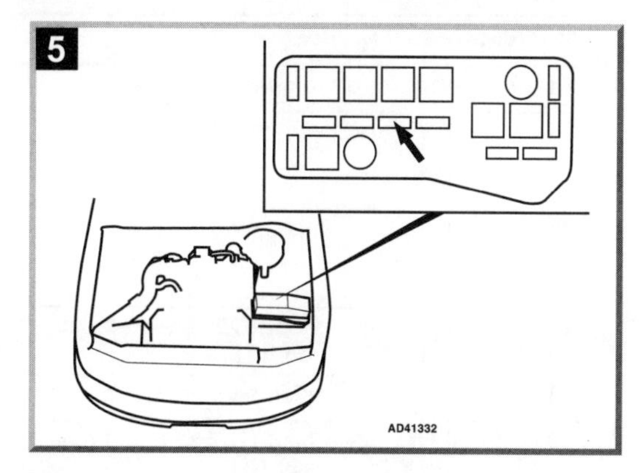

Erasing

- Ensure ignition switched OFF.
- LS400 – remove EFi 20A fuse from underhood fusebox 5.
- SC400 – remove EFi 30A fuse from underhood fusebox 6.
- Wait 10 seconds minimum.
- Reinstall fuse.
- Repeat checking procedure to ensure no data remains in ECM fault memory.

NOTE: ***Disconnecting battery ground lead will also erase trouble codes but may erase memory from electronic units such as radio.***

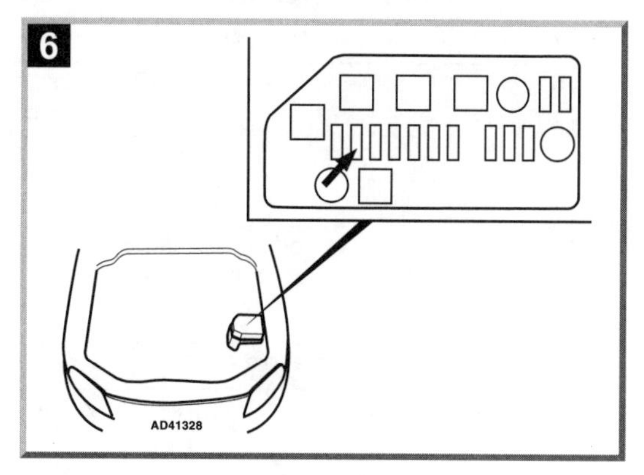

Trouble code identification

Flash code	Fault location	Probable cause
12	Crankshaft position (CKP) sensor	Wiring, CKP sensor, starter motor, ECM
12	Camshaft position (CMP) sensor – LH/RH	Wiring, CMP sensor, starter motor, ECM
13	Crankshaft position (CKP) sensor	Wiring, valve timing, CKP sensor, ECM
13	Camshaft position sensor – LH/RH	Valve timing, CMP sensor, ECM
14	Ignition module/ignition primary confirmation signal – LH bank	Wiring, ignition module, ECM
15	Ignition module/ignition primary confirmation signal – RH bank	Wiring, ignition module, ECM
16	Automatic transmission signal	ECM
21	Heated oxygen sensor (HO2S) – front LH	Wiring, HO2S, ECM
22	Engine coolant temperature (ECT) sensor	Wiring, ECT sensor, ECM

Flash code	Fault location	Probable cause
24	Intake air temperature (IAT) sensor	Wiring, IAT sensor, ECM
25	Mixture control – lean	Intake/fuel/ignition system, wiring, front LH/RH HO2S, ECT/VAF sensor, injector, valve timing, mechanical fault
26	Mixture control – rich	Intake/fuel/ignition system, wiring, VAF sensor, injector, valve timing, mechanical fault
27	Oxygen sensor (O2S)/heated oxygen sensor (HO2S) – rear LH	Wiring, O2S/HO2S, ECM
28	Heated oxygen sensor (HO2S) – front RH	Wiring, HO2S, ECM
29	Oxygen sensor (O2S)/heated oxygen sensor (HO2S) – rear RH	Wiring, O2S/HO2S, ECM
31	Volume air flow (VAF) sensor – incorrect signal	Wiring, VAF sensor, ECM
35	Barometric pressure (BARO) sensor	ECM
38	Transmission fluid temperature (TFT) sensor – AT	Wiring, TFT sensor, ECM
41	Throttle position (TP) sensor 1	Wiring, TP sensor, ECM
42	Vehicle speed sensor (VSS) 1	Wiring, VSS, ECM
43	Cranking signal	Wiring, ignition switch, starter motor relay, ECM
46	AT shift solenoid 4	Wiring, shift solenoid 4, ECM
47	Throttle position (TP) sensor 2 – traction control system	Wiring, TP sensor, ECM
48	Secondary air injection (AIR) solenoid	Wiring, AIR solenoid, ECM
51	Switch signal – A/C switch ON during diagnosis	Wiring, A/C switch, A/C amplifier, ECM
51	Switch signal – closed throttle position (CTP) switch OFF during diagnosis	Wiring, CTP switch, ECM
51	Switch signal – park/neutral position (PNP) switch not in P or N during diagnosis	Wiring, PNP switch, ECM
52	Knock sensor (KS) – LH	Wiring, KS, ECM
53	Knock sensor (KS) control signal	ECM
55	Knock sensor (KS) – RH	Wiring, KS, ECM
61	Vehicle speed sensor (VSS) 2 – AT	Wiring, VSS, ECM
62	AT shift solenoid 1	Wiring, shift solenoid, ECM
63	AT shift solenoid 2	Wiring, shift solenoid, ECM
64	AT shift solenoid 3	Wiring, shift solenoid, ECM
67	Overdrive (OD) clutch speed sensor	Wiring, OD clutch speed sensor, ECM
71	Exhaust gas recirculation (EGR) system	Hose/pipe leak or blockage, wiring, EGR valve, EGRT sensor, ECM
78	Fuel pump control module (FPCM)	Wiring, FPCM, fuel pump, ECM

MAZDA

Model:	Engine identification:	Year:
B2200 Pickup	**F2**	**1990-94**
B2600 Pickup	**G6**	**1989-93**
626/MX-6 2.2L	**F2**	**1991-92**
626/MX-6 2.2L Turbo	**F2**	**1991-92**

System: **Mazda EGI**

Self-diagnosis

General information

- Refer to the front of this manual for general test conditions, terminology, detailed descriptions of wiring faults and a general trouble shooter for electrical and mechanical faults.
- Engine control module (ECM) incorporates self-diagnosis function.
- ECM operates in backup mode if sensors fail, to enable vehicle to be driven to workshop.
- Malfunction indicator lamp (MIL) will illuminate if certain faults are recorded.
- Trouble codes can be displayed with an LED test lamp or accessed with a suitable code reader connected to data link connector (DLC) **1** [**2**].

Accessing

- Ensure ignition switched OFF.
- Jump test connector (green, 1 pin) to ground **1** [**1**].
- Connect LED test lamp between data link connector (DLC) terminals 1 and 2 **1** [**2**].

NOTE: ***Connect LED test lamp positive connection to DLC terminal 2.***

- Switch ignition ON.
- Count LED flashes and compare with trouble code table.
  - 1.2 second flashes indicate the 'tens' of the trouble code **2** [**A**].
  - 0.4 second flashes indicate the 'units' of the trouble code **2** [**D**].
  - A 1.6 second pause (LED OFF) separates 'TENS' and 'UNITS' **2** [**C**].
  - A 0.4 second pause (LED OFF) separates each flash **2** [**B**].
  - A 4 second pause separates each trouble code **2** [**E**].
  - For example: Trouble code 22 displayed **2**.
- Start engine.
- Allow to idle.
- Check for further trouble codes.
- Switch ignition OFF.
- Remove jump lead and LED test lamp.

1

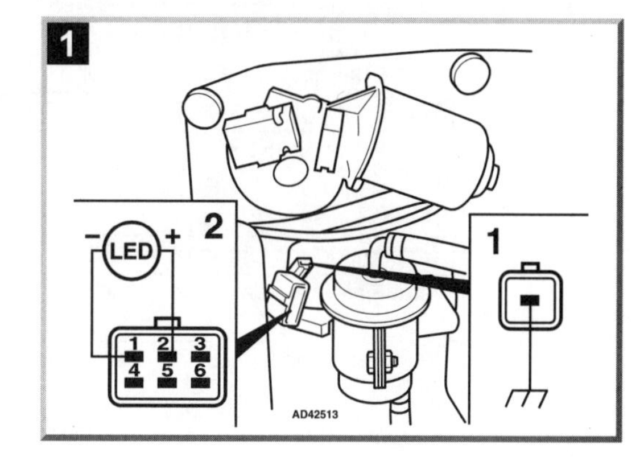

2

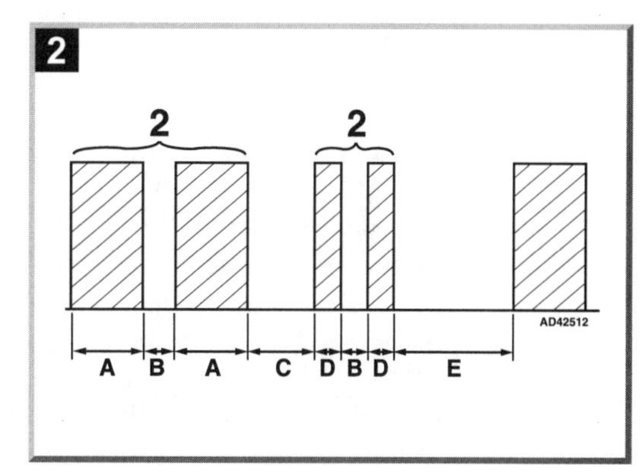

Erasing

- Ensure ignition switched OFF.
- Disconnect battery ground lead.
- Depress the brake pedal for a minimum of 5 seconds.
- Reconnect battery ground lead.
- Jump test connector (green, 1 pin) to ground **1** [**1**].
- Connect LED test lamp between DLC terminals **1** [**2**].
- Switch ignition ON. Do not start engine.
- Wait 6 seconds.
- Start engine.
- Ensure engine is at normal operating temperature
- Run engine at 2000 rpm for 3 minutes.
- Check that no trouble codes are displayed.
- Switch ignition OFF.
- Remove jump lead and LED tester.

NOTE: ***Disconnecting battery ground lead will also erase trouble codes but may erase memory from electronic units such as radio.***

Trouble code identification

Flash code	Fault location	Probable cause
01	Ignition signal	**Engine fault, distributor, ignition coil wiring, ECM**
02	Engine speed (RPM) sensor/ crankshaft position (CKP) sensor	**Wiring, distributor, ECM**
03	Camshaft position (CMP) sensor 1	**Wiring, distributor, ECM**
04	Camshaft position (CMP) sensor 2	**Wiring, distributor, ECM**
05	Knock sensor (KS)	**Wiring, KS sensor, ECM**
08	Volume air flow (VAF) sensor/mass air flow (MAF) sensor	**Wiring, VAF/MAF sensor, ECM**
09	Engine coolant temperature (ECT) sensor	**Wiring, ECT sensor, ECM**
10	Intake air temperature (IAT) sensor – in VAF sensor	**Wiring, VAF sensor, ECM**
11	Intake air temperature (IAT) sensor – in manifold	**Wiring, IAT sensor, ECM**
12	Throttle position (TP) sensor	**Wiring, TP sensor, ECM**
14	Barometric pressure (BARO) sensor	**ECM**
15	Oxygen sensor (O2S)/heated oxygen sensor (HO2S)	**Spark plugs, air leaks, fuel pressure, O2S/HO2S, ECM**
16	Exhaust gas recirculation (EGR) valve position sensor	**Wiring, EGR sensor, ECM**
17	Oxygen sensor (O2S)/heated oxygen sensor (HO2S) feedback system	**Spark plugs, air leaks, O2S/HO2S, ECM**
23	Heated oxygen sensor (HO2S)	**Wiring, fuel system, HO2S, ECM**
24	Heated oxygen sensor (HO2S) feedback system	**Spark plugs, air leaks, HO2S, ECM**
25	Fuel pressure (FP) control solenoid	**Wiring, FP control solenoid, ECM**
26	Evaporative emission (EVAP) canister purge valve	**Wiring, EVAP solenoid, ECM**
28	Exhaust gas recirculation (EGR) solenoid	**Wiring, EGR solenoid, ECM**
29	Exhaust gas recirculation (EGR) solenoid	**Wiring, EGR solenoid, ECM**
34	Idle air control (IAC) valve	**Wiring, IAC solenoid, ECM**
40	Oxygen sensor heater relay	**Wiring, HO2S**
41	Intake manifold air control solenoid	**Wiring, air control solenoid, ECM**

Model:	Engine identification:	Year:
B2300	VIN code digit 8 = A	1995-97
B2300	VIN code digit 8 = D	2001-04
B2500	VIN code digit 8 = C	1998-00
B3000	VIN code digit 8 = U/V	1995-04
B4000	VIN code digit 8 = E/X	1995-04
Mazda 6 2.0L/3.0L	L3/AJ	2003-04
Millenia 2.3L/2.5L	KJ/KL	1995-02
MPV 2.5L	GY	2000-02
MPV 3.0L	JE	1996-99
MPV 3.0L	AJ	2002-04
MX-5 Miata 1.8L	BP	1996-04
Protegé 1.5/1.8L	Z5/BP	1995-98
Protegé/Protegé 5 1.6/1.8/2.0L	ZM/FP/FS	1999-03
Tribute 2.0L/3.0L	VIN code digit 8 = B/1	2001-04
626/MX-6 2.0L/2.5L	FS/KL	1996-02

System: **Mazda EGI/Ford EEC**

Self-diagnosis

General information

- Refer to the front of this manual for general test conditions, terminology, detailed descriptions of wiring faults and a general trouble shooter for electrical and mechanical faults.
- Engine control module (ECM) incorporates self-diagnosis function.
- Malfunction indicator lamp (MIL) will illuminate if certain faults are recorded.
- ECM operates in backup mode if sensors fail, to enable vehicle to be driven to workshop.

Accessing

- Trouble codes can be displayed by using a Scan Tool connected to the data link connector (DLC) **1**.

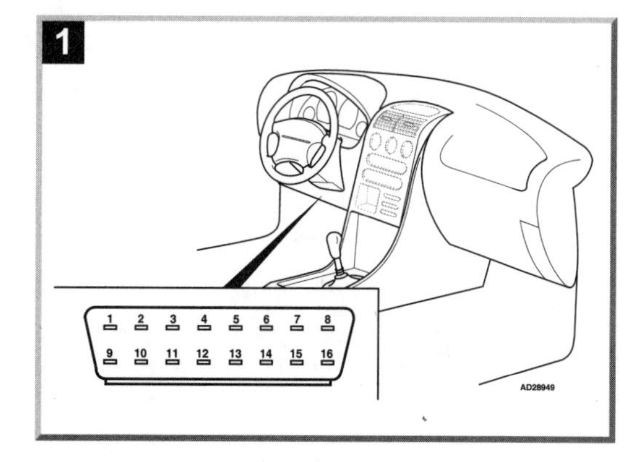

Erasing

- After the faults have been rectified, trouble codes can be erased by using a Scan Tool connected to the data link connector (DLC).

Trouble code identification

OBD-II code	Fault location	Probable cause
All P0, P2 and U0 codes	Refer to OBD-II trouble code tables at the front of this manual	–
P1000	OBD-II monitor testing not complete	**Wiring, VSS sensor, mechanical fault, ECM**
P1001	Data link connector (DLC) – self-test terminated	**Incorrect operator input, ECM**
P1100	Mass air flow (MAF) sensor – intermittent sensing	**Wiring, MAF sensor, ECM**
P1101	Mass air flow (MAF) sensor – out of self test range	**Wiring, MAF sensor, ECM**
P1102	Mass air flow (MAF) sensor – incomparable with TP sensor signal – signal low	**Wiring, MAF sensor, TP sensor**
P1103	Mass air flow (MAF) sensor – incomparable with RPM signal – signal high	**Wiring, MAF sensor**
P1110	Intake air temperature (IAT) sensor (engine) – short/open circuit	**Wiring, IAT sensor, ECM**
P1112	Intake air temperature (IAT) sensor – intermittent sensing	**Wiring, IAT sensor, ECM**
P1113	Intake air temperature (IAT) sensor (supercharger) – short/open circuit	**Wiring, IAT sensor, ECM**
P1114	Intake air temperature (IAT) sensor – voltage low	**Wiring, IAT sensor, ECM**
P1115	Intake air temperature (IAT) sensor – voltage high	**Wiring, IAT sensor, ECM**
P1116	Engine coolant temperature (ECT) sensor – out of self test range	**Wiring, ECT sensor, ECM**
P1117	Engine coolant temperature (ECT) sensor – intermittent sensing	**Wiring, ECT sensor, ECM**
P1120	Throttle position (TP) sensor – out of range "low"	**Wiring, TP sensor, ECM**
P1121	Throttle position (TP) sensor – incomparable with MAF sensor signal	**Wiring, TP sensor, ECM**
P1122	Throttle position (TP) sensor – stuck closed	**Wiring, TP sensor, ECM**
P1123	Throttle position (TP) sensor – stuck open	**Wiring, TP sensor, ECM**
P1124	Throttle position (TP) sensor – out of self test range	**Wiring, TP sensor, ECM**
P1125	Throttle position (TP) sensor – intermittent circuit fault	**Wiring, TP sensor, ECM**
P1126	Throttle position (TP) sensor – circuit fault	**Wiring, TP sensor, ECM**
P1127	Heated oxygen sensor (HO2S) – RH or rear bank – front (6 cyl) – heater circuit malfunction	**Wiring, HO2S**
P1128	Heated oxygen sensor (HO2S) – front – bank 1/bank 2 sensors transposed	**Wiring**
P1129	Heated oxygen sensor (HO2S) – rear – bank 1/bank 2 sensors transposed	**Wiring**

OBD-II code	Fault location	Probable cause
P1130	Heated oxygen sensor (HO2S) – front (4 cyl) – adaptive limit achieved	Wiring, HO2S, fuel pressure, fuel system, mechanical fault, ECM
P1130	Heated oxygen sensor (HO2S) – RH or rear bank – front (6 cyl) – adaptive limit achieved	Wiring, HO2S, fuel pressure, fuel system, mechanical fault, ECM
P1131	Heated oxygen sensor (HO2S) – front (4 cyl) – lean mixture indicated	Wiring, HO2S, fuel pressure, fuel system, mechanical fault, ECM
P1131	Heated oxygen sensor (HO2S) – RH or rear bank – front (6 cyl) – lean mixture indicated	Wiring, HO2S, fuel pressure, fuel system, mechanical fault, ECM
P1132	Heated oxygen sensor (HO2S) – front (4 cyl) – rich mixture indicated	Wiring, HO2S, fuel pressure, fuel system, mechanical fault, ECM
P1132	Heated oxygen sensor (HO2S) – RH or rear bank – front (6 cyl) – rich mixture indicated	Wiring, HO2S, fuel pressure, fuel system, mechanical fault, ECM
P1133	Fuel trim (FT) lean mixture – bank 1	Fuel pressure, injectors, MAF sensor/MAP sensor, intake leak
P1134	Fuel trim (FT) rich mixture – bank 1	Fuel pressure, injectors, MAF sensor/MAP sensor, EVAP system
P1135	Heated oxygen sensor (HO2S) 1, bank 1, heater control – circuit low	Wiring, HO2S, ECM
P1136	Heated oxygen sensor (HO2S) 1, bank 1, heater control – circuit high	Wiring, HO2S, ECM
P1137	Heated oxygen sensor (HO2S) – rear – adaptive limit achieved	Wiring, HO2S, exhaust system, ECM
P1138	Heated oxygen sensor (HO2S) – rear – adaptive limit achieved	Wiring, HO2S, exhaust system, ECM
P1141	Heated oxygen sensor (HO2S) 2, bank 1, heater control – circuit low	Wiring, HO2S, ECM
P1142	Heated oxygen sensor (HO2S) 2, bank 1, heater control – circuit high	Wiring, HO2S, ECM
P1143	Heated oxygen sensor (HO2S) 2, bank 1 – lean mixture indicated	Wiring, HO2S, fuel pressure, fuel system, mechanical fault, ECM
P1144	Heated oxygen sensor (HO2S) 2, bank 1 – rich mixture indicated	Wiring, HO2S, fuel pressure, fuel system, mechanical fault, ECM
P1150	Heated oxygen sensor (HO2S) – LH or front bank – front (6 cyl) – adaptive limit achieved	Wiring, HO2S, fuel pressure, fuel system, mechanical fault, ECM
P1151	Heated oxygen sensor (HO2S) – LH or front bank – front (6 cyl) – lean mixture indicated	Wiring, HO2S, fuel pressure, fuel system, mechanical fault, ECM
P1152	Heated oxygen sensor (HO2S) – LH or front bank – front (6 cyl) – rich mixture indicated	Wiring, HO2S, fuel pressure, fuel system, mechanical fault, ECM
P1153	Fuel trim (FT) lean mixture – bank 2	Fuel pressure, injectors, MAF sensor/MAP sensor, intake leak
P1154	Fuel trim (FT) rich mixture – bank 2	Fuel pressure, injectors, MAF sensor/MAP sensor, EVAP system
P1157	Heated oxygen sensor (HO2S) 2, bank 2 – lean mixture indicated	Wiring, HO2S, fuel pressure, fuel system, mechanical fault, ECM
P1158	Heated oxygen sensor (HO2S) 2, bank 2 – rich mixture indicated	Wiring, HO2S, fuel pressure, fuel system, mechanical fault, ECM

OBD-II code	Fault location	Probable cause
P1169	Heated oxygen sensor (HO2S) – rear (6 cyl) – voltage high	**Wiring, HO2S, fuel pressure, fuel system, mechanical fault, ECM**
P1170	Heated oxygen sensor (HO2S) – front (4 cyl) – voltage high	**Wiring, HO2S, fuel pressure, fuel system, mechanical fault, ECM**
P1170	Heated oxygen sensor (HO2S) LH or front bank – front (6 cyl) – voltage high	**Wiring, HO2S, fuel pressure, fuel system, mechanical fault, ECM**
P1173	Heated oxygen sensor (HO2S) RH or rear bank – front (6 cyl) – voltage high	**Wiring, HO2S, fuel pressure, fuel system, mechanical fault, ECM**
P1195	Exhaust gas recirculation (EGR) system sensor – malfunction	**Wiring, EGR sensor, ECM**
P1196	Starter signal – circuit malfunction	**Wiring, ignition switch, mechanical fault**
P1244	Generator/regulator system – load input out of range	**Wiring, generator, regulator**
P1245	Generator/regulator system – load input out of range	**Wiring, generator, regulator**
P1246	Generator/regulator system – load input out of range	**Wiring, generator, regulator**
P1250	Fuel pressure (FP) control solenoid valve 1 – malfunction	**Wiring, FP solenoid, ECM**
P1251	Idle speed control (ISC) system – rpm lower than expected	**Wiring, IAC valve, throttle valve tight/ sticking, ECM**
P1252	Fuel pressure (FP) control solenoid valve 2 – malfunction	**Wiring, FP solenoid, ECM**
P1260	Theft detected – engine immobilized	**immobilizer fault**
P1270	Maximum engine RPM or vehicle speed obtained	**ECM**
P1285	Cylinder head temperature (CHT) sensor – overheating detected	**Wiring, CHT sensor, engine overheating, ECM**
P1288	Cylinder head temperature (CHT) sensor – out of self test range	**Wiring, CHT sensor, ECM**
P1289	Cylinder head temperature (CHT) sensor – circuit malfunction	**Wiring, CHT sensor, ECM**
P1290	Cylinder head temperature (CHT) sensor – circuit malfunction	**Wiring, CHT sensor, ECM**
P1299	Cylinder head temperature (CHT) sensor – fail safe cooling activated	**Wiring, CHT sensor, engine overheating, ECM**
P1309	Misfire detection monitor – not enabled	**CMP not syncronized, CMP, ECM**
P1336	Crankshaft position (CKP) sensor/camshaft position (CMP) sensor – circuit range/ performance	**Wiring, CKP sensor, CMP sensor, mechanical fault**
P1345	Camshaft position (CMP) sensor – signal malfunction	**Wiring, CMP sensor, ECM**
P1351	Engine control module (ECM) – loss of ignition diagnostic monitor	**ECM**

OBD-II code	Fault location	Probable cause
P1352	Ignition coil A – primary circuit malfunction	**Wiring, ignition coil**
P1353	Ignition coil B – primary circuit malfunction	**Wiring, ignition coil**
P1354	Ignition coil C – primary circuit malfunction	**Wiring, ignition coil**
P1358	Engine control module (ECM) – loss of ignition diagnostic monitor	**ECM**
P1359	Engine control module (ECM) – loss of ignition output signal	**ECM**
P1360	Ignition coil A – secondary circuit malfunction	**Wiring, ignition coil**
P1361	Ignition coil B – secondary circuit malfunction	**Wiring, ignition coil**
P1362	Ignition coil C – secondary circuit malfunction	**Wiring, ignition coil**
P1364	Ignition coil – primary circuit malfunction	**Wiring, ignition coil**
P1365	Ignition coil – secondary circuit malfunction	**Wiring, ignition coil**
P1382	Camshaft position (CMP) sensor – malfunction	**Insecure sensor/rotor, air gap, wiring, CMP sensor, ECM**
P1387	Camshaft position (CMP) sensor – malfunction	**Insecure sensor/rotor, air gap, wiring, CMP sensor, ECM**
P1390	Octane adjust – out of self test range	**Wiring, octane adjuster, ECM**
P1400	Differential pressure feedback sensor – low voltage detected	**Wiring, pressure feedback sensor, ECM**
P1401	Differential pressure feedback sensor – high voltage detected	**Wiring, pressure feedback sensor, ECM**
P1402	Exhaust gas recirculation (EGR) valve position sensor – voltage low/high	**Wiring, EGR sensor, ECM**
P1403	Differential pressure feedback sensor – hoses reversed	**Mechanical fault**
P1405	Differential pressure feedback sensor – upstream hose off or plugged	**Mechanical fault**
P1406	Differential pressure feedback sensor – downstream hose off or plugged	**Mechanical fault**
P1407	Exhaust gas recirculation (EGR) system – no flow detected	**Wiring, EGR solenoid valve, mechanical fault**
P1408	Exhaust gas recirculation (EGR) system – flow out of self test range	**Wiring, EGR solenoid valve, mechanical fault**
P1409	Electronic vacuum regulator control – circuit malfunction	**Wiring, ECM**
P1410	Intake air flap control system (Mazda 6) – circuit malfunction	**Intake blocked, wiring, intake air flap control actuator, ECM**
P1410	Exhaust gas recirculation (EGR) boost sensor/ solenoid – stuck	**Wiring, EGR boost sensor, EGR boost solenoid, hoses blocked/leaking**
P1412	Exhaust gas recirculation (EGR) solenoid – malfunction	**Wiring, EGR solenoid, ECM**
P1432	Engine coolant thermostat – circuit malfunction	**Wiring, engine coolant thermostat, ECM**

OBD-II code	Fault location	Probable cause
P1443	Evaporative emission (EVAP) canister purge control system	Wiring, EVAP valve, ECM
P1444	Purge flow sensor – low input	Wiring, purge flow sensor
P1445	Purge flow sensor – high input	Wiring, purge flow sensor
P1446	Evaporative emission (EVAP) system	Hose connection(s), intake leak, EVAP canister, EVAP canister purge valve, ECM
P1449	Evaporative emission (EVAP) canister purge system – tank pressure problem	Wiring, EVAP valve, ECM
P1450	Evaporative emission (EVAP) canister purge system – leak	Wiring, EVAP valve, ECM
P1451	Evaporative emission (EVAP) canister vent valve – circuit malfunction	Wiring, EVAP vent valve, fuel tank pressure sensor, ECM
P1455	Fuel tank level sensor – short/open circuit	Wiring, fuel tank level sensor
P1456	Evaporative emission (EVAP) system	Hose connection(s), intake leak, EVAP canister, EVAP canister purge valve, ECM
P1457	Evaporative emission (EVAP) system	Hose connection(s), intake leak, EVAP canister, EVAP canister purge valve, ECM
P1460	Wide open throttle (WOT) air conditioning cut off – circuit malfunction	Wiring, TP sensor
P1461	A/C refrigerant pressure sensor – high voltage	Wiring, A/C refrigerant pressure sensor, ECM
P1462	A/C refrigerant pressure sensor – low voltage	Wiring, A/C refrigerant pressure sensor, ECM
P1463	A/C refrigerant pressure sensor – insufficient pressure change	Wiring, A/C system – mechanical fault, A/C refrigerant pressure sensor, A/C compressor clutch
P1464	A/C control signal – voltage high during self-test	Wiring, AC master switch, AC compressor clutch relay, ECM
P1465	A/C control signal – malfunction	Wiring, AC master switch, AC compressor clutch relay, ECM
P1469	Frequent A/C compressor clutch cycling	Wiring, A/C system – mechanical fault
P1473	Cooling fan – circuit malfunction	Wiring, cooling fan relay(s), cooling fan(s), ECM
P1474	A/C control signal – voltage high during self-test	Wiring
P1474	Cooling fan "low" (Tribute) – circuit malfunction	Wiring, cooling fan relay, cooling fan, ECM
P1475	Cooling fan – circuit malfunction	Wiring, cooling fan relay(s), cooling fan(s), ECM
P1476	Cooling fan – circuit malfunction	Wiring, cooling fan relay(s), cooling fan(s), ECM
P1477	Cooling fan "medium" (Tribute) – circuit malfunction	Wiring, cooling fan relay, cooling fan, ECM
P1479	A/C condenser blower motor relay – circuit malfunction	Wiring, A/C condenser blower motor relay, ECM

OBD-II code	Fault location	Probable cause
P1479	Cooling fan "high" (Tribute) – circuit malfunction	**Wiring, cooling fan relay, cooling fan, ECM**
P1480	Cooling fan – circuit malfunction	**Wiring, cooling fan relay(s), cooling fan(s), ECM**
P1481	Cooling fan – circuit malfunction	**Wiring, cooling fan relay(s), cooling fan(s), ECM**
P1485	Exhaust gas recirculation (EGR) valve – malfunction	**Wiring, EGR valve, ECM**
P1486	Exhaust gas recirculation (EGR) valve – malfunction	**Wiring, EGR valve, ECM**
P1487	Exhaust gas recirculation (EGR) valve – malfunction	**Wiring, EGR valve, ECM**
P1496	Exhaust gas recirculation (EGR) valve coil 1 – short/open circuit	**Wiring, EGR valve, ECM**
P1497	Exhaust gas recirculation (EGR) valve coil 2 – short/open circuit	**Wiring, EGR valve, ECM**
P1498	Exhaust gas recirculation (EGR) valve coil 3 – short/open circuit	**Wiring, EGR valve, ECM**
P1499	Exhaust gas recirculation (EGR) valve coil 4 – short/open circuit	**Wiring, EGR valve, ECM**
P1500	Vehicle speed sensor (VSS) – intermittent fault	**Wiring, VSS sensor, ECM**
P1501	Vehicle speed sensor (VSS) – out of self test range	**Wiring, VSS sensor, ECM**
P1502	Vehicle speed sensor (VSS) – signal malfunction	**Wiring, VSS sensor, ABS control module, multifunction control module, ECM**
P1504	Idle air control (IAC) system – circuit malfunction	**Wiring, IAC valve, ECM**
P1505	Idle air control (IAC) system	**Wiring, IAC valve, mechanical fault, ECM**
P1506	Idle air control (IAC) system – overspeed error	**Wiring, IAC valve, mechanical fault, ECM**
P1507	Idle air control (IAC) system – underspeed error	**Wiring, IAC valve, mechanical fault, ECM**
P1508	Idle air control (IAC) valve 1 – short/open circuit	**Wiring, IAC valve, ECM**
P1509	Idle air control (IAC) valve 2 – short/open circuit	**Wiring, IAC valve, ECM**
P1510	Idle air control (IAC) system – malfunction	**Wiring, IAC valve, mechanical fault, ECM**
P1511	Idle air control (IAC) system – malfunction	**Wiring, IAC valve, mechanical fault, ECM**
P1512	Intake manifold air control – stuck close	**Wiring, air control solenoid, ECM**
P1516	Neutral position (NP) switch – circuit malfunction	**Wiring, NP switch, ECM**
P1518	Intake manifold air control – stuck open	**Wiring, air control solenoid, ECM**
P1519	Intake manifold air control – stuck closed	**Wiring, air control solenoid, ECM**
P1520	Intake manifold air control – circuit malfunction	**Wiring, air control solenoid, ECM**
P1521	Intake manifold air control solenoid 1 – short/open circuit	**Wiring, air control solenoid, ECM**

OBD-II code	Fault location	Probable cause
P1522	Intake manifold air control solenoid 2 – short/open circuit	**Wiring, air control solenoid, ECM**
P1523	Intake manifold air control solenoid – circuit malfunction	**Wiring, air control solenoid, ECM**
P1524	Charge air cooler bypass solenoid valve – malfunction	**Wiring, bypass solenoid, ECM**
P1525	Supercharger air control solenoid (vacuum) – short/open circuit	**Wiring, air control solenoid, ECM**
P1526	Supercharger air control solenoid (vent) – short/open circuit	**Wiring, air control solenoid, ECM**
P1527	Bypass air solenoid valve – short/open circuit	**Wiring, air control solenoid, ECM**
P1528	Idle air control (IAC) system – malfunction	**Wiring, IAC valve, mechanical fault, ECM**
P1529	L/C atmospheric balance air control solenoid valve – malfunction	**Wiring, air control solenoid, ECM**
P1537	Intake manifold air control solenoid, bank 1 – solenoid stuck open	**Wiring, intake manifold air control solenoid, mechanical fault**
P1538	Intake manifold air control solenoid, bank 2 – solenoid stuck open	**Wiring, intake manifold air control solenoid, mechanical fault**
P1540	Supercharger air control solenoid – short/open circuit	**Wiring, valve, ECM**
P1550	Power steering pressure (PSP) switch – malfunction	**Wiring, PSP switch, ECM**
P1562	Engine control module (ECM) – battery voltage low	**Wiring, ECM**
P1565	Cruise control system	**Wiring, CPP switch, BPP switch, cruise control module, cruise control master switch, VSS**
P1566	Cruise control system (Tribute)	**Wiring, CPP switch, BPP switch, cruise control module, cruise control master switch, VSS**
P1566	Transmission control module (TCM) – battery voltage low	**Wiring, fuse, TCM**
P1567	Cruise control system	**Wiring, CPP switch, BPP switch, cruise control module, cruise control master switch, VSS**
P1568	Cruise control system	**Wiring, CPP switch, BPP switch, cruise control module, cruise control master switch, VSS**
P1569	Intake manifold air control – circuit low	**Wiring, air control solenoid, ECM**
P1570	Intake manifold air control – circuit high	**Wiring, air control solenoid, ECM**
P1572	Brake pedal position (BPP) switch – circuit malfunction	**Wiring, BPP switch, ECM**
P1600	Engine control module (ECM) – KAM error	**ECM**
P1601	ECM to AT communication – signal error	**Wiring, ECM**

OBD-II code	Fault location	Probable cause
P1602	Immobilizer control module – communication error	**Wiring, incorrect key, immobilizer control module, ECM**
P1603	Engine control module (ECM) – incorrect immobilizer code	**Immobilizer word code not programmed**
P1604	Engine control module (ECM) – incorrect immobilizer code	**Immobilizer key code not programmed**
P1605	Engine control module (ECM) – keep alive test error	**ECM**
P1608	Engine control module (ECM) – knock function malfunction	**ECM**
P1609	Engine control module (ECM) – supply voltage	**Wiring, fuses, ECM**
P1621	Engine control module (ECM) – incorrect immobilizer code after cranking	**Immobilizer word code not programmed**
P1622	Engine control module (ECM) – incorrect immobilizer key code	**Immobilizer key code not programmed**
P1623	Engine control module (ECM) – immobilizer read/write error, EEPROM damaged	**ECM**
P1624	Engine control module (ECM) – immobilizer communication error	**ECM**
P1627	ECM to ABS/TCS communication – signal error	**Wiring, ECM**
P1628	ECM to ABS/TCS communication – signal error	**Wiring, ECM**
P1630	Generator voltage – circuit malfunction	**Wiring, mechanical fault**
P1631	Generator voltage – no charge	**Wiring, mechanical fault**
P1632	Battery voltage monitor – circuit malfunction	**Wiring, mechanical fault**
P1633	Battery voltage – overcharging	**Wiring, mechanical fault**
P1634	Generator terminal B – open circuit	**Wiring, mechanical fault**
P1635	Engine control module (ECM) – processor fault	**ECM**
P1639	Engine control module (ECM) – keep alive test error	**Wiring, ECM not programmed, ECM**
P1641	Fuel pump – circuit malfunction	**Wiring, fuel pump control module, fuel pump, ECM**
P1645	Fuel pump – circuit malfunction	**Wiring, fuel pump control module, fuel pump, ECM**
P1650	Power steering pressure (PSP) switch – out of self test range	**Wiring, PSP switch, ECM**
P1651	Power steering pressure (PSP) switch – input malfunction	**Wiring, PSP switch, ECM**
P1652	Power steering pressure (PSP) switch – circuit malfunction	**Wiring, PSP switch, ECM**
P1700	Transmission system – malfunction	**Transmission**
P1701	Transmission solenoid malfunction – reverse gear engagement error	**Wiring, solenoid valve, mechanical fault**

OBD-II code	Fault location	Probable cause
P1702	Transmission range (TR) sensor – circuit malfunction	Wiring, TR sensor
P1703	Brake pedal position (BPP) switch – torque converter – out of self test range	Wiring, switch
P1704	Transmission range (TR) sensor – malfunction	Wiring, TR sensor
P1705	Transmission range (TR) sensor – out of self test range	Wiring, TR sensor
P1708	Clutch pedal position (CPP) switch – malfunction	Wiring, CPP switch
P1709	Park/Neutral position switch (except B series Pickup) – out of self test range	Wiring, switch
P1709	Clutch pedal position (CPP) switch (B series Pickup/Tribute) – malfunction	Wiring, CPP switch
P1711	Transmission fluid temperature sensor – out of self test range	Wiring, temperature sensor, mechanical fault
P1713	Transmission fluid temperature sensor – performance	Wiring, temperature sensor
P1714	AT – shift solenoid 1 – mechanical failure	Wiring, mechanical fault
P1715	AT – shift solenoid 2 – mechanical failure	Wiring, mechanical fault
P1716	AT – shift solenoid 3 – mechanical failure	Wiring, mechanical fault
P1717	AT – shift solenoid 4 – mechanical failure	Wiring, mechanical fault
P1718	Transmission fluid temperature sensor – performance	Wiring, temperature sensor
P1719	Overdrive system – loss of signal	Wiring, mechanical fault
P1720	Vehicle speed sensor (VSS) – no signal	Wiring, VSS, mechanical fault
P1721	Shift solenoid (SS) 1 – malfunction	Wiring, shift solenoid, ECM
P1722	Shift solenoid (SS) 2 – malfunction	Wiring, shift solenoid, ECM
P1723	Shift solenoid (SS) 3 – malfunction	Wiring, shift solenoid, ECM
P1724	Shift solenoid (SS) 4 – malfunction	Wiring, shift solenoid, ECM
P1729	4x4 low range switch – switch error	Wiring, switch
P1735	Gear selection, 1-2 – shift malfunction	Wiring, shift solenoids, transmission mechanical fault
P1736	Gear selection, 2-3 – shift malfunction	Wiring, shift solenoids, transmission mechanical fault
P1737	Torque converter clutch (TCC) solenoid valve – malfunction	Wiring, TCC solenoid valve, mechanical fault
P1738	Shift time error	Mechanical fault
P1740	Torque converter clutch (TCC) control – malfunction	Wiring, mechanical fault
P1741	Torque converter clutch (TCC) control – malfunction	Wiring, mechanical fault

OBD-II code	Fault location	Probable cause
P1742	Torque converter clutch (TCC) solenoid valve – malfunction	**Wiring, TCC solenoid valve, mechanical fault**
P1743	Torque converter clutch (TCC) solenoid valve – malfunction	**Wiring, TCC solenoid valve, mechanical fault**
P1744	Torque converter clutch (TCC) system – malfunction	**Wiring, mechanical fault**
P1745	AT – electronic pressure control solenoid – circuit malfunction	**Wiring, mechanical fault, pressure control solenoid**
P1746	AT – electronic pressure control solenoid – low input	**Wiring, mechanical fault, pressure control solenoid**
P1747	AT – electronic pressure control solenoid – high input	**Wiring, mechanical fault, pressure control solenoid**
P1748	AT – electronic pressure control solenoid – malfunction	**Wiring, mechanical fault, pressure control solenoid**
P1749	AT – electronic pressure control solenoid system	**Wiring, mechanical fault, pressure control solenoid**
P1751	AT – shift solenoid 1 – open circuit	**Wiring, shift solenoid**
P1752	AT – shift solenoid 1 – short circuit	**Wiring, shift solenoid**
P1754	AT – freewheeling clutch solenoid – circuit malfunction	**Wiring, mechanical fault, freewheeling clutch solenoid**
P1756	AT – shift solenoid 2 – open circuit	**Wiring, shift solenoid**
P1757	AT – shift solenoid 2 – short circuit	**Wiring, shift solenoid**
P1760	AT – electronic pressure control solenoid – short to ground	**Wiring, mechanical fault, pressure control solenoid**
P1761	AT – shift solenoid 3 – performance	**Wiring, mechanical fault, shift solenoid**
P1762	Transmission – band failure	**Mechanical fault**
P1765	AT – gear 3-2 timing solenoid – malfunction	**Wiring, mechanical fault, timing solenoid**
P1767	AT – torque converter clutch (TCC) solenoid – circuit malfunction	**Wiring, TCC solenoid**
P1770	AT – overrun clutch solenoid valve – malfunction	**Wiring, mechanical fault, overrun clutch solenoid**
P1771	AT – throttle position (TP) sensor – open circuit	**Wiring, TP sensor**
P1772	AT – throttle position (TP) sensor – short circuit	**Wiring, TP sensor**
P1775	AT – torque reduction signal 1 – signal problem	**Wiring**
P1776	AT – torque reduction signal 2 – signal problem	**Wiring**
P1777	AT – torque reduction signal (water temperature signal) – malfunction	**Wiring**
P1780	Transmission control switch – out of self test range	**Wiring, switch, mechanical fault**
P1781	4x4 low range switch – out of self test range	**Wiring, switch**
P1783	Transmission – over temperature condition	**Mechanical fault**

OBD-II code	Fault location	Probable cause
P1788	AT – 3-2 timing/coasting clutch solenoid – open circuit during self test	Wiring, mechanical fault, engine performance, ECM
P1789	AT – 3-2 timing/coasting clutch solenoid – short circuit during self test	Wiring, mechanical fault, engine performance, ECM
P1790	Throttle position (TP) sensor – malfunction	Wiring, TP sensor, ECM
P1792	Barometric pressure (BARO) sensor – circuit malfunction	ECM
P1793	Mass air flow (MAF) sensor/intake air temperature (IAT) sensor – circuit malfunction	Wiring, MAF sensor, IAT sensor, ECM
P1794	Engine control module (ECM) – battery voltage low	Wiring, fuses
P1795	Closed throttle position (CTP) switch – malfunction	Wiring, CTP switch, ECM
P1797	Park/neutral position (PNP) switch/clutch pedal position (CPP) switch	Wiring, PNP switch, CPP switch, mechanical fault, ECM
P1798	Engine coolant temperature (ECT) sensor – circuit intermittent	Wiring, ECT sensor, cooling system, ECM
P1799	Torque converter clutch (TCC) solenoid valve – malfunction	Wiring, TCC solenoid valve, mechanical fault
P1900	Turbine shaft speed (TSS) sensor – intermittent	Wiring, TSS sensor, ECM/PCM/TCM
P1901	Turbine shaft speed (TSS) sensor – intermittent	Wiring, TSS sensor, ECM/PCM/TCM
U1020	CAN data bus – A/C clutch signal malfunction	Wiring, CAN data bus, ECM
U1039	CAN data bus – VSS signal malfunction	Wiring, CAN data bus, ECM
U1041	CAN data bus – VSS signal malfunction	Wiring, CAN data bus, ECM
U1051	CAN data bus – brake signal malfunction	Wiring, CAN data bus, ECM
U1131	CAN data bus – fuel system signal malfunction	Wiring, CAN data bus, ECM
U1147	CAN data bus – vehicle security signal malfunction	Wiring, CAN data bus, ECM
U1262	CAN data bus – circuit malfunction	Wiring, CAN data bus, ECM
U1451	CAN data bus – alarm system control module signal malfunction	Wiring, CAN data bus, ECM
U2243	CAN data bus – external environment signal malfunction	Wiring, CAN data bus, ECM

Model:	Engine identification:	Year:
B2300 Pickup 2.3L	VIN code digit 8 = A	1994
B3000 Pickup 3.0L	VIN code digit 8 = U	1994
B4000 Pickup 4.0L	VIN code digit 8 = X	1994
Navajo 4.0L	VIN code digit 8 = X	1991-94

System: **Ford EEC-IV**

Self-diagnosis

General information

- Refer to the front of this manual for general test conditions, terminology, detailed descriptions of wiring faults and a general trouble shooter for electrical and mechanical faults.
- Engine control module (ECM) incorporates self-diagnosis function.
- Malfunction indicator lamp (MIL) will illuminate if certain faults are recorded.
- ECM operates in backup mode if sensors fail, to enable car to be driven to workshop.
- Trouble codes can be accessed with a voltmeter or a suitable code reader connected to the data link connector (DLC) **1**.

Accessing – key ON engine OFF test

- Start engine.
- Ensure engine is at normal operating temperature.
- Switch ignition OFF.
- Wait 10 seconds.
- Connect an analogue voltmeter between data link connector (DLC) terminal 4 and battery (12 volt) positive **1**.
- Jump DLC 1 terminal **1** [**1**] to DLC 2 terminal 2 **1** [**2**].
- Switch ignition ON.
- Count voltmeter pulses and compare with trouble code table.
  - A group of 0.5 second pulses indicate each of the 2 (1991), or 3 (1992-94) trouble code digits **2** [**A**].
  - A 0.5 second pause separates each pulse **2** [**B**].
  - A 2 second pause separates each trouble code digit **2** [**C**].
  - A 4 second pause separates each trouble code **2** [**D**].
  - For example: Trouble code 231 displayed **2**.

NOTE: ***The trouble codes refer to present faults.***

- When all present trouble codes have been displayed:
  - A 6-9 second pause **3** [**A**], followed by a 0.5 second pulse **3** [**B**], and another 6-9 second pause **3** [**C**], indicate the start of the memorised trouble codes **3** [**D**].
- Trouble codes are indicated in the same manner as above.

NOTE: ***Memorised trouble codes refer to faults that have occurred in the last 80 ignition key cycles.***

- Count voltmeter pulses and compare with trouble code table.
- Switch ignition OFF.
- Remove jump lead and voltmeter.

NOTE: ***The malfunction indicator lamp (MIL) will also flash to display the trouble code.***

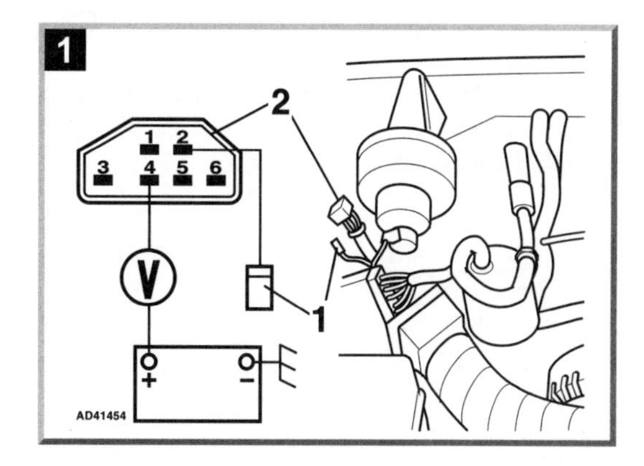

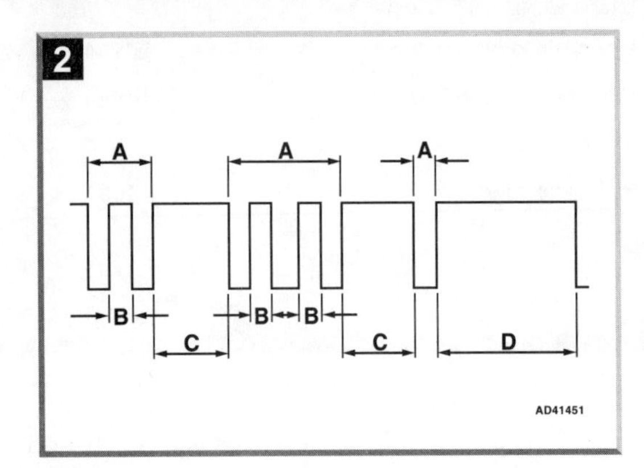

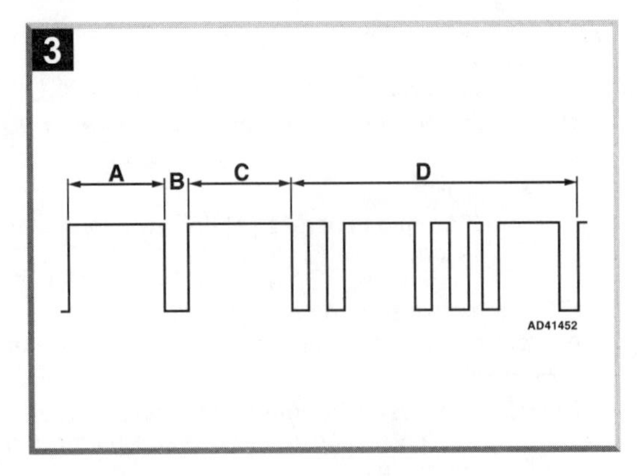

Accessing – key ON engine ON test

- Start engine.
- Ensure engine is at normal operating temperature.
- Switch ignition OFF.
- Wait 10 seconds.
- Connect an analogue voltmeter between data link connector (DLC) terminal 4 and battery (12 volt) positive **1**.
- Jump DLC 1 terminal **1** [**1**] to DLC 2 terminal 2 **1** [**2**].
- Start engine.
- After a group of 3 0.5 second pulses (ID code) are displayed:
- Press and release the brake pedal.
- If after 6-20 seconds, a single 0.5 second pulse is displayed (dynamic response signal):
- Increase engine speed sharply.

NOTE: ***Do not depress throttle during the test without the dynamic response signal.***

- Trouble codes are indicated in the same manner as above.
- Count voltmeter pulses and compare with trouble code table.
- Switch ignition OFF.
- Remove jump lead and voltmeter.

Erasing

- Ensure ignition switched OFF.
- Connect an analogue voltmeter between data link connector (DLC) terminal 4 and battery (12 volt) positive **1**.
- Jump DLC 1 terminal **1** [**1**] to DLC 2 terminal 2 **1** [**2**].
- Switch ignition ON.
- When trouble codes start to be displayed:
- Disconnect the jumper lead between DLC 1 terminal **1** [**1**] to DLC 2 terminal 2 **1** [**2**].
- If system components have been replaced:
- Disconnect battery ground lead for 5 minutes to erase adaptive memory from engine control module (ECM).
- Switch ignition OFF.
- Remove the voltmeter.
- Repeat checking procedure to ensure no data remains in ECM fault memory.

NOTE: ***The vehicle may suffer with driveability problems until the adaptive memory returns.***

NOTE: ***Disconnecting battery ground lead may erase memory from electronic units such as the radio.***

Trouble code identification – 1991

Trouble code	Fault location	Probable cause
11	No fault	–
12	Idle air control (IAC) valve – high rpm function during diagnostic test	**Wiring, IAC valve, mechanical fault**
13	Idle air control (IAC) valve – low rpm function during diagnostic test	**Wiring, IAC valve, mechanical fault**
14	Camshaft position (CMP) sensor – signal absent	**Wiring, CMP sensor, ECM**
15	Engine control module (ECM) – memory function during diagnostic test	**ECM – ROM failure**
15	Engine control module (ECM) – supply voltage	**Wiring, fuses**
16	Ignition diagnostic monitor – ICM – signal absent	**Wiring, ignition control module**
18	Ignition diagnostic monitor – ICM – circuit failure	**Wiring, ignition control module**
18	Spark output – open circuit	**Wiring, ignition control module**
19	Engine control module (ECM)	**ECM internal fault**
21	Engine coolant temperature (ECT) sensor – abnormal signal	**Wiring, ECT sensor, ECM**
22	Barometric pressure (BARO) sensor – abnormal signal	**Wiring, BARO sensor, ECM**
23	Throttle position (TP) sensor – abnormal signal	**Wiring, TP sensor, ECM**
24	Intake air temperature (IAT) sensor – abnormal signal	**Wiring, IAT sensor, ECM**
26	Mass air flow (MAF) sensor – abnormal signal	**Wiring, MAF sensor, ECM**
29	Vehicle speed sensor (VSS) – signal absent	**Wiring, VSS, ECM**
41	Heated oxygen sensor (HO2S) – signal low	**Wiring, HO2S, fuel pressure, fuel system, mechanical fault, ECM**
42	Heated oxygen sensor (HO2S) – signal high	**Wiring, HO2S, fuel pressure, fuel system, mechanical fault, ECM**
45	Ignition coil – Nos. 1, 2 & 3 primary – open or short circuit	**Wiring, coil**
51	Engine coolant temperature (ECT) sensor – signal high	**Wiring, ECT sensor, ECM**
53	Throttle position (TP) sensor – signal high	**Wiring, TP sensor, ECM**
54	Intake air temperature (IAT) sensor – signal high	**Wiring, IAT sensor, ECM**
56	Mass air flow (MAF) sensor – signal high	**Wiring, MAF sensor, ECM**
61	Engine coolant temperature (ECT) sensor – signal low	**Wiring, ECT sensor, ECM**
63	Throttle position (TP) sensor – signal low	**Wiring, TP sensor, ECM**

Trouble code	Fault location	Probable cause
64	Intake air temperature (IAT) sensor – signal low	**Wiring, IAT sensor, ECM**
66	Mass air flow (MAF) sensor – signal low	**Wiring, MAF sensor, ECM**
67	Park neutral position (PNP) switch – AT	**Wiring, switch**
67	Clutch pedal position (CPP) switch/neutral position (NP) switch – MT – open or short circuit	**Wiring, CPP switch, NP switch**
72	Mass air flow (MAF) sensor – abnormal signal	**Wiring, MAF sensor, ECM**
73	Throttle position (TP) sensor – abnormal signal – open throttle	**Wiring, TP sensor, ECM**
74	Brake pedal position (BPP) switch – open or short circuit	**Wiring, BPP switch**
77	Dynamic response test – incorrect rpm change during diagnostic test	**Incorrect operator input, ECM**
79	A/C switch	**Wiring, A/C switch**
86	AT – shift solenoid	**Wiring, shift solenoid**
87	Fuel pump – open or short circuit	**Wiring, fuel pump, relay**
89	AT – clutch converter override	**Wiring**
95	Fuel pump – open or short circuit	**Wiring, fuel pump, relay**
96	Fuel pump – open circuit	**Wiring, fuel pump, relay**
98	Engine control module (ECM)	**ECM**

Trouble code identification – 1992-95

Flash code	Fault location	Probable cause
111	No fault	–
112	Intake air temperature (IAT) sensor – signal low	**Wiring, IAT sensor, ECM**
113	Intake air temperature (IAT) sensor – signal high	**Wiring, IAT sensor, ECM**
114	Intake air temperature (IAT) sensor – abnormal signal	**Wiring, IAT sensor, ECM**
116	Engine coolant temperature (ECT) sensor – abnormal signal	**Wiring, ECT sensor, ECM**
117	Engine coolant temperature (ECT) sensor – signal low	**Wiring, ECT sensor, ECM**
118	Engine coolant temperature (ECT) sensor – signal high	**Wiring, ECT sensor, ECM**
121	Throttle position (TP) sensor – abnormal signal	**Wiring, TP sensor, ECM**

Flash code	Fault location	Probable cause
122	Throttle position (TP) sensor – signal low	**Wiring, TP sensor, ECM**
123	Throttle position (TP) sensor – signal high	**Wiring, TP sensor, ECM**
124	Throttle position (TP) sensor – signal high	**Wiring, TP sensor, ECM**
129	Mass air flow (MAF) sensor – abnormal	**Wiring, MAF sensor, ECM**
136	Heated oxygen sensor (HO2S) – LH – signal low	**Wiring, HO2S, fuel pressure, fuel system, mechanical fault, ECM**
137	Heated oxygen sensor (HO2S) – LH – signal high	**Wiring, HO2S, fuel pressure, fuel system, mechanical fault, ECM**
139	Heated oxygen sensor (HO2S) – LH – open or short circuit	**Wiring**
144	Heated oxygen sensor (HO2S) – RH – open or short circuit	**Wiring**
157	Mass air flow (MAF) sensor – signal low	**Wiring, MAF sensor, ECM**
158	Mass air flow (MAF) sensor – signal high	**Wiring, MAF sensor, ECM**
159	Mass air flow (MAF) sensor – abnormal signal	**Wiring, MAF sensor, ECM**
167	Throttle position (TP) sensor – abnormal signal	**Wiring, TP sensor, ECM**
171	Adaptive fuel control – RH – exceeds limits	**HO2S, fuel pressure, fuel system, mechanical fault, ECM**
172	Heated oxygen sensor (HO2S) – RH – signal low	**Wiring, HO2S, fuel pressure, fuel system, mechanical fault, ECM**
173	Heated oxygen sensor (HO2S) – RH – signal high	**Wiring, HO2S, fuel pressure, fuel system, mechanical fault, ECM**
175	Adaptive fuel control – LH – exceeds limits	**HO2S, fuel pressure, fuel system, mechanical fault, ECM**
176	Heated oxygen sensor (HO2S) – LH – signal low	**Wiring, HO2S, fuel pressure, fuel system, mechanical fault, ECM**
177	Heated oxygen sensor (HO2S) – LH – signal high	**Wiring, HO2S, fuel pressure, fuel system, mechanical fault, ECM**
179	Adaptive fuel control – RH – signal low – open throttle	**HO2S, fuel pressure, fuel system, mechanical fault, ECM**
181	Adaptive fuel control – RH – signal high – open throttle	**HO2S, fuel pressure, fuel system, mechanical fault, ECM**
182	Adaptive fuel control – RH – signal low – idle	**HO2S, fuel pressure, fuel system, mechanical fault, ECM**
183	Adaptive fuel control – RH – signal high – idle	**HO2S, fuel pressure, fuel system, mechanical fault, ECM**
184	Mass air flow (MAF) sensor – signal high	**Wiring, MAF sensor, ECM**
185	Mass air flow (MAF) sensor – signal low	**Wiring, MAF sensor, ECM**
186	Injector – pulse width long	**Wiring, injector**
187	Injector – pulse width short	**Wiring, injector**
188	Adaptive fuel control – LH – signal low – open throttle	**HO2S, fuel pressure, fuel system, mechanical fault, ECM**

Flash code	Fault location	Probable cause
189	Adaptive fuel control – LH – signal high – open throttle	**HO2S, fuel pressure, fuel system, mechanical fault, ECM**
191	Adaptive fuel control – LH – signal low – idle	**HO2S, fuel pressure, fuel system, mechanical fault, ECM**
192	Adaptive fuel control – LH – signal high – idle	**HO2S, fuel pressure, fuel system, mechanical fault, ECM**
211	Camshaft position (CMP) sensor – signal absent	**Wiring, CMP sensor, ECM**
212	Ignition diagnostic monitor – ICM – signal absent	**Wiring, ignition control module**
213	Spark output – open circuit	**Wiring, ignition control module**
214	Camshaft position (CMP) sensor – open or short circuit	**Wiring, CMP sensor**
215	Ignition coil – No.1 primary – open or short circuit	**Wiring, coil**
216	Ignition coil – No.2 primary – open or short circuit	**Wiring, coil**
217	Ignition coil – No.3 primary – open or short circuit	**Wiring, coil**
226	Ignition diagnostic monitor – ICM – signal absent	**Wiring, ignition control module**
232	Ignition coil – Nos.1, 2 & 3 primary – open or short circuit	**Wiring, coil**
326	Exhaust gas pressure sensor – signal low	**Wiring, pressure sensor, mechanical fault**
327	Exhaust gas pressure sensor – signal low	**Wiring, pressure sensor, mechanical fault**
332	Exhaust gas pressure sensor – insufficient EGR flow	**Wiring, pressure sensor, mechanical fault**
335	Exhaust gas pressure sensor – abnormal signal during diagnostic test	**Wiring, pressure sensor, mechanical fault**
336	Exhaust gas pressure sensor – signal high	**Wiring, pressure sensor, mechanical fault**
337	Exhaust gas pressure sensor – signal high	**Wiring, pressure sensor, mechanical fault**
341	Octane adjust circuit – open circuit	**Wiring, octane adjuster, ECM**
411	Idle air control (IAC) valve – low rpm function during diagnostic test	**Wiring, IAC valve, mechanical fault**
412	Idle air control (IAC) valve – high rpm function during diagnostic test	**Wiring, IAC valve, mechanical fault**
415	Idle air control (IAC) valve – minimum limit	**Wiring, IAC valve, ECM**
416	Idle air control (IAC) valve – maximum limit	**Wiring, IAC valve, ECM**
452	Vehicle speed sensor (VSS) – signal absent	**Wiring, VSS, ECM**
511	Engine control module (ECM) – memory function during diagnostic test	**ECM**
512	Engine control module (ECM) – supply voltage	**Wiring, fuses**

Flash code	Fault location	Probable cause
513	Engine control module (ECM)	ECM
522	Park/neutral position (PNP) switch	Wiring, PNP switch, mechanical fault
525	A/C switch	Wiring, switch
528	Clutch pedal position (CPP) switch/neutral position (NP) switch – MT – open or short circuit	Wiring, CPP switch, NP switch, mechanical fault
536	Brake pedal position (BPP) switch – open or short circuit	Wiring, BPP switch, mechanical fault
538	Dynamic response test – incorrect rpm change during diagnostic test	Incorrect operator input, ECM
539	A/C switch	Wiring, A/C switch
542	Fuel pump – open or short circuit	Wiring, fuel pump, relay
543	Fuel pump – open circuit	Wiring, fuel pump, relay
556	Fuel pump – open or short circuit	Wiring, fuel pump, relay
558	Exhaust gas recirculation (EGR) solenoid – open or short circuit	Wiring, EGR solenoid
565	Evaporative emission (EVAP) canister purge valve – open or short circuit	Wiring, EVAP solenoid
566	Shift solenoid 3-4 – AT – open or short circuit	Wiring, shift solenoid
629	Torque converter clutch solenoid – AT – open or short circuit	Wiring, clutch solenoid
998	Engine control module (ECM)	ECM

MAZDA

Model:	Engine identification:	Year:
MPV 2.6/3.0L	**G6/JE**	**1994-95**
MX-3 1.6/1.8L	**B6/K8**	**1992-95**
MX-5 Miata 1.6/1.8L	**B6/BP**	**1990-95**
Protegé 1.8L	**BP**	**1990-94**
323 1.6/1.8L	**B6/BP**	**1991-94**

System: **EGI**

Self-diagnosis

General information

- Refer to the front of this manual for general test conditions, terminology, detailed descriptions of wiring faults and a general trouble shooter for electrical and mechanical faults.
- Engine control module (ECM) incorporates self-diagnosis function.
- ECM operates in backup mode if sensors fail, to enable vehicle to be driven to workshop.
- Malfunction indicator lamp (MIL) will illuminate if certain faults are recorded.
- Trouble codes can be displayed with an LED test lamp or accessed with a suitable code reader connected to the data link connector (DLC) **1**.

Accessing

- Ensure ignition switched OFF.
- Connect LED tester between data link connector (DLC) terminal FEN and battery (12 volt) positive **2**.
- Jump DLC terminals TEN and GND **2**.
- Switch ignition ON.
- Count LED flashes and compare with trouble code table.
  - 1.2 second flashes indicate the 'tens' of the trouble code **3** [**A**].
  - 0.4 second flashes indicate the 'units' of the trouble code **3** [**D**].
  - A 1.6 second pause (LED OFF) separates 'TENS' and 'UNITS' **3** [**C**].
  - A 0.4 second pause (LED OFF) separates each flash **3** [**B**].
  - A 4 second pause separates each trouble code **3** [**E**].
  - For example: Trouble code 22 displayed **3**.
- Switch ignition OFF.

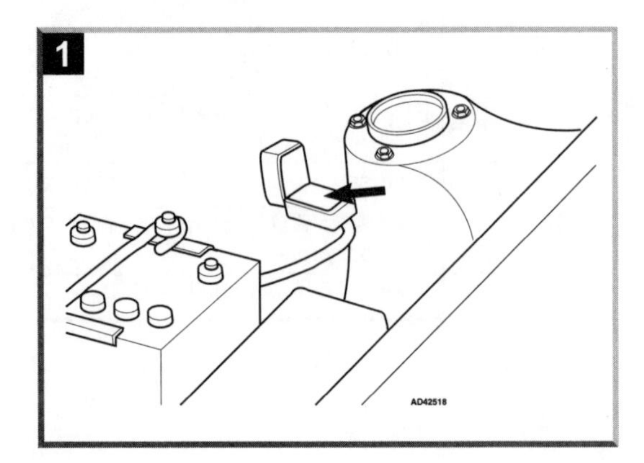

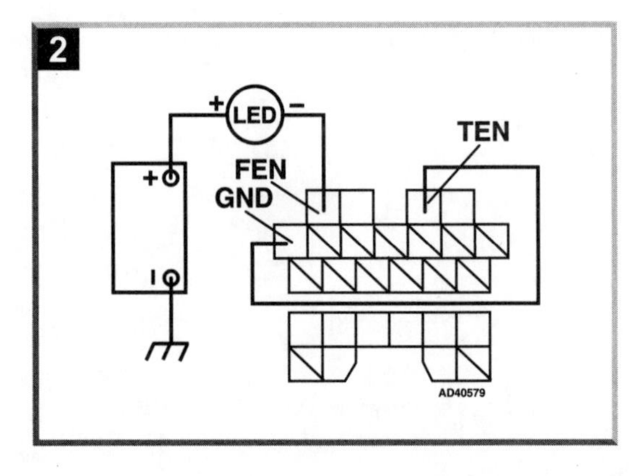

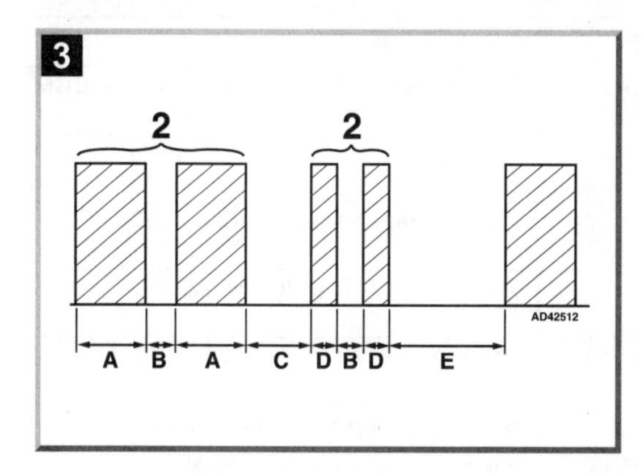

Erasing

- Ensure ignition switched OFF.
- Disconnect battery ground lead.
- Depress the brake pedal for a minimum of 5 seconds.
- Reconnect battery ground lead.
- Connect LED tester between DLC terminal FEN and battery (12 volt) positive **2**.
- Jump DLC terminal TEN and GND **2**.
- Switch ignition ON. Do not start engine.
- Wait 6 seconds.
- Start engine.
- Ensure engine is at normal operating temperature.
- Run engine at 2000 rpm for 3 minutes.
- Check that no trouble codes are displayed.
- Switch ignition OFF.
- Remove jump lead and LED tester.

NOTE: ***Disconnecting battery ground lead will also erase trouble codes but may erase memory from electronic units such as radio.***

Trouble code identification

Flash code	Fault location	Probable cause
01	Ignition signal	**Engine fault, distributor, ignition coil wiring, ECM**
02	Crankshaft position (CKP) sensor	**Wiring, distributor, ECM**
03	Camshaft position (CMP) sensor 1	**Wiring, distributor, ECM**
04	Camshaft position (CMP) sensor 2	**Wiring, distributor, ECM**
05	Knock sensor (KS)	**Wiring, KS sensor, ECM**
06	Vehicle speed sensor (VSS)	**Wiring, VSS, ECM**
08	Volume air flow (VAF) sensor	**Wiring, VAF sensor, ECM**
09	Engine coolant temperature (ECT) sensor	**Wiring, ECT sensor, ECM**
10	Intake air temperature (IAT) sensor – in VAF sensor	**Wiring, VAF sensor, ECM**
11	Intake air temperature (IAT) sensor – in manifold	**Wiring, IAT sensor, ECM**
12	Throttle position (TP) sensor	**Wiring, TP sensor, ECM**
14	Barometric pressure (BARO) sensor	**ECM**
15	Oxygen sensor (O2S) – 4 cyl	**Spark plugs, air leaks, fuel pressure, O2S, ECM**
15	Heated oxygen sensor (HO2S) – LH or front bank (6 cyl)	**Spark plugs, air leaks, fuel pressure, HO2S, ECM**

Flash code	Fault location	Probable cause
16	Exhaust gas recirculation (EGR) valve position sensor – short/open circuit	**Wiring, EGR position sensor, valve, ECM**
17	Oxygen sensor (O2S) – feedback system – 4 cyl	**Spark plugs, air leaks, O2S, ECM**
17	Heated oxygen sensor (HO2S) – LH or front bank (6 cyl)	**Spark plugs, air leaks, HO2S, ECM**
23	Heated oxygen sensor (HO2S) – RH or rear bank (6 cyl)	**Spark plugs, air leaks, HO2S, ECM**
24	Heated oxygen sensor (HO2S) – RH or rear bank (6 cyl)	**Spark plugs, air leaks, HO2S, ECM**
25	Fuel pressure (FP) control solenoid	**Wiring, FP control solenoid, ECM**
26	Evaporative emission (EVAP) canister purge valve	**Wiring, EVAP solenoid, ECM**
28	Exhaust gas recirculation (EGR) solenoid	**Wiring, EGR solenoid, ECM**
29	Exhaust gas recirculation (EGR) solenoid	**Wiring, EGR solenoid, ECM**
34	Idle air control (IAC) valve	**Wiring, IAC solenoid, ECM**
35	Fuel pressure (FP) control solenoid	**Wiring, FP control solenoid, ECM**
36	Heated oxygen sensor (HO2S) heater – (6 cyl)	**Wiring, HO2S**
41	Intake manifold air control solenoid 1	**Wiring, air control solenoid, ECM**
46	Intake manifold air control solenoid 2	**Wiring, air control solenoid, ECM**
55	Pulse generator – MX-3 AT	**Wiring**
56	AT – fluid temperature sensor	**Wiring, temperature sensor**
60	1-2 shift solenoid valve – AT	**Wiring, mechanical fault**
61	2-3 shift solenoid valve – AT	**Wiring, mechanical fault**
62	3-4 shift solenoid valve – AT	**Wiring, mechanical fault**
63	Solenoid valve (lockup control) – AT	**Wiring, mechanical fault**

MAZDA

Model:	Engine identification:	Year:
626/MX-6 2.0L	FS-DOHC	1993-95
626/MX-6 2.5L	KL V6	1993-95
929 3.0L	JE	1993-95

System: **EGI**

Self-diagnosis

General information

- Refer to the front of this manual for general test conditions, terminology, detailed descriptions of wiring faults and a general trouble shooter for electrical and mechanical faults.
- Engine control module (ECM) incorporates self-diagnosis function.
- Malfunction indicator lamp (MIL) will illuminate if certain faults are recorded.
- ECM operates in backup mode if sensors fail, to enable car to be driven to workshop.
- Trouble codes can be accessed with an LED test lamp or suitable code reader connected to the data link connector (DLC):
  - except 929 – **1**.
  - 929 – **2**.

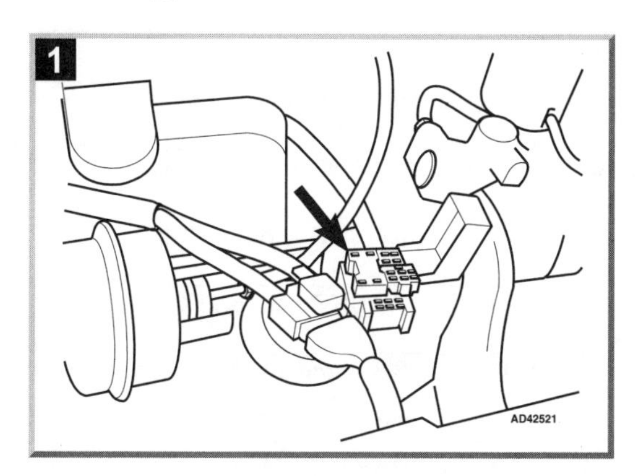

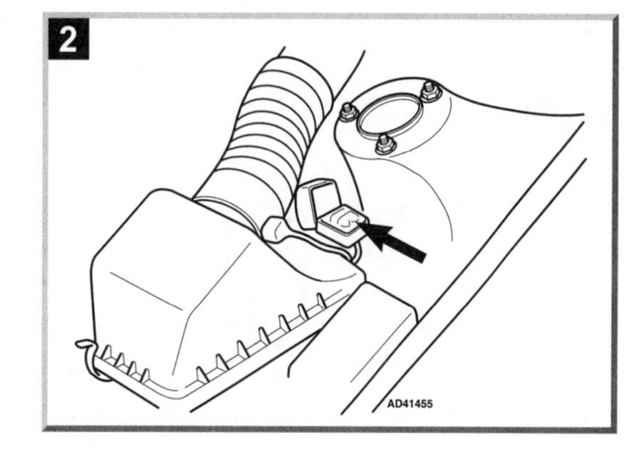

Accessing

- Ensure ignition switched OFF.
- Connect LED tester between data link connector (DLC) terminal FEN and battery (12 volt) positive **3**.
- Jump DLC terminals TEN and GND **3**.
- Switch ignition ON.
- Count LED flashes and compare with trouble code table.
  - 1.2 second flashes indicate the 'tens' of the trouble code **4** [**A**].
  - 0.4 second flashes indicate the 'units' of the trouble code **4** [**D**].
  - A 1.6 second pause (LED OFF) separates 'TENS' and 'UNITS' **4** [**C**].
  - A 0.4 second pause (LED OFF) separates each flash **4** [**B**].
  - A 4 second pause separates each trouble code **4** [**E**].
  - For example: Trouble code 22 displayed **4**.
- Switch ignition OFF.
- Remove jump lead and LED tester.

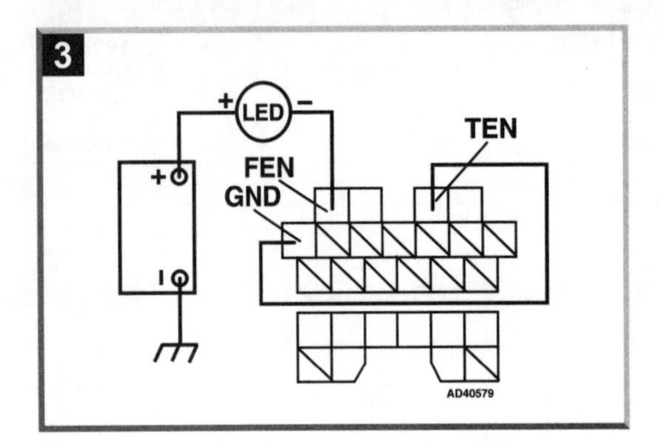

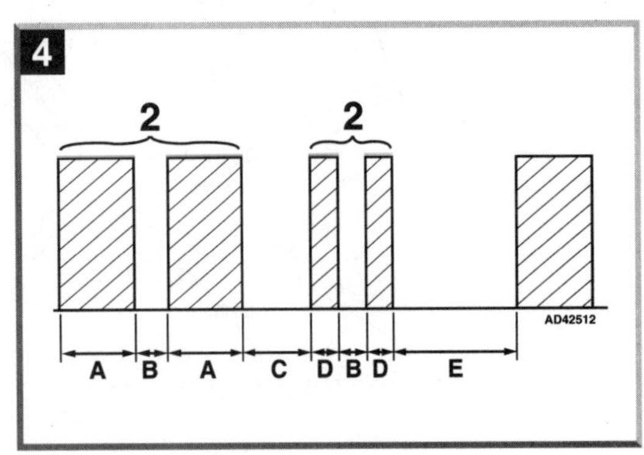

Erasing

- Ensure ignition switched OFF.
- Disconnect battery ground lead.
- Depress the brake pedal for a minimum of 5 seconds.
- Reconnect battery ground lead.
- Connect LED tester between DLC terminal FEN and battery (12 volt) positive **3**.
- Jump DLC terminal TEN and GND **3**.
- Switch ignition ON. Do not start engine.
- Wait 6 seconds.
- Start engine.
- Ensure engine is at normal operating temperature.
- Run engine at 2000 rpm for 3 minutes.
- Check that no trouble codes are displayed.
- Switch ignition OFF.
- Remove jump lead and LED tester.

NOTE: ***Disconnecting battery ground lead will also erase trouble codes but may erase memory from electronic units such as radio.***

Trouble code identification

Flash code	Fault location	Probable cause
01	Ignition signal	Engine fault, distributor, ignition coil wiring, ECM
02	Crankshaft position (CKP) sensor/camshaft position (CMP) sensor	Wiring, CKP/CMP sensor, ECM
03	Crankshaft position (CKP) sensor/camshaft position (CMP) sensor	Wiring, CKP/CMP sensor, ECM
04	Crankshaft position (CKP) sensor/camshaft position (CMP) sensor	Wiring, CKP/CMP sensor, ECM
05	Knock sensor (KS) – LH or single	Wiring, KS sensor, ECM
06	Vehicle speed sensor (VSS)	Wiring, VSS sensor, ECM
07	Knock sensor (KS) – RH	Wiring, KS sensor, ECM
08	Mass air flow (MAF)/volume air flow (VAF) sensor	Wiring, MAF/VAF sensor, ECM
09	Engine coolant temperature (ECT) sensor	Wiring, ECT sensor, ECM
10	Intake air temperature (IAT) sensor – in VAF sensor	Wiring, VAF sensor, ECM

Flash code	Fault location	Probable cause
11	Intake air temperature (IAT) sensor – in manifold	**Wiring, IAT sensor, ECM**
12	Throttle position (TP) sensor	**Wiring, TP sensor, ECM**
14	Barometric pressure (BARO) sensor	**ECM**
15	Heated oxygen sensor (HO2S)/oxygen sensor (O2S) – LH or single	**Spark plugs, air leaks, fuel pressure, HO2S/O2S, ECM**
16	Exhaust gas recirculation (EGR) valve position sensor	**Wiring, EGR sensor, valve, ECM**
17	Heated oxygen sensor (HO2S)/oxygen sensor (O2S) – LH or single	**Spark plugs, air leaks, HO2S/O2S, ECM**
23	Heated oxygen sensor (HO2S) – RH	**Spark plugs, air leaks, HO2S, ECM**
24	Heated oxygen sensor (HO2S) feedback system – RH	**Spark plugs, air leaks, HO2S, ECM**
25	Fuel pressure (FP) control solenoid	**Wiring, FP control solenoid, ECM**
26	Evaporative emission (EVAP) canister purge valve	**Wiring, EVAP solenoid, ECM**
28	Exhaust gas recirculation (EGR) solenoid	**Wiring, EGR solenoid, ECM**
29	Exhaust gas recirculation (EGR) solenoid	**Wiring, EGR solenoid, ECM**
30	Cold start injector relay	**Wiring, relay**
34	Idle speed control (ISC) actuator/idle air control (IAC) valve	**Wiring, ISC/IAC, ECM**
36	Oxygen sensor (O2S) heaters	**Wiring, O2S**
37	Oxygen sensor (O2S) heaters	**Wiring, O2S**
41	Intake manifold air control solenoid 1	**Wiring, air control solenoid, ECM**
46	Intake manifold air control solenoid 2	**Wiring, air control solenoid, ECM**
55	Pulse generator – AT	**Wiring**
56	AT – fluid temperature sensor	**Wiring, fluid temperature sensor**
60	1-2 shift solenoid valve – AT	**Wiring, mechanical fault**
61	2-3 shift solenoid valve – AT	**Wiring, mechanical fault**
62	3-4 shift solenoid valve – AT	**Wiring, mechanical fault**
63	Solenoid valve (lockup control) – AT	**Wiring, mechanical fault**
64	Solenoid valve (3-2 timing) – AT	**Wiring**
65	A/C control signal – 929	**Wiring**
65	Solenoid valve (lockup) – AT – except 929	**Wiring**
66	Solenoid valve (line pressure) – AT	**Wiring, mechanical fault**
67	Coolant fan relay No.1	**Wiring, relay**
68	Coolant fan relay Nos.2 & 3	**Wiring, relay**
69	Engine coolant temperature (ECT) sensor (fan)	**Wiring, ECT sensor**

MERCEDES-BENZ

Model:	**C-Class (202) • C-Class (203) • E-Class (210) • M-Class (163) S-Class (220) • CL (215) • CLK (208) • CLK (209) • SL (129) • SL (230) SLK (170)**
Year:	**2000-04**
Engine identification:	**111, 112, 113**
System:	**Bosch ME 2.0/2.1/2.8 • Siemens ME SIM4**

Data link connector (DLC) locations

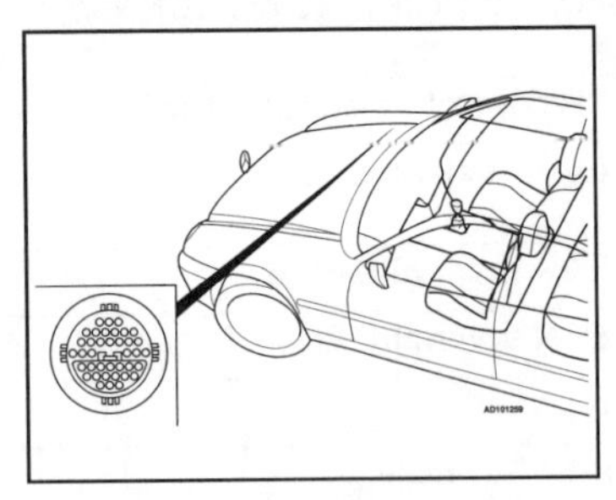

38-pin – C-Class (202)

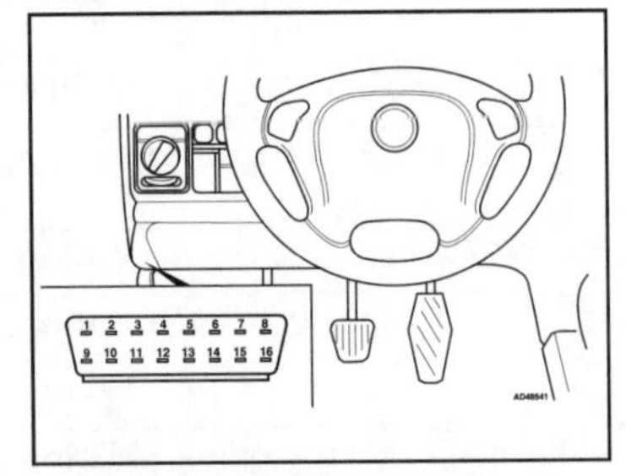

16-pin – C-Class (203), E-Class (210), M-Class (163), S-Class (220), CL (215), CLK (208/209), SL (129/230), SLK (170)

Self-diagnosis

General information

- Refer to the front of this manual for general test conditions, terminology, detailed descriptions of wiring faults and a general trouble shooter for electrical and mechanical faults.

Accessing and erasing

- The engine control module (ECM) fault memory can only be accessed and erased using diagnostic equipment connected to the data link connector (DLC).

Trouble code identification

EOBD type	Fault location	Probable cause
All P0, P2 and U0 codes **1**	Refer to OBD-II trouble code tables at the front of this manual	–
P0801	Engine coolant blower motor/AC condenser blower motor – malfunction	**Wiring, engine coolant blower motor/ AC condenser blower motor**

EOBD type	Fault location	Probable cause
P0802	Intake manifold air control solenoid – malfunction	Wiring, intake manifold air control solenoid
P0803	Supercharger (SC) bypass valve motor – malfunction	Wiring, SC bypass valve motor, ECM
P0805	Supercharger (SC) pressure – incorrect	Supercharger (SC) bypass valve motor sticking, intake leak/blockage, SC, ECM
P0806	Supercharger (SC) clutch – malfunction	Wiring, SC clutch, ECM
P0809	Crankshaft position (CKP) sensor/camshaft position (CMP) sensor – signal incompatibility	Incorrect valve timing
P0811	CAN data bus, ignition switch control module – malfunction	Wiring, ignition switch control module, ECM
P1000	Transmission shift lever control module – defective	Transmission shift lever control module
P1031	Heated oxygen sensor (HO2S) 1 – malfunction	Wiring, connector, HO2S
P1146	Mass air flow (MAF) sensor(s) – malfunction	Wiring, intake leak, MAF sensor(s)
P1147	Engine coolant temperature (ECT) sensor(s) – malfunction	Wiring, ECT sensor(s)
P1148	Intake air temperature (IAT) sensor(s) – malfunction	Wiring, IAT sensor(s)
P1149	Manifold absolute pressure (MAP) sensor(s) – malfunction	Wiring, hose(s), MAP sensor(s), ECM
P1162	Throttle position (TP) sensor(s) – malfunction	Wiring, TP sensor(s)
P1163	Engine oil level switch – malfunction	Wiring, engine oil level switch, ECM
P1177	Engine oil sensor, quality/level/temperature – malfunction	Wiring, engine oil sensor
P1178	Engine oil sensor, level – implausible signal	Wiring, oil level, engine oil sensor
P1179	Engine oil sensor, quality – implausible signal	Wiring, engine oil sensor
P1180	Engine oil sensor, temperature – too high	Wiring, engine oil sensor
P1181	Engine coolant blower motor/AC condenser blower motor – malfunction	Wiring, engine coolant blower motor/ AC condenser blower motor, ECM
P1182	Starter motor relay – defective	Wiring, starter motor relay, ignition switch, ECM
P1183	Cylinder cut-out solenoid, bank 1 – defective	Wiring, cylinder cut-out solenoid
P1184	Cylinder cut-out solenoid, bank 2 – defective	Wiring, cylinder cut-out solenoid
P1185	Engine oil sensor, quality – water in oil	Engine oil
P1186	Fuel shut-off – recognised	Wiring, cruise control actuator, mechanical fault, ECM
P1225	Intake manifold air control solenoid – malfunction	Wiring, intake manifold air control solenoid, ECM
P1233	Throttle actuator – mechanical fault	Throttle actuator tight/sticking
P1235	Supercharger (SC) pressure – incorrect	Supercharger (SC) bypass valve motor sticking, intake leak/blockage, SC, ECM

EOBD type	Fault location	Probable cause
P1236	Supercharger (SC) clutch – malfunction	**Wiring, SC clutch, ECM**
P1300	Crankshaft position (CKP) sensor(s) – malfunction	**Wiring, flywheel ring gear damaged, CKP sensor(s)**
P1355	Cylinder cut-out solenoid, bank 1 – solenoid stuck open	**Wiring, cylinder cut-out solenoid**
P1356	Cylinder cut-out solenoid, bank 2 – solenoid stuck open	**Wiring, cylinder cut-out solenoid**
P1357	Cylinder cut-out system ON, cylinder 2, 3, 5 or 8 – inlet valve still opening	**Mechanical fault**
P1358	Cylinder cut-out system OFF, cylinder 5 – exhaust valve not opening	**Mechanical fault**
P1359	Cylinder cut-out system OFF, cylinder 2 – exhaust valve not opening	**Mechanical fault**
P1360	Cylinder cut-out system OFF, cylinder 3 – exhaust valve not opening	**Mechanical fault**
P1361	Cylinder cut-out system OFF, cylinder 8 – exhaust valve not opening	**Mechanical fault**
P1366	Cylinder cut-out system, throttle switch-over valve – malfunction	**Wiring, throttle switch-over valve**
P1380	Cylinder cut-out system OFF – one inlet valve of a cylinder is not opening	**Mechanical fault**
P1384	Knock sensor (KS) 1, bank 2 – malfunction	**Wiring, KS**
P1385	Knock sensor (KS) 2, bank 2 – malfunction	**Wiring, KS**
P1386	Engine control module (ECM), knock control – defective	**Fuel contamination, mechanical fault, ECM**
P1386	Engine control module (ECM), bank 1, knock control – defective	**Fuel contamination, mechanical fault, ECM**
P1397	Camshaft position (CMP) sensor(s) – malfunction	**Wiring, CMP sensor(s)**
P1400	Exhaust gas recirculation (EGR) solenoid – circuit malfunction	**Wiring, EGR solenoid, hose(s), ECM**
P1420	Secondary air injection (AIR) valve/solenoid – defective	**Wiring, fuse, AIR valve/solenoid, ECM**
P1437	Catalytic converter temperature sensor, bank 1 – defective	**Wiring, catalytic converter temperature sensor, ECM**
P1443	Evaporative emission (EVAP) canister purge valve(s) – defective	**Wiring, hoses, EVAP canister purge valve(s)**
P1444	Catalytic converter temperature sensor, bank 2 – defective	**Wiring, catalytic converter temperature sensor, ECM**
P1453	Secondary air injection (AIR) pump relay – defective	**Wiring, fuse, AIR pump relay, ECM**
P1463	Secondary air injection (AIR) system – defective	**Wiring, hoses, AIR pump relay, AIR valve/ solenoid**
P1490	Evaporative emission (EVAP) canister purge valve(s) – defective	**Wiring, hoses, EVAP canister purge valve(s)**

EOBD type	Fault location	Probable cause
P1491	AC system – pressure too high	**AC control module trouble code(s) stored**
P1519	Camshaft position (CMP) control, bank 1 – mechanical fault	**Camshaft position control system**
P1522	Camshaft position (CMP) control, bank 2 – mechanical fault	**Camshaft position control system**
P1525	Camshaft position (CMP) actuator, bank 1 – defective	**Wiring, CMP actuator, ECM**
P1533	Camshaft position (CMP) actuator, bank 2 – defective	**Wiring, CMP actuator, ECM**
P1542	Accelerator pedal position (APP) sensor – signal	**Wiring, APP sensor, ECM**
P1551	AC compressor clutch cut-off	**Wiring, incorrect AC control module, ECM**
P1570	Ignition switch control module/ECM – attempted theft/data bus/not compatible	**Wiring, trouble code(s) stored in other system(s), ignition switch control module, ECM**
P1580	Throttle motor, bank 1, TPM/ISC – circuit malfunction	**Wiring, throttle motor, ECM**
P1581	Throttle motor, bank 2, TPM/ISC – circuit malfunction	**Wiring, throttle motor, ECM**
P1584	Brake pedal position (BPP) switch – malfunction	**Wiring, BPP switch, trouble code(s) stored in other system(s)**
P1587	Engine control module (ECM), bank 1 & 2 – supply voltage	**Wiring, connectors, ECM**
P1588	CAN data bus, central locking control module – malfunction	**Wiring**
P1589	Engine control module (ECM), bank 2, knock control – defective	**Fuel contamination, mechanical fault, ECM**
P1603	CAN data bus, ignition switch control module – malfunction	**Wiring**
P1605	Acceleration sensor, ESP/ASR system – rough road signal, comparison of wheel speeds	**Trouble code(s) stored in other system(s)**
P1632	Engine control module (ECM), bank 2 – internal fault	**ECM**
P1641	Engine control module (ECM), bank 1 & 2 – TP signal	**Wiring, ECM**
P1642	Engine control module (ECM) – incorrectly coded	**ECM coded for MT with AT fitted**
P1643	Engine control module (ECM) – incorrectly coded/data bus	**ECM coded for AT with MT fitted, wiring**
P1644	Transmission control module (TCM) – supply voltage low	**Wiring, TCM trouble code(s) stored**
P1666	Cylinder cut-out solenoid, bank 1 or 2 – does not open	**Wiring, cylinder cut-out solenoid**
P1681	Engine control module (ECM), crash signal – plausibility	**Wiring, connectors, SRS control module trouble code(s) stored, ECM**

EOBD type	Fault location	Probable cause
P1747	CAN data bus, TCM/instrument panel – malfunction	**Wiring, trouble code(s) stored in other system(s)**
P1750	Transmission shift lever control module – supply voltage low	**Wiring, multifunction control module, transmission shift lever control module**
P1817	Reverse lamps – short to ground	**Wiring short to ground, reverse lamps switch**
P1817	Reverse lamps – short to positive	**Wiring short to positive, reverse lamps switch**
P1817	Reverse lamps – supply voltage, open circuit	**Wiring, reverse lamps switch**
P1832	Transmission shift lever control module – short circuit	**Wiring, connector, transmission shift lever control module**
P1833	Transmission shift lever control module – open circuit	**Wiring, connector, transmission shift lever control module**
P1856	Transmission shift lever recognition – malfunction	**Transmission shift lever control module**
P1860	CAN data bus, VSS signal, rear axle – malfunction	**Wiring, trouble code(s) stored in other system(s)**
P1861	CAN data bus, VSS signal, rear axle – malfunction	**Wiring, trouble code(s) stored in other system(s)**
P1875	CAN data bus, transmission shift lever control module – malfunction	**Wiring, transmission shift lever control module**
P1876	CAN data bus, VSS signal – malfunction	**Wiring, trouble code(s) stored in other system(s), electronic traction control (ETC) module**
P1904	CAN data bus, TCM/instrument panel – malfunction	**Wiring, trouble code(s) stored in other system(s)**
P1906	CAN data bus, stop lamp switch – malfunction	**Wiring, trouble code(s) stored in other system(s), electronic ignition switch**
P1910	Transmission shift lever control module – supply voltage high	**Wiring, multifunction control module, transmission shift lever control module**
P1911	Transmission shift lever illumination – open circuit	**Transmission shift lever control module**
P1912	Transmission shift lever control module, range switch – implausible signal	**Stored trouble code(s), transmission shift lever control module**
P1925	Transmission shift lever control module, immobilizer signal – plausibility	**Wiring, trouble code(s) stored in other system(s), electronic ignition switch, transmission shift lever control module**

1 Except those P0 codes listed

MERCEDES-BENZ

Model:	**C230K/240/280 (202) • CLK320 (208) • E320 (210) • S320/420/500/600 SL500/600 (129) • CL500/600 (140) • SLK230 (170)**
Year:	**1993-02**
Engine identification:	**104.941/944/994, 111.943/947/973/975, 112.910/911/920/921/940/941, 119.980/981/982/985, 120.982/983**
System:	**Bosch ME 1.0/2.0/ME-SFI 2.1**

Data link connector (DLC) locations

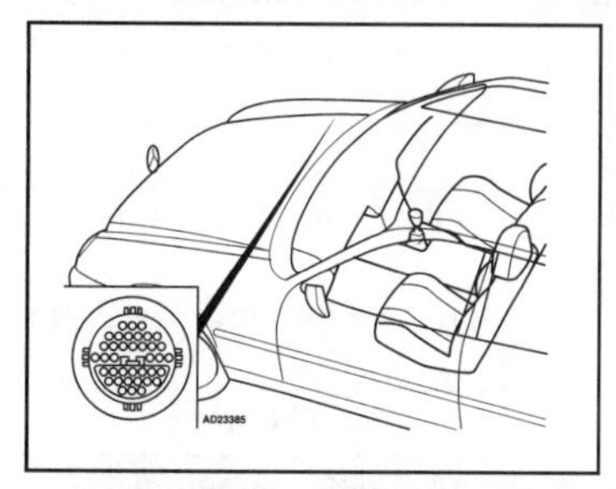

C-Class, CLK, S-Class, SL, SLK – RH engine bay

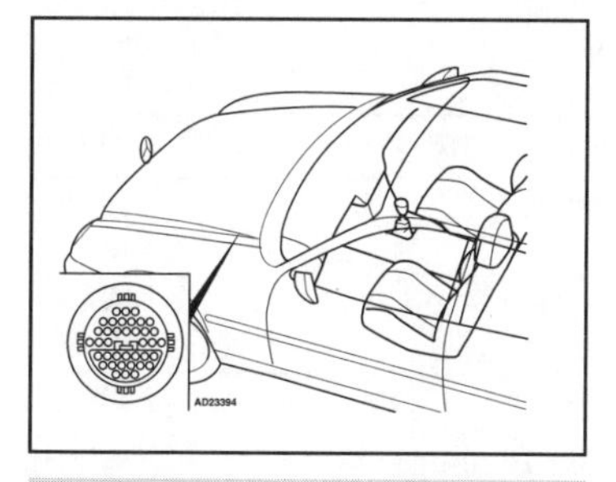

E-Class (210) – LH engine bay

Self-diagnosis

General information

- Refer to the front of this manual for general test conditions, terminology, detailed descriptions of wiring faults and a general trouble shooter for electrical and mechanical faults.

Accessing and erasing

- The engine control module (ECM) fault memory can only be accessed and erased using diagnostic equipment connected to the data link connector (DLC).

Trouble code identification

OBD-II code	Fault location	Probable cause
All P0, P2 and U0 codes	Refer to OBD-II trouble code tables at the front of this manual	–
P1031	Heated oxygen sensor (HO2S) 1 – malfunction	**Wiring, connector, HO2S**
P1146	Mass air flow (MAF) sensor(s) – malfunction	**Wiring, intake leak, MAF sensor(s)**
P1147	Engine coolant temperature (ECT) sensor(s) – malfunction	**Wiring, ECT sensor(s)**
P1148	Intake air temperature (IAT) sensor(s) – malfunction	**Wiring, IAT sensor(s)**

OBD-II code	Fault location	Probable cause
P1149	Manifold absolute pressure (MAP) sensor(s) – malfunction	**Wiring, hose(s), MAP sensor(s), ECM**
P1162	Throttle position (TP) sensor(s) – malfunction	**Wiring, TP sensor(s)**
P1163	Engine oil level switch – malfunction	**Wiring, engine oil level switch, ECM**
P1177	Engine oil sensor, quality/level/temperature – malfunction	**Wiring, engine oil sensor**
P1178	Engine oil sensor, level – implausible signal	**Wiring, oil level, engine oil sensor**
P1179	Engine oil sensor, quality – implausible signal	**Wiring, engine oil sensor**
P1180	Engine oil sensor, temperature – too high	**Wiring, engine oil sensor**
P1181	Engine coolant blower motor/AC condenser blower motor – malfunction	**Wiring, engine coolant blower motor/ AC condenser blower motor, ECM**
P1182	C-Class 06/97 →: Starter motor relay – defective	**Wiring, starter motor relay, ignition switch, ECM**
P1185	Engine oil sensor, quality – water in oil	**Engine oil**
P1186	Fuel shut-off – recognised	**Wiring, cruise control actuator, mechanical fault, ECM**
P1225	Intake manifold air control solenoid – malfunction	**Wiring, intake manifold air control solenoid, ECM**
P1233	Throttle actuator – mechanical fault	**Throttle actuator tight/sticking**
P1300	Crankshaft position (CKP) sensor(s) – malfunction	**Wiring, flywheel ring gear damaged, CKP sensor(s)**
P1384	Knock sensor (KS) 1, bank 2 – malfunction	**Wiring, KS**
P1385	Knock sensor (KS) 2, bank 2 – malfunction	**Wiring, KS**
P1386	Engine control module (ECM), knock control – defective	**Fuel contamination, mechanical fault, ECM**
P1386	Engine control module (ECM), bank 1, knock control – defective	**Fuel contamination, mechanical fault, ECM**
P1397	Camshaft position (CMP) sensor(s) – malfunction	**Wiring, CMP sensor(s)**
P1400	Exhaust gas recirculation (EGR) solenoid – circuit malfunction	**Wiring, EGR solenoid, hose(s), ECM**
P1420	Secondary air injection (AIR) valve/solenoid – defective	**Wiring, fuse, AIR valve/solenoid, ECM**
P1443	Evaporative emission (EVAP) canister purge valve(s) – defective	**Wiring, hoses, EVAP canister purge valve(s)**
P1453	Secondary air injection (AIR) pump relay – defective	**Wiring, fuse, AIR pump relay, ECM**
P1463	Secondary air injection (AIR) system – defective	**Wiring, hoses, AIR pump relay, AIR valve/solenoid**
P1490	Evaporative emission (EVAP) canister purge valve(s) – defective	**Wiring, hoses, EVAP canister purge valve(s)**
P1491	AC system – pressure too high	**AC control module trouble code(s) stored**

OBD-II code	Fault location	Probable cause
P1519	Camshaft position (CMP) control, bank 1 – mechanical fault	**Camshaft position control system**
P1522	Camshaft position (CMP) control, bank 2 – mechanical fault	**Camshaft position control system**
P1525	Camshaft position (CMP) actuator, bank 1 – defective	**Wiring, CMP actuator, ECM**
P1533	Camshaft position (CMP) actuator, bank 2 – defective	**Wiring, CMP actuator, ECM**
P1542	Accelerator pedal position (APP) sensor – signal	**Wiring, APP sensor, ECM**
P1570	Ignition switch control module/ECM – attempted theft/data bus/not compatible	**Wiring, trouble code(s) stored in other system(s), ignition switch control module, ECM**
P1580	Throttle motor, bank 1, TPM/ISC – circuit malfunction	**Wiring, throttle motor, ECM**
P1581	Throttle motor, bank 2, TPM/ISC – circuit malfunction	**Wiring, throttle motor, ECM**
P1584	Brake pedal position (BPP) switch – malfunction	**Wiring, BPP switch, trouble code(s) stored in other system(s)**
P1587	Engine control module (ECM), bank 1 & 2 – supply voltage	**Wiring, connectors, ECM**
P1588	CAN data bus, central locking control module – malfunction	**Wiring**
P1589	Engine control module (ECM), bank 2, knock control – defective	**Fuel contamination, mechanical fault, ECM**
P1603	CAN data bus, ignition switch control module – malfunction	**Wiring**
P1605	Acceleration sensor, ESP/ASR system – rough road signal, comparison of wheel speeds	**Trouble code(s) stored in other system(s)**
P1632	Engine control module (ECM), bank 2 – internal fault	**ECM**
P1641	Engine control module (ECM), bank 1 & 2 – TP signal	**Wiring, ECM**
P1642	Engine control module (ECM) – incorrectly coded	**ECM coded for MT with AT fitted**
P1643	Engine control module (ECM) – incorrectly coded/data bus	**ECM coded for AT with MT fitted, wiring**
P1644	Transmission control module (TCM) – supply voltage low	**Wiring, TCM trouble code(s) stored**
P1681	Engine control module (ECM), crash signal – plausibility	**Wiring, connectors, SRS control module trouble code(s) stored, ECM**
P1747	CAN data bus, TCM/instrument panel – malfunction	**Wiring, trouble code(s) stored in other system(s)**

MERCEDES-BENZ

Model:	**C220/280 (202) • E280/320 (124) • S320 (140) • SL320 (129)**
Year:	**1993-02/95**
Engine identification:	**104.941/942/943/944/945/991/994/995/111.920, 921/940/941/942/944/945/946/960/961/970/974/975**
System:	**Bosch HFM • Bosch HFM-SFI • Siemens PMS**

Data link connector (DLC) locations

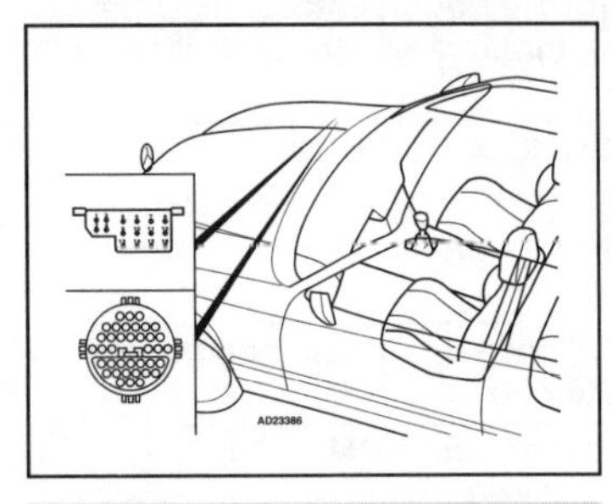

C-Class, E-Class (124), S-Class, SL

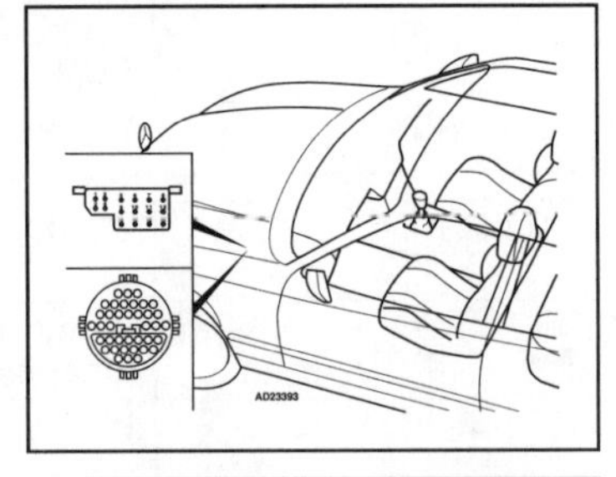

E-Class (210) – LH engine bay

Self-diagnosis

General information

- Refer to the front of this manual for general test conditions, terminology, detailed descriptions of wiring faults and a general trouble shooter for electrical and mechanical faults.
- Trouble codes are displayed by using an LED connected to the data link connector (DLC).
- The ECM fault memory can also be checked using diagnostic equipment connected to the data link connector (DLC).
- Diagnostic equipment can also be used to erase data from ECM fault memory.
- More detailed fault diagnosis information can be obtained using diagnostic equipment connected to the data link connector (DLC).

HFM/HFM/SFI/PMS system

Accessing – flash codes →02/95

- Ensure ignition switched OFF.
- Bridge data link connector (DLC) terminals 1 and 4 or 1 and 8 with a switched lead – contacts normally open **1** or **2**.
- Connect LED test lamp between terminals 3 and 4 or 8 and 16 **1** or **2**.

NOTE: ***Connect LED test lamp positive connection to terminal 3 or 16.***

- Switch ignition ON.
- Operate switch for 2-4 seconds.
- Count LED flashes. Note trouble code.
- Trouble codes consist of short flashes **3**.
- For example: Trouble code 3 displayed **3**.
- Repeat operation. Note trouble codes. Compare with trouble code table.
- Switch ignition OFF.

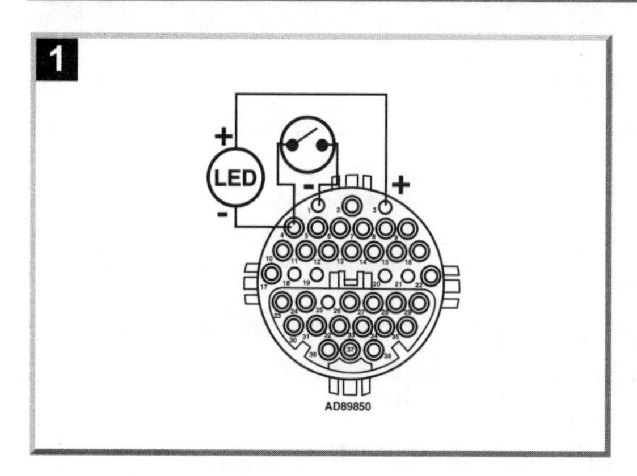

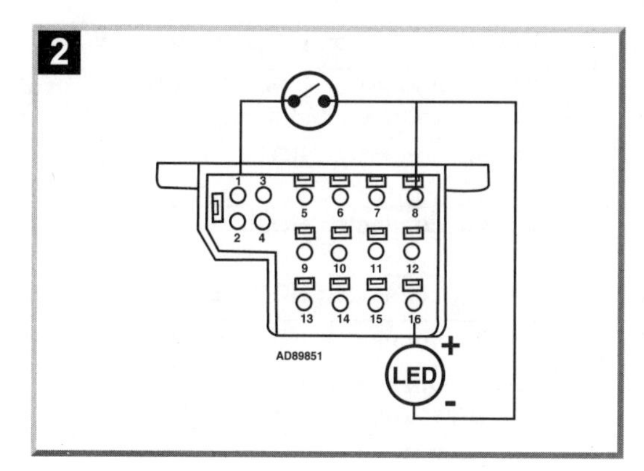

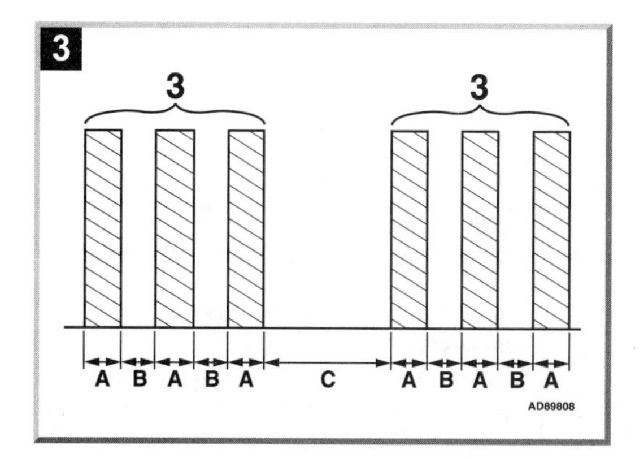

Erasing – flash codes

- Ensure ignition switched OFF.
- Bridge data link connector (DLC) terminals 1 and 4 or 1 and 8 with a switched lead – contacts normally open **1** or **2**.
- Connect LED test lamp between terminals 3 and 4 or 8 and 16 **1** or **2**.

NOTE: ***Connect LED test lamp positive connection to terminal 3 or 16.***

- Switch ignition ON.
- Operate switch for 2-4 seconds.
- Operate switch for 5-6 seconds – Bosch ECM →08/93.
- Allow LED to display trouble code.
- Operate switch for 6-8 seconds.
- Operate switch for 8-9 seconds – Bosch ECM →08/93.
- Repeat operation to erase all stored trouble codes.
- After erasing trouble code 1: Switch ignition OFF.
- Wait 2 seconds.
- Switch ignition ON.
- Wait 10 seconds minimum.
- Start engine.

Trouble code identification

Flash code	Fault location	Probable cause
1	No fault found	–
2	Engine coolant temperature (ECT) sensor – electrical fault	**Wiring, connector, ECT sensor, ECM**
3	Intake air temperature (IAT) sensor – electrical fault	**Wiring, connector, IAT sensor, ECM**
4	Bosch HFM: Mass air flow (MAF) sensor – range/performance problem	**Wiring, connector, intake leak, MAF sensor, ECM**
	Siemens PMS: Manifold absolute pressure (MAP) sensor – vacuum leak	**Hoses, ECM**
5	Idle speed control (ISC) actuator – malfunction	**Wiring, connector, ISC actuator**
6	Throttle position (TP) sensor – range/performance problem	**Wiring, connector, TP sensor, ECM**
7	Idle speed control (ISC) actuator – malfunction	**Wiring, connector, ISC actuator, ECM**
8	Idle speed control (ISC) – malfunction	**Wiring, connector, intake leak, MAF sensor, throttle valve tight/sticking**
9 [1]	Heated oxygen sensor (HO2S) 1 – range/performance problem	**Wiring, connector, HO2S**
10	Heated oxygen sensor (HO2S) 2 – range/performance problem	**Wiring, connector, HO2S**
11	Oxygen sensor heater 1 – electrical fault	**Wiring, connector, HO2S, ECM**
12	Oxygen sensor heater 2 – electrical fault	**Wiring, connector, HO2S, ECM**
13	Lambda regulation – system too lean/rich	**Wiring, intake/exhaust leak, MAF sensor, fuel pressure, injector(s), ECM**
14	Bosch HFM: Injector 1 – circuit malfunction	**Wiring, injector, ECM**
	Siemens PMS: Injector 1 & 4 – circuit malfunction	**Wiring, injector, ECM**
15	Bosch HFM: Injector 2 – circuit malfunction	**Wiring, injector, ECM**
	Siemens PMS: Injector 2 & 3 – circuit malfunction	**Wiring, injector, ECM**
16	Injector 3 – circuit malfunction	**Wiring, injector, ECM**
17	Injector 4 – circuit malfunction	**Wiring, injector, ECM**
18	Injector 5 – circuit malfunction	**Wiring, injector, ECM**
19	Injector 6 – circuit malfunction	**Wiring, injector, ECM**
20	Bosch HFM: Fuel trim (FT), idle speed – range/performance problem	**Intake/exhaust leak, injector(s), fuel pressure, HO2S, ECM**
	Siemens PMS: Fuel trim (FT) – range/performance problem	**Intake/exhaust leak, injector(s), fuel pressure, HO2S, EVAP system, ECM**
21	111: Ignition amplifier/ignition coil, cylinder 1 & 4 – defective	**Wiring, ignition amplifier, ignition coil, ECM**
	104: Ignition amplifier/ignition coil, cylinder 1 & 6 – defective	**Wiring, ignition amplifier, ignition coil, ECM**

Flash code	Fault location	Probable cause
22	Bosch HFM: Ignition amplifier 1/ignition coil, cylinder 1 – defective	**Wiring, ignition amplifier, ignition coil, ECM**
	Siemens PMS: Ignition amplifier/ignition coil, cylinder 2 & 3 – defective	**Wiring, ignition amplifier, ignition coil, ECM**
	104: Ignition amplifier/ignition coil, cylinder 2 & 5 – defective	**Wiring, ignition amplifier, ignition coil, ECM**
23	111: Ignition amplifier 2/ignition coil, cylinder 2 & 3 – defective	**Wiring, ignition amplifier, ignition coil, ECM**
	104: Ignition amplifier/ignition coil, cylinder 3 & 4 – defective	**Wiring, ignition amplifier, ignition coil, ECM**
24	Crankshaft position (CKP) sensor – malfunction	**Wiring, air gap, flywheel ring gear damaged, CKP sensor**
25	Camshaft position (CMP) sensor – incorrect signal	**Wiring, CMP sensor**
26	Engine control module (ECM) coding plug – electrical fault	**Wiring, ECM**
27	Engine control module (ECM), RPM signal – output	**Wiring, RPM sensor, ECM**
28	ABS control module/engine control module (ECM) communication – VSS signal	**Wiring, VSS, ABS control module trouble code(s) stored**
29 2	Intake manifold heater relay – malfunction	**Wiring, intake manifold heater relay, ECM**
30	Fuel pump relay – defective	**Wiring, fuel pump relay**
31	Without cat: CO adjustment resistor – malfunction	**Wiring, CO adjustment resistor, ECM**
32	Knock sensor(s) (KS) – defective	**Wiring, KS**
33	Ignition timing, knocking – maximum retardation reached	**Fuel contamination, mechanical fault**
34	Engine control module (ECM), knock control – internal fault	**ECM**
35	Secondary air injection (AIR) pump/relay – malfunction	**Wiring, AIR pump/relay, ECM**
36	Evaporative emission (EVAP) canister purge valve – electrical fault	**Wiring, EVAP canister purge valve, hose(s), ECM**
37	Transmission shift control valve – electrical fault	**Wiring, transmission shift control valve**
38	Camshaft position (CMP) actuator – malfunction	**Wiring, mechanical fault, CMP actuator, ECM**
39	Exhaust gas recirculation (EGR) solenoid – malfunction	**Wiring, EGR solenoid, hose(s)**
40	Transmission overload protection switch – malfunction	**Wiring, transmission overload protection switch**
41	CAN data bus, ECM – malfunction	**Wiring, ECM**

Flash code	Fault location	Probable cause
42	111: CAN data bus, diagnostic module – malfunction	Wiring
	104: CAN data bus – malfunction	Wiring
43	Engine control module (ECM), pin 50 – starter signal	Wiring
45	With cruise control system: Cruise control, inertia fuel shut-off (IFS) switch – active	Wiring, TP sensor, idle speed control (ISC) actuator
46	Intake manifold air control solenoid – malfunction	Wiring, intake manifold air control solenoid, ECM
48	Oxygen sensor heater relay 2 – electrical fault	Wiring, oxygen sensor heater relay, ECM
49	Engine control module (ECM) – supply voltage	Wiring, ignition switch, overvoltage protection relay
50	Engine control module (ECM) – internal fault	ECM

1 →12/92: Code may be displayed although no fault exists
2 E-Class (124) →07/93

HFM/HFM/SFI/PMS system

Accessing and erasing – scanner codes

- The engine control module (ECM) fault memory can only be accessed and erased using diagnostic equipment connected to the data link connector (DLC).

Trouble code identification

Scan code	Fault location	Probable cause
–	No fault found	–
002	Engine coolant temperature (ECT) sensor – short circuit	Wiring, connector, ECT sensor, ECM
003	Engine coolant temperature (ECT) sensor – open circuit	Wiring, connector, ECT sensor, ECM
004	Engine coolant temperature (ECT) sensor – range/performance problem	–
005	Engine coolant temperature (ECT) sensor – plug connection(s)	Multi-plug(s)
006	Intake air temperature (IAT) sensor – short circuit	Wiring, connector, IAT sensor, ECM
007	Intake air temperature (IAT) sensor – open circuit	Wiring, connector, IAT sensor, ECM

Scan code	Fault location	Probable cause
008	Intake air temperature (IAT) sensor – plug connection(s)	**Multi-plug(s)**
009	Bosch HFM: Mass air flow (MAF) sensor – range/performance problem	**Wiring, connector, intake leak, MAF sensor**
	Siemens PMS: Manifold absolute pressure (MAP) sensor – vacuum leak	**Hoses, ECM**
010	Bosch HFM: Mass air flow (MAF) sensor – open circuit	**Wiring, connector, MAF sensor, ECM**
	Siemens PMS: Manifold absolute pressure (MAP) sensor – vacuum leak	**Hoses, ECM**
011	Idle speed control (ISC) actuator – range/performance problem	**Wiring, connector, ISC actuator**
012	Idle speed control (ISC) actuator – circuit malfunction	**Wiring, connector, ISC actuator**
013	Bosch HFM: Idle speed control (ISC) actuator – plug connection(s)	**Multi-plug(s)**
	Siemens PMS: Throttle position (TP) sensor – range/performance problem	**Wiring, connector, TP sensor, ECM**
014	Throttle position (TP) sensor – signal high	**Wiring, connector, TP sensor, ECM**
015	Throttle position (TP) sensor – signal low	**Wiring, connector, TP sensor, ECM**
016	Bosch HFM: Throttle position (TP) sensor – plug connection(s)	**Multi-plug(s)**
	Siemens PMS: Idle speed control (ISC) actuator – malfunction	**Wiring, connector, ISC actuator, ECM**
017	Idle speed control (ISC) actuator – signal high	**Wiring, connector, ISC actuator, ECM**
018	Idle speed control (ISC) actuator – signal low	**Wiring, connector, ISC actuator, ECM**
019	Bosch HFM: Idle speed control (ISC) actuator – plug connection(s)	**Multi-plug(s)**
	Siemens PMS: Idle speed control (ISC) – malfunction	**Wiring, connector, intake leak, MAF sensor, throttle valve tight/sticking**
020	Idle speed control (ISC) – lower limit reached	**Wiring, connector, intake leak, MAF sensor, throttle valve tight/sticking**
021	Idle speed control (ISC) – upper limit reached	**Wiring, connector, intake leak, MAF sensor, throttle valve tight/sticking**
022	Cruise control/idle speed control (ISC) actuator – limp-home mode	**Wiring, ISC actuator, throttle valve tight/sticking**
022	Siemens PMS: Heated oxygen sensor (HO2S) 1 – range/performance problem	**Wiring, connector, HO2S**
023	Heated oxygen sensor (HO2S) 1 – voltage high	**Wiring, connector, HO2S**
024	Heated oxygen sensor (HO2S) 1 – open circuit	**Wiring, connector, HO2S**
025	Bosch HFM: Heated oxygen sensor (HO2S) 1 – range/performance problem	**Wiring, connector, HO2S**
	Siemens PMS: Oxygen sensor heater 1 – electrical fault	**Wiring, connector, HO2S, ECM**

Scan code	Fault location	Probable cause
026	Bosch HFM: Heated oxygen sensor (HO2S) 2 – voltage high	**Wiring, connector, HO2S**
	Siemens PMS: Oxygen sensor heater 1 – electrical fault	**Wiring, connector, HO2S, ECM**
027	Bosch HFM: Heated oxygen sensor (HO2S) 2 – open circuit	**Wiring, connector, HO2S**
	Siemens PMS: Oxygen sensor heater 1 – electrical fault	**Wiring, connector, HO2S, ECM**
028	Bosch HFM: Heated oxygen sensor (HO2S) 2 – range/performance problem	**Wiring, connector, HO2S**
	Siemens PMS: Lambda regulation – system too lean	**Wiring, intake/exhaust leak, MAF sensor, fuel pressure, injector(s), ECM**
029	Bosch HFM: Oxygen sensor heater – current low	**Wiring, connector, HO2S, ECM**
	Siemens PMS: Lambda regulation – system too rich	**Wiring, intake/exhaust leak, MAF sensor, fuel pressure, injector(s), ECM**
030	Bosch HFM: Oxygen sensor heater – current high	**Wiring, connector, HO2S, ECM**
	Siemens PMS: Injector 1 & 4 – circuit malfunction	**Wiring, injector, ECM**
031	Bosch HFM: Oxygen sensor heater – short circuit	**Wiring, connector, HO2S, ECM**
	Siemens PMS: Injector 1 & 4 – circuit malfunction	**Wiring, injector, ECM**
032	Bosch HFM: Oxygen sensor heater 2 – current low	**Wiring, connector, HO2S, ECM**
	Siemens PMS: Injector 2 & 3 – circuit malfunction	**Wiring, injector, ECM**
033	Bosch HFM: Oxygen sensor heater 2 – current high	**Wiring, connector, HO2S, ECM**
	Siemens PMS: Injector 2 & 3 – circuit malfunction	**Wiring, injector, ECM**
034	Bosch HFM: Oxygen sensor heater 2 – short circuit	**Wiring, connector, HO2S, ECM**
	Siemens PMS: Fuel trim (FT) – range/performance problem	**Intake/exhaust leak, injector(s), fuel pressure, HO2S, EVAP system, ECM**
035	Lambda regulation – system too lean	**Wiring, intake/exhaust leak, MAF sensor, fuel pressure, injector(s), ECM**
036	Lambda regulation – system too rich	**Wiring, intake/exhaust leak, MAF sensor, fuel pressure, injector(s), ECM**
037	Injector 1 – short to positive	**Wiring short to positive, injector, ECM**
038	Bosch HFM: Injector 1 – open circuit/short to ground	**Wiring open circuit/short to ground, injector, ECM**
	Siemens PMS: Ignition amplifier/ignition coil, cylinder 1 & 4 – defective	**Wiring, ignition amplifier, ignition coil, ECM**

Scan code	Fault location	Probable cause
039	Bosch HFM: Injector 2 – short to positive	**Wiring short to positive, injector, ECM**
	Siemens PMS: Ignition amplifier/ignition coil, cylinder 1 & 4 – defective	**Wiring, ignition amplifier, ignition coil, ECM**
040	Bosch HFM: Injector 2 – open circuit/short to ground	**Wiring open circuit/short to ground, injector, ECM**
	Siemens PMS: Ignition amplifier/ignition coil, cylinder 1 & 4 – defective	**Wiring, ignition amplifier, ignition coil, ECM**
041	Bosch HFM: Injector 3 – short to positive	**Wiring short to positive, injector, ECM**
	Siemens PMS: Ignition amplifier/ignition coil, cylinder 2 & 3 – defective	**Wiring, ignition amplifier, ignition coil, ECM**
042	Bosch HFM: Injector 3 – open circuit/short to ground	**Wiring open circuit/short to ground, injector, ECM**
	Siemens PMS: Ignition amplifier/ignition coil, cylinder 2 & 3 – defective	**Wiring, ignition amplifier, ignition coil, ECM**
043	Bosch HFM: Injector 4 – short to positive	**Wiring short to positive, injector, ECM**
	Siemens PMS: Ignition amplifier/ignition coil, cylinder 2 & 3 – defective	**Wiring, ignition amplifier, ignition coil, ECM**
044	Bosch HFM: Injector 4 – open circuit/short to ground	**Wiring open circuit/short to ground, injector, ECM**
	Siemens PMS: Crankshaft position (CKP) sensor – malfunction	**Wiring, air gap, flywheel ring gear damaged, CKP sensor**
045	111: Crankshaft position (CKP) sensor – malfunction	**Wiring, air gap, flywheel ring gear damaged, CKP sensor**
	104: Injector 5 – short to positive	**Wiring short to positive, injector, ECM**
046	111: Crankshaft position (CKP) sensor – malfunction	**Wiring, air gap, flywheel ring gear damaged, CKP sensor**
	104: Injector 5 – open circuit/short to ground	**Wiring open circuit/short to ground, injector, ECM**
047	111: Engine control module (ECM) coding plug – electrical fault	**Wiring, ECM**
	104: Injector 6 – short to positive	**Wiring short to positive, injector, ECM**
048	111: Engine control module (ECM) coding plug – electrical fault	**Wiring, ECM**
	104: Injector 6 – open circuit/short to ground	**Wiring open circuit/short to ground, injector, ECM**
049	Bosch HFM: Fuel trim (FT), idle speed – mixture rich	**Intake/exhaust leak, injector(s), fuel pressure, HO2S, ECM**
	Siemens PMS: Engine control module (ECM), RPM signal – output/short to ground	**Wiring short to ground, RPM sensor**
050	Bosch HFM: Fuel trim (FT), idle speed – mixture lean	**Intake/exhaust leak, injector(s), fuel pressure, HO2S, ECM**
	Siemens PMS: Engine control module (ECM), RPM signal – output/short to positive	**Wiring short to positive, RPM sensor**

Scan code	Fault location	Probable cause
051	Bosch HFM: Fuel trim (FT), lower part load – mixture rich	**Intake/exhaust leak, injector(s), fuel pressure, HO2S, ECM**
	Siemens PMS: ABS control module/engine control module (ECM) communication – VSS signal	**Wiring, VSS, ABS control module trouble code(s) stored**
052	Bosch HFM: Fuel trim (FT), lower part load – mixture lean	**Intake/exhaust leak, injector(s), fuel pressure, HO2S, ECM**
	Siemens PMS: ABS control module/engine control module (ECM) communication – VSS signal	**Wiring, VSS, ABS control module trouble code(s) stored**
053	Fuel trim (FT), upper part load – mixture rich	**Intake/exhaust leak, injector(s), fuel pressure, HO2S, ECM**
053 1	Intake manifold heater relay – malfunction	**Wiring, intake manifold heater relay, ECM**
054	Fuel trim (FT), upper part load – mixture lean	**Intake/exhaust leak, injector(s), fuel pressure, HO2S, ECM**
054 1	Intake manifold heater relay – malfunction	**Wiring, intake manifold heater relay, ECM**
055	Bosch HFM: Ignition amplifier 1/ignition coil, cylinder 1 – misfire	**Wiring, ignition amplifier, ignition coil, ECM**
	Siemens PMS: Fuel pump relay – defective	**Wiring, fuel pump relay**
	104: Ignition amplifier 1/ignition coil, cylinder 2 – misfire	**Wiring, ignition amplifier, ignition coil, ECM**
056	Bosch HFM: Ignition amplifier 1/ignition coil, cylinder 4 – misfire	**Wiring, ignition amplifier, ignition coil, ECM**
	Siemens PMS: Fuel pump relay – defective	**Wiring, fuel pump relay**
	104: Ignition amplifier 1/ignition coil, cylinder 5 – misfire	**Wiring, ignition amplifier, ignition coil, ECM**
057	Bosch HFM: Ignition amplifier 1/ignition coil, cylinder 1 & 4 – defective	**Wiring, ignition amplifier, ignition coil, ECM**
	Siemens PMS: CO adjustment resistor – malfunction	**Wiring, CO adjustment resistor, ECM**
	104: Ignition amplifier 3/ignition coil, cylinder 2 & 5 – defective	**Wiring, ignition amplifier, ignition coil, ECM**
058	Bosch HFM: Ignition amplifier 2/ignition coil, cylinder 2 – misfire	**Wiring, ignition amplifier, ignition coil, ECM**
	Siemens PMS: CO adjustment resistor – malfunction	**Wiring, CO adjustment resistor, ECM**
	104: Ignition amplifier 2/ignition coil, cylinder 3 – misfire	**Wiring, ignition amplifier, ignition coil, ECM**
059	Bosch HFM: Ignition amplifier 2/ignition coil, cylinder 3 – misfire	**Wiring, ignition amplifier, ignition coil, ECM**
	Siemens PMS: Evaporative emission (EVAP) canister purge valve – electrical fault	**Wiring, EVAP canister purge valve, hose(s), ECM**
	104: Ignition amplifier 2/ignition coil, cylinder 4 – misfire	**Wiring, ignition amplifier, ignition coil, ECM**

Scan code	Fault location	Probable cause
060	Bosch HFM: Ignition amplifier 2/ignition coil, cylinder 2 & 3 – defective	**Wiring, ignition amplifier, ignition coil, ECM**
	Siemens PMS: Evaporative emission (EVAP) canister purge valve – electrical fault	**Wiring, EVAP canister purge valve, hose(s), ECM**
	104: Ignition amplifier 2/ignition coil, cylinder 3 & 4 – defective	**Wiring, ignition amplifier, ignition coil, ECM**
061	111: Short term fuel trim (FT), idle speed/part load – malfunction	**Fuel system**
	104: Ignition amplifier 3/ignition coil, cylinder 1 – misfire	**Wiring, ignition amplifier, ignition coil, ECM**
062	111: Transmission shift control valve – electrical fault	**Wiring, transmission shift control valve**
	104: Ignition amplifier 3/ignition coil, cylinder 6 – misfire	**Wiring, ignition amplifier, ignition coil, ECM**
063	111: Engine control module (ECM) – supply voltage	**Wiring, connector, fuse, ignition switch, central locking control module, engine control relay**
	104: Ignition amplifier 3/ignition coil, cylinder 1 & 6 – defective	**Wiring, ignition amplifier, ignition coil, ECM**
064	Bosch HFM: Crankshaft position (CKP) sensor – signal	**Wiring, air gap, flywheel ring gear damaged, CKP sensor**
	Siemens PMS: Central locking control module, immobilizer signal – open circuit/short to positive	**Wiring, central locking control module, trouble code(s) stored in other system(s)**
065	Bosch HFM: Crankshaft position (CKP) sensor – defective	**Wiring, air gap, flywheel ring gear damaged, CKP sensor**
	Siemens PMS: Central locking control module, immobilizer signal – open circuit/short to ground	**Wiring, central locking control module, trouble code(s) stored in other system(s)**
066	Bosch HFM: Crankshaft position (CKP) sensor – range/performance problem	**Wiring, air gap, flywheel ring gear damaged, CKP sensor**
	Siemens PMS: Attempt to start engine while central locking still activated	**Erase fault memory, trouble code(s) stored in other system(s)**
067	Bosch HFM: Camshaft position (CMP) sensor – incorrect signal	**Wiring, CMP sensor**
	Siemens PMS: Central locking control module, immobilizer signal – plausibility	**Wrong central locking control module**
068	Engine control module (ECM) coding plug – short to ground	**Wiring short to ground, ECM**
069	Engine control module (ECM) coding plug – open circuit/short to positive	**Wiring open circuit/short to positive, ECM**
070	Engine control module (ECM), RPM signal – output/short to ground	**Wiring short to ground, RPM sensor**
071	Engine control module (ECM), RPM signal – output/short to positive	**Wiring short to positive, RPM sensor**

Scan code	Fault location	Probable cause
072	ABS control module/engine control module (ECM) communication, VSS signal – signal not recognised	**Wiring, VSS, ABS control module trouble code(s) stored**
073	ABS control module/engine control module (ECM) communication, VSS signal – signal high	**Wiring, VSS, ABS control module trouble code(s) stored**
074 [2]	Intake manifold heater relay – short to positive	**Wiring short to positive, intake manifold heater relay, ECM**
075 [2]	Intake manifold heater relay – open circuit/short to ground	**Wiring open circuit/short to ground, intake manifold heater relay, ECM**
075 [3]	Engine control module (ECM), ETC signal – not recognised/short to positive	**Wiring, electronic traction control (ETC) module**
076	Fuel pump relay – defective	**Wiring, fuel pump relay**
077	Without cat: CO adjustment resistor – short to positive	**Wiring short to positive, CO adjustment resistor, ECM**
078	Without cat: CO adjustment resistor – malfunction	**Wiring, connector(s), CO adjustment resistor, ECM**
079	Knock sensor(s) (KS) 1 – open circuit	**Wiring, KS**
080	Knock sensor(s) (KS) 2 – open circuit	**Wiring, KS**
081	Ignition timing, knocking – maximum retardation reached	**Fuel contamination, mechanical fault**
082	Ignition timing, knocking – firing angle between cylinders	**Fuel contamination, mechanical fault**
083	111: Secondary air injection (AIR) pump/relay – malfunction	**Wiring, AIR pump/relay, ECM**
	104: Engine control module (ECM), knock control – internal fault	**ECM**
084	Short term fuel trim (FT), max. value – idle speed/part load	**Fuel system, ECM**
085	Secondary air injection (AIR) pump/relay – malfunction	**Wiring, AIR pump/relay, ECM**
086	Evaporative emission (EVAP) canister purge valve – open circuit/short circuit	**Wiring, AIR pump/relay, ECM**
087	Evaporative emission (EVAP) canister purge valve – short to positive	**Wiring short to positive, AIR pump/relay, ECM**
088	Transmission shift control valve – electrical fault	**Wiring, transmission shift control valve**
089	Camshaft position (CMP) actuator – short to positive	**Wiring, CMP actuator, ECM**
090	Camshaft position (CMP) actuator – open circuit/short to ground	**Wiring, CMP actuator, ECM**
091	Exhaust gas recirculation (EGR) solenoid – short to positive	**Wiring short to positive, EGR solenoid, hose(s)**
092	Exhaust gas recirculation (EGR) solenoid – open circuit/short to ground	**Wiring open circuit/short to ground, EGR solenoid, hose(s)**

Scan code	Fault location	Probable cause
093	Transmission overload protection switch – short to ground	**Wiring short to ground, transmission overload protection switch**
094	Transmission overload protection switch – switch closed	**Wiring, transmission overload protection switch**
095	Transmission overload protection switch – switch open	**Wiring, transmission overload protection switch**
096	Transmission overload protection switch – plausibility	**Wiring, transmission overload protection switch**
097	CAN data bus, ECM – malfunction	**Wiring, ECM**
098	CAN data bus, ASR system – malfunction	**Wiring**
099	CAN data bus, cruise control system/TP control module – malfunction	**Wiring**
100	CAN data bus, diagnostic module – malfunction	**Wiring**
101	Engine control module (ECM), pin 50 – starter signal	**Wiring**
104	With AT/cruise control system: Cruise control, inertia fuel shut-off (IFS) switch – active	**Wiring, TP sensor, idle speed control (ISC) actuator**
105	Intake manifold air control solenoid – short to positive	**Wiring, intake manifold air control solenoid, ECM**
106	Intake manifold air control solenoid – open circuit/short to ground	**Wiring, intake manifold air control solenoid, ECM**
107	Ignition amplifier – malfunction	**Wiring, ignition amplifier, ignition coil(s), ECM**
108	Oxygen sensor heater relay 2 – short to positive	**Wiring short to positive, oxygen sensor heater relay, ECM**
109	Oxygen sensor heater relay 2 – open circuit/short to ground	**Wiring open circuit/short to ground, oxygen sensor heater relay, ECM**
110	Engine control module (ECM) – supply voltage, plausibility	**Wiring, ignition switch, overvoltage protection relay**
111	Engine control module (ECM) – supply voltage low	**Wiring, ignition switch, overvoltage protection relay**
112	Engine control module (ECM) – internal fault	**ECM**
113	Engine control module (ECM) – coding	**Not coded**
114	Engine control module (ECM) – identification	**ECM not coded, ECM**
115	Engine control module (ECM) – coding	**Not coded, ECM**
116	CAN data bus, remote central locking – malfunction	**Wiring**
	Bosch HFM 06/93→: Engine control module (ECM) – supply voltage low	**Wiring, connector, fuse, overvoltage protection relay**
117	Attempt to start engine while central locking still activated	**Erase fault memory, trouble code(s) stored in other system(s)**
118	Supercharger (SC) – malfunction	**Wiring, mechanical fault, hose(s), intake air flap control actuator, ECM**

Scan code	Fault location	Probable cause
119	Supercharger (SC) clutch – open circuit/short to ground	**Wiring open circuit/short to ground, SC clutch, ECM**
120	Electronic traction control (ETC) – signal	**Wiring short to ground, ETC module**
123	Intake air flap control actuator, supercharger (SC) – malfunction	**Wiring open circuit/short to ground, intake air flap control actuator, ECM**
124	Intake air flap control actuator, supercharger (SC) – malfunction	**Wiring short to positive, intake air flap control actuator, ECM**
125	Engine control module (ECM) – internal fault	**ECM**
126	Engine control module (ECM) – internal fault	**ECM**
127	Idle speed control (ISC) actuator/ISC actuator, cruise control system – malfunction	**Wiring, ISC actuator, ISC actuator – cruise control system**
128	Engine control module (ECM) – internal fault	**ECM**
129	Engine control module (ECM) – internal fault	**ECM**
130	Throttle position (TP) sensor – malfunction	**Wiring, TP sensor, ECM**
131	Engine control module (ECM) – internal fault	**ECM**
132	Engine control module (ECM) – internal fault	**ECM**
133	Idle speed control (ISC) actuator – defective	**Wiring, ISC actuator**
134	Engine control module (ECM) – internal fault	**ECM**
135	Idle speed control (ISC) actuator – supply voltage	**Wiring, ECM**
136	Idle speed control (ISC) actuator – defective	**Wiring, ISC actuator, ECM**
137	Engine control module (ECM) – internal fault	**ECM**
138	Idle speed control (ISC) actuator – defective	**Wiring, ISC actuator**
139	Cruise control selector switch – defective	**Wiring, cruise control selector switch**
140	Engine control module (ECM) – internal fault	**ECM**
141	Engine control module (ECM) – internal fault	**ECM**
142	Engine control module (ECM) – internal fault	**ECM**
143	Brake pedal position (BPP) switch, cruise control system – defective	**Wiring, BPP switch**
144	Engine control module (ECM) – internal fault	**ECM**
145	CAN data bus, VSS signal, rear axle – malfunction	**Wiring, trouble code(s) stored in other system(s)**
146	CAN data bus, VSS signal, front axle – malfunction	**Wiring, trouble code(s) stored in other system(s)**
147	CAN data bus, AC – malfunction	**Wiring, AC control module trouble code(s) stored**
148	Starter motor relay – short to positive	**Wiring, ignition switch, ECM**
149	Starter motor relay – short to ground	**Wiring, ignition switch, ECM**
150	CAN data bus, AT – malfunction	**Wiring, TCM trouble code(s) stored**

Scan code	Fault location	Probable cause
151	CAN data bus, AC – malfunction	**Wiring, AC control module trouble code(s) stored**
152	CAN data bus – malfunction	**Wiring**
153	Engine coolant blower motor/AC condenser blower motor – short to positive	**Wiring, ECT sensor, AC control module, ECM**
154	Engine coolant blower motor/AC condenser blower motor – short to ground	**Wiring, ECT sensor, AC control module, ECM**
155	Transmission control module (TCM) – coding	**Incorrectly coded**
156	CAN data bus, ABS/ETC – plausibility	**Wiring, trouble code(s) stored in other system(s)**
157	CAN data bus, ABS/ETC – plausibility	**Wiring, trouble code(s) stored in other system(s)**
158	CAN data bus, ABS – plausibility	**Wiring, ABS control module trouble code(s) stored**

1 E-Class (124) →07/93, Siemens PMS
2 E-Class (124) →07/93

Accessing and erasing – OBD-II codes

- The engine control module (ECM) fault memory can only be accessed and erased using diagnostic equipment connected to the data link connector (DLC).

Trouble code identification

OBD-II code	Fault location	Probable cause
All P0, P2 and U0 codes	Refer to OBD-II trouble code tables at the front of this manual	–
P1131	Heated oxygen sensor (HO2S) 1 – signal/electrical fault	**Wiring, HO2S**
P1132	Heated oxygen sensor (HO2S) control 1 – system too rich	**Intake/exhaust leak, HO2S**
P1137	Heated oxygen sensor (HO2S) 2 – signal/electrical fault	**Wiring, HO2S**
P1138	Heated oxygen sensor (HO2S) 2 – range/performance problem	**Wiring, HO2S**
P1170	Short term fuel trim (FT) – malfunction	**Intake/exhaust leak, injector(s), fuel pressure, MAF sensor**
P1335	Engine speed (RPM) sensor – signal	**Wiring, ECM**
P1336	Crankshaft position (CKP) sensor – signal	**Wiring, air gap, flywheel ring gear damaged, CKP sensor**

OBD-II code	Fault location	Probable cause
P1337	Engine control module (ECM), RPM signal – output	**Wiring, RPM sensor, ECM**
P1340	Camshaft position (CMP) sensor – signal	**Wiring, CMP sensor, ECM**
P1341	Camshaft position (CMP) actuator – mechanical fault	**Camshaft timing mechanism, CMP actuator**
P1342	Camshaft position (CMP) actuator – electrical fault	**Wiring, connector, CMP actuator, ECM**
P1400	Exhaust gas recirculation (EGR) valve/ solenoid – circuit malfunction	**Wiring, connector, EGR valve/solenoid, ECM**
P1411	Secondary air injection (AIR) system – malfunction	**Wiring, connector, hoses, AIR pump, AIR solenoid**
P1443	Evaporative emission (EVAP) canister purge valve – malfunction	**Wiring, connector, EVAP canister purge valve**
P1444	Evaporative emission (EVAP) system – malfunction	**Wiring, connector, hose(s), EVAP canister purge valve, ECM**
P1700	Transmission shift control – malfunction	**Wiring, transmission shift control valve, TCM trouble code(s) stored**
P1701	Transmission shift control valve – electrical fault	**Wiring, transmission shift control valve**
P1711	Intake manifold air control solenoid – malfunction	**Wiring, connectors, intake manifold air control solenoid, ECM**
P1740	Engine control module (ECM), load signal – plausibility	**Wiring, electronic throttle system (ETS), MAF sensor, TP sensor, ECM**
P1741	Throttle position (TP) sensor, load signal – plausibility	**Wiring, electronic throttle system (ETS), MAF sensor, TP sensor, ECM**
P1750	Diagnostic module, pin 30 – supply voltage low	**Wiring, connector(s), fuse(s)**

MINI

Model:	Engine identification:	Year:
Cooper 1.6L	W10B	2002-04
Cooper Convertible 1.6L	W10B	2004
Cooper S 1.6L	W11B	2002-04
Cooper S Convertible 1.6L	W11B	2004

System: **Siemens EMS 2000**

Self-diagnosis

General information

- Refer to the front of this manual for general test conditions, terminology, detailed descriptions of wiring faults and a general trouble shooter for electrical and mechanical faults.
- Engine control module (ECM) incorporates self-diagnosis function.
- Malfunction indicator lamp (MIL) will illuminate if certain faults are recorded.
- ECM operates in backup mode if sensors fail, to enable vehicle to be driven to workshop.

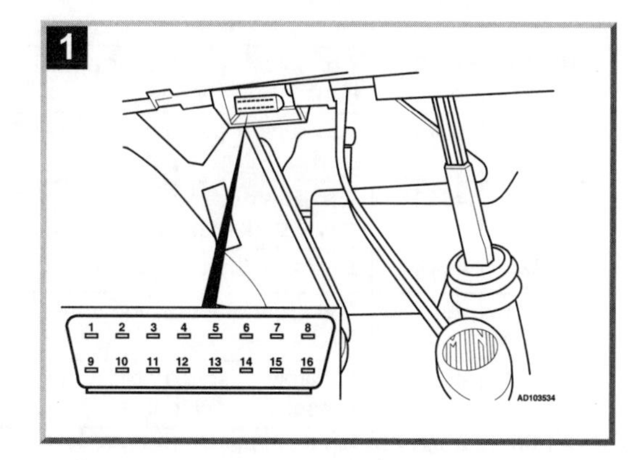

Accessing

- The engine control module (ECM) fault memory can only be accessed and erased using diagnostic equipment connected to the data link connector (DLC) 1.

Erasing

- After the faults have been rectified, trouble codes can only be erased by using a Scan Tool connected to the data link connector (DLC) 1.

Trouble code identification

OBD-II code	BMW type	Fault location	Probable cause
All P0, P2 and U0 codes	–	Refer to OBD-II trouble code tables at the front of this manual	–
P1106	4358	Manifold absolute pressure (MAP) sensor – signal too low at engine stop	**Wiring short to ground, poor connection, sensor supply, MAP sensor**
P1107	4359	Manifold absolute pressure (MAP) sensor – signal too low at idle	**Wiring short to ground, poor connection, MAP sensor**
P1108	4360	Manifold absolute pressure (MAP) sensor, signal at low engine RPM too low at full load	**Wiring short to ground, MAP sensor**
P1109	4361	Manifold absolute pressure – pressure too high under deceleration	**Throttle valve sticking, air leak**
P1122	4386	Accelerator pedal position (APP) sensor, track 1 – low input	**Wiring short to ground, APP sensor**

OBD-II code	BMW type	Fault location	Probable cause
P1123	4387	Accelerator pedal position (APP) sensor, track 1 – high input	**Wiring short to positive, APP sensor**
P1125	4389	Throttle motor position sensor, A/B – small range/performance problem	**Wiring, throttle motor position sensor**
P1126	4390	Throttle motor position sensor, A/B – large range/performance problem	**Wiring, throttle motor position sensor**
P1143	4419	Heated oxygen sensor (HO2S) 2, bank 1 – signal too high	**Wiring short to positive, catalytic converter, fuel pressure high, HO2S**
P1144	4420	Heated oxygen sensor (HO2S) 2, bank 1 – signal too low	**Wiring short to ground, poor connection, exhaust leak, fuel pressure low, HO2S**
P1222	4642	Accelerator pedal position (APP) sensor, track 2 – low input	**Wiring short to ground, APP sensor**
P1223	4643	Accelerator pedal position (APP) sensor, track 2 – high input	**Wiring short to positive, APP sensor**
P1224	4644	Accelerator pedal position (APP) sensor – range/performance problem	**Wiring, APP sensor**
P1226	4646	Electronic throttle system (ETS) – malfunction	**Throttle valve tight/sticking, ETS**
P1229	4649	Throttle motor position sensor – adaptation failure	**ETS**
P1237	4663	Manifold absolute pressure (MAP) sensor 2 – low input	**Wiring short to ground, MAP sensor**
P1238	4664	Manifold absolute pressure (MAP) sensor 2 – high input	**Wiring short to positive, MAP sensor**
P1239	4665	Manifold absolute pressure (MAP) sensor 2 – signal too low at engine stop	**Wiring short to ground, poor connection, sensor supply, MAP sensor**
P1240	4672	Manifold absolute pressure (MAP) sensor 2 – signal too low at idle	**Wiring short to ground, poor connection, MAP sensor**
P1241	4673	Manifold absolute pressure (MAP) sensor 2 – signal at low engine RPM too low at full load	**Wiring short to ground, MAP sensor**
P1242	4674	Manifold absolute pressure (MAP) sensor 2 – pressure too high under deceleration	**Throttle valve sticking, air leak**
P1320	4896	Flywheel adaptation for misfire detection – range	–
P1321	4897	Flywheel adaptation for misfire detection – performance problem	–
P1366	4966	Ignition coil A, primary/secondary – signal low	**Wiring short to ground, ignition coil**
P1367	4967	Ignition coil B, primary/secondary – signal low	**Wiring short to ground, ignition coil**
P1570	5488	Engine control module (ECM), sensor supply circuit A – low output	**Wiring short to ground, ECM**
P1571	5489	Engine control module (ECM), sensor supply circuit A – high output	**Wiring short to positive, ECM**

OBD-II code	BMW type	Fault location	Probable cause
P1572	5490	Engine control module (ECM), sensor supply circuit A – noisy signal	Wiring, interference
P1573	5491	Engine control module (ECM), sensor supply circuit B – low output	Wiring short to ground, ECM
P1574	5492	Engine control module (ECM), sensor supply circuit B – high output	Wiring short to positive, ECM
P1575	5493	Engine control module (ECM), sensor supply circuit B – noisy signal	Wiring, interference
P1600	5632	Engine control module (ECM) – RAM error	–
P1607	5639	CAN data bus version	Programming
P1612	5650	CAN data bus, instrument panel	Wiring, instrument panel
P1613	5651	CAN data bus, automatic stability control (ASC) system	Wiring, ASC
P1615	5653	Engine control module (ECM), processor fault – SPI-bus failure	ECM
P1617	5655	Engine control module (ECM), controller – malfunction	ECM
P1679	5753	Electronic throttle system (ETS), monitor level 2/3 – low engine torque detected	ETS, ECM
P1680	5760	Electronic throttle system (ETS), monitor level 2/3 – processor error	ECM
P1681	5761	Electronic throttle system (ETS), monitor level 2/3 – engine RPM fault	ETS, ECM
P1682	5762	Electronic throttle system (ETS), monitor level 2/3 – idle speed fault	ETS, ECM
P1683	5763	Electronic throttle system (ETS), monitor level 2/3 – idle speed fault	ETS, ECM
P1684	5764	Electronic throttle system (ETS), monitor level 2/3 – malfunction	ETS, ECM
P1685	5765	Electronic throttle system (ETS), monitor level 2/3 – malfunction	ETS, ECM
P1686	5766	Electronic throttle system (ETS), monitor level 2/3 – APP sensor diagnostic error	Wiring, APP sensor, ECM
P1687	5767	Electronic throttle system (ETS), monitor level 2/3 – throttle motor position sensor diagnostic error	Wiring, throttle motor position sensor, ECM
P1688	5768	Electronic throttle system (ETS), monitor level 2/3 – MAF calculation	Wiring, intake system, MAF sensor, ECM
P1689	5769	Electronic throttle system (ETS), monitor level 2/3 – incorrect engine torque	ETS, ECM
P1691	5777	Electronic throttle system (ETS), monitor level 2/3 – throttle motor control malfunction	ETS, ECM
P1692	5778	Electronic throttle system (ETS), monitor level 2/3 – throttle motor control/injection cut-off	ETS, ECM

MITSUBISHI

Model:	Engine identification:	Year:
Diamante 3.0L	**6G72**	**1992-93**
Expo/LRV 2.4L	**4G64**	**1992-93**
Expo/LRV 1.8L	**4G93**	**1992-93**
3000GT Turbo	**6G72 DOHC**	**1991-93**
3000GT	**6G72 DOHC**	**1991-93**

System: **MPI**

Self-diagnosis

General information

- Refer to the front of this manual for general test conditions, terminology, detailed descriptions of wiring faults and a general trouble shooter for electrical and mechanical faults.
- Engine control module (ECM) incorporates self-diagnosis function.
- Malfunction indicator lamp (MIL) will illuminate if certain faults are recorded.
- ECM operates in backup mode if certain sensors fail, to enable car to be driven to workshop.
- Trouble codes can be accessed with a voltmeter or a suitable code reader connected to the data link connector (DLC) **1**.

Accessing

- Ensure ignition switched OFF.
- Connect an analogue voltmeter between data link connector (DLC) terminals 1 and 12 **1**.
- Switch ignition ON. Do NOT start engine.
- Voltmeter should be pulsing.
  - 1.5 second pulses indicate the 'tens' of the trouble code **2** [**A**].
  - 0.5 second pulses indicate the 'units' of the trouble code **2** [**B**].
  - A 0.5 second pause separates each pulse **2** [**C**].
  - A 2 second pause separates 'tens' and 'units' **2** [**D**].
  - A 3 second pause separates each trouble code **2** [**E**].
  - For example: Trouble code 24 displayed **2**.
- If no trouble codes are stored, short pulses will continue to be displayed.
- If continuous voltage is displayed: Suspect faulty engine control module (ECM).
- Count voltmeter pulses and compare with trouble code table.
- Switch ignition OFF.
- Remove voltmeter.

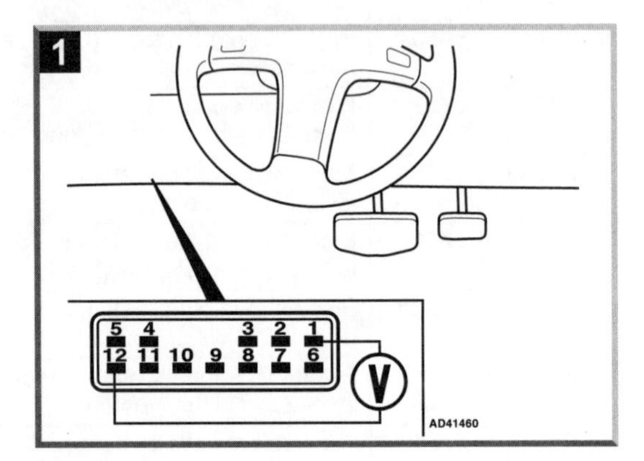

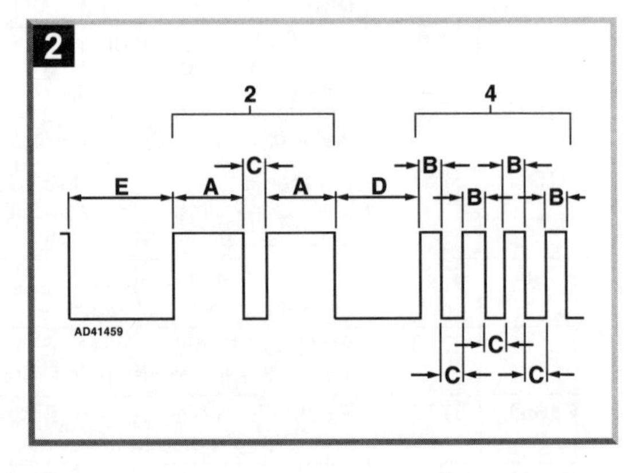

Erasing

- Ensure ignition switched OFF.
- Disconnect battery ground lead for 15 seconds minimum.
- Reconnect battery ground lead.
- Repeat checking procedure to ensure no data remains in ECM fault memory.

NOTE: ***Disconnecting battery ground lead may erase memory from electronic units such as radio.***

Trouble code identification

Trouble code	Fault location	Probable cause
11	Oxygen sensor (O2S)/heated oxygen sensor (HO2S) – front – malfunction	**Wiring, O2S/HO2S, fuel system, injectors, ECM**
12	Volume air flow (VAF) sensor – malfunction	**Wiring, VAF sensor, ECM**
13	Intake air temperature (IAT) sensor – malfunction	**Wiring, IAT sensor, ECM**
14	Throttle position (TP) sensor – malfunction	**Wiring, TP sensor, CTP switch, ECM**
15	Idle speed control (ISC) actuator position sensor – malfunction	**Wiring, ISC sensor, ECM**
21	Engine coolant temperature (ECT) sensor – malfunction	**Wiring, ECT sensor, ECM**
22	Crankshaft position (CKP) sensor – malfunction	**Wiring, CKP sensor, ECM**
23	Camshaft position (CMP) sensor – malfunction	**Wiring, CMP sensor, ECM**
24	Vehicle speed sensor (VSS) – malfunction	**Wiring, VSS, ECM**
25	Barometric pressure (BARO) sensor – malfunction	**Wiring, BARO sensor, ECM**
31	Knock sensor (KS) – malfunction	**Wiring, KS, ECM**
36	Ignition timing adjustment	**Wiring short to ground**
39	Heated oxygen sensor (HO2S) – front/RH front – malfunction	**Wiring, HO2S, fuel system, injectors, ECM**
41	Injectors – malfunction	**Wiring, injectors, ECM**
42	Fuel pump – malfunction	**Wiring, relay module, fuel pump, ECM**
43	Exhaust gas recirculation (EGR) system – malfunction	**Wiring, EGR, EGRT, EGRTVV, hose leak, ECM**

Trouble code	Fault location	Probable cause
44	Ignition coil/ignition module – cylinders 1 and 4 – malfunction	**Wiring, ignition coil/module, ECM**
52	Ignition coil/ignition module – cylinders 2 and 5 – malfunction	**Wiring, ignition coil/module, ECM**
53	Ignition coil/ignition module – cylinders 3 and 6 – malfunction	**Wiring, ignition coil/module, ECM**
55	Idle speed control (ISC) actuator position sensor – malfunction	**Wiring, ISC sensor, ECM**
59	Heated oxygen sensor (HO2S) – rear/LH rear – malfunction	**Wiring, HO2S, ECM**
61	AT – torque reduction signal – malfunction	**Wiring, TCM**
62	Intake manifold air control actuator position sensor – malfunction	**Wiring, position sensor, actuator, ECM**
69	Heated oxygen sensor (HO2S) – RH rear – malfunction	**Wiring, HO2S, ECM**
71	Traction control vacuum solenoid – malfunction	**Wiring, solenoid, ECM**
72	Traction control ventilation solenoid – malfunction	**Wiring, solenoid, ECM**

MITSUBISHI

Model:	Engine identification:	Year:
Diamante 3.0L	**6G72**	**1994-95**
Galant 2.4L	**4G64**	**1994-95**
Mirage 1.8L – California	**4G93**	**1994**
Mirage 1.8L – Federal	**4G93**	**1994-95**
3000GT 3.0L/Turbo	**6G72 DOHC**	**1994-95**

System: **MPI**

Self-diagnosis

General information

- Refer to the front of this manual for general test conditions, terminology, detailed descriptions of wiring faults and a general trouble shooter for electrical and mechanical faults.
- Engine control module (ECM) incorporates self-diagnosis function.
- Malfunction indicator lamp (MIL) will illuminate if certain faults are recorded.
- ECM operates in backup mode if certain sensors fail, to enable car to be driven to workshop.
- Trouble codes can be accessed with a voltmeter or a suitable code reader connected to the data link connector (DLC).

Accessing

- Ensure ignition switched OFF.
- Access data link connector (DLC):
  - Mirage – **1**.
  - Galant/3000GT – **2**.
  - Diamante – **3**.
- Jump DLC terminal 1 and ground **4**.
- Switch ignition ON. Do NOT start engine.
- MIL should be flashing.
  - 1.5 second flashes indicate the 'tens' of the trouble code **5** [**A**].
  - 0.5 second flashes indicate the 'units' of the trouble code **5** [**B**].
  - A 0.5 second pause separates each flash **5** [**C**].
  - A 2 second pause separates 'tens' and 'units' **5** [**D**].
  - A 3 second pause separates each trouble code **5** [**E**].
  - For example: Trouble code 24 displayed **5**.
- If no trouble codes are stored, short flashes will continue to be displayed.
- If MIL remains ON: Suspect faulty engine control module (ECM).
- Count MIL flashes and compare with trouble code table.
- Switch ignition OFF.
- Remove jump lead.

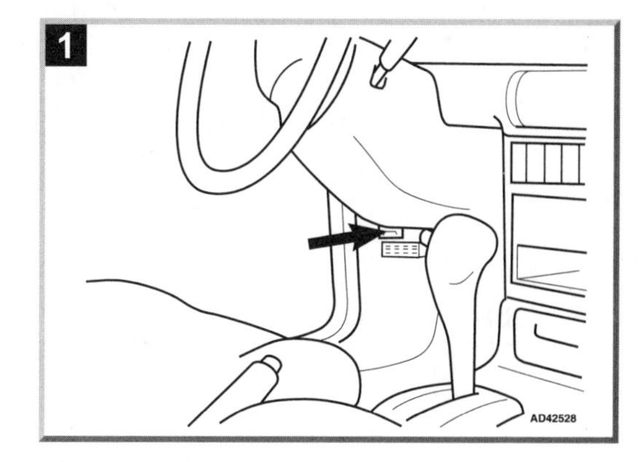

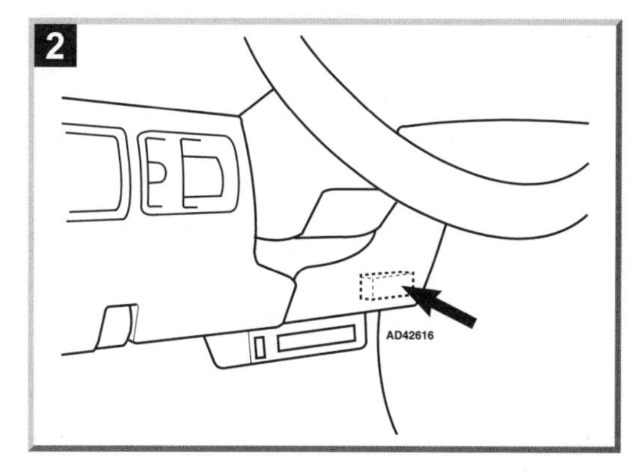

3

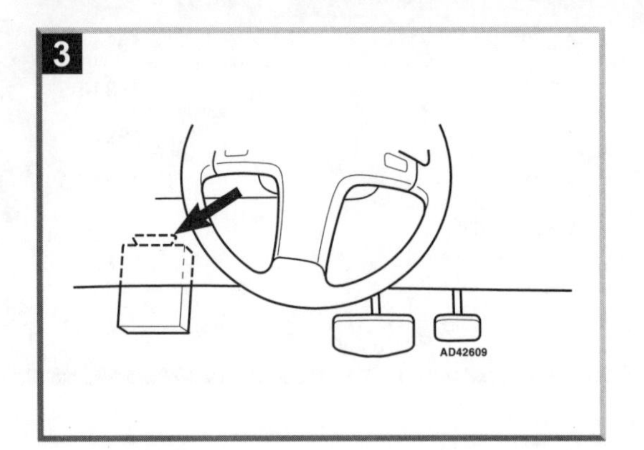

5

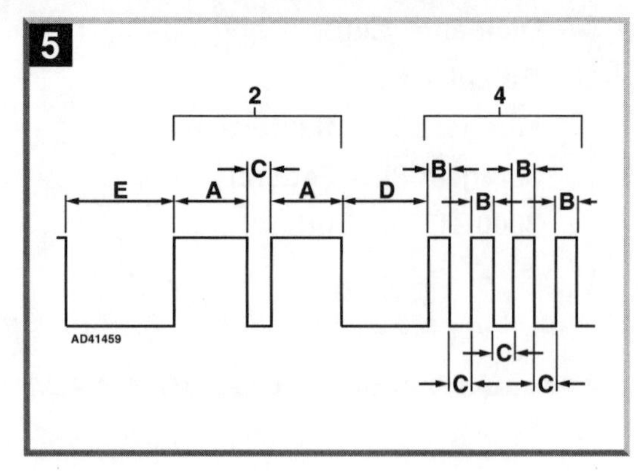

4

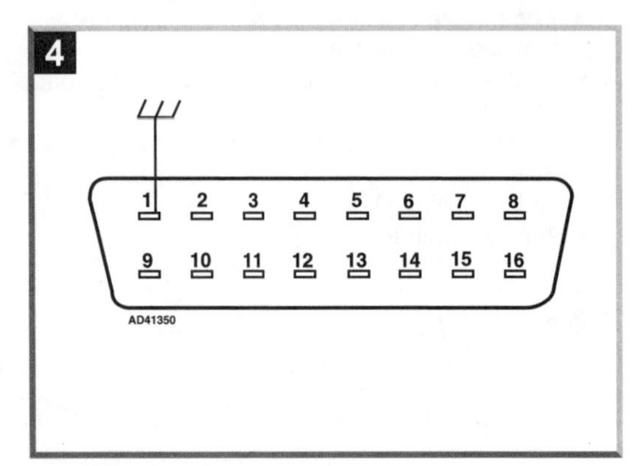

Erasing

- Ensure ignition switched OFF.
- Disconnect battery ground lead for 15 seconds minimum.
- Reconnect battery ground lead.
- Repeat checking procedure to ensure no data remains in ECM fault memory.

NOTE: ***Disconnecting battery ground lead may erase memory from electronic units such as radio.***

Trouble code identification

Flash code	Fault location	Probable cause
11	Oxygen sensor (O2S)/heated oxygen sensor (HO2S) – front – malfunction	**Wiring, O2S/HO2S, fuel system, injectors, ECM**
12	Volume air flow (VAF) sensor – malfunction	**Wiring, VAF sensor, ECM**
13	Intake air temperature (IAT) sensor – malfunction	**Wiring, IAT sensor, ECM**
14	Throttle position (TP) sensor – malfunction	**Wiring, TP sensor, CTP switch, ECM**
21	Engine coolant temperature (ECT) sensor – malfunction	**Wiring, ECT sensor, ECM**
22	Crankshaft position (CKP) sensor – malfunction	**Wiring, CKP sensor, ECM**
23	Camshaft position (CMP) sensor – malfunction	**Wiring, CMP sensor, ECM**
24	Vehicle speed sensor (VSS) – malfunction	**Wiring, VSS, ECM**

Flash code	Fault location	Probable cause
25	Barometric pressure (BARO) sensor – malfunction	**Wiring, BARO sensor, ECM**
31	Knock sensor (KS) – malfunction	**Wiring, KS, ECM**
36	Ignition timing adjustment	**Wiring short to ground**
39	Heated oxygen sensor (HO2S) – front/RH front – malfunction	**Wiring, HO2S, fuel system, injectors, ECM**
41	Injectors – malfunction	**Wiring, injectors, ECM**
42	Fuel pump – malfunction	**Wiring, relay module, fuel pump, ECM**
43	Exhaust gas recirculation (EGR) system – malfunction	**Wiring, EGR, EGRT, EGRTVV, hose leak, ECM**
44	Ignition coil/ignition module – cylinders 1 and 4 – malfunction	**Wiring, ignition coil/module, ECM**
52	Ignition coil/ignition module – cylinders 2 and 5 – malfunction	**Wiring, ignition coil/module, ECM**
53	Ignition coil/ignition module – cylinders 3 and 6 – malfunction	**Wiring, ignition coil/module, ECM**
55	Idle speed control (ISC) actuator position sensor – malfunction	**Wiring, ISC sensor, ECM**
59	Heated oxygen sensor (HO2S) – rear/LH rear – malfunction	**Wiring, HO2S, ECM**
61	AT – torque reduction signal – malfunction	**Wiring, TCM**
62	Intake manifold air control actuator position sensor – malfunction	**Wiring, position sensor, actuator, ECM**
69	Heated oxygen sensor (HO2S) – RH rear – malfunction	**Wiring, HO2S, ECM**
71	Traction control vacuum solenoid – malfunction	**Wiring, solenoid, ECM**
72	Traction control ventilation solenoid – malfunction	**Wiring, solenoid, ECM**

MITSUBISHI

Model:	Engine identification:	Year:
Diamante 3.0/3.5L	6G72/6G74	1996-04
Eclipse 2.0L Turbo	4G63	1995-99
Eclipse 2.4L	4G64	1996-04
Eclipse 3.0L	6G72	2000-04
Endeavor 3.8L	6G75	2004
Galant 2.4/3.0/3.8L	4G64/6G72/6G75	1996-04
Lancer Evolution/Sportback 2.0/2.4L/Turbo	4G94/4G63/4G69	2002-04
Montero Sport 2.4L	4G64	1997-99
Montero/Montero Sport 3.0/3.5/3.8L	6G72/6G74/6G75	1996-04
Mighty Max Pickup 2.4L	4G64	1996
Outlander 2.4L	4G69	2003-04
3000GT 3.0L/Turbo	6G72	1996-99

System: **MPI**

Self-diagnosis

General information

- Refer to the front of this manual for general test conditions, terminology, detailed descriptions of wiring faults and a general trouble shooter for electrical and mechanical faults.
- Engine control module (ECM) incorporates self-diagnosis function.
- Malfunction indicator lamp (MIL) will illuminate if certain faults are recorded.
- ECM operates in backup mode if sensors fail, to enable vehicle to be driven to workshop.

Accessing

- Trouble codes can be displayed by using a Scan Tool connected to the data link connector (DLC):
  - Diamante/Eclipse/Galant/3000GT – **1**.
  - Montero/Montero Sport/Mighty Max Pickup – **2**.
  - Endeavor/Outlander/Lancer Evolution/Sportback – **3**.

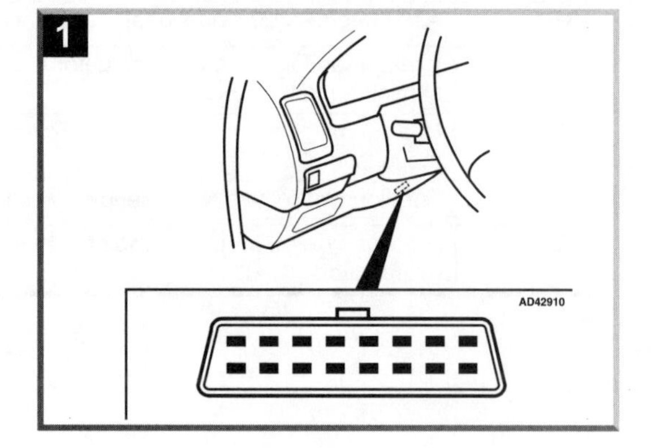

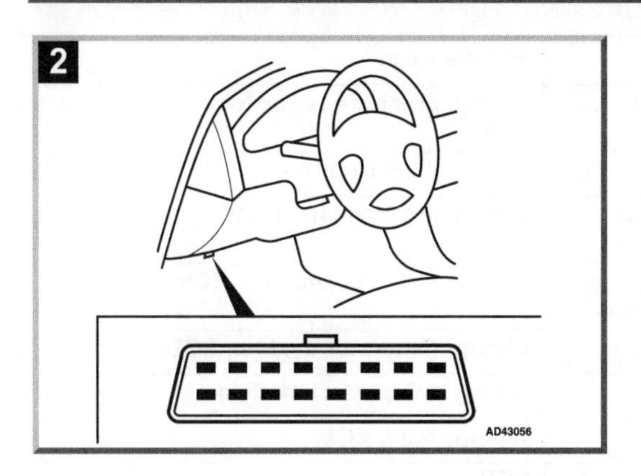

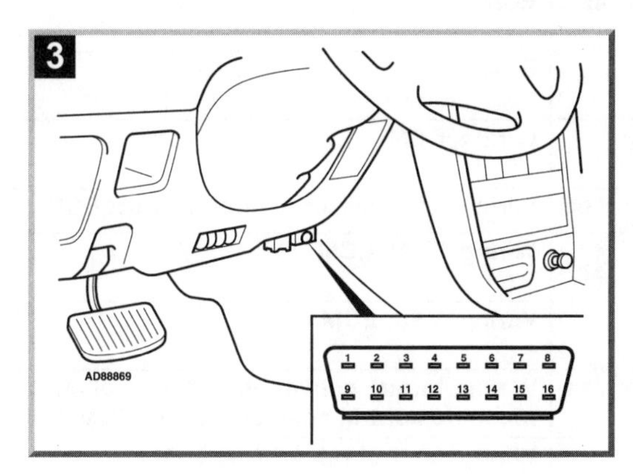

Erasing

- Trouble codes can be erased by using a Scan Tool connected to the data link connector (DLC) or as follows:
- Ensure ignition switched OFF.
- Disconnect battery ground cable for 30 seconds minimum.
- Reconnect battery ground cable.
- Repeat checking procedure to ensure no data remains in ECM fault memory.

NOTE: ***Disconnecting battery ground lead may erase memory from electronic units such as radio.***

Trouble code identification

OBD-II code	Fault location	Probable cause
All P0, P2 and U0 codes	Refer to OBD-II trouble code tables at the front of this manual	–
P1020	Variable valve timing control – performance problem	**Wiring, engine oil pressure switch, valve timing control valve, actuator, ECM**
P1021	Valve timing control valve – circuit malfunction	**Wiring, valve timing control valve, ECM**
P1100	Intake manifold air control actuator position sensor – circuit malfunction	**Wiring, position sensor, actuator, ECM**
P1101	Traction control vacuum solenoid – circuit malfunction	**Wiring, vacuum solenoid, ECM**
P1102	Traction control vent solenoid – circuit malfunction	**Wiring, vent solenoid, ECM**
P1103	Turbocharger (TC) wastegate actuator – malfunction	**Wiring, VAF sensor, ECM**
P1104	Turbocharger (TC) wastegate regulating valve – malfunction	**Wiring, TC wastegate regulating valve, ECM**

OBD-II code	Fault location	Probable cause
P1105	Fuel pressure regulator control solenoid – malfunction	**Wiring, pressure regulator control solenoid, ECM**
P1300	Ignition timing adjustment – circuit malfunction	**Wiring short to ground, ECM**
P1400	Manifold absolute pressure (MAP) sensor – circuit malfunction	**Wiring, MAP sensor, ECM**
P1443	Evaporative emission (EVAP) canister purge system – control valve 2 circuit malfunction	**Wiring, EVAP solenoid 2, ECM**
P1500	Generator system – circuit malfunction	**Wiring, generator system, ECM**
P1530	AC control module – circuit malfunction	**Wiring, AC control module**
P1600	AT to ECM – signal failure	**Wiring, TCM, ECM**
P1601	ECM to throttle control unit – communication malfunction	**ECM**
P1602	ECM to system SLI – communication malfunction	**ECM**
P1603	Engine control module (ECM) – battery supply voltage	**Wiring, ECM**
P1610	Immobilizer control module – circuit malfunction	**Wiring, immobilizer control module, ECM**
P1715	Automatic transmission speed sensors A and B – malfunction	**Wiring, TCM, ECM**
P1720	Vehicle speed sensor (VSS) – circuit malfunction	**Wiring, VSS, ECM**
P1750	AT – torque converter/shift control and pressure control solenoid valves – malfunction	**Wiring, torque converter/shift control and pressure control solenoid valves, TCM, ECM**
P1751	Automatic transmission – control relay malfunction	**Wiring, control relay, TCM, ECM**
P1791	Engine coolant temperature to TCM – signal malfunction	**Wiring, TCM, ECM**
P1795	Throttle position (TP) switch to TCM – signal malfunction	**Wiring, TCM, ECM**
U1073	CAN data bus – bus off, malfunction detected	**Wiring, ECM**
U1102	Antilock brake system (ABS)/traction control system (TCS), CAN data bus – time out	**Wiring, ABS control module, ECM**
U1108	Instrumentation control module, CAN data bus – time out	**Wiring, instrumentation control module, ECM**
U1110	AC, CAN data bus – time out	**Wiring, AC control module, ECM**

MITSUBISHI

Model:	Engine identification:	Year:
Eclipse 1.8L	**4G37**	**1990-94**
Eclipse 2.0L/Turbo	**4G63**	**1992-94**
Galant 2.0L 8V SOHC	**4G63**	**1989-92**
Galant 2.0L 16V DOHC	**4G63**	**1989-93**
Mirage 1.5L	**4G15**	**1989-94**
Precis 1.5L	**G4AJ**	**1990-93**

System: **MPI**

Self-diagnosis

General information

- Refer to the front of this manual for general test conditions, terminology, detailed descriptions of wiring faults and a general trouble shooter for electrical and mechanical faults.
- Engine control module (ECM) incorporates self-diagnosis function.
- Malfunction indicator lamp (MIL) will illuminate if certain faults are recorded.
- ECM operates in backup mode if sensors fail, to enable vehicle to be driven to workshop.
- Trouble codes can be displayed using a voltmeter connected to the data link connector (DLC) **3**.

Accessing

- Ensure ignition switched OFF.
- Access data link connector (DLC):
  - Galant/Eclipse/Mirage – **1**.
  - Precis – **2**.
- Connect analogue type voltmeter between data link connector (DLC) terminals 1 and 12 **3**.
- Switch ignition ON.
- Count voltmeter needle deflections and compare with trouble code table.
- Trouble codes will be displayed by voltmeter deflections of 1.5 seconds, indicating 'TENS' **4** [**A**], followed by voltmeter deflections of 0.5 seconds, indicating 'UNITS' **4** [**D**].
- A 0.5 second pause separates each deflection **4** [**B**].
- A 2 second pause separates 'TENS' and 'UNITS' **4** [**C**].
- A 3 second pause separates each trouble code **4** [**E**].
- For example: Voltmeter deflects for 1.5 seconds twice, followed by deflections of 0.5 seconds twice, indicates trouble code 22 **4**.
- Switch ignition OFF.
- Disconnect voltmeter.

NOTE: ***If a constant deflection is displayed, suspect engine control module (ECM) fault.***

NOTE: ***If constant short pulses are displayed, no trouble code is stored.***

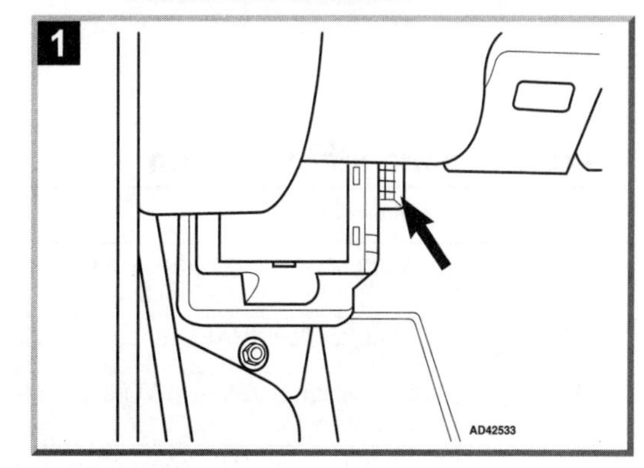

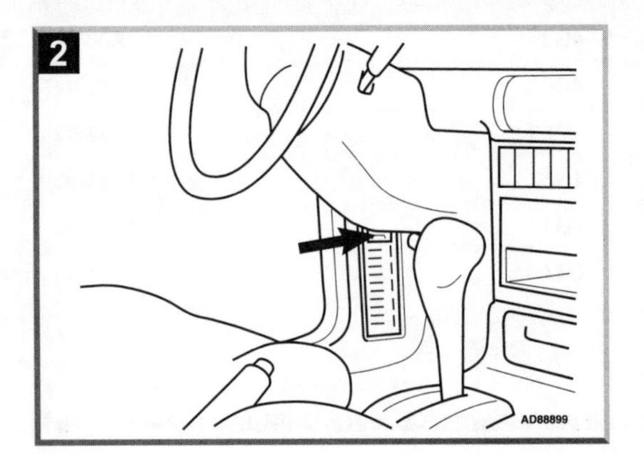

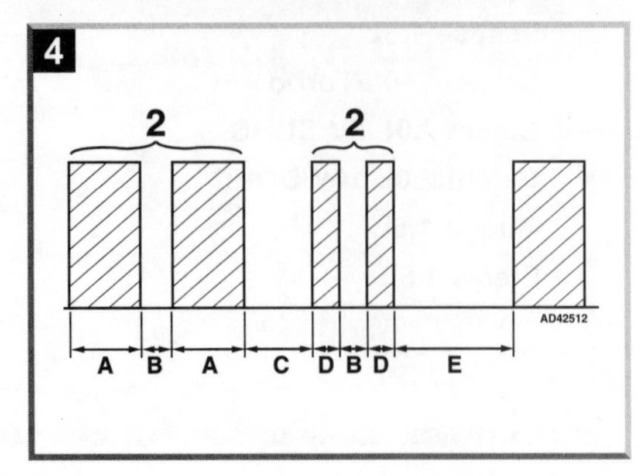

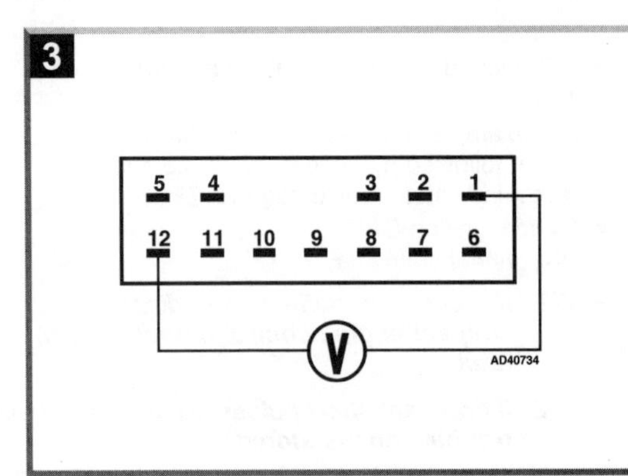

Erasing

- Ensure ignition switched OFF.
- Disconnect battery ground lead.
- Wait 15 seconds minimum.
- Reconnect battery ground lead.
- Repeat checking procedure to ensure no data remains in ECM fault memory.

NOTE: ***Disconnecting battery ground lead may erase memory from electronic units such as radio.***

Trouble code identification

Trouble code	Fault location	Probable cause
11	Oxygen sensor (O2S)/heated oxygen sensor (HO2S) – front – malfunction	**Wiring, O2S/HO2S, fuel system, injectors, ECM**
12	Volume air flow (VAF) sensor – malfunction	**Wiring, VAF sensor, ECM**
13	Intake air temperature (IAT) sensor – malfunction	**Wiring, IAT sensor, ECM**
14	Throttle position (TP) sensor – malfunction	**Wiring, TP sensor, CTP switch, ECM**
15	Idle speed control (ISC) actuator position sensor – malfunction	**Wiring, ISC sensor, ECM**
21	Engine coolant temperature (ECT) sensor – malfunction	**Wiring, ECT sensor, ECM**

Trouble code	Fault location	Probable cause
22	Crankshaft position (CKP) sensor – malfunction	**Wiring, CKP sensor, ECM**
23	Camshaft position (CMP) sensor – malfunction	**Wiring, CMP sensor, ECM**
24	Vehicle speed sensor (VSS) – malfunction	**Wiring, VSS, ECM**
25	Barometric pressure (BARO) sensor – malfunction	**Wiring, BARO sensor, ECM**
31	Knock sensor (KS) – malfunction	**Wiring, KS, ECM**
36	Ignition timing adjustment	**Wiring short to ground**
41	Injectors – malfunction	**Wiring, injectors, ECM**
42	Fuel pump – malfunction	**Wiring, relay module, fuel pump, ECM**
43	Exhaust gas recirculation (EGR) system – malfunction	**Wiring, EGR, EGRT, EGRTVV, hose leak, ECM**
44	Ignition coil/module – malfunction	**Wiring, ignition coil/module, ECM**
55	Idle speed control (ISC) actuator position sensor – malfunction	**Wiring, ISC sensor, ECM**
59	Heated oxygen sensor (HO2S) – rear – malfunction	**Wiring, HO2S, ECM**

MITSUBISHI

Model:	Engine identification:	Year:
Eclipse 2.0L	420A	1995-99

System: **MPI**

Self-diagnosis

General information

- Refer to the front of this manual for general test conditions, terminology, detailed descriptions of wiring faults and a general trouble shooter for electrical and mechanical faults.
- Engine control module (ECM) incorporates self-diagnosis function.
- Malfunction indicator lamp (MIL) will illuminate if certain faults are recorded.
- ECM operates in backup mode if sensors fail, to enable vehicle to be driven to workshop.
- Trouble codes can be displayed by using a Scan Tool connected to the data link connector (DLC) **1**.
- Trouble codes can also be displayed by the MIL.

NOTE: ***The use of a Scan Tool is essential to obtain full diagnostic information.***

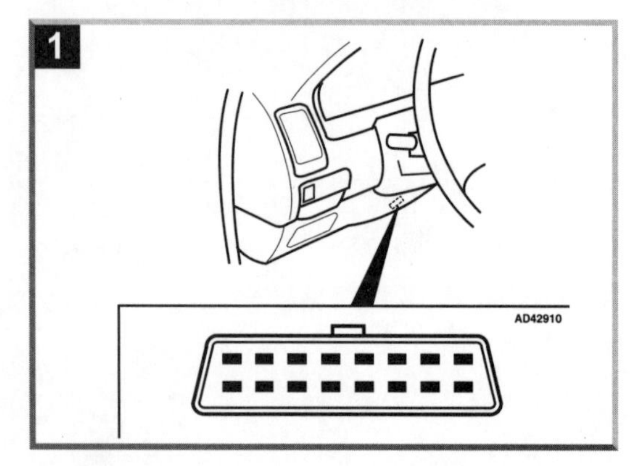

Accessing

- Ensure ignition switched OFF.
- Within 5 seconds turn ignition key as follows:
  - Ignition ON – OFF – ON – OFF – ON.
- MIL should be flashing.
- Count MIL flashes and compare with trouble code table.
  - The first group of short flashes indicate the 'TENS' **2** [**A**] of the trouble code.
  - The second group of short flashes indicates the 'UNITS' **2** [**D**] of the trouble code.
  - A short pause separates each flash **2** [**B**].
  - A long pause separates each set of flashes **2** [**C**].
  - A 4 second pause separates each trouble code **2** [**E**].
  - For example: Trouble code 32 displayed **2**.
- Switch ignition OFF.

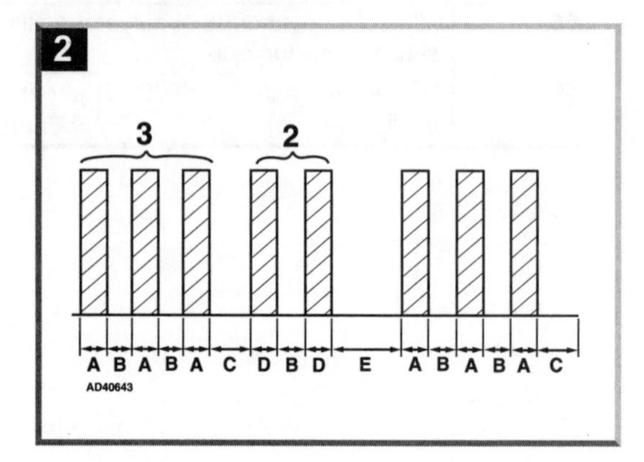

Erasing

- Trouble codes can be erased by using a Scan Tool connected to the data link connector (DLC) or as follows:
- Ensure ignition switched OFF.
- Disconnect battery ground cable for 30 seconds minimum.
- Reconnect battery ground cable.
- Repeat checking procedure to ensure no data remains in ECM fault memory.

NOTE: ***Disconnecting battery ground lead will erase trouble codes but may also erase memory from electronic units such as radio.***

Trouble code identification

Flash code – OBD-II code	Fault location	Probable cause	
–	*All P0, P2 and U0 codes*	Refer to OBD-II trouble code tables at the front of this manual	–
11	–	Crankshaft position (CKP) sensor – circuit malfunction	**Wiring, CKP sensor, ECM**
11	P1390	CKP/CMP sensor – timing incorrect	**Wiring, timing belt, CKP/CMP sensor, ECM**
11	P1391	CKP/CMP sensor – signal failure	**Wiring, CKP/CMP sensor, ECM**
11	P1398	Adaptive speed for misfire detection – at limit	**Wiring, CKP sensor, mechanical fault, ECM**
13	P1297	Manifold absolute pressure (MAP) sensor – circuit malfunction	**Wiring, MAP sensor, ECM**
14	–	Manifold absolute pressure (MAP) sensor – voltage low/high	**Wiring, MAP sensor, ECM**
14	P1496	Manifold absolute pressure (MAP) sensor – voltage supply too low	**Wiring short to ground, MAP sensor, ECM**
14	P1296	Manifold absolute pressure (MAP) sensor – no voltage supply	**Wiring, MAP sensor, ECM**
15	–	Vehicle speed sensor (VSS) – circuit malfunction	**Wiring, VSS, ECM**
16	–	Knock sensor (KS) – circuit malfunction	**Wiring, KS, ECM**
17	–	Engine coolant temperature – voltage low	**Wiring, ECT sensor, cooling system, ECM**
21	–	Heated oxygen sensor (HO2S) – front/rear	**Wiring, fuel system, exhaust system, HO2S, ECM**
22	–	Engine coolant temperature (ECT) sensor – circuit/voltage low/high	**Wiring, ECT sensor, ECM**
23	–	Intake air temperature (IAT) sensor – voltage low/high	**Wiring, IAT sensor, ECM**
24	–	Throttle position (TP) sensor – voltage low/ high/ circuit performance problem	**Wiring, TP sensor, MAP sensor, ECM**
24	P1295	Throttle position (TP) sensor – no voltage supply	**Wiring, TP sensor, ECM**
25	–	Idle air (IAC) control system – malfunction	**Wiring, IAC system, ECM**
25	P1294	Idle air (IAC) control system – incorrect idle speed	**Wiring, IAC motor, ECM**
27	–	Injector – circuit malfunction	**Wiring, injectors, ECM**
31	–	Evaporative emission (EVAP) canister purge system	**Wiring, Fuel tank/cap/vent valve/hoses, EVAP solenoid, EVAP canister, ECM**
31	P1495	Evaporative emission (EVAP) canister purge system – vent solenoid circuit malfunction	**Wiring, EVAP solenoid, ECM**

Flash code – OBD-II code	Fault location	Probable cause	
31	P1494	Evaporative emission (EVAP) canister purge system – vent control malfunction	**Wiring, EVAP solenoid, hose blockage, ECM**
31	P1486	Evaporative emission (EVAP) canister purge system – leak monitor hose malfunction	**Hose blockage, EVAP solenoid, ECM**
32	–	Exhaust gas recirculation (EGR) system – malfunction	**Wiring, EGR solenoid, hose blockage, ECM**
33	–	A/C clutch relay – circuit malfunction	**Wiring, A/C clutch relay, ECM**
35	P1490	Engine coolant blower motor relay – low speed/circuit malfunction	**Wiring, blower motor relay, ECM**
35	P1489	AC condenser blower motor relay – high speed/circuit malfunction	**Wiring, blower motor relay, ECM**
35	P1487	Engine coolant blower motor relay – high speed/circuit malfunction	**Wiring, blower motor relay, ECM**
36	–	Pulsed secondary air injection (PAIR) system – malfunction	**Wiring, PAIR valve, hose blockage, ECM**
37	P1899	Park/neutral position (PNP) switch – circuit malfunction	**Wiring, PNP switch, ECM**
41	–	Generator system – field insufficient switching	**Wiring, generator, ECM**
42	–	Relay module – circuit malfunction	**Wiring, relay module, ECM**
42	–	Fuel gauge tank sensor – circuit malfunction	**Wiring, tank sensor, ECM**
43	–	Misfire	**Wiring, ignition system, CKP sensor, mechanical fault, fuel system, injectors, ECM**
43	–	Ignition coil 1/2 – circuit malfunction	**Wiring, ignition coil, ECM**
44	–	Battery temperature sensor – circuit malfunction	**Wiring, temperature sensor, ECM**
44	P1493	Battery temperature sensor – voltage low	**Wiring, temperature sensor, ECM**
44	P1492	Battery temperature sensor – voltage high	**Wiring, temperature sensor, ECM**
45	–	Transmission control module – trouble code detected	**Various faults**
46	–	Generator system – voltage high	**Wiring, generator, ECM**
47	–	Generator system – voltage low	**Wiring, generator, ECM**
51	–	Fuel trim – mixture too lean	**Wiring, ECT/IAT sensor, mechanical fault, injector, MAP sensor, fuel system, HO2S, ECM**
52	–	Fuel trim – mixture too rich	**Wiring, ECT/IAT sensor, mechanical fault, injector, MAP sensor, fuel system, HO2S, ECM**
53	–	Engine control module (ECM) – malfunction	**ECM**

Flash code – OBD-II code	Fault location	Probable cause	
54	–	Camshaft position (CMP) sensor – circuit malfunction	**Wiring, CMP sensor, ECM**
61	–	Barometric pressure (BARO) sensor – range/performance	**Wiring, MAP sensor, ECM**
62	P1697	Engine control module (ECM) – service indicator malfunction	**ECM**
63	P1696	Engine control module (ECM) – EEPROM write unsuccessful	**ECM**
64	–	Catalytic converter – efficiency below limit	**Wiring, catalytic converter, HO2S, ECM**
65	–	Power steering pressure (PSP) sensor – range/performance malfunction	**Wiring, PSP sensor, ECM**
66	P1698	AT to ECM – signal failure	**Wiring, TCM, ECM**

MITSUBISHI

Model:	Engine identification:	Year:
Expo 2.4L	**4G64**	**1994-95**
Expo LRV 2.4L	**4G64**	**1994-95**
Expo 2.4L FWD – California	**4G64**	**1994**
Expo LRV 2.4L FWD – California	**4G64**	**1994**
Expo LRV 1.8L – Federal	**4G93**	**1994-95**
Expo LRV 1.8L – California	**4G93**	**1994**

System: **MPI**

Self-diagnosis

General information

- Refer to the front of this manual for general test conditions, terminology, detailed descriptions of wiring faults and a general trouble shooter for electrical and mechanical faults.
- Engine control module (ECM) incorporates self-diagnosis function.
- Malfunction indicator lamp (MIL) will illuminate if certain faults are recorded.
- ECM operates in backup mode if sensors fail, to enable car to be driven to workshop.
- Trouble codes can be displayed with the MIL or a suitable code reader connected to the data link connector (DLC) **1**.

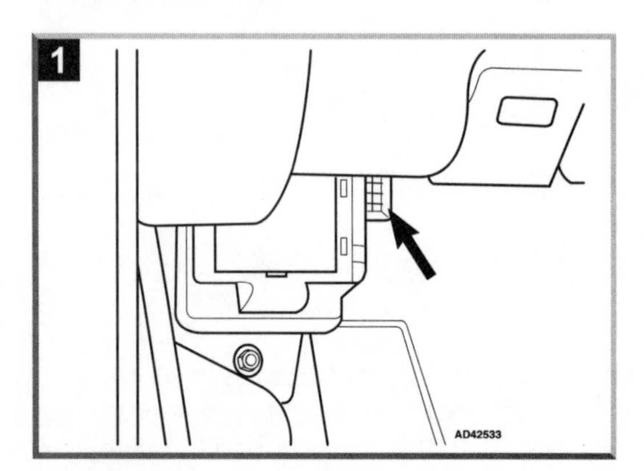

Accessing

- Ensure ignition switched OFF.
- Access data link connector (DLC) **1**.
- Jump DLC terminal 1 and ground **2**.
- Switch ignition ON. Do NOT start engine.
- MIL should be flashing.
  - 1.5 second flashes indicate the 'tens' of the trouble code **3** [**A**].
  - 0.5 second flashes indicate the 'units' of the trouble code **3** [**B**].
  - A 0.5 second pause separates each flash **3** [**C**].
  - A 2 second pause separates 'tens' and 'units' **3** [**D**].
  - A 3 second pause separates each trouble code **3** [**E**].
  - For example: Trouble code 24 displayed **3**.
- If no trouble codes are stored, short flashes will continue to be displayed.
- If MIL remains ON: Suspect faulty engine control module (ECM).
- Count MIL flashes and compare with trouble code table.
- Switch ignition OFF.
- Remove jump lead.

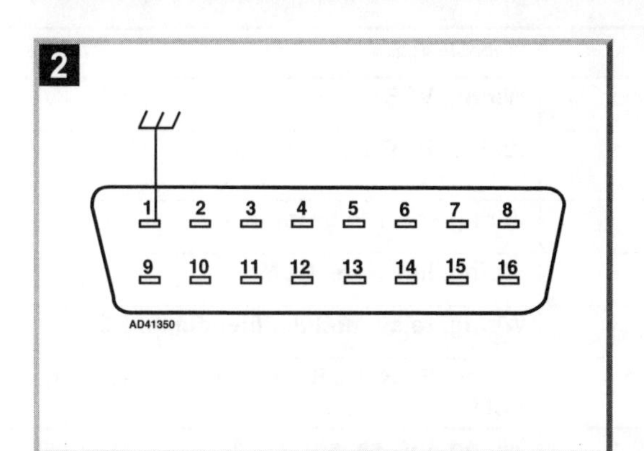

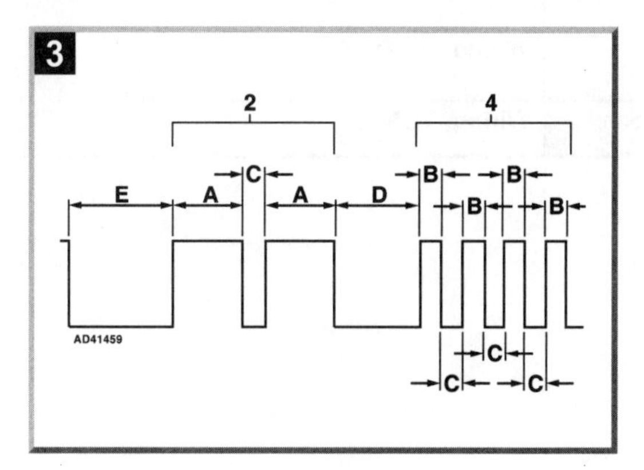

Erasing

- Ensure ignition switched OFF.
- Disconnect battery ground lead for 15 seconds minimum.
- Reconnect battery ground lead.
- Repeat checking procedure to ensure no data remains in ECM fault memory.

NOTE: ***Disconnecting battery ground lead may erase memory from electronic units such as radio.***

Trouble code identification

Flash code	Fault location	Probable cause
11	Oxygen sensor (O2S)/heated oxygen sensor (HO2S) – front – malfunction	**Wiring, O2S/HO2S, fuel system, injectors, ECM**
12	Volume air flow (VAF) sensor – malfunction	**Wiring, VAF sensor, ECM**
13	Intake air temperature (IAT) sensor – malfunction	**Wiring, IAT sensor, ECM**
14	Throttle position (TP) sensor – malfunction	**Wiring, TP sensor, CTP switch, ECM**
21	Engine coolant temperature (ECT) sensor – malfunction	**Wiring, ECT sensor, ECM**
22	Crankshaft position (CKP) sensor – malfunction	**Wiring, CKP sensor, ECM**
23	Camshaft position (CMP) sensor – malfunction	**Wiring, CMP sensor, ECM**

Flash code	Fault location	Probable cause
24	Vehicle speed sensor (VSS) – malfunction	Wiring, VSS, ECM
25	Barometric pressure (BARO) sensor – malfunction	Wiring, BARO sensor, ECM
36	Ignition timing adjustment	Wiring short to ground
41	Injectors – malfunction	Wiring, injectors, ECM
42	Fuel pump – malfunction	Wiring, relay module, fuel pump, ECM
43	Exhaust gas recirculation (EGR) system – malfunction	Wiring, EGR, EGRT, EGRTVV, hose leak, ECM
55	Idle speed control (ISC) actuator position sensor – malfunction	Wiring, ISC sensor, ECM
59	Heated oxygen sensor (HO2S) – rear – malfunction	Wiring, HO2S, ECM
61	AT – torque reduction signal – malfunction	Wiring, TCM

MITSUBISHI

Model:	Engine identification:	Year:
Expo/LRV 2.4L – Federal	**4G64**	**1996**
Expo/LRV 2.4L FWD – California	**4G64**	**1995-96**
Expo 2.4L 4X4 – California	**4G64**	**1996**
Expo LRV 1.8L – Federal	**4G93**	**1996**
Expo LRV 1.8L – California	**4G93**	**1995-96**
Mirage 1.5/1.8L	**4G15/4G93**	**1995-02**

System: **MPI**

Self-diagnosis

General information

- Refer to the front of this manual for general test conditions, terminology, detailed descriptions of wiring faults and a general trouble shooter for electrical and mechanical faults.
- Engine control module (ECM) incorporates self-diagnosis function.
- Malfunction indicator lamp (MIL) will illuminate if certain faults are recorded.
- ECM operates in backup mode if sensors fail, to enable vehicle to be driven to workshop.

Accessing

- Trouble codes can be displayed by using a Scan Tool connected to the data link connector (DLC):
  - Expo/LRV – **1**.
  - Mirage – **2**.

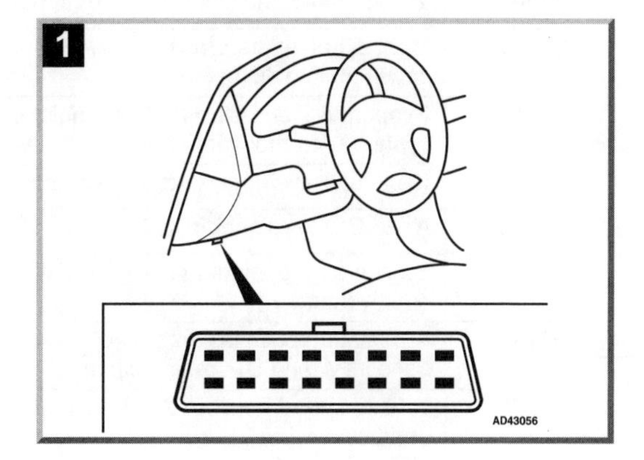

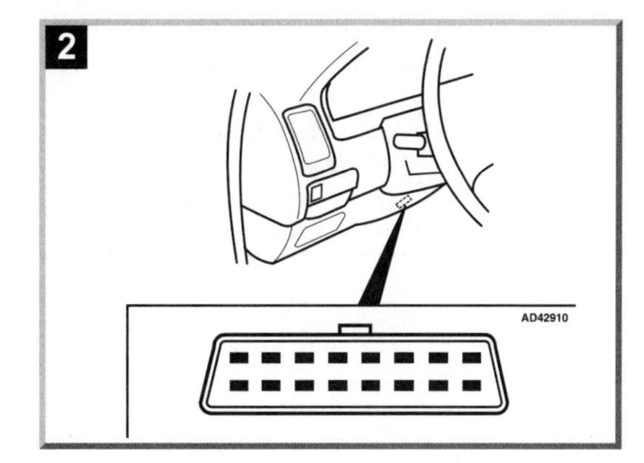

Erasing

- Trouble codes can be erased by using a Scan Tool connected to the data link connector (DLC) or as follows:
- Ensure ignition switched OFF.
- Disconnect battery ground cable for 30 seconds minimum.
- Reconnect battery ground cable.
- Repeat checking procedure to ensure no data remains in ECM fault memory.

NOTE: ***Disconnecting battery ground lead will erase trouble codes but may also erase memory from electronic units such as radio.***

Trouble code identification

OBD-II code	Fault location	Probable cause
All P0, P2 and U0 codes	Refer to OBD-II trouble code tables at the front of this manual	–
P1300	Ignition timing adjustment – circuit malfunction	**Wiring short to ground, ECM**
P1400	Manifold absolute pressure (MAP) sensor – circuit malfunction	**Wiring, MAP sensor, ECM**
P1443	Evaporative emission (EVAP) canister purge system – control valve 2 circuit malfunction	**Wiring, EVAP solenoid 2, ECM**
P1500	Generator system – circuit malfunction	**Wiring, generator system, ECM**
P1600	AT to ECM – signal failure	**Wiring, TCM, ECM**
P1603	Engine control module (ECM) – battery supply voltage	**Wiring, ECM**
P1715	Automatic transmission speed sensors A and B – malfunction	**Wiring, TCM, ECM**
P1720	vehicle speed sensor (VSS) – circuit malfunction	**Wiring, VSS, ECM**
P1750	AT – torque converter/shift control and pressure control solenoid valves – malfunction	**Wiring, torque converter/shift control and pressure control solenoid valves, TCM, ECM**
P1751	Automatic transmission – control relay malfunction	**Wiring, control relay, TCM, ECM**
P1791	Engine coolant temperature to TCM – signal malfunction	**Wiring, TCM, ECM**
P1795	Throttle position (TP) switch to TCM – signal malfunction	**Wiring, TCM, ECM**

MITSUBISHI

Model:	Engine identification:	Year:
Mighty Max Pickup 2.4L	**4G64**	**1990-95**
Mighty Max Pickup 3.0L	**6G72**	**1990-96**
Montero 3.0L	**6G72**	**1989-93**

System: **MPI**

Self-diagnosis

General information

- Refer to the front of this manual for general test conditions, terminology, detailed descriptions of wiring faults and a general trouble shooter for electrical and mechanical faults.
- Engine control module (ECM) incorporates self-diagnosis function.
- Malfunction indicator lamp (MIL) will illuminate if certain faults are recorded.
- ECM operates in backup mode if sensors fail, to enable vehicle to be driven to workshop.
- Trouble codes can be displayed using a voltmeter connected to the data link connector (DLC) **3**.

Accessing

- Ensure ignition switched OFF.
- Access data link connector (DLC) **1** Mighty Max Pickup, or **2** Montero.
- Connect analogue type voltmeter between data link connector (DLC) terminals 1 and 12 **3**.
- Switch ignition ON.
- Count voltmeter needle deflections and compare with trouble code table.
- Trouble codes will be displayed by voltmeter deflections of 1.5 seconds, indicating 'TENS' **4** [**A**], followed by voltmeter deflections of 0.5 seconds, indicating 'UNITS' **4** [**D**].
- A 0.5 second pause separates each deflection **4** [**B**].
- A 2 second pause separates 'TENS' and 'UNITS' **4** [**C**].
- A 3 second pause separates each trouble code **4** [**E**].
- For example: Voltmeter deflects for 1.5 seconds twice, followed by deflections of 0.5 seconds twice, indicates trouble code 22 **4**.
- Switch ignition OFF.
- Disconnect voltmeter.

NOTE: ***If a constant deflection is displayed, suspect engine control module (ECM) fault.***

NOTE: ***If constant short pulses are displayed, no trouble code is stored.***

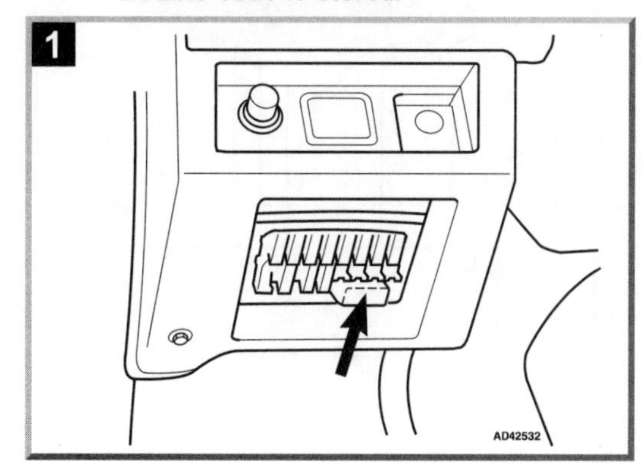

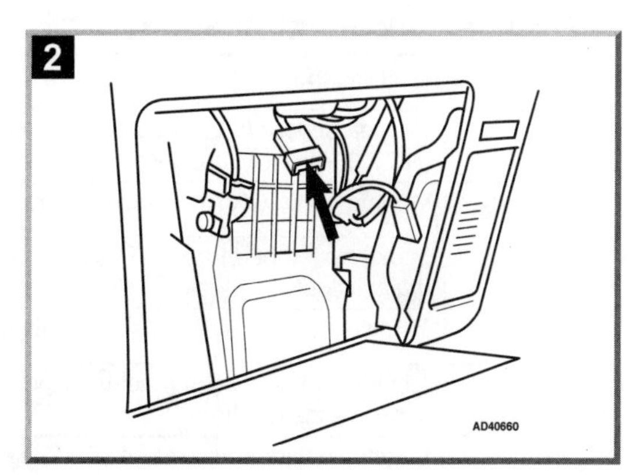

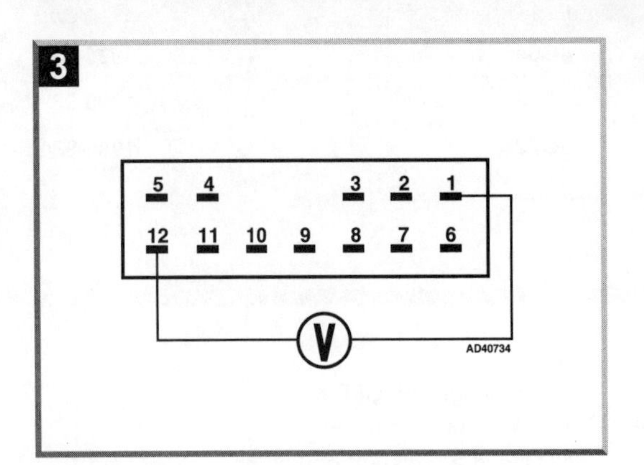

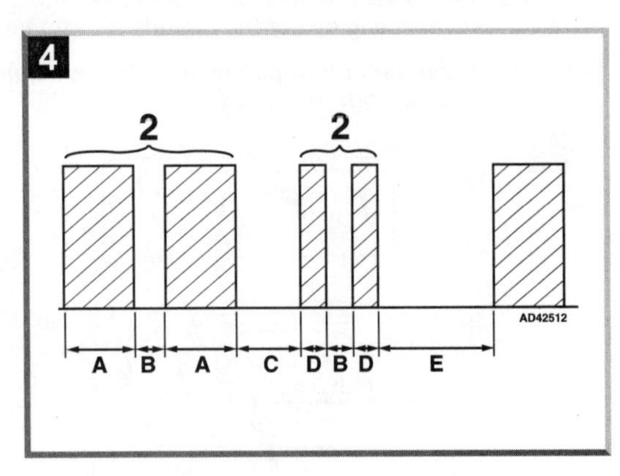

Erasing

- Ensure ignition switched OFF.
- Disconnect battery ground lead.
- Wait 15 seconds minimum.
- Reconnect battery ground lead.
- Repeat checking procedure to ensure no data remains in ECM fault memory.

NOTE: ***Disconnecting battery ground lead may erase memory from electronic units such as radio.***

Trouble code identification

Trouble code	Fault location	Probable cause
11	Oxygen sensor (O2S)/heated oxygen sensor (HO2S) – front – malfunction	**Wiring, O2S/HO2S, fuel system, injectors, ECM**
12	Volume air flow (VAF) sensor – malfunction	**Wiring, VAF sensor, ECM**
13	Intake air temperature (IAT) sensor – malfunction	**Wiring, IAT sensor, ECM**
14	Throttle position (TP) sensor – malfunction	**Wiring, TP sensor, CTP switch, ECM**
15	Idle speed control (ISC) actuator position sensor – malfunction	**Wiring, ISC sensor, ECM**
21	Engine coolant temperature (ECT) sensor – malfunction	**Wiring, ECT sensor, ECM**
22	Crankshaft position (CKP) sensor – malfunction	**Wiring, CKP sensor, ECM**

Trouble code	Fault location	Probable cause
23	Camshaft position (CMP) sensor – malfunction	**Wiring, CMP sensor, ECM**
24	Vehicle speed sensor (VSS) – malfunction	**Wiring, VSS, ECM**
25	Barometric pressure (BARO) sensor – malfunction	**Wiring, BARO sensor, ECM**
36	Ignition timing adjustment	**Wiring short to ground**
41	Injectors – malfunction	**Wiring, injectors, ECM**
42	Fuel pump – malfunction	**Wiring, relay module, fuel pump, ECM**
43	Exhaust gas recirculation (EGR) system – malfunction	**Wiring, EGR, EGRT, EGRTVV, hose leak, ECM**
55	Idle speed control (ISC) actuator position sensor – malfunction	**Wiring, ISC sensor, ECM**
59	Heated oxygen sensor (HO2S) – rear – malfunction	**Wiring, HO2S, fuel system, injectors, ECM**

MITSUBISHI

Model:	Engine identification:	Year:
Montero 3.0L 12V	**6G72**	**1994-95**

System: **MPI**

Self-diagnosis

General information

- Refer to the front of this manual for general test conditions, terminology, detailed descriptions of wiring faults and a general trouble shooter for electrical and mechanical faults.
- Engine control module (ECM) incorporates self-diagnosis function.
- Malfunction indicator lamp (MIL) will illuminate if certain faults are recorded.
- ECM operates in backup mode if certain sensors fail, to enable car to be driven to workshop.
- Trouble codes can be accessed with a voltmeter or a suitable code reader connected to the data link connector (DLC).

Accessing

- Ensure ignition switched OFF.
- Connect an analogue voltmeter between data link connector (DLC) 1 **1** [**2**] terminal 4 or 5 and DLC 2 **1** [**1**] terminal 25.
- Switch ignition ON. Do NOT start engine.
- Voltmeter should be pulsing.
  - 1.5 second pulses indicate the 'tens' of the trouble code **2** [**A**].
  - 0.5 second pulses indicate the 'units' of the trouble code **2** [**B**].
  - A 0.5 second pause separates each pulse **2** [**C**].
  - A 2 second pause separates 'tens' and 'units' **2** [**D**].
  - A 3 second pause separates each trouble code **2** [**E**].
  - For example: Trouble code 24 displayed **2**.
- If no trouble codes are stored, short pulses will continue to be displayed.
- If continuous voltage is displayed: Suspect faulty engine control module (ECM).
- Count voltmeter pulses and compare with trouble code table.
- Switch ignition OFF.
- Remove voltmeter.

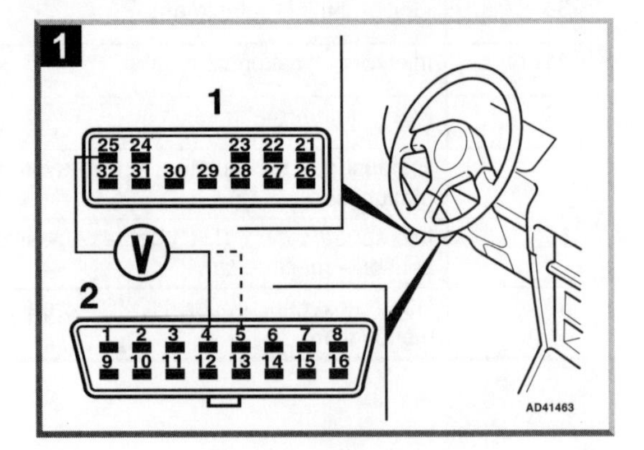

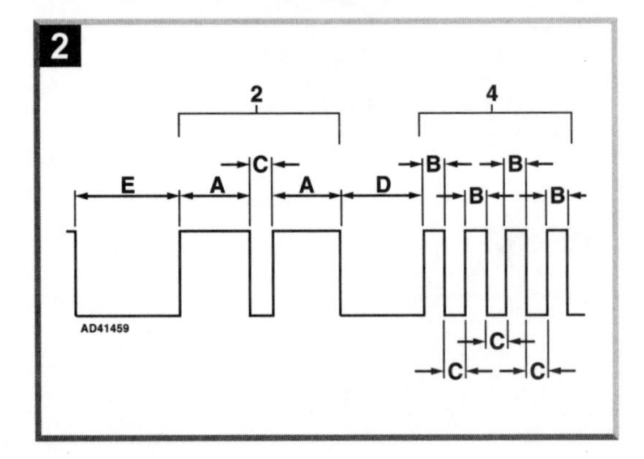

Erasing

- Ensure ignition switched OFF.
- Disconnect battery ground lead for 15 seconds minimum.
- Reconnect battery ground lead.
- Repeat checking procedure to ensure no data remains in ECM fault memory.

NOTE: ***Disconnecting battery ground lead may erase memory from electronic units such as radio.***

Trouble code identification

Trouble code	Fault location	Probable cause
11	Heated oxygen sensor (HO2S) – front – malfunction	**Wiring, HO2S, fuel system, injectors, ECM**
12	Volume air flow (VAF) sensor – malfunction	**Wiring, VAF sensor, ECM**
13	Intake air temperature (IAT) sensor – malfunction	**Wiring, IAT sensor, ECM**
14	Throttle position (TP) sensor – malfunction	**Wiring, TP sensor, CTP switch, ECM**
21	Engine coolant temperature (ECT) sensor – malfunction	**Wiring, ECT sensor, ECM**
22	Crankshaft position (CKP) sensor – malfunction	**Wiring, CKP sensor, ECM**
23	Camshaft position (CMP) sensor – malfunction	**Wiring, CMP sensor, ECM**
24	Vehicle speed sensor (VSS) – malfunction	**Wiring, VSS, ECM**
25	Barometric pressure (BARO) sensor – malfunction	**Wiring, BARO sensor, ECM**
36	Ignition timing adjustment	**Wiring short to ground**
41	Injectors – malfunction	**Wiring, injectors, ECM**
42	Fuel pump – malfunction	**Wiring, relay module, fuel pump, ECM**
43	Exhaust gas recirculation (EGR) system – malfunction	**Wiring, EGR, EGRT, EGRTVV, hose leak, ECM**

NISSAN

Model:	Engine identification:	Year:
Altima 2.4L	**KA24DE**	**1993-94**
Sentra/NX 1.6L	**GA16DE**	**1991-94**
Sentra/NX 2.0L	**SR20DE**	**1991-94**
Stanza 2.4L	**KA24E**	**1990-92**
240SX 2.4L	**KA24DE**	**1991-94**

System: **ECCS**

Self-diagnosis

General information

- Refer to the front of this manual for general test conditions, terminology, detailed descriptions of wiring faults and a general trouble shooter for electrical and mechanical faults.
- Engine control module (ECM) incorporates self-diagnosis function.
- Malfunction indicator lamp (MIL) will illuminate if certain faults are recorded.
- ECM operates in backup mode if sensors fail, to enable vehicle to be driven to workshop.
- Trouble codes can be displayed by the red LED 3 [**1**] in the engine control module (ECM) and the MIL lamp in diagnostic mode 2, or using a suitable code reader.

Accessing – Mode 1

- Access engine control module (ECM):
  - Except 240SX – center fascia – 1.
  - 240SX – RH kick panel – 2.
- Switch ignition ON.
- Check MIL and red LED in ECM illuminate.
- Start engine. Allow to idle.
- If MIL and LED extinguish, no trouble codes have been recorded.
- If MIL and LED remain illuminated, access trouble codes.

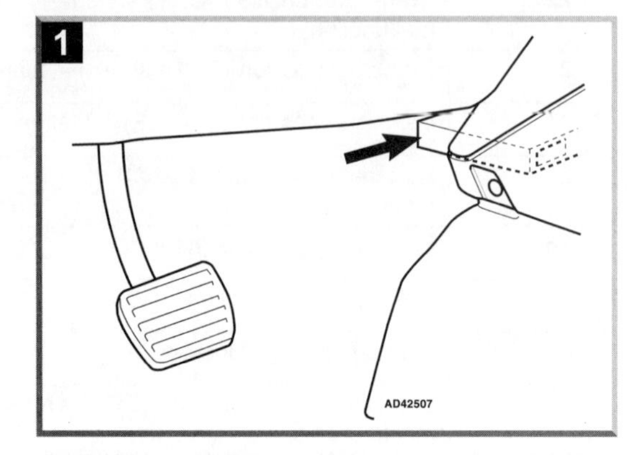

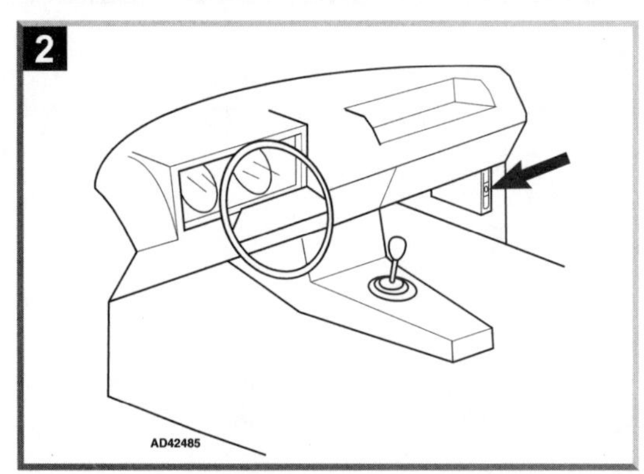

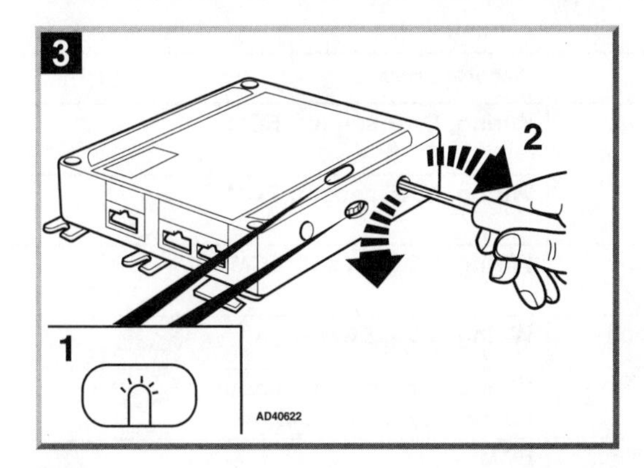

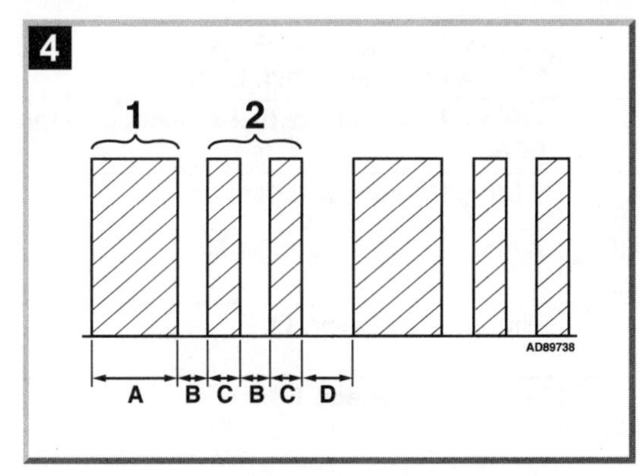

Accessing – Mode 2

- Switch ignition ON. Do NOT start engine.
- Turn diagnostic mode selector on ECM fully clockwise **3** [**2**].
- Wait at least 2 seconds.
- Turn diagnostic mode selector on ECM fully counterclockwise **3** [**2**].
- Self-diagnosis now in Mode 2.
- MIL and red LED in ECM should flash.
- Count flashes and compare with trouble code table.
  - Long flashes indicate the 'tens' of the trouble code **4** [**A**].
  - Short flashes indicate the 'units' of the trouble code **4** [**C**].
  - A short pause separates each flash **4** [**B**].
  - A long pause separates each trouble code **4** [**D**].
  - For example: Trouble code 12 displayed **4** – mass air flow (MAF) sensor.
- Switch ignition OFF.
- Self-diagnosis will return to Mode 1.

NOTE: ***Starting engine in Self-diagnosis Mode 2 will activate heated oxygen sensor (HO2S) diagnosis.***

Erasing

- Switch ignition ON. Do NOT start engine.
- Turn diagnostic mode selector on ECM fully clockwise **3** [**2**].
- Wait at least 2 seconds.
- Turn diagnostic mode selector on ECM fully counterclockwise **3** [**2**].
- Self-diagnosis now in Mode 2.
- Turn diagnostic mode selector on ECM fully clockwise **3** [**2**].
- Wait at least 2 seconds.
- Turn diagnostic mode selector on ECM fully counterclockwise **3** [**2**].
- Self-diagnosis will return to Mode 1.
- Trouble codes now erased.
- Switch ignition OFF.
- Repeat checking procedure to ensure no data remains in ECM fault memory.

NOTE: ***Disconnecting battery ground lead will also erase trouble codes but may erase memory from electronic units such as radio.***

Trouble code identification

Flash code	Fault location	Probable cause
11	Crankshaft position (CKP) sensor – circuit malfunction	**Wiring, CKP sensor, ECM**
12	Mass air flow (MAF) sensor – range/ performance problem	**Wiring, MAF sensor, ECM**
13	Engine coolant temperature (ECT) sensor – range/performance problem	**Wiring, ECT sensor, ECM**
14	Vehicle speed sensor (VSS) – malfunction	**Wiring, VSS, ECM**
21	Ignition coil – no primary signal	**Wiring, ignition coil/module, ignition system, ECM**
31	Engine control module (ECM) – malfunction	**ECM**
32	Exhaust gas recirculation (EGR) system – control valve malfunction	**Wiring, hose leak/blockage, EGR cut OFF solenoid, ECM**
33	Heated oxygen sensor (HO2S) – circuit malfunction	**Wiring, HO2S, Intake/fuel system, injector, ECM**
34	Knock sensor (KS) – circuit malfunction	**Wiring, KS sensor, ECM**
35	Exhaust gas recirculation temperature (EGRT) sensor – circuit malfunction	**Wiring, EGRT sensor, ECM**
41	Intake air temperature (IAT) sensor – circuit malfunction	**Wiring, IAT sensor, ECM**
43	Throttle position (TP) sensor – circuit malfunction	**Wiring, TP sensor, ECM**
45	Injector leak – malfunction	**Injector, fuel system, ECM**
54	ECM/TCM – communication malfunction	**Wiring, ECM, TCM**
55	No faults detected	–

NISSAN

Model:	Engine identification:	Year:
Altima 2.4L	KA24DE	1995
Maxima 3.0L	VQ30DE	1995
Sentra 1.6/2.0L	GA16DE/SR20DE	1995
200SX 1.6L	GA16DE	1995
200SX 2.0L	SR20DE	1995
240SX 2.4L	KA24DE	1995

System: **ECCS**

Self-diagnosis

General information

- Refer to the front of this manual for general test conditions, terminology, detailed descriptions of wiring faults and a general trouble shooter for electrical and mechanical faults.
- Engine control module (ECM) incorporates self-diagnosis function.
- Malfunction indicator lamp (MIL) will illuminate if certain faults are recorded.
- ECM operates in backup mode if sensors fail, to enable vehicle to be driven to workshop.
- Trouble codes can be displayed by the malfunction indicator lamp (MIL) or by using a Scan Tool connected to the data link connector (DLC):
  - Except Maxima – **1** [1].
  - Maxima – **3** [1].

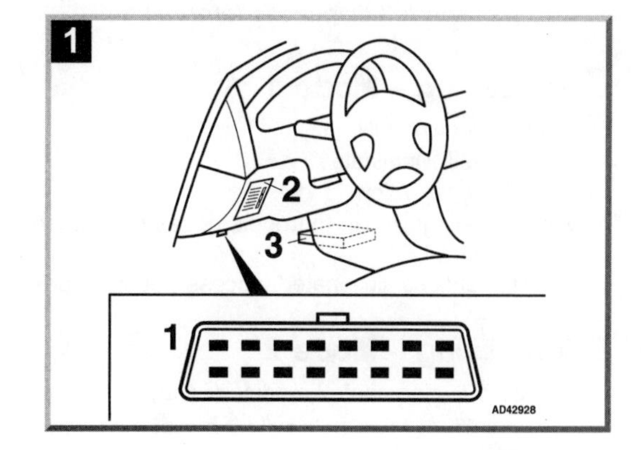

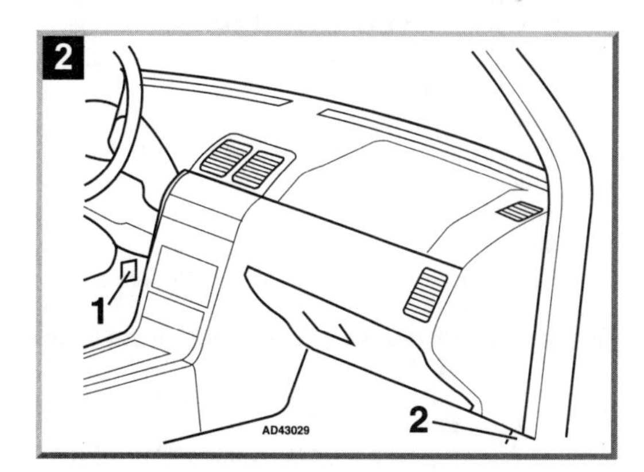

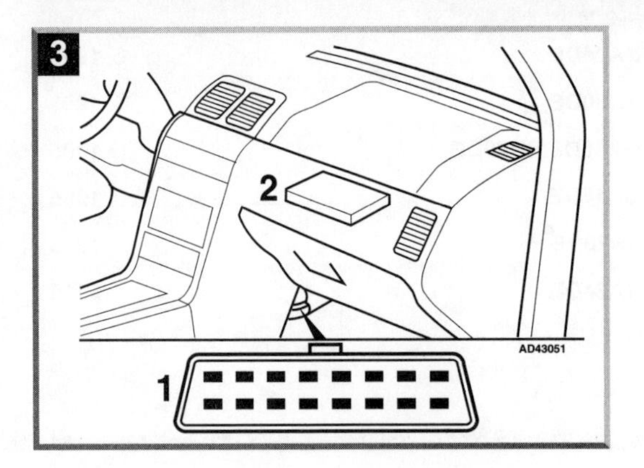

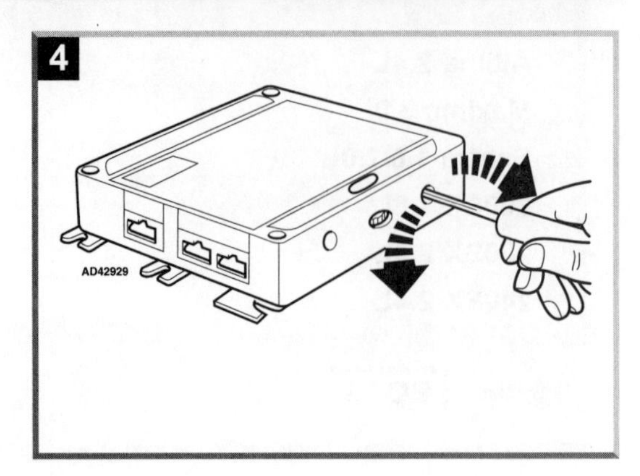

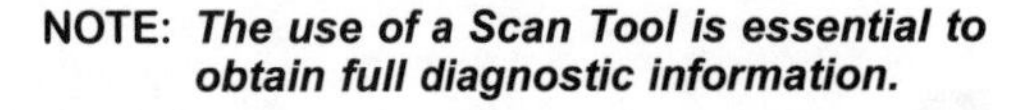

NOTE: ***The use of a Scan Tool is essential to obtain full diagnostic information.***

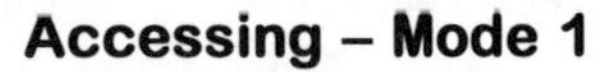

Accessing – Mode 1

- Switch ignition ON.
- Check MIL illuminates.
- Start engine. Allow to idle.
- If MIL extinguishes, no trouble codes have been recorded.
- If MIL remains illuminated, access trouble codes.

Accessing – Mode 2

- Access engine control module (ECM):
  - Altima/Maxima/Sentra/200SX – **1** [3].
  - 240SX – **2** [2].
- Switch ignition ON. Do NOT start engine.
- Turn diagnostic mode selector on ECM fully clockwise **4**.
- Wait at least 2 seconds.
- Turn diagnostic mode selector on ECM fully counterclockwise **4**.
- Self-diagnosis now in Mode 2.
- MIL should flash.
- Count flashes and compare with trouble code table.
  - Long flashes indicate the 'tens' of the trouble code **5** [**A**].
  - Short flashes indicate the 'units' of the trouble code **5** [**C**].
  - A short pause separates each flash **5** [**B**].
  - A long pause separates each trouble code **5** [**D**].
  - For example: Trouble code 12 displayed **5**.
- Switch ignition OFF.
- Self-diagnosis will return to Mode 1.

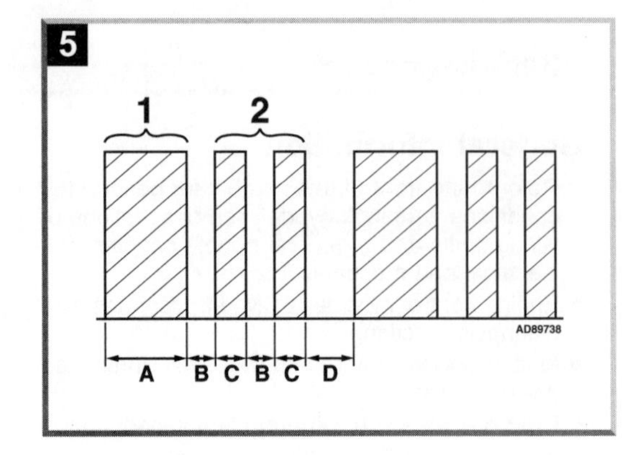

Erasing

- Trouble codes can be erased by using a Scan Tool connected to the data link connector (DLC) or as follows:
- With the ignition switch ON and the engine control module (ECM) in diagnosis test mode 2.
- Turn diagnostic mode selector on ECM fully clockwise **4**.
- Trouble codes are erased.
- Switch ignition OFF.
- Turn diagnostic mode selector on ECM fully counterclockwise **4**.
- Repeat checking procedure to ensure no data remains in ECM fault memory.

Trouble code identification

Flash code	OBD-II code	Fault location	Probable cause
–	*All P0, P2 and U0 codes*	Refer to OBD-II trouble code tables at the front of this manual	–
11	–	Camshaft position (CMP) sensor – circuit malfunction	**Wiring, CMP sensor, Starter motor, ECM**
12	–	Mass air flow (MAF) sensor – circuit malfunction	**Wiring, MAF sensor, ECM**
13	–	Engine coolant temperature (ECT) sensor – circuit malfunction	**Wiring, ECT sensor, ECM**
14	–	Vehicle speed sensor (VSS) – circuit malfunction	**Wiring, VSS, ECM**
21	P1320	Ignition signal – circuit malfunction	**Wiring, ignition coil/module, ignition system, CKP sensor, ECM**
25	–	Idle air control (IAC) system – malfunction	**Wiring, IAC, ECM**
28	P1900	Engine coolant blower motor – circuit malfunction	**Wiring, blower motor, cooling system, ECM**
31	–	Engine control module (ECM) – malfunction	**ECM**
32	–	Exhaust gas recirculation (EGR) system – flow excessively high or low	**Wiring, hose leak/blockage, EGR valve, EGR/EVAP valve, ECM**
33	–	Heated oxygen sensor (HO2S) – LH/front – circuit malfunction	**Wiring, HO2S, intake/fuel system, injector, ECM**
34	–	Knock sensor (KS) – circuit malfunction	**Wiring, KS, ECM**
35	P1401	Exhaust gas recirculation temperature (EGRT) sensor – circuit malfunction	**Wiring, EGRT sensor, ECM**
36	–	Exhaust gas recirculation (EGR) system – valve malfunction	**Wiring, EGR/EVAP valve, hose leak/ blockage, ECM**
37	–	Closed loop control – except Maxima – inoperative	**Wiring, HO2S, intake/fuel system, ECM**
37	–	Closed loop control – Maxima – LH – inoperative	**Wiring, HO2S, intake/fuel system, ECM**
38	–	Closed loop control – RH – inoperative	**Wiring, HO2S, intake/fuel system, ECM**
41	–	Intake air temperature (IAT) sensor – circuit malfunction	**Wiring, IAT sensor, ECM**
42	–	Fuel temperature sensor – voltage high or low	**Wiring, fuel temperature sensor, ECM**
43	–	Throttle position (TP) sensor – circuit malfunction	**Wiring, TP sensor, ECM**
47	P1335	Crankshaft position (CKP) sensor 2 – Maxima – circuit malfunction	**Wiring, CKP sensor, starter motor, ECM**
53	–	Heated oxygen sensor (HO2S) – RH front – voltage high	**Wiring, HO2S, intake/fuel system, injector, ECM**
54	–	ECM/TCM – communication malfunction	**Wiring, ECM, TCM**
55	–	No self diagnostic failure indicated	–

Flash code	OBD-II code	Fault location	Probable cause
63	–	Cylinder No.6 – misfire	**Wiring, mechanical fault, EGR valve, injector, intake/ignition system, CKP sensor, ECM**
64	–	Cylinder No.5 – misfire	**Wiring, mechanical fault, EGR valve, injector, intake/ignition system, CKP sensor, ECM**
65	–	Cylinder No.4 – misfire	**Wiring, mechanical fault, EGR valve, injector, intake/ignition system, CKP sensor, ECM**
66	–	Cylinder No.3 – misfire	**Wiring, mechanical fault, EGR valve, injector, intake/ignition system, CKP sensor, ECM**
67	–	Cylinder No.2 – misfire	**Wiring, mechanical fault, EGR valve, injector, intake/ignition system, CKP sensor, ECM**
68	–	Cylinder No.1 – misfire	**Wiring, mechanical fault, EGR valve, injector, intake/ignition system, CKP sensor, ECM**
71	–	Random misfire	**Wiring, mechanical fault, EGR valve, injector, intake/ignition system, CKP sensor, ECM**
72	–	Catalytic converter – except Maxima – efficiency below limit	**Catalytic converter, wiring, intake/ exhaust system, injector, ECM**
72	–	Catalytic converter – Maxima – RH – efficiency below limit	**Catalytic converter, wiring, intake/ exhaust system, injector, ECM**
73	–	Catalytic converter – LH – efficiency below limit	**Catalytic converter, wiring, intake/ exhaust system, injector, ECM**
76	–	Mixture control – except Maxima	**Wiring, HO2S front, injector, exhaust/fuel system, MAF sensor, ECM**
76	–	Mixture control – Maxima – RH	**Wiring, HO2S front, injector, exhaust/fuel system, MAF sensor, ECM**
77	–	Heated oxygen sensor (HO2S) – rear – circuit malfunction	**Wiring, HO2S, intake/fuel system, injector, ECM**
82	–	Crankshaft position (CKP) sensor – except Maxima – circuit malfunction	**Wiring, CKP sensor, starter motor, ECM**
82	–	Crankshaft position (CKP) sensor 1 – Maxima – circuit malfunction	**Wiring, CKP sensor, starter motor, ECM**
84	P1605	AT diagnosis communication line – communication malfunction	**Wiring, TCM**
85	P1110	Camshaft position (CMP) actuator – circuit malfunction	**Wiring, CMP actuator, ECM**
86	–	Mixture control – LH	**Wiring, HO2S front, injector, exhaust/fuel system, MAF sensor, ECM**
87	–	Evaporative emission (EVAP) canister purge system – control valve circuit malfunction	**Wiring, EVAP valve, ECM**

Flash code	OBD-II code	Fault location	Probable cause
91	–	Heated oxygen sensor (HO2S) – RH/front – heater circuit malfunction	**Wiring, HO2S, ECM**
94	P1550	Torque converter clutch solenoid – voltage low	**Wiring, clutch solenoid, TCM, ECM**
95	P1336	Crankshaft position (CKP) sensor – rotor teeth damage	**Wiring, CKP sensor, mechanical fault, ECM**
98	–	Engine coolant temperature (ECT) sensor – malfunction	**Wiring, ECT sensor, cooling system, ECM**
101	–	Heated oxygen sensor (HO2S) – LH front – heater circuit malfunction	**Wiring, HO2S, ECM**
103	–	Park/neutral position (PNP) switch – circuit malfunction	**Wiring, PNP switch, ECM**
103	–	Neutral position (NP) switch – MT – circuit malfunction	**Wiring, NP switch, ECM**
105	P1400	Exhaust gas recirculation (EGR) system/ evaporative emission (EVAP) canister purge system – valve malfunction	**Wiring, EGR/EVAP valve, ECM**
108	–	Evaporative emission (EVAP) canister purge system – control valve circuit malfunction	**Wiring, EVAP valve, ECM**
111	–	Park/neutral position (PNP) switch – circuit malfunction	**Wiring, PNP switch, TCM, ECM**
112	–	AT – vehicle speed sensor (VSS)	**Wiring, VSS, TCM**
113	–	AT – first gear selection – malfunction	**Valve block**
114	–	AT – second gear selection – malfunction	**Valve block**
115	–	AT – third gear selection – malfunction	**Valve block**
116	–	AT – fourth gear selection – malfunction	**Valve block**
118	–	AT – shift solenoid A – voltage low	**Wiring, solenoid A, TCM**
121	–	AT – shift solenoid B – voltage low	**Wiring, solenoid B, TCM**
123	P1760	AT – overrun clutch solenoid – voltage low	**Wiring, clutch solenoid, TCM**
124	–	Torque converter clutch solenoid – voltage low	**Wiring, clutch solenoid, TCM**
125	–	AT – Line pressure solenoid – voltage low	**Wiring, pressure solenoid, TCM**
126	P1705	AT – throttle position (TP) sensor – voltage high or low	**Wiring, TP sensor, TCM, ECM**
127	–	Engine speed (RPM) – incorrect voltage signal ECM to TCM	**Wiring, RPM sensor, TCM, ECM**
128	–	Transmission fluid temperature sensor – excessively high or low voltage	**Wiring, temperature sensor, TCM, ECM**

NISSAN

Model:	Engine identification:	Year:
Altima 2.4L	KA24DE	1996-01
Sentra 1.6L	GA16DE	1996-00
Sentra 1.8L	QG18DE	2002
Sentra 2.0L	SR20DE	1996-00
Sentra 2.5L	QR25DE	2002
Pick-up 2.4L	KA24DE	1996-97
Frontier 2.4L	KA24DE	1998-02
200SX 1.6/2.0L	GA16DE/SR20DE	1996-98
240SX 2.4L	KA24DE	1996
Xterra 2.4L	KA24DE	2000-01

System: **ECCS**

Self-diagnosis

General information

- Refer to the front of this manual for general test conditions, terminology, detailed descriptions of wiring faults and a general trouble shooter for electrical and mechanical faults.
- Engine control module (ECM) incorporates self-diagnosis function.
- Malfunction indicator lamp (MIL) will illuminate if certain faults are recorded.
- ECM operates in backup mode if sensors fail, to enable vehicle to be driven to workshop.
- Except Sentra →1999/Altima, Frontier & Xterra →2000: Trouble codes can be displayed by the malfunction indicator lamp (MIL) or by using a Scan Tool connected to the data link connector (DLC) **1** [1].
- Sentra 2000→/Altima, Frontier & Xterra 2001→: Trouble codes can only be displayed using a Scan Tool connected to the data link connector (DLC) **1** [1].

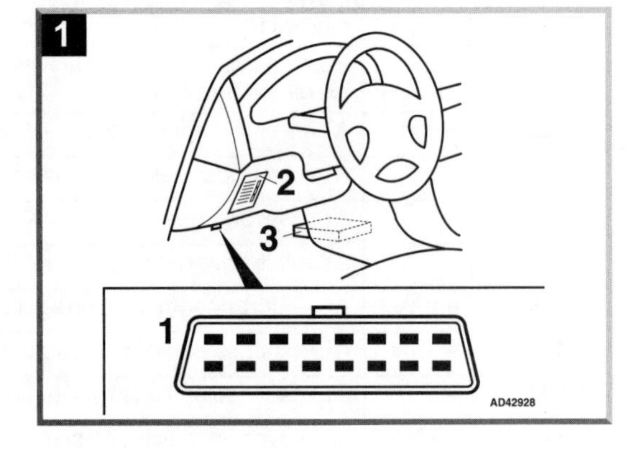

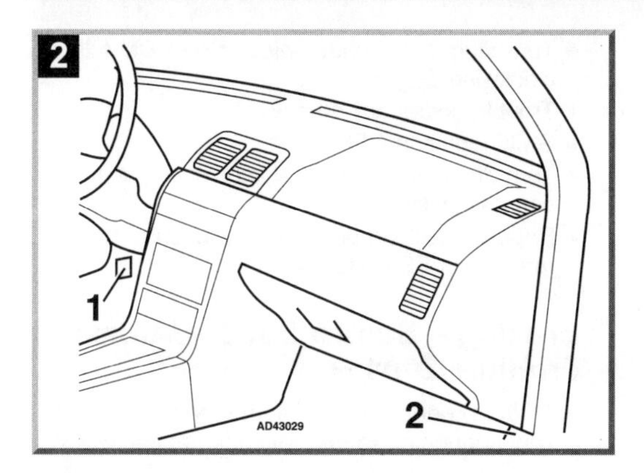

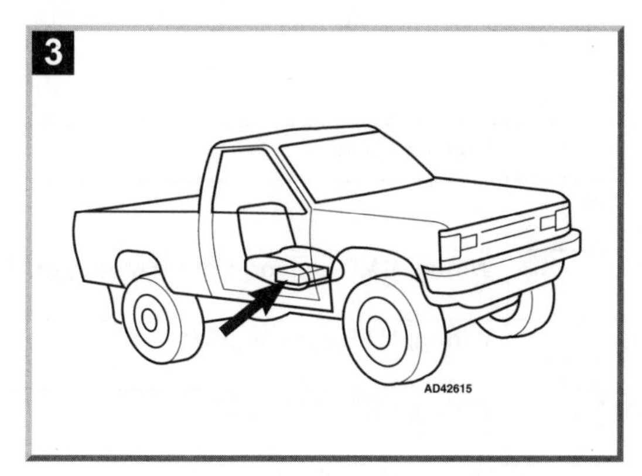

NOTE: ***The use of a Scan Tool is essential to obtain full diagnostic information.***

Accessing – Mode 1

- Switch ignition ON.
- Check MIL illuminates.
- Start engine. Allow to idle.
- If MIL extinguishes, no trouble codes have been recorded.
- If MIL remains illuminated, access trouble codes.

Accessing – Mode 2 – Sentra →1999/ Altima, Frontier & Xterra →2000

- Access engine control module (ECM):
  - Altima/Sentra/200SX – 1 [3].
  - 240SX – 2 [2].
  - Pick-up – 3.
- Switch ignition ON. Do NOT start engine.
- Turn diagnostic mode selector on ECM fully clockwise 4.
- Wait at least 2 seconds.
- Turn diagnostic mode selector on ECM fully counterclockwise 4.
- Self-diagnosis now in Mode 2.
- MIL should flash.
- Count flashes and compare with trouble code table.
  - Long flashes indicate the 'hundreds' of the trouble code 5 [A].
  - Short flashes indicate the 'units' of the trouble code 5 [C].
  - A short pause separates each flash 5 [B].
  - A long pause separates each trouble code 5 [D].
  - For example: Trouble code 102 displayed 5.
- Switch ignition OFF.
- Self-diagnosis will return to Mode 1.

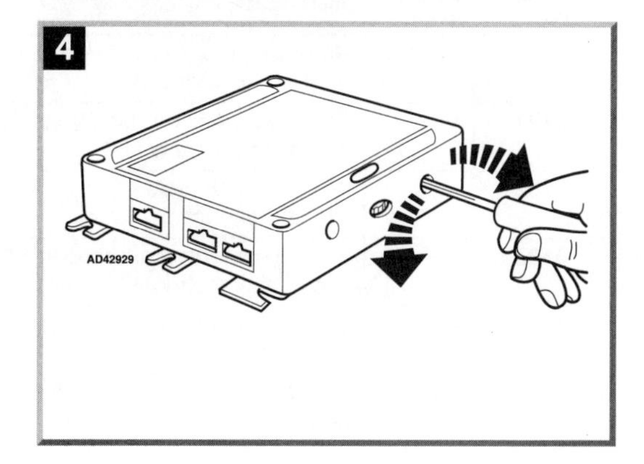

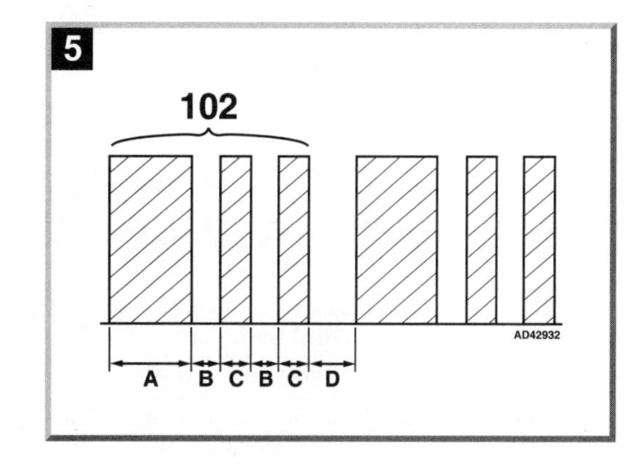

Accessing – Mode 2 – Sentra 2000→/ Altima, Frontier & Xterra 2001→

- Trouble codes can only be displayed using a Scan Tool connected to the data link connector (DLC) **1** [1].

Erasing – Sentra →1999/Altima, Frontier & Xterra →2000

- Trouble codes can be erased by using a Scan Tool connected to the data link connector (DLC) or as follows:
- With the ignition switch ON and the engine control module (ECM) in diagnosis test mode 2.
- Turn diagnostic mode selector on ECM fully clockwise **4**.
- Trouble codes are erased.
- Switch ignition OFF.
- Turn diagnostic mode selector on ECM fully counterclockwise **4**.
- Repeat checking procedure to ensure no data remains in ECM fault memory.

Erasing – Sentra 2000→/Altima & Frontier 2001→

- Trouble codes can only be erased by using a Scan Tool connected to the data link connector (DLC).

Trouble code identification

Flash code	OBD-II code	Fault location	Probable cause
–	*All P0, P2 and U0 codes*	Refer to OBD-II trouble code tables at the front of this manual	–
0101	–	Camshaft position (CMP) sensor – circuit malfunction	**Wiring, CMP sensor, Starter motor, ECM**
0102	–	Mass air flow (MAF) sensor – circuit malfunction	**Wiring, MAF sensor, ECM**
0103	–	Engine coolant temperature (ECT) sensor – circuit malfunction	**Wiring, ECT sensor, ECM**
0104	–	Vehicle speed sensor (VSS) – circuit malfunction	**Wiring, VSS, ECM**
0111	**P1447**	Evaporative emission (EVAP) canister purge system – flow malfunction	**Wiring, EVAP valve, hose leak/blockage, canister, BARO sensor, MAP/BARO switching valve, ECM**
0112	**P1165**	Intake manifold air control system – vacuum check switch circuit malfunction	**Wiring, vacuum check switch, hose leak/ blockage, air control solenoid, ECM**
0114	–	Fuel trim – mixture too rich	**Wiring, intake/exhaust system, HO2S front, fuel system, injector, MAF sensor, ECM**
0115	–	Fuel trim – mixture too lean	**Wiring, exhaust system, HO2S front, fuel system, injector, MAF sensor, ECM**
0201	**P1320**	Ignition signal – circuit malfunction	**Wiring, ignition coil/module, ignition system, CKP sensor, ECM**
0203	–	Closed throttle position (CTP) switch – circuit malfunction	**Wiring, CTP switch, ECM**
0205	–	Idle air control (IAC) system – malfunction	**Wiring, IAC system, ECM**
0208	**P1900**	Engine coolant blower motor – circuit malfunction	**Wiring, blower motor, cooling system, ECM**

Flash code	OBD-II code	Fault location	Probable cause
0213	P1440	Evaporative emission (EVAP) canister purge system – leaking/malfunction	**Wiring, fuel tank/cap, hose leak/ blockage, EVAP valve/canister, EVAP vent valve, BARO sensor, MAP/BARO switching valve, ECM**
0214	P1444	Evaporative emission (EVAP) canister purge system – control valve malfunction	**Wiring, EVAP pressure sensor, hose leak/blockage, EVAP valve/canister, ECM**
0215	P1446	Evaporative emission (EVAP) canister purge system – vent control valve malfunction	**Wiring, EVAP pressure sensor, hose leak/blockage, EVAP vent valve, EVAP canister contaminated, ECM**
0301	–	Engine control module (ECM) – malfunction	**ECM**
0302	–	Exhaust gas recirculation (EGR) system – flow excessively high or low	**Wiring, hose leak/blockage, EGR valve, EGR/EVAP valve, ECM**
0303	–	Heated oxygen sensor (HO2S) – front – circuit malfunction	**Wiring, HO2S, intake/fuel system, injector, ECM**
0304	–	Knock sensor (KS) – circuit malfunction	**Wiring, KS, ECM**
0305	P1401	Exhaust gas recirculation temperature (EGRT) sensor – circuit malfunction	**Wiring, EGRT sensor, ECM**
0306	–	Exhaust gas recirculation (EGR) system – valve malfunction	**Wiring, EGR valve, EGR back pressure transducer, hose leak/blockage, CMP sensor, MAF sensor, EGR/EVAP valve, ECM**
0307	–	Closed loop control – 1996-97 – inoperative	**Wiring, HO2S front, ECM**
0307	P1148	Closed loop control bank 1 – inoperative	**Wiring, HO2S front, ECM**
0309	P1448	Evaporative emission (EVAP) canister purge system – vent control valve malfunction	**Wiring, EVAP vent valve, EVAP pressure sensor, hose leak/blockage, EVAP canister contaminated, EVAP vacuum cut valve, ECM**
0311	P1491	Evaporative emission (EVAP) canister purge system – bypass vacuum valve malfunction	**Wiring, EVAP vacuum cut/bypass valve, hose leak/blockage, EVAP pressure sensor/vent valve, EVAP canister contaminated, ECM**
0312	P1493	Evaporative emission (EVAP) canister purge system – control valve malfunction	**Wiring, EVAP valve, hose leak/blockage, EVAP pressure sensor/vent valve, EVAP canister contaminated, ECM**
0401	–	Intake air temperature (IAT) sensor – circuit malfunction	**Wiring, IAT sensor, ECM**
0402	–	Fuel temperature sensor – circuit malfunction	**Wiring, fuel temperature sensor, ECM**
0403	–	Throttle position (TP) sensor – circuit malfunction	**Wiring, TP sensor, ECM**
0409	–	Heated oxygen sensor (HO2S) – front – slow response	**Wiring, HO2S, fuel/intake system, injector, exhaust system, MAF sensor, ECM**
0410	–	Heated oxygen sensor (HO2S) – front – voltage high	**Wiring, HO2S, fuel system, injector, ECM**

Flash code	OBD-II code	Fault location	Probable cause
0411	–	Heated oxygen sensor (HO2S) – front – voltage low	Wiring, HO2S, fuel/intake system, injector, ECM
0412	–	Heated oxygen sensor (HO2S) – front – excessively high voltage	Wiring, HO2S, ECM
0505	–	No self diagnostic failure indicated	–
0510	–	Heated oxygen sensor (HO2S) – rear – voltage high	Wiring, HO2S, fuel/intake system, injector, ECM
0511	–	Heated oxygen sensor (HO2S) – rear – voltage low	Wiring, HO2S, fuel system, injector, ECM
0512	–	Heated oxygen sensor (HO2S) – rear – excessively high voltage	Wiring, HO2S, ECM
0513	P1776	Torque converter clutch solenoid – converter slip during lock condition	Wiring, clutch solenoid, torque converter, TCM, ECM
0514	P1402	Exhaust gas recirculation (EGR) system – excessive flow	Wiring, EGR/EVAP valve, EGR valve, EGRT sensor, EGR back pressure transducer, ECM
0605	–	Cylinder No.4 – misfire	Wiring, mechanical fault, fuel system, EGR valve, injector, intake/ignition system, CKP sensor, HO2S front, ECM
0606	–	Cylinder No.3 – misfire	Wiring, mechanical fault, fuel system, EGR valve, injector, intake/ignition system, CKP sensor, HO2S front, ECM
0607	–	Cylinder No.2 – misfire	Wiring, mechanical fault, fuel system, EGR valve, injector, intake/ignition system, CKP sensor, HO2S front, ECM
0608	–	Cylinder No.1 – misfire	Wiring, mechanical fault, fuel system, EGR valve, injector, intake/ignition system, CKP sensor, HO2S front, ECM
0701	–	Random misfire	Wiring, mechanical fault, fuel system, EGR valve, injector, intake/ignition system, CKP sensor, HO2S front, ECM
0702	–	Catalytic converter – efficiency below limit	Catalytic converter, wiring, intake/ exhaust system, injector, ECM
0704	–	Evaporative emission (EVAP) canister purge system – pressure sensor malfunction	Wiring, EVAP pressure sensor, hose leak/blockage, EVAP vent/control valve, EVAP canister, ECM
0705	–	Evaporative emission (EVAP) canister purge system – leaking/malfunction	Wiring, fuel tank/cap, hose leak/ blockage, EVAP valve/canister, EVAP vent valve, BARO sensor, MAP/BARO switching valve, ECM
0706	–	Mixture control – malfunction	Wiring, intake/exhaust system, injector, fuel system, MAF sensor, ECM
0707	–	Heated oxygen sensor (HO2S) – rear – circuit malfunction	Wiring, HO2S, intake/fuel system, injector, ECM
0715	–	Evaporative emission (EVAP) canister purge system – large leak	EVAP canister purge system, hose leak, EVAP pressure sensor/vent valve, fuel filler cap, ECM

Flash code	OBD-II code	Fault location	Probable cause
0801	P1490	Evaporative emission (EVAP) canister purge system – bypass vacuum valve malfunction	**Wiring, EVAP bypass/cut valve, hose leak/blockage, ECM**
0801	P1441	Evaporative emission (EVAP) canister purge system – bypass vacuum cut valve malfunction	**Wiring, EVAP bypass vacuum cut valve, EVAP cut valve/pressure sensor, hose leak/blockage, ECM**
0802	–	Crankshaft position (CKP) sensor – circuit malfunction	**Wiring, CKP sensor, ECM**
0803	–	Manifold absolute pressure (MAP) sensor – circuit malfunction	**Wiring, MAP sensor, ECM**
0804	P1605	TCM diagnosis communication line – malfunction	**Wiring, TCM**
0805	P1110	Camshaft position (CMP) actuator – circuit malfunction	**Wiring, CMP actuator, ECM**
0807	–	Evaporative emission (EVAP) canister purge system – 1996-97 – control valve malfunction	**Wiring, EVAP control valve, EVAP valve/ pressure sensor, hose leak/blockage, ECM**
0807	P1492	Evaporative emission (EVAP) canister purge system – 1998 – control valve malfunction	**Wiring, EVAP control valve, EVAP valve/ pressure sensor, hose leak/blockage, ECM**
0901	–	Heated oxygen sensor (HO2S) – front – heater circuit malfunction	**Wiring, HO2S, ECM**
0902	–	Heated oxygen sensor (HO2S) – rear – heater circuit malfunction	**Wiring, HO2S, ECM**
0903	–	Evaporative emission (EVAP) canister purge system – vent control valve circuit malfunction	**Wiring, EVAP vent control valve, EVAP breather valve/pressure sensor, hose leak/blockage, ECM**
0904	P1550	Torque converter clutch solenoid – 1996-97 – voltage low	**Wiring, clutch solenoid, TCM, ECM**
0904	P1775	Torque converter clutch solenoid – 1998 – voltage low	**Wiring, clutch solenoid, TCM, ECM**
0905	P1336	Crankshaft position (CKP) sensor – rotor teeth damage	**Wiring, CKP sensor, ECM**
0908	–	Engine coolant temperature – too low for closed loop control	**Wiring, ECT sensor, cooling system, ECM**
1003	–	Park/neutral position (PNP) switch – 1996-97 – circuit malfunction	**Wiring, PNP switch, ECM**
1003	P1706	Park/neutral position (PNP) switch – 1998 – circuit malfunction	**Wiring, PNP switch, ECM**
1004	P1130	Intake manifold air control solenoid – malfunction	**Wiring, control solenoid, intake system, hose leak/blockage, control vacuum check switch, ECM**
1005	P1400	Exhaust gas recirculation (EGR) system/ evaporative emission (EVAP) canister purge system – valve malfunction	**Wiring, EGR/EVAP valve, ECM**

Flash code	OBD-II code	Fault location	Probable cause
1008	**P1445**	Evaporative emission (EVAP) canister purge system – 1996-97 – volume control valve malfunction	**Wiring, EVAP valve, ECM**
1008	**–**	Evaporative emission (EVAP) canister purge system – 1998 – control valve malfunction	**Wiring, EVAP valve, ECM**
1101	**–**	Park/neutral position (PNP) switch – circuit malfunction	**Wiring, PNP switch, ECM**
1102	**–**	AT – vehicle speed sensor (VSS)	**Wiring, VSS, TCM**
1103	**–**	AT – first gear selection – malfunction	**Valve block**
1104	**–**	AT – second gear selection – malfunction	**Valve block**
1105	**–**	AT – third gear selection – malfunction	**Valve block**
1106	**–**	AT – fourth gear selection – malfunction	**Valve block**
1107	**–**	Torque converter lock-up solenoid – circuit malfunction	**Wiring, clutch solenoid, TCM**
1108	**–**	AT – shift solenoid A – voltage low	**Wiring, solenoid A, TCM**
1201	**–**	AT – shift solenoid B – voltage low	**Wiring, solenoid B, TCM**
1203	**P1760**	AT – overrun clutch solenoid – voltage low	**Wiring, clutch solenoid, TCM**
1204	**–**	Torque converter clutch solenoid – voltage low	**Wiring, clutch solenoid, TCM**
1205	**–**	AT – Line pressure solenoid – voltage low	**Wiring, pressure solenoid, TCM**
1206	**P1705**	AT – throttle position (TP) sensor – voltage high or low	**Wiring, TP sensor, TCM, ECM**
1207	**–**	Engine speed (RPM) – incorrect voltage signal ECM to TCM	**Wiring, RPM sensor, TCM, ECM**
1208	**–**	Transmission fluid temperature sensor – excessively high or low voltage	**Wiring, temperature sensor, TCM, ECM**
1302	**P1105**	MAP/BARO switching valve – circuit malfunction	**Wiring, MAP/BARO switching valve, ECM**
1305	**P1220**	Fuel pump control module – circuit malfunction	**Wiring, fuel pump speed resistor, fuel pump control module**
1308	**P1900**	Engine coolant blower motor – circuit malfunction	**Wiring, coolant blower motor, cooling system, ECM**
–	**P1031**	Air fuel (A/F) ratio sensor 1, heater – circuit malfunction	**Wiring, A/F sensor**
–	**P1032**	Air fuel (A/F) ratio sensor 1, heater – circuit malfunction	**Wiring, A/F sensor**
–	**P1065**	Engine control module (ECM) – battery voltage supply	**Wiring, fuse, ECM**
–	**P1111**	Camshaft position (CMP) actuator, intake – circuit malfunction	**Wiring, CMP actuator, ECM**
–	**P1121**	Throttle motor – malfunction	**Throttle motor, ECM**

Flash code	OBD-II code	Fault location	Probable cause
–	P1122	Throttle motor – circuit malfunction	**Wiring, throttle motor, ECM**
–	P1124	Throttle motor – short circuit	**Wiring, throttle motor, throttle motor relay, ECM**
–	P1126	Throttle motor – open circuit – Sentra 2002	**Wiring, throttle motor, throttle motor relay, ECM**
–	P1126	Thermostat – stuck open – except Sentra 2002	**Wiring, thermostat, ECT sensor**
–	P1128	Throttle motor – short circuit	**Wiring, throttle motor, ECM**
–	P1132	Intake manifold air control solenoid – circuit malfunction	**Wiring, intake manifold air control solenoid, ECM**
–	P1137	Intake manifold air control actuator position sensor – circuit malfunction	**Wiring, intake manifold air control actuator position sensor, ECM**
–	P1138	Intake manifold air control solenoid – circuit malfunction	**Wiring, intake manifold air control solenoid, intake manifold air control actuator position sensor, ECM**
–	P1140	Camshaft position (CMP)	**Wiring, mixture, HO2S, ECM**
–	P1143	Heated oxygen sensor (HO2S) 1 – voltage low	
–	P1144	Heated oxygen sensor (HO2S) 1 – voltage high	**Wiring, mixture, HO2S, ECM**
–	P1146	Heated oxygen sensor (HO2S) 2 – voltage low	**Wiring, mixture, HO2S, ECM**
–	P1147	Heated oxygen sensor (HO2S) 2 – voltage low	**Wiring, injectors, fuel pressure, air leak, HO2S, ECM**
–	P1163	Heated oxygen sensor (HO2S)	**Wiring, HO2S, ECM**
–	P1164	Heated oxygen sensor (HO2S) – voltage high	**Wiring, mixture, HO2S, ECM**
–	P1166	Heated oxygen sensor (HO2S) 2 – voltage low	**Wiring, mixture, HO2S, ECM**
–	P1167	Heated oxygen sensor (HO2S) 2 – voltage low	**Wiring, injectors, fuel pressure, air leak, HO2S, ECM**
–	P1168	Closed loop control bank 2 – inoperative	**Wiring, HO2S front, ECM**
–	P1169	Heated oxygen sensor (HO2S) 3 – voltage low	**Wiring, fuel pressure, injectors, HO2S, ECM**
–	P1170	Heated oxygen sensor (HO2S) 3 – voltage low	**Wiring, fuel pressure, injectors, HO2S, ECM**
–	P1217	Engine over temperature	**Cooling system, engine coolant blower motor**
–	P1223	Throttle position (TP) sensor 2 – voltage low	**Wiring, TP sensor**
–	P1224	Throttle position (TP) sensor 2 – voltage high	**Wiring, TP sensor**
–	P1225	Throttle motor position sensor – learning problem, voltage low	**Wiring, TP sensor 1, TP sensor 2**

Flash code	OBD-II code	Fault location	Probable cause
–	P1226	Throttle motor position sensor – learning problem	**Wiring, TP sensor 1, TP sensor 2**
–	P1227	Accelerator pedal position (APP) sensor 2 – voltage low	**Wiring, APP sensor**
–	P1228	Accelerator pedal position (APP) sensor 2 – voltage high	**Wiring, APP sensor**
–	P1229	Engine control module (ECM) – sensor supply voltage low/high	**Wiring, APP sensor, throttle motor position sensor, MAF sensor, fuel tank pressure sensor, TP sensor(s), refrigerant pressure sensor**
–	P1271	Air fuel (A/F) ratio sensor 1 – voltage low	**Wiring, A/F sensor, ECM**
–	P1272	Air fuel (A/F) ratio sensor 1 – voltage high	**Wiring, A/F sensor, ECM**
–	P1273	Air fuel (A/F) ratio sensor 1 – lean mixture	**A/F sensor, fuel pressure, injectors, intake air leak**
–	P1274	Air fuel (A/F) ratio sensor 1 – rich mixture	**HO2S, A/F sensor, fuel pressure, injectors, intake air leak**
–	P1275	Air fuel (A/F) ratio sensor 1 – response time	**HO2S, A/F sensor, fuel pressure, injectors, intake air leak, exhaust leak, MAF sensor**
–	P1276	Air fuel (A/F) ratio sensor 1 – activity	**Wiring, A/F sensor**
–	P1277	Air fuel (A/F) ratio sensor 1 – slow response	**Wiring, A/F sensor**
–	P1278	Air fuel (A/F) ratio sensor 1 – slow response	**Wiring, fuel pressure, injectors, MAF sensor, PCV, A/F sensor**
–	P1279	Air fuel (A/F) ratio sensor 1 – slow response	**Wiring, fuel pressure, injectors, MAF sensor, PCV, A/F sensor**
–	P1442	Evaporative emission (EVAP) canister purge system – small leak	**EVAP canister purge system, hose leak, EVAP pressure sensor/vent valve, fuel filler cap, ECM**
–	P1446	Evaporative emission (EVAP) canister purge system – very small leak	**EVAP canister purge system, hose leak, EVAP pressure sensor/vent valve, fuel filler cap, ECM**
–	P1456	Evaporative emission (EVAP) canister purge system – very small leak	**EVAP canister purge system, hose leak, EVAP pressure sensor/vent valve, fuel filler cap, ECM**
–	P1464	Fuel gauge tank sensor – voltage high	**Wiring, fuel gauge tank sensor, ECM**
–	P1564	Cruise control master switch – circuit malfunction	**Wiring, cruise control master switch, ECM**
–	P1572	Cruise control brake pedal switch – circuit malfunction	**Wiring, cruise control brake pedal switch, clutch pedal position (CPP) switch, ECM**
–	P1574	Cruise control vehicle speed sensor – signal malfunction	**Wiring, cruise control vehicle speed sensor, ECM**
–	P1610	Ignition key/engine control module (ECM) – malfunction	**Incorrect ignition key, ECM**

Flash code	OBD-II code	Fault location	Probable cause
–	P1611	Immobilizer control module/engine control module (ECM) – coding	**Immobilizer control module/ECM incorrectly coded**
–	P1612	Immobilizer control module/engine control module (ECM) communication – malfunction	**Wiring, immobilizer control module, ECM**
–	P1613	Engine control module (ECM), immobilizer function – internal failure	**ECM**
–	P1614	Immobilizer control module/module coding plug communication – no signal	**Wiring, immobilizer control module, module coding plug**
–	P1615	Ignition key/immobilizer control module communication – malfunction	**Incorrect ignition key, immobilizer control module**
–	P1805	Brake pedal position (BPP) switch – circuit malfunction	**Wiring, BPP switch**

NISSAN

Model:	Engine identification:	Year:
Altima 2.5L	QR25DE	2002
Altima 3.5L	VQ35DE	2002
Pathfinder 3.5L	VQ35DE	2002
Xterra 2.4L	KA24DE	2002
Xterra 3.3L	VG33E/VG33ER	2002

System: **ECCS**

Self-diagnosis

General information

- Refer to the front of this manual for general test conditions, terminology, detailed descriptions of wiring faults and a general trouble shooter for electrical and mechanical faults.
- Engine control module (ECM) incorporates self-diagnosis function.
- Malfunction indicator lamp (MIL) will illuminate if certain faults are recorded.
- ECM operates in backup mode if sensors fail, to enable vehicle to be driven to workshop.
- Trouble codes can be displayed by using a Scan Tool connected to the data link connector (DLC) **1** or by the malfunction indicator lamp (MIL).

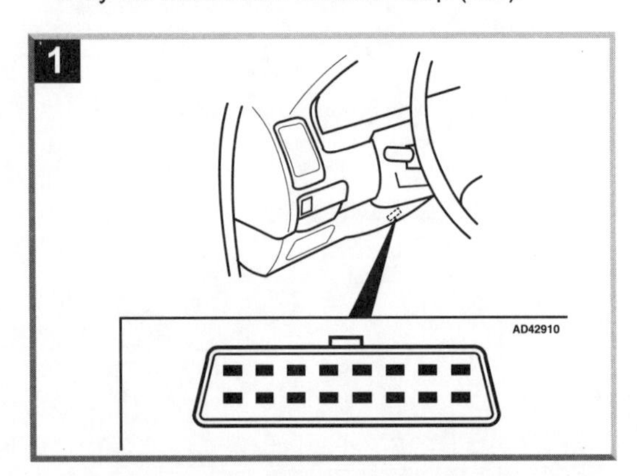

NOTE: ***The use of a Scan Tool is essential to obtain full diagnostic information.***

Accessing – Mode 1

- Switch ignition ON.
- Check MIL illuminates.
- Start engine. Allow to idle.
- If MIL extinguishes, no trouble codes have been recorded.
- If MIL remains illuminated, access trouble codes.

Accessing – Mode 2

- Ensure accelerator pedal is fully released.
- Switch ignition ON. Do NOT start engine.
- Wait 3 seconds.
- Within 5 seconds, repeat the following 5 times:
  - Fully depress accelerator pedal.
  - Fully release accelerator pedal.
- Wait 7 seconds.
- Fully depress accelerator pedal.
- After approximately 10 seconds MIL will start flashing.
- Fully release accelerator pedal.
- Self-diagnosis now in Mode 2.
- Count flashes and compare with trouble code table.
- Each trouble code consists of four groups of one or more flashes.
- The first group of flashes indicate the 'thousands' of the trouble code **2** [A].
- The second group of flashes indicate the 'hundreds' of the trouble code **2** [B].
- The third group of flashes indicate the 'tens' of the trouble code **2** [C].
- The fourth group of flashes indicate the 'units' of the trouble code **2** [D].
- Ten flashes in a group indicate '0'.
- A short pause separates each trouble code group **2** [E].

- A long pause separates each trouble code 2 [F].
- For example: Trouble code P1130 displayed 2.
- Switch ignition OFF.
- Self-diagnosis will return to Mode 1.

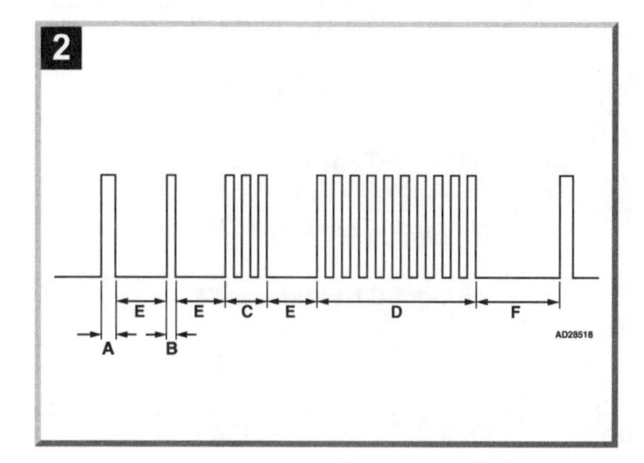

Erasing

- Trouble codes can be erased by using a Scan Tool connected to the data link connector (DLC) or as follows:
- Ensure accelerator pedal is fully released.
- Switch ignition ON. Do NOT start engine.
- Wait 3 seconds.
- Within 5 seconds, repeat the following 5 times:
  - Fully depress accelerator pedal.
  - Fully release accelerator pedal.
- Wait 7 seconds.
- Fully depress accelerator pedal.
- After approximately 10 seconds MIL will start flashing.
- Fully release accelerator pedal.
- Self-diagnosis now in Mode 2.
  - Fully depress accelerator pedal for more than 10 seconds.
  - Fully release accelerator pedal.
  - Trouble code 0000 should be displayed.

Trouble code identification

Flash code	OBD-II code	Fault location	Probable cause
–	*All P0, P2 and U0 codes*	Refer to OBD-II trouble code tables at the front of this manual	–
0011	P0011	Camshaft variable timing control – bank 1 – performance	**Wiring, CMP sensor, CKP sensor, camshaft valve timing sensor**
0021	P0021	Camshaft variable timing control – bank 2 – performance	**Wiring, CMP sensor, CKP sensor, camshaft variable valve timing sensor**
0031	P0031	Heated oxygen sensor (HO2S) 1 bank 1 – heater voltage low	**Wiring, HO2S heater, ECM**
0032	P0032	Heated oxygen sensor (HO2S) 1 bank 1 – heater voltage high	**Wiring, HO2S heater, ECM**
0037	P0037	Heated oxygen sensor (HO2S) 2 bank 1 – heater voltage low	**Wiring, HO2S heater, ECM**
0038	P0038	Heated oxygen sensor (HO2S) 2 bank 1 – heater voltage high	**Wiring, HO2S heater, ECM**
0051	P0051	Heated oxygen sensor (HO2S) 1 bank 2 – heater voltage low	**Wiring, HO2S heater, ECM**
0052	P0052	Heated oxygen sensor (HO2S) 1 bank 2 – heater voltage high	**Wiring, HO2S heater, ECM**
0057	P0057	Heated oxygen sensor (HO2S) 2 bank 2 – heater voltage low	**Wiring, HO2S heater, ECM**
0058	P0058	Heated oxygen sensor (HO2S) 2 bank 2 – heater voltage high	**Wiring, HO2S heater, ECM**
0101	P0101	Mass air flow (MAF) sensor – range/ performance problem	**Wiring, MAF sensor, air leak, ECM**

Flash code	OBD-II code	Fault location	Probable cause
0102	P0102	Mass air flow (MAF) sensor – voltage low	**Wiring, MAF sensor, air leak, ECM**
0103	P0103	Mass air flow (MAF) sensor – voltage high	**Wiring, MAF sensor, ECM**
0107	P0107	Manifold absolute pressure (MAP) sensor – voltage low	**Wiring, MAP sensor, ECM**
0108	P0108	Manifold absolute pressure (MAP) sensor – voltage high	**Wiring, MAP sensor, ECM**
0112	P0112	Intake air temperature (IAT) sensor – voltage low	**Wiring, IAT sensor, ECM**
0113	P0113	Intake air temperature (IAT) sensor – voltage high	**Wiring, IAT sensor, ECM**
0117	P0117	Engine coolant temperature (ECT) sensor – voltage low	**Wiring, ECT sensor, ECM**
0118	P0118	Engine coolant temperature (ECT) sensor – voltage high	**Wiring, ECT sensor, ECM**
0121	P0121	Throttle position (TP) sensor – range/ performance problem	**Wiring, TP sensor, ECM**
0122	P0122	Throttle position (TP) sensor – voltage low	**Wiring, TP sensor, ECM**
0123	P0123	Throttle position (TP) sensor – voltage high	**Wiring, TP sensor, ECM**
0125	P0125	Engine coolant temperature – too low for closed loop control	**Wiring, ECT sensor, cooling system, ECM**
0127	P0127	Intake air temperature (IAT) sensor – temperature too high	**Wiring short circuit, IAT sensor, ECM**
0128	P0128	Engine coolant temperature – too low	**Thermostat, ECT sensor, cooling system, ECM**
0132	P0132	Heated oxygen sensor (HO2S) 1, bank 1 – voltage high	**Wiring, HO2S, fuel system, injector, ECM**
0133	P0133	Heated oxygen sensor (HO2S) 1, bank 1 – slow response	**Wiring, HO2S, fuel/intake system, injector, exhaust system, MAF sensor, ECM**
0134	P0134	Heated oxygen sensor (HO2S) 1, bank 1 – excessively high voltage	**Wiring, HO2S, ECM**
0138	P0138	Heated oxygen sensor (HO2S) 2, bank 1 – voltage high	**Wiring, HO2S, fuel/intake system, injector, ECM**
0139	P0139	Heated oxygen sensor (HO2S) 2, bank 1 – circuit malfunction	**Wiring, HO2S, intake/fuel system, injector, ECM**
0152	P0152	Heated oxygen sensor (HO2S) 1, bank 2 – voltage high	**Wiring, HO2S, fuel system, injector, ECM**
0153	P0153	Heated oxygen sensor (HO2S) 1, bank 2 – slow response	**Wiring, HO2S, fuel/intake system, injector, exhaust system, crankcase vent valve, MAF sensor, ECM**
0154	P0154	Heated oxygen sensor (HO2S) 1, bank 2 – excessively high voltage	**Wiring, HO2S, ECM**
0158	P0158	Heated oxygen sensor (HO2S) 2, bank 2 – voltage high	**Wiring, HO2S, fuel/intake system, injector, ECM**

Flash code	OBD-II code	Fault location	Probable cause
0159	P0159	Heated oxygen sensor (HO2S) 2, bank 2 – circuit malfunction	Wiring, HO2S, intake/fuel system, injector, ECM
0171	P0171	Fuel trim – bank 1 – mixture too lean	Wiring, exhaust system, HO2S front, fuel system, injector, MAF sensor, ECM
0172	P0172	Fuel trim – bank 1 – mixture too rich	Wiring, intake/exhaust system, HO2S front, fuel system, injector, MAF sensor, ECM
0174	P0174	Fuel trim – bank 2 – mixture too lean	Wiring, exhaust system, HO2S front, fuel system, injector, MAF sensor, ECM
0175	P0175	Fuel trim – bank 2 – mixture too rich	Wiring, exhaust system, HO2S front, fuel system, injector, MAF sensor, ECM
0181	P0181	Fuel tank temperature sensor – range/ performance	Wiring, fuel tank temperature sensor, ECM
0182	P0182	Fuel tank temperature sensor – voltage low	Wiring, fuel tank temperature sensor, ECM
0183	P0183	Fuel tank temperature sensor – voltage high	Wiring, fuel tank temperature sensor, ECM
0217	P0217	Engine coolant over temperature condition	Wiring, coolant fan, coolant hoses, thermostat, radiator, coolant temperature switch
0221	P0221	Throttle position (TP) sensor 1 & 2 – range/performance problem	Wiring, TP sensor, ECM
0222	P0222	Throttle position (TP) sensor 1 – voltage low	Wiring, TP sensor, ECM
0223	P0223	Throttle position (TP) sensor 1 – voltage high	Wiring, TP sensor, ECM
0226	P0226	Accelerator pedal position (APP) sensor 1/2 – range/performance problem	Wiring, APP sensor, ECM
0227	P0227	Accelerator pedal position (APP) sensor 1 – voltage low	Wiring, APP sensor, ECM
0228	P0228	Accelerator pedal position (APP) sensor 1 – voltage high	Wiring, APP sensor, ECM
0245	P0245	Supercharger (SC) bypass solenoid valve – malfunction	Wiring, SC bypass valve, hose leak/ blockage
0300	P0300	Random misfire	Wiring, mechanical fault, fuel system, EGR valve, injector, intake/ignition system, CKP sensor, HO2S front, ECM
0301	P0301	Cylinder No.1 – misfire	Wiring, mechanical fault, fuel system, EGR valve, injector, intake/ignition system, CKP sensor, HO2S front, ECM
0302	P0302	Cylinder No.2 – misfire	Wiring, mechanical fault, fuel system, EGR valve, injector, intake/ignition system, CKP sensor, HO2S front, ECM
0303	P0303	Cylinder No.3 – misfire	Wiring, mechanical fault, fuel system, EGR valve, injector, intake/ignition system, CKP sensor, HO2S front, ECM

Flash code	OBD-II code	Fault location	Probable cause
0304	P0304	Cylinder No.4 – misfire	**Wiring, mechanical fault, fuel system, EGR valve, injector, intake/ignition system, CKP sensor, HO2S front, ECM**
0305	P0305	Cylinder No.5 – misfire	**Wiring, mechanical fault, fuel system, EGR valve, injector, intake/ignition system, CKP sensor, HO2S front, ECM**
0306	P0306	Cylinder No.6 – misfire	**Wiring, mechanical fault, fuel system, EGR valve, injector, intake/ignition system, CKP sensor, HO2S front, ECM**
0327	P0327	Knock sensor (KS) – voltage low	**Wiring, KS, ECM**
0328	P0328	Knock sensor (KS) – voltage high	**Wiring, KS, ECM**
0335	P0335	Crankshaft position (CKP) sensor 1 – circuit malfunction	**Wiring, CKP sensor, ECM**
0340	P0340	Camshaft position (CMP) sensor – circuit malfunction	**Wiring, CMP sensor, Starter motor, ECM**
0400	P0400	Exhaust gas recirculation (EGR) system – flow excessively high or low	**Wiring, hose leak/blockage, EGR valve, EGR/EVAP valve, ECM**
0402	P0402	Exhaust gas recirculation (EGR) system – valve malfunction	**Wiring, EGR valve, EGR back pressure transducer, hose leak/blockage, CMP sensor, MAF sensor, EGR/EVAP valve, ECM**
0405	P0405	Exhaust gas recirculation temperature (EGRT) sensor – low voltage	**Wiring, EGRT sensor, ECM**
0406	P0406	Exhaust gas recirculation temperature (EGRT) sensor – high voltage	**Wiring, EGRT sensor, ECM**
0420	P0420	Catalytic converter – bank 1 – efficiency below limit	**Catalytic converter, wiring, intake/ exhaust system, injector, ECM**
0430	P0430	Catalytic converter – bank 2 – efficiency below limit	**Catalytic converter, wiring, intake/ exhaust system, injector, ECM**
0441	P0441	Evaporative emission (EVAP) canister purge system – flow malfunction	**Wiring, EVAP valve, hose leak/blockage, canister, BARO sensor, MAP/BARO switching valve, ECM**
0442	P0442	Evaporative emission (EVAP) canister purge system – small leak	**Wiring, fuel tank/cap, hose leak/ blockage, EVAP valve/canister, EVAP vent valve, BARO sensor, MAP/BARO switching valve, ECM**
0444	P0444	Evaporative emission (EVAP) canister purge system – control valve malfunction	**Wiring, EVAP control valve, EVAP valve/ pressure sensor, hose leak/blockage, ECM**
0445	P0445	Evaporative emission (EVAP) canister purge system – control valve malfunction	**Wiring, EVAP control valve, EVAP valve/ pressure sensor, hose leak/blockage, ECM**
0447	P0447	Evaporative emission (EVAP) canister purge system – vent control valve malfunction	**Wiring, EVAP vent control valve, EVAP breather valve/pressure sensor, hose leak/blockage, ECM**

Flash code	OBD-II code	Fault location	Probable cause
0452	P0452	Evaporative emission (EVAP) canister purge system – pressure sensor malfunction	**Wiring, EVAP pressure sensor, hose leak/blockage, EVAP vent/control valve, EVAP canister, ECM**
0453	P0453	Evaporative emission (EVAP) canister purge system – pressure sensor malfunction	**Wiring, EVAP pressure sensor, hose leak/blockage, EVAP vent/control valve, EVAP canister, ECM**
0455	P0455	Evaporative emission (EVAP) canister purge system – gross leak	**Wiring, fuel tank/cap, hose leak/ blockage, EVAP valve/canister, EVAP vent valve, BARO sensor, MAP/BARO switching valve, ECM**
0456	P0456	Evaporative emission (EVAP) canister purge system – very small leak	**Wiring, fuel tank/cap, hose leak/ blockage, EVAP valve/canister, EVAP vent valve, BARO sensor, MAP/BARO switching valve, ECM**
0460	P0460	Fuel tank level sensor – signal malfunction	**Wiring open/short circuit, fuel tank level sensor**
0461	P0461	Fuel tank level sensor – range performance problem	**Wiring, fuel tank level sensor**
0462	P0462	Fuel tank level sensor – voltage low	**Wiring, fuel tank level sensor**
0463	P0463	Fuel tank level sensor – voltage high	**Wiring, fuel tank level sensor**
0500	P0500	Vehicle speed sensor (VSS) – circuit malfunction	**Wiring, VSS, ECM**
0505	P0505	Idle speed control system – malfunction	**Wiring, ISC system, ECM**
0506	P0506	Idle speed control system – RPM too low	**Wiring, ISC system, ECM**
0507	P0507	Idle speed control system – RPM too high	**Wiring, ISC system, ECM**
0510	P0510	Closed throttle position (CTP) switch – circuit malfunction	**Wiring, CTP switch, ECM**
0600	P0600	ECM/TCM – communication malfunction	**Wiring, ECM, TCM**
0605	P0605	Engine control module (ECM) – malfunction	**ECM**
0650	P0650	Malfunction indicator lamp (MIL) – circuit malfunction	**Wiring, ECM**
0705	P0705	Park/neutral position (PNP) switch – circuit malfunction	**Wiring, PNP switch, ECM**
0710	P0710	Transmission fluid temperature sensor – excessively high or low voltage	**Wiring, temperature sensor, TCM, ECM**
0720	P0720	AT – vehicle speed sensor (VSS)	**Wiring, VSS, TCM**
0725	P0725	Engine speed (RPM) – incorrect voltage signal ECM to TCM	**Wiring, RPM sensor, TCM, ECM**
0731	P0731	AT – first gear selection – malfunction	**Valve block**
0732	P0732	AT – second gear selection – malfunction	**Valve block**
0733	P0733	AT – third gear selection – malfunction	**Valve block**
0734	P0734	AT – fourth gear selection – malfunction	**Valve block**

Flash code	OBD-II code	Fault location	Probable cause
0740	P0740	Torque converter clutch solenoid – voltage low	Wiring, clutch solenoid, TCM
0744	P0744	Torque converter lock-up solenoid – circuit malfunction	Wiring, clutch solenoid, TCM
0745	P0745	AT – Line pressure solenoid – voltage low	Wiring, pressure solenoid, TCM
0750	P0750	AT – shift solenoid A – voltage low	Wiring, solenoid A, TCM
0755	P0755	AT – shift solenoid B – voltage low	Wiring, solenoid B, TCM
1102	P1102	Mass air flow (MAF) sensor – range/performance problem	Wiring, MAF sensor, air leak, ECM
1111	P1111	Camshaft position (CMP) actuator – bank 1 – malfunction	Wiring, CMP actuator, ECM
1121	P1121	Throttle valve position motor – malfunction	Throttle valve position motor
1122	P1122	Throttle valve position motor – range/performance problem	Wiring, throttle valve position motor
1124	P1124	Throttle valve position motor relay – short circuit	Wiring, throttle valve position motor relay
1126	P1126	Throttle valve position motor relay – open circuit	Wiring, throttle valve position motor relay
1128	P1128	Throttle valve position motor – short circuit	Wiring, throttle valve position motor
1130	P1130	Intake manifold air control solenoid – malfunction	Wiring, intake manifold air control solenoid, ECM
1131	P1131	Intake manifold air control solenoid – malfunction	Wiring, intake manifold air control solenoid, ECM
1136	P1136	Camshaft position (CMP) actuator – bank 2 – malfunction	Wiring, CMP actuator, ECM
1140	P1140	Camshaft variable valve timing sensor – bank 1 – range/performance problem	Wiring, camshaft variable valve timing sensor, ECM
1143	P1143	Heated oxygen sensor (HO2S) 1, bank 1 – lean shift monitoring	Wiring, HO2S, fuel system, injector, air leak, ECM
1144	P1144	Heated oxygen sensor (HO2S) 1, bank 1 – rich shift monitoring	Wiring, HO2S, fuel system, injector, ECM
1145	P1145	Camshaft variable valve timing sensor – bank 2 – range/performance problem	Wiring, camshaft variable valve timing sensor, ECM
1146	P1146	Heated oxygen sensor (HO2S) 2, bank 1 – minimum voltage monitoring	Wiring, HO2S, fuel system, injector, ECM
1147	P1147	Heated oxygen sensor (HO2S) 2, bank 1 – maximum voltage monitoring	Wiring, HO2S, fuel system, injector, ECM
1148	P1148	Closed loop control – bank 1 – inoperative	Wiring, HO2S front, ECM
1163	P1163	Heated oxygen sensor (HO2S) 1, bank 2 – lean shift monitoring	Wiring, HO2S, fuel system, injector, air leak, ECM
1164	P1164	Heated oxygen sensor (HO2S) 1, bank 2 – rich shift monitoring	Wiring, HO2S, fuel system, injector, ECM

Flash code	OBD-II code	Fault location	Probable cause
1165	P1165	Intake manifold air control vacuum check switch – malfunction	**Wiring, intake manifold air control solenoid, intake manifold air control vacuum check switch, ECM**
1166	P1166	Heated oxygen sensor (HO2S) 2, bank 2 – minimum voltage monitoring	**Wiring, HO2S, fuel system, injector, ECM**
1167	P1167	Heated oxygen sensor (HO2S) 2, bank 2 – maximum voltage monitoring	**Wiring, HO2S, fuel system, injector, ECM**
1168	P1168	Closed loop control – bank 2 – inoperative	**Wiring, HO2S front, ECM**
1211	P1211	ABS control module – malfunction	**Wiring, ABS control module**
1212	P1212	ABS control module/traction control module – communication failure	**Wiring, CAN data bus, ABS control module**
1217	P1217	Engine coolant over temperature condition	**Wiring, coolant fan, coolant hoses, thermostat, radiator, coolant temperature switch**
1223	P1223	Throttle position (TP) sensor 2 – voltage low	**Wiring, TP sensor, ECM**
1224	P1224	Throttle position (TP) sensor 2 – voltage high	**Wiring, TP sensor, ECM**
1225	P1225	Closed throttle position, learning – voltage too low	**Throttle valve position motor, TP sensor 1 & 2,**
1226	P1226	Closed throttle position, learning – unsuccessful	**Throttle valve position motor, TP sensor 1 & 2,**
1227	P1227	Accelerator pedal position (APP) sensor 2 – voltage low	**Wiring, APP sensor, ECM**
1228	P1228	Accelerator pedal position (APP) sensor 2 – voltage high	**Wiring, APP sensor, ECM**
1229	P1229	Sensor supply voltage – short circuit	**Wiring, APP sensor, TP sensor, MAF sensor, EVAP sensor, PS sensor, refrigerant sensor, ECM**
1335	P1335	Crankshaft position (CKP) sensor 2 – circuit malfunction	**Wiring, CKP sensor, ECM**
1336	P1336	Crankshaft position (CKP) sensor 1 – rotor teeth damage	**Wiring, CKP sensor, ECM**
1400	P1400	Exhaust gas recirculation (EGR) system/ evaporative emission (EVAP) canister purge system – valve malfunction	**Wiring, EGR/EVAP valve, ECM**
1402	P1402	Exhaust gas recirculation (EGR) system – excessive flow	**Wiring, EGR/EVAP valve, EGR valve, EGRT sensor, EGR back pressure transducer, ECM**
1442	P1442	Evaporative emission (EVAP) canister purge system – small leak	**Wiring, fuel tank/cap, hose leak/ blockage, EVAP valve/canister, EVAP vent valve, BARO sensor, MAP/BARO switching valve, ECM**
1444	P1444	Evaporative emission (EVAP) canister purge system – control valve malfunction	**Wiring, EVAP pressure sensor, hose leak/blockage, EVAP valve/canister, ECM**

Flash code	OBD-II code	Fault location	Probable cause
1446	P1446	Evaporative emission (EVAP) canister purge system – vent control valve malfunction	**Wiring, EVAP pressure sensor, hose leak/blockage, EVAP vent valve, EVAP canister contaminated, ECM**
1448	P1448	Evaporative emission (EVAP) canister purge system – vent control valve malfunction	**Wiring, EVAP vent valve, EVAP pressure sensor, hose leak/blockage, EVAP canister contaminated, EVAP vacuum cut valve, ECM**
1456	P1456	Evaporative emission (EVAP) canister purge system – very small leak	**Wiring, fuel tank/cap, hose leak/ blockage, EVAP valve/canister, EVAP vent valve, BARO sensor, MAP/BARO switching valve, ECM**
1464	P1464	Fuel tank level sensor – voltage high	**Wiring, fuel tank level sensor**
1490	P1490	Evaporative emission (EVAP) canister purge system – bypass vacuum valve malfunction	**Wiring, EVAP bypass/cut valve, hose leak/blockage, ECM**
1491	P1491	Evaporative emission (EVAP) canister purge system – bypass vacuum valve malfunction	**Wiring, EVAP vacuum cut/bypass valve, hose leak/blockage, EVAP pressure sensor/vent valve, EVAP canister contaminated, ECM**
1564	P1564	Cruise control master switch – malfunction	**Wiring, cruise control master switch, ECM**
1572	P1572	Cruise control brake switch – circuit malfunction	**Wiring, cruise control brake switch, stop lamp switch, clutch pedal switch, ECM**
1574	P1574	Cruise control vehicle speed sensor – signal variation between two vehicle speed sensors	**Wiring, CAN data bus, vehicle speed sensor, instrument panel, ECM**
1605	P1605	TCM diagnosis communication line – malfunction	**Wiring, TCM**
1705	P1705	AT – throttle position (TP) sensor – voltage high or low	**Wiring, TP sensor, TCM, ECM**
1706	P1706	Park/neutral position (PNP) switch – 1998 – circuit malfunction	**Wiring, PNP switch, ECM**
1760	P1760	AT – overrun clutch solenoid – voltage low	**Wiring, clutch solenoid, TCM**
1800	P1800	Intake manifold air control solenoid – malfunction	**Wiring, intake manifold air control solenoid, ECM**
1805	P1805	Brake pedal position (BPP) switch – circuit malfunction	**Wiring, BPP switch, ECM**

NISSAN

Model:	Engine identification:	Year:
Armada 5.6L	**VK56DE**	**2004**
Murano 3.5L	**VQ35DE**	**2003-04**
Titan 5.6L	**VK56DE**	**2004**
350Z 3.5L	**VQ35DE**	**2003-04**

System: **ECCS**

Self-diagnosis

General information

- Refer to the front of this manual for general test conditions, terminology, detailed descriptions of wiring faults and a general trouble shooter for electrical and mechanical faults.
- Engine control module (ECM) incorporates self-diagnosis function.
- Malfunction indicator lamp (MIL) will illuminate if certain faults are recorded.
- ECM operates in backup mode if sensors fail, to enable vehicle to be driven to workshop.
- Trouble codes can be displayed by using a Scan Tool connected to the data link connector (DLC) or by the malfunction indicator lamp (MIL):
  - Armada/Murano/Titan – **1**
  - 350Z – **2**

1

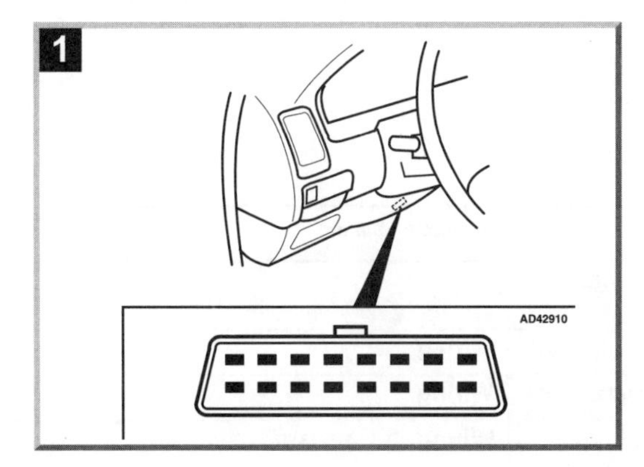

2

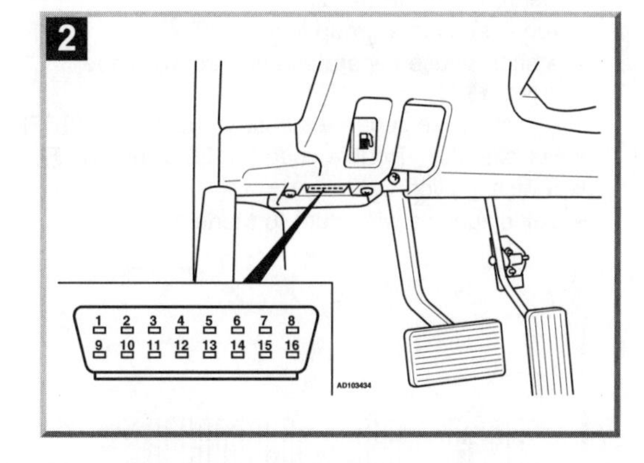

NOTE: ***The use of a Scan Tool is essential to obtain full diagnostic information.***

Accessing – Mode 1

- Switch ignition ON.
- Check MIL illuminates.
- Start engine. Allow to idle.
- If MIL extinguishes, no trouble codes have been recorded.
- If MIL remains illuminated, access trouble codes.

Accessing – Mode 2

- Ensure accelerator pedal is fully released.
- Switch ignition ON. Do NOT start engine.
- Wait 3 seconds.
- Within 5 seconds, repeat the following 5 times:
  - Fully depress accelerator pedal.
  - Fully release accelerator pedal.
- Wait 7 seconds.
- Fully depress accelerator pedal.
- After approximately 10 seconds MIL will start flashing.

- Fully release accelerator pedal.
- Self-diagnosis now in Mode 2.
- Count flashes and compare with trouble code table.
- Each trouble code consists of four groups of one or more flashes.
- The first group of flashes indicate the 'thousands' of the trouble code **3** [**A**].
- The second group of flashes indicate the 'hundreds' of the trouble code **3** [**B**].
- The third group of flashes indicate the 'tens' of the trouble code **3** [**C**].
- The fourth group of flashes indicate the 'units' of the trouble code **3** [**D**].
- Ten flashes in a group indicate '0'.
- A short pause separates each trouble code group **3** [**E**].
- A long pause separates each trouble code **3** [**F**].
- For example: Trouble code P1130 displayed **3**.
- Switch ignition OFF.
- Self-diagnosis will return to Mode 1.

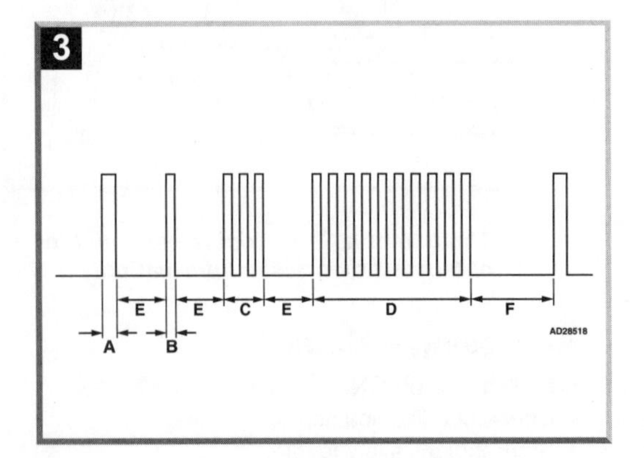

Erasing

- Trouble codes can be erased by using a Scan Tool connected to the data link connector (DLC) or as follows:
- Ensure accelerator pedal is fully released.
- Switch ignition ON. Do NOT start engine.
- Wait 3 seconds.
- Within 5 seconds, repeat the following 5 times:
  - Fully depress accelerator pedal.
  - Fully release accelerator pedal.
- Wait 7 seconds.
- Fully depress accelerator pedal.
- After approximately 10 seconds MIL will start flashing.
- Fully release accelerator pedal.
- Self-diagnosis now in Mode 2.
  - Fully depress accelerator pedal for more than 10 seconds.
  - Fully release accelerator pedal.
  - Trouble code 0000 should be displayed.

Trouble code identification

Flash code	OBD-II code	Fault location	Probable cause
–	*All P0, P2 and U0 codes*	Refer to OBD-II trouble code tables at the front of this manual	–
1000	**U1000**	CAN communication line – signal malfunction	**Wiring**
1001	**U1001**	CAN communication line – signal malfunction	**Wiring**
1031	**P1031**	Air fuel (A/F) ratio sensor bank 1, heater – voltage low	**Wiring, A/F sensor**
1032	**P1032**	Air fuel (A/F) ratio sensor bank 1, heater – voltage high	**Wiring, A/F sensor**
1051	**P1051**	Air fuel (A/F) ratio sensor bank 2, heater – voltage low	**Wiring, A/F sensor**

Flash code	OBD-II code	Fault location	Probable cause
1052	P1052	Air fuel (A/F) ratio sensor bank 2, heater – voltage high	**Wiring, A/F sensor**
1065	P1065	Engine control module (ECM) – battery voltage supply	**Wiring, fuse, ECM**
1111	P1111	Intake valve timing control solenoid bank 1 – circuit malfunction	**Wiring, intake valve control solenoid, ECM**
1121	P1121	Throttle valve position motor – malfunction	**Throttle valve position motor**
1122	P1122	Throttle valve position motor – range/ performance problem	**Wiring, throttle valve position motor**
1124	P1124	Throttle valve position motor relay – short circuit	**Wiring, throttle valve position motor relay**
1126	P1126	Throttle valve position motor relay – open circuit	**Wiring, throttle valve position motor relay**
1128	P1128	Throttle valve position motor – short circuit	**Wiring, throttle valve position motor**
1136	P1136	Intake valve timing control solenoid bank 2 – circuit malfunction	**Wiring, intake valve control solenoid, ECM**
1143	P1143	Heated oxygen sensor (HO2S) 1, bank 1 – voltage low	**Wiring, mixture, HO2S, ECM**
1144	P1144	Heated oxygen sensor (HO2S) 1, bank 1 – voltage high	**HO2S, fuel pressure, fuel system, ECM**
1146	P1146	Heated oxygen sensor (HO2S) 2, bank 1 – minimum voltage monitoring	**Wiring, HO2S, fuel system, injector, ECM**
1147	P1147	Heated oxygen sensor (HO2S) 2, bank 1 – maximum voltage monitoring	**Wiring, HO2S, fuel system, injector, intake air leaks, ECM**
1148	P1148	Closed loop control – bank 1 – inoperative	**Wiring, HO2S front, ECM**
1163	P1163	Heated oxygen sensor (HO2S) 1, bank 2 – voltage low	**Wiring, mixture, HO2S, ECM**
1164	P1164	Heated oxygen sensor (HO2S) 1, bank 2 – voltage high	**HO2S, fuel pressure, fuel system, ECM**
1166	P1166	Heated oxygen sensor (HO2S) 2, bank 2 – minimum voltage monitoring	**Wiring, HO2S, fuel system, injector, ECM**
1167	P1167	Heated oxygen sensor (HO2S) 2, bank 2 – maximum voltage monitoring	**Wiring, HO2S, fuel system, injector, intake air leaks, ECM**
1168	P1168	Closed loop control – bank 2 – inoperative	**Wiring, HO2S front, ECM**
1211	P1211	ABS/TCS control module – malfunction	**Wiring, ABS/TCS control module**
1212	P1212	ABS control module/traction control module – communication failure	**Wiring, CAN data bus, ABS control module**
1217	P1217	Engine coolant over temperature condition	**Wiring, coolant fan, coolant hoses, thermostat, radiator, coolant temperature switch**
1225	P1225	Closed throttle position, learning – voltage too low	**Throttle valve position motor, TP sensor 1 & 2,**
1226	P1226	Closed throttle position, learning – unsuccessful	**Throttle valve position motor, TP sensor 1 & 2,**

Flash code	OBD-II code	Fault location	Probable cause
1229	P1229	Sensor supply voltage – short circuit	Wiring, APP sensor, TP sensor, MAF sensor, EVAP sensor, PS sensor, refrigerant sensor, ECM
1271	P1271	Air fuel (A/F) ratio sensor bank 1 – voltage low	Wiring, A/F sensor
1272	P1272	Air fuel (A/F) ratio sensor bank 1 – voltage high	Wiring, A/F sensor
1273	P1273	Air fuel (A/F) ratio sensor bank 1 – lean mixture	A/F sensor, fuel pressure, injectors, intake air leak
1274	P1274	Air fuel (A/F) ratio sensor bank 1 – rich mixture	A/F sensor, fuel pressure, injectors
1276	P1276	Air fuel (A/F) ratio sensor bank 1 – activity	Wiring, A/F sensor
1278	P1278	Air fuel (A/F) ratio sensor bank 1 – slow response	Wiring, fuel pressure, injectors, MAF sensor, PCV, A/F sensor, intake air leaks, exhaust leaks
1279	P1279	Air fuel (A/F) ratio sensor bank 1 – slow response	Wiring, fuel pressure, injectors, MAF sensor, PCV, A/F sensor, intake air leaks, exhaust leaks
1281	P1281	Air fuel (A/F) ratio sensor bank 2 – voltage low	Wiring, A/F sensor
1282	P1282	Air fuel (A/F) ratio sensor bank 2 – voltage high	Wiring, A/F sensor
1283	P1283	Air fuel (A/F) ratio sensor bank 2 – lean mixture	A/F sensor, fuel pressure, injectors, intake air leak
1284	P1284	Air fuel (A/F) ratio sensor bank 2 – rich mixture	A/F sensor, fuel pressure, injectors
1286	P1286	Air fuel (A/F) ratio sensor bank 2 – activity	Wiring, A/F sensor
1288	P1288	Air fuel (A/F) ratio sensor bank 2 – slow response	Wiring, fuel pressure, injectors, MAF sensor, PCV, A/F sensor, intake air leaks, exhaust leaks
1289	P1289	Air fuel (A/F) ratio sensor bank 2 – slow response	Wiring, fuel pressure, injectors, MAF sensor, PCV, A/F sensor, intake air leaks, exhaust leaks
1444	P1444	Evaporative emission (EVAP) canister purge system – control valve malfunction	Wiring, EVAP pressure sensor, hose leak/blockage, EVAP valve/canister, ECM
1446	P1446	Evaporative emission (EVAP) canister purge system – vent control valve malfunction	Wiring, EVAP pressure sensor, hose leak/blockage, EVAP vent valve, EVAP canister contaminated, ECM
1564	P1564	Cruise control master switch – malfunction	Wiring, cruise control master switch, ECM
1572	P1572	Cruise control brake switch – circuit malfunction	Wiring, cruise control brake switch, stop lamp switch, clutch pedal switch, ECM
1574	P1574	Cruise control vehicle speed sensor – signal malfunction	Wiring, TCM – dtc's stored, ABS/TCS control module – dtc's stored, instrument panel, ECM

Flash code	OBD-II code	Fault location	Probable cause
1610	P1610	Ignition key/engine control module (ECM) – malfunction	**Incorrect ignition key, ECM**
1615	P1615	Ignition key/immobilizer control module communication – malfunction	**Incorrect ignition key, immobilizer control module**
1700	P1700	CVT shift control – malfunction	**TCM trouble code(s) stored**
1705	P1705	AT – throttle position (TP) sensor – voltage high or low	**Wiring, TP sensor, mechanical fault, ECM**
1706	P1706	Park/neutral position (PNP) switch – circuit malfunction	**Wiring, PNP switch, instrument panel, TCM**
1715	P1715	Automatic transmission input speed sensor – malfunction	**Wiring, TCM**
1716	P1716	Automatic transmission turbine revolution sensor – circuit malfunction	**Wiring, turbine sensor 1, turbine sensor 2, ECM**
1720	P1720	Vehicle speed sensor (VSS) – signal variation	**Wiring, mechanical fault, ABS/TCS control module, TCM**
1730	P1730	Automatic transmission fail safe function – circuit malfunction	**Wiring, transmission clutches, transmission solenoid valves, transmission pressure switches**
1740	P1740	CVT lock-up select solenoid valve – malfunction	**Wiring, lock-up select solenoid valve**
1752	P1752	Automatic transmission input clutch – solenoid valve malfunction	**Wiring, input clutch solenoid valve**
1754	P1754	Automatic transmission input clutch – solenoid valve malfunction	**Wiring, input clutch solenoid valve, transmission pressure switch 3**
1757	P1757	Automatic transmission front brake – solenoid valve malfunction	**Wiring, front brake solenoid valve**
1759	P1759	Automatic transmission front brake – solenoid valve malfunction	**Wiring, front brake solenoid valve, transmission pressure switch 1**
1762	P1762	Automatic transmission direct clutch – solenoid valve malfunction	**Wiring, direct clutch solenoid valve**
1764	P1764	Automatic transmission direct clutch – solenoid valve malfunction	**Wiring, direct clutch solenoid valve, transmission pressure switch 5**
1767	P1767	Automatic transmission high/low reverse – solenoid valve malfunction	**Wiring, high/low reverse solenoid valve**
1769	P1769	Automatic transmission high/low reverse – solenoid valve malfunction	**Wiring, high/low reverse solenoid valve, transmission pressure switch 6**
1772	P1772	Automatic transmission low coast brake – solenoid valve malfunction	**Wiring, low coast brake solenoid valve**
1774	P1774	Automatic transmission low coast brake – solenoid valve malfunction	**Wiring, low coast brake solenoid valve, transmission pressure switch 2**
1777	P1777	CVT step motor – circuit malfunction	**Wiring, step motor**
1778	P1778	CVT step motor – mechanical malfunction	**Step motor**
1800	P1800	VIAS control solenoid – circuit malfunction	**Wiring, VIAS control solenoid**
1805	P1805	Brake pedal position (BPP) switch – circuit malfunction	**Wiring, BPP switch, ECM**

NISSAN

Model:	Engine identification:	Year:
Axxess 2.4L	**KA24E**	**1990**
Pickup 2.4L	**KA24E**	**1990-95**
Pickup/Pathfinder 3.0L	**VG30E**	**1990-95**
Pulsar 1.6L	**E16i/GA16i**	**1988-90**
Pulsar 1.8L	**CA18DE**	**1988-89**
Sentra 1.6L	**E16i/GA16i**	**1988-90**

System: **ECCS**

Self-diagnosis

General information

- Refer to the front of this manual for general test conditions, terminology, detailed descriptions of wiring faults and a general trouble shooter for electrical and mechanical faults.
- Engine control module (ECM) incorporates self-diagnosis function.
- Malfunction indicator lamp (MIL) will illuminate if certain faults are recorded – California 1988 on.
- ECM operates in backup mode if certain sensors fail, to enable vehicle to be driven to workshop.
- Trouble codes are displayed by the red and green LED's in the engine control module (ECM), in diagnostic test mode 3, or using a suitable code reader.
- The red LED flashes indicate 'tens', followed by the green which indicate 'units'.

Accessing

Mode selection

- Access engine control module (ECM):
  - Sentra/Pulsar – **1**.
  - Axxess – **2**.
  - Pickup/Pathfinder – **3**.
- Switch ignition ON.
- Turn ECM diagnosis mode selector fully clockwise **4** [**1**].
  - Wait until the LED's flash **4** [**2**].
  - After LED's flash required mode number, turn mode selector fully counterclockwise **4** [**1**].
  - For example 3 flashes indicates Mode 3.

NOTE: ***If the ignition is switched OFF during self-diagnosis, it will return to mode 1 when It Is turned ON again.***

1

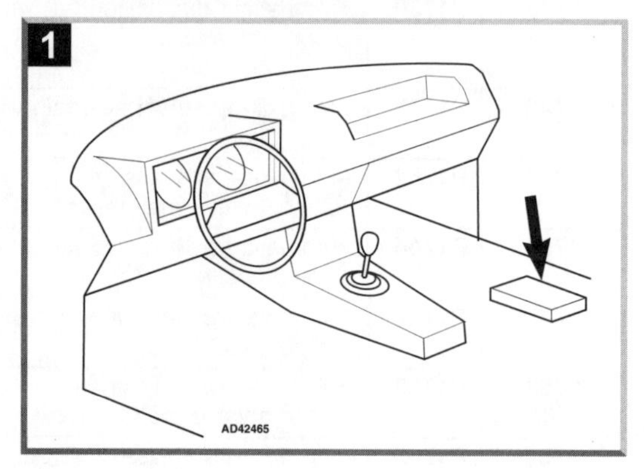

2

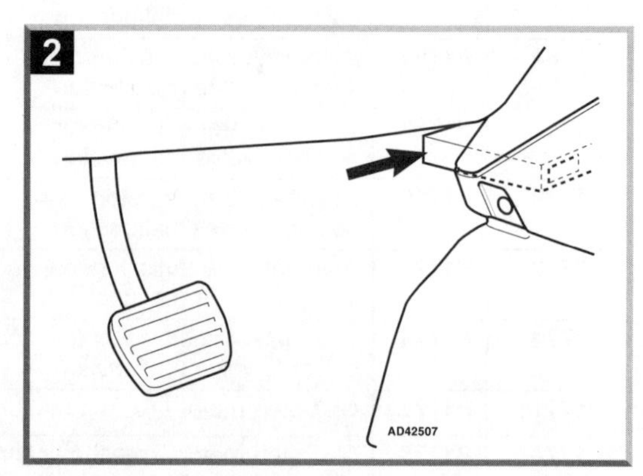

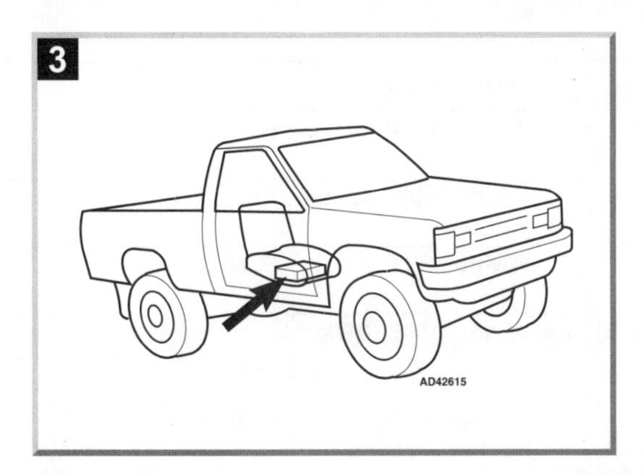

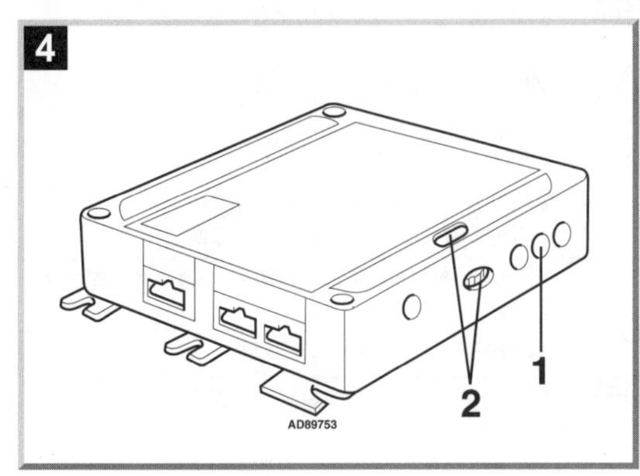

Mode description

- Mode 1: Mixture ratio feedback – heated oxygen sensor (HO2S)/oxygen sensor (O2S).
- Mode 2: Mixture ratio feedback – CO adjustment section.
- Mode 3: Code display.
- Mode 4: Tests diagnosis ON/OFF switching.
- Mode 5: Real time diagnosis, codes displayed as they occur.

Mode 3

- Access engine control module (ECM).
- Switch ignition ON. Do NOT start engine.
- Select mode 3.
- Count LED flashes and compare with trouble code table.

Mode 4

- Access engine control module (ECM).
- Switch ignition ON. Do NOT start engine.
- Select mode 4.
- Red LED goes off.
- Crank engine, red LED should come ON.
- Depress accelerator pedal, red LED goes OFF.
- Raise driven wheels of vehicle and drive vehicle, green LED should come ON when speed reaches 12 mph (20 kph).

Mode 5

- Access engine control module (ECM).
- Start engine.
- Select mode 5.
- Test drive vehicle.
- LED's will flash any codes occurring during test.

Erasing

NOTE: ***Trouble codes are automatically erased once the starter is operated 50 times after the last trouble code was stored.***

Method 1

- Select mode 3.
- Read flash codes.
- Select mode 4.
- Switch ignition OFF.

Method 2

- Ensure ignition switched OFF.
- Disconnect battery ground cable.
- Wait at least 15 seconds.
- Reconnect battery ground cable.
- Repeat checking procedure to ensure no data remains in ECM fault memory.

NOTE: ***Disconnecting battery ground lead may also erase memory from electronic units such as radio.***

Trouble code identification

Flash code	Fault location	Probable cause
11	Crankshaft position (CKP) sensor – circuit malfunction	**Wiring, CKP sensor, ECM**
12	Mass air flow (MAF) sensor – range/ performance problem	**Wiring, MAF sensor, ECM**
13	Engine coolant temperature (ECT) sensor – range/performance problem	**Wiring, ECT sensor, ECM**
14	Vehicle speed sensor (VSS) – malfunction	**Wiring, VSS, ECM**
15	Injector – mixture lean – 1988	**Injector, fuel system, ECM**
21	Ignition coil – no primary signal	**Wiring, ignition coil/module, ignition system, ECM**
23	Closed throttle position (CTP) switch – circuit malfunction	**Wiring, CTP switch, ECM**
25	Idle air control (IAC) valve – idle speed higher than specified	**Wiring, IAC valve, ECM**
31	Engine control module (ECM) – malfunction	**ECM**
32	Exhaust gas recirculation (EGR) system – control valve malfunction	**Wiring, hose leak/blockage, EGR cut OFF solenoid, ECM**
33	Heated oxygen sensor (HO2S) – circuit malfunction	**Wiring, HO2S, intake/fuel system, injector, ECM**
34	Knock sensor (KS) – circuit malfunction	**Wiring, KS sensor, ECM**
35	Exhaust gas recirculation temperature (EGRT) sensor – circuit malfunction	**Wiring, EGRT sensor, ECM**
41	Intake air temperature (IAT) sensor – circuit malfunction	**Wiring, IAT sensor, ECM**
43	Throttle position (TP) sensor – circuit malfunction	**Wiring, TP sensor, ECM**
45	Injector leak – malfunction	**Injector, fuel system, ECM**
51	Injector – circuit malfunction	**Wiring, injector, ECM**
55	No faults detected	–

NISSAN

Model:	Engine identification:	Year:
Maxima 3.0L	**VQ30DE**	**1995-02**
Pathfinder 3.3L	**VG33E**	**1996-00**
Pathfinder 3.5L	**VQ35DE**	**2001**
Quest 3.0L	**VG30E**	**1996-98**
Quest 3.3L	**VG33E**	**1999-02**
300ZX 3.0L	**VG30DE**	**1996**
300ZX 3.0L Turbo	**VG30DETT**	**1996**
Frontier 3.3L	**VG33E/VG33ER**	**1999-02**
Xterra 3.3L	**VG33E**	**2000-01**

System: **ECCS**

Self-diagnosis

General information

- Refer to the front of this manual for general test conditions, terminology, detailed descriptions of wiring faults and a general trouble shooter for electrical and mechanical faults.
- Engine control module (ECM) incorporates self-diagnosis function.
- Malfunction indicator lamp (MIL) will illuminate if certain faults are recorded.
- ECM operates in backup mode if sensors fail, to enable vehicle to be driven to workshop.
- Maxima & Quest 2000→/Pathfinder & Xterra 2001: Trouble codes can only be displayed using a Scan Tool connected to the data link connector (DLC).
- Except Maxima & Quest 2000→/Pathfinder & Xterra →2001: Trouble codes can be displayed by the malfunction indicator lamp (MIL) or by using a Scan Tool connected to the data link connector (DLC).
  - Maxima →1995: **1** [1].
  - Maxima 1996→/300ZX/Quest/Pathfinder/Xterra: **2** [1].

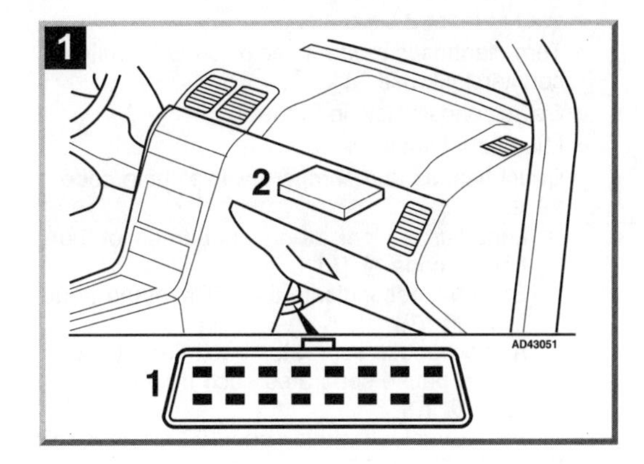

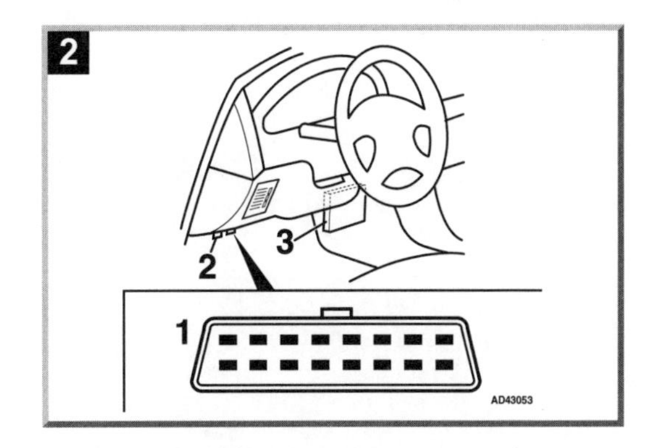

NOTE: ***The use of a Scan Tool is essential to obtain full diagnostic information.***

Accessing – Mode 1

- Switch ignition ON.
- Check MIL illuminates.
- Start engine. Allow to idle.
- If MIL extinguishes, no trouble codes have been recorded.
- If MIL remains illuminated, access trouble codes.

Accessing – Mode 2 – except Maxima & Quest 2000 →/ Pathfinder & Xterra 2001

- Access engine control module (ECM):
  - Maxima – 3 [2].
  - Quest – 1 [2].
  - Pathfinder – 2 [3].
  - 300ZX – 4 [1].
- Switch ignition ON. Do NOT start engine.
- Turn diagnostic mode selector on ECM fully clockwise 5.
- Wait at least 2 seconds.
- Turn diagnostic mode selector on ECM fully counterclockwise 5.
- Self-diagnosis now in Mode 2.
- MIL should flash.
- Count flashes and compare with trouble code table.
  - Long flashes indicate the 'hundreds' of the trouble code 6 [A].
  - Short flashes indicate the 'units' of the trouble code 6 [C].
  - A short pause separates each flash 6 [B].
  - A long pause separates each trouble code 6 [D].
  - For example: Trouble code 102 displayed 6.
- Switch ignition OFF.
- Self-diagnosis will return to Mode 1.

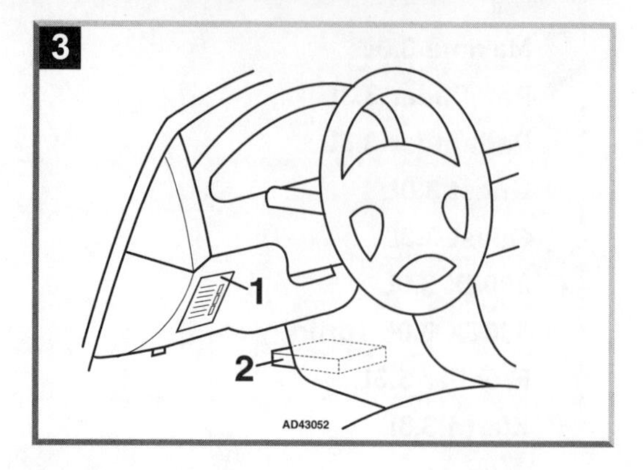

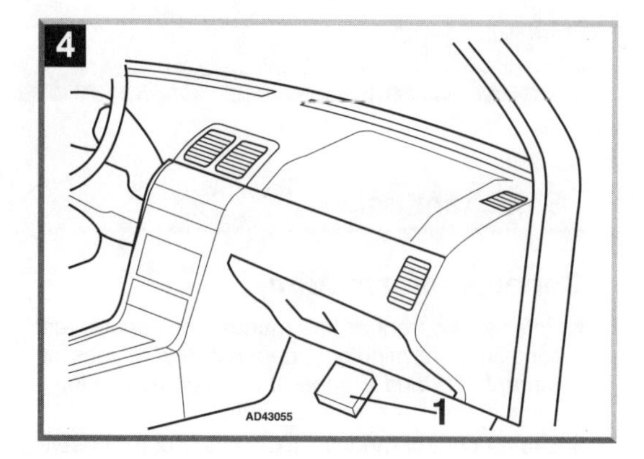

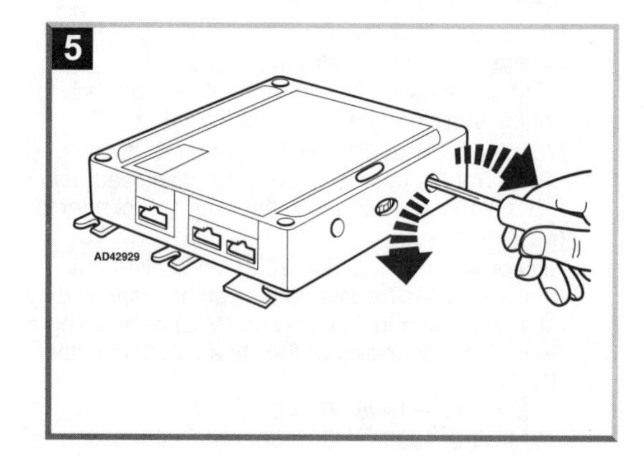

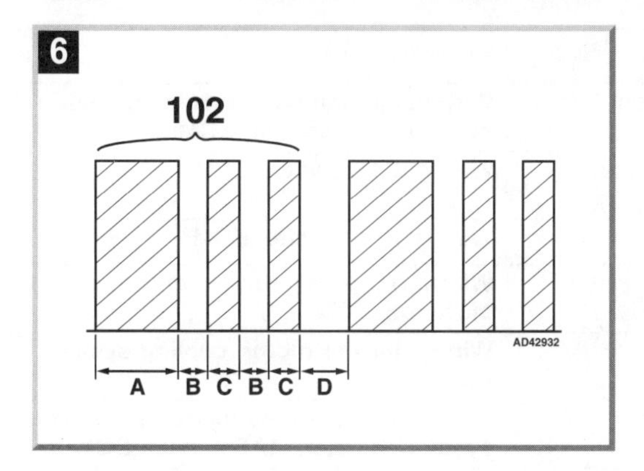

Accessing – Mode 2 – Maxima & Quest 2000 →/ Pathfinder & Xterra 2001

- Trouble codes can only be displayed using a Scan Tool connected to the data link connector (DLC).

Erasing – except Maxima & Quest 2000 →/Pathfinder & Xterra 2001

- Trouble codes can be erased by using a Scan Tool connected to the data link connector (DLC) or as follows:
- With the ignition switch ON and the engine control module (ECM) in diagnosis test mode 2.
- Turn diagnostic mode selector on ECM fully clockwise **5**.
- Trouble codes are erased.
- Switch ignition OFF.
- Turn diagnostic mode selector on ECM fully counterclockwise **5**.
- Repeat checking procedure to ensure no data remains in ECM fault memory.

Erasing – Maxima & Quest 2000 →/ Pathfinder & Xterra 2001

- Trouble codes can only be erased by using a Scan Tool connected to the data link connector (DLC).

Trouble code identification

Flash code	OBD-II code	Fault location	Probable cause
–	*All P0, P2 and U0 codes*	Refer to OBD-II trouble code tables at the front of this manual	–
0101	–	Camshaft position (CMP) sensor – circuit malfunction	**Wiring, CMP sensor, Starter motor, ECM**
0102	–	Mass air flow (MAF) sensor – circuit malfunction	**Wiring, MAF sensor, ECM**
0103	–	Engine coolant temperature (ECT) sensor – circuit malfunction	**Wiring, ECT sensor, ECM**
0104	–	Vehicle speed sensor (VSS) – circuit malfunction	**Wiring, VSS, ECM**
0111	P1447	Evaporative emission (EVAP) canister purge system – flow malfunction	**Wiring, EVAP valve, hose leak/blockage, canister, BARO sensor, MAP/BARO switching valve, ECM**
0112	P1165	Intake manifold air control system – vacuum check switch circuit malfunction	**Wiring, vacuum check switch, hose leak/ blockage, air control solenoid, ECM**
0113	P1443	Evaporative emission (EVAP) canister purge system – vacuum check switch circuit malfunction	**Wiring, EVAP vacuum check switch, hose leak/blockage, TP/ECT sensor, EVAP valve, ECM**
0114	–	Fuel trim – RH – mixture too rich	**Wiring, intake/exhaust system, HO2S front, fuel system, injector, MAF sensor, ECM**
0115	–	Fuel trim – RH – mixture too lean	**Wiring, exhaust system, HO2S front, fuel system, injector, MAF sensor, ECM**

Flash code	OBD-II code	Fault location	Probable cause
0201	P1320	Ignition signal – circuit malfunction	Wiring, ignition coil/module, ignition system, CKP sensor, ECM
0203	–	Closed throttle position (CTP) switch – circuit malfunction	Wiring, CTP switch, ECM
0205	–	Idle control system – malfunction	Wiring, IAC system, ECM
0206	P1160	Turbocharger (TC) wastegate pressure sensor – circuit malfunction	Wiring, TC pressure sensor, hose leak/ blockage, ECM
0208	P1900	Engine coolant blower motor – circuit malfunction	Wiring, blower motor, cooling system, ECM
0209	–	Fuel trim – LH – mixture too rich	Wiring, exhaust system, HO2S front, fuel system, injector, MAF sensor, ECM
0210	–	Fuel trim – LH – mixture too lean	Wiring, exhaust system, HO2S front, fuel system, injector, MAF sensor, ECM
0211	–	Engine over temperature	Cooling system, engine coolant blower motor
0213	P1440	Evaporative emission (EVAP) canister purge system – leaking/malfunction	Wiring, fuel tank/cap, hose leak/ blockage, EVAP valve/canister, EVAP vent valve, BARO sensor, MAP/BARO switching valve, ECM
0214	P1444	Evaporative emission (EVAP) canister purge system – control valve malfunction	Wiring, EVAP pressure sensor, hose leak/blockage, EVAP valve/canister, ECM
0215	P1446	Evaporative emission (EVAP) canister purge system – vent control valve malfunction	Wiring, EVAP pressure sensor, hose leak/blockage, EVAP vent valve, EVAP canister contaminated, ECM
0301	–	Engine control module (ECM) – malfunction	ECM
0302	–	Exhaust gas recirculation (EGR) system – flow excessively high or low	Wiring, hose leak/blockage, EGR valve, EGR/EVAP valve, ECM
0303	–	Heated oxygen sensor (HO2S) – LH front – circuit malfunction	Wiring, HO2S, intake/fuel system, injector, ECM
0304	–	Knock sensor (KS) – circuit malfunction	Wiring, KS, ECM
0305	P1401	Exhaust gas recirculation temperature (EGRT) sensor – circuit malfunction	Wiring, EGRT sensor, ECM
0306	–	Exhaust gas recirculation (EGR) system – valve malfunction	Wiring, EGR valve, EGR back pressure transducer, hose leak/blockage, CMP sensor, MAF sensor, EGR/EVAP valve, ECM
0307	–	Closed loop control – RH 1996-97 – inoperative	Wiring, HO2S front, ECM
0307	P1148	Closed loop control – RH 1998 – inoperative	Wiring, HO2S front, ECM
0308	–	Closed loop control – LH 1996-97 – inoperative	Wiring, HO2S front, ECM
0308	P1168	Closed loop control – LH 1998 – inoperative	Wiring, HO2S front, ECM

Flash code	OBD-II code	Fault location	Probable cause
0309	P1448	Evaporative emission (EVAP) canister purge system – vent control valve malfunction	**Wiring, EVAP vent valve, EVAP pressure sensor, hose leak/blockage, EVAP canister contaminated, EVAP vacuum cut valve, ECM**
0311	P1491	Evaporative emission (EVAP) canister purge system – bypass vacuum valve malfunction	**Wiring, EVAP vacuum cut/bypass valve, hose leak/blockage, EVAP pressure sensor/vent valve, EVAP canister contaminated, ECM**
0312	P1493	Evaporative emission (EVAP) canister purge system – control valve malfunction	**Wiring, EVAP valve, hose leak/blockage, EVAP pressure sensor/vent valve, EVAP canister contaminated, ECM**
0313	–	Heated oxygen sensor (HO2S) – LH rear – voltage high	**Wiring, HO2S, fuel/intake system, injector, ECM**
0314	–	Heated oxygen sensor (HO2S) – LH rear – voltage low	**Wiring, HO2S, fuel system, injector, ECM**
0315	–	Heated oxygen sensor (HO2S) – LH rear – excessively high voltage	**Wiring, HO2S, ECM**
0401	–	Intake air temperature (IAT) sensor – circuit malfunction	**Wiring, IAT sensor, ECM**
0402	–	Fuel temperature sensor – circuit malfunction	**Wiring, fuel temperature sensor, ECM**
0403	–	Throttle position (TP) sensor – circuit malfunction	**Wiring, TP sensor, ECM**
0407	P1335	Crankshaft position (CKP) sensor 2 – circuit malfunction	**Wiring, CKP sensor, ECM**
0409	–	Heated oxygen sensor (HO2S) – RH front – slow response	**Wiring, HO2S, fuel/intake system, injector, exhaust system, MAF sensor, ECM**
0410	–	Heated oxygen sensor (HO2S) – RH front – voltage high	**Wiring, HO2S, fuel system, injector, ECM**
0411	–	Heated oxygen sensor (HO2S) – RH front – voltage low	**Wiring, HO2S, fuel/intake system, injector, ECM**
0412	–	Heated oxygen sensor (HO2S) – RH front – excessively high voltage	**Wiring, HO2S, ECM**
0413	–	Heated oxygen sensor (HO2S) – LH front – slow response	**Wiring, HO2S, fuel/intake system, injector, exhaust system, crankcase vent valve, MAF sensor, ECM**
0414	–	Heated oxygen sensor (HO2S) – LH front – voltage high	**Wiring, HO2S, fuel system, injector, ECM**
0415	–	Heated oxygen sensor (HO2S) – LH front – voltage low	**Wiring, HO2S, fuel/intake system, injector, ECM**
0503	–	Heated oxygen sensor (HO2S) – RH front – circuit malfunction	**Wiring, HO2S, ECM**
0504	–	ECM/TCM – communication malfunction	**Wiring, ECM, TCM**
0505	–	No self diagnostic failure indicated	–

Flash code	OBD-II code	Fault location	Probable cause
0509	–	Heated oxygen sensor (HO2S) – LH front – excessively high voltage	Wiring, HO2S, ECM
0510	–	Heated oxygen sensor (HO2S) – rear – voltage high	Wiring, HO2S, fuel/intake system, injector, ECM
0511	–	Heated oxygen sensor (HO2S) – rear – voltage low	Wiring, HO2S, fuel system, injector, ECM
0512	–	Heated oxygen sensor (HO2S) – rear – excessively high voltage	Wiring, HO2S, ECM
0514	P1402	Exhaust gas recirculation (EGR) system – excessive flow	Wiring, EGR/EVAP valve, EGR valve, EGRT sensor, EGR back pressure transducer, ECM
0515	–	Exhaust gas recirculation (EGR) system – circuit malfunction	Wiring, EGR valve, ECM
0603	–	Cylinder No.6 – misfire	Wiring, mechanical fault, fuel system, EGR valve, injector, intake/ignition system, CKP sensor, HO2S front, ECM
0604	–	Cylinder No.5 – misfire	Wiring, mechanical fault, fuel system, EGR valve, injector, intake/ignition system, CKP sensor, HO2S front, ECM
0605	–	Cylinder No.4 – misfire	Wiring, mechanical fault, fuel system, EGR valve, injector, intake/ignition system, CKP sensor, HO2S front, ECM
0606	–	Cylinder No.3 – misfire	Wiring, mechanical fault, fuel system, EGR valve, injector, intake/ignition system, CKP sensor, HO2S front, ECM
0607	–	Cylinder No.2 – misfire	Wiring, mechanical fault, fuel system, EGR valve, injector, intake/ignition system, CKP sensor, HO2S front, ECM
0608	–	Cylinder No.1 – misfire	Wiring, mechanical fault, fuel system, EGR valve, injector, intake/ignition system, CKP sensor, HO2S front, ECM
0701	–	Random misfire	Wiring, mechanical fault, fuel system, EGR valve, injector, intake/ignition system, CKP sensor, HO2S front, ECM
0702	–	Catalytic converter – RH – efficiency below limit	Catalytic converter, wiring, intake/ exhaust system, injector, ECM
0703	–	Catalytic converter – LH – efficiency below limit	Catalytic converter, wiring, intake/ exhaust system, injector, ECM
0704	–	Evaporative emission (EVAP) canister purge system – pressure sensor malfunction	Wiring, EVAP pressure sensor, hose leak/blockage, EVAP vent/control valve, EVAP canister, ECM
0705	–	Evaporative emission (EVAP) canister purge system – leaking/malfunction	Wiring, fuel tank/cap, hose leak/ blockage, EVAP valve/canister, EVAP vent valve, BARO sensor, MAP/BARO switching valve, ECM
0706	–	Mixture control	Wiring, intake/exhaust system, injector, fuel system, MAF sensor, ECM

Flash code	OBD-II code	Fault location	Probable cause
0707	–	Heated oxygen sensor (HO2S) – Maxima 1996-97 rear – circuit malfunction	**Wiring, HO2S, intake/fuel system, injector, ECM**
0707	–	Heated oxygen sensor (HO2S) – except Maxima 1996-97 RH rear – circuit malfunction	**Wiring, HO2S, intake/fuel system, injector, ECM**
0707	–	Heated oxygen sensor (HO2S) – 1998 rear – circuit malfunction	**Wiring, HO2S, intake/fuel system, injector, ECM**
0708	–	Heated oxygen sensor (HO2S) – LH rear – circuit malfunction	**Wiring, HO2S, intake/fuel system, injector, ECM**
0715	–	Evaporative emission (EVAP) canister purge system – large leak	**EVAP canister purge system, hose leak, EVAP pressure sensor/vent valve, fuel filler cap, ECM**
0801	P1441	Evaporative emission (EVAP) canister purge system – bypass vacuum valve malfunction	**Wiring, EVAP bypass/cut valve, hose leak/blockage, ECM**
0801	P1490	Evaporative emission (EVAP) canister purge system – bypass vacuum valve malfunction	**Wiring, EVAP bypass/cut valve, hose leak/blockage, ECM**
0802	–	Crankshaft position (CKP) sensor 1 – circuit malfunction	**Wiring, CKP sensor, ECM**
0803	–	Manifold absolute pressure (MAP) sensor – circuit malfunction	**Wiring, MAP sensor, ECM**
0804	P1605	TCM diagnosis communication line – malfunction	**Wiring, TCM**
0805	P1110	Camshaft position (CMP) actuator – circuit malfunction	**Wiring, CMP actuator, ECM**
0807	–	Evaporative emission (EVAP) canister purge system – 1996-97 – control valve malfunction	**Wiring, EVAP control valve, EVAP valve/ pressure sensor, hose leak/blockage, ECM**
0807	P1492	Evaporative emission (EVAP) canister purge system – 1998 – control valve malfunction	**Wiring, EVAP control valve, EVAP valve/ pressure sensor, hose leak/blockage, ECM**
0901	–	Heated oxygen sensor (HO2S) – RH front – heater circuit malfunction	**Wiring, HO2S, ECM**
0902	–	Heated oxygen sensor (HO2S) – rear – Maxima – heater circuit malfunction	**Wiring, HO2S, ECM**
0902	–	Heated oxygen sensor (HO2S) – RH rear – 300ZX/Pathfinder – heater circuit malfunction	**Wiring, HO2S, ECM**
0903	–	Evaporative emission (EVAP) canister purge system – vent control valve circuit malfunction	**Wiring, EVAP vent control valve, EVAP breather valve/pressure sensor, hose leak/blockage, ECM**
0904	P1550	Torque converter clutch solenoid – 1996-97 – voltage low	**Wiring, clutch solenoid, TCM, ECM**
0904	P1775	Torque converter clutch solenoid – 1998 – voltage low	**Wiring, clutch solenoid, TCM, ECM**

Flash code	OBD-II code	Fault location	Probable cause
0905	P1336	Crankshaft position (CKP) sensor 1 – rotor teeth damage	Wiring, CKP sensor, ECM
0908	–	Engine coolant temperature – too low for closed loop control	Wiring, ECT sensor, cooling system, ECM
1001	–	Heated oxygen sensor (HO2S) – LH front – heater circuit malfunction	Wiring, HO2S, ECM
1002	–	Heated oxygen sensor (HO2S) – LH rear – heater circuit malfunction	Wiring, HO2S, ECM
1003	–	Park/neutral position (PNP) switch – 1996-97 – circuit malfunction	Wiring, PNP switch, ECM
1003	P1706	Park/neutral position (PNP) switch – 1998 – circuit malfunction	Wiring, PNP switch, ECM
1004	P1130	Intake manifold air control solenoid – malfunction	Wiring, control solenoid, intake system, hose leak/blockage, control vacuum check switch, ECM
1005	P1400	Exhaust gas recirculation (EGR) system/ evaporative emission (EVAP) canister purge system – valve malfunction	Wiring, EGR/EVAP valve, ECM
1008	P1445	Evaporative emission (EVAP) canister purge system – 1996-97 – volume control valve malfunction	Wiring, EVAP valve, ECM
1008	–	Evaporative emission (EVAP) canister purge system – 1998 – control valve malfunction	Wiring, EVAP valve, ECM
1101	–	Park/neutral position (PNP) switch – circuit malfunction	Wiring, PNP switch, ECM
1102	–	AT – vehicle speed sensor (VSS)	Wiring, VSS, TCM
1103	–	AT – first gear selection – malfunction	Valve block
1104	–	AT – second gear selection – malfunction	Valve block
1105	–	AT – third gear selection – malfunction	Valve block
1106	–	AT – fourth gear selection – malfunction	Valve block
1107	–	Torque converter lock-up solenoid – circuit malfunction	Wiring, clutch solenoid, TCM
1108	–	AT – shift solenoid A – voltage low	Wiring, solenoid A, TCM
1201	–	AT – shift solenoid B – voltage low	Wiring, solenoid B, TCM
1203	P1760	AT – overrun clutch solenoid – voltage low	Wiring, clutch solenoid, TCM
1204	–	Torque converter clutch solenoid – voltage low	Wiring, clutch solenoid, TCM
1205	–	AT – Line pressure solenoid – voltage low	Wiring, pressure solenoid, TCM
1206	P1705	AT – throttle position (TP) sensor – voltage high or low	Wiring, TP sensor, TCM, ECM
1207	–	Engine speed (RPM) – incorrect voltage signal ECM to TCM	Wiring, RPM sensor, TCM, ECM
1208	–	Transmission fluid temperature sensor – excessively high or low voltage	Wiring, temperature sensor, TCM, ECM

Flash code	OBD-II code	Fault location	Probable cause
1302	P1105	MAP – BARO switching valve – circuit malfunction	**Wiring, MAP/BARO switching valve, ECM**
1305	P1220	Fuel pump (FP) control module – circuit malfunction	**Wiring, FP control module, ECM**
1306	P1150	Turbocharger (TC) wastegate regulating valve – LH – circuit malfunction	**Wiring, TC wastegate regulating valve, ECM**
1307	P1155	Turbocharger (TC) wastegate regulating valve – RH – circuit malfunction	**Wiring, TC wastegate regulating valve, ECM**
1308	P1900	Engine coolant blower motor – circuit malfunction	**Wiring, coolant blower motor, cooling system, ECM**
–	P1065	Engine control module (ECM) – battery voltage supply	**Wiring, fuse, ECM**
–	P1121	Throttle motor – malfunction	**Throttle motor, ECM**
–	P1122	Throttle motor – circuit malfunction	**Wiring, throttle motor, ECM**
–	P1124	Throttle motor – short circuit	**Wiring, throttle motor, throttle motor relay, ECM**
–	P1126	Throttle motor – open circuit – Maxima 2002	**Wiring, throttle motor, throttle motor relay, ECM**
–	P1126	Thermostat – stuck open – except Maxima 2002	**Wiring, thermostat, ECT sensor**
–	P1128	Throttle motor – short circuit	**Wiring, throttle motor, ECM**
–	P1143	Heated oxygen sensor (HO2S) 1 – voltage low	**Wiring, mixture, HO2S, ECM**
–	P1144	Heated oxygen sensor (HO2S) 1 – voltage high	**Wiring, mixture, HO2S, ECM**
–	P1146	Heated oxygen sensor (HO2S) 2 – voltage low	**Wiring, mixture, HO2S, ECM**
–	P1147	Heated oxygen sensor (HO2S) 2 – voltage low	**Wiring, injectors, fuel pressure, air leak, HO2S, ECM**
–	P1163	Heated oxygen sensor (HO2S)	**Wiring, HO2S, ECM**
–	P1164	Heated oxygen sensor (HO2S) – voltage high	**Wiring, mixture, HO2S, ECM**
–	P1166	Heated oxygen sensor (HO2S) 2 – voltage low	**Wiring, mixture, HO2S, ECM**
–	P1167	Heated oxygen sensor (HO2S) 2 – voltage low	**Wiring, injectors, fuel pressure, air leak, HO2S, ECM**
–	P1211	ABS/TCS control unit – communication malfunction	**Wiring, ABS/TCS control unit, ECM**
–	P1212	ABS/TCS control unit – circuit malfunction	**Wiring, ABS/TCS control unit, ECM**
–	P1217	Engine over temperature	**Cooling system, engine coolant blower motor**
–	P1223	Throttle position (TP) sensor 2 – voltage low	**Wiring, TP sensor**
–	P1224	Throttle position (TP) sensor 2 – voltage high	**Wiring, TP sensor**

Flash code	OBD-II code	Fault location	Probable cause
–	P1225	Throttle motor position sensor – learning problem, voltage low	Wiring, TP sensor 1, TP sensor 2
–	P1226	Throttle motor position sensor – learning problem	Wiring, TP sensor 1, TP sensor 2
–	P1227	Accelerator pedal position (APP) sensor 2 – voltage low	Wiring, APP sensor
–	P1228	Accelerator pedal position (APP) sensor 2 – voltage high	Wiring, APP sensor
–	P1229	Engine control module (ECM) – sensor supply voltage low/high	Wiring, APP sensor, throttle motor position sensor, MAF sensor, fuel tank pressure sensor, TP sensor(s), refrigerant pressure sensor
–	P1243	Supercharger (SC) bypass valve – circuit malfunction	Wiring, SC bypass valve
–	P1442	Evaporative emission (EVAP) canister purge system – small leak	EVAP canister purge system, hose leak, EVAP pressure sensor/vent valve, fuel filler cap, ECM
–	P1456	Evaporative emission (EVAP) canister purge system – very small leak	EVAP canister purge system, hose leak, EVAP pressure sensor/vent valve, fuel filler cap, ECM
–	P1464	Fuel gauge tank sensor – voltage high	Wiring, fuel gauge tank sensor, ECM
–	P1572	Cruise control brake pedal switch – circuit malfunction	Wiring, cruise control brake pedal switch, clutch pedal position (CPP) switch, ECM
–	P1574	Cruise control vehicle speed sensor – signal malfunction	Wiring, cruise control vehicle speed sensor, ECM
–	P1610	Ignition key/engine control module (ECM) – malfunction	Incorrect ignition key, ECM
–	P1611	Immobilizer control module/engine control module (ECM) – coding	Immobilizer control module/ECM incorrectly coded
–	P1612	Immobilizer control module/engine control module (ECM) communication – malfunction	Wiring, immobilizer control module, ECM
–	P1613	Engine control module (ECM), immobilizer function – internal failure	ECM
–	P1614	Immobilizer control module/module coding plug communication – no signal	Wiring, immobilizer control module, module coding plug
–	P1615	Ignition key/immobilizer control module communication – malfunction	Incorrect ignition key, immobilizer control module
–	P1800	Intake manifold air control solenoid – circuit malfunction	Wiring, intake manifold air control solenoid
–	P1805	Brake pedal position (BPP) switch – circuit malfunction	Wiring, BPP switch actuator position sensor, intake – circuit malfunction Wiring, CMP actuator position sensor, ECM

NISSAN

Model:	Engine identification:	Year:
Maxima 3.0L	VG30E	1987-94
Stanza 2.0L	CA20E	1987-89
200SX 1.8L/2.0L	CA18ET/CA20E	1986-88
200SX 3.0L	VG30E	1987-88
240SX 2.4L	KA24E	1989-90
300ZX 3.0L/Turbo	VG30E/VG30ET	1987-89

System: **ECCS**

Self-diagnosis

General information

- Refer to the front of this manual for general test conditions, terminology, detailed descriptions of wiring faults and a general trouble shooter for electrical and mechanical faults.
- Engine control module (ECM) incorporates self-diagnosis function.
- Malfunction indicator lamp (MIL) will illuminate if certain faults are recorded – California 1988 on.
- ECM operates in backup mode if certain sensors fail, to enable vehicle to be driven to workshop.
- Trouble codes are displayed by the red and green LED's in the engine control module (ECM), in diagnostic test mode 3, or using a suitable code reader.
- The red LED flashes indicate 'tens', followed by the green which indicate 'units'.

Accessing

Mode selection

- Access engine control module (ECM):
  - Stanza/Maxima 87-88 – **1**.
  - Maxima 89-94 – **2**.
  - 200SX – **3**.
  - 240SX/300ZX – **4**.
- Switch ignition ON.
- Turn ECM diagnosis mode selector fully clockwise **5** [**1**].
  - Wait until the LED's flash **5** [**2**].
  - After LED's flash required mode number, turn mode selector fully counterclockwise **5** [**1**].
  - For example 3 flashes indicates Mode 3.

NOTE: ***If the ignition is switched OFF during self-diagnosis, it will return to mode 1 when it is turned ON again.***

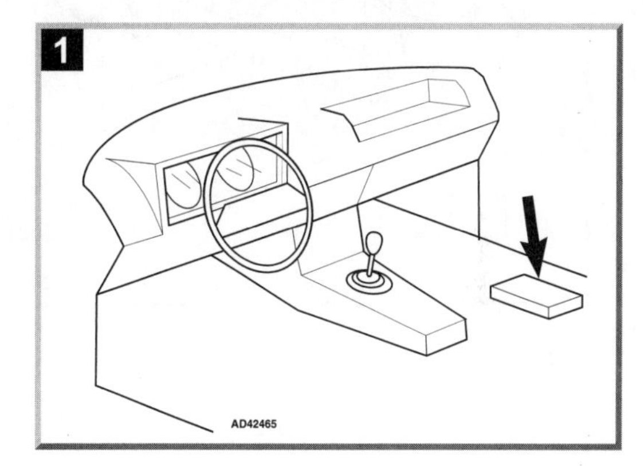

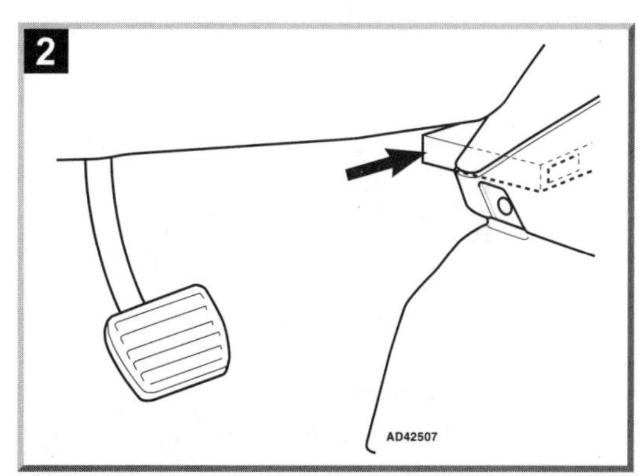

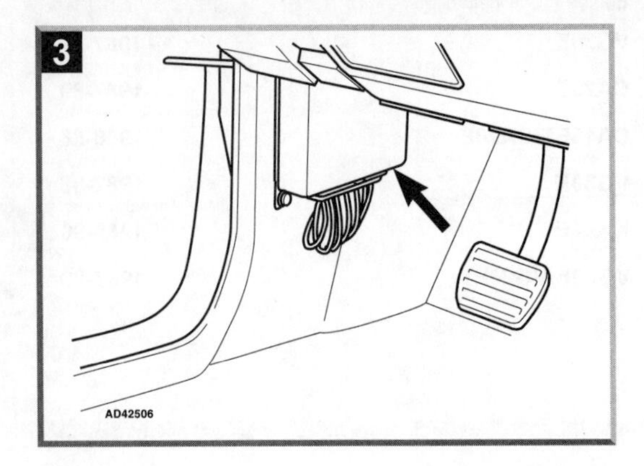

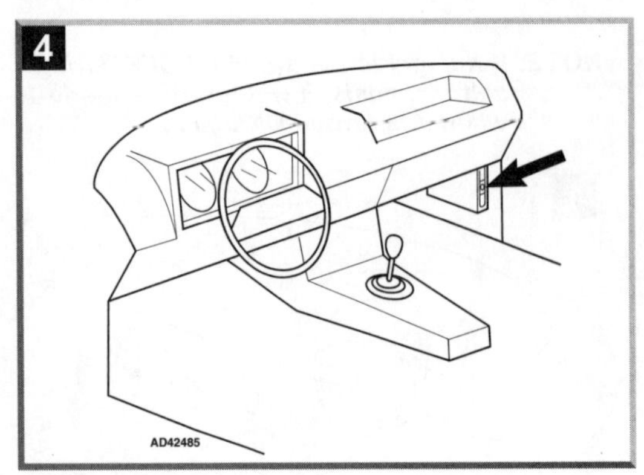

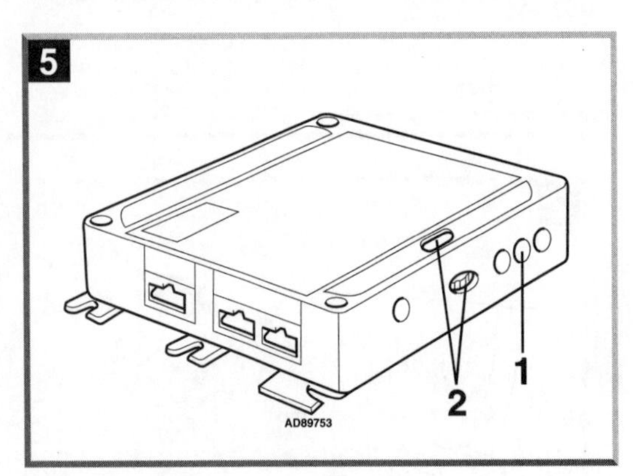

Mode description

- Mode 1: Mixture ratio feedback – heated oxygen sensor (HO2S)/oxygen sensor (O2S) diagnosis.
- Mode 2: Mixture ratio feedback – CO adjustment.
- Mode 3: Code display.
- Mode 4: Tests diagnosis ON/OFF switching.
- Mode 5: Real time diagnosis, codes displayed as they occur.

Mode 3

- Access engine control module (ECM).
- Switch ignition ON. Do NOT start engine.
- Select mode 3.
- Count LED flashes and compare with trouble code table.

Mode 4

- Access engine control module (ECM).
- Switch ignition ON. Do NOT start engine.
- Select mode 4.
- Red LED goes off.
- Crank engine, red LED should come ON.
- Depress accelerator pedal, red LED goes OFF.
- Raise driven wheels of vehicle and drive vehicle, green LED should come ON when speed reaches 12 mph (20 kph).

Mode 5

- Access engine control module (ECM).
- Start engine.
- Select mode 5.
- Test drive vehicle.
- LED's will flash any codes occurring during test.

Erasing

NOTE: ***Trouble codes are automatically erased once the starter is operated 50 times after the last trouble code was stored.***

Method 1

- Select mode 3.
- Read flash codes.
- Select mode 4.
- Switch ignition OFF.

Method 2

- Ensure ignition switched OFF.
- Disconnect battery ground cable.
- Wait at least 15 seconds.
- Reconnect battery ground cable.
- Repeat checking procedure to ensure no data remains in ECM fault memory.

NOTE: ***Disconnecting battery ground lead may also erase memory from electronic units such as radio.***

Trouble code identification

Flash code	Fault location	Probable cause
11	Crankshaft position (CKP) sensor – circuit malfunction	**Wiring, CKP sensor, ECM**
12	Mass air flow (MAF) sensor – range/ performance problem	**Wiring, MAF sensor, ECM**
12	Volume air flow (VAF) sensor – range/ performance problem	**Wiring, VAF sensor, ECM**
13	Engine coolant temperature (ECT) sensor – range/performance problem	**Wiring, ECT sensor, ECM**
14	Vehicle speed sensor (VSS) – malfunction	**Wiring, VSS, ECM**
21	Ignition coil – No primary signal	**Wiring, ignition coil/module, ignition system, ECM**
22	Fuel pump – circuit malfunction	**Wiring, fuel pump/relay, ignition switch, ECM**
23	Closed throttle position (CTP) switch – circuit malfunction	**Wiring, CTP switch, ECM**
24	Wide open throttle (WOT) switch – circuit malfunction	**Wiring, WOT switch, ECM**
31	Engine control module (ECM) – malfunction	**ECM**
32	Exhaust gas recirculation (EGR) system – control valve malfunction	**Wiring, hose leak/blockage, EGR cut OFF solenoid, ECM**
33	Heated oxygen sensor (HO2S) – circuit malfunction	**Wiring, HO2S, intake/fuel system, injector, ECM**
34	Knock sensor (KS) – circuit malfunction	**Wiring, KS sensor, ECM**
35	Exhaust gas recirculation temperature (EGRT) sensor – circuit malfunction	**Wiring, EGRT sensor, ECM**
41	Intake air temperature (IAT) sensor – 1.8/2.0L – circuit malfunction	**Wiring, IAT sensor, ECM**
41	Fuel temperature sensor – 3.0L – circuit malfunction	**Wiring, temperature sensor, ECM**
42	Fuel temperature sensor – circuit malfunction	**Wiring, temperature sensor, ECM**
43	Throttle position (TP) sensor – circuit malfunction	**Wiring, TP sensor, ECM**
44	No faults detected	–
45	Injector leak – malfunction	**Injector, fuel system, ECM**
51	Injector – circuit malfunction	**Wiring, injector, ECM**
54	ECM/TCM – communication malfunction	**Wiring, ECM, TCM**
55	No faults detected	–

NISSAN

Model:	Engine identification:	Year:
Maxima 3.0L	**VE30DE**	**1992-94**
Quest 3.0L	**VG30E**	**1993-95**
300ZX 3.0L	**VG30DE**	**1990-95**
300ZX 3.0L Turbo	**VG30DETT**	**1990-95**

System: **ECCS**

Self-diagnosis

General information

- Refer to the front of this manual for general test conditions, terminology, detailed descriptions of wiring faults and a general trouble shooter for electrical and mechanical faults.
- Engine control module (ECM) incorporates self-diagnosis function.
- Malfunction indicator lamp (MIL) will illuminate if certain faults are recorded.
- ECM operates in backup mode if sensors fail, to enable vehicle to be driven to workshop.
- Trouble codes can be displayed by the red LED 4 [1] in the engine control module (ECM) and the MIL lamp in diagnostic mode 2, or using a suitable code reader.

Accessing – Mode 1

- Access engine control module (ECM):
  - Maxima – center fascia – 1.
  - 300ZX – RH footwell – 2.
  - Quest – behind glove box – 3.
- Switch ignition ON.
- Check MIL and red LED in ECM illuminate.
- Start engine. Allow to idle.
- If MIL and LED extinguish, no trouble codes have been recorded.
- If MIL and LED remain illuminated, access trouble codes.

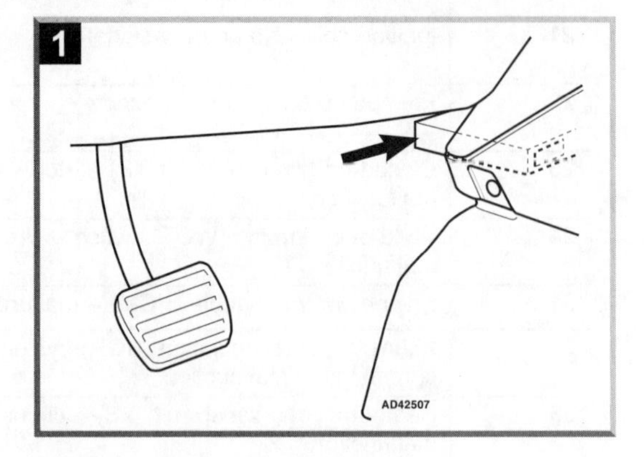

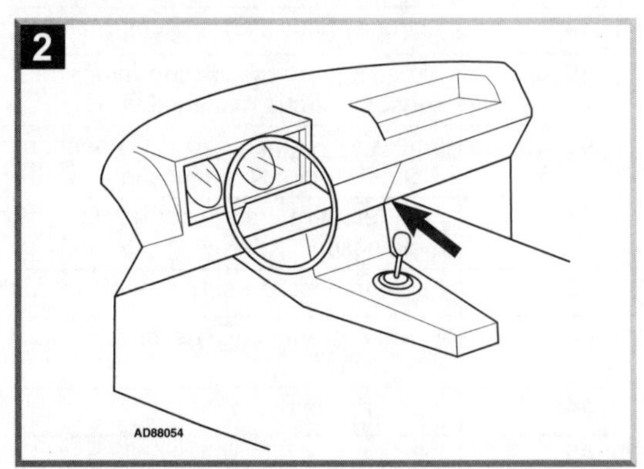

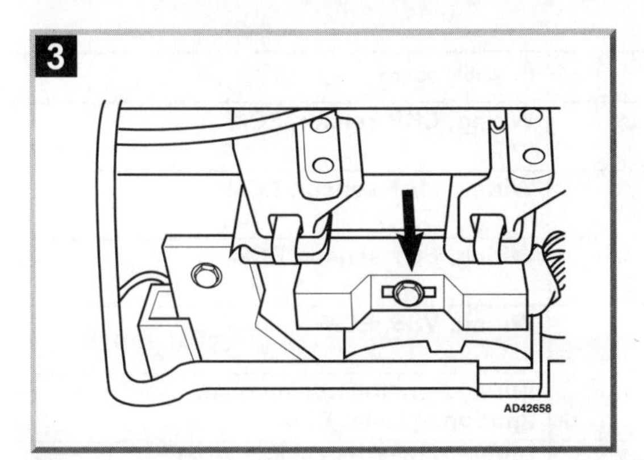

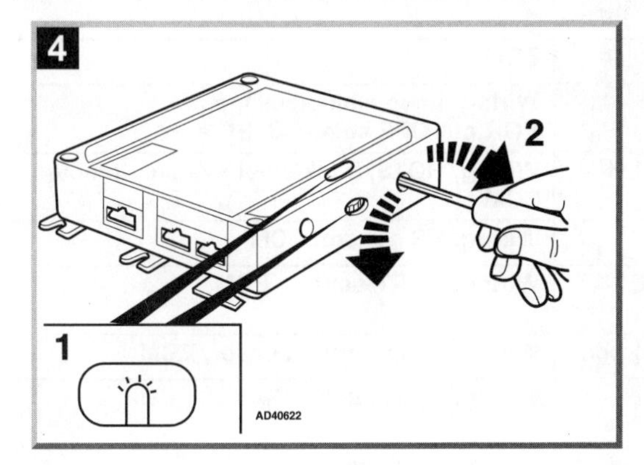

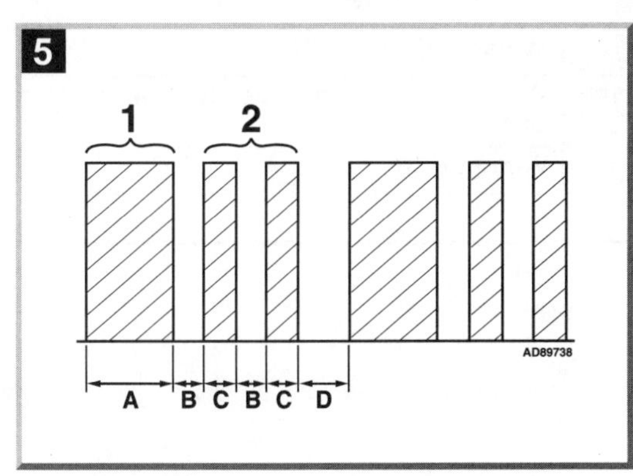

Accessing – Mode 2

- Switch ignition ON. Do NOT start engine.
- Turn diagnostic mode selector on ECM fully clockwise **4** [**2**].
- Wait at least 2 seconds.
- Turn diagnostic mode selector on ECM fully counterclockwise **4** [**2**].
- Self-diagnosis now in Mode 2.
- MIL and red LED in ECM should flash.
- Count flashes and compare with trouble code table.
  - Long flashes indicate the 'tens' of the trouble code **5** [**A**].
  - Short flashes indicate the 'units' of the trouble code **5** [**C**].
  - A short pause separates each flash **5** [**B**].
  - A long pause separates each trouble code **5** [**D**].
  - For example: Trouble code 12 displayed **5**.
- Switch ignition OFF.
- Self-diagnosis will return to Mode 1.

NOTE: ***Starting engine in Self-diagnosis Mode 2 will activate heated oxygen sensor (HO2S) diagnosis.***

Erasing

- Switch ignition ON. Do NOT start engine.
- Turn diagnostic mode selector on ECM fully clockwise **4** [**2**].
- Wait at least 2 seconds.
- Turn diagnostic mode selector on ECM fully counterclockwise **4** [**2**].
- Self-diagnosis now in Mode 2.
- Turn diagnostic mode selector on ECM fully clockwise **4** [**2**].
- Wait at least 2 seconds.
- Turn diagnostic mode selector on ECM fully counterclockwise **4** [**2**].
- Self-diagnosis will return to Mode 1.
- Trouble codes now erased.
- Switch ignition OFF.
- Repeat checking procedure to ensure no data remains in ECM fault memory.

NOTE: ***Disconnecting battery ground lead will also erase trouble codes but may erase memory from electronic units such as radio.***

Trouble code identification

Flash code	Fault location	Probable cause
11	Crankshaft position (CKP) sensor – circuit malfunction	**Wiring, CKP sensor, ECM**
12	Mass air flow (MAF) sensor – range/ performance problem	**Wiring, MAF sensor, ECM**
13	Engine coolant temperature (ECT) sensor – range/performance problem	**Wiring, ECT sensor, ECM**
14	Vehicle speed sensor (VSS) – circuit malfunction	**Wiring, VSS, ECM**
21	Ignition coil – no primary signal	**Wiring, ignition coil/module, ignition system, ECM**
26	Turbocharger (TC) boost pressure sensor – circuit malfunction	**Wiring, pressure sensor, ECM**
31	Engine control module (ECM) – malfunction	**ECM**
32	Exhaust gas recirculation (EGR) system – control valve malfunction	**Wiring, hose leak/blockage, EGR cut OFF solenoid, ECM**
33	Heated oxygen sensor (HO2S) – single/LH – circuit malfunction	**Wiring, HO2S, intake/fuel system, injector, ECM**
34	Knock sensor (KS) – circuit malfunction	**Wiring, KS sensor, ECM**
35	Exhaust gas recirculation temperature (EGRT) sensor – circuit malfunction	**Wiring, EGRT sensor, ECM**
42	Fuel temperature sensor – circuit malfunction	**Wiring, temperature sensor, ECM**
43	Throttle position (TP) sensor – circuit malfunction	**Wiring, TP sensor, ECM**
45	Injector leak – malfunction	**Injector, fuel system, ECM**
51	Injector – circuit malfunction	**Wiring, injector, ECM**
53	Heated oxygen sensor (HO2S) – RH – circuit malfunction	**Wiring, HO2S, intake/fuel system, injector, ECM**
54	ECM/TCM – communication malfunction	**Wiring, ECM, TCM**
55	No faults detected	–

SAAB

Model:	9-3 2.0L Turbo • 9-3 2.3L Turbo • 9-5 2.3L Turbo • 9-5 3.0L
Year:	1999-02
Engine identification:	B204E, B204L, B204R, B205E, B205F, B205L, B205R, B235E, B235R, B308E,
System:	Trionic T7 • Trionic T5

Data link connector (DLC) locations

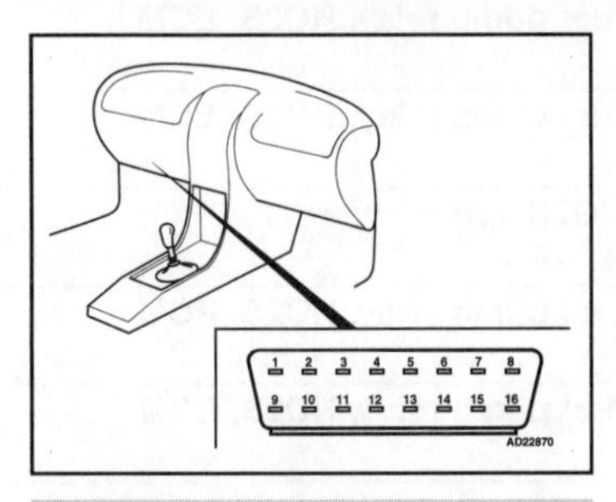

Under steering column

Self-diagnosis

General information

- Refer to the front of this manual for general test conditions, terminology, detailed descriptions of wiring faults and a general trouble shooter for electrical and mechanical faults.

Accessing and erasing

- The engine control module (ECM) fault memory can only be accessed and erased using diagnostic equipment connected to the data link connector (DLC).

Trouble code identification

OBD-II code	Fault location	Probable cause
All P0, P2 and U0 codes	Refer to OBD-II trouble code tables at the front of this manual	–
P1105	Manifold absolute pressure (MAP) sensor, TC system – leak detected	Intake leak, hose(s)
P1106	Manifold absolute pressure (MAP) sensor, TC system – range/performance problem	MAP sensor – TC system
P1107	Manifold absolute pressure (MAP) sensor, TC system – open circuit/short to ground	Wiring open circuit/short to ground, ECM

OBD-II code	Fault location	Probable cause
P1108	Manifold absolute pressure (MAP) sensor, TC system – short to positive	**Wiring short to positive, ECM**
P1110	Turbocharger (TC) bypass valve – range/ performance problem	**Hose(s), TC bypass solenoid, TC bypass valve**
P1131	Heated oxygen sensor (HO2S) 1, bank 1 – malfunction	**Wiring, HO2S, ECM**
P1132	Heated oxygen sensor (HO2S) 1, bank 1 – malfunction	**Wiring, HO2S, ECM**
P1133	Heated oxygen sensor (HO2S) 1, bank 1 – voltage low	**Wiring, HO2S**
P1135	Heated oxygen sensor (HO2S) 1, bank 1 – heater low input	**Wiring, fuel pump relay, HO2S, ECM**
P1136	Heated oxygen sensor (HO2S) 1, bank 1 – heater high input	**Wiring, fuel pump relay, HO2S, ECM**
P1137	Heated oxygen sensor (HO2S) 2, bank 1 – voltage low	**Wiring, HO2S**
P1141	Heated oxygen sensor (HO2S) 2, bank 1 – heater low input	**Wiring, fuel pump relay, HO2S, ECM**
P1142	Heated oxygen sensor (HO2S) 2, bank 1 – heater high input	**Wiring, fuel pump relay, HO2S, ECM**
P1171	Short term fuel trim (FT), max. value – weak mixture	**Fuel pressure, injector(s), intake system, mass air flow (MAF) sensor**
P1172	Short term fuel trim (FT), min. value – rich mixture	**Fuel pressure, injector(s), evaporative emission (EVAP) canister purge valve**
P1181	Long term fuel trim (FT), max. value – weak mixture	**Fuel pressure, injector(s), intake system, mass air flow (MAF) sensor**
P1182	Long term fuel trim (FT), min. value – rich mixture	**Fuel pressure, injector(s), evaporative emission (EVAP) canister purge valve**
P1230	Throttle position (TP) sensor 1 & 2 – range/performance problem	**Wiring, connectors, throttle control unit, ECM**
P1240	Throttle position motor – short circuit	**Wiring, throttle control unit, ECM**
P1251	Throttle position motor – fully closed signal	**Wiring, mechanical fault, throttle control unit, ECM**
P1252	Throttle position motor – fully open signal/ no current	**Wiring, throttle control unit, ECM**
P1253	Throttle position motor – fully open signal during cranking	**Wiring, mechanical fault, throttle control unit, ECM**
P1260	Throttle return spring – weak spring force	**Mechanical fault, throttle control unit**
P1261	Throttle valve – tight/sticking	**Mechanical fault, throttle control unit, ECM**
P1263	Throttle valve – moved manually while engine running	**Throttle control unit, mechanical fault**
P1264	Throttle valve open – accelerator depressed	**Throttle control unit, mechanical fault**

OBD-II code	Fault location	Probable cause
P1300	Random/multiple cylinder(s) – misfire detected	**Mechanical fault, wiring, ignition/fuel system, mass air flow (MAF) sensor, fuel pump relay, manifold absolute pressure (MAP) sensor, intake leak**
P1301	Cylinder 1 – misfire detected	**Mechanical fault, ignition/fuel system, injector**
P1302	Cylinder 2 – misfire detected	**Mechanical fault, ignition/fuel system, injector**
P1303	Cylinder 3 – misfire detected	**Mechanical fault, ignition/fuel system, injector**
P1304	Cylinder 4 – misfire detected	**Mechanical fault, ignition/fuel system, injector**
P1310	Ignition coils, bank 1 – no supply voltage	**Wiring, ignition coil assembly, ECM**
P1312	Ignition coil – knock detection, cylinder 1 & 2	**Wiring open circuit/short to positive, ignition coil assembly, ECM**
P1320	Ignition coils, bank 2 – no supply voltage	**Wiring, ignition coil assembly, ECM**
P1334	Ignition coil – knock detection, cylinder 3 & 4	**Wiring open circuit/short to positive, ignition coil assembly, ECM**
P1390	Misfire detected, random – fuel level low	**Wiring, connectors, injector(s), fuel pump relay, fuel pressure, mass air flow (MAF) sensor, intake leak, fuel level**
P1391	Cylinder 1, misfire detected – fuel level low	**Wiring, connectors, injector, fuel level**
P1392	Cylinder 2, misfire detected – fuel level low	**Wiring, connectors, injector, fuel level**
P1393	Cylinder 3, misfire detected – fuel level low	**Wiring, connectors, injector, fuel level**
P1394	Cylinder 4, misfire detected – fuel level low	**Wiring, connectors, injector, fuel level**
P1395	Cylinder 5, misfire detected – fuel level low	**Wiring, connectors, injector, fuel level**
P1396	Cylinder 6, misfire detected – fuel level low	**Wiring, connectors, injector, fuel level**
P1413	Secondary air injection (AIR) pump relay – open circuit/short to ground	**Wiring open circuit/short to ground, AIR pump relay**
P1414	Secondary air injection (AIR) pump relay – short to positive	**Wiring short to positive, AIR pump relay**
P1416	Fuel system/misfire detected – fuel level low	**Ignition/fuel system, injector(s), fuel level**
P1441	Evaporative emission (EVAP) canister purge valve/fuel tank level sensor – malfunction	**Wiring, connectors, EVAP canister purge valve, fuel tank level sensor, ECM**
P1442	Evaporative emission (EVAP) system/fuel tank level sensor – small leak detected/malfunction	**Hose connection(s), intake leak, EVAP canister, EVAP canister purge valve, wiring, fuel tank level sensor**
P1444	Evaporative emission (EVAP) system – open circuit/short to ground	**Wiring open circuit/short to ground, connectors, EVAP canister purge valve**
P1445	Evaporative emission (EVAP) system – open circuit/short to positive	**Wiring open circuit/short to positive, connectors, EVAP canister purge valve**
P1451	Evaporative emission (EVAP) pressure sensor, in fuel tank – range/performance problem	**Wiring, connectors, EVAP pressure sensor**

OBD-II code	Fault location	Probable cause
P1452	Evaporative emission (EVAP) pressure sensor, in fuel tank – signal too low when depressurised	**Wiring, EVAP canister purge valve, EVAP system leak, EVAP pressure sensor**
P1453	Evaporative emission (EVAP) pressure sensor, in fuel tank – signal too high when depressurised	**Wiring, EVAP canister purge valve, hose(s), EVAP pressure sensor**
P1455	Evaporative emission (EVAP) pressure sensor/ fuel tank level sensor, in fuel tank – large leak detected/malfunction	**Hose connection(s), wiring, EVAP pressure sensor, fuel tank level sensor**
P1460	Immobilizer activated – incorrectly programmed/not programmed	**Trouble code(s) stored in other system(s), ECM/instrumentation control module/ ignition key not matched to immobilizer control module, immobilizer control module**
P1491	Evaporative emission (EVAP) pressure sensor/ fuel tank level sensor, in fuel tank – range/performance problem/malfunction	**Wiring, connector(s), EVAP pressure sensor, fuel tank level sensor**
P1492	Evaporative emission (EVAP) pressure sensor/ fuel tank level sensor, in fuel tank – signal too low when depressurised/malfunction	**Wiring, connector(s), EVAP pressure sensor, fuel tank level sensor**
P1493	Evaporative emission (EVAP) pressure sensor/ fuel tank level sensor, in fuel tank – signal too high when depressurised/malfunction	**Wiring, connector(s), EVAP pressure sensor, fuel tank level sensor**
P1500	Battery voltage – too low/high	**Wiring, alternator, battery**
P1530	Accelerator pedal position (APP) sensor 1 & 2 – range/performance problem	**Wiring, throttle control unit, ECM**
P1531	Accelerator pedal position (APP) sensor 1 & 2 – adapted value out of range	**Wiring, throttle control unit, ECM**
P1532	Accelerator pedal position (APP) sensor 1 & 2 – short circuit	**Wiring, throttle control unit, ECM**
P1549	Turbocharger (TC) system – malfunction	**Mechanical fault, hose(s), TC wastegate regulating valve**
P1576	Brake pedal position (BPP) switch – open circuit/short to positive	**Wiring open circuit/short to positive, BPP switch**
P1577	Brake pedal position (BPP) switch – short to ground	**Wiring short to ground, BPP switch**
P1601	Engine control module (ECM) – defective	**ECM**
P1602	Engine control module (ECM) – defective	**ECM**
P1603	Engine control module (ECM) – defective	**ECM**
P1604	Engine control module (ECM) – defective	**ECM**
P1605	Engine control module (ECM) – defective	**ECM**
P1606	Engine control module (ECM) – defective	**ECM**
P1607	Engine control module (ECM) – defective	**ECM**
P1608	Engine control module (ECM) – defective	**ECM**
P1609	Engine control module (ECM) – defective	**ECM**

OBD-II code	Fault location	Probable cause
P1610	Engine control module (ECM) – defective	**ECM**
P1611	Engine control module (ECM) – defective	**ECM**
P1611 1	Malfunction indicator lamp (MIL) – short to ground	**Wiring short to ground, TCM trouble code(s) stored**
P1613	Engine control module (ECM) – defective	**ECM**
P1614	Engine control module (ECM) – defective	**ECM**
P1621	Engine control module (ECM) – defective	**ECM**
P1623	CAN data bus, TCM	**Wiring**
P1624	CAN data bus, TCM	**TCM trouble code(s) stored**
P1625	CAN data bus, TCS/ABS	**Wiring**
P1631	Engine control module (ECM) – defective	**ECM**
P1632	Engine control module (ECM) – defective	**ECM**
P1633	Engine control module (ECM) – defective	**ECM**
P1640	Engine control relay – no voltage to ECM pin 1	**Wiring, connectors, engine control relay**
P1640 1	Immobilizer code – not programmed	**Code must be programmed into multifunction control module 2 after replacing ECM, instrument panel, multifunction control module 2**
P1641	Fuel pump relay – no voltage to fuel pump/ oxygen sensor heater	**Wiring, connectors, fuel pump relay**
P1641 1	Immobilizer code – incorrect	**Multifunction control module 2 incorrectly coded**
P1652	Engine control relay – open circuit/short to ground	**Wiring open circuit/short to ground, connectors, engine control relay, ECM**
P1653	Engine control relay – short to positive	**Wiring short to positive, connectors, engine control relay, ECM**
P1654	Fuel pump relay – open circuit/short to ground	**Wiring open circuit/short to ground, connectors, fuel pump relay**
P1655	Fuel pump relay – short to positive	**Wiring short to positive, connectors, fuel pump relay**
P1656	AC relay – open circuit/short to ground	**Wiring open circuit/short to ground, connectors, AC relay**
P1657	AC relay – short to positive	**Wiring short to positive, connectors, AC relay**
P1658	Turbocharger (TC) bypass valve – open circuit/ short to ground	**Wiring open circuit/short to ground, connectors, TC bypass valve**
P1659	Turbocharger (TC) bypass valve – short to positive	**Wiring short to positive, connectors, TC bypass valve**
P1662	Turbocharger (TC) wastegate regulating valve – open circuit/short to ground	**Wiring open circuit/short to ground, connectors, TC wastegate regulating valve**
P1663	Turbocharger (TC) wastegate regulating valve – short to positive	**Wiring short to positive, connectors, TC wastegate regulating valve**

OBD-II code	Fault location	Probable cause
P1670	Throttle actuator limp home relay – open circuit/short to ground	**Wiring open circuit/short to ground, connectors, throttle actuator limp home relay**
P1671	Throttle actuator limp home relay – short to positive	**Wiring short to positive, connectors, throttle actuator limp home relay**
P1676	Injector circuits – open circuit/short to ground/positive	**Wiring, fuse, injector(s), ECM**
P1901 [2]	CAN data bus, multifunction control module 1	**Wiring**
P1902 [3]	CAN data bus, multifunction control module 2	**Wiring**
P1908	CAN data bus, instrument panel	**Wiring**
P1921	CAN data bus, ECM	**Wiring**
P1923	CAN data bus, TCM	**Wiring**
P1925	CAN data bus, TCS/ABS	**Wiring**

[1] Vehicle fitted with TRIONIC T5

[2] Multifunction control module 1 controls: headlamps, interior lamps, instrument illumination, intermittent wiper, door mirror heater, heated rear window

[3] Multifunction control module 2 controls: central locking, alarm system, immobilizer, rear seat heater, seat belt warning, electrically adjustable passenger's seat

Model:	**900 2.0L Turbo • 9000 2.3L/Turbo**
Year:	**1992-98**
Engine identification:	**B204E, B204I, B204L, B204S, B234, B234E, B234I, B234L, B234R**
System:	**Trionic**

Data link connector (DLC) locations

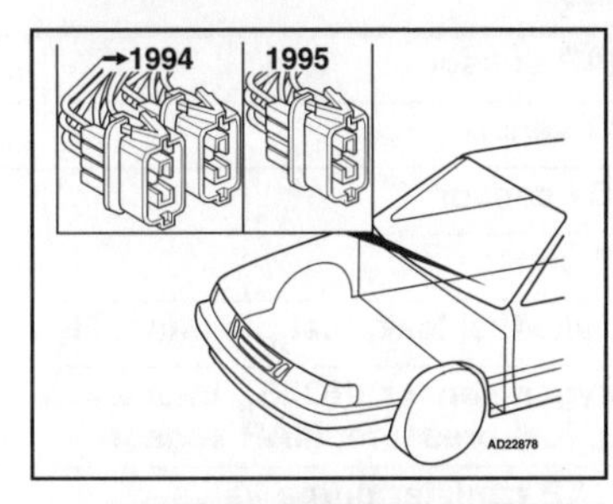

→1994 black DLC/9000 →1995 – under RH front seat

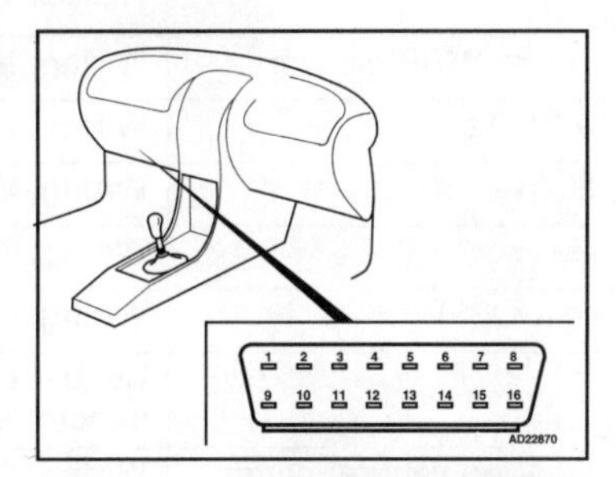

16-pin – under steering column

Self-diagnosis

General information

- Refer to the front of this manual for general test conditions, terminology, detailed descriptions of wiring faults and a general trouble shooter for electrical and mechanical faults.
- Trouble codes are displayed by the malfunction indicator lamp (MIL).

Accessing – flash codes

- Ensure ignition switched OFF.
- Switch ignition ON – DO NOT start engine.
- MIL will illuminate for 3 seconds **1** [**A**].
- MIL will extinguish for 3 seconds **1** [**B**].
- MIL will illuminate for 3 seconds **1** [**A**]
- MIL will extinguish for 2 seconds **1** [**C**].
- The last MIL ON/OFF indicates start of trouble code display.
- Trouble codes are displayed as a single digit number with a 3 second interval between each separate code.
- Count MIL flashes. Note trouble codes. Compare with trouble code table.
- For example: Trouble code 2 displayed – manifold absolute pressure (MAP) sensor **1**.
- After the last trouble code has been displayed, MIL extinguishes for 3 seconds **1** [**B**].
- MIL illuminates for 3 seconds **1** [**A**].
- MIL will now repeat all recorded trouble codes.
- Switch ignition OFF.

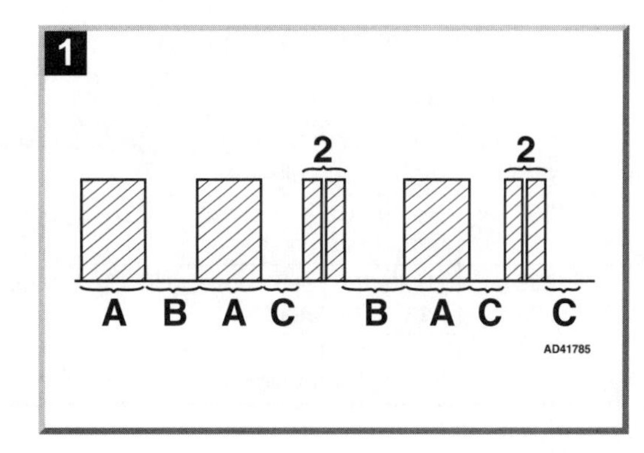

Erasing – flash codes

- Ensure ignition switched OFF.
- Disconnect ECM multi-plug or battery ground lead for at least 5 minutes.

NOTE: ***ECM adaptive memory may be erased causing erratic running faults. Carry out road test to allow ECM to re-learn basic values.***

- Diagnostic equipment can also be used to erase data from ECM fault memory.

WARNING: ***Disconnecting the battery may erase memory from electronic units (e.g. radio, clock).***

Trouble code identification

Flash code	Fault location	Probable cause
2	Manifold absolute pressure (MAP) sensor	**Wiring, MAP sensor**
3	Intake air temperature (IAT) sensor	**Wiring, IAT sensor**
4	Engine coolant temperature (ECT) sensor	**Wiring, ECT sensor**
5	Throttle position (TP) sensor	**Wiring, TP sensor**
6	Heated oxygen sensor (HO2S)	**Wiring, intake/fuel leak, fuel pressure, HO2S**
7	Mixture rich/lean	**Heated oxygen sensor (HO2S), intake leak, injector(s), fuel pressure, MAP sensor**
8	Evaporative emission (EVAP) canister purge valve	**Wiring, EVAP canister purge valve**
9	Engine control module (ECM) – internal fault	**ECM**

Accessing and erasing – OBD-II codes

- The engine control module (ECM) fault memory can only be accessed and erased using diagnostic equipment connected to the data link connector (DLC).

Trouble code identification

OBD-II code	Fault location	Probable cause
All P0, P2 and U0 codes	Refer to OBD-II trouble code tables at the front of this manual	–
P1130	Heated oxygen sensor (HO2S) 1, bank 1 – heater high input	**Wiring, fuses, HO2S**
P1135	Heated oxygen sensor (HO2S) 1, bank 1 – heater low input	**Wiring, fuses, HO2S**
P1322	Electronic throttle system (ETS) – incorrect rpm signal	**Wiring, engine control module (ECM)/ETS control module communication**

OBD-II code	Fault location	Probable cause
P1443	Evaporative emission (EVAP) canister purge valve – malfunction	**Wiring, connectors, EVAP canister purge valve**
P1444	Evaporative emission (EVAP) canister purge valve – high input	**Wiring, connectors, EVAP canister purge valve**
P1445	Evaporative emission (EVAP) canister purge valve – low input	**Wiring, connectors, EVAP canister purge valve**
P1500	Battery voltage – too low/high	**Wiring, alternator, battery**
P1651	Engine control module (ECM) – RAM error	**ECM**
P1652	Engine control module (ECM) – ROM error	**ECM**

SAAB

Model:	**9000 2.3L/Turbo**
Year:	**1992-94**
Engine identification:	**B202I, B202L, B234, B202S**
System:	**Bosch LH-Jetronic 2.4/2.4.2 (35-pin ECM)**

Data link connector (DLC) locations

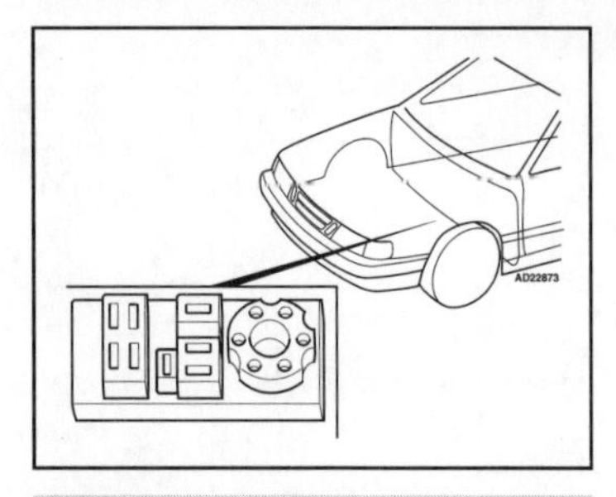

9000

Self-diagnosis

General information

- Refer to the front of this manual for general test conditions, terminology, detailed descriptions of wiring faults and a general trouble shooter for electrical and mechanical faults.
- Trouble codes are displayed by the malfunction indicator lamp (MIL).
- The ECM fault memory can also be checked using diagnostic equipment connected to the data link connector (DLC).

Accessing

- Ensure ignition switched OFF.
- Bridge data link connector (DLC) terminal 1 (grey/red wire) and ground with a switched lead – contacts normally open **1**.
- Switch ignition ON.
- Check that MIL illuminates.
- Set switch to ON **1**.
- Check MIL extinguishes.
- Wait 2.5 seconds approx.
- Check MIL flashes and set switch to OFF immediately **1**.
- MIL will flash code 12112 continuously **2**.

NOTE: ***MIL will display long flash at start and finish of each trouble code.***

NOTE: ***If engine switched OFF during test, MIL will flash code 12231 continuously*** **3**. ***Crank engine for 5 seconds. If ignition signal satisfactory, flash code 12231 will be extinguished when ignition key released.***

- Set switch to ON **1**.
- The first trouble code will then be displayed.
- When MIL displays short flash, set switch to OFF immediately.
- The second trouble code will then be displayed.
- Count MIL flashes. Note trouble codes. Compare with trouble code table.
- Repeat operation. Note trouble codes.
- A series of long flashes indicates no further faults.

- If necessary, repeat test procedure as follows:
  - Set switch to ON **1**.
  - After two short flashes, set switch to OFF immediately **1**.
  - Note trouble codes. Compare with trouble code table.

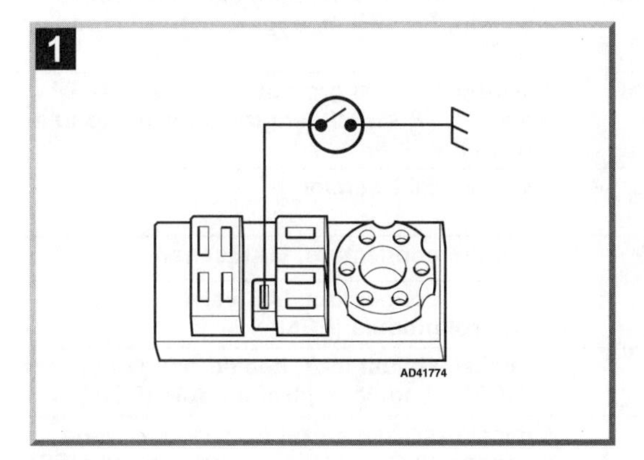

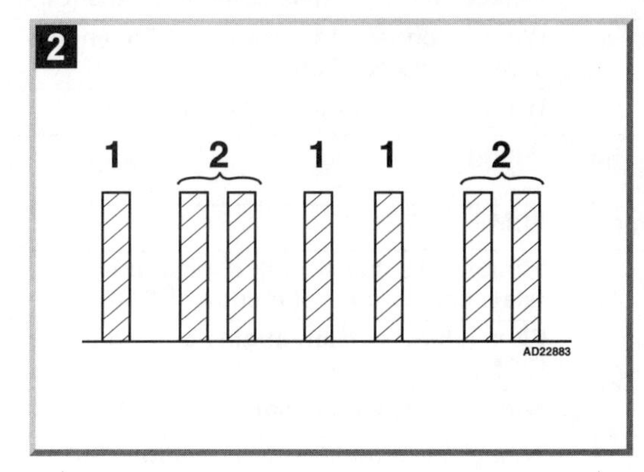

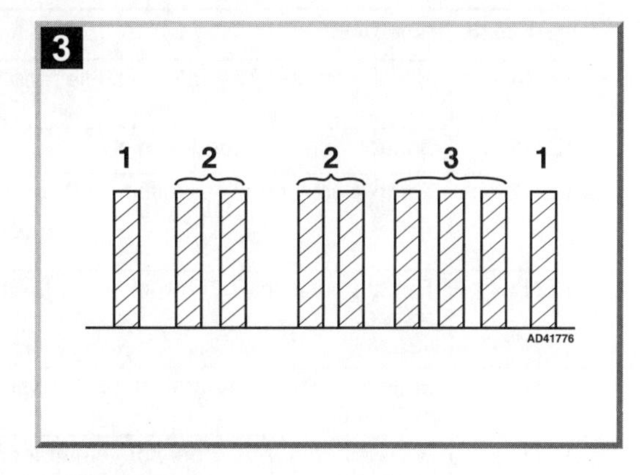

Erasing

NOTE: ***Trouble codes cannot be erased until MIL has displayed trouble code 00000 indicating end of trouble codes.***

- Ensure ignition switched OFF.
- Bridge data link connector (DLC) terminal 1 (grey/red wire) and ground with a switched lead – contacts normally open **1**.
- Switch ignition ON.
- Set switch to ON **1**.
- After MIL displays three short flashes, set switch to OFF immediately.
- Trouble codes now erased.

Trouble code identification

Flash code	Fault location	Probable cause
00000	End of trouble codes/no fault found	–
12111	Mixture adaption – during idling	**Intake leak, heated oxygen sensor (HO2S), injector(s), fuel pressure, engine control module (ECM)**
12112	Mixture adaption – throttle open	**Intake leak, heated oxygen sensor (HO2S), injector(s), fuel pressure, engine control module (ECM)**
12113	Idle air control (IAC) valve – pulse ratio too low	**Check/adjust throttle valve, intake leak, IAC valve, engine control module (ECM)**

Flash code	Fault location	Probable cause
12114	Idle air control (IAC) valve – pulse ratio too high	**IAC valve tight/sticking, mechanical fault, engine control module (ECM)**
12211	Battery voltage – too low/high	**Wiring, alternator, battery**
12212	Throttle position (TP) switch – idling contacts	**Wiring short to ground, check/adjust TP switch, TP switch, engine control module (ECM)**
12213	Throttle position (TP) switch – full throttle contacts	**Wiring short to ground, check/adjust TP switch, TP switch, engine control module (ECM)**
12214	Engine coolant temperature (ECT) sensor – malfunction	**Wiring, ECT sensor**
12221	Mass air flow (MAF) sensor – malfunction	**Wiring, connectors, MAF sensor**
12222	Idle air control (IAC) valve – malfunction	**Wiring, connectors, IAC valve, engine control module (ECM)**
12223	System too lean	**Intake/exhaust leak, heated oxygen sensor (HO2S), engine control module (ECM)**
12224	System too rich	**Injector(s), fuel pressure, heated oxygen sensor (HO2S), engine control module (ECM)**
12225	Heated oxygen sensor (HO2S) – malfunction	**Wiring, connectors, fuse(s), HO2S, engine control module (ECM)**
12231	Ignition signal – no signal	**Wiring, connectors, ignition system**
12232	Engine control module (ECM) – supply voltage, memory	**Wiring**
12233	Engine control module (ECM) – ROM error	**ECM**
12241	Injector(s) – malfunction	**Wiring, fuse, injector(s), engine control relay, engine control module (ECM)**
12242	Mass air flow (MAF) sensor – filament burn-off	**Wiring, MAF sensor, engine control module (ECM)**
12243	Vehicle speed sensor (VSS) – no signal	**Wiring, VSS, engine control module (ECM)**
12244	No 'DRIVE' signal (automatic transmission)	**Wiring, fuse, park/neutral position (PNP) switch**
12245	Exhaust gas recirculation (EGR) system – malfunction	**Wiring, EGR solenoid**
12251 1	Throttle position (TP) sensor – malfunction	**Wiring, TP sensor**
12252	Evaporative emission (EVAP) canister purge valve – malfunction	**Wiring, EVAP canister purge valve**
12253	Pre-ignition – signal more than 20 seconds	**Wiring, ignition control module (ICM)/engine control module (ECM) communication, ignition control module (ICM), engine control module (ECM)**
12254	Crankshaft position (CKP) sensor/engine speed (RPM) sensor – no signal	**Wiring, crankshaft position (CKP) sensor/ engine speed (RPM) sensor**

1 Vehicle fitted with LH-Jetronic 2.4.2

Model:	900 2.3L/2.5L
Year:	1993-98
Engine identification:	B204I, B206I, B234I, B258I
System:	Bosch Motronic 2.8.1 • Bosch Motronic 2.10.2/3

Data link connector (DLC) locations

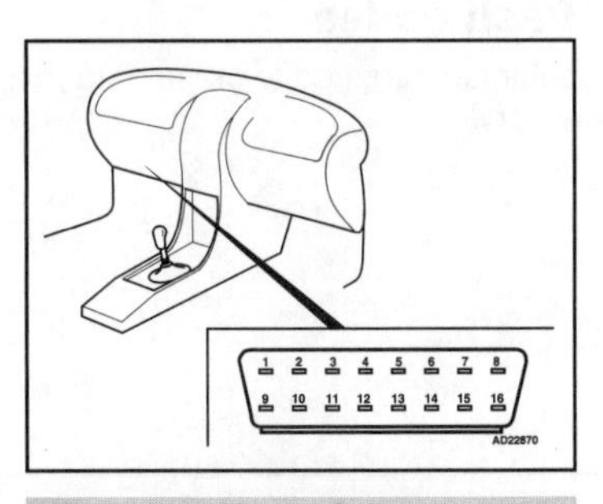

Under steering column

Self-diagnosis

General information

- Refer to the front of this manual for general test conditions, terminology, detailed descriptions of wiring faults and a general trouble shooter for electrical and mechanical faults.
- Trouble codes are displayed by the malfunction indicator lamp (MIL).
- More detailed fault diagnosis information can be obtained using diagnostic equipment connected to the data link connector (DLC).

Accessing – flash codes

- Ensure ignition switched OFF.
- Bridge data link connector (DLC) terminal 6 and ground with a switched lead – contacts normally open **1**.
- Switch ignition ON.
- Set switch to ON for 4 seconds max. **1**.
- Trouble codes are displayed as a two digit number with a long flash between each separate code.
- For example: Trouble code 12 displayed – no fault found **2**.
- Count MIL flashes. Note trouble codes. Compare with trouble code table.

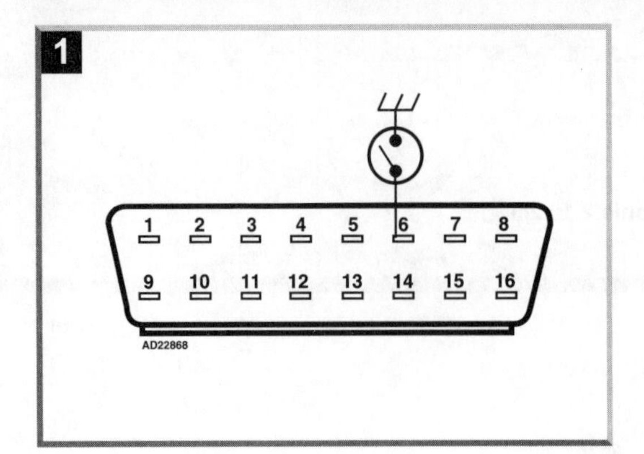

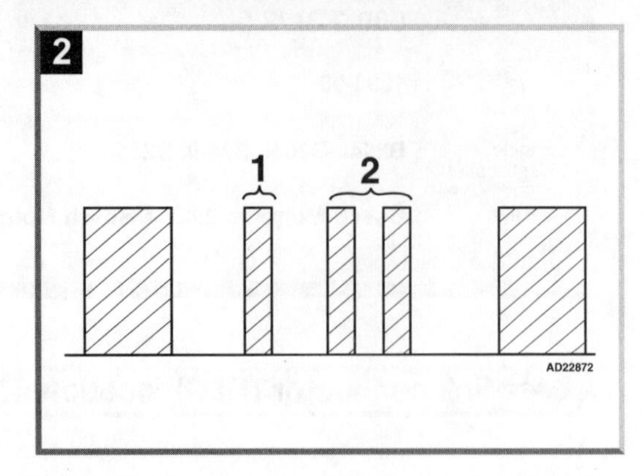

Erasing – flash codes

- Diagnostic equipment required to erase data from ECM fault memory.

Trouble code identification

Flash code	Fault location	Probable cause
11	Secondary air injection (AIR) pump relay, short to ground/positive	**Wiring, relay**
12	No fault found	–
21	Mass air flow (MAF) sensor, short to ground/ positive	**Wiring, MAF sensor**
31	Intake air temperature (IAT) sensor	**Wiring, IAT sensor**
41	Engine coolant temperature (ECT) sensor	**Wiring, ECT sensor**
51	Throttle position (TP) sensor	**Wiring, TP sensor**
61	Heated oxygen sensor (HO2S)	**Wiring, HO2S, intake/fuel leak, fuel pressure**
61	Heated oxygen sensor (HO2S) 2	**Wiring, HO2S, intake/fuel leak, fuel pressure**
62	Heated oxygen sensor (HO2S) 1	**Wiring, HO2S, intake/fuel leak, fuel pressure**
71	Mixture rich/lean, cylinder 1-3-5	**Heated oxygen sensor (HO2S), intake leak, injector(s), fuel pressure**
72	Mixture rich/lean, cylinder 2-4-6	**Heated oxygen sensor (HO2S), intake leak, injector(s), fuel pressure**
72 1	Mixture rich/lean – long time	**Wiring, heated oxygen sensor (HO2S), intake leak, exhaust leak, injector(s), fuel pressure**
73	Mixture rich/lean – short time	**Wiring, heated oxygen sensor (HO2S), intake leak, exhaust leak, injector(s), fuel pressure**

Flash code	Fault location	Probable cause
81	Evaporative emission (EVAP) canister purge valve	**Wiring, EVAP canister purge valve**
91	Engine control module (ECM) – RAM error	**ECM**
92	Engine control module (ECM) – ROM error	**ECM**

1 Motronic 2.10.3

Accessing and erasing – OBD-II codes

- The engine control module (ECM) fault memory can only be accessed and erased using diagnostic equipment connected to the data link connector (DLC).

Trouble code identification

OBD-II code	Fault location	Probable cause
All P0, P2 and U0 codes	Refer to OBD-II trouble code tables at the front of this manual	–
P1001	Evaporative emission (EVAP) canister purge valve – low output	**Wiring open circuit/short to ground, EVAP canister purge valve**
P1002	Evaporative emission (EVAP) canister purge valve – high output	**Wiring short to positive, EVAP canister purge valve**
P1011	Injector 1 – low output	**Wiring open circuit/short to ground, injector**
P1012	Injector 1 – high output	**Wiring short to positive, injector**
P1021	Injector 2 – low output	**Wiring open circuit/short to ground, injector**
P1022	Injector 2 – high output	**Wiring short to positive, injector**
P1031	Injector 3 – low output	**Wiring open circuit/short to ground, injector**
P1032	Injector 3 – high output	**Wiring short to positive, injector**
P1041	Injector 4 – low output	**Wiring open circuit/short to ground, injector**
P1042	Injector 4 – high output	**Wiring short to positive, injector**
P1051	Injector 5 – low output	**Wiring open circuit/short to ground, injector**
P1052	Injector 5 – high output	**Wiring short to positive, injector**
P1061	Injector 6 – low output	**Wiring open circuit/short to ground, injector**
P1062	Injector 6 – high output	**Wiring short to positive, injector**
P1171 **1**	Intake manifold air control solenoid, outer flap – short to ground	**Wiring, intake manifold air control solenoid**
P1172	System too lean, during idling	**Intake leak, battery**

OBD-II code	Fault location	Probable cause
P1172 [1]	Intake manifold air control solenoid, outer flap – short to positive	**Wiring, intake manifold air control solenoid**
P1173	System too rich, during idling	**Fuel pressure, battery, exhaust leak, injector(s)**
P1206	Secondary air injection (AIR) pump relay – low output	**Wiring open circuit/short to ground, secondary air injection (AIR) pump, relay**
P1207	Secondary air injection (AIR) pump relay – high output	**Wiring short to positive, secondary air injection (AIR) pump, relay**
P1211	Idle air control (IAC) valve – low output	**Wiring open circuit/short to ground, IAC valve**
P1212	Idle air control (IAC) valve – high output	**Wiring short to positive, IAC valve**
P1251	Malfunction indicator lamp (MIL) – low output	**Wiring short to ground**
P1252	Malfunction indicator lamp (MIL) – high output	**Wiring short to positive**
P1300	Engine control module (ECM) – AT torque reduction signal	**Wiring, transmission control module (TCM)**
P1301	Engine control module (ECM) – AT torque reduction signal	**Wiring, transmission control module (TCM)**
P1455	AC compressor clutch relay – low output	**Wiring open circuit/short to ground, relay**
P1456	AC compressor clutch relay – high output	**Wiring short to positive, relay**
P1500	Battery voltage – too low/high	**Wiring, alternator, battery**
P1560	Traction control module/ECM communication – signal low	**Wiring open circuit/short to ground, throttle position (TP) sensor**
P1561	Traction control module/ECM communication – signal high	**Wiring short to positive, throttle position (TP) sensor**
P1601	Fuel pump relay – high output	**Wiring short to positive, relay**
P1602	Fuel pump relay – low output	**Wiring open circuit/short to ground, relay**
P1621	AC compressor clutch relay – high output	**Wiring short to positive, AC compressor clutch relay**
P1622	AC compressor clutch relay – low output	**Wiring open circuit/short to ground, AC compressor clutch relay**
P1628	Battery voltage – too low/high	**Wiring, alternator, battery**
P1628 [1]	Engine control module (ECM) – supply voltage missing	**Wiring, connector, fuse**
P1630	AC compressor clutch relay – high output	**Wiring short to positive, relay**
P1631	AC compressor clutch relay – low output	**Wiring open circuit/short to ground, relay**

[1] Motronic 2.10.3

Model:	**9000 3.0L V6**
Year:	**1995-97**
Engine identification:	**B308I**
System:	**Motronic 2.8.1**

Data link connector (DLC) locations

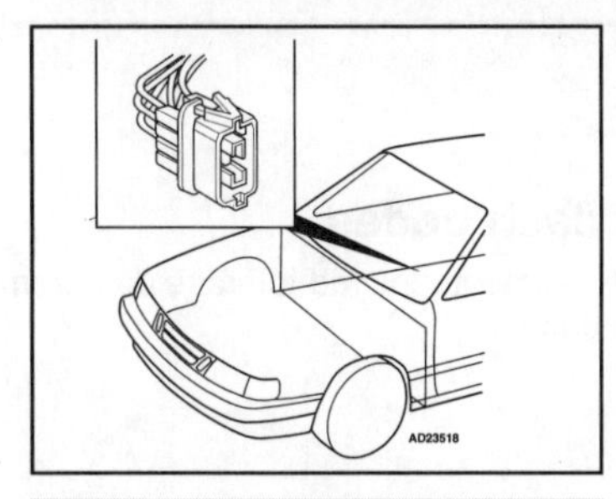

1995 – under RH front seat

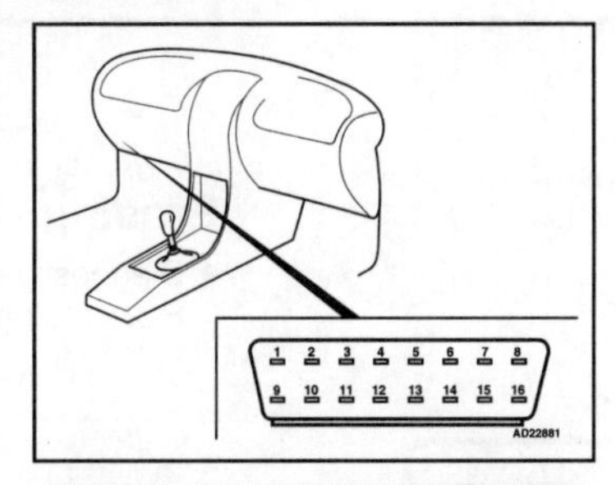

1996→ – under fascia, driver's side

Self-diagnosis

General information

- Refer to the front of this manual for general test conditions, terminology, detailed descriptions of wiring faults and a general trouble shooter for electrical and mechanical faults.
- Trouble codes are displayed by the malfunction indicator lamp (MIL).
- The ECM fault memory can also be checked using diagnostic equipment connected to the data link connector (DLC).

Accessing – flash codes

- Ensure ignition switched OFF.
- Bridge data link connector (DLC) terminal 2 (1995) [1] or terminal 6 (1996→) [2] and ground with a switched lead – contacts normally open.
- Switch ignition ON.
- Set switch to ON [1] or [2].
- Check that MIL flashes.
- Trouble codes are displayed as a two digit number with a 3 second interval between each separate code.
- Count MIL flashes. Note trouble codes. Compare with trouble code table.
- For example: Trouble code displayed: 31 – intake air temperature (IAT) sensor [3].
- Trouble codes will continue to be displayed if data link connector (DLC) terminal 2 or terminal 6 switched to ground.

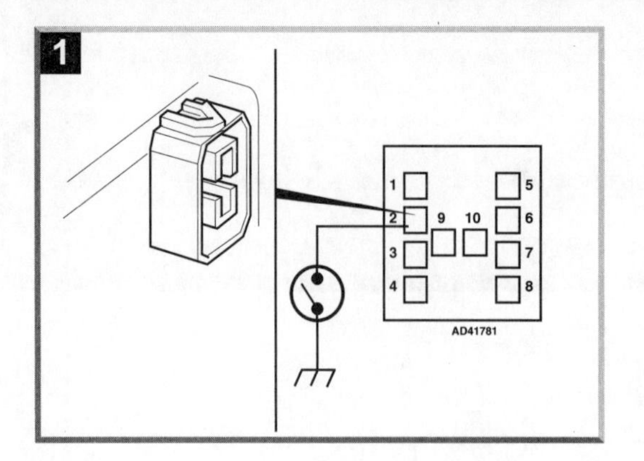

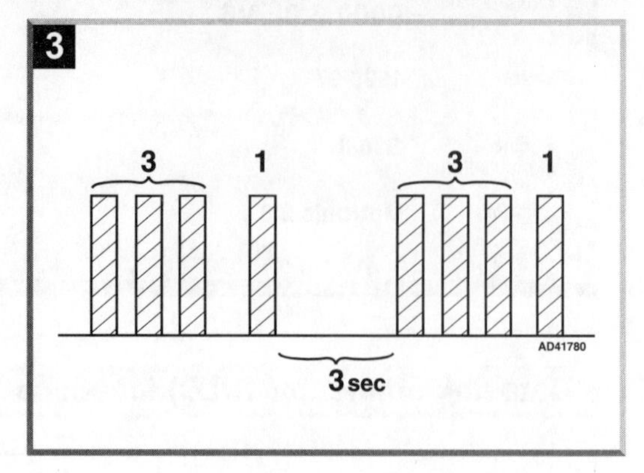

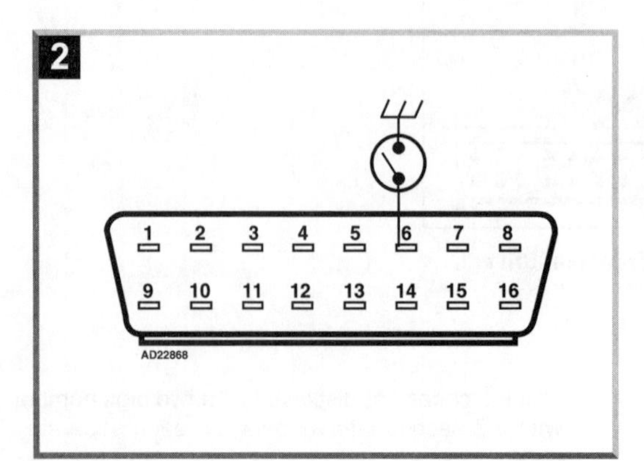

Erasing – flash codes

- Diagnostic equipment required to erase data from ECM fault memory.

Trouble code identification

Flash code	Fault location	Probable cause
11	Secondary air injection (AIR) pump relay, short to ground/positive	Wiring, relay
12	No fault found	–
21	Mass air flow (MAF) sensor, short to ground/ positive	Wiring, MAF sensor
31	Intake air temperature (IAT) sensor, short to ground/positive	Wiring, IAT sensor
41	Engine coolant temperature (ECT) sensor, short to ground/positive	Wiring, ECT sensor
51	Throttle position (TP) sensor, short to ground/ positive	Wiring, TP sensor

Flash code	Fault location	Probable cause
61	Heated oxygen sensor (HO2S) 2	**Wiring, fuse, connectors, fuel pump relay, HO2S**
62	Heated oxygen sensor (HO2S) 1	**Wiring, fuse, connectors, fuel pump relay, HO2S**
71	Mixture rich/lean, cylinder 1-3-5	**Heated oxygen sensor (HO2S), intake leak, injector(s), fuel pressure, evaporative emission (EVAP) canister purge valve**
72	Mixture rich/lean, cylinder 2-4-6	**Heated oxygen sensor (HO2S), intake leak, injector(s), fuel pressure, evaporative emission (EVAP) canister purge valve**
81	Evaporative emission (EVAP) canister purge valve	**Wiring, EVAP canister purge valve**
91	Engine control module (ECM)	**Wiring, connector(s), ECM**

Accessing and erasing – OBD-II codes

- The engine control module (ECM) fault memory can only be accessed and erased using diagnostic equipment connected to the data link connector (DLC).

Trouble code identification

OBD-II code	Fault location	Probable cause
All P0, P2 and U0 codes	Refer to OBD-II trouble code tables at the front of this manual	–
P1001	Evaporative emission (EVAP) canister purge valve – low output	**Wiring open circuit/short to ground, EVAP canister purge valve**
P1002	Evaporative emission (EVAP) canister purge valve – high output	**Wiring short to positive, EVAP canister purge valve**
P1011	Injector 1 – low output	**Wiring open circuit/short to ground, injector**
P1012	Injector 1 – high output	**Wiring short to positive, injector**
P1021	Injector 2 – low output	**Wiring open circuit/short to ground, injector**
P1022	Injector 2 – high output	**Wiring short to positive, injector**
P1031	Injector 3 – low output	**Wiring open circuit/short to ground, injector**
P1032	Injector 3 – high output	**Wiring short to positive, injector**
P1041	Injector 4 – low output	**Wiring open circuit/short to ground, injector**
P1042	Injector 4 – high output	**Wiring short to positive, injector**
P1051	Injector 5 – low output	**Wiring open circuit/short to ground, injector**
P1052	Injector 5 – high output	**Wiring short to positive, injector**

OBD-II code	Fault location	Probable cause
P1061	Injector 6 – low output	Wiring open circuit/short to ground, injector
P1062	Injector 6 – high output	Wiring short to positive, injector
P1206	Secondary air injection (AIR) pump relay – low output	Wiring open circuit/short to ground, secondary air injection (AIR) pump, relay
P1207	Secondary air injection (AIR) pump relay – high output	Wiring short to positive, secondary air injection (AIR) pump, relay
P1211	Idle air control (IAC) valve – low output	Wiring open circuit/short to ground, IAC valve
P1212	Idle air control (IAC) valve – high output	Wiring short to positive, IAC valve
P1236	Intake manifold air control solenoid, outer flap – open circuit/short to ground	Wiring, intake manifold air control solenoid
P1237	Intake manifold air control solenoid, outer flap – short to positive	Wiring, intake manifold air control solenoid
P1246	Intake manifold air control solenoid, inner flap – open circuit/short to ground	Wiring, intake manifold air control solenoid
P1247	Intake manifold air control solenoid, inner flap – short to positive	Wiring, intake manifold air control solenoid
P1251	Malfunction indicator lamp (MIL) – low output	Wiring short to ground
P1252	Malfunction indicator lamp (MIL) – high output	Wiring short to positive
P1450	AC relay – low output	Wiring, AC relay
P1451	AC relay – high output	Wiring, AC relay
P1500	Battery voltage – too low/high	Wiring, alternator, battery
P1601	Fuel pump relay – high output	Wiring short to positive, relay
P1602	Fuel pump relay – low output	Wiring open circuit/short to ground, relay
P1630	Traction control system (TCS) – signal low	Wiring open circuit/short to ground, traction control module/engine control module (ECM) communication
P1631	Traction control system (TCS) – signal high	Wiring short to positive, traction control module/engine control module (ECM) communication

SCION

Model:	Engine identification:	Year:
xA 1.5L	1NZ-FE	2004
xB 1.5L	1NZ-FE	2004

System: **SFI**

Self-diagnosis

General information

- Refer to the front of this manual for general test conditions, terminology, detailed descriptions of wiring faults and a general trouble shooter for electrical and mechanical faults.
- Engine control module (ECM) incorporates self-diagnosis function.
- Malfunction indicator lamp (MIL) will illuminate if certain faults are recorded.
- ECM operates in backup mode if sensors fail, to enable vehicle to be driven to workshop.

Accessing

- Trouble codes can be displayed by using a Scan Tool connected to the data link connector (DLC):
  - xA – 1.
  - xB – 2.

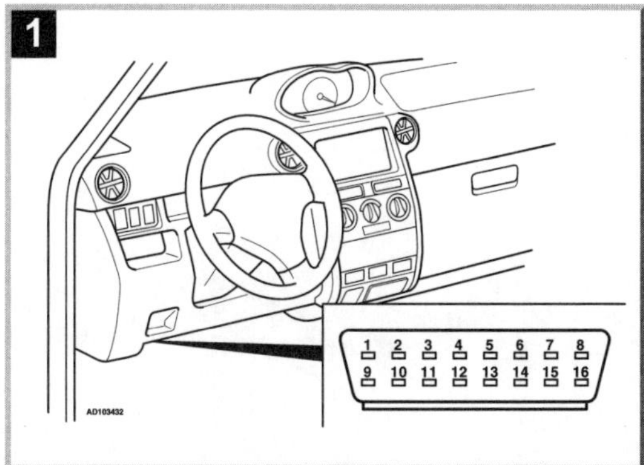

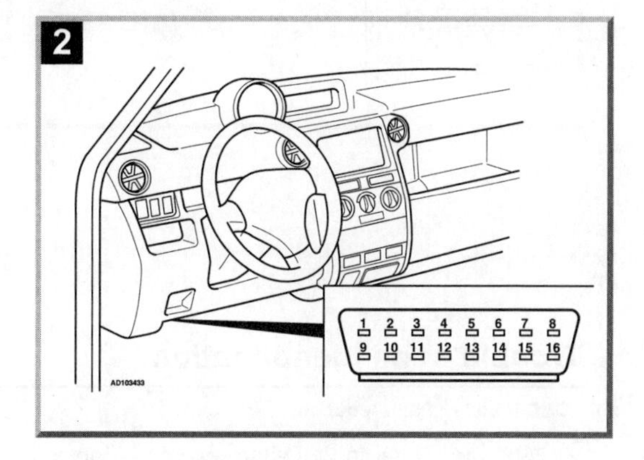

Erasing

- After the faults have been rectified, trouble codes can be erased by using a Scan Tool connected to the data link connector (DLC) or as follows:
- Ensure ignition switched OFF.
- xA – remove EFI fuse (15A) from underhood fusebox 3. Wait 60 seconds minimum.
- xB – remove EFI fuse (15A) from underhood fusebox 4. Wait 60 seconds minimum.
- Reinstall fuse.
- Repeat checking procedure to ensure no data remains in ECM fault memory.

NOTE: ***Disconnecting battery ground lead will also erase trouble codes but may also erase memory from electronic units such as radio.***

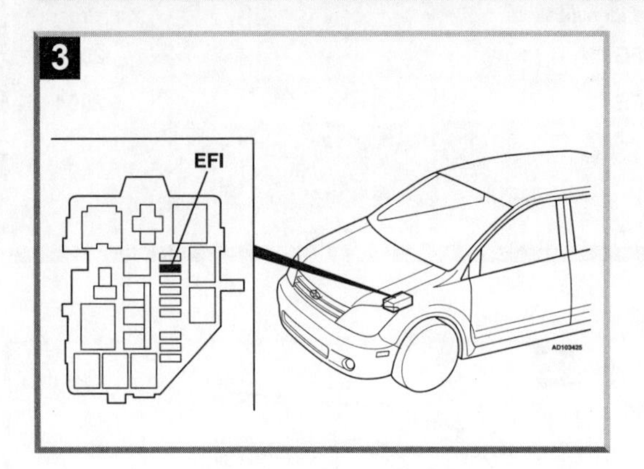

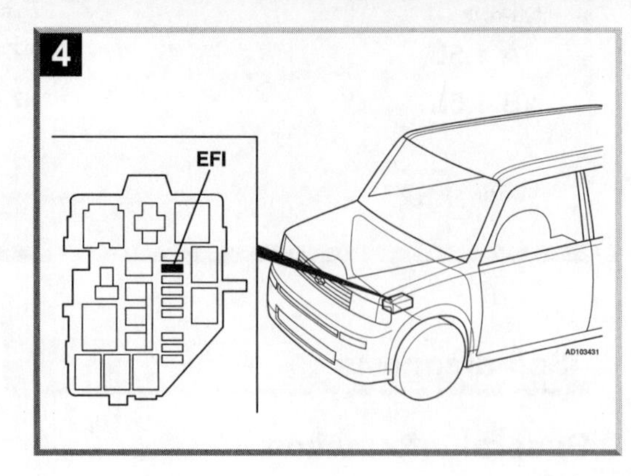

Trouble code identification.

OBD-II code	Fault location	Probable cause
All P0, P2 and U0 codes	Refer to OBD-II trouble code tables at the front of this manual	– NOTE: ***No P1 codes applicable to this model***

Model:	Engine identification:	Year:
Baja 2.5L 16V	**VIN code digit 6 = 6**	**2003-04**
Baja 2.5L Turbo	**VIN code digit 6 = 6**	**2004**
Forester 2.5L 16V	**VIN code digit 6 = 6**	**1998-04**
Forester 2.5L Turbo	**VIN code digit 6 = 6**	**2004**
Impreza 1.8L	**VIN code digit 6 = 2**	**1995-97**
Impreza 2.0L Turbo	**VIN code digit 6 = 2**	**2002-04**
Impreza 2.2L 16V	**VIN code digit 6 = 4**	**1996-01**
Impreza 2.5L 16V	**VIN code digit 6 = 6**	**1998-04**
Impreza 2.5L Turbo	**VIN code digit 6 = 7**	**2004**
Legacy 2.2L 16V	**VIN code digit 6 = 4**	**1995-99**
Legacy/Outback 2.5L 16V	**VIN code digit 6 = 6**	**1996-04**
Outback 3.0L	**VIN code digit 6 = 8**	**2002-04**

System: **MFI/MPI**

Self-diagnosis

General information

- Refer to the front of this manual for general test conditions, terminology, detailed descriptions of wiring faults and a general trouble shooter for electrical and mechanical faults.
- Engine control module (ECM) incorporates self-diagnosis function.
- Malfunction indicator lamp (MIL) will illuminate if certain faults are recorded.
- ECM operates in backup mode if sensors fail, to enable vehicle to be driven to workshop.

Accessing

- Trouble codes can be displayed by using a Scan Tool connected to the data link connector (DLC):
  - Impreza →1997/Legacy →1999/Outback →1999 **1**
  - Forester/Impreza →2001 **2**.
  - Impreza 2002→ **3**.
  - Baja/Legacy 2000→/Outback 2000→ **4**.

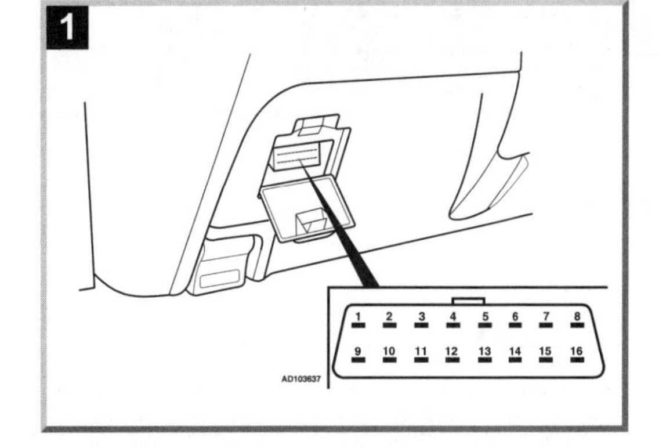

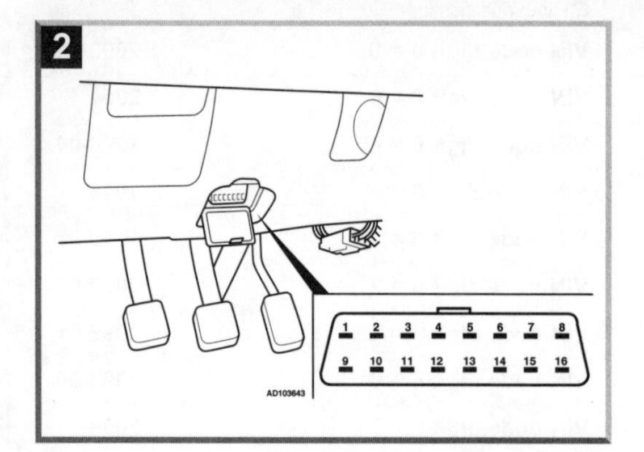

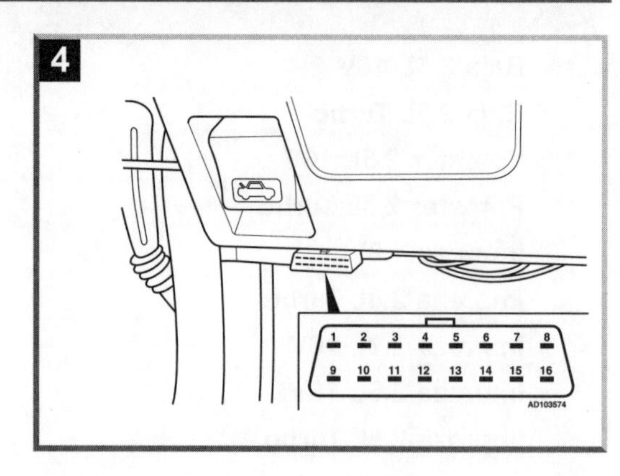

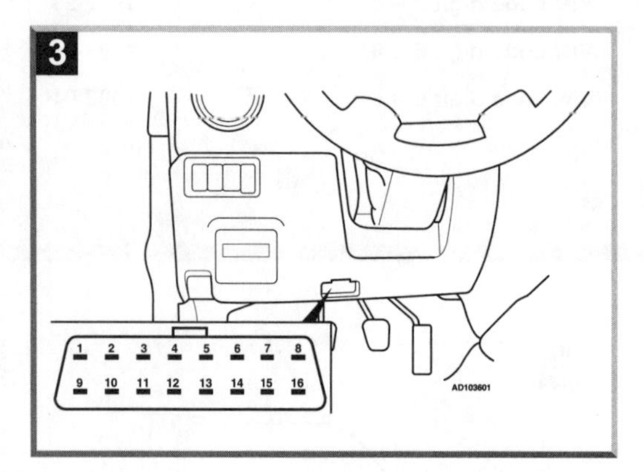

Erasing - models →2002

- Trouble codes can be erased by using a Scan Tool connected to the data link connector (DLC) or as follows:
- Ensure ignition switched OFF.
- Disconnect battery ground cable for 30 seconds minimum.
- Reconnect battery ground cable.

NOTE: ***Disconnecting battery ground lead will erase trouble codes but may also erase memory from electronic units such as radio.***

Erasing - models 2003→

- Trouble codes can only be erased by using a Scan Tool connected to the data link connector (DLC).

Trouble code identification

OBD-II code	Fault location	Probable cause
All P0, P2 and U0 codes	Refer to OBD-II trouble code tables at the front of this manual	–
P1086	Intake manifold air control actuator position sensor, LH – voltage low	**Wiring, intake manifold air control actuator position sensor, ECM**
P1087	Intake manifold air control actuator position sensor, LH – voltage high	**Wiring, intake manifold air control actuator position sensor, ECM**
P1088	Intake manifold air control actuator position sensor, RH – voltage low	**Wiring, intake manifold air control actuator position sensor, ECM**
P1089	Intake manifold air control actuator position sensor, RH – voltage high	**Wiring, intake manifold air control actuator position sensor, ECM**
P1090	Intake manifold air control actuator, RH – stuck open	**Wiring, intake manifold air control actuator position sensor, ECM**

OBD-II code	Fault location	Probable cause
P1091	Intake manifold air control actuator, RH – stuck closed	**Wiring, intake manifold air control actuator position sensor, ECM**
P1092	Intake manifold air control actuator, LH – stuck open	**Wiring, intake manifold air control actuator position sensor, ECM**
P1093	Intake manifold air control actuator, LH – stuck closed	**Wiring, intake manifold air control actuator position sensor, ECM**
P1094	Intake manifold air control actuator, RH – open circuit	**Wiring, intake manifold air control actuator position sensor, ECM**
P1095	Intake manifold air control actuator, RH – over current/short circuit	**Wiring, intake manifold air control actuator position sensor, ECM**
P1096	Intake manifold air control actuator, LH – open circuit	**Wiring, intake manifold air control actuator position sensor, ECM**
P1097	Intake manifold air control actuator, LH – over current/short circuit	**Wiring, intake manifold air control actuator position sensor, ECM**
P1100	Starter signal – circuit malfunction	**Starter motor circuit, NP/TR switch, CPP switch, ECM**
P1101	Neutral position (NP) switch – MT – circuit malfunction	**Wiring, NP switch, ECM**
P1101	Park/neutral position (PNP) switch – circuit malfunction	**Wiring, PNP switch, ECM**
P1102	MAP/BARO sensor – switching valve – circuit malfunction	**Wiring, MAP/BARO sensor switching valve, ECM**
P1103	Engine torque control signal 1 – circuit malfunction	**Wiring, ECM, TCM**
P1104	Traction control signal – circuit malfunction	**Wiring, traction control module**
P1106	Engine torque control signal 2 – circuit malfunction	**Wiring, ECM, TCM**
P1107	Secondary air injection (AIR) diagnosis solenoid – circuit malfunction	**Wiring, AIR diagnosis solenoid, ECM**
P1110	Barometric pressure (BARO) sensor – voltage low	**Wiring, BARO sensor, ECM**
P1111	Barometric pressure (BARO) sensor – voltage high	**Wiring, BARO sensor, ECM**
P1112	Barometric pressure (BARO) sensor – range/performance problem	**Wiring, BARO sensor, ECM**
P1115	Engine torque control cut signal – voltage high	**Wiring, ECM, TCM**
P1116	Engine torque control cut signal – voltage low	**Wiring, ECM, TCM**
P1120	Starter signal – voltage high	**Starter motor circuit, wiring short to positive, ECM**
P1121	Park/neutral position (PNP) switch/neutral position (NP) switch – voltage low	**Wiring short to ground, selector cable, PNP switch, ECM**
P1122	MAP/BARO sensor switching valve – voltage high	**Wiring open/short to positive, MAP/BARO switching valve, ECM**
P1124	Traction control signal – voltage high	**Wiring open/short to positive**

OBD-II code	Fault location	Probable cause
P1130	Heated oxygen sensor (HO2S) 1 – open circuit	Wiring, HO2S, ECM
P1131	Heated oxygen sensor (HO2S) 1 – short circuit	Wiring, HO2S, ECM
P1132	Heated oxygen sensor (HO2S) 1 – voltage low	Wiring, HO2S, ECM
P1133	Heated oxygen sensor (HO2S) 1 – voltage high	Wiring, HO2S, ECM
P1134	Engine control module (ECM)	ECM
P1137	Heated oxygen sensor (HO2S) 1 – range/performance problem	Wiring, HO2S, ECM
P1139	Heated oxygen sensor (HO2S) 1, heater control – range/performance problem	Wiring, HO2S, ECM
P1140	Heated oxygen sensor (HO2S) 1, bank 2, heater – range/performance problem	Wiring, HO2S, ECM
P1141	Mass air flow (MAF) sensor – range/performance problem voltage high	MAF sensor
P1142	Throttle position (TP) sensor – range/performance problem voltage low	TP sensor
P1143	MAP/BARO sensor – range/performance problem voltage low	Hose leak/blockage, MAP/BARO sensor, MAP/BARO sensor switching solenoid
P1144	Manifold absolute pressure (MAP) sensor/ barometric pressure (BARO) sensor – range/performance problem voltage high	MAP/BARO sensor
P1146	Manifold absolute pressure (MAP) sensor – range/performance problem voltage high	Wiring, MAP sensor, TP sensor
P1150	Heated oxygen sensor (HO2S) 1, heater – voltage high	Wiring, HO2S, ECM
P1151	Heated oxygen sensor (HO2S) 2 – voltage high	Wiring, HO2S, ECM
P1152	Heated oxygen sensor (HO2S) 1, bank 1 – range/performance problem voltage low	Wiring, HO2S, ECM
P1153	Heated oxygen sensor (HO2S) 1, bank 1 – range/performance problem voltage high	Wiring, HO2S, ECM
P1154	Heated oxygen sensor (HO2S) 1, bank 2 – range/performance problem voltage low	Wiring, HO2S, ECM
P1155	Heated oxygen sensor (HO2S) 1, bank 2 – range/performance problem voltage high	Wiring, HO2S, ECM
P1160	Throttle actuator control (TAC) – return spring failure	Wiring, TAC motor
P1207	Injector air control solenoid – voltage low	Wiring, Injector air control solenoid, ECM
P1208	Injector air control solenoid – voltage high	Wiring, Injector air control solenoid, ECM
P1208	Knock sensor (KS) – voltage low	Wiring, KS, ECM
P1230	Fuel pump (FP) control module – circuit malfunction	Wiring, FP control module, ECM
P1244	Turbocharger (TC) wastegate regulating valve – voltage low	Wiring, TC wastegate regulating valve, ECM

OBD-II code	Fault location	Probable cause
P1245	Turbocharger (TC) wastegate regulating valve – failsafe	**Wiring, TC wastegate regulating valve, ECM**
P1301	Exhaust gas temperature sensor – misfire/fire due to increased temperature	**Repair other DTC's, replace pre-CAT**
P1312	Exhaust gas temperature sensor	**Wiring, exhaust gas temperature sensor**
P1325	Knock sensor (KS) – voltage low	**Wiring, KS, ECM**
P1400	Fuel tank pressure control solenoid – circuit malfunction voltage low	**Wiring open/short to ground, fuel tank pressure control solenoid, ECM**
P1401	Fuel tank pressure control system – malfunction	**EVAP system, fuel tank pressure sensor**
P1402	Fuel level sensor – circuit malfunction	**Wiring, instrument cluster, fuel level sensor, ECM**
P1420	Fuel tank pressure control solenoid – circuit malfunction voltage high	**Wiring open/short to positive, fuel tank pressure control solenoid, ECM**
P1421	Exhaust gas recirculation (EGR) system – voltage high	**Wiring open/short to positive, EGR solenoid, ECM**
P1422	Evaporative emission (EVAP) canister purge valve – voltage high	**Wiring open/short to positive, EVAP valve, ECM**
P1423	Evaporative emission (EVAP) canister purge system – vent control voltage high	**Wiring open/short to positive, EVAP vent valve, ECM**
P1440	Fuel tank pressure control system – voltage low	**Fuel tank pressure control solenoid, fuel filler cap, EVAP drain/vent valve, EVAP canister/valve, fuel tank, hose/pipe leak**
P1441	Fuel tank pressure control system – voltage high	**Fuel tank pressure control solenoid, fuel filler cap, EVAP drain/vent valve, EVAP valve, fuel tank, hose/pipe blockage**
P1442	Fuel level sensor – range/performance problem	**Fuel level sensor**
P1443	Evaporative emission (EVAP) canister purge system – vent control malfunction	**Hose blockage, EVAP drain valve**
P1445	Injector air control solenoid – circuit malfunction	**Wiring, Injector air control solenoid, ECM**
P1446	Evaporative emission (EVAP) atmospheric pressure solenoid – voltage low	**Wiring, atmospheric pressure solenoid, ECM**
P1447	Evaporative emission (EVAP) atmospheric pressure solenoid – voltage high	**Wiring, atmospheric pressure solenoid, ECM**
P1448	Evaporative emission (EVAP) atmospheric pressure solenoid – range/performance problem	**Wiring, atmospheric pressure solenoid, ECM**
P1480	Engine coolant blower motor relay 1 – voltage high	**Wiring, blower motor relay, ECM**
P1490	Thermostat – malfunction	**Thermostat, cooling system, engine coolant blower motor**
P1491	Positive crankcase ventilation – malfunction	**Wiring, hose blocked/leaking**
P1492	Exhaust gas recirculation (EGR) solenoid valve – signal 1 – voltage low	**Wiring, EGR solenoid valve**

OBD-II code	Fault location	Probable cause
P1493	Exhaust gas recirculation (EGR) solenoid valve – signal 1 – voltage high	**Wiring, EGR solenoid valve, ECM**
P1494	Exhaust gas recirculation (EGR) solenoid valve – signal 2 – voltage low	**Wiring, EGR solenoid valve**
P1495	Exhaust gas recirculation (EGR) solenoid valve – signal 2 – voltage high	**Wiring, EGR solenoid valve, ECM**
P1496	Exhaust gas recirculation (EGR) solenoid valve – signal 3 – voltage low	**Wiring, EGR solenoid valve**
P1497	Exhaust gas recirculation (EGR) solenoid valve – signal 3 – voltage high	**Wiring, EGR solenoid valve, ECM**
P1498	Exhaust gas recirculation (EGR) solenoid valve – signal 4 – voltage low	**Wiring, EGR solenoid valve**
P1499	Exhaust gas recirculation (EGR) solenoid valve – signal 4 – voltage high	**Wiring, EGR solenoid valve, ECM**
P1500	Engine coolant blower motor relay 1 – circuit malfunction	**Wiring, blower motor relay, ECM**
P1502	Engine coolant blower motor – malfunction	**Cooling system, radiator ventilation obstructed**
P1505	Idle air (IAC) control valve – voltage high	**Wiring, IAC valve, ECM**
P1507	Idle air (IAC) control system – malfunction	**Intake system, IAC valve**
P1510	Idle air (IAC) control valve circuit 1 – voltage low	**Wiring, IAC valve, ECM**
P1511	Idle air (IAC) control valve circuit 1 – voltage high	**Wiring, IAC valve, ECM**
P1512	Idle air (IAC) control valve circuit 2 – voltage low	**Wiring, IAC valve, ECM**
P1513	Idle air (IAC) control valve circuit 2 – voltage high	**Wiring, IAC valve, ECM**
P1514	Idle air (IAC) control valve circuit 3 – voltage low	**Wiring, IAC valve, ECM**
P1515	Idle air (IAC) control valve circuit 3 – voltage high	**Wiring, IAC valve, ECM**
P1516	Idle air (IAC) control valve circuit 4 – voltage low	**Wiring, IAC valve, ECM**
P1517	Idle air (IAC) control valve circuit 4 – voltage high	**Wiring, IAC valve, ECM**
P1518	Starter motor switch – circuit malfunction	**Wiring**
P1520	Engine coolant blower motor relay 1 – voltage high	**Wiring open/short to positive, blower motor relay, ECM**
P1540	Vehicle speed sensor (VSS) – malfunction 2	**Wiring, instrument cluster, VSS, ECM**
P1544	Exhaust gas temperature sensor – increased temperature detected	**Repair other DTC's, exhaust system, replace pre-CAT**
P1560	Engine control module (ECM) – battery voltage supply	**Wiring, fuse, ECM**

OBD-II code	Fault location	Probable cause
P1590	Park neutral position (PNP) switch – voltage high	**Wiring, PNP switch, ECM**
P1591	Park neutral position (PNP) switch – voltage low	**Wiring, PNP switch, ECM**
P1592	Park neutral position (PNP) switch – circuit malfunction	**Wiring, PNP switch, ECM**
P1594	Transmission control module (TCM) – communication circuit malfunction	**Wiring, TCM, ECM**
P1595	Transmission control module (TCM) – communication circuit voltage low	**Wiring, TCM, ECM**
P1596	Transmission control module (TCM) – communication circuit voltage high	**Wiring, TCM, ECM**
P1698	Engine torque control cut signal – voltage low	**Wiring, ECM, TCM**
P1699	Engine torque control cut signal – voltage high	**Wiring, ECM, TCM**
P1700	AT – throttle position (TP) sensor – circuit malfunction	**Wiring, TP sensor**
P1701	Cruise control set signal – AT – circuit malfunction	**Wiring, cruise control switch, TCM**
P1702	AT diagnosis signal – voltage low	**Wiring short circuit, TCM**
P1703	Shift solenoid (SS) low clutch timing – circuit malfunction	**Wiring, shift solenoid, TCM**
P1704	Shift solenoid (SS) 2-4 brake timing – circuit malfunction	**Wiring, shift solenoid, TCM**
P1705	Shift solenoid (SS) 2-4 brake pressure control – circuit malfunction	**Wiring, shift solenoid, TCM**
P1711	Engine torque control signal 1 – circuit malfunction	**Wiring, ECM, TCM**
P1712	Engine torque control signal 2 – circuit malfunction	**Wiring, ECM, TCM**
P1722	AT diagnosis signal – voltage high	**Wiring short to positive, TCM, ECM**
P1742	Automatic transmission diagnosis signal – circuit malfunction	**Wiring, TCM**

SUBARU

Model:	Engine identification:	Year:
Impreza 1.8L	**VIN digit 6 = 2**	**1993-94**

System: **MPFI**

Self-diagnosis

General information

- Refer to the front of this manual for general test conditions, terminology, detailed descriptions of wiring faults and a general trouble shooter for electrical and mechanical faults.
- Engine control module (ECM) incorporates self-diagnosis function.
- Malfunction indicator lamp (MIL) will illuminate if certain faults are recorded.
- ECM operates in backup mode if sensors fail, to enable car to be driven to workshop.
- Trouble codes can be accessed with suitable code reader or displayed with the malfunction indicator lamp (MIL).

Accessing – Read memory mode

- Access engine diagnostic link connectors **1**.
- Ensure ignition switched OFF.
- Connect together engine diagnostic link connectors 3 and 4 (black) **1**.
- Switch ignition ON. Do NOT start engine.
- MIL should be flashing.
  - 1.3 second flashes indicate 'tens' of the trouble code **2** [**A**].
  - 0.2 second flashes indicate 'units' of the trouble code **2** [**C**].
  - A 0.2 second pause separates each flash **2** [**B**].
  - A 1.8 second pause separates each trouble code **2** [**D**].
  - For example: Trouble code 12 displayed **2**.
- Count MIL flashes and compare with trouble code table.
- Disconnect engine diagnostic link connectors 3 and 4 (black) **1**.
- Carry out self-diagnosis – D-Check mode.

NOTE: ***If no trouble codes are recorded the MIL will flash ON and OFF every 0.5 seconds.***

- Switch ignition OFF.

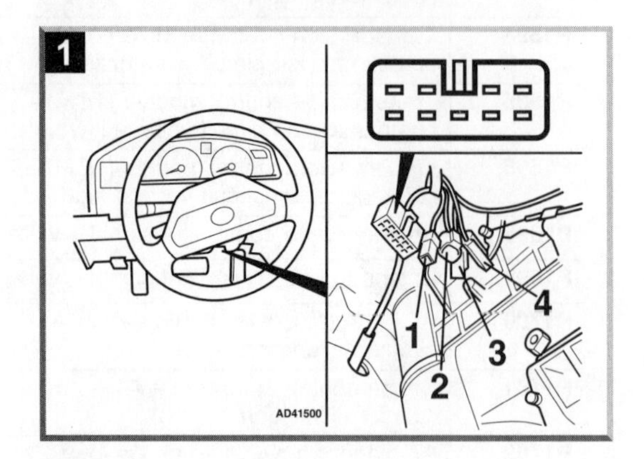

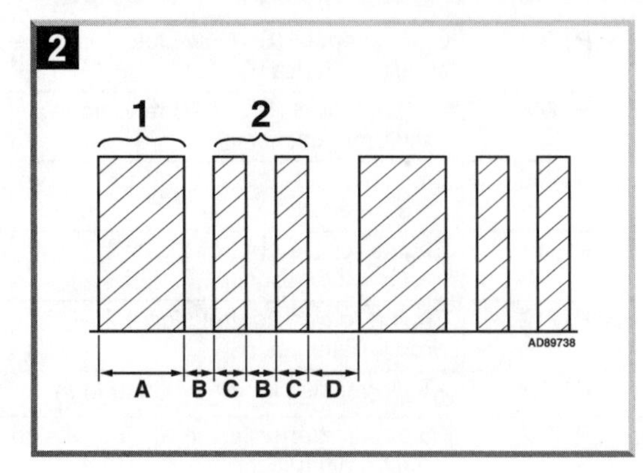

Accessing – D-Check mode

- Carry out self-diagnosis – Read memory mode.
- Start engine.
- Ensure engine is at normal operating temperature.
- Switch ignition OFF.
- Connect together engine diagnostic link connectors 1 and 2 (green) **1**.
- Switch ignition ON. Do NOT start engine.
- Check MIL illuminates.
- If not: Check wiring, fuses, relay, ECM and MIL lamp.
- Start engine.
- MIL lamp should be flashing.
- Count MIL flashes and compare with trouble code table.
- If MIL does not flash:
- Drive vehicle at 7 mph minimum for 1 minute minimum.

NOTE: ***Ensure that 4th gear is engaged whilst driving (MT).***

- Run engine at 2000 rpm.
- Ensure engine is at normal operating temperature.

NOTE: ***If no trouble codes are recorded the MIL will flash ON and OFF every 0.5 seconds.***

- Switch ignition OFF.
- Disconnect engine diagnostic link connectors 1 and 2 (green) **1**.

Erasing

- Start engine.
- Ensure engine is at normal operating temperature.
- Switch ignition OFF.
- Connect together engine diagnostic link connectors 1 and 2 (green) **1**.
- Connect together engine diagnostic link connectors 3 and 4 (black) **1**.
- Switch ignition ON. Do NOT start engine.
- Check MIL illuminates.
- If not: Check wiring, fuses, relay, ECM and MIL lamp.
- Start engine.
- Drive vehicle at 7 mph minimum for 1 minute minimum.

NOTE: ***Ensure that 4th gear is engaged whilst driving (MT).***

- Run engine at 2000 rpm.
- Check that MIL flashes.
- Check that no trouble codes are recorded.

NOTE: ***If no trouble codes are recorded the MIL will flash ON and OFF every 0.5 seconds.***

- Switch ignition OFF.
- Disconnect engine diagnostic link connectors 1 and 2 (green) **1**.
- Disconnect engine diagnostic link connectors 3 and 4 (black) **1**.

Trouble code identification

Flash code	Fault location	Probable cause
11	Crankshaft position (CKP) sensor – no signal	**Wiring, CKP sensor, ECM**
12	Starter signal – circuit	**Wiring, ignition switch, NP switch, PNP switch, ECM**
13	Camshaft position (CMP) sensor – no signal	**Wiring, CMP sensor, ECM**
14	Injector No.1 – signal out of limits	**Wiring, injector, ECM**
15	Injector No.2 – signal out of limits	**Wiring, injector, ECM**
16	Injector No.3 – signal out of limits	**Wiring, injector, ECM**
17	Injector No.4 – signal out of limits	**Wiring, injector, ECM**
21	Engine coolant temperature (ECT) sensor	**Wiring, ECT sensor, ECM**
23	Mass air flow (MAF) sensor	**Wiring, MAF sensor, ECM**

Flash code	Fault location	Probable cause
24	Idle air control (IAC) valve	**Wiring, IAC valve, ECM**
31	Throttle position (TP) sensor	**Wiring, TP sensor, ECM**
32	Heated oxygen sensor (HO2S)	**Wiring, HO2S, ECM**
33	Vehicle speed sensor (VSS) – no signal	**Wiring, VSS, instrument cluster, ECM**
34	Exhaust gas recirculation (EGR) solenoid	**Wiring, EGR valve, ECM**
35	Evaporative emission (EVAP) canister purge valve	**Wiring, EVAP valve, ECM**
36	Secondary air injection (AIR) solenoid	**Wiring, AIR solenoid, ECM**
41	Fuel trim (mixture)	**Injector, MAF sensor, ECT/TP sensor, HO2S, fuel pressure, ECM**
51	Neutral position (NP) switch – MT	**Wiring, NP switch, ECM**
51	Park/neutral position switch (PNP)	**Selector cable incorrectly adjusted, wiring, PNP switch, ECM**
55	Exhaust gas recirculation temperature (EGRT) sensor	**Wiring, EGRT sensor, ECM**
56	Exhaust gas recirculation (EGR) system	**Hose leak/blockage, EGR solenoid**

SUBARU

Model:	Engine identification:	Year:
Legacy 2.2L	VIN digit 6 = 6	1990-94
Legacy 2.2L Turbo	VIN digit 6 = 6	1991-94

System: **MPFI**

Self-diagnosis

General information

- Refer to the front of this manual for general test conditions, terminology, detailed descriptions of wiring faults and a general trouble shooter for electrical and mechanical faults.
- Engine control module (ECM) incorporates self-diagnosis function.
- Malfunction indicator lamp (MIL) will illuminate if certain faults are recorded.
- ECM operates in backup mode if sensors fail, to enable vehicle to be driven to workshop.
- Trouble codes can be accessed with suitable code reader connected to the data link connectors (DLC) **1** and **2**, under left hand fascia.
- Trouble codes can be displayed by the MIL using the engine diagnostic links.

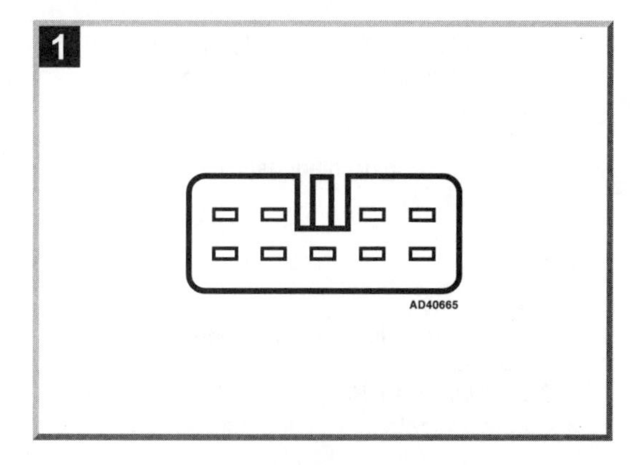

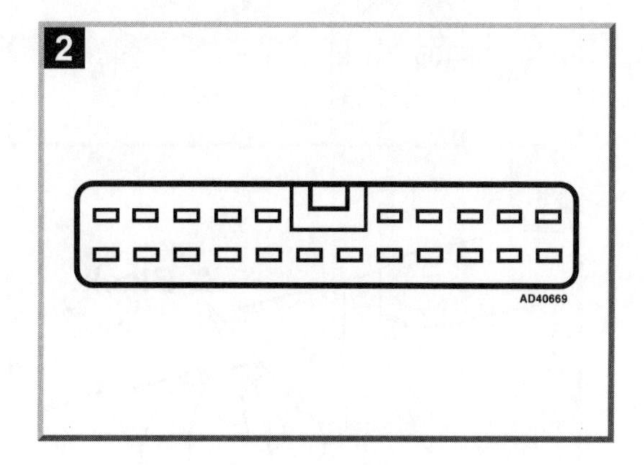

Accessing – Read memory mode

- Access engine diagnostic link connectors **3**.
- Ensure ignition switched OFF.
- Connect together engine diagnostic link connectors (black) **4** [**2**].
- Switch ignition ON. Do NOT start engine.
- MIL should be flashing.
- Count MIL flashes and compare with trouble code table.
  - Long flashes indicate the 'tens' of the trouble code **5** [**A**].
  - Short flashes indicate the 'units' of the trouble code **5** [**C**].
  - A short pause separates each flash **5** [**B**].
  - A long pause separates each trouble code **5** [**D**].
  - For example: Trouble code 12 displayed **5**.
- Disconnect engine diagnostic link connectors (black) **4** [**2**].
- Carry out self-diagnosis – D-Check mode.

NOTE: ***If no trouble codes are recorded the vehicle specification code is displayed.***

3

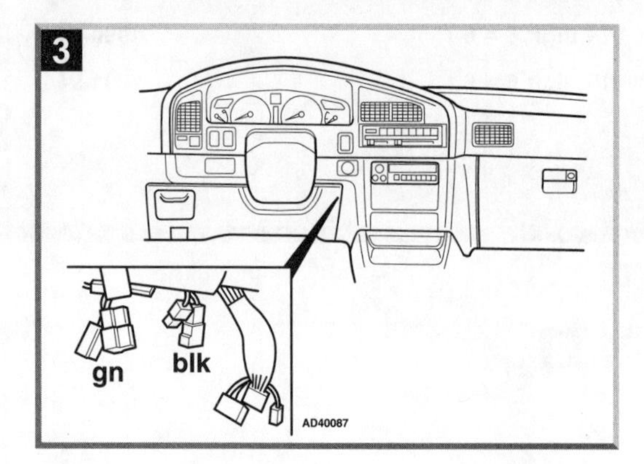

4

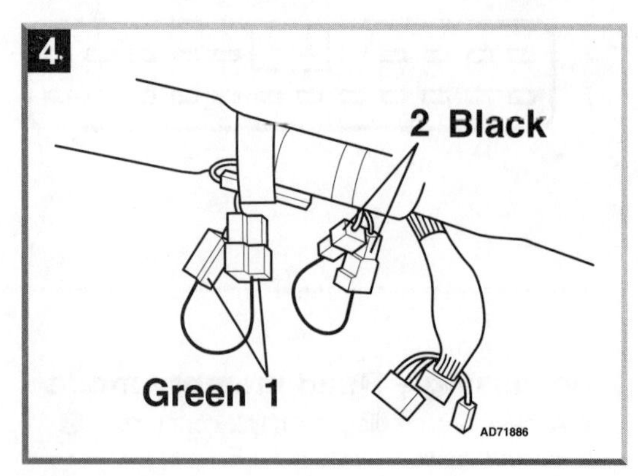

5

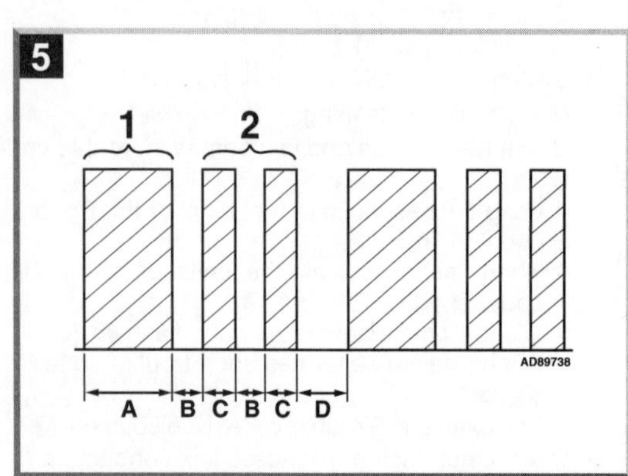

Accessing – D-Check mode

- Carry out self-diagnosis – Read memory mode.
- Start engine.
- Ensure engine is at normal operating temperature.
- Switch ignition OFF.
- Connect together engine diagnostic link connectors (green) 4 [**1**].
- Switch ignition ON. Do NOT start engine.
- Check MIL illuminates.
- If not: Check wiring, fuses, relay, ECM and MIL lamp.
- Fully depress accelerator pedal, then release to one half throttle position.
- Wait two seconds.
- Fully release accelerator pedal.
- Start engine.
- Drive vehicle at 7 mph minimum and shift up to 4th gear (MT).
- Increase engine speed to over 2000 rpm.
- MIL should be flashing.
- Count MIL flashes and compare with trouble code table.
  - Long flashes indicate the 'tens' of the trouble code 5 [**A**].
  - Short flashes indicate the 'units' of the trouble code 5 [**C**].
  - A short pause separates each flash 5 [**B**].
  - A long pause separates each trouble code 5 [**D**].
  - For example: Trouble code 12 displayed 5.
- If MIL extinguishes: No trouble codes are recorded.
- Switch ignition OFF.
- Disconnect engine diagnostic link connectors (green) 4 [**1**].

Erasing

- Start engine.
- Ensure engine is at normal operating temperature.
- Switch ignition OFF.
- Connect together engine diagnostic link connectors (green) 4 [**1**].
- Connect together engine diagnostic link connectors (black) 4 [**2**].
- Switch ignition ON. Do NOT start engine.
- Check MIL illuminates.
- Fully depress accelerator pedal, then release to one half throttle position.
- Wait two seconds.

- Fully release accelerator pedal.
- Start engine.
- Drive vehicle at 7 mph minimum for at least 1 minute and shift up to 4th gear (MT).
- Increase engine speed to over 2000 rpm.
- Check that MIL flashes.

NOTE: ***If MIL remains ON, compare recorded codes with trouble code table and rectify faults as necessary.***

- Switch ignition OFF.
- Disconnect engine diagnostic link connectors (green) **4** [**1**].
- Disconnect engine diagnostic link connectors (black) **4** [**2**].

Trouble code identification

Flash code	Fault location	Probable cause
11	Crankshaft position (CKP) sensor – no signal	Wiring, CKP sensor, ECM
12	Starter signal – circuit	Wiring, ignition switch, ECM
13	Camshaft position (CMP) sensor – no signal	Wiring, CMP sensor, ECM
14	Injector No.1 – signal out of limits	Wiring, injector, ECM
15	Injector No.2 – signal out of limits	Wiring, injector, ECM
16	Injector No.3 – signal out of limits	Wiring, injector, ECM
17	Injector No.4 – signal out of limits	Wiring, injector, ECM
21	Engine coolant temperature (ECT) sensor	Wiring, ECT sensor, ECM
22	Knock sensor (KS) – circuit	Wiring, KS, ECM
23	Mass air flow (MAF) sensor	Wiring, MAF sensor, ECM
24	Idle air control (IAC) valve	Wiring, IAC valve, ECM
31	Throttle position (TP) sensor	Wiring, TP sensor, ECM
32	Heated oxygen sensor (HO2S)	Wiring, HO2S, ECM
33	Vehicle speed sensor (VSS) – no signal	Wiring, VSS, ECM
35	Evaporative emission (EVAP) canister purge valve	Wiring, EVAP valve, ECM
41	Fuel trim (mixture)	Injector, MAF sensor, ECT/TP sensor, HO2S, fuel pressure, ECM
42	Throttle position (TP) sensor – CTP switch contacts	Wiring, TP sensor, ECM
44	Turbocharger (TC) wastegate regulating valve	Wiring, TC wastegate regulating valve, ECM
45	Barometric pressure (BARO) sensor	ECM
49	Mass air flow (MAF) sensor	Incorrect MAF sensor, ECM
51	Neutral position (NP) switch – MT	Wiring, NP switch, ECM
51	Park/neutral position switch (PNP) – AT in N position	Wiring, PNP switch, ECM
52	Park/neutral position switch (PNP) – AT in P position	Wiring, PNP switch, ECM

SUBARU

Model:	Engine identification:	Year:
Loyale 1.8L	**VIN digit 6 = 4 & 5**	**1989-94**
Loyale 1.8L Turbo	**VIN digit 6 = 4 & 5**	**1989-94**

System: **SPFI/MPFI**

Self-diagnosis

General information

- Refer to the front of this manual for general test conditions, terminology, detailed descriptions of wiring faults and a general trouble shooter for electrical and mechanical faults.
- Engine control module (ECM) incorporates self-diagnosis function.
- Malfunction indicator lamp (MIL) will illuminate if certain faults are recorded.
- ECM operates in backup mode if sensors fail, to enable vehicle to be driven to workshop.
- Trouble codes can be accessed with suitable code reader connected to the data link connectors (DLC) **1**, **2** and **3**, under left hand fascia.
- Trouble codes can be displayed by the ECM LED or MIL using the engine diagnostic links.

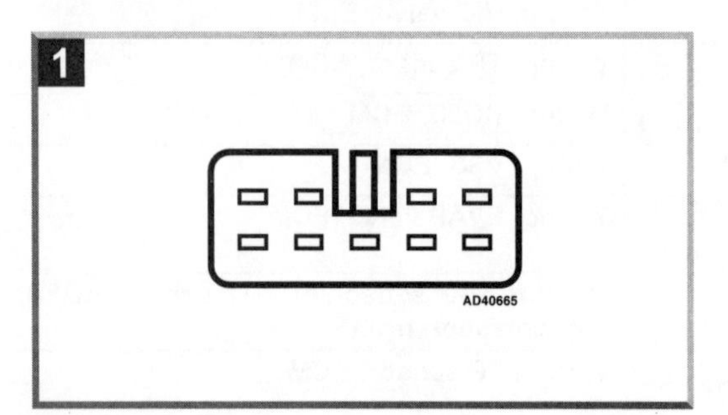

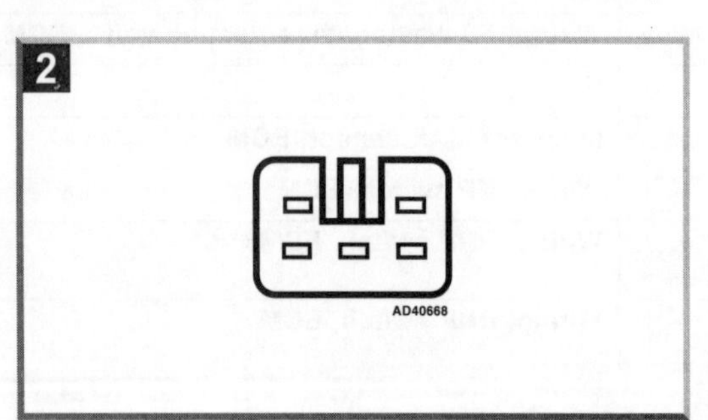

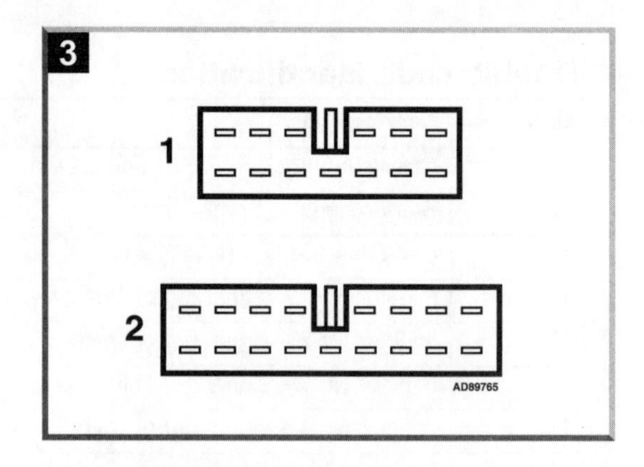

Accessing – Read memory mode

- Access engine control module (ECM) **4** [**5**].
- Ensure ignition switched OFF.
- Connect together engine diagnostic link connectors 3 and 4 (black) **4**.
- Switch ignition ON. Do NOT start engine.
- MIL should be ON.
- LED in ECM should be flashing **4**.
- Count LED flashes and compare with trouble code table.
  - Long flashes indicate the 'tens' of the trouble code **5** [**A**].
  - Short flashes indicate the 'units' of the trouble code **5** [**C**].
  - A short pause separates each flash **5** [**B**].
  - A long pause separates each trouble code **5** [**D**].
  - For example: Trouble code 12 displayed **5**.
- Disconnect engine diagnostic link connectors 3 and 4 (black) **4**.
- Carry out self-diagnosis – D-Check mode.

NOTE: ***If no trouble codes are recorded the vehicle specification code is displayed.***

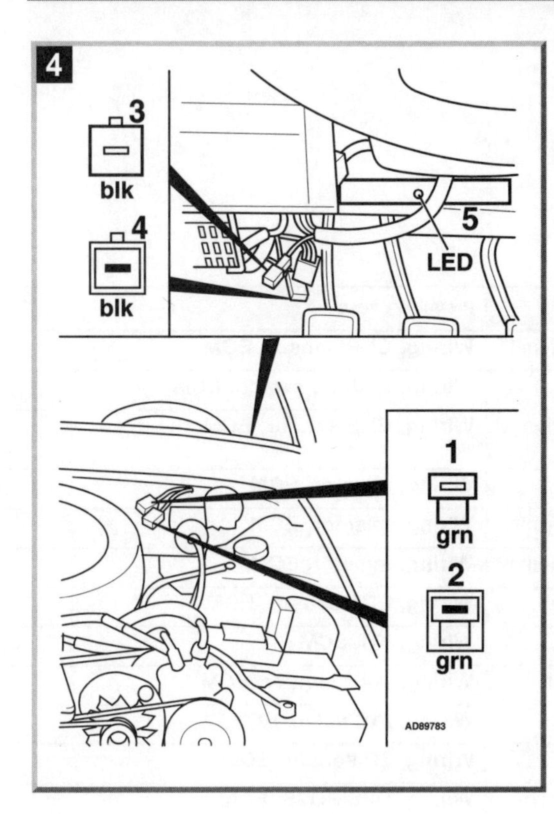

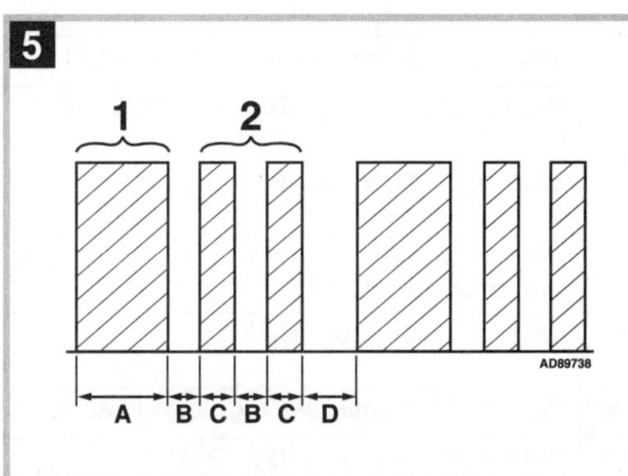

Accessing – D-Check mode

- Carry out self-diagnosis – Read memory mode.
- Start engine.
- Ensure engine is at normal operating temperature.
- Switch ignition OFF.
- Connect together engine diagnostic link connectors 1 and 2 (green) 4.
- Switch ignition ON. Do NOT start engine.
- Check MIL illuminates.
- If not: Check wiring, fuses, relay, ECM and MIL lamp.
- Fully depress accelerator pedal, then release to one half throttle position.
- Wait two seconds.
- Fully release accelerator pedal.
- Start engine.
- Drive vehicle at 7 mph minimum and shift up to 4th gear (MT).
- Increase engine speed to over 2000 rpm.
- LED should be flashing.
- Count LED flashes and compare with trouble code table.
  - Long flashes indicate the 'tens' of the trouble code 5 [**A**].
  - Short flashes indicate the 'units' of the trouble code 5 [**C**].
  - A short pause separates each flash 5 [**B**].
  - A long pause separates each trouble code 5 [**D**].
  - For example: Trouble code 12 displayed 5.
- If MIL extinguishes: No trouble codes are recorded.
- Switch ignition OFF.
- Disconnect engine diagnostic link connectors 1 and 2 (green) 4.

Erasing

- Start engine.
- Ensure engine is at normal operating temperature.
- Switch ignition OFF.
- Connect together engine diagnostic link connectors 1 and 2 (green) 4.
- Connect together engine diagnostic link connectors 3 and 4 (black) 4.
- Switch ignition ON. Do NOT start engine.
- Check MIL illuminates.
- Fully depress accelerator pedal, then release to one half throttle position.
- Wait two seconds.
- Fully release accelerator pedal.
- Start engine.
- Drive vehicle at 7 mph minimum for at least 1 minute and shift up to 4th gear (MT).
- Increase engine speed to over 2000 rpm.
- Check that MIL flashes.

NOTE: ***If MIL remains ON, compare recorded codes with trouble code table and rectify faults as necessary.***

- Switch ignition OFF.
- Disconnect engine diagnostic link connectors 1 and 2 (green) 4.
- Disconnect engine diagnostic link connectors 3 and 4 (black) 4.

Trouble code identification

Flash code	Fault location	Probable cause
11	Crankshaft position (CKP) sensor – no signal	**Wiring, CKP sensor, ECM**
12	Starter signal – circuit	**Wiring, ignition switch, ECM**
13	Crankshaft position (CKP) sensor – position signal	**Wiring, CKP sensor, ECM**
14	Injector – non turbo – signal out of limits	**Wiring, injector, ECM**
14	Injector Nos.1 & 2 – turbo – signal out of limits	**Wiring, injector, ECM**
15	Injector Nos.3 & 4 – turbo – signal out of limits	**Wiring, injector, ECM**
21	Engine coolant temperature (ECT) sensor	**Wiring, ECT sensor, ECM**
22	Knock sensor (KS) – circuit	**Wiring, KS, ECM**
23	Manifold absolute pressure (MAP) sensor	**Wiring, MAP sensor, ECM**
24	Idle air control (IAC) valve	**Wiring, IAC valve, ECM**
31	Throttle position (TP) sensor	**Wiring, TP sensor, ECM**
32	Oxygen sensor (O2S)/heated oxygen sensor (HO2S)	**Wiring, O2S/HO2S, ECM**
33	Vehicle speed sensor (VSS) – no signal	**Wiring, VSS, instrument cluster, ECM**
34	Exhaust gas recirculation (EGR) solenoid – continuously ON or OFF	**Blown fuse, wiring, EGR solenoid, ECM**
35	Evaporative emission (EVAP) canister purge valve – continuously ON or OFF	**Wiring, EVAP valve, ECM**
41	Fuel trim (mixture)	**Injector, MAF sensor, ECT/TP sensor, HO2S, fuel pressure, ECM**
42	Throttle position (TP) sensor – CTP switch contacts	**Wiring, TP sensor, ECM**
44	Turbocharger (TC) wastegate regulating valve	**Wiring, TC wastegate regulating valve, ECM**
45	Transmission kick-down relay – continuously ON or OFF	**Wiring, kick-down relay, ECM**
51	Neutral position (NP) switch – MT	**Wiring, NP switch, ECM**
51	Park/neutral position switch (PNP) – AT in N position	**Wiring, PNP switch, ECM**
55	Exhaust gas recirculation temperature (EGRT) sensor	**Wiring, EGRT sensor, ECM**
61	Park/neutral position switch (PNP) – AT in P position	**Wiring, PNP switch, ECM**

SUZUKI

Model:	Engine identification:	Year:
Aerio/Aerio SX 2.0L	J20	2003
Esteem 1.6L 16V	G16	1996-98
Grand Vitara 2.5L	H25	1999-03
Sidekick 1.6L 16V	G16	1996-98
Sidekick 1.8L 16V	J18	1996-98
Swift 1.3L	G13	1996-98
Vitara 1.6L/2.0L	G16/J20	1999-03
X-90 1.6L 16V	G16	1996-98
XL-7 2.7L	H27	2001-03

System: **EPI**

Self-diagnosis

General information

- Refer to the front of this manual for general test conditions, terminology, detailed descriptions of wiring faults and a general trouble shooter for electrical and mechanical faults.
- Engine control module (ECM) incorporates self-diagnosis function.
- Malfunction indicator lamp (MIL) will illuminate if certain faults are recorded.
- ECM operates in backup mode if sensors fail, to enable vehicle to be driven to workshop.

Accessing

- Trouble codes can be displayed by using a Scan Tool connected to the data link connector (DLC) **1**.

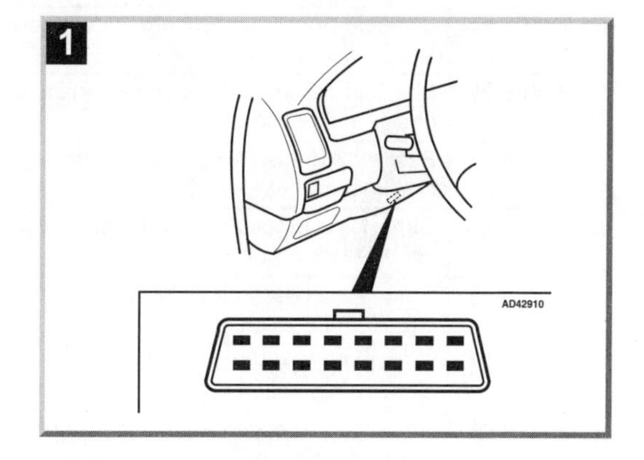

Erasing – except Aerio

- After the faults have been rectified, trouble codes can be erased by using a Scan Tool connected to the data link connector (DLC) or as follows:
- Ensure ignition switched OFF.
- Disconnect battery ground cable for 30 seconds minimum.
- Reconnect battery ground cable.
- Repeat checking procedure to ensure no data remains in ECM fault memory.

NOTE: ***Disconnecting battery ground lead will erase trouble codes but may also erase memory from electronic units such as radio.***

Erasing – Aerio

- After the faults have been rectified, trouble codes can only be erased by using a Scan Tool connected to the data link connector (DLC) **1**.

Trouble code identification

OBD-II code	Fault location	Probable cause
All P0, P2 and U0 codes	Refer to OBD-II trouble code tables at the front of this manual	–
P1132	Heated oxygen sensor (HO2S) 1, bank 1 – heater control – circuit high	**Wiring short to positive, HO2S**
P1152	Heated oxygen sensor (HO2S) 1, bank 2 – heater control – circuit high	**Wiring short to positive, HO2S**
P1250	Intake manifold heater – circuit malfunction	**Wiring, intake manifold heater/relay, ECM**
P1408 **1**	Manifold absolute pressure (MAP) sensor – EGR system – circuit malfunction	**Hose leak/blockage, MAP sensor, ECM**
P1408 **2**	Manifold absolute pressure (MAP) sensor – circuit malfunction	**Hose leak/blockage, MAP sensor, ECM**
P1408 **3**	Manifold differential pressure sensor – circuit malfunction	**Wiring, Manifold differential pressure sensor, ECM**
P1410	Fuel tank pressure control system – malfunction	**Wiring, fuel tank pressure control solenoid, ECM**
P1450 **4** **7**	Barometric pressure (BARO) sensor – circuit malfunction	**ECM**
P1450 **5**	Barometric pressure (BARO) sensor – circuit malfunction	**Manifold differential pressure sensor, ECM**
P1450 **6**	Barometric pressure (BARO) sensor – circuit malfunction	**MAP sensor, ECM**
P1451 **4**	Barometric pressure (BARO) sensor – range/performance problem	**MAP sensor, ECM**
P1451 **5**	Barometric pressure (BARO) sensor – range/performance problem	**Manifold differential pressure sensor, ECM**
P1460	Engine coolant blower motor – system malfunction	**Wiring, blower motor relay, ECM**
P1500	Starter signal – circuit malfunction	**Wiring, ECM**
P1510	Engine control module (ECM) – supply voltage	**Wiring, ECM**
P1530	Ignition timing adjustment connector – circuit malfunction	**Wiring, ECM**
P1600	TCM to ECM – signal failure	**Wiring, TCM, ECM**
P1700	AT – throttle position (TP) switch – signal malfunction	**Wiring, TP sensor, TCM, ECM**
P1705	AT – engine coolant temperature (ECT) sensor – signal malfunction	**Wiring, ECT sensor, TCM, ECM**

OBD-II code	Fault location	Probable cause
P1710	Back-up vehicle speed (VSS) sensor – circuit malfunction	**Wiring, VSS, Speedometer, TCM, ECM**
P1715	Park/neutral position (PNP) switch – circuit malfunction	**Wiring, PNP switch, ECM**
P1717	Park/neutral position (PNP) switch – circuit malfunction	**Wiring, PNP switch, ECM**
P1875	4WD low gear position switch – circuit malfunction	**Wiring, low gear position switch, TCM or ECM**
P1895	AT – ECM torque reduction signal – circuit malfunction	**Wiring, TCM, ECM**

1 Except XL-7/Vitara/Grand Vitara
2 XL-7/2001 → Vitara/Grand Vitara
3 →2000 Vitara/Grand Vitara
4 Except →2000 Vitara/Grand Vitara
5 →2000 Vitara/Grand Vitara
6 2000 → Grand Vitara
7 Vitara

SUZUKI

Model:	Engine identification:	Year:
Esteem 1.6L	**G16**	**1995**
Sidekick 1.6L	**G16**	**1991-95**
Sidekick 1.6L 16V	**G16**	**1992-95**

System: **EPI**

Self-diagnosis

General information

- Refer to the front of this manual for general test conditions, terminology, detailed descriptions of wiring faults and a general trouble shooter for electrical and mechanical faults.
- Engine control module (ECM) incorporates self-diagnosis function.
- Malfunction indicator lamp (MIL) will illuminate if certain faults are recorded.
- ECM operates in backup mode if sensors fail, to enable vehicle to be driven to workshop.
- Trouble codes can be displayed by the MIL.

Accessing

- Access data link connector (DLC) [1] Sidekick and [2] Esteem.
- Ensure ignition switched OFF.
- Jump data link connector (DLC) terminals B and C (4 pins) or B and D (6 pins) [3].
- Switch ignition ON.

NOTE: ***If engine will not start, crank for at least 2 seconds and without turning the ignition OFF jump DLC terminals*** **[3].**

- Count MIL flashes and compare with trouble code table.
  - First flashes of 0.3 second intervals indicate the 'tens' of the trouble code [4].
  - After a 1 second pause, the next flashes of 0.3 second intervals indicate the 'units' of the trouble code [4].
  - A 3 second pause separates each trouble code [4].
  - For example: Trouble code 21 displayed [4].
- Switch ignition OFF.

NOTE: ***If MIL permanently illuminated, suspect engine control module (ECM) fault.***

- Remove jump lead.

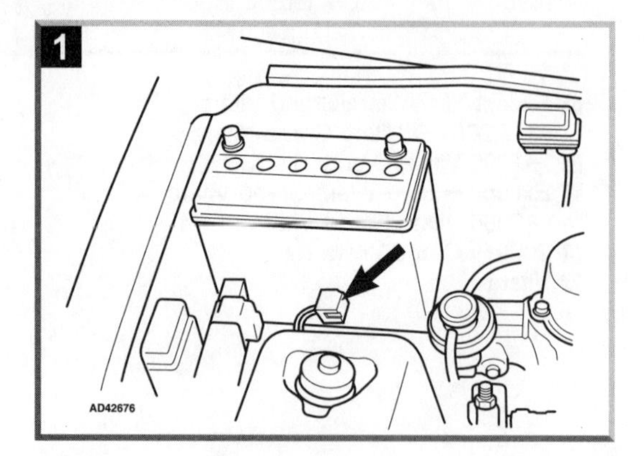

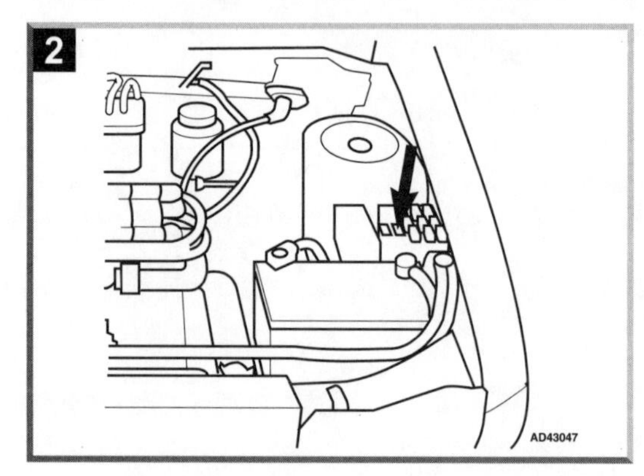

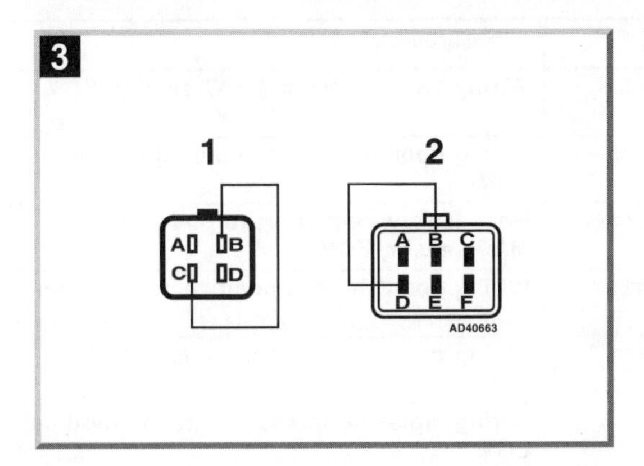

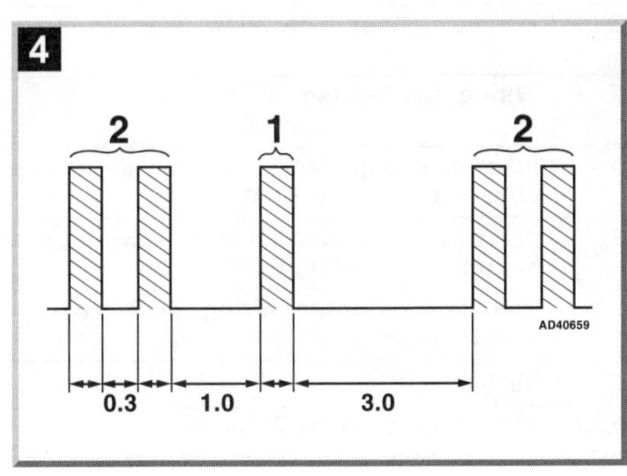

Erasing

- Ensure ignition switched OFF.
- Disconnect battery ground lead.
- Wait 20 seconds minimum.
- Reconnect battery ground lead.
- Start engine.
- Ensure engine is at normal operating temperature.
- Switch ignition OFF.
- Switch ignition ON.
- Jump data link connector (DLC) terminals B and C (4 pins) or B and D (6 pins) **3**.
- Check that trouble code 12 is displayed.
- Switch ignition OFF.
- Remove jump lead.

NOTE: ***Disconnecting battery ground lead will also erase trouble codes but may erase memory from electronic units such as radio.***

Trouble code identification

Flash code	Fault location	Probable cause
12	No fault	–
13	Oxygen sensor (O2S)/heated oxygen sensor (HO2S) – signal not fluctuating	**Wiring, O2S/HO2S, ECM**
14	Engine coolant temperature (ECT) sensor – high voltage signal	**Wiring open/short to positive, ECT sensor, ECM**
15	Engine coolant temperature (ECT) sensor – low voltage signal	**Wiring short to ground, ECT sensor, ECM**
21	Throttle position (TP) sensor – high voltage signal	**Wiring, TP sensor, ECM**
22	Throttle position (TP) sensor – low voltage signal	**Wiring, TP sensor, ECM**
23	Intake air temperature (IAT) sensor – high voltage signal	**Wiring open/short to positive, IAT sensor, ECM**
24	Vehicle speed sensor (VSS) – circuit	**Wiring, VSS, ECM**

Flash code	Fault location	Probable cause
25	Intake air temperature (IAT) sensor – low voltage signal	**Wiring short to ground, IAT sensor, ECM**
31	Manifold absolute pressure (MAP) sensor – low voltage signal	**Wiring open/short to ground, MAP sensor, ECM**
32	Manifold absolute pressure (MAP) sensor – high voltage signal	**Hose leak, wiring short to positive, MAP sensor, ECM**
33	Mass air flow (MAF) sensor – high voltage signal	**Wiring open/short to positive, MAF sensor, ECM**
34	Mass air flow (MAF) sensor – low voltage signal	**Wiring open circuit, MAF sensor, ECM**
41	Ignition fail safe signal	**Wiring, noise suppressor, ignition module, ECM**
42	Camshaft position (CMP) sensor – 1.6L 16V – circuit	**Wiring, CMP sensor, ECM**
42	Crankshaft position (CKP) sensor – 1.6L – circuit	**Wiring, CKP sensor, ECM**
44	Throttle position (TP) sensor – CTP switch contacts – circuit	**Incorrectly adjusted TP sensor, wiring open circuit, TP sensor, ECM**
45	Throttle position (TP) sensor – CTP switch contacts – circuit	**Incorrectly adjusted TP sensor, wiring short circuit, CTP switch, ECM**
51	Exhaust gas recirculation (EGR) system – circuit	**Hose leak/blockage, wiring, EGR valve, EGRT sensor, ECM**
52	Mixture control – high or low	**Injectors, ECM**
53	Engine control module (ECM) – ground circuit	**Wiring open circuit, ECM**

SUZUKI

Model:	Engine identification:	Year:
Samurai 1.3L	G13	1990-94
Swift 1.3L	G13	1989-95
Swift 1.3L 16V DOHC	G13	1989-94

System: **EPI**

Self-diagnosis

General information

- Refer to the front of this manual for general test conditions, terminology, detailed descriptions of wiring faults and a general trouble shooter for electrical and mechanical faults.
- Engine control module (ECM) incorporates self-diagnosis function.
- Malfunction indicator lamp (MIL) will illuminate if certain faults are recorded.
- ECM operates in backup mode if sensors fail, to enable vehicle to be driven to workshop.
- Trouble codes can be displayed by the MIL.

Accessing

- Swift – access data link connector (DLC) **1**.
- Samurai – access data link connector (DLC) **2**.
- Ensure ignition switched OFF.
- Jump data link connector (DLC) terminals B and C (4 pins) or B and D (6 pins) **3**.
- Switch ignition ON.

NOTE: ***If engine will not start, crank for at least 2 seconds and without turning the ignition OFF jump DLC terminals 3.***

- Count MIL flashes and compare with trouble code table.
  - First flashes of 0.3 second intervals indicate the 'tens' of the trouble code **4**.
  - After a 1 second pause, the next flashes of 0.3 second intervals indicate the 'units' of the trouble code **4**.
  - A 3 second pause separates each trouble code **4**.
  - For example: Trouble code 21 displayed **4**.
- Switch ignition OFF.

NOTE: ***If MIL permanently illuminated, suspect engine control module (ECM) fault.***

- Remove jump lead.

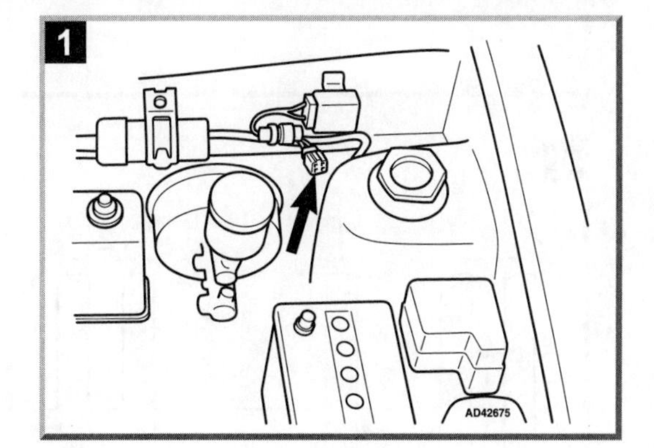

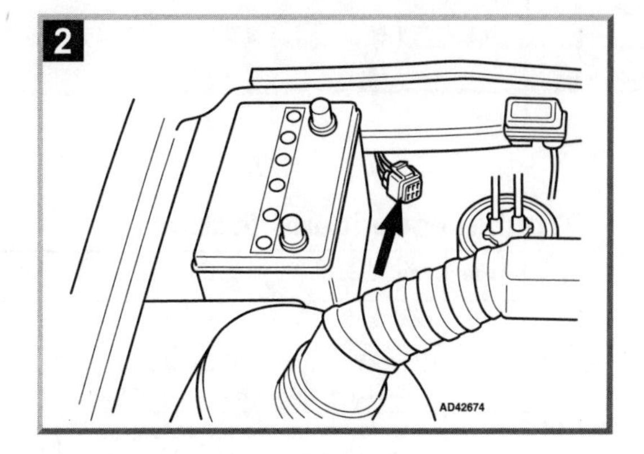

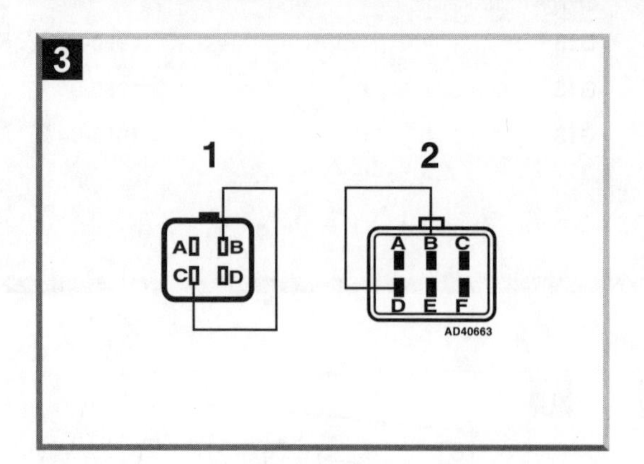

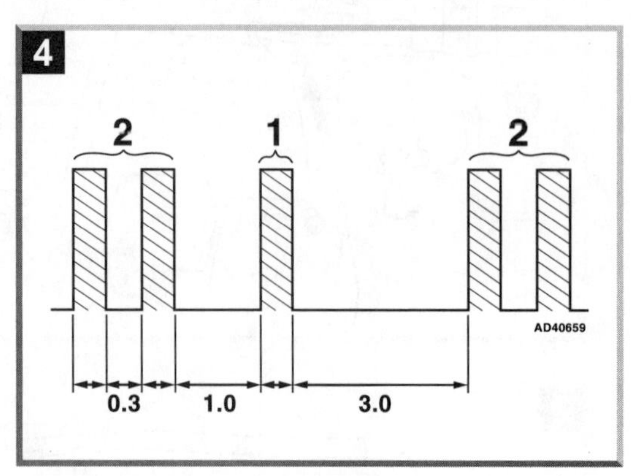

Erasing

- Ensure ignition switched OFF.
- Disconnect battery ground lead.
- Wait 60 seconds minimum.
- Reconnect battery ground lead.
- Start engine.
- Ensure engine is at normal operating temperature.
- Switch ignition OFF.
- Switch ignition ON.
- Jump data link connector (DLC) terminals B and C (4 pins) or B and D (6 pins) **3**.
- Check that trouble code 12 is displayed.
- Switch ignition OFF.
- Remove jump lead.

NOTE: ***Disconnecting battery ground lead may erase memory from electronic units such as radio.***

Trouble code identification

Flash code	Fault location	Probable cause
12	No fault	–
13	Oxygen sensor (O2S)/heated oxygen sensor (HO2S) – signal not fluctuating	**Wiring, O2S/HO2S, ECM**
14	Engine coolant temperature (ECT) sensor – high voltage signal	**Wiring open/short to positive, ECT sensor, ECM**
15	Engine coolant temperature (ECT) sensor – low voltage signal	**Wiring short to ground, ECT sensor, ECM**
21	Throttle position (TP) switch – Swift 1.3L MT	**Wiring short circuit, TP switch, ECM**
21	Throttle position (TP) sensor – Swift 1.3L AT/ 1.3L 16V – high voltage signal	**Wiring, TP sensor, ECM**
21	Throttle position (TP) sensor – Samurai – high voltage signal	**Wiring, TP sensor, ECM**
22	Throttle position (TP) sensor – Swift 1.3L AT/ 1.3L 16V – low voltage signal	**Wiring, TP sensor, ECM**

Flash code	Fault location	Probable cause
22	Throttle position (TP) sensor – Samurai – low voltage signal	**Wiring, TP sensor, ECM**
23	Intake air temperature (IAT) sensor – high voltage signal	**Wiring open/short to positive, IAT sensor, ECM**
24	Vehicle speed sensor (VSS) – circuit	**Wiring, VSS, ECM**
25	Intake air temperature (IAT) sensor – low voltage signal	**Wiring short to ground, IAT sensor, ECM**
31	Manifold absolute pressure (MAP) sensor – low voltage signal	**Wiring open/short to ground, MAP sensor, ECM**
32	Manifold absolute pressure (MAP) sensor – high voltage signal	**Hose leak, wiring short to positive, MAP sensor, ECM**
33	Mass air flow (MAF) sensor – high voltage signal	**Wiring open/short to positive, MAF sensor, ECM**
34	Mass air flow (MAF) sensor – low voltage signal	**Wiring open circuit, MAF sensor, ECM**
41	Ignition fail safe signal – circuit	**Wiring, noise suppressor, ignition module, ECM**
42	Camshaft position (CMP) sensor – Swift 1.3L – circuit	**Air gap, wiring, CMP sensor, ECM**
42	Crankshaft position (CKP) sensor – Swift 1.3L 16V/Samurai – circuit	**Air gap, wiring, CKP sensor, ECM**
44	Throttle position (TP) sensor – CTP switch contacts – circuit	**Incorrectly adjusted TP sensor, wiring open circuit, TP sensor, ECM**
45	Throttle position (TP) sensor – CTP switch contacts – circuit	**Incorrectly adjusted TP sensor, wiring short circuit, CTP switch, ECM**
46	Idle speed control (ISC) system – circuit	**Wiring, ISC actuator/relay, CTP switch, ECM**
51	Exhaust gas recirculation (EGR) system – circuit	**Hose leak/blockage, wiring, EGR valve/solenoid, ECM**
52	Injector – fuel leak	**Injector, ECM**
53	Engine control module (ECM) – ground circuit	**Wiring open circuit, ECM**
54	5th gear switch – continuously ON	**Wiring, 5th gear switch, ECM**
71	Diagnosis terminal – grounded	**DLC terminals jumped, wiring short to ground, ECM**

TOYOTA

Model:	Engine identification:	Year:
Avalon 3.0L	1MZ-FE	1995-04
Camry/Camry Solara 2.4L	2AZ-FE	2002-04
Camry/Camry Solara 3.0L	1MZ-FE	1994-04
Camry/Camry Solara 3.3L	3MZ-FE	2004
Echo 1.5L	1NZ-FE	2000-04
Sienna 3.0L	1MZ-FE	1998-03
Sienna 3.3L	3MZ-FE	2004

System: **SFI**

Self-diagnosis

General information

- Refer to the front of this manual for general test conditions, terminology, detailed descriptions of wiring faults and a general trouble shooter for electrical and mechanical faults.
- Engine control module (ECM) incorporates self-diagnosis function.
- Malfunction indicator lamp (MIL) will illuminate if certain faults are recorded.
- ECM operates in backup mode if sensors fail, to enable vehicle to be driven to workshop.

Accessing

- Trouble codes can be displayed by using a Scan Tool connected to the data link connector (DLC) **1**.

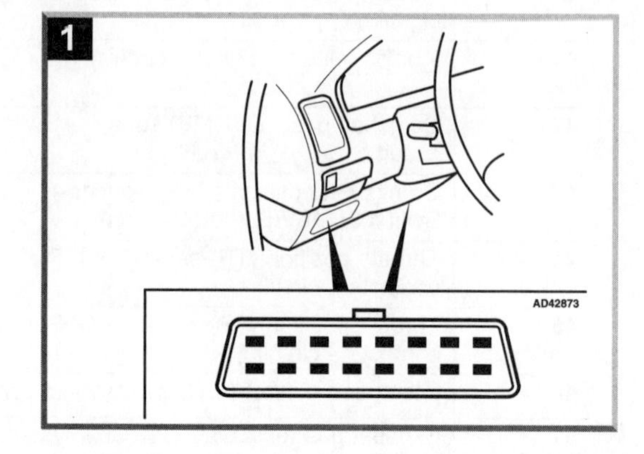

Erasing

- After the faults have been rectified, trouble codes can be erased by using a Scan Tool connected to the data link connector (DLC) or as follows:
- Ensure ignition switched OFF.
- Disconnect battery ground cable for 60 seconds minimum.
- Reconnect battery ground cable.
- Repeat checking procedure to ensure no data remains in ECM fault memory.

NOTE: ***Disconnecting battery ground lead will erase trouble codes but may also erase memory from electronic units such as radio.***

Trouble code identification

OBD-II code	Fault location	Probable cause
All P0, P2 and U0 codes	Refer to OBD-II trouble code tables at the front of this manual	–
P1100	Barometric pressure (BARO) sensor – circuit malfunction	**Wiring, BARO sensor, ECM**
P1120	Accelerator pedal position (APP) sensor – circuit malfunction	**Wiring, APP sensor, ECM**
P1121	Accelerator pedal position (APP) sensor – range/performance problem	**APP sensor, ECM**
P1125	Throttle actuator control (TAC) motor – circuit malfunction	**Wiring, TAC motor, ECM**
P1126	Throttle actuator control (TAC) clutch – circuit malfunction	**Wiring, magnetic clutch, ECM**
P1127	Throttle actuator control (TAC) power source – circuit malfunction	**Wiring, ECM**
P1128	Throttle actuator control (TAC) lock – malfunction	**TAC motor, throttle body, ECM**
P1129	Throttle actuator control (TAC) system – malfunction	**TAC system, ECM**
P1130	Air fuel (A/F) ratio sensor – LH front – range/ performance malfunction	**Wiring, air fuel (A/F) ratio sensor, air induction system, fuel pressure, injector(s), ECM**
P1133	Air fuel (A/F) ratio sensor – LH front – response malfunction	**Wiring, air fuel (A/F) ratio sensor, air induction system, fuel pressure, injector(s), ECM**
P1135	Air fuel (A/F) sensor – LH front – heater circuit malfunction	**Wiring, air fuel (A/F) ratio sensor, air induction system, fuel pressure, injector(s), ECM**
P1150	Air fuel (A/F) sensor – RH front – performance malfunction	**Wiring, air fuel (A/F) ratio sensor, ECM**
P1153	Air fuel (A/F) sensor – RH front – response malfunction	**Air fuel (A/F) ratio sensor**
P1155	Air fuel (A/F) sensor – RH front – heater circuit malfunction	**Wiring, air fuel (A/F) ratio sensor, ECM**
P1300	Ignition control – circuit malfunction – Sienna →00/Solara 3.0L	**Wiring, ignition module, ECM**
P1300	Ignition control – cylinder No.1 – circuit malfunction	**Wiring, ignition coil/module, ECM**
P1305	Ignition control – cylinder No.2 – circuit malfunction	**Wiring, ignition coil/module, ECM**
P1310	Ignition control – cylinder No.3 – circuit malfunction	**Wiring, ignition coil/module, ECM**
P1315	Ignition control – cylinder No.4 – circuit malfunction	**Wiring, ignition coil/module, ECM**
P1320	Ignition control – cylinder No.5 – circuit malfunction	**Wiring, ignition coil/module, ECM**

OBD-II code	Fault location	Probable cause
P1325	Ignition control – cylinder No.6 – circuit malfunction	**Wiring, ignition coil/module, ECM**
P1335	Crankshaft position (CKP) sensor – no signal	**Wiring, CKP sensor, starter motor, ECM**
P1345	Variable valve timing sensor – LH bank – circuit malfunction	**Wiring, valve timing sensor, ECM**
P1346	Variable valve timing sensor – LH bank – range/performance problem	**Valve timing, ECM**
P1349	Variable valve timing system – LH bank – malfunction	**Valve timing, oil control valve, valve timing control module, ECM**
P1350	Variable valve timing sensor – RH bank – circuit malfunction	**Wiring, valve timing sensor, ECM**
P1351	Variable valve timing sensor – RH bank – range/performance problem	**Valve timing, ECM**
P1354	Variable valve timing system – RH bank – malfunction	**Valve timing, oil control valve, valve timing control module, ECM**
P1400	Throttle position (TP) sensor 2 – circuit malfunction	**Wiring, TP sensor, ECM**
P1401	Throttle position (TP) sensor 2 – range/ performance problem	**TP sensor**
P1410	Exhaust gas recirculation (EGR) valve position sensor – malfunction	**Wiring, EGR valve position sensor, ECM**
P1411	Exhaust gas recirculation (EGR) valve position sensor – range/performance problem	**EGR valve position sensor**
P1500	Starter signal – circuit malfunction	**Wiring, ignition switch, engine control relay, ECM**
P1520	Stop lamp switch – signal malfunction	**Wiring, stop lamp switch, ECM**
P1566	Cruise control switch – circuit malfunction	**Wiring, cruise control switch, ECM**
P1570	Cruise control laser radar sensor – malfunction	**Cruise control laser radar sensor, cruise control module, ECM**
P1572	Cruise control laser radar sensor – improper aiming of beam axis	**Cruise control laser radar sensor alignment, cruise control laser radar sensor**
P1575	Cruise control/skid control warning buzzer – malfunction	**Wiring, warning buzzer, ABS control module**
P1578	Brake system – malfunction	**ABS system, ABS control module**
P1600	Engine control module (ECM) – supply voltage	**Wiring, ECM**
P1605	Knock control – malfunction	**ECM**
P1615	Data bus, cruise control module – communication error	**CAN data bus, cruise control module, ECM**
P1616	Data bus, cruise control module – communication error	**CAN data bus, ECM, cruise control module**
P1633	Throttle actuator control (TAC) system – malfunction	**ECM**
P1645	Body control module – malfunction	**Body control module, A/C control module, wiring**

OBD-II code	Fault location	Probable cause
P1647	Transmission control module (TCM) – system malfunction	Trouble code(s) stored
P1656	Oil control valve/variable valve timing – LH bank – malfunction	Wiring, oil control valve, ECM
P1663	Oil control valve/variable valve timing – RH bank – malfunction	Wiring, oil control valve, ECM
P1690	Oil control valve/variable valve timing – malfunction	Wiring, oil control valve, ECM
P1692	Variable valve timing – malfunction	Wiring, valve timing sensor, ECM
P1693	Variable valve timing – malfunction	Wiring, valve timing sensor, ECM
P1700	Output shaft speed (OSS) sensor	Wiring, OSS sensor, ECM
P1705	AT – direct clutch speed sensor	Wiring, direct clutch speed sensor, ECM
P1715	Rear wheel speed sensor(s) – malfunction	Wiring, rear wheel speed sensor(s), ECM
P1725	Input shaft speed (ISS) sensor/turbine shaft speed (TSS) sensor – circuit malfunction	Wiring, ISS sensor/TSS sensor, ECM
P1730	Intermediate shaft speed sensor – AT	Wiring, intermediate shaft speed sensor, ECM
P1755	Torque converter clutch (TCC) solenoid – circuit malfunction	Wiring, torque converter, TCC solenoid, ECM
P1760	AT – pressure control shift solenoid	Wiring, shift solenoid, ECM
P1765	AT – pressure control shift solenoid	Wiring, shift solenoid, ECM
P1770	Differential lock solenoid – circuit malfunction	Wiring, differential lock solenoid, ECM
P1780	Park/neutral position (PNP) switch/transmission range (TR) sensor – circuit malfunction	Wiring, PNP switch/TR sensor, ECM
P1781	Shift lever switch – malfunction	Wiring, shift lever switch, ECM
P1790	Shift/timing solenoid	Wiring, shift/timing solenoid, ECM
P1810	CAN data bus malfunction – malfunction	Wiring
P1815	Shift control solenoid – malfunction	Wiring, shift control solenoid, ECM
P1818	Shift control solenoid – circuit malfunction	Wiring, shift control solenoid, ECM
B2785	Ignition switch ON – malfunction	Immobilizer system
B2786	Ignition switch OFF – malfunction	Immobilizer system
B2795	Unrecognized key code	Immobilizer system
B2796	Immobilizer system – no communication	Immobilizer system
B2797	Communication – malfunction 1	Immobilizer system
B2798	Communication – malfunction 2	Immobilizer system
B2799	Immobilizer – malfunction	Wiring, immobilizer system, ECM

TOYOTA

Model:	Engine identification:	Year:
Camry 2.2L	5S-FE	1992-95
Cressida 3.0L	7M-GE	1990-92

System: **TCCS**

Self-diagnosis

General information

- Refer to the front of this manual for general test conditions, terminology, detailed descriptions of wiring faults and a general trouble shooter for electrical and mechanical faults.
- Engine control module (ECM) incorporates self-diagnosis function.
- Malfunction indicator lamp (MIL) will illuminate if certain faults are recorded.
- ECM operates in backup mode if sensors fail, to enable vehicle to be driven to workshop.
- ECM trouble codes can be accessed with suitable code reader connected to the data link connector (DLC) **1** or displayed by the MIL.
- A second diagnostic link connector is fitted which can also be used to output diagnostic codes:
  - Cressida – **2**.
  - Camry – **3**

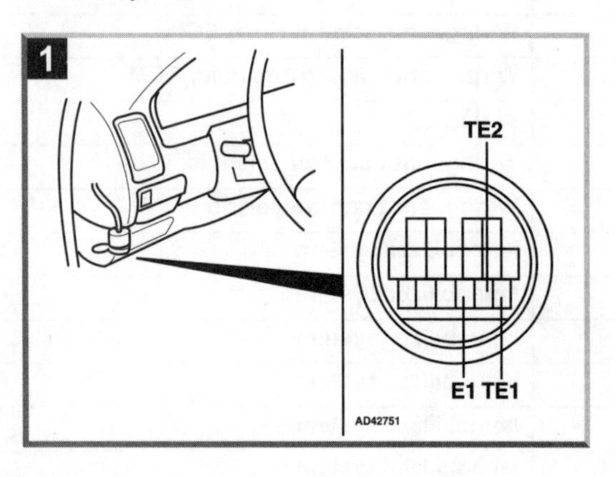

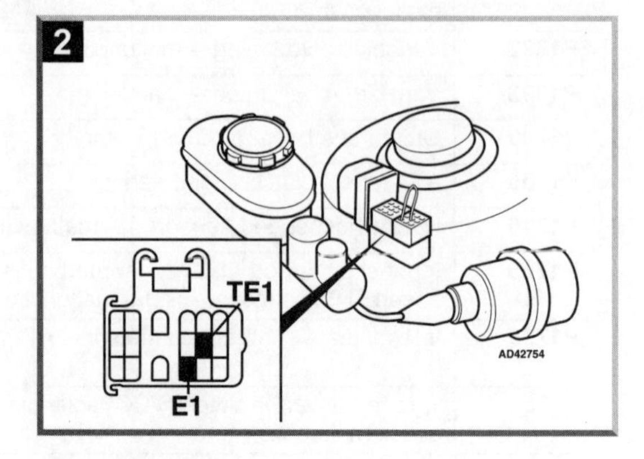

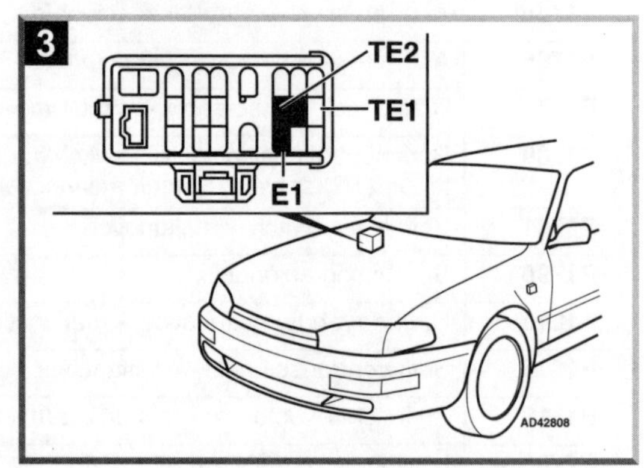

Preparatory conditions

- Battery voltage 11 V minimum.
- Closed throttle position (CTP) switch fully closed.
- Transmission in neutral position.
- All auxiliary equipment, including air conditioning, switched OFF.
- Engine at normal operating temperature.

Accessing – Normal mode

- Switch ignition ON. Do NOT start engine.
- Check MIL illuminates.
- Switch ignition OFF.
- Cressida – jump DLC terminals TE1 and E1 with a short lead 1 or 2.
- Camry – jump DLC terminals TE1 and E1 with a short lead 1 or 3.
- Switch ignition ON.
- Count MIL flashes and compare with trouble code table.
- First flashes of 0.5 second duration indicate 'tens', after a 1.5 second pause, next flashes of 0.5 second duration indicate 'units' 4.
- A 2.5 second pause occurs between codes.
- After a 4.5 second pause, trouble codes will be repeated.

NOTE: ***The MIL flashes once every 0.25 seconds if the system is correct 5.***

- Switch ignition OFF.
- Remove jump lead.

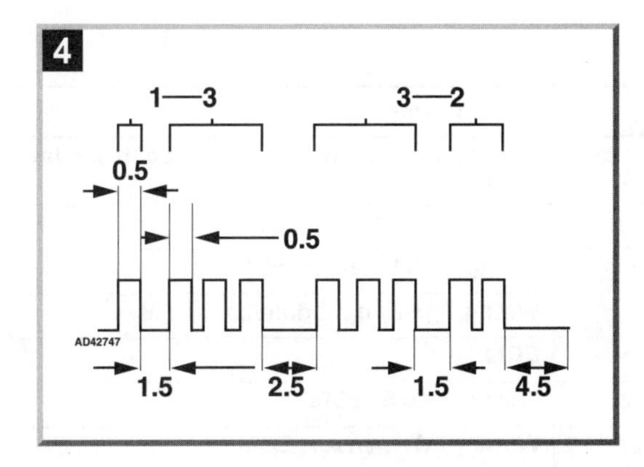

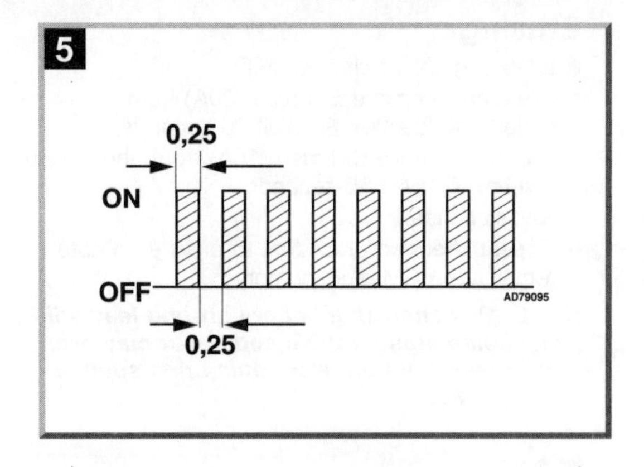

Accessing – Test mode

- Carry out preparatory conditions as in normal mode.
- Jump DLC terminals TE2 and E1 with a short lead 1.
- Switch ignition ON. Do NOT start engine.
- Check MIL is flashing to confirm test mode is operating.

NOTE: ***Test mode will not be activated if the DLC terminals are connected after the ignition switch is turned ON.***

- Start the engine.
- Drive vehicle over 10 mph to simulate possible fault conditions.
- Stop vehicle.
- Jump DLC terminals TE1 and E1 with a short lead 1.
- Count MIL flashes and compare with trouble code table.
- Switch ignition OFF.
- Remove jump leads.

Erasing

- Ensure ignition switched OFF.
- Cressida – remove EFI fuse (20A) from underhood fusebox 6. Wait 20 seconds.
- Camry – remove EFI fuse (15A) from underhood fusebox 7. Wait 30 seconds.
- Reinstall fuse.
- Repeat checking procedure to ensure no data remains in ECM fault memory.

NOTE: ***Disconnecting battery ground lead will also erase trouble codes but may erase memory from electronic units such as radio.***

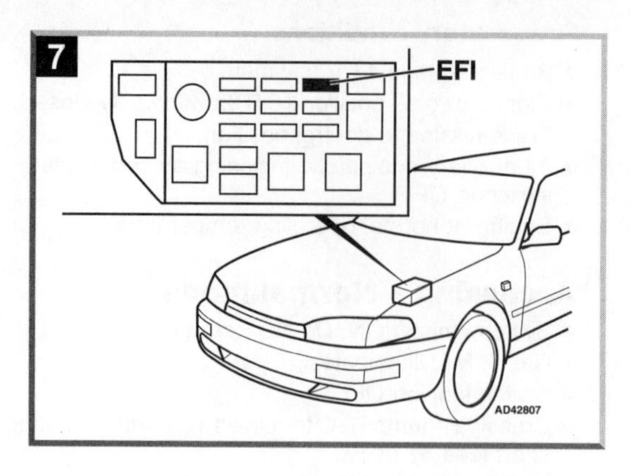

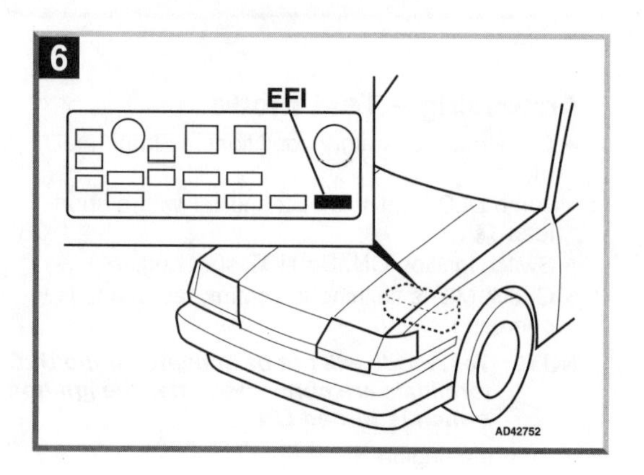

Trouble code identification

Flash code	Fault location	Probable cause
11	Engine control module (ECM) – supply voltage	**Wiring, ignition switch, engine control relay, ECM**
12	Engine speed (RPM) sensor – circuit	**Wiring, RPM sensor, ECM**
13	Engine speed (RPM) sensor – above 1000 rpm	**Wiring, RPM sensor, ECM**
14	Ignition reference signal – no signal	**Wiring, ignition module/coil, ECM**
16	Transmission control signal – malfunction	**ECM**
21	Heated oxygen sensor (HO2S) – front – circuit	**Wiring, HO2S, ECM**
22	Engine coolant temperature (ECT) sensor – circuit	**Wiring, ECT sensor, ECM**
24	Intake air temperature (IAT) sensor – circuit	**Wiring, IAT sensor, ECM**
25	Mixture control – continuously lean	**Wiring, injector, HO2S, ECT/MAP/VAF sensor, intake/fuel/ignition system, ECM**

Flash code	Fault location	Probable cause
26	Mixture control – continuously rich	**Wiring, injector, fuel system, cold start injector, HO2S, ECT/MAP/VAF sensor, ECM**
27	Heated oxygen sensor (HO2S) – rear – circuit	**Wiring, HO2S, ECM**
31	Volume air flow (VAF) sensor – Cressida – circuit	**Wiring, VAF sensor, ECM**
31	Manifold absolute pressure (MAP) sensor – Camry – circuit	**Wiring, MAP sensor, ECM**
32	Volume air flow (VAF) sensor – Cressida – circuit	**Wiring, VAF sensor, ECM**
41	Throttle position (TP) sensor – circuit	**Wiring, TP sensor, ECM**
42	Vehicle speed sensor (VSS) – circuit	**Wiring, VSS, ECM**
43	Starter signal – circuit	**Wiring, ignition switch, engine control relay, ECM**
51	Switch signal – A/C switch ON during diagnosis	**Wiring, A/C switch, A/C amplifier, ECM**
51	Switch signal – closed throttle position (CTP) switch OFF during diagnosis	**Wiring, CTP switch, ECM**
51	Switch signal – park/neutral position (PNP) switch not in P or N during diagnosis	**Wiring, PNP switch, ECM**
52	Knock sensor (KS) – circuit	**Wiring, KS, ECM**
53	Knock control – malfunction	**ECM**
71	Exhaust gas recirculation (EGR) system – malfunction	**Hose leak/blockage, wiring, EGRT sensor, EGR solenoid, ECM**

TOYOTA

Model:	Engine identification:	Year:
Camry/Celica 2.2L	5S-FE	1996-01
Celica 1.8L	7A-FE	1996-97
Celica 1.8L	1ZZ-FE/2ZZ-GE	2001-04
Corolla 1.6/1.8L	4A-FE/7A-FE	1996-97
Corolla 1.8L	1ZZ-FE	1998-04
Matrix 1.8L	1ZZ-FE/2ZZ-GE	2003-04
MR2 1.8L	1ZZ-FE	2000-04
Paseo 1.5L	5E-FE	1995-98
RAV4 2.0L	3S-FE	1996-02
Tercel 1.5L	5E-FE	1995-98

System: **SFI**

Self-diagnosis

General information

- Refer to the front of this manual for general test conditions, terminology, detailed descriptions of wiring faults and a general trouble shooter for electrical and mechanical faults.
- Engine control module (ECM) incorporates self-diagnosis function.
- Malfunction indicator lamp (MIL) will illuminate if certain faults are recorded.
- ECM operates in backup mode if sensors fail, to enable vehicle to be driven to workshop.

Accessing

- Trouble codes can be displayed by using a Scan Tool connected to the data link connector (DLC):
  - Except Celica →97 – **1**.
  - Celica →97 – **2**.

1

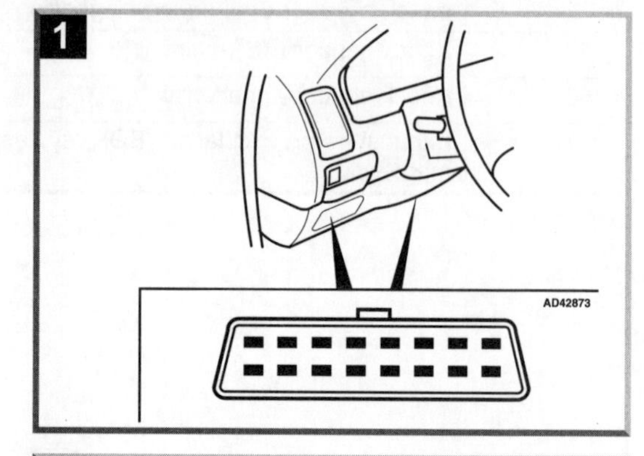

2

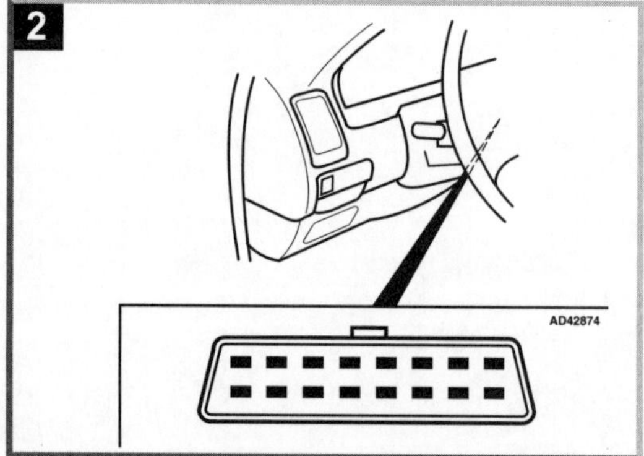

Erasing

- After the faults have been rectified, trouble codes can be erased by using a Scan Tool connected to the data link connector (DLC) or as follows:
- Ensure ignition switched OFF.
- Disconnect battery ground cable for 60 seconds minimum.
- Reconnect battery ground cable.
- Repeat checking procedure to ensure no data remains in ECM fault memory.

NOTE: ***Disconnecting battery ground lead will erase trouble codes but may also erase memory from electronic units such as radio.***

Trouble code identification

OBD-II code	Fault location	Probable cause
All P0, P2 and U0 codes	Refer to OBD-II trouble code tables at the front of this manual	–
P1010	Oil control valve/variable valve lift – LH bank – circuit malfunction	**Wiring, oil control valve, variable valve lift oil pressure switch, ECM**
P1011	Oil control valve/variable valve lift – LH bank – open malfunction	**Wiring, oil control valve, variable valve lift oil pressure switch, ECM**
P1012	Oil control valve/variable valve lift – LH bank – close malfunction	**Wiring, oil control valve, variable valve lift oil pressure switch, ECM**
P1100	Barometric pressure (BARO) sensor – circuit malfunction	**Wiring, BARO sensor, ECM**
P1120	Accelerator pedal position (APP) sensor – circuit malfunction	**Wiring, APP sensor, ECM**
P1121	Accelerator pedal position (APP) sensor – range/performance problem	**APP sensor, ECM**
P1125	Throttle actuator control (TAC) motor – circuit malfunction	**Wiring, TAC motor, ECM**
P1126	Throttle actuator control (TAC) clutch – circuit malfunction	**Wiring, magnetic clutch, ECM**
P1127	Throttle actuator control (TAC) power source – circuit malfunction	**Wiring, ECM**
P1128	Throttle actuator control (TAC) lock – malfunction	**TAC motor, throttle body, ECM**
P1129	Throttle actuator control (TAC) system – malfunction	**TAC system, ECM**
P1130	Air fuel (A/F) ratio sensor – performance malfunction	**Wiring, air fuel (A/F) ratio sensor, ECM**
P1133	Air fuel (A/F) ratio sensor – response malfunction	**Air fuel (A/F) ratio sensor**
P1135	Air fuel (A/F) ratio sensor – heater circuit malfunction	**Wiring, air fuel (A/F) ratio sensor**
P1150	Air fuel (A/F) sensor – RH front – performance malfunction	**Wiring, air fuel (A/F) ratio sensor, ECM**
P1153	Air fuel (A/F) sensor – RH front – response malfunction	**Air fuel (A/F) ratio sensor**

OBD-II code	Fault location	Probable cause
P1155	Air fuel (A/F) sensor – RH front – heater circuit malfunction	**Wiring, air fuel (A/F) ratio sensor, ECM**
P1300	Ignition control 1 – circuit malfunction	**Wiring, ignition coil 1, ECM**
P1310	Ignition control 2 – circuit malfunction	**Wiring, ignition coil 2, ECM**
P1300	Ignition control – cylinder No.1 – circuit malfunction – Corolla (1998→)	**Wiring, ignition coil/module, ECM**
P1305	Ignition control – cylinder No.2 – circuit malfunction	**Wiring, ignition coil/module, ECM**
P1310	Ignition control – cylinder No.3 – circuit malfunction	**Wiring, ignition coil/module, ECM**
P1315	Ignition control – cylinder No.4 – circuit malfunction	**Wiring, ignition coil/module, ECM**
P1320	Ignition control – cylinder No.5 – circuit malfunction	**Wiring, ignition coil/module, ECM**
P1325	Ignition control – cylinder No.6 – circuit malfunction	**Wiring, ignition coil/module, ECM**
P1335	Crankshaft position (CKP) sensor – circuit malfunction	**Wiring, CKP sensor, ECM**
P1345	Variable valve timing sensor – LH bank – circuit malfunction	**Wiring, valve timing sensor, ECM**
P1346	Variable valve timing sensor – LH bank – range/performance problem	**Valve timing, ECM**
P1349	Variable valve timing system – LH bank – malfunction	**Valve timing, oil control valve, valve timing control module, ECM**
P1350	Variable valve timing sensor – RH bank – circuit malfunction	**Wiring, valve timing sensor, ECM**
P1351	Variable valve timing sensor – RH bank – range/performance problem	**Valve timing, ECM**
P1354	Variable valve timing system – RH bank – malfunction	**Valve timing, oil control valve, valve timing control module, ECM**
P1400	Throttle position (TP) sensor 2 – circuit malfunction	**Wiring, TP sensor, ECM**
P1401	Throttle position (TP) sensor 2 – range/performance problem	**TP sensor**
P1410	Exhaust gas recirculation (EGR) valve position sensor – malfunction	**Wiring, EGR valve position sensor, ECM**
P1411	Exhaust gas recirculation (EGR) valve position sensor – range/performance problem	**EGR valve position sensor**
P1500	Starter signal – circuit malfunction	**Wiring, ignition switch, engine control relay, ECM**
P1520	Stop lamp switch – signal malfunction	**Wiring, stop lamp switch, ECM**
P1566	Cruise control switch – circuit malfunction	**Wiring, cruise control switch, ECM**
P1600	Engine control module (ECM) – supply voltage	**Wiring, ECM**
P1605	Knock control – malfunction	**ECM**

OBD-II code	Fault location	Probable cause
P1633	Throttle actuator control (TAC) system – malfunction	ECM
P1645	Body control module – malfunction	Body control module, instruments, A/C control module, ABS control module, SRS control module, wiring
P1646	Transmission control module (TCM) – system malfunction	Wiring, TCM, ECM
P1647	Transmission control module (TCM) – system malfunction	Trouble code(s) stored
P1656	Oil control valve/variable valve timing – LH bank – malfunction	Wiring, oil control valve, ECM
P1663	Oil control valve/variable valve timing – RH bank – malfunction	Wiring, oil control valve, ECM
P1690	Oil control valve/variable valve timing – malfunction	Wiring, oil control valve, ECM
P1692	Variable valve timing – malfunction	Wiring, valve timing sensor, ECM
P1693	Variable valve timing – malfunction	Wiring, valve timing sensor, ECM
P1700	Output shaft speed (OSS) sensor	Wiring, OSS sensor, ECM
P1705	AT – direct clutch speed sensor	Wiring, direct clutch speed sensor, ECM
P1715	Rear wheel speed sensor(s) – malfunction	Wiring, rear wheel speed sensor(s), ECM
P1725	Input shaft speed (ISS) sensor/turbine shaft speed (TSS) sensor – circuit malfunction	Wiring, ISS sensor/TSS sensor, ECM
P1730	Intermediate shaft speed sensor – AT	Wiring, intermediate shaft speed sensor, ECM
P1755	Torque converter clutch (TCC) solenoid – circuit malfunction	Wiring, torque converter, TCC solenoid, ECM
P1760	AT – pressure control shift solenoid	Wiring, shift solenoid, ECM
P1765	AT – pressure control shift solenoid	Wiring, shift solenoid, ECM
P1770	Differential lock solenoid – circuit malfunction	Wiring, differential lock solenoid, ECM
P1780	Park/neutral position (PNP) switch/transmission range (TR) sensor – circuit malfunction	Wiring, PNP switch/TR sensor, ECM
P1781	Shift lever switch – malfunction	Wiring, shift lever switch, ECM
P1790	Shift/timing solenoid	Wiring, shift/timing solenoid, ECM
P1810	CAN data bus malfunction – malfunction	Wiring
P1815	Shift control solenoid – malfunction	Wiring, shift control solenoid, ECM
P1818	Shift control solenoid – circuit malfunction	Wiring, shift control solenoid, ECM
B2795	Unrecognized key code	Immobilizer system
B2796	Immobilizer system – no communication	Immobilizer system
B2797	Communication – malfunction 1	Immobilizer system
B2798	Communication – malfunction 2	Immobilizer system
B2799	Immobilizer – malfunction	Wiring, immobilizer system, ECM

TOYOTA

Model:	Engine identification:	Year:
Camry 3.0L	**3VZ-FE**	**1992-93**
Camry Wagon 3.0L	**3VZ-FE**	**1992-93**

System: **TCCS**

Self-diagnosis

General information

- Refer to the front of this manual for general test conditions, terminology, detailed descriptions of wiring faults and a general trouble shooter for electrical and mechanical faults.
- Engine control module (ECM) incorporates self-diagnosis function.
- Malfunction indicator lamp (MIL) will illuminate if certain faults are recorded.
- ECM operates in backup mode if sensors fail, to enable vehicle to be driven to workshop.
- ECM trouble codes can be accessed with suitable code reader connected to the data link connector (DLC) **1** or displayed by the MIL.

NOTE: ***A second diagnostic link connector is fitted* 2*, which can also be used to output diagnostic codes.***

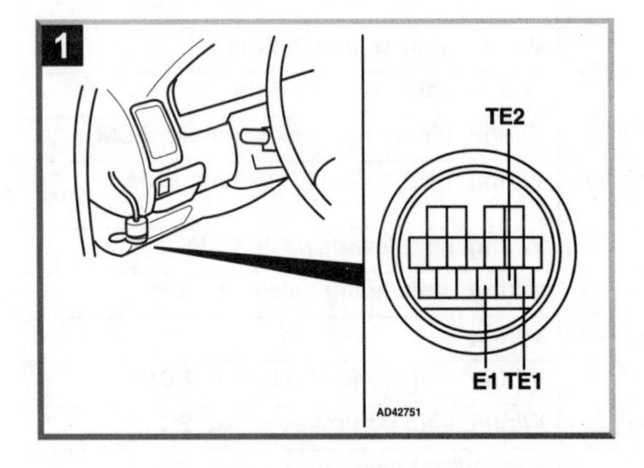

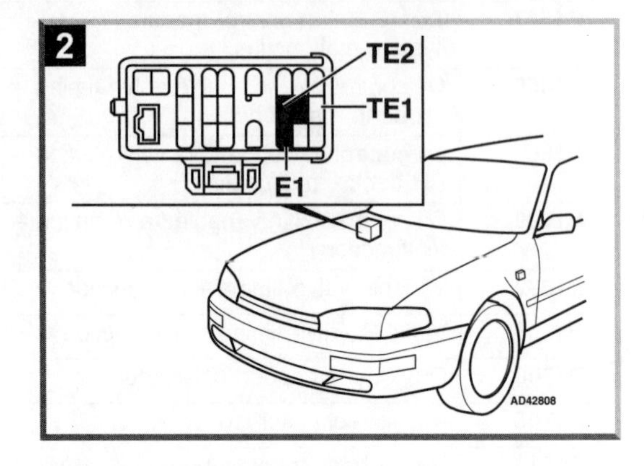

Preparatory conditions

- Battery voltage 11 V minimum.
- Closed throttle position (CTP) switch fully closed.
- Transmission in neutral position.
- All auxiliary equipment, including air conditioning, switched OFF.
- Engine at normal operating temperature.

Accessing – Normal mode

- Switch ignition ON. Do NOT start engine.
- Check MIL illuminates.
- Switch ignition OFF.
- Jump DLC terminals TE1 and E1 with a short lead **1**.
- Switch ignition ON.
- Count MIL flashes and compare with trouble code table.
- First flashes of 0.5 second duration indicate 'tens', after a 1.5 second pause, next flashes of 0.5 second duration indicate 'units' **3**.
- A 2.5 second pause occurs between codes.
- After a 4.5 second pause, trouble codes will be repeated.

NOTE: ***The MIL flashes once every 0.25 seconds if the system is correct* 4*.***

- Switch ignition OFF.
- Remove jump lead.

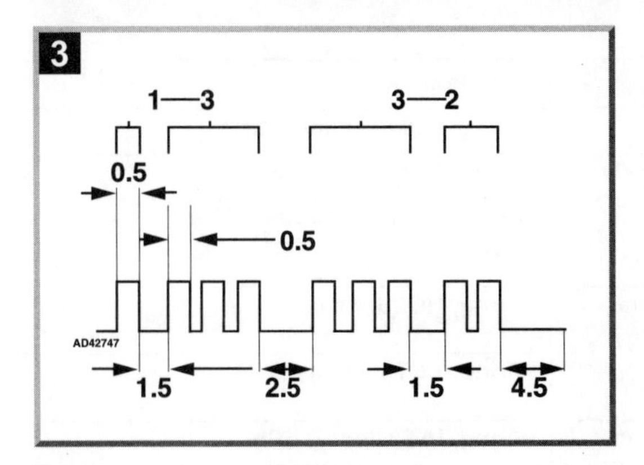

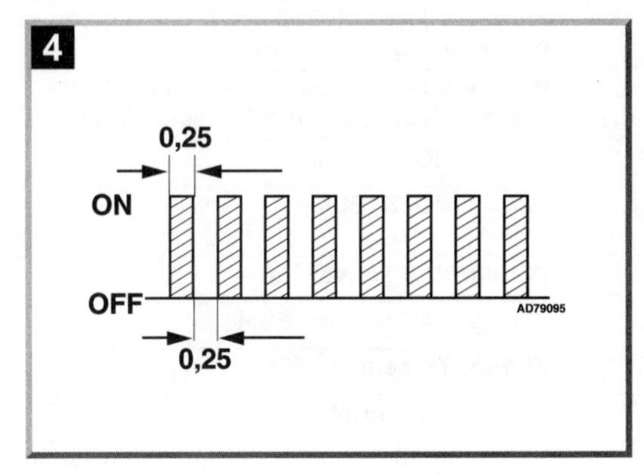

Accessing – Test mode

- Carry out preparatory conditions as in normal mode.
- Jump DLC terminals TE2 and E1 with a short lead **1**.
- Switch ignition ON. Do NOT start engine.
- Check MIL is flashing to confirm test mode is operating.

NOTE: ***Test mode will not be activated if the DLC terminals are connected after the ignition switch is turned ON.***

- Start the engine.
- Drive vehicle over 10 mph to simulate possible fault conditions.
- Stop vehicle.
- Jump DLC terminals TE1 and E1 with a short lead **1**.
- Count MIL flashes and compare with trouble code table.
- Switch ignition OFF.
- Remove jump leads.

Erasing

- Ensure ignition switched OFF.
- Remove EFI fuse (15A) from underhood fusebox **5**.
- Wait 30 seconds.
- Reinstall fuse.
- Repeat checking procedure to ensure no data remains in ECM fault memory.

NOTE: ***Disconnecting battery ground lead will also erase trouble codes but may erase memory from electronic units such as radio.***

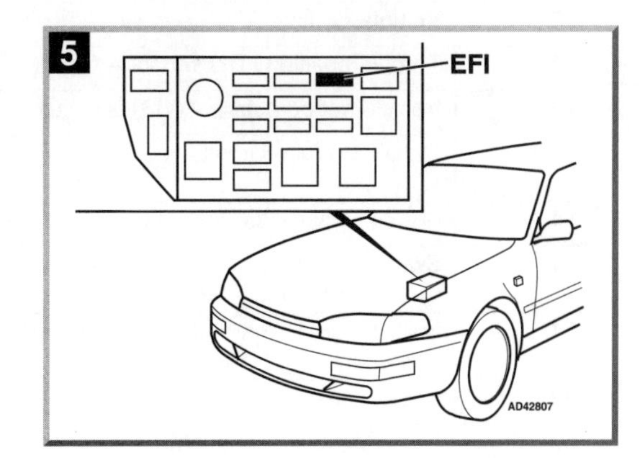

Trouble code identification

Flash code	Fault location	Probable cause
12	Engine speed (RPM) sensor – circuit	**Wiring, RPM sensor, ECM**
13	Engine speed (RPM) sensor – above 1000 rpm	**Wiring, RPM sensor, ECM**
14	Ignition reference signal – no signal	**Wiring, ignition module, ECM**
16	Transmission control signal – malfunction	**ECM**
21	Heated oxygen sensor (HO2S) – LH front – circuit	**Wiring, HO2S, ECM**
22	Engine coolant temperature (ECT) sensor – circuit	**Wiring, ECT sensor, ECM**
24	Intake air temperature (IAT) sensor – circuit	**Wiring, IAT sensor, ECM**
25	Mixture control – continuously lean	**Wiring, injector, HO2S, ECT/VAF sensor, intake/fuel/ignition system, ECM**
26	Mixture control – continuously rich	**Wiring, injector, fuel system, cold start injector, HO2S, ECT/VAF sensor, ECM**
27	Heated oxygen sensor (HO2S) – rear – circuit	**Wiring, HO2S, ECM**
28	Heated oxygen sensor (HO2S) – RH front – circuit	**Wiring, HO2S, ECM**
31	Volume air flow (VAF) sensor – circuit	**Wiring, VAF sensor, ECM**
32	Volume air flow (VAF) sensor – circuit	**Wiring, VAF sensor, ECM**
41	Throttle position (TP) sensor – circuit	**Wiring, TP sensor, ECM**
42	Vehicle speed sensor (VSS) – circuit	**Wiring, VSS, ECM**
43	Starter signal – circuit	**Wiring, ignition switch, engine control relay, ECM**
51	Switch signal – A/C switch ON during diagnosis	**Wiring, A/C switch, A/C amplifier, ECM**
51	Switch signal – closed throttle position (CTP) switch OFF during diagnosis	**Wiring, CTP switch, ECM**
51	Switch signal – park/neutral position (PNP) switch not in P or N during diagnosis	**Wiring, PNP switch, ECM**
52	Knock sensor (KS) – LH – circuit	**Wiring, KS, ECM**
53	Knock control – malfunction	**ECM**
55	Knock sensor (KS) – RH – circuit	**Wiring, KS, ECM**
71	Exhaust gas recirculation (EGR) system – malfunction	**Hose leak/blockage, wiring, EGRT sensor, EGR solenoid, ECM**

TOYOTA

Model:	Engine identification:	Year:
Celica 1.6L	**4A-FE**	**1990-93**
Celica 2.2L	**5S-FE**	**1990-93**
Corolla 1.6L	**4A-FE**	**1989-92**

System: **TCCS**

Self-diagnosis

General information

- Refer to the front of this manual for general test conditions, terminology, detailed descriptions of wiring faults and a general trouble shooter for electrical and mechanical faults.
- Engine control module (ECM) incorporates self-diagnosis function.
- Malfunction indicator lamp (MIL) will illuminate if certain faults are recorded.
- ECM operates in backup mode if sensors fail, to enable vehicle to be driven to workshop.
- Trouble codes can be displayed by the MIL or accessed with suitable code reader connected to the data link connector (DLC):
  - Corolla – **1**.
  - Celica – **2**.

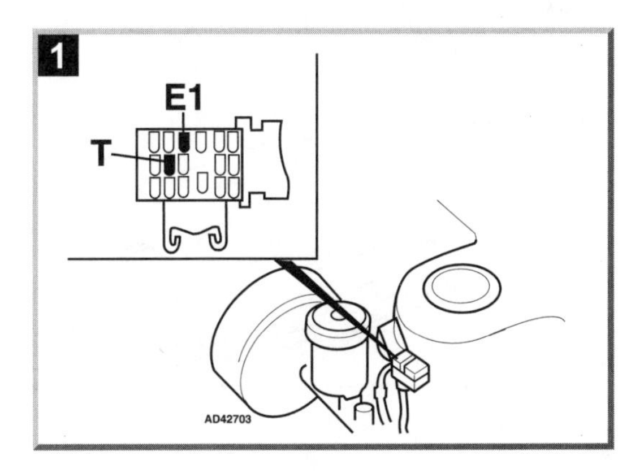

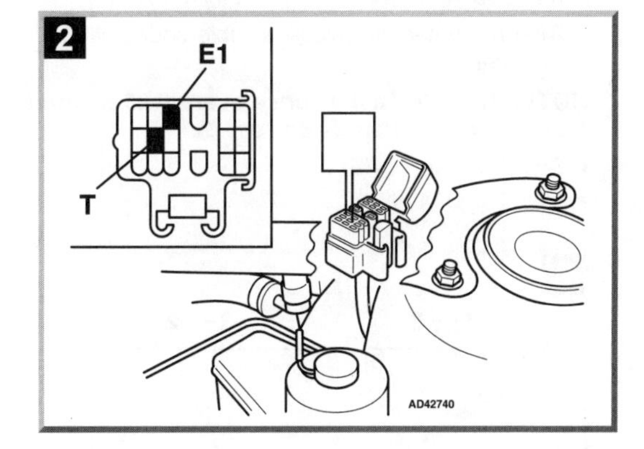

Preparatory conditions

- Battery voltage 11 V minimum.
- Closed throttle position (CTP) switch fully closed.
- Transmission in neutral position.
- All auxiliary equipment, including air conditioning, switched OFF.
- Engine at normal operating temperature.

Accessing

- Switch ignition ON. Do NOT start engine.
- Corolla – jump DLC terminals T and E1 with a short lead [1].
- Celica – jump DLC terminals T and E1 with a short lead [2].
- Count MIL flashes and compare with trouble code table.
- First flashes of 0.5 second duration indicate 'tens', after a 1.5 second pause, next flashes of 0.5 second duration indicate 'units' [3].
- A 2.5 second pause occurs between codes.
- After a 4.5 second pause, trouble codes will be repeated.

NOTE: ***The MIL flashes once every 0.25 seconds if the system is correct*** **[4].**

- Switch ignition OFF.
- Remove jump lead.

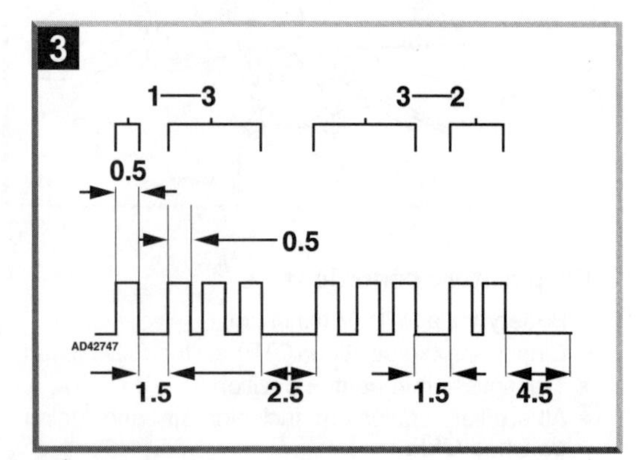

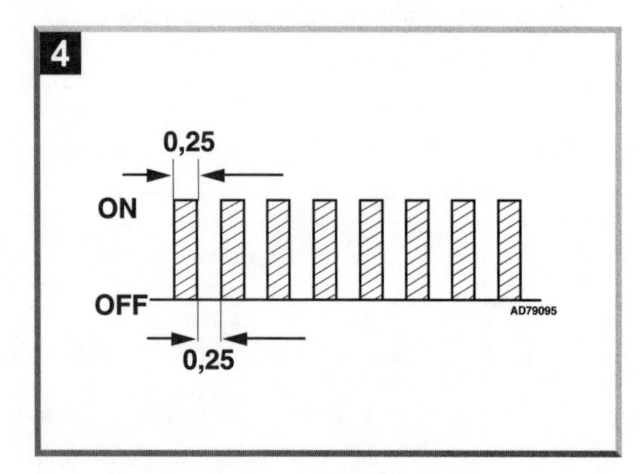

Erasing

- Ensure ignition switched OFF.
- Corolla – remove STOP fuse (15A) from LH kick panel fusebox [5]. Wait 60 seconds.
- Celica – remove EFI fuse (15A) from underhood fusebox [6]. Wait 20 seconds.
- Reinstall fuse.
- Repeat checking procedure to ensure no data remains in ECM fault memory.

NOTE: ***Disconnecting battery ground lead will also erase trouble codes but may erase memory from electronic units such as radio.***

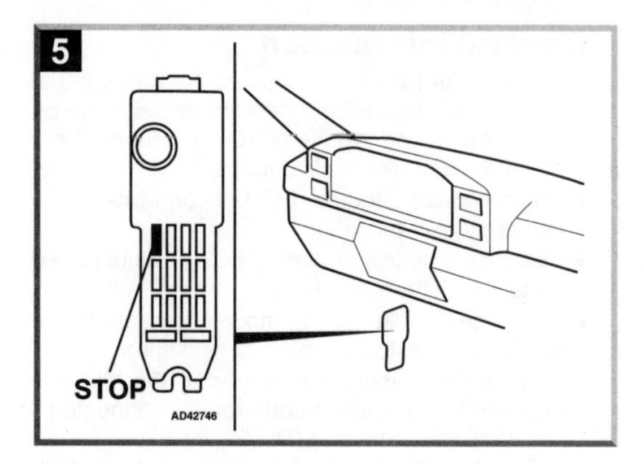

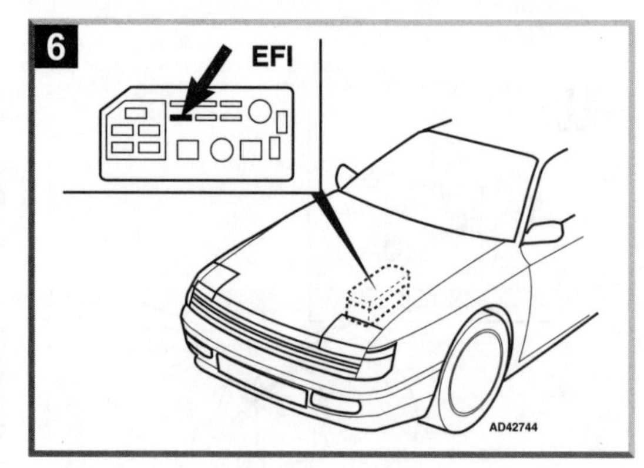

Trouble code identification

Flash code	Fault location	Probable cause
11	Engine control module (ECM) – supply voltage	**Wiring, ignition switch, engine control relay, ECM**
12	Engine speed (RPM) sensor – circuit	**Wiring, RPM sensor, ECM**
13	Engine speed (RPM) sensor – above 1000 rpm	**Wiring, RPM sensor, ECM**
14	Ignition reference signal – no signal	**Wiring, ignition module/coil, ECM**
16	Transmission control signal – malfunction	**ECM**
21	Heated oxygen sensor (HO2S) – circuit	**Wiring, HO2S, ECM**
22	Engine coolant temperature (ECT) sensor – circuit	**Wiring, ECT sensor, ECM**
24	Intake air temperature (IAT) sensor – circuit	**Wiring, IAT sensor, ECM**
25	Mixture control – continuously lean	**Wiring, injector, HO2S, ECT/MAP sensor, intake/fuel/ignition system, ECM**
26	Mixture control – continuously rich	**Wiring, injector, fuel system, cold start injector, HO2S, ECT/MAP sensor, ECM**
27	Heated oxygen sensor (HO2S) – rear – circuit	**Wiring, HO2S, ECM**
31	Manifold absolute pressure (MAP) sensor – circuit	**Wiring, MAP sensor, ECM**
41	Throttle position (TP) sensor – circuit	**Wiring, TP sensor, ECM**
42	Vehicle speed sensor (VSS) – circuit	**Wiring, VSS, ECM**
43	Starter signal – circuit	**Wiring, ignition switch, ECM**
51	Switch signal – A/C switch ON during diagnosis	**Wiring, A/C switch, A/C amplifier, ECM**
51	Switch signal – closed throttle position (CTP) switch OFF during diagnosis	**Wiring, CTP switch, ECM**
51	Switch signal – park/neutral position (PNP) switch not in P or N during diagnosis	**Wiring, PNP switch, ECM**
52	Knock sensor (KS) – circuit	**Wiring, KS, ECM**
53	Knock control – malfunction	**ECM**
71	Exhaust gas recirculation (EGR) system – malfunction	**Hose leak/blockage, wiring, EGRT sensor, EGR solenoid, ECM**

TOYOTA

Model:	Engine identification:	Year:
Celica 1.8L	7A-FE	1994-95
Celica 2.2L	5S-FE	1994-95
Corolla 1.6L	4A-FE	1993-95
Corolla 1.8L	7A-FE	1993-95

System: **TCCS**

Self-diagnosis

General information

- Refer to the front of this manual for general test conditions, terminology, detailed descriptions of wiring faults and a general trouble shooter for electrical and mechanical faults.
- Engine control module (ECM) incorporates self-diagnosis function.
- Malfunction indicator lamp (MIL) will illuminate if certain faults are recorded.
- ECM operates in backup mode if sensors fail, to enable vehicle to be driven to workshop.
- ECM trouble codes can be displayed by the MIL or accessed with suitable code reader connected to the data link connector (DLC):
  - Corolla – **1**.
  - Celica – **2**

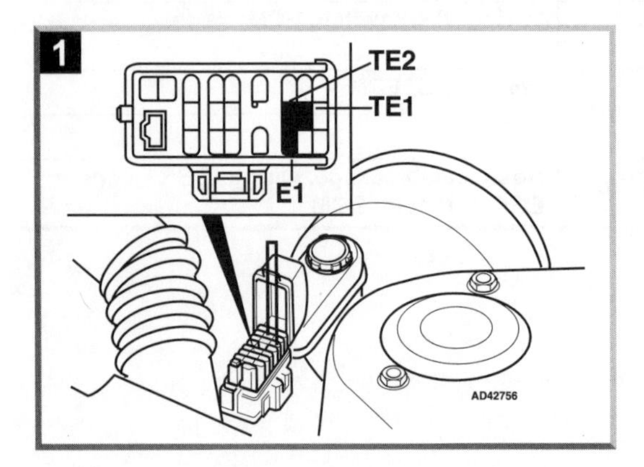

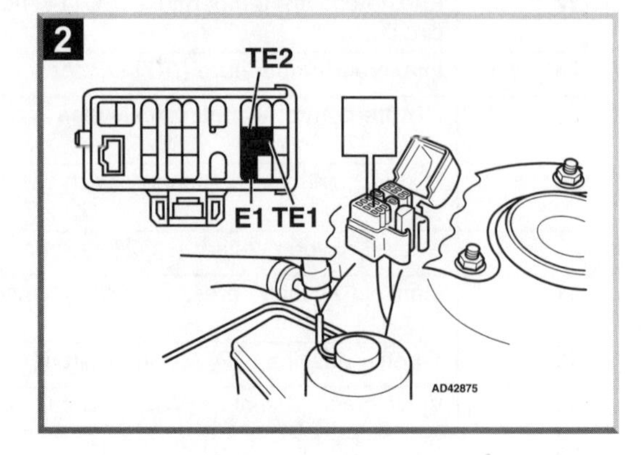

Preparatory conditions

- Battery voltage 11 V minimum.
- Closed throttle position (CTP) switch fully closed.
- Transmission in neutral position.
- All auxiliary equipment, including air conditioning, switched OFF.
- Engine at normal operating temperature.

Accessing – Normal mode

- Switch ignition ON. Do NOT start engine.
- Check MIL illuminates.
- Switch ignition OFF.
- Corolla – jump DLC terminals TE1 and E1 with a short lead **1**.
- Celica – jump DLC terminals TE1 and E1 with a short lead **2**.
- Switch ignition ON.
- Count MIL flashes and compare with trouble code table.

- First flashes of 0.5 second duration indicate 'tens', after a 1.5 second pause, next flashes of 0.5 second duration indicate 'units' **3**.
- A 2.5 second pause occurs between codes.
- After a 4.5 second pause, trouble codes will be repeated.

NOTE: ***The MIL flashes once every 0.25 seconds if the system is correct 4.***

- Switch ignition OFF.
- Remove jump lead.

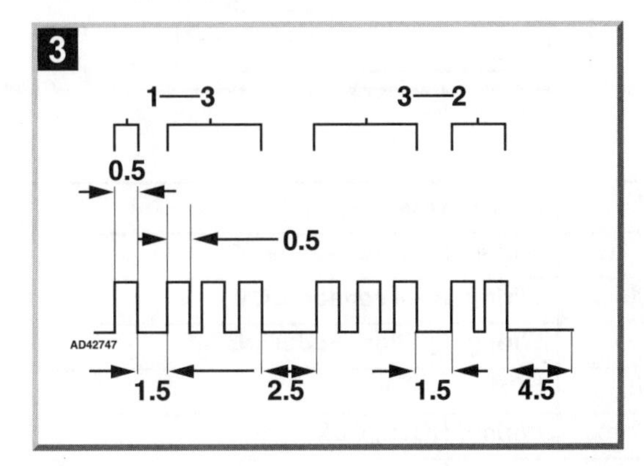

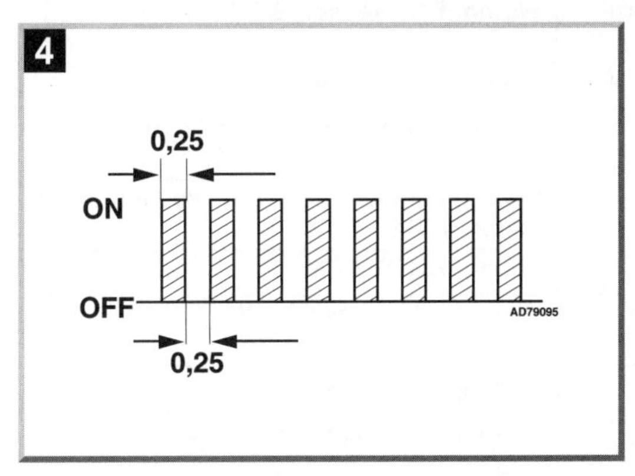

Accessing – Test mode

- Carry out preparatory conditions as in normal mode.
- Corolla – jump DLC terminals TE2 and E1 with a short lead **1**.
- Celica – jump DLC terminals TE2 and E1 with a short lead **2**.
- Switch ignition ON. Do NOT start engine.
- Check MIL is flashing to confirm test mode is operating.

NOTE: ***Test mode will not be activated if the DLC terminals are connected after the ignition switch is turned ON.***

- Start the engine.
- Drive vehicle over 10 mph to simulate possible fault conditions.
- Stop vehicle.
- Jump DLC terminals TE1 and E1 with a short lead **1** or **2**.
- Count MIL flashes and compare with trouble code table.
- Switch ignition OFF.
- Remove jump leads.

Erasing

- Ensure ignition switched OFF.
- Corolla – remove EFI fuse (15A) from underhood fusebox **5**.
- Celica – remove EFI fuse (15A) from underhood fusebox **6**.
- Wait 30 seconds.
- Reinstall fuse.
- Repeat checking procedure to ensure no data remains in ECM fault memory.

NOTE: ***Disconnecting battery ground lead will also erase trouble codes but may erase memory from electronic units such as radio.***

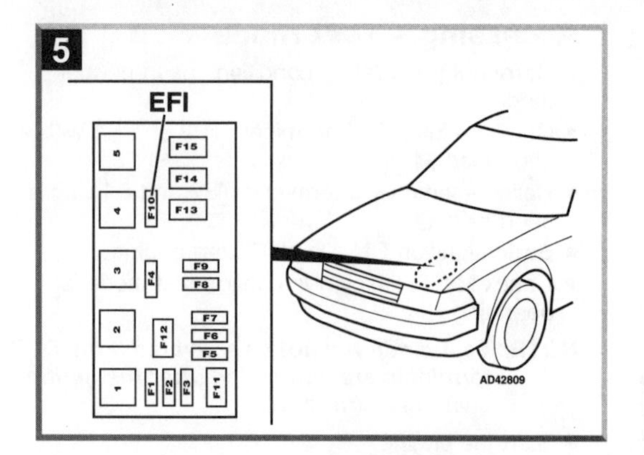

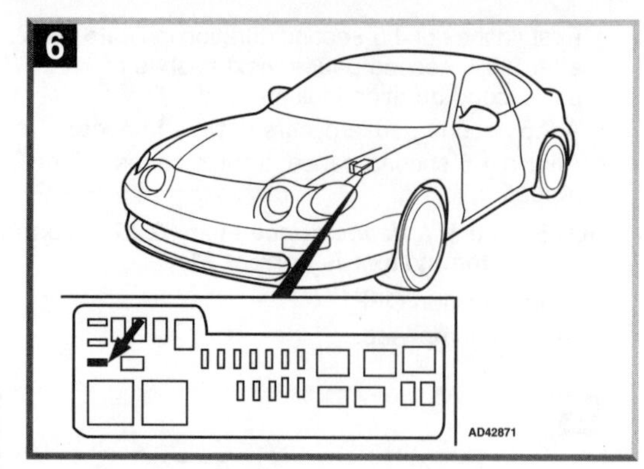

Trouble code identification

Flash code	Fault location	Probable cause
12	Engine speed (RPM) sensor – circuit	**Wiring, RPM sensor, ECM**
13	Engine speed (RPM) sensor – above 1500 rpm	**Wiring, RPM sensor, ECM**
14	Ignition reference signal – no signal	**Wiring, ignition module, ECM**
16	Transmission control signal – malfunction	**ECM**
21	Heated oxygen sensor (HO2S) – front – circuit	**Wiring, HO2S, ECM**
22	Engine coolant temperature (ECT) sensor – circuit	**Wiring, ECT sensor, ECM**
24	Intake air temperature (IAT) sensor – circuit	**Wiring, IAT sensor, ECM**
25	Mixture control – continuously lean	**Wiring, injector, HO2S, intake/fuel/ignition system, valve timing, ECM**
26	Mixture control – continuously rich	**Wiring, injector, intake/fuel system, valve timing, ECM**
27	Heated oxygen sensor (HO2S) – rear – circuit	**Wiring, HO2S, ECM**
31	Manifold absolute pressure (MAP) sensor – circuit	**Wiring, MAP sensor, ECM**
41	Throttle position (TP) sensor – circuit	**Wiring, TP sensor, ECM**
42	Vehicle speed sensor (VSS) – circuit	**Wiring, VSS, instrument cluster, ECM**
43	Starter signal – circuit	**Wiring, ignition switch, engine control relay, ECM**
51	Switch signal – A/C switch ON during diagnosis	**Wiring, A/C switch, A/C amplifier, ECM**
51	Switch signal – closed throttle position (CTP) switch OFF during diagnosis	**Wiring, CTP switch, ECM**
51	Switch signal – park/neutral position (PNP) switch not in P or N during diagnosis	**Wiring, PNP switch, ECM**
52	Knock sensor (KS) – circuit	**Wiring, KS, ECM**
71	Exhaust gas recirculation (EGR) system – malfunction	**Hose leak/blockage, wiring, EGRT sensor, EGR solenoid, ECM**

TOYOTA

Model:	Engine identification:	Year:
Highlander 2.4L	2AZ-FE	2001-04
Highlander 3.0L	1MZ-FE	2001-03
Highlander 3.3L	3MZ-FE	2004
RAV4 2.0L	1AZ-FE	2002-03
RAV4 2.4L	2AZ-FE	2004
Tacoma 3.4L	5VZ -FE	1995-04
Tundra Pickup 3.4L	5VZ-FE	2000-04
T100 Pickup 3.4L	5VZ-FE	1995-98
4Runner 3.4L	5VZ-FE	1996-02
4Runner 4.0L	1GR-FE	2003-04

System: **SFI**

Self-diagnosis

General information

- Refer to the front of this manual for general test conditions, terminology, detailed descriptions of wiring faults and a general trouble shooter for electrical and mechanical faults.
- Engine control module (ECM) incorporates self-diagnosis function.
- Malfunction indicator lamp (MIL) will illuminate if certain faults are recorded.
- ECM operates in backup mode if sensors fail, to enable vehicle to be driven to workshop.

Accessing

- Trouble codes can be displayed by using a Scan Tool connected to the data link connector (DLC) **1**.

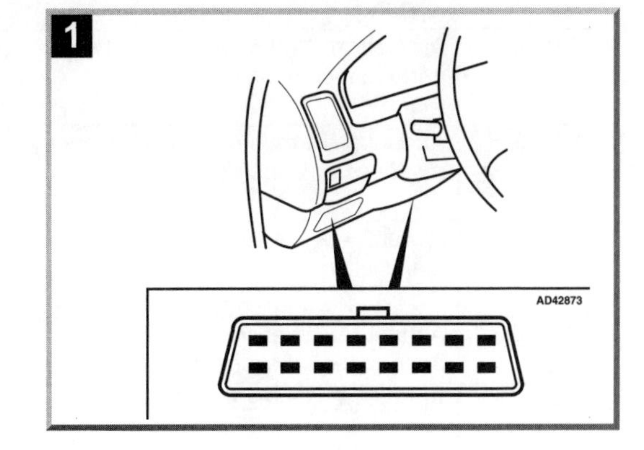

Erasing

- After the faults have been rectified, trouble codes can be erased by using a Scan Tool connected to the data link connector (DLC) or as follows:
- Ensure ignition switched OFF.
- Disconnect battery ground cable for 60 seconds minimum.
- Reconnect battery ground cable.
- Repeat checking procedure to ensure no data remains in ECM fault memory.

NOTE: ***Disconnecting battery ground lead will erase trouble codes but may also erase memory from electronic units such as radio.***

Trouble code identification

OBD-II code	Fault location	Probable cause
All P0, P2 and U0 codes	Refer to OBD-II trouble code tables at the front of this manual	–
P1100	Barometric pressure (BARO) sensor – circuit malfunction	**Wiring, BARO sensor, ECM**
P1120	Accelerator pedal position (APP) sensor – circuit malfunction	**Wiring, APP sensor, ECM**
P1121	Accelerator pedal position (APP) sensor – range/performance problem	**APP sensor, ECM**
P1125	Throttle actuator control (TAC) motor – circuit malfunction	**Wiring, TAC motor, ECM**
P1126	Throttle actuator control (TAC) clutch – circuit malfunction	**Wiring, magnetic clutch, ECM**
P1127	Throttle actuator control (TAC) power source – circuit malfunction	**Wiring, ECM**
P1128	Throttle actuator control (TAC) lock – malfunction	**TAC motor, throttle body, ECM**
P1129	Throttle actuator control (TAC) system – malfunction	**TAC system, ECM**
P1130	Air fuel (A/F) ratio sensor – LH front – range/ performance malfunction	**Wiring, air fuel (A/F) ratio sensor, air induction system, fuel pressure, injector(s), ECM**
P1133	Air fuel (A/F) ratio sensor – LH front – response malfunction	**Wiring, air fuel (A/F) ratio sensor, air induction system, fuel pressure, injector(s), ECM**
P1135	Air fuel (A/F) sensor – LH front – heater circuit malfunction	**Wiring, air fuel (A/F) ratio sensor, air induction system, fuel pressure, injector(s), ECM**
P1150	Air fuel (A/F) sensor – RH front – performance malfunction	**Wiring, air fuel (A/F) ratio sensor, ECM**
P1153	Air fuel (A/F) sensor – RH front – response malfunction	**Air fuel (A/F) ratio sensor**
P1155	Air fuel (A/F) sensor – RH front – heater circuit malfunction	**Wiring, air fuel (A/F) ratio sensor, ECM**
P1300	Ignition control – circuit malfunction – T100/4Runner/4Runner/Tacoma/Tundra	**Wiring, ignition module, ECM**
P1300	Ignition control – cylinder No.1 – circuit malfunction	**Wiring, ignition coil/module, ECM**
P1305	Ignition control – cylinder No.2 – circuit malfunction	**Wiring, ignition coil/module, ECM**
P1310	Ignition control – cylinder No.3 – circuit malfunction	**Wiring, ignition coil/module, ECM**
P1315	Ignition control – cylinder No.4 – circuit malfunction	**Wiring, ignition coil/module, ECM**
P1320	Ignition control – cylinder No.5 – circuit malfunction	**Wiring, ignition coil/module, ECM**

OBD-II code	Fault location	Probable cause
P1325	Ignition control – cylinder No.6 – circuit malfunction	Wiring, ignition coil/module, ECM
P1335	Crankshaft position (CKP) sensor – no signal	Wiring, CKP sensor, starter motor, ECM
P1345	Variable valve timing sensor – LH bank – circuit malfunction	Wiring, valve timing sensor, ECM
P1346	Variable valve timing sensor – LH bank – range/performance problem	Valve timing, ECM
P1349	Variable valve timing system – LH bank – malfunction	Valve timing, oil control valve, valve timing control module, ECM
P1350	Variable valve timing sensor – RH bank – circuit malfunction	Wiring, valve timing sensor, ECM
P1351	Variable valve timing sensor – RH bank – range/performance problem	Valve timing, ECM
P1354	Variable valve timing system – RH bank – malfunction	Valve timing, oil control valve, valve timing control module, ECM
P1400	Throttle position (TP) sensor 2 – circuit malfunction	Wiring, TP sensor, ECM
P1401	Throttle position (TP) sensor 2 – range/ performance problem	TP sensor
P1410	Exhaust gas recirculation (EGR) valve position sensor – malfunction	Wiring, EGR valve position sensor, ECM
P1411	Exhaust gas recirculation (EGR) valve position sensor – range/performance problem	EGR valve position sensor
P1500	Starter signal – circuit malfunction	Wiring, ignition switch, engine control relay, ECM
P1520	Stop lamp switch – signal malfunction	Wiring, stop lamp switch, ECM
P1566	Cruise control switch – circuit malfunction	Wiring, cruise control switch, ECM
P1570	Cruise control laser radar sensor – malfunction	Cruise control laser radar sensor, cruise control module, ECM
P1572	Cruise control laser radar sensor – improper aiming of beam axis	Cruise control laser radar sensor alignment, cruise control laser radar sensor
P1575	Cruise control/skid control warning buzzer – malfunction	Wiring, warning buzzer, ABS control module
P1578	Brake system – malfunction	ABS system, ABS control module
P1600	Engine control module (ECM) – supply voltage	Wiring, ECM
P1605	Knock control – malfunction	ECM
P1615	Data bus, cruise control module – communication error	CAN data bus, cruise control module, ECM
P1616	Data bus, cruise control module – communication error	CAN data bus, ECM, cruise control module
P1633	Throttle actuator control (TAC) system – malfunction	ECM
P1645	Body control module – malfunction	Body control module, A/C control module, wiring

OBD-II code	Fault location	Probable cause
P1647	Transmission control module (TCM) – system malfunction	**Trouble code(s) stored**
P1656	Oil control valve/variable valve timing – LH bank – malfunction	**Wiring, oil control valve, ECM**
P1663	Oil control valve/variable valve timing – RH bank – malfunction	**Wiring, oil control valve, ECM**
P1690	Oil control valve/variable valve timing – malfunction	**Wiring, oil control valve, ECM**
P1692	Variable valve timing – malfunction	**Wiring, valve timing sensor, ECM**
P1693	Variable valve timing – malfunction	**Wiring, valve timing sensor, ECM**
P1700	Output shaft speed (OSS) sensor	**Wiring, OSS sensor, ECM**
P1705	AT – direct clutch speed sensor	**Wiring, direct clutch speed sensor, ECM**
P1715	Rear wheel speed sensor(s) – malfunction	**Wiring, rear wheel speed sensor(s), ECM**
P1725	Input shaft speed (ISS) sensor/turbine shaft speed (TSS) sensor – circuit malfunction	**Wiring, ISS sensor/TSS sensor, ECM**
P1730	Intermediate shaft speed sensor – AT	**Wiring, intermediate shaft speed sensor, ECM**
P1755	Torque converter clutch (TCC) solenoid – circuit malfunction	**Wiring, torque converter, TCC solenoid, ECM**
P1760	AT – pressure control shift solenoid	**Wiring, shift solenoid, ECM**
P1765	AT – pressure control shift solenoid	**Wiring, shift solenoid, ECM**
P1770	Differential lock solenoid – circuit malfunction	**Wiring, differential lock solenoid, ECM**
P1780	Park/neutral position (PNP) switch/transmission range (TR) sensor – circuit malfunction	**Wiring, PNP switch/TR sensor, ECM**
P1781	Shift lever switch – malfunction	**Wiring, shift lever switch, ECM**
P1782	Transfer box low ratio switch – circuit malfunction	**Wiring, transfer box low ratio switch, ECM**
P1783	Transfer box neutral switch – circuit malfunction	**Wiring, transfer box neutral switch, ECM**
P1790	Shift/timing solenoid	**Wiring, shift/timing solenoid, ECM**
P1810	CAN data bus malfunction – malfunction	**Wiring**
P1815	Shift control solenoid – malfunction	**Wiring, shift control solenoid, ECM**
P1818	Shift control solenoid – circuit malfunction	**Wiring, shift control solenoid, ECM**
B2785	Ignition switch ON – malfunction	**Immobilizer system**
B2786	Ignition switch OFF – malfunction	**Immobilizer system**
B2795	Unrecognized key code	**Immobilizer system**
B2796	Immobilizer system – no communication	**Immobilizer system**
B2797	Communication – malfunction 1	**Immobilizer system**
B2798	Communication – malfunction 2	**Immobilizer system**
B2799	Immobilizer – malfunction	**Wiring, immobilizer system, ECM**

TOYOTA

Model:	Engine identification:	Year:
Land Cruiser 4.0L	3F-E	1988-90
Land Cruiser 4.0L	3F-E	1991-94

System: **TCCS**

Self-diagnosis

General information

- Refer to the front of this manual for general test conditions, terminology, detailed descriptions of wiring faults and a general trouble shooter for electrical and mechanical faults.
- Engine control module (ECM) incorporates self-diagnosis function.
- Malfunction indicator lamp (MIL) will illuminate if certain faults are recorded.
- ECM operates in backup mode if sensors fail, to enable vehicle to be driven to workshop.
- Trouble codes can be displayed by the MIL or accessed with suitable code reader connected to the data link connector (DLC):
  - 1988-90 – **1**.
  - 1991-94 – **2**.

1

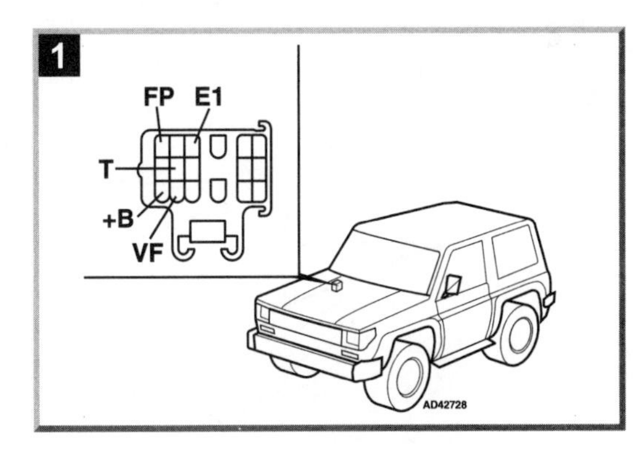

2

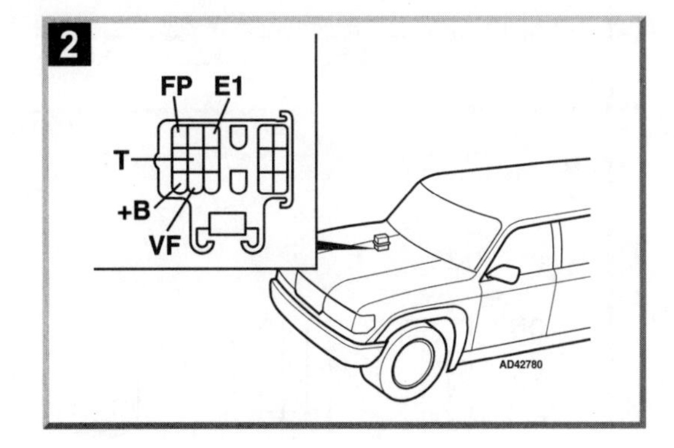

Accessing – MIL

- Switch ignition ON.
- Check MIL illuminates.
- Start engine. Allow to idle.
- If MIL extinguishes, no trouble codes have been recorded.
- If MIL remains illuminated, access trouble codes.

Accessing – trouble codes

- Switch ignition ON. Do NOT start engine.
- 1988-90 – jump DLC terminals T and E1 with a short lead **1**.
- 1991-94 – jump DLC terminals T and E1 with a short lead **2**.
- Count MIL flashes and compare with trouble code table.
- First flashes of 0.5 second duration indicate 'tens', after a 1.5 second pause, next flashes of 0.5 second duration indicate 'units' **3**.
- A 2.5 second pause occurs between codes.
- After a 4.5 second pause, trouble codes will be repeated.

NOTE: ***The MIL flashes once every 0.25 seconds if the system is correct* 4.**

- Switch ignition OFF.
- Remove jump lead.

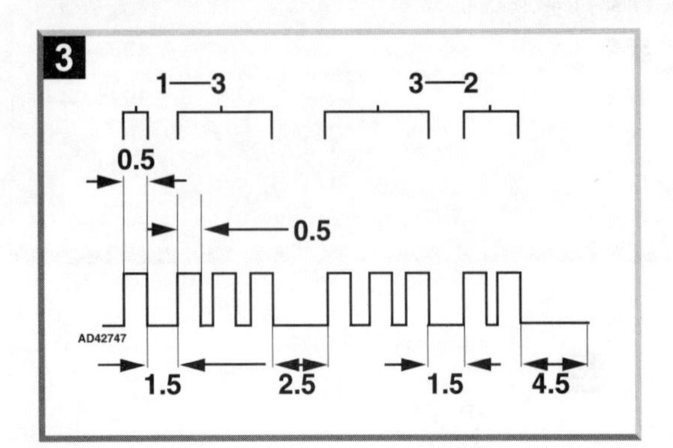

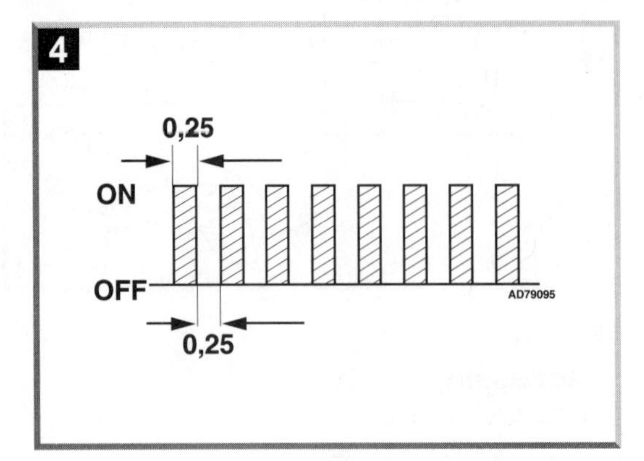

Erasing

- After the faults have been rectified, erase the trouble codes as follows:
- 1988-90 – remove EFI fuse (15A) from LH fascia panel fusebox for 30 seconds minimum **5**.
- 1991-94 – remove EFI fuse (15A) from underhood fusebox for 30 seconds minimum **6**.
- Reinstall fuse.
- Repeat checking procedure to ensure no data remains in ECM fault memory.

NOTE: ***Disconnecting battery ground lead will also erase trouble codes but may erase memory from electronic units such as radio.***

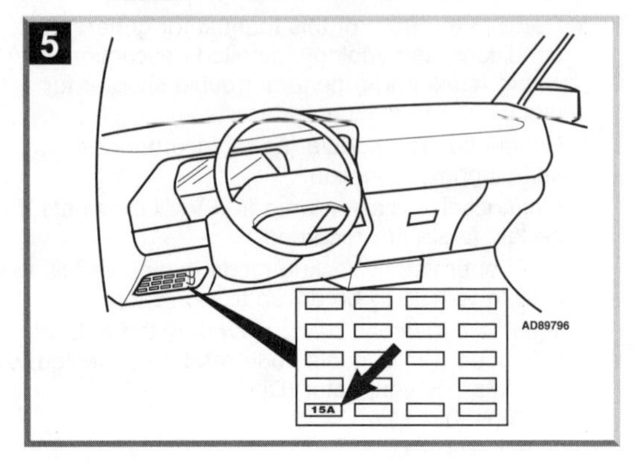

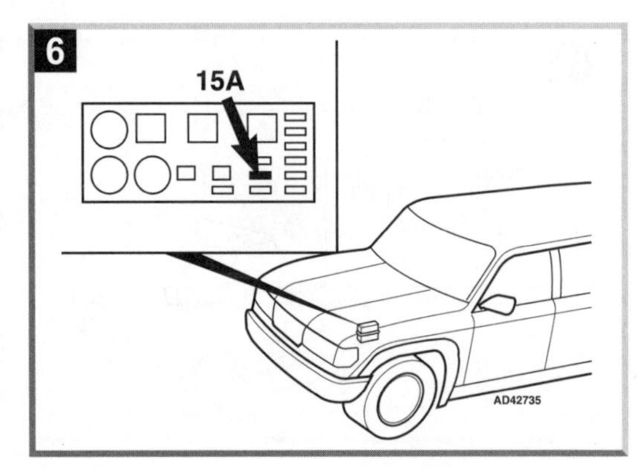

Trouble code identification

Flash code	Fault location	Probable cause
11	Engine control module (ECM) power supply	**Wiring, ignition switch, engine control relay, ECM**
12	Engine speed (RPM) sensor – circuit	**Wiring, RPM sensor, ECM**
13	Engine speed (RPM) sensor – above 1000 rpm	**Wiring, RPM sensor, ECM**
14	Ignition reference signal – no signal	**Wiring, ignition module/coil, ECM**
21	Heated oxygen sensor (HO2S) 1 – circuit	**Wiring, HO2S, ECM**
22	Engine coolant temperature (ECT) sensor – circuit	**Wiring, ECT sensor, ECM**
24	Intake air temperature (IAT) sensor – circuit	**Wiring, IAT sensor, ECM**
25	Mixture control – continuously lean	**Wiring, injector, HO2S, ECT/VAF sensor, intake/fuel/ignition system, ECM**
26	Mixture control – continuously rich	**Wiring, injector, fuel system, cold start injector, HO2S, ECT/VAF sensor, ECM**
28	Heated oxygen sensor (HO2S) 2 – circuit	**Wiring, HO2S, ECM**
31	Volume air flow (VAF) sensor – circuit	**Wiring, VAF sensor, ECM**
32	Volume air flow (VAF) sensor – circuit	**Wiring, VAF sensor, ECM**
35	Barometric pressure (BARO) sensor – circuit	**ECM**
41	Throttle position (TP) sensor – circuit	**Wiring, TP sensor, ECM**
42	Vehicle speed sensor (VSS) – circuit	**Wiring, VSS, ECM**
43	Starter signal – circuit	**Wiring, ignition switch, ECM**
51	Switch signal – A/C switch ON during diagnosis	**Wiring, A/C switch, A/C amplifier, ECM**
51	Switch signal – closed throttle position (CTP) switch OFF during diagnosis	**Wiring, CTP switch, ECM**
51	Switch signal – park/neutral position (PNP) switch not in P or N during diagnosis	**Wiring, PNP switch, ECM**
52	Knock sensor (KS) 1 – circuit	**Wiring, KS, ECM**
53	Knock control – malfunction	**ECM**
55	Knock sensor (KS) 2 – circuit	**Wiring, KS, ECM**
71	Exhaust gas recirculation (EGR) system – malfunction	**Hose leak/blockage, wiring, EGRT sensor, EGR solenoid, ECM**
81	Transmission control module (TCM) – communication	**Wiring**
83	Transmission control module (TCM) – communication	**Wiring**
84	Transmission control module (TCM) – communication	**Wiring**
85	Transmission control module (TCM) – communication	**Wiring**

TOYOTA

Model: **Land Cruiser 4.5L**

Engine identification: **1FZ-FE**

Year: **1993-94**

System: **TCCS**

Self-diagnosis

General information

- Refer to the front of this manual for general test conditions, terminology, detailed descriptions of wiring faults and a general trouble shooter for electrical and mechanical faults.
- Engine control module (ECM) incorporates self-diagnosis function.
- Malfunction indicator lamp (MIL) will illuminate if certain faults are recorded.
- ECM operates in backup mode if sensors fail, to enable vehicle to be driven to workshop.
- ECM trouble codes can be accessed with suitable code reader connected to the data link connector (DLC) **1** or displayed by the MIL.

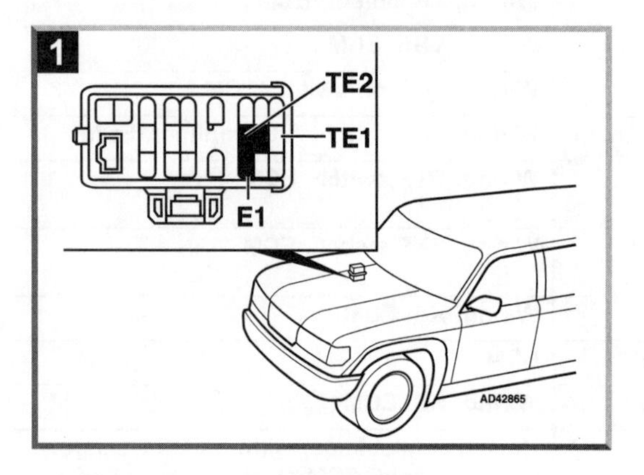

Preparatory conditions

- Battery voltage 11 V minimum.
- Closed throttle position (CTP) switch fully closed.
- Transmission in neutral position.
- All auxiliary equipment, including air conditioning, switched OFF.
- Engine at normal operating temperature.

Accessing – Normal mode

- Switch ignition ON. Do NOT start engine.
- Check MIL illuminates.
- Switch ignition OFF.
- Jump DLC terminals TE1 and E1 with a short lead **1**.
- Switch ignition ON.
- Count MIL flashes and compare with trouble code table.
- First flashes of 0.5 second duration indicate 'tens', after a 1.5 second pause, next flashes of 0.5 second duration indicate 'units' **2**.
- A 2.5 second pause occurs between codes.
- After a 4.5 second pause, trouble codes will be repeated.

NOTE: ***The MIL flashes once every 0.25 seconds if the system is correct* 3.**

- Switch ignition OFF.
- Remove jump lead.

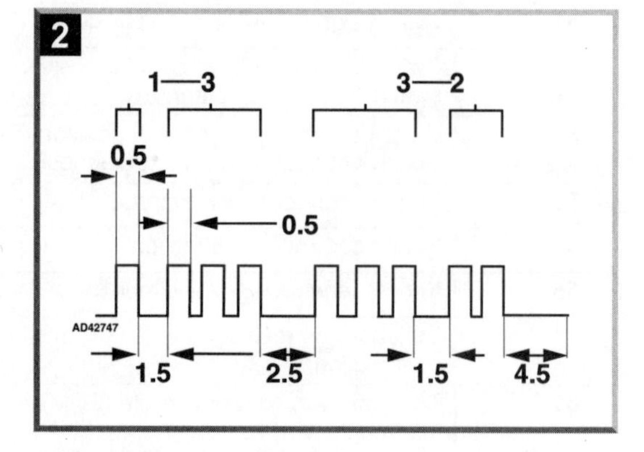

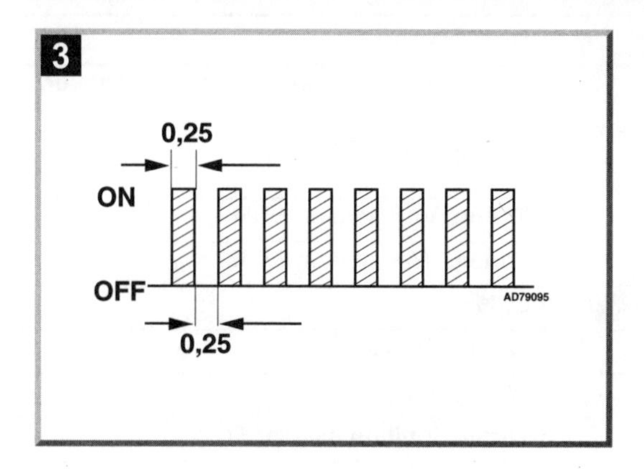

Accessing – Test mode

- Carry out preparatory conditions as in normal mode.
- Jump DLC terminals TE2 and E1 with a short lead 1.
- Switch ignition ON. Do NOT start engine.
- Check MIL is flashing to confirm test mode is operating.

NOTE: ***Test mode will not be activated if the DLC terminals are connected after the ignition switch is turned ON.***

- Start the engine.
- Drive vehicle over 10 mph to simulate possible fault conditions.
- Stop vehicle.
- Jump DLC terminals TE1 and E1 with a short lead 1.
- Count MIL flashes and compare with trouble code table.
- Switch ignition OFF.
- Remove jump leads.

Erasing

- Ensure ignition switched OFF.
- Remove EFI fuse (15A) from underhood fusebox 4.
- Wait 30 seconds.
- Reinstall fuse.
- Repeat checking procedure to ensure no data remains in ECM fault memory.

NOTE: ***Disconnecting battery ground lead will also erase trouble codes but may erase memory from electronic units such as radio.***

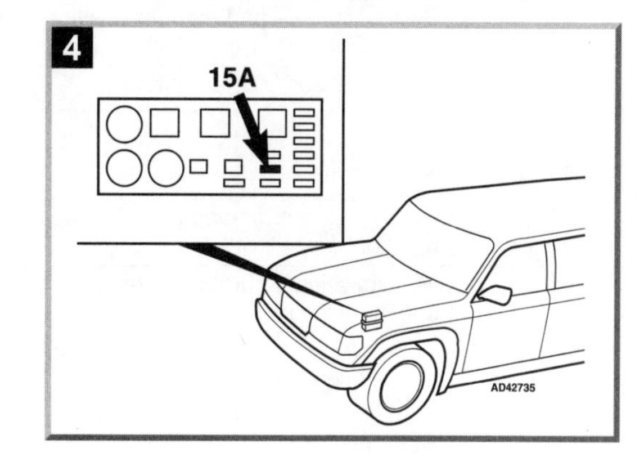

Trouble code identification

Flash code	Fault location	Probable cause
12	Engine speed (RPM) sensor – circuit	**Wiring, RPM sensor, ECM**
13	Engine speed (RPM) sensor – above 1000 rpm	**Wiring, RPM sensor, ECM**
14	Ignition reference signal – no signal	**Wiring, ignition module/coil, ECM**
21	Heated oxygen sensor (HO2S) – cylinders 1, 2 & 3 – circuit	**Wiring, HO2S, ECM**
22	Engine coolant temperature (ECT) sensor – circuit	**Wiring, ECT sensor, ECM**
24	Intake air temperature (IAT) sensor – circuit	**Wiring, IAT sensor, ECM**
25	Mixture control – continuously lean	**Wiring, injector, HO2S, VAF sensor, intake/fuel/ignition/PAIR system, ECM**

Flash code	Fault location	Probable cause
26	Mixture control – continuously rich	**Wiring, injector, fuel system, ECT sensor, ECM**
28	Heated oxygen sensor (HO2S) – cylinders 4, 5 & 6 – circuit	**Wiring, HO2S, ECM**
31	Volume air flow (VAF) sensor – circuit	**Wiring, VAF sensor, ECM**
32	Volume air flow (VAF) sensor – circuit	**Wiring, VAF sensor, ECM**
35	Barometric pressure (BARO) sensor – circuit	**ECM**
41	Throttle position (TP) sensor – circuit	**Wiring, TP sensor, ECM**
42	Vehicle speed sensor (VSS) – circuit	**Wiring, VSS, ECM**
43	Starter signal – circuit	**Wiring, ignition switch, ECM**
51	Switch signal – A/C switch ON during diagnosis	**Wiring, A/C switch, A/C amplifier, ECM**
51	Switch signal – closed throttle position (CTP) switch OFF during diagnosis	**Wiring, CTP switch, ECM**
51	Switch signal – park/neutral position (PNP) switch not in P or N during diagnosis	**Wiring, PNP switch, ECM**
52	Knock sensor (KS) 1 – circuit	**Wiring, KS, ECM**
53	Knock control – malfunction	**ECM**
55	Knock sensor (KS) 2 – circuit	**Wiring, KS, ECM**
71	Exhaust gas recirculation (EGR) system – malfunction	**Hose leak/blockage, wiring, EGRT sensor, EGR solenoid, ECM**
81	Transmission control module (TCM) – communication	**Wiring**
83	Transmission control module (TCM) – communication	**Wiring**
84	Transmission control module (TCM) – communication	**Wiring**
85	Transmission control module (TCM) – communication	**Wiring**

Model:	Engine identification:	Year:
Land Cruiser 4.5L	**1FZ-FE**	**1995-97**
Supra 3.0L	**2JZ-GE**	**1996-98**
Supra 3.0L Turbo	**2JZ-GTE**	**1996-98**

System: **SFI**

Self-diagnosis

General information

- Refer to the front of this manual for general test conditions, terminology, detailed descriptions of wiring faults and a general trouble shooter for electrical and mechanical faults.
- Engine control module (ECM) incorporates self-diagnosis function.
- Malfunction indicator lamp (MIL) will illuminate if certain faults are recorded.
- ECM operates in backup mode if sensors fail, to enable vehicle to be driven to workshop.

Accessing

- Trouble codes can be displayed by using a Scan Tool connected to the data link connector (DLC) **1**.

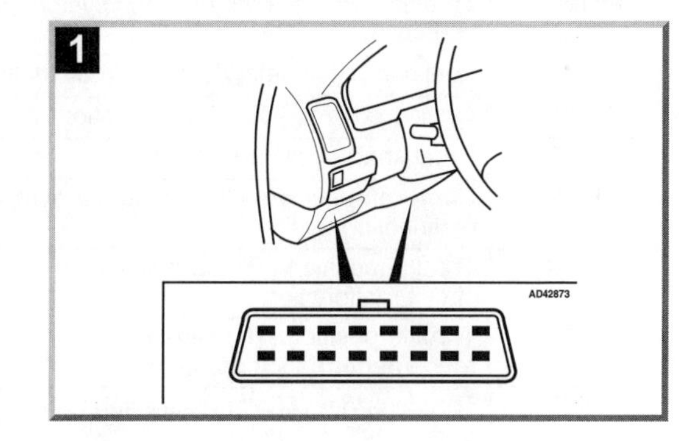

Erasing

- After the faults have been rectified, trouble codes can be erased by using a Scan Tool connected to the data link connector (DLC) or as follows:
- Ensure ignition switched OFF.
- Disconnect battery ground cable for 30 seconds minimum.
- Reconnect battery ground cable.
- Repeat checking procedure to ensure no data remains in ECM fault memory.

NOTE: ***Disconnecting battery ground lead will erase trouble codes but may also erase memory from electronic units such as radio.***

Trouble code identification

OBD-II code	Fault location	Probable cause
All P0, P2 and U0 codes	Refer to OBD-II trouble code tables at the front of this manual	–
P1100	Barometric pressure (BARO) sensor – circuit malfunction	**ECM**
P1120	Accelerator pedal position (APP) sensor – circuit malfunction	**Wiring, APP sensor, ECM**

OBD-II code	Fault location	Probable cause
P1121	Accelerator pedal position (APP) sensor – range/performance problem	**APP sensor, ECM**
P1125	Throttle actuator control (TAC) motor – circuit malfunction	**Wiring, TAC motor, ECM**
P1126	Magnetic clutch – circuit malfunction	**Wiring, magnetic clutch, ECM**
P1127	Traction control system – circuit malfunction	**Wiring, ECM**
P1128	Throttle actuator control (TAC) motor – lock malfunction	**TAC motor, throttle body, ECM**
P1129	Throttle actuator control (TAC) system – malfunction	**Wiring, ECM**
P1200	Fuel pump (FP) relay/ECM – circuit malfunction	**Wiring, fuel pump/control module, ECM**
P1300	Ignition control – circuit malfunction	**Wiring, ignition module, ECM**
P1335	Crankshaft position (CKP) sensor – no signal	**Wiring, CKP sensor, starter motor, ECM**
P1349	Camshaft position (CMP) actuator system – malfunction	**Valve timing, CMP actuator, ECM**
P1400	Throttle position (TP) sensor 2 – circuit malfunction	**Wiring, TP sensor, ECM**
P1401	Throttle position (TP) sensor 2 – range/performance problem	**TP sensor**
P1405	Turbocharger (TC) pressure sensor – circuit malfunction	**Wiring, TC pressure sensor, ECM**
P1406	Turbocharger (TC) pressure sensor – range/performance problem	**TC pressure sensor**
P1500	Starter signal – circuit malfunction	**Wiring, ignition switch, engine control relay, ECM**
P1511	Turbocharger (TC) pressure – too low	**Intake system/air control valve, wiring, TC wastegate regulating valve/actuator, ECM**
P1512	Turbocharger (TC) pressure – too high	**Wiring, TC wastegate regulating valve/ actuator, ECM**
P1520	Stop lamp switch – signal malfunction	**Wiring, stop lamp switch, ECM**
P1565	Cruise control switch – signal malfunction	**Wiring, cruise control switch, ECM**
P1600	Engine control module (ECM) – supply voltage	**Wiring, ECM**
P1605	Knock control – malfunction	**ECM**
P1630	Traction control system – malfunction	**Wiring, throttle control module, ECM**
P1633	Engine control module (ECM) – throttle actuator control (TAC) system – malfunction	**ECM**
P1652	Turbocharger (TC) 2 air control solenoid – circuit malfunction	**Wiring, air control solenoid, ECM**
P1656	Oil control valve – circuit malfunction	**Wiring, oil control valve, ECM**
P1658	Turbocharger (TC) wastegate regulating valve – circuit malfunction	**Wiring, TC wastegate regulating valve, ECM**
P1661	Turbocharger (TC) 2 exhaust gas control valve – circuit malfunction	**Wiring, exhaust gas control valve, ECM**
P1662	Exhaust bypass valve – circuit malfunction	**Wiring, exhaust bypass valve, ECM**
P1780	Park/neutral position (PNP) switch – circuit malfunction	**Wiring, PNP switch, ECM**

TOYOTA

Model:	Engine identification:	Year:
Land Cruiser 4.7L	2UZ-FE	1998-04
Sequoia 4.7L	2UZ-FE	2001-04
Tundra 4.7L	2UZ-FE	2000-04
4Runnner 4.7L	2UZ-FE	2003-04

System: **SFI**

Self-diagnosis

General information

- Refer to the front of this manual for general test conditions, terminology, detailed descriptions of wiring faults and a general trouble shooter for electrical and mechanical faults.
- Engine control module (ECM) incorporates self-diagnosis function.
- Malfunction indicator lamp (MIL) will illuminate if certain faults are recorded.
- ECM operates in backup mode if sensors fail, to enable vehicle to be driven to workshop.

Accessing

- Trouble codes can be displayed by using a Scan Tool connected to the data link connector (DLC) **1**.

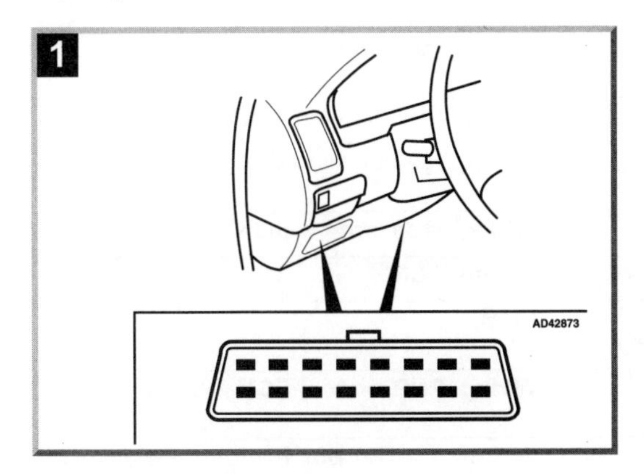

Erasing

- After the faults have been rectified, trouble codes can be erased by using a Scan Tool connected to the data link connector (DLC) or as follows:
- Ensure ignition switched OFF.
- Disconnect battery ground cable for 60 seconds minimum.
- Reconnect battery ground cable.
- Repeat checking procedure to ensure no data remains in ECM fault memory.

NOTE: ***Disconnecting battery ground lead will erase trouble codes but may also erase memory from electronic units such as radio.***

Trouble code identification

OBD-II code	Fault location	Probable cause
All P0, P2 and U0 codes	Refer to OBD-II trouble code tables at the front of this manual	–
P1120	Accelerator pedal position (APP) sensor – circuit malfunction	**Wiring, APP sensor, ECM**
P1121	Accelerator pedal position (APP) sensor – range/performance problem	**APP sensor, ECM**
P1125	Throttle actuator control (TAC) motor – circuit malfunction	**Wiring, TAC motor, ECM**
P1126	Magnetic clutch – circuit malfunction	**Wiring, magnetic clutch, ECM**
P1127	Traction control system – circuit malfunction	**Wiring, ECM**
P1128	Throttle actuator control (TAC) motor – lock malfunction	**TAC motor, throttle body, ECM**
P1129	Throttle actuator control (TAC) system – malfunction	**Wiring, ECM**
P1200	Fuel pump (FP) relay/ECM – circuit malfunction	**Wiring, fuel pump/control module, ECM**
P1300	Ignition control – cylinder No.1 – circuit malfunction	**Wiring, ignition module/coil, ECM**
P1305	Ignition control – cylinder No.2 – circuit malfunction	**Wiring, ignition module/coil, ECM**
P1310	Ignition control – cylinder No.3 – circuit malfunction	**Wiring, ignition module/coil, ECM**
P1315	Ignition control – cylinder No.4 – circuit malfunction	**Wiring, ignition module/coil, ECM**
P1320	Ignition control – cylinder No.5 – circuit malfunction	**Wiring, ignition module/coil, ECM**
P1325	Ignition control – cylinder No.6 – circuit malfunction	**Wiring, ignition module/coil, ECM**
P1330	Ignition control – cylinder No.7 – circuit malfunction	**Wiring, ignition module/coil, ECM**
P1335	Crankshaft position (CKP) sensor – no signal	**Wiring, CKP sensor, starter motor, ECM**
P1340	Ignition control – cylinder No.8 – circuit malfunction	**Wiring, ignition module/coil, ECM**
P1520	Stop lamp switch – signal malfunction	**Wiring, stop lamp switch, ECM**
P1600	Engine control module (ECM) – supply voltage	**Wiring, ECM**
P1633	Engine control module (ECM) – throttle actuator control (TAC) system – malfunction	**ECM**
P1780	Park/neutral position (PNP) switch – circuit malfunction	**Wiring, PNP switch, ECM**
P1782	Transfer box low ratio switch – circuit malfunction	**Wiring, transfer box low ratio switch, ECM**
P1783	Transfer box neutral switch – circuit malfunction	**Wiring, transfer box neutral switch, ECM**
B2785	Ignition switch ON – malfunction	**Wiring, ignition switch, engine control relay**

OBD-II code	Fault location	Probable cause
B2786	Ignition switch OFF – malfunction	Wiring, ignition switch, engine control relay
B2791	Key unlock warning switch – malfunction	Wiring, engine control relay, key unlock warning switch
B2795	Unrecognized key code	Key, incorrect key
B2796	Immobilizer system – no communication	Key, transponder, wiring, ECM
B2797	Communication malfunction 1 – interference	Incorrect key has been used, wiring, ignition module, ECM
B2798	Communication malfunction 2	Key, transponder, ignition module, wiring, ECM
B2799	Immobilizer – malfunction	Wiring, immobilizer system, ECM

TOYOTA

Model:	Engine identification:	Year:
MR2 2.2L	5S-FE	1991-95
Paseo 1.5L	5E-FE	1992-95

System: **TCCS**

Self-diagnosis

General information

- Refer to the front of this manual for general test conditions, terminology, detailed descriptions of wiring faults and a general trouble shooter for electrical and mechanical faults.
- Engine control module (ECM) incorporates self-diagnosis function.
- Malfunction indicator lamp (MIL) will illuminate if certain faults are recorded.
- ECM operates in backup mode if sensors fail, to enable vehicle to be driven to workshop.
- ECM trouble codes can be displayed by the MIL or accessed with suitable code reader connected to the data link connector (DLC):
  - Paseo – **1**.
  - MR2 – **2**.

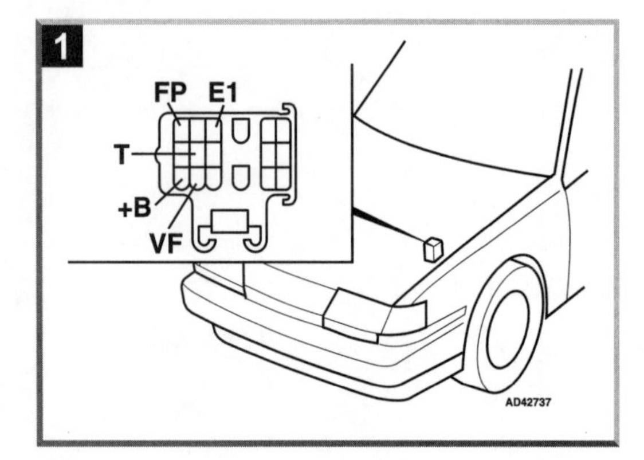

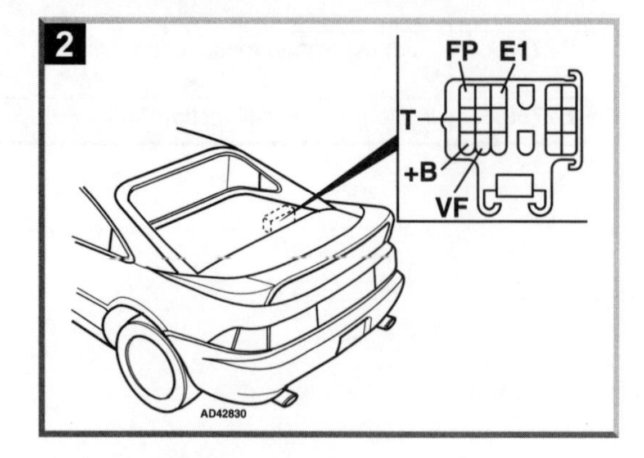

Preparatory conditions

- Battery voltage 11 V minimum.
- Closed throttle position (CTP) switch fully closed.
- Transmission in neutral position.
- All auxiliary equipment, including air conditioning, switched OFF.
- Engine at normal operating temperature.

Accessing

- Switch ignition ON. Do NOT start engine.
- Check MIL illuminates.
- Switch ignition OFF.
- Paseo – jump DLC terminals TE1 and E1 with a short lead **1**.
- MR2 – jump DLC terminals TE1 and E1 with a short lead **2**.
- Switch ignition ON.
- Count MIL flashes and compare with trouble code table.
- First flashes of 0.5 second duration indicate 'tens', after a 1.5 second pause, next flashes of 0.5 second duration indicate 'units' **3**.
- A 2.5 second pause occurs between codes.

- After a 4.5 second pause, trouble codes will be repeated.

NOTE: ***The MIL flashes once every 0.25 seconds if the system is correct 4.***

- Switch ignition OFF.
- Remove jump lead.

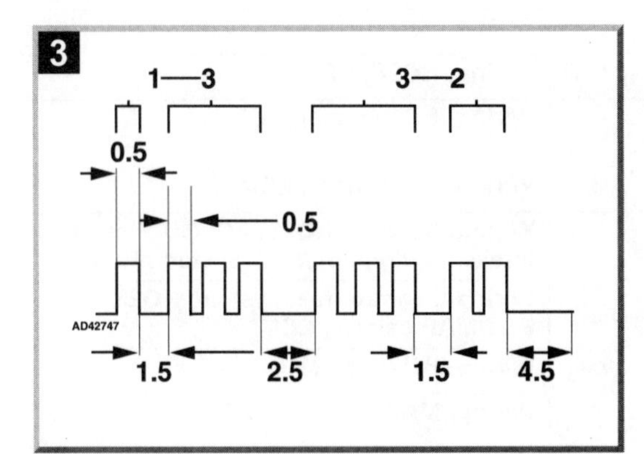

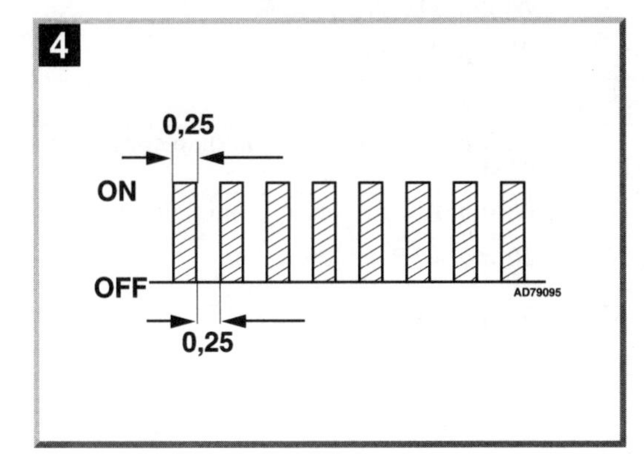

Erasing

- Ensure ignition switched OFF.
- Paseo – remove EFI fuse (15A) from underhood fusebox 5.
- MR2 – remove EFI fuse (15A) from underhood fusebox 6.
- Wait 30 seconds.
- Reinstall fuse.
- Repeat checking procedure to ensure no data remains in ECM fault memory.

NOTE: ***Disconnecting battery ground lead will also erase trouble codes but may erase memory from electronic units such as radio.***

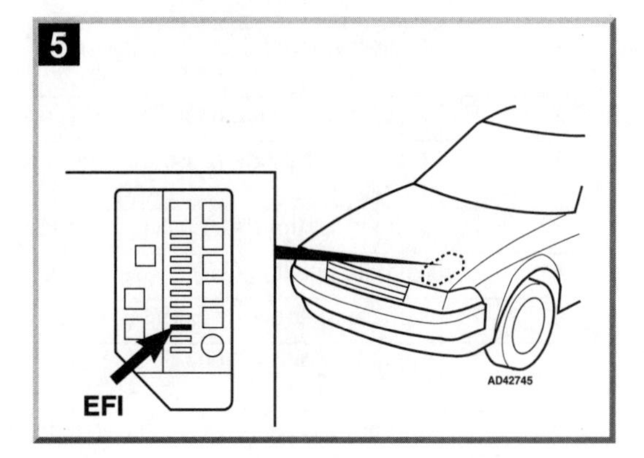

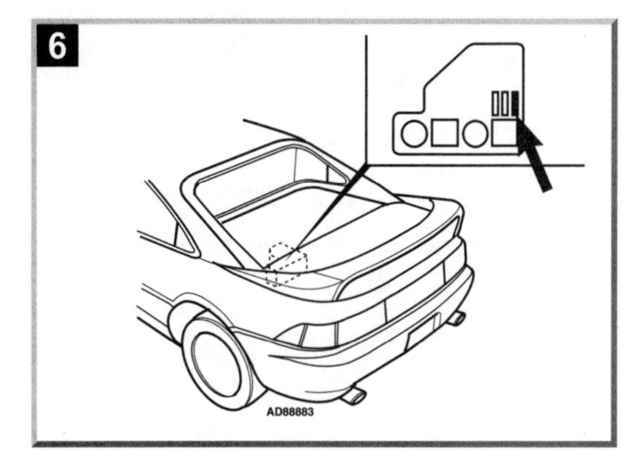

Trouble code identification

Flash code	Fault location	Probable cause
12	Engine speed (RPM) sensor – circuit	**Wiring, RPM sensor, ECM**
13	Engine speed (RPM) sensor – above 1000 rpm	**Wiring, RPM sensor, ECM**
14	Ignition reference signal – no signal	**Wiring, ignition module, ECM**
16	Transmission control signal – malfunction	**ECM**
21	Heated oxygen sensor (HO2S) – front – circuit	**Wiring, HO2S, ECM**
22	Engine coolant temperature (ECT) sensor – circuit	**Wiring, ECT sensor, ECM**
24	Intake air temperature (IAT) sensor – circuit	**Wiring, IAT sensor, ECM**
25	Mixture control – continuously lean	**Wiring, injector, HO2S, ECT/MAP sensor, intake/fuel/ignition/EGR system, ECM**
26	Mixture control – continuously rich	**Wiring, injector, fuel system, HO2S, ECT/MAP sensor, ECM**
27	Heated oxygen sensor (HO2S) – rear – circuit	**Wiring, HO2S, ECM**
31	Manifold absolute pressure (MAP) sensor – circuit	**Wiring, MAP sensor, ECM**
33	Idle air control (IAC) valve – circuit	**Wiring, IAC valve, ECM**
41	Throttle position (TP) sensor – circuit	**Wiring, TP sensor, ECM**
42	Vehicle speed sensor (VSS) – circuit	**Wiring, VSS, ECM**
43	Starter signal – circuit	**Wiring, ignition switch, engine control relay, ECM**
51	Switch signal – A/C switch ON during diagnosis	**Wiring, A/C switch, A/C amplifier, ECM**
51	Switch signal – closed throttle position (CTP) switch OFF during diagnosis	**Wiring, CTP switch, ECM**
51	Switch signal – park/neutral position (PNP) switch not in P or N during diagnosis	**Wiring, PNP switch, ECM**
52	Knock sensor (KS) – circuit	**Wiring, KS, ECM**
71	Exhaust gas recirculation (EGR) system – malfunction	**Hose leak/blockage, wiring, EGRT sensor, EGR solenoid, ECM**

Model:	Engine identification:	Year:
Pickup 3.0L	3VZ-E	1992-95
T100 Pickup 3.0L	3VZ-E	1993-94
4Runner 3.0L	3VZ-E	1992-95

System: **TCCS**

Self-diagnosis

General information

- Refer to the front of this manual for general test conditions, terminology, detailed descriptions of wiring faults and a general trouble shooter for electrical and mechanical faults.
- Engine control module (ECM) incorporates self-diagnosis function.
- Malfunction indicator lamp (MIL) will illuminate if certain faults are recorded.
- ECM operates in backup mode if sensors fail, to enable vehicle to be driven to workshop.
- ECM trouble codes can be displayed by the MIL or accessed with suitable code reader connected to the data link connector (DLC):
  - 4Runner/Pickup – **1**.
  - T100 Pickup – **2**

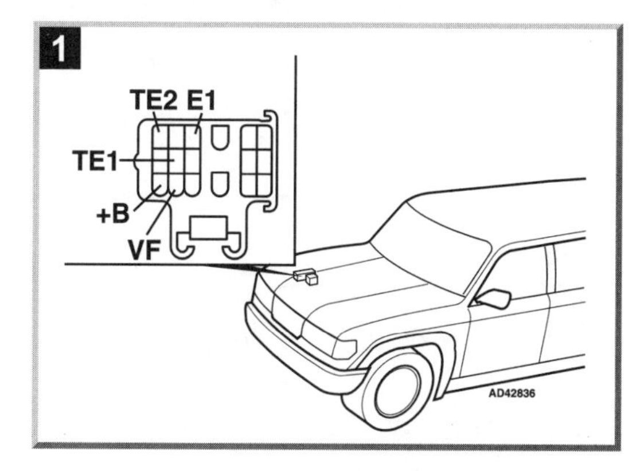

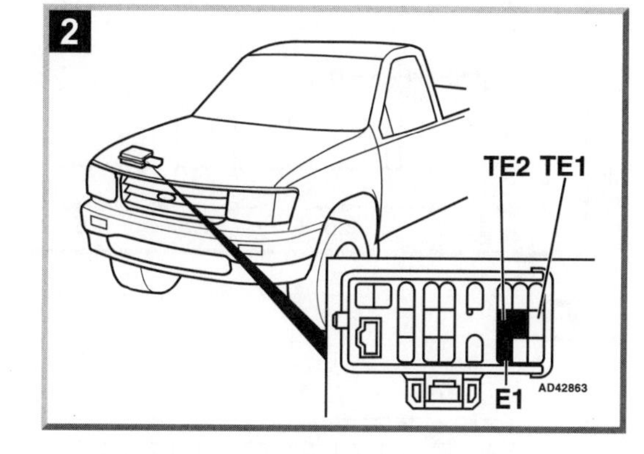

Preparatory conditions

- Battery voltage 11 V minimum.
- Closed throttle position (CTP) switch fully closed.
- Transmission in neutral position.
- All auxiliary equipment, including air conditioning, switched OFF.
- Engine at normal operating temperature.

Accessing – Normal mode

- Switch ignition ON. Do NOT start engine.
- Check MIL illuminates.
- Switch ignition OFF.
- 4Runner/Pickup – jump DLC terminals TE1 and E1 with a short lead **1**.
- T100 Pickup – jump DLC terminals TE1 and E1 with a short lead **2**.
- Switch ignition ON.
- Count MIL flashes and compare with trouble code table.
- First flashes of 0.5 second duration indicate 'tens', after a 1.5 second pause, next flashes of 0.5 second duration indicate 'units' **3**.
- A 2.5 second pause occurs between codes.

- After a 4.5 second pause, trouble codes will be repeated.

NOTE: ***The MIL flashes once every 0.25 seconds if the system is correct 4.***

- Switch ignition OFF.
- Remove jump lead.

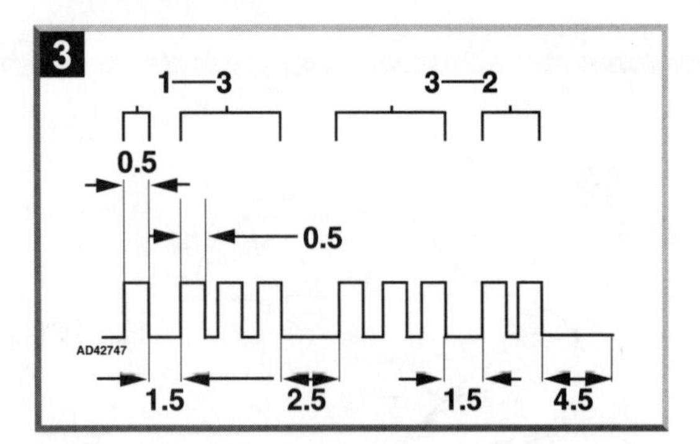

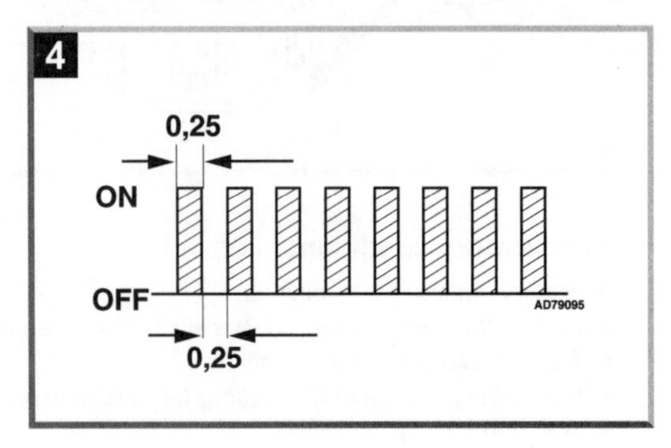

Accessing – Test mode

- Carry out preparatory conditions as in normal mode.
- 4Runner/Pickup – jump DLC terminals TE2 and E1 with a short lead 1.
- T100 Pickup – jump DLC terminals TE2 and E1 with a short lead 2.
- Switch ignition ON. Do NOT start engine.
- Check MIL is flashing to confirm test mode is operating.

NOTE: ***Test mode will not be activated if the DLC terminals are connected after the ignition switch is turned ON.***

- Start the engine.
- Drive vehicle over 10 mph to simulate possible fault conditions.
- Stop vehicle.
- Jump DLC terminals TE1 and E1 with a short lead 1 or 2.
- Count MIL flashes and compare with trouble code table.
- Switch ignition OFF.
- Remove jump leads.

Erasing

- Ensure ignition switched OFF.
- 4Runner/Pickup – remove EFI fuse (15A) from underhood fusebox 5.
- T100 Pickup – remove EFI fuse (15A) from underhood fusebox 6.
- Wait 30 seconds.
- Reinstall fuse.
- Repeat checking procedure to ensure no data remains in ECM fault memory.

NOTE: ***Disconnecting battery ground lead will also erase trouble codes but may erase memory from electronic units such as radio.***

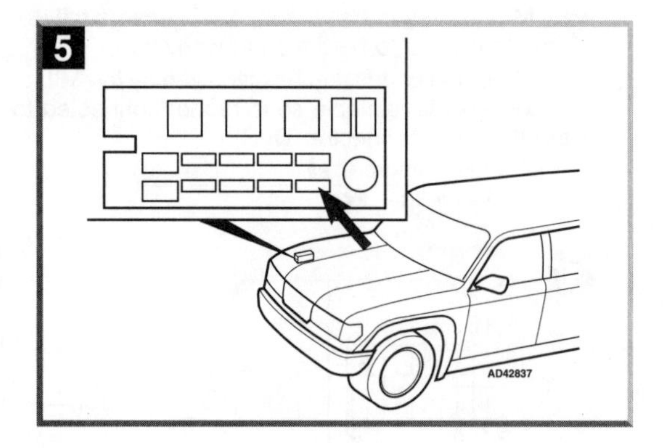

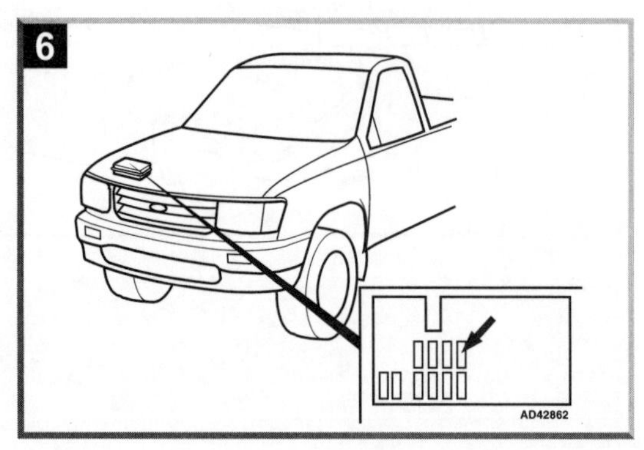

Trouble code identification

Flash code	Fault location	Probable cause
12	Engine speed (RPM) sensor – circuit	**Wiring, RPM sensor, ECM**
13	Engine speed (RPM) sensor – above 1000 rpm	**Wiring, RPM sensor, ECM**
14	Ignition reference signal – no signal	**Wiring, ignition module, ECM**
16	Transmission control signal – malfunction	**ECM**
21	Heated oxygen sensor (HO2S) – front – circuit	**Wiring, HO2S, ECM**
22	Engine coolant temperature (ECT) sensor – circuit	**Wiring, ECT sensor, ECM**
24	Intake air temperature (IAT) sensor – circuit	**Wiring, IAT sensor, ECM**
25	Mixture control – continuously lean	**Wiring, injector, HO2S, ECT/VAF sensor, intake/fuel/ignition system, ECM**
26	Mixture control – continuously rich	**Wiring, injector, fuel system, cold start injector, HO2S, ECT/VAF sensor, ECM**
27	Heated oxygen sensor (HO2S) – rear – circuit	**Wiring, HO2S, ECM**
31	Volume air flow (VAF) sensor – circuit	**Wiring, VAF sensor, ECM**
32	Volume air flow (VAF) sensor – circuit	**Wiring, VAF sensor, ECM**
35	Barometric pressure (BARO) sensor – circuit	**ECM**
41	Throttle position (TP) sensor – circuit	**Wiring, TP sensor, ECM**
42	Vehicle speed sensor (VSS) – circuit	**Wiring, VSS, ECM**
43	Starter signal – circuit	**Wiring, ignition switch, engine control relay, ECM**
51	Switch signal – A/C switch ON during diagnosis	**Wiring, A/C switch, A/C amplifier, ECM**
51	Switch signal – closed throttle position (CTP) switch OFF during diagnosis	**Wiring, CTP switch, ECM**
51	Switch signal – park/neutral position (PNP) switch not in P or N during diagnosis	**Wiring, PNP switch, ECM**
52	Knock sensor (KS) – circuit	**Wiring, KS, ECM**
53	Knock control – malfunction	**ECM**
71	Exhaust gas recirculation (EGR) system – malfunction	**Hose leak/blockage, wiring, EGRT sensor, EGR solenoid, ECM**

TOYOTA

Model: **Previa 2.4L**
Engine identification: **2TZ-FE**
Year: **1991-95**

System: **TCCS**

Self-diagnosis

General information

- Refer to the front of this manual for general test conditions, terminology, detailed descriptions of wiring faults and a general trouble shooter for electrical and mechanical faults.
- Engine control module (ECM) incorporates self-diagnosis function.
- Malfunction indicator lamp (MIL) will illuminate if certain faults are recorded.
- ECM operates in backup mode if sensors fail, to enable vehicle to be driven to workshop.
- ECM trouble codes can be accessed with suitable code reader connected to the data link connector (DLC) **1** or displayed by the MIL.

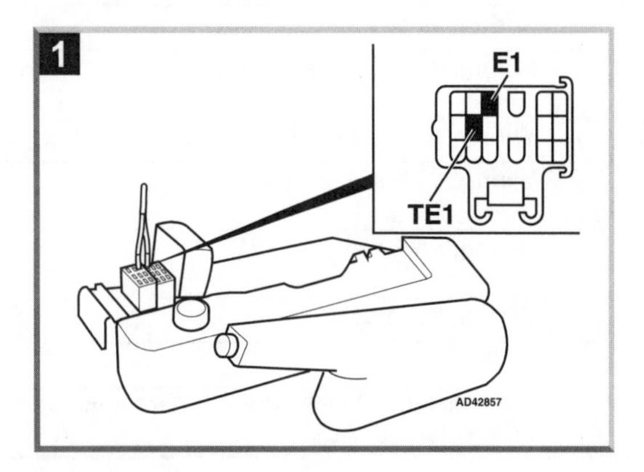

Preparatory conditions

- Battery voltage 11 V minimum.
- Closed throttle position (CTP) switch fully closed.
- Transmission in neutral position.
- All auxiliary equipment, including air conditioning, switched OFF.
- Engine at normal operating temperature.

Accessing

- Switch ignition ON. Do NOT start engine.
- Check MIL illuminates.
- Switch ignition OFF.
- Jump DLC terminals TE1 and E1 **1**.
- Switch ignition ON.
- Count MIL flashes and compare with trouble code table.
- First flashes of 0.5 second duration indicate 'tens', after a 1.5 second pause, next flashes of 0.5 second duration indicate 'units' **2**.
- A 2.5 second pause occurs between codes.
- After a 4.5 second pause, trouble codes will be repeated.

NOTE: ***The MIL flashes once every 0.25 seconds if the system is correct*** **3**.

- Switch ignition OFF.
- Remove jump lead.

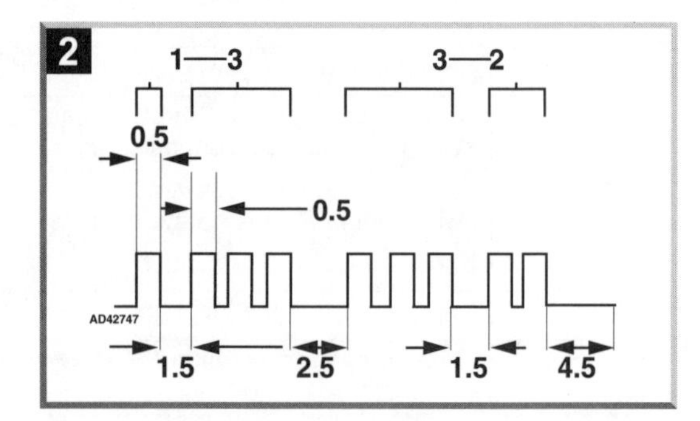

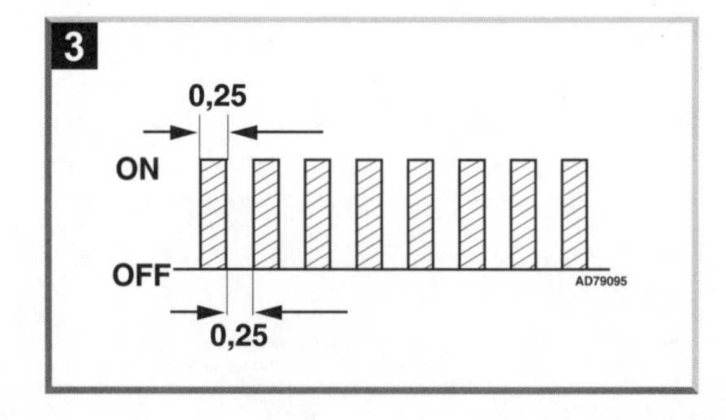

Erasing

- Ensure ignition switched OFF.
- Remove EFI fuse (15A) from fascia fusebox **4**.
- Wait 30 seconds.
- Reinstall fuse.
- Repeat checking procedure to ensure no data remains in ECM fault memory.

NOTE: ***Disconnecting battery ground lead will also erase trouble codes but may erase memory from electronic units such as radio.***

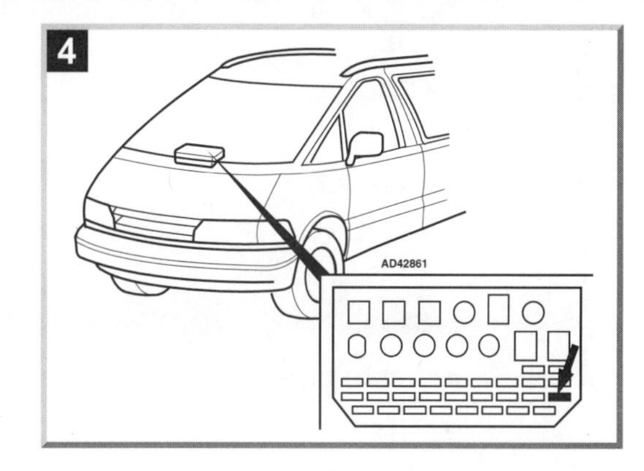

Trouble code identification

Flash code	Fault location	Probable cause
12	Engine speed (RPM) sensor – circuit	**Wiring, RPM sensor, ignition module, ECM**
13	Engine speed (RPM) sensor – above 1000 rpm	**Wiring, RPM sensor, ECM**
14	Ignition reference signal – no signal	**Wiring, ignition module, ECM**
21	Heated oxygen sensor (HO2S) – front – circuit	**Wiring, HO2S, ECM**
22	Engine coolant temperature (ECT) sensor – circuit	**Wiring, ECT sensor, ECM**
24	Intake air temperature (IAT) sensor – circuit	**Wiring, IAT sensor, ECM**
25	Mixture control – continuously lean	**Wiring, injector, HO2S, ECT/VAF sensor, intake/fuel/ignition system, ECM**
26	Mixture control – continuously rich	**Wiring, injector, fuel system, cold start injector, HO2S, ECT/VAF sensor, ECM**
27	Heated oxygen sensor (HO2S) – rear – circuit	**Wiring, HO2S, ECM**
31	Volume air flow (VAF) sensor – circuit	**Wiring, VAF sensor, ECM**
32	Volume air flow (VAF) sensor – circuit	**Wiring, VAF sensor, ECM**
41	Throttle position (TP) sensor – circuit	**Wiring, TP sensor, ECM**
42	Vehicle speed sensor (VSS) – circuit	**Wiring, VSS, ECM**
43	Starter signal – circuit	**Wiring, ignition switch, engine control relay, ECM**
51	Switch signal – closed throttle position (CTP) switch OFF during diagnosis	**Wiring, CTP switch, ECM**
51	Switch signal – park/neutral position (PNP) switch not in P or N during diagnosis	**Wiring, PNP switch, ECM**
52	Knock sensor (KS) – circuit	**Wiring, KS, ECM**
53	Knock control – malfunction	**ECM**
71	Exhaust gas recirculation (EGR) system – malfunction	**Hose leak/blockage, wiring, EGRT sensor, EGR solenoid, ECM**

TOYOTA

Model:	Engine identification:	Year:
Previa 2.4L	2TZ-FZE	1995-97
Tacoma 2.4L	2RZ-FE	1995-04
Tacoma 2.7L	3RZ-FE	1995-04
T100 Pickup 2.7L	3RZ-FE	1994-98
4Runner 2.7L	3RZ-FE	1995-00

System: **SFI**

Self-diagnosis

General information

- Refer to the front of this manual for general test conditions, terminology, detailed descriptions of wiring faults and a general trouble shooter for electrical and mechanical faults.
- Engine control module (ECM) incorporates self-diagnosis function.
- Malfunction indicator lamp (MIL) will illuminate if certain faults are recorded.
- ECM operates in backup mode if sensors fail, to enable vehicle to be driven to workshop.

Accessing

- Trouble codes can be displayed by using a Scan Tool connected to the data link connector (DLC)
  - Previa – 1
  - Except Previa – 2

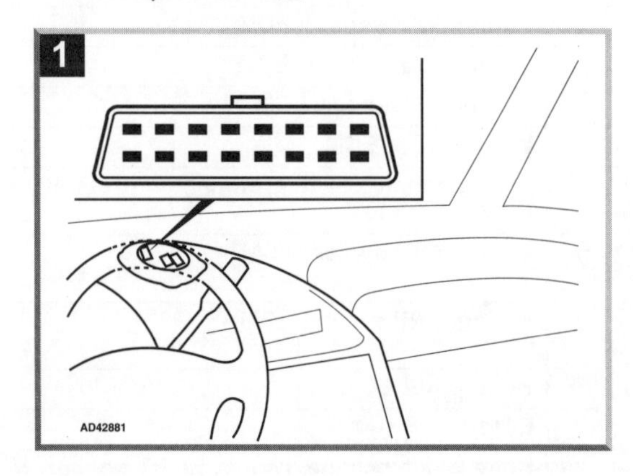

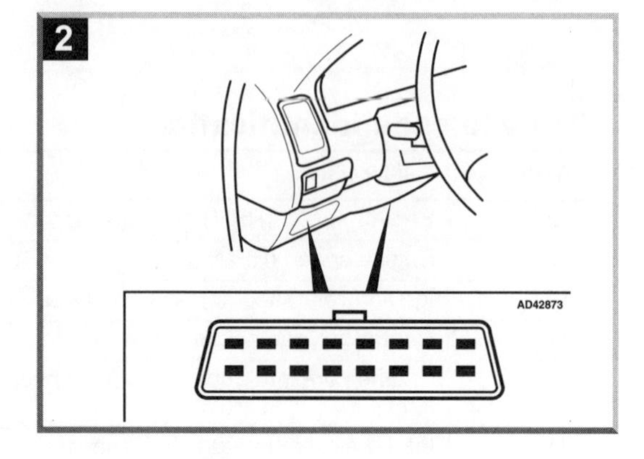

Erasing

- After the faults have been rectified, trouble codes can be erased by using a Scan Tool connected to the data link connector (DLC) or as follows:
- Ensure ignition switched OFF.
- Disconnect battery ground cable for 60 seconds minimum.
- Reconnect battery ground cable.
- Repeat checking procedure to ensure no data remains in ECM fault memory.

NOTE: ***Disconnecting battery ground lead will erase trouble codes but may also erase memory from electronic units such as radio.***

Trouble code identification

OBD-II code	Fault location	Probable cause
All P0, P2 and U0 codes	Refer to OBD-II trouble code tables at the front of this manual	–
P1130	Air fuel (A/F) ratio sensor – LH front – range/performance malfunction	**Wiring, air fuel (A/F) ratio sensor, air induction system, fuel pressure, injector(s), ECM**
P1133	Air fuel (A/F) ratio sensor – LH front – response malfunction	**Air fuel (A/F) ratio sensor, air induction system, fuel pressure, injector(s), ECM**
P1135	Air fuel (A/F) sensor – LH front – heater circuit malfunction	**Wiring, air fuel (A/F) ratio sensor, ECM**
P1300	Ignition control 1 – circuit malfunction	**Wiring, ignition module, ECM**
P1310	Ignition control 2 – circuit malfunction	**Wiring, ignition module, ECM**
P1300	Ignition control – cylinder No.1 – circuit malfunction – Tacoma (2000→)	**Wiring, ignition coil/module, ECM**
P1305	Ignition control – cylinder No.2 – circuit malfunction – Tacoma (2000→)	**Wiring, ignition coil/module, ECM**
P1310	Ignition control – cylinder No.3 – circuit malfunction – Tacoma (2000→)	**Wiring, ignition coil/module, ECM**
P1315	Ignition control – cylinder No.4 – circuit malfunction – Tacoma (2000→)	**Wiring, ignition coil/module, ECM**
P1335	Crankshaft position (CKP) sensor – no signal	**Wiring, CKP sensor, ECM**
P1500	Starter signal – circuit malfunction	**Wiring, ignition switch, engine control relay, ECM**
P1510	Supercharger (SC) pressure control – circuit malfunction	**Wiring, SC magnetic clutch/relay, SC bypass valve, intake system, ECM**
P1520	Stop lamp switch – signal malfunction	**Wiring, stop lamp switch, ECM**
P1600	Engine control module (ECM) – supply voltage	**Wiring, ECM**
P1605	Knock control – malfunction	**ECM**
P1780	Park/neutral position (PNP) switch – circuit malfunction	**Wiring, PNP switch, ECM**
P1782	Transfer box low ratio switch – circuit malfunction	**Wiring, transfer box low ratio switch, ECM**
P1783	Transfer box neutral switch – circuit malfunction	**Wiring, transfer box neutral switch, ECM**

TOYOTA

Model:	Engine identification:	Year:
Supra 3.0L	**2JZ-GE**	**1993-95**
Supra 3.0L Turbo	**2JZ-GTE**	**1993-95**

System: **TCCS**

Self-diagnosis

General information

- Refer to the front of this manual for general test conditions, terminology, detailed descriptions of wiring faults and a general trouble shooter for electrical and mechanical faults.
- Engine control module (ECM) incorporates self-diagnosis function.
- Malfunction indicator lamp (MIL) will illuminate if certain faults are recorded.
- ECM operates in backup mode if sensors fail, to enable vehicle to be driven to workshop.
- ECM trouble codes can be accessed with suitable code reader connected to the data link connector (DLC) **1** or displayed by the MIL.

NOTE: ***A second diagnostic link connector is fitted 2, which can also be used to output diagnostic codes.***

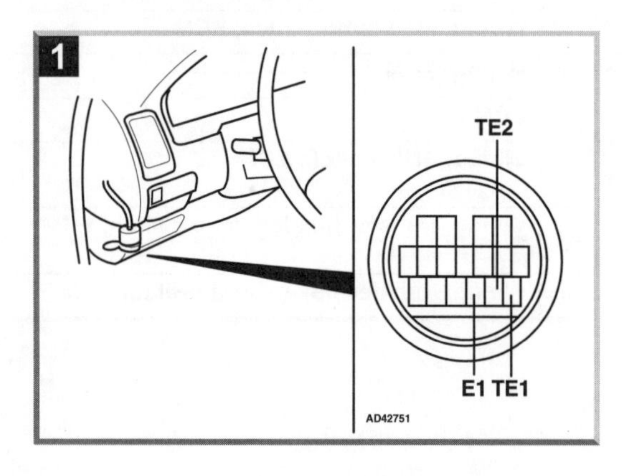

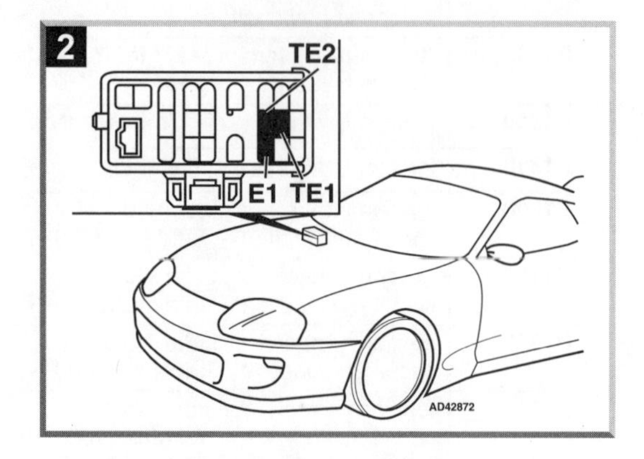

Preparatory conditions

- Battery voltage 11 V minimum.
- Closed throttle position (CTP) switch fully closed.
- Transmission in neutral position.
- All auxiliary equipment, including air conditioning, switched OFF.
- Engine at normal operating temperature.

Accessing – Normal mode

- Switch ignition ON. Do NOT start engine.
- Check MIL illuminates.
- Switch ignition OFF.
- Jump DLC terminals TE1 and E1 with a short lead [1] or [2].
- Switch ignition ON.
- Count MIL flashes and compare with trouble code table.
- First flashes of 0.5 second duration indicate 'tens', after a 1.5 second pause, next flashes of 0.5 second duration indicate 'units' [3].
- A 2.5 second pause occurs between codes.
- After a 4.5 second pause, trouble codes will be repeated.

NOTE: ***The MIL flashes once every 0.25 seconds if the system is correct [4].***

- Switch ignition OFF.
- Remove jump lead.

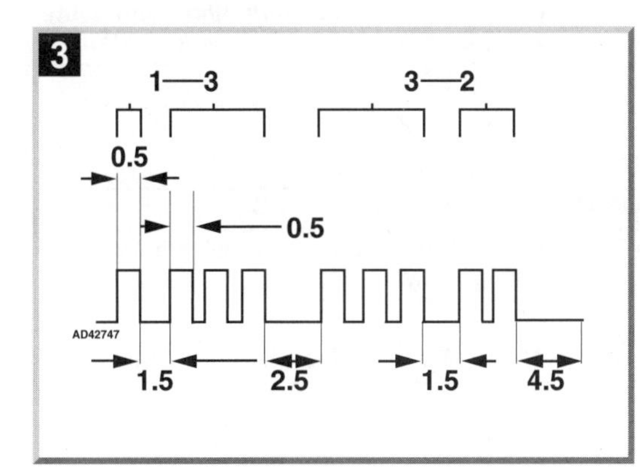

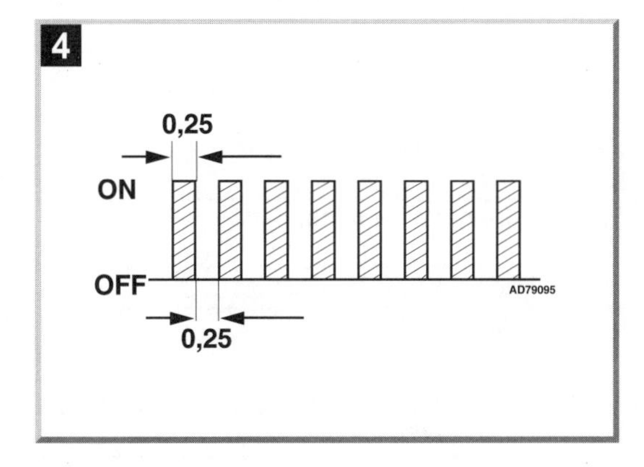

Accessing – Test mode

- Carry out preparatory conditions as in normal mode.
- Jump DLC terminals TE2 and E1 with a short lead [1].
- Switch ignition ON. Do NOT start engine.
- Check MIL is flashing to confirm test mode is operating.

NOTE: ***Test mode will not be activated if the DLC terminals are connected after the ignition switch is turned ON.***

- Start the engine.
- Drive vehicle over 10 mph to simulate possible fault conditions.
- Stop vehicle.
- Jump DLC terminals TE1 and E1 with a short lead [1].
- Count MIL flashes and compare with trouble code table.
- Switch ignition OFF.
- Remove jump leads.

Erasing

- Ensure ignition switched OFF.
- Remove EFI fuse (30A) from underhood fusebox [5].
- Wait 30 seconds.
- Reinstall fuse.
- Repeat checking procedure to ensure no data remains in ECM fault memory.

NOTE: ***Disconnecting battery ground lead will also erase trouble codes but may erase memory from electronic units such as radio.***

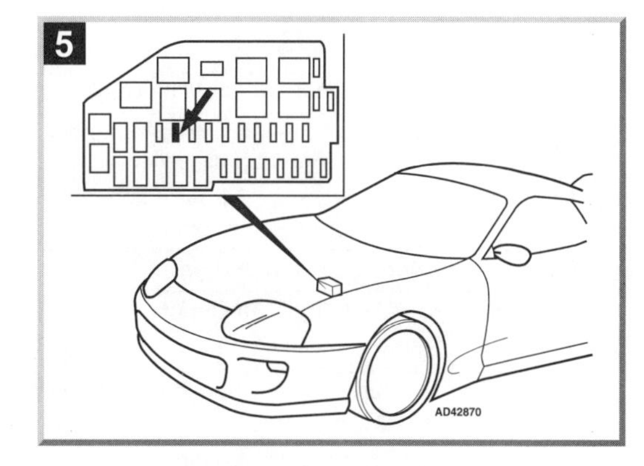

Trouble code identification

Flash code	Fault location	Probable cause
12	Crankshaft position (CKP) sensor – circuit	**Wiring, CKP sensor, starter motor, ECM**
12	Camshaft position (CMP) sensor 1 & 2 – circuit	**Wiring, CMP sensor, starter motor, ECM**
13	Crankshaft position (CKP) sensor – above 1000 rpm	**Wiring, valve timing, CKP sensor, ECM**
13	Camshaft position (CMP) sensor – above 1000 rpm	**Wiring, valve timing, CMP sensor, ECM**
14	Ignition reference signal – no signal	**Wiring, ignition module, ECM**
16	Transmission control signal – malfunction	**ECM**
21	Heated oxygen sensor (HO2S) 1 – front – circuit	**Wiring, HO2S, ECM**
22	Engine coolant temperature (ECT) sensor – circuit	**Wiring, ECT sensor, ECM**
24	Intake air temperature (IAT) sensor – circuit	**Wiring, IAT sensor, ECM**
25	Mixture control – continuously lean	**Wiring, injector, HO2S, MAF/VAF sensor, valve timing, intake/fuel/ignition, ECM**
26	Mixture control – continuously rich	**Wiring, injector, intake/fuel/ignition system, valve timing, MAF/VAF sensor, ECM**
27	Heated oxygen sensor (HO2S) – rear – circuit	**Wiring, HO2S, ECM**
28	Heated oxygen sensor (HO2S) 2 – front – circuit	**Wiring, HO2S, ECM**
31	Volume air flow (VAF) sensor – 2JZ-GE – circuit	**Wiring, VAF sensor, ECM**
31	Mass air flow (MAF) sensor – 2JZ-GTE – circuit	**Wiring, MAF sensor, ECM**
34	Turbocharger (TC) pressure – too high	**TC wastegate regulating valve/actuator, wiring, ECM**
35	Barometric pressure (BARO) sensor – circuit	**ECM**
35	Turbocharger (TC) pressure sensor – circuit	**Wiring, TC pressure sensor, ECM**
41	Throttle position (TP) sensor – circuit	**Wiring, TP sensor, ECM**
42	Vehicle speed sensor (VSS) – circuit	**Wiring, instrument cluster, VSS, ECM**
43	Starter signal – circuit	**Wiring, ignition switch, engine control relay, ECM**
47	Throttle position (TP) sensor 2 – circuit	**Wiring, TP sensor, ECM**
51	Switch signal – A/C switch ON during diagnosis	**Wiring, A/C switch, ECM**
51	Switch signal – closed throttle position (CTP) switch OFF during diagnosis	**Wiring, CTP switch, ECM**
51	Switch signal – park neutral position (PNP) switch not in P or N during diagnosis	**Wiring, PNP switch, ECM**
52	Knock sensor (KS) – front – circuit	**Wiring, KS, ECM**
53	Knock control – malfunction	**ECM**
55	Knock sensor (KS) – rear – circuit	**Wiring, KS, ECM**
71	Exhaust gas recirculation (EGR) system – circuit	**Wiring, EGRT sensor, EGR valve/solenoid, hose leak/blockage, ECM**
78	Fuel pump control – circuit	**Wiring, fuel pump/control module, ECM/ supply**

TOYOTA

Model: **Tercel 1.5L**

Engine identification: **3E-E**

Year: **1990-94**

System: **TCCS**

Self-diagnosis

General information

- Refer to the front of this manual for general test conditions, terminology, detailed descriptions of wiring faults and a general trouble shooter for electrical and mechanical faults.
- Engine control module (ECM) incorporates self-diagnosis function.
- Malfunction indicator lamp (MIL) will illuminate if certain faults are recorded.
- ECM operates in backup mode if sensors fail, to enable vehicle to be driven to workshop.
- Trouble codes can be accessed with suitable code reader connected to the data link connector (DLC) **1** or displayed by the MIL.

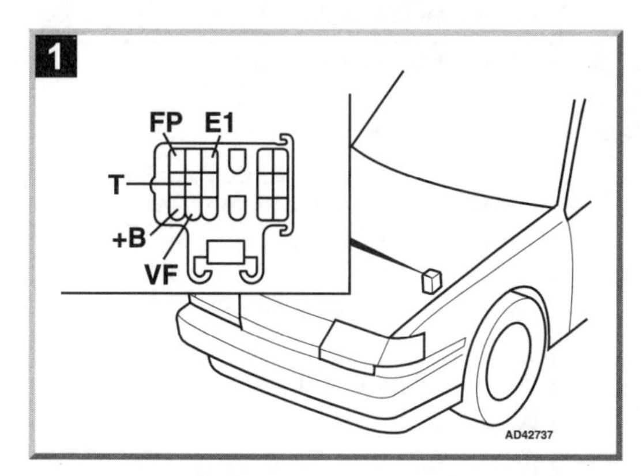

Preparatory conditions

- Battery voltage 11 V minimum.
- Closed throttle position (CTP) switch fully closed.
- Transmission in neutral position.
- All auxiliary equipment, including air conditioning, switched OFF.
- Engine at normal operating temperature.

Accessing

- Switch ignition ON. Do NOT start engine.
- Jump DLC terminals T and E1 with a short lead **1**.
- Count MIL flashes and compare with trouble code table.
- First flashes of 0.5 second duration indicate 'tens', after a 1.5 second pause, next flashes of 0.5 second duration indicate 'units' **2**.
- A 2.5 second pause occurs between codes.
- After a 4.5 second pause, trouble codes will be repeated.

NOTE: ***The MIL flashes once every 0.25 seconds if the system is correct*** **3**.

- Switch ignition OFF.
- Remove jump lead.

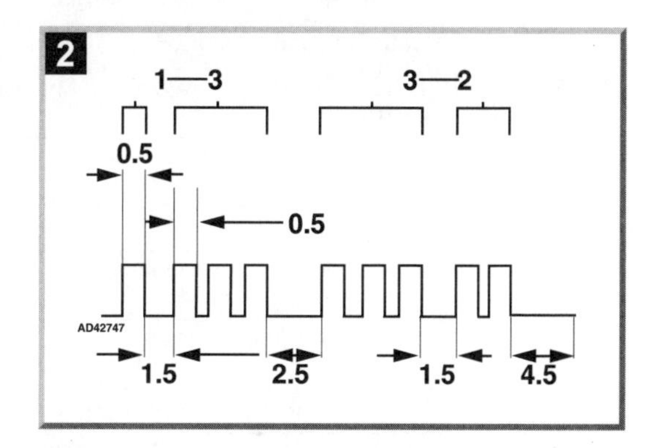

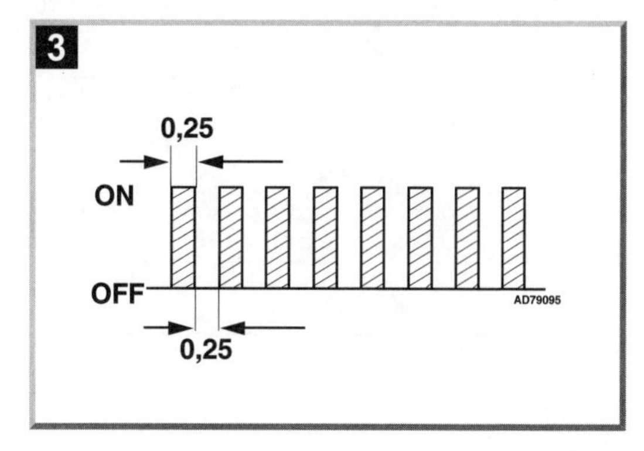

Erasing

- Ensure ignition switched OFF.
- Remove EFI fuse (15A) from underhood fusebox 4.
- Wait 60 seconds.
- Reinstall fuse.
- Repeat checking procedure to ensure no data remains in ECM fault memory.

NOTE: ***Disconnecting battery ground lead will also erase trouble codes but may erase memory from electronic units such as radio.***

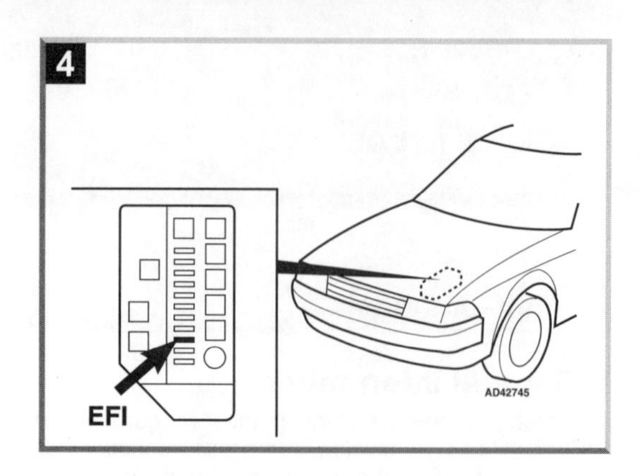

Trouble code identification

Flash code	Fault location	Probable cause
12	Engine speed (RPM) sensor – circuit	**Wiring, RPM sensor, ECM**
13	Engine speed (RPM) sensor – above 1000 rpm	**Wiring, RPM sensor, ECM**
14	Ignition reference signal – no signal	**Wiring, ignition module/coil, ECM**
21	Heated oxygen sensor (HO2S) – circuit	**Wiring, HO2S, ECM**
22	Engine coolant temperature (ECT) sensor – circuit	**Wiring, ECT sensor, ECM**
24	Intake air temperature (IAT) sensor – circuit	**Wiring, IAT sensor, ECM**
25	Mixture control – continuously lean	**Wiring, injector, HO2S, ECT/MAP sensor, intake/fuel/ignition system, ECM**
26	Mixture control – continuously rich	**Wiring, injector, fuel system, cold start injector, HO2S, ECT/MAP sensor, ECM**
31	Manifold absolute pressure (MAP) sensor – circuit	**Wiring, MAP sensor, ECM**
41	Throttle position (TP) sensor – circuit	**Wiring, TP sensor, ECM**
42	Vehicle speed sensor (VSS) – circuit	**Wiring, VSS, ECM**
43	Starter signal – circuit	**Wiring, ignition switch, ECM**
51	Switch signal – A/C switch ON during diagnosis	**Wiring, A/C switch, A/C amplifier, ECM**
51	Switch signal – closed throttle position (CTP) switch OFF during diagnosis	**Wiring, CTP switch, ECM**
51	Switch signal – park/neutral position (PNP) switch not in P or N during diagnosis	**Wiring, PNP switch, ECM**
71	Exhaust gas recirculation (EGR) system – malfunction	**Hose leak/blockage, wiring, EGRT sensor, EGR solenoid, ECM**

VOLKSWAGEN

Model:	**Golf • Jetta • Cabrio • New Beetle • New Beetle Cabriolet • Passat Corrado • Touareg • Euro Van**
Year:	**1991-04**
Engine identification:	**1Y, 1Z, 2E, 3F, AAA, AAC, AAF, AAZ, ABF, ABV, ACK, ACU, ACV, ADP, ADR, ADY, AEB, AEE, AEF, AEG, AEH, AEK, AER, AES, AET, AEX, AEY, AFB, AFH, AFK, AFN, AFP, AFT, AGD, AGG, AGN, AGP, AGR, AGU, AGX, AGZ, AHA, AHB, AHD, AHF, AHG, AHH, AHL, AHT, AHU, AHW, AHY, AJH, AJM, AJT, AJV, AKK, AKL, AKN, AKP, AKQ, AKR, AKS, AKT, AKU, AKV, AKW, ALD, ALE, ALG, ALH, ALL, ALM, ALZ, AMF, AMV, AMX, AMY, ANA, ANB, ANJ, ANU, ANV, ANW, ANX, ANY, APA, APE, APF, APH, APK, APL, APQ, APR, APT, APU, AQA, AQD, AQM, AQN, AQP, AQQ, AQY, ARC, ARG, ARL, ARM, ARR, ARZ, ASV, ASX, ASZ, ATD, ATJ, ATM, ATN, ATQ, ATU, ATW, AUA, AUB, AUC, AUD, AUE, AUF, AUG, AUM, AUR, AUS, AUY, AVB, AVC, AVF, AVG, AVH, AVT, AVU, AVY, AWC, AWD, AWF, AWG, AWH, AWM, AWP, AWT, AWU, AWV, AWW, AXK, AXQ, AWX, AXG, AXP, AYL, AYZ, AZD, AZG, AZH, AZM, AZX, BAA, BBW, BDC, BDF, BDP, BEV, BEW, BGD, BHW, BHX, BJS, BKF, BKW, BMX, BNU, PG, PY**
System:	**Bosch EDC/1.3.3/1.4 • Bosch EDC 15M/P/V • Bosch EDC 16 • Bosch Motronic M2.7/2.9/2.9.1 • Bosch Motronic M3.2 • Bosch Motronic M3.8.1/2/3/4/5 • Bosch Motronic M5.9/5.9.2 • Bosch Motronic ME7.1/7.1.1/7.5/7.5.10 • Bosch Motronic MP9.0 Bosch MSA 12/15/15.5 • Magneti Marelli 1AV/4AV/4CV/4LV/4MV • Siemens Simos VAG Digifant 3.0/3.2 • VAG Digifant ML5.4/5.5/5.9 • VAG Digifant MP4.1/2/3/4/7**

Data link connector (DLC) locations

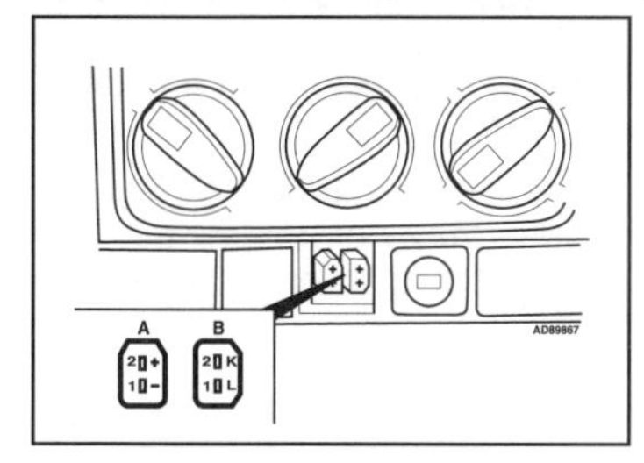

Golf/Jetta →07/93 – below heater controls

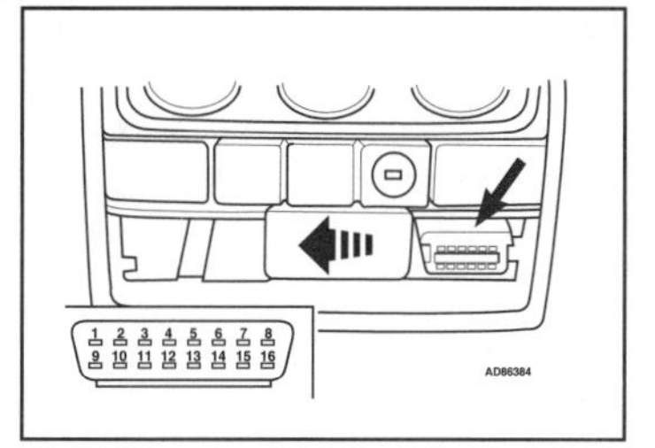

Golf/Jetta 08/93-97, Cabrio – adjacent to ashtray

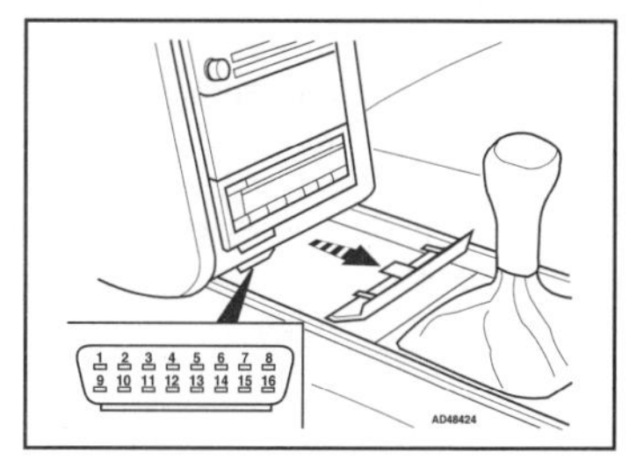

Golf/Jetta →1999 – center console

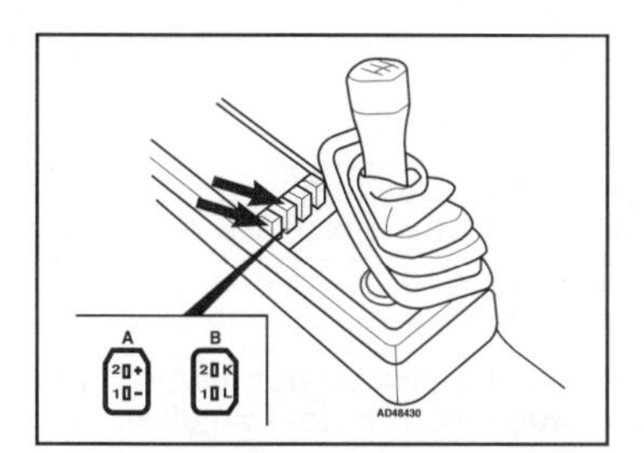

Passat →1993, Corrado →07/93 – center console

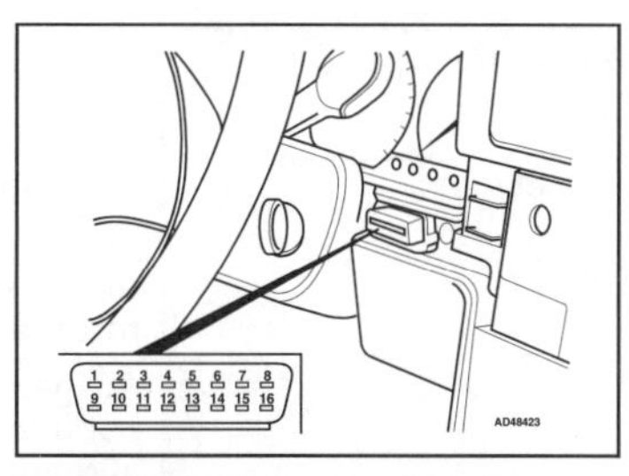

Passat 1994-96 – adjacent to steering column

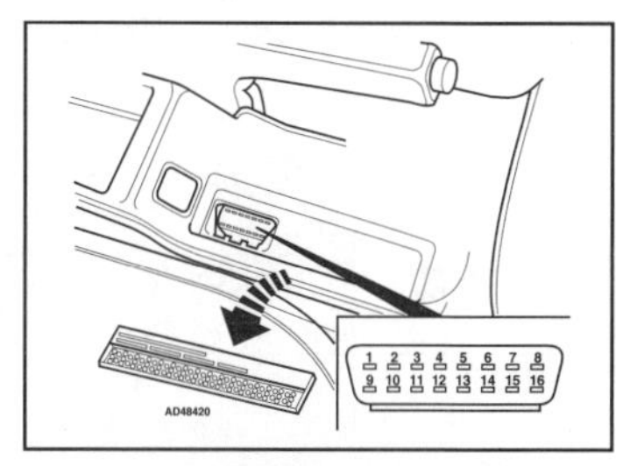

Passat 1996-09/00 – near handbrake, below rubber mat

VOLKSWAGEN

Golf • Jetta • Cabrio • New Beetle • New Beetle Cabriolet • Passat Corrado • Touareg • Euro Van

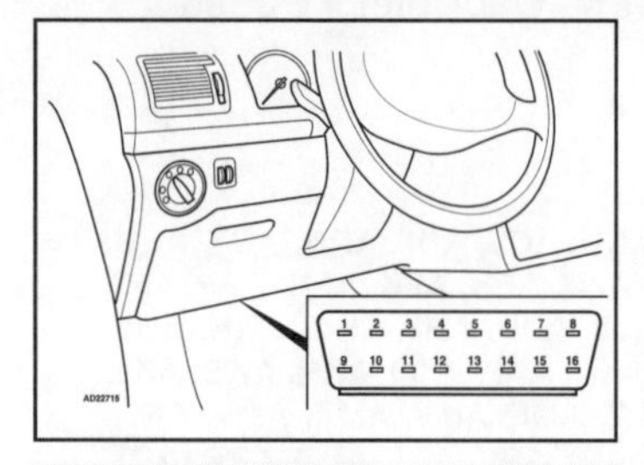

Beetle, Golf 1999→, Jetta 1999→, Passat 10/00→, Touareg, – fascia, driver's side

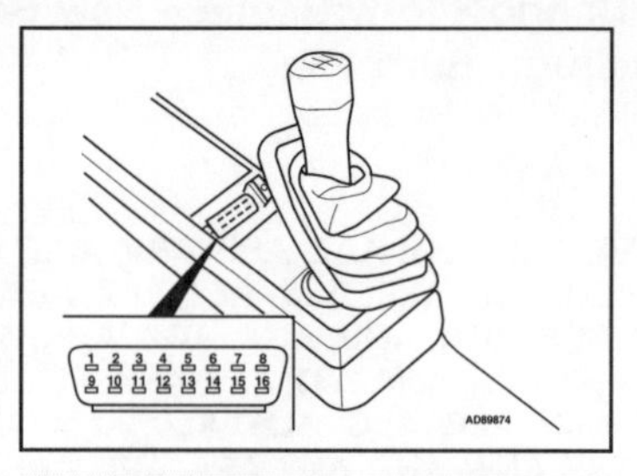

Corrado 08/93→ – center console

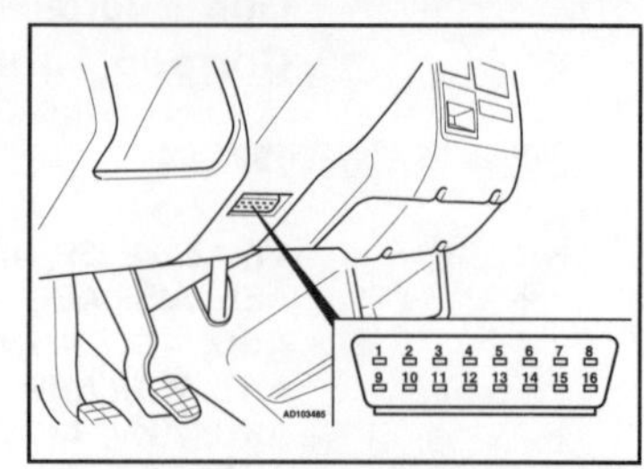

Euro Van 1999-03 – below steering column

Self-diagnosis

General information

- Refer to the front of this manual for general test conditions, terminology, detailed descriptions of wiring faults and a general trouble shooter for electrical and mechanical faults.
- Carry out road test for at least 10 minutes.
- Except 2.8/2.9: Ensure accelerator pedal is briefly fully depressed and engine speed exceeds 3500 rpm.
- 2.8/2.9: Ensure accelerator pedal is briefly fully depressed and engine speed exceeds 4600 rpm.
- Automatic transmission in 'P' or 'N'.
- If engine does not start: Crank engine for 6 seconds. Leave ignition switched ON.
- Data link connector (DLC) – 2-pin: (A) black, (B) brown/white.

Accessing and erasing

- The engine control module (ECM) fault memory can only be accessed and erased using diagnostic equipment connected to the data link connector (DLC).
- Passat, engine code 2E: Self-diagnosis capability only available from 08/92.
- Engine code PG: Self-diagnosis capability only available from 08/92.
- Engine code 3F: Self-diagnosis capability only available from 10/90.
- Engine code PY: Self-diagnosis capability only available from 1991.
- Engine code 1Y/AAZ: Self-diagnosis capability only available from 10/94.

NOTE: ***Self-diagnosis output using 4-digit trouble codes may not display all available diagnostic information (early models).***

Trouble code identification

Scan code 4-digit	Fault location	Probable cause
1111	Engine control module (ECM) – defective	ECM
1231	Vehicle speed sensor (VSS)	Wiring, speedometer, VSS
1232	Idle speed control (ISC) actuator	Throttle valve tight/sticking, wiring, multi-plug incorrectly wired, ISC actuator
2111 [1]	Crankshaft position (CKP) sensor	Air gap, metal particles, insecure sensor/rotor, wiring, CKP sensor
2113	Camshaft position (CMP) sensor	Air gap, insecure sensor/rotor, wiring, poor connection, fuse, distributor/camshaft alignment, CMP sensor

Scan code 4-digit	Fault location	Probable cause
2121	Closed throttle position (CTP) switch	**Accelerator cable adjustment, CTP switch adjustment, wiring, CTP switch**
2141	Knock control – control limit exceeded	**ECM**
2142	Knock sensor (KS) 1	**Wiring, KS, ECM**
2144	Knock sensor (KS) 2	**Wiring, KS, ECM**
2212	Throttle position (TP) sensor	**Wiring, poor connection, TP sensor**
2214	Maximum engine RPM exceeded	**Incorrect gear shift, CKP/RPM sensor**
2222	Manifold absolute pressure (MAP) sensor	**TC wastegate regulating valve, hose connection(s), oil contamination, valve timing, poor connection, wiring, MAP sensor, ECM**
2231	Idle speed control (ISC)	**Intake leak/blockage, throttle valve tight/sticking, IAC valve or ISC actuator/ position sensor**
2234	Engine control module (ECM) – supply voltage	**Fuse, alternator, battery, current draw with ignition OFF, engine control relay, wiring**
2242	Mixture adjustment resistor	**Wiring, mixture adjustment resistor**
2243	Instrument panel, fuel consumption signal	**Wiring short to positive, instrument panel**
2312	Engine coolant temperature (ECT) sensor	**Wiring, poor connection, ECT sensor**
2314	Engine/gearbox electrical connection	**Wiring, transmission fault**
2322	Intake air temperature (IAT) sensor	**Wiring, poor connection, IAT sensor**
2323	Volume air flow (VAF) sensor	**Intake leak, wiring, VAF sensor**
2324	Mass air flow (MAF) sensor	**Intake leak, wiring, MAF sensor**
2341	Heated oxygen sensor (HO2S) – lambda control	**Heating inoperative, intake/exhaust leak, misfire, fuel level low, fuel pressure/pump, injector(s), EVAP canister purge valve, MAF sensor filament burn-off, wiring, HO2S**
2342	Heated oxygen sensor (HO2S)	**Wiring, HO2S, heating inoperative, fuel level low**
2411	Exhaust gas recirculation temperature (EGRT) sensor	**Wiring, EGRT sensor**
2412	Intake air temperature (IAT) sensor	**Wiring, poor connection, IAT sensor**
2413	Mixture control (MC)	**Fuel level low, fuel pressure/pump, intake/exhaust leak, misfire, MAF sensor filament burn-off, HO2S, EVAP canister purge valve, injector(s), excessive fuel in engine oil**
3434	Oxygen sensor heater relay	**Wiring, oxygen sensor heater relay**
4312	Exhaust gas recirculation (EGR) solenoid	**Wiring, EGR solenoid**
4332	Engine control module (ECM) – output stages	**Wiring, ECM controlled components**
4343	Evaporative emission (EVAP) canister purge valve	**Wiring, fuse, EVAP canister purge valve**
4411	Injector 1	**Wiring, fuse, injector**
4412	Injector 2	**Wiring, fuse, injector**

Scan code 4-digit	Fault location	Probable cause
4413	Injector 3	**Wiring, fuse, injector**
4414	Injector 4	**Wiring, fuse, injector**
4421	Injector 5	**Wiring, injector**
4422	Injector 6	**Wiring, injector**
4431	Idle air control (IAC) valve	**Wiring, IAC valve**
4433	Fuel pump relay	**Wiring, fuse, fuel pump relay**
4444	No fault found	–

[1] Trouble code may be displayed if engine is not idling during self-diagnosis due to missing CKP sensor signal. Ignore trouble code if engine starts.

Scan code 5-digit	OBD-II code	Fault location	Probable cause
–	*All P0, P2 and U0 codes*	Refer to OBD-II trouble code tables at the front of this manual	–
00000	–	No fault found	–
00263	–	Transmission control module (TCM) – incorrect signal	**Wiring short to ground, TCM trouble code(s) stored, ECM**
00268	–	Idle speed control (ISC) actuator	**Wiring, ISC actuator**
00281	–	Vehicle speed sensor (VSS)	**Wiring, speedometer, VSS**
00282	–	Idle speed control (ISC) actuator/throttle motor	**Throttle valve tight/sticking, wiring, multi-plug incorrectly wired, ISC actuator/throttle motor**
00305	–	Instrument panel, fuel consumption signal	**Wiring, instrument panel**
00513 [1]	–	Crankshaft position (CKP) sensor	**Air gap, metal particles, insecure sensor/rotor, wiring, CKP sensor**
00514	–	Crankshaft position (CKP) sensor	**Air gap, insecure sensor/rotor, wiring, poor connection, CKP sensor**
00515	–	Except ACU/ADL: Camshaft position (CMP) sensor	**Air gap, insecure sensor/rotor, wiring, poor connection, fuse, distributor/ camshaft alignment, CMP sensor**
00515	–	ACU/ADL: Crankshaft position (CKP) sensor	**Air gap, insecure sensor/rotor, wiring, poor connection, fuse, distributor/ camshaft alignment, CKP sensor**
00516	–	Closed throttle position (CTP) switch	**Accelerator cable adjustment, CTP switch adjustment, throttle valve tight/ sticking, wiring, CTP switch**
00518	–	Throttle position (TP) sensor	**Wiring, poor connection, TP sensor**
00519	–	Manifold absolute pressure (MAP) sensor	**TC wastegate regulating valve, hose connection(s), oil contamination, valve timing, poor connection, wiring, MAP sensor, ECM**
00520	–	Mass air flow (MAF) sensor	**Wiring, MAF sensor**

Scan code 5-digit	OBD-II code	Fault location	Probable cause
00521	–	Mixture adjustment resistor	Wiring, mixture adjustment resistor
00522	–	Engine coolant temperature (ECT) sensor	Wiring, poor connection, ECT sensor
00523	–	Intake air temperature (IAT) sensor	Wiring, poor connection, IAT sensor
00524	–	Knock sensor (KS) 1	Wiring, KS, ECM
00525	–	Heated oxygen sensor (HO2S)	Wiring, HO2S, heating inoperative, fuel level low
00526	–	Stop lamp switch, cruise control	Wiring, stop lamp switch
00527	–	Intake air temperature (IAT) sensor	Wiring, poor connection, IAT sensor
00528	–	Barometric pressure (BARO) sensor	Wiring, BARO sensor
00529	–	Crankshaft position (CKP) sensor	Wiring, CKP sensor
00530	–	Idle speed control (ISC) actuator/position sensor	Wiring, poor connection, multi-plug incorrectly wired, ISC actuator/position sensor
00532	–	Engine control module (ECM) – supply voltage	Fuse, alternator, battery, current draw with ignition OFF, engine control relay, wiring
00533	–	Idle speed control (ISC)	Intake leak/blockage, throttle valve tight/sticking, IAC valve or ISC actuator/ position sensor
00534	–	Engine oil temperature (EOT) sensor	Wiring, EOT sensor
00535	–	Engine control module (ECM) – knock control 1	Wiring, KS, ECM
00536	–	Engine control module (ECM) – knock control 2	Wiring, KS, ECM
00537	–	Heated oxygen sensor (HO2S) – lambda control	CO adjustment, heating inoperative, intake/exhaust leak, misfire, fuel level low, fuel pressure/pump, injector(s), EVAP canister purge valve, MAF sensor filament burn-off, wiring, HO2S
00539	–	Fuel temperature sensor	Wiring, fuel temperature sensor
00540	–	Knock sensor (KS) 2	Wiring, KS, ECM
00542	–	Injector needle lift sensor	Air in fuel system, fuel level low, wiring, injector needle lift sensor
00543	–	Maximum engine RPM exceeded	Incorrect gear shift, CKP/RPM sensor, AT fault, ECM
00544	–	Maximum boost pressure exceeded	Hoses interchanged/not connected, hoses blocked/leaking, TC wastegate actuator/regulating valve, MAP sensor
00545	–	ECM/TCM electrical connection	TCM trouble code(s) stored, wiring, transmission fault
00546	–	Data link connector (DLC) – defective	Wiring
00549	–	Instrument panel, fuel consumption signal	Wiring short to positive, instrument panel
00550	–	Start of injection – control	Wiring, fuel injection timing solenoid, injector needle lift sensor, air in fuel system, fuel level low, pump timing

Scan code 5-digit	OBD-II code	Fault location	Probable cause
00552	–	Volume air flow (VAF) sensor	**Intake leak, wiring, VAF sensor**
00553	–	Mass air flow (MAF) sensor	**Intake leak, wiring, MAF sensor**
00557	–	Power steering pressure (PSP) switch – short to ground	**Wiring short to ground, PSP switch**
00560	–	Exhaust gas recirculation (EGR) – control difference	**Intake leak, EGR valve/solenoid**
00561	–	Mixture control (MC)	**Trouble codes 00525/00533 stored, fuel level low, fuel pressure/pump, intake/exhaust leak, misfire, MAF sensor filament burn-off, HO2S, EVAP canister purge valve, injector(s), excessive fuel in engine oil**
00575	–	Intake manifold pressure	**Intake leak/blockage, MAF sensor, TC wastegate actuator/regulating valve, hoses interchanged/not connected, wiring**
00577	–	Knock control, cylinder 1 – control limit exceeded	**Fuel pressure, misfire, intake leak, injector(s)**
00578	–	Knock control, cylinder 2 – control limit exceeded	**Fuel pressure, misfire, intake leak, injector(s)**
00579	–	Knock control, cylinder 3 – control limit exceeded	**Fuel pressure, misfire, intake leak, injector(s)**
00580	–	Knock control, cylinder 4 – control limit exceeded	**Fuel pressure, misfire, intake leak, injector(s)**
00581	–	Knock control, cylinder 5 – control limit exceeded	**Fuel pressure, misfire, intake leak, injector(s)**
00582	–	Knock control, cylinder 6 – control limit exceeded	**Fuel pressure, misfire, intake leak, injector(s)**
00585	–	Exhaust gas recirculation temperature (EGRT) sensor	**Wiring, EGRT sensor**
00586	–	Exhaust gas recirculation (EGR) system – control	**EGR solenoid**
00609	–	Ignition amplifier, primary circuit 1	**Wiring, ignition amplifier**
00610	–	Ignition amplifier, primary circuit 2	**Wiring, ignition amplifier**
00611	–	Ignition amplifier, primary circuit 3	**Wiring, ignition amplifier**
00624	–	AC signal – compressor cut-in	**Wiring, AC system**
00625	–	Vehicle speed signal	**Wiring, speedometer, VSS**
00626	–	Glow plug warning lamp	**Wiring, glow plug warning lamp**
00627	–	Fuel filter water level sensor	**Water in filter, wiring, fuel filter water level sensor**
00628	–	Fuel injection pump control module – engine stop malfunction	**Wiring, fuel injection pump**
00635	–	Heated oxygen sensor (HO2S) 1, bank 1 – heater circuit malfunction	**Wiring, HO2S**
00638	–	ECM/TCM electrical connection	**Wiring, transmission fault**
00640	–	Oxygen sensor heater relay	**Wiring, oxygen sensor heater relay**

Scan code 5-digit	OBD-II code	Fault location	Probable cause
00650	–	Clutch pedal position (CPP) switch – short to positive	**Wiring short to positive, CPP switch**
00653	–	Transmission control module (TCM)/ TR sensor – implausible TR signal	**Wiring, transmission fault, poor connection**
00667	–	Outside air temperature signal	**Wiring, instrument panel, AC system, outside air temperature sensor**
00668	–	Engine control module (ECM) – supply voltage low	**Battery, wiring, engine control relay**
00670	–	Idle speed control (ISC) actuator position sensor	**Wiring, throttle valve, ISC actuator position sensor**
00671	–	Cruise control master switch	**Wiring, cruise control master switch**
00740	–	Camshaft position (CMP) sensor	**Wiring, CMP sensor**
00741	–	Stop lamp switch/brake pedal position (BPP) switch – implausible ratio	**Wiring, both switch positions not synchronised, stop lamp switch, BPP switch**
00750	–	Malfunction indicator lamp (MIL) – circuit malfunction	**ECM incorrectly coded, wiring, MIL**
00758	–	Secondary air injection (AIR) system	**AIR solenoid/relay, wiring**
00765	–	Fuel quantity adjuster position sensor	**Wiring, fuel injection pump**
00777	–	Accelerator pedal position (APP) sensor	**Incorrectly adjusted, wiring, APP sensor**
00792	–	AC pressure switch	**Wiring, AC pressure switch**
01013	–	AC compressor clutch, load signal	**Wiring, AC system**
01025	–	Malfunction indicator lamp (MIL)	**Wiring, MIL**
01028	–	Engine coolant blower motor relay	**Wiring, engine coolant blower motor relay**
01044	–	Engine control module (ECM) – coding	**Incorrectly coded**
01050	–	Glow plug monitoring	**Fuse, wiring, glow plug relay, glow plugs**
01052	–	Fuel lever position sensor	**Wiring, fuel lever position sensor**
01087	–	Engine control module (ECM) – basic setting	**Basic setting not completed, throttle valve tight/sticking**
01088	–	Mixture control (MC)	**Fuel level low, fuel pressure/pump, MAF sensor, intake/exhaust leak, EVAP canister purge valve, excessive fuel in engine oil, injector(s)**
01117	–	Alternator load signal	**Wiring, alternator**
01119	–	Gear recognition signal – AT	**Wiring**
01120	–	Camshaft position (CMP) control, bank 1 – mechanical fault	**Wiring, fuse, CMP actuator**
01121	–	Camshaft position (CMP) control, bank 2 – mechanical fault	**Wiring, fuse, CMP actuator**
01126	–	Engine RPM signal	**Wiring, CKP/RPM sensor, instrument panel**
01163	–	Backfire	**Intake leak, wiring, ignition amplifier, injector(s)**

Scan code 5-digit	OBD-II code	Fault location	Probable cause
01165	–	Idle speed control (ISC) actuator/throttle motor	**Accelerator cable adjustment, throttle valve, wiring, ISC actuator/throttle motor, basic setting not carried out**
01167	–	Full throttle stop solenoid	**Wiring, full throttle stop valve**
01168	–	Idle speed boost solenoid	**Wiring, full throttle stop valve**
01169	–	Door contact switch, driver's	**Wiring, door contact switch**
01170	–	Fuel injection timing sensor	**Wiring, fuel injection timing sensor**
01177	–	Engine control module (ECM) – defective	**ECM**
01180	–	Engine/AC electrical connection	**Wiring**
01182	–	Mass air flow (MAF) sensor/throttle position (TP) sensor – signal variation exceeded	**Throttle valve sticking, ISC actuator/ throttle motor sticking/mechanically damaged, incorrect throttle control unit, intake leak between MAF sensor and throttle valve, air filter blocked**
01183	–	Malfunction indicator lamp (MIL) – circuit malfunction	**ECM incorrectly coded, wiring, MIL**
01193	–	Engine coolant heater relay 1, low output	**Wiring, engine coolant heater relay**
01194	–	Engine coolant heater relay 2, high output	**Wiring, engine coolant heater relay**
01196	–	CAN data bus, TCM – incorrect signal	**Wiring**
01204	–	Engine coolant temperature (ECT) sensor	**Wiring, ECT sensor**
01208	–	Engine control module (ECM) – data changed	**ECM**
01209	–	Alternator speed signal	**Wiring, alternator**
01235	–	Secondary air injection (AIR) solenoid	**Wiring, AIR solenoid**
01237	–	Fuel shut-off solenoid	**Wiring, fuel shut-off solenoid**
01242	–	Engine control module (ECM) – output stages	**Wiring, ECM controlled components**
01243	–	Intake manifold air control solenoid	**Wiring, intake manifold air control solenoid**
01247	–	Evaporative emission (EVAP) canister purge valve	**Wiring, fuse, EVAP canister purge valve**
01249	–	Injector 1	**Wiring, fuse, injector**
01250	–	Injector 2	**Wiring, fuse, injector**
01251	–	Injector 3	**Wiring, fuse, injector**
01252	–	Injector 4	**Wiring, fuse, injector**
01253	–	Injector 5	**Wiring, injector**
01254	–	Injector 6	**Wiring, injector**
01257	–	Idle air control (IAC) valve	**Wiring, IAC valve**
01259	–	Fuel pump relay	**Wiring, fuse, fuel pump relay**
01262	–	Turbocharger (TC) wastegate regulating valve	**Wiring, fuse, TC wastegate regulating valve**
01265	–	Exhaust gas recirculation (EGR) solenoid	**Wiring, EGR solenoid**

Scan code 5-digit	OBD-II code	Fault location	Probable cause
01266	–	Glow plug relay	Wiring, glow plug relay
01268	–	Fuel quantity adjuster	Incorrectly set, wiring, fuel injection pump
01269	–	Fuel injection timing solenoid	Wiring, fuel injection timing solenoid
01282	–	Intake manifold air control solenoid	Wiring, intake manifold air control solenoid
01283	–	Intake manifold air control actuator	Wiring, intake manifold flap, intake manifold air control actuator
01312	–	CAN data bus – defective	Trouble code(s) stored in other system(s), wiring
01314	–	Engine control module (ECM), cruise control system – defective	ECM
01315	–	CAN data bus, TCM – no signal	TCM trouble code(s) stored, wiring, matching resistor in ECM
01316	–	CAN data bus, ABS – defective	ABS control module incorrectly coded, wiring
01317	–	CAN data bus, instrumentation	Trouble code(s) stored in other system(s), wiring, instrumentation control module
01318	–	Fuel injection pump control module	Data bus wiring, fuel injection pump
01321	–	CAN data bus, SRS	SRS control module trouble code(s) stored, wiring
01375	–	Engine mounting control solenoid, bank 1 & 2	Wiring, engine mounting control solenoid
01376	–	Fuel injection pump position sensor	Wiring, fuel injection pump position sensor
01437	–	Throttle control unit – basic setting	Basic setting not carried out, CTP switch, ISC actuator/position sensor, TP sensor
01440	–	Fuel level signal	Wiring, instrument panel, fuel gauge tank sensor
01441	–	Fuel low level sensor	Wiring, fuel transfer pump, fuel level sensor
01442	–	Engine misfire – fuel pump housing empty	Fuel level low, fuel transfer pump
01575	–	Auxiliary drive – switched OFF	Auxiliary drive overloaded
01613	–	Fuel cooling pump relay – circuit malfunction	Wiring, fuel cooling pump relay
01656	–	SRS control module – crash signal	Wiring
01686	–	Engine coolant blower motor run-on relay	Wiring, engine coolant blower motor run-on relay
01695	–	Fuel temperature sensor – circuit malfunction	Wiring, fuel temperature sensor
16394	–	Camshaft position (CMP) actuator, intake/left/front, bank 1 – circuit malfunction	Wiring, CMP actuator
16395	–	Camshaft position (CMP), intake/left/front, bank 1 – timing over advanced/system performance	Valve timing, camshaft position (CMP) actuator

Scan code 5-digit	OBD-II code	Fault location	Probable cause
16396	–	Camshaft position (CMP), intake/left/front, bank 1 – timing over retarded	**Valve timing, camshaft position (CMP) actuator**
16398	–	Camshaft position (CMP) actuator, exhaust/right/rear, bank 1 – timing over advanced/system performance	**Valve timing, camshaft position (CMP) actuator**
16399	–	Camshaft position (CMP) actuator, exhaust/right/rear, bank 1 – timing over retarded	**Valve timing, camshaft position (CMP) actuator**
16414	–	Heated oxygen sensor (HO2S) 1, bank 1, heater control – circuit malfunction	**Wiring open circuit, HO2S**
16415	–	Heated oxygen sensor (HO2S) 1, bank 1, heater control – circuit low	**Wiring short to ground, HO2S**
16416	–	Heated oxygen sensor (HO2S) 1, bank 1, heater control – circuit high	**Wiring short to positive, HO2S**
16474	–	Fuel metering solenoid – open circuit	**Wiring open circuit, fuel metering solenoid**
16475	–	Fuel metering solenoid – short to ground	**Wiring short to ground, fuel metering solenoid**
16476	–	Fuel metering solenoid – short to positive	**Wiring short to ground, fuel metering solenoid**
16485	–	Mass air flow (MAF) sensor, bank 1 – range/performance problem	**Intake leak, wiring, MAF sensor**
16486	–	Mass air flow (MAF) sensor, bank 1 – low input	**Intake leak, air filter blocked, wiring short to ground, fuse, MAF sensor**
16487	–	Mass air flow (MAF) sensor, bank 1 – high input	**Wiring short to positive, ground wire defective, MAF sensor**
16490	–	Manifold absolute pressure (MAP) sensor/ barometric pressure (BARO) sensor – range/performance problem	**Intake/exhaust leak, wiring, MAP sensor, BARO sensor**
16491	–	Manifold absolute pressure (MAP) sensor/ barometric pressure (BARO) sensor – low input	**Wiring short to ground, MAP sensor, BARO sensor**
16492	–	Manifold absolute pressure (MAP) sensor/ barometric pressure (BARO) sensor – high input	**Wiring short to positive, MAP sensor, BARO sensor**
16496	–	Intake air temperature (IAT) sensor – low input	**Wiring short to ground, IAT sensor**
16497	–	Intake air temperature (IAT) sensor – high input	**Wiring open circuit/short to positive, ground wire defective, IAT sensor**
16500	–	Engine coolant temperature (ECT) sensor – range/performance problem	**Coolant thermostat, poor connection, wiring, ECT sensor**
16501	–	Engine coolant temperature (ECT) sensor – low input	**Coolant thermostat, wiring short to ground, ECT sensor**
16502	–	Engine coolant temperature (ECT) sensor – high input	**Coolant thermostat, wiring open circuit/ short to positive, ground wire defective, ECT sensor**
16504	–	Throttle position (TP) sensor – circuit malfunction	**Poor connection, wiring, TP sensor**

Scan code 5-digit	OBD-II code	Fault location	Probable cause
16505	–	Throttle position (TP) sensor – range/performance problem	**Poor connection, TP sensor**
16506	–	Throttle position (TP) sensor – low input	**Signal wire short to ground, supply wire defective, TP sensor**
16507	–	Throttle position (TP) sensor – high input	**Signal wire open circuit/short to positive, ground wire defective, TP sensor**
16514	–	Heated oxygen sensor (HO2S) 1, bank 1 – circuit malfunction	**Heating inoperative, poor connection, wiring, HO2S**
16515	–	Heated oxygen sensor (HO2S) 1, bank 1 – voltage low	**Wiring short to ground, HO2S**
16516	–	Heated oxygen sensor (HO2S) 1, bank 1 – voltage high	**Wiring short to positive, HO2S**
16517	–	Heated oxygen sensor (HO2S) 1, bank 1 – slow response	**Heating inoperative, wiring, HO2S**
16518	–	Heated oxygen sensor (HO2S) 1, bank 1 – no activity detected	**Wiring open circuit, heating inoperative, HO2S**
16519	–	Heated oxygen sensor (HO2S) 1, bank 1 – heater circuit malfunction	**Fuse, wiring, HO2S**
16520	–	Heated oxygen sensor (HO2S) 2, bank 1 – circuit malfunction	**Heating inoperative, wiring, HO2S**
16521	–	Heated oxygen sensor (HO2S) 2, bank 1 – low voltage	**Wiring short to ground, exhaust leak, HO2S**
16522	–	Heated oxygen sensor (HO2S) 2, bank 1 – high voltage	**Wiring short to positive, HO2S**
16523	–	Heated oxygen sensor (HO2S) 2, bank 1 – slow response	**Heating inoperative, wiring, HO2S**
16524	–	Heated oxygen sensor (HO2S) 2, bank 1 – no activity detected	**Wiring, HO2S**
16525	–	Heated oxygen sensor (HO2S) 2, bank 1 – heater circuit malfunction	**Wiring, HO2S**
16534	–	Heated oxygen sensor (HO2S) 1, bank 2 – circuit malfunction	**Wiring, HO2S**
16535	–	Heated oxygen sensor (HO2S) 1, bank 2 – low voltage	**Wiring short to ground, HO2S**
16536	–	Heated oxygen sensor (HO2S) 1, bank 2 – high voltage	**Wiring short to positive, HO2S**
16537	–	Heated oxygen sensor (HO2S) 1, bank 2 – slow response	**Heating inoperative, wiring, HO2S**
16538	–	Heated oxygen sensor (HO2S) 1, bank 2 – no activity detected	**Wiring, HO2S**
16540	–	Heated oxygen sensor (HO2S) 2, bank 2 – circuit malfunction	**Heating inoperative, wiring, HO2S**
16541	–	Heated oxygen sensor (HO2S) 2, bank 2 – low voltage	**Wiring short to ground, exhaust leak, HO2S**
16542	–	Heated oxygen sensor (HO2S) 2, bank 2 – high voltage	**Wiring short to positive, HO2S**
16543	–	Heated oxygen sensor (HO2S) 2, bank 2 – slow response	**Heating inoperative, wiring, HO2S**

Scan code 5-digit	OBD-II code	Fault location	Probable cause
16544	–	Heated oxygen sensor (HO2S) 2, bank 2 – no activity detected	**Wiring, HO2S**
16545	–	Heated oxygen sensor (HO2S) 2, bank 2 – heater circuit malfunction	**Wiring, HO2S**
16554	–	Fuel trim (FT), bank 1 – malfunction	**Intake leak, AIR system, fuel pressure/ pump, injector(s), EVAP canister purge valve, HO2S**
16555	–	System too lean, bank 1	**Intake/exhaust leak, AIR system, MAF sensor, fuel pressure/pump, injector(s), EVAP canister purge valve, HO2S**
16556	–	System too rich, bank 1	**EVAP canister purge valve, fuel pressure, injector(s), HO2S**
16557	–	Fuel trim (FT), bank 2 – malfunction	**Fuel pressure/pump, injector(s), AIR system, hose connection(s), intake leak**
16558	–	System too lean, bank 2	**Fuel pressure/pump, injector(s), intake/ exhaust leak, AIR system, hose connection(s)**
16559	–	System too rich, bank 2	**Fuel pressure, injector(s), EVAP canister purge valve**
16575	–	Fuel rail pressure (FRP) sensor – range/ performance problem	**Wiring, FRP sensor**
16576	–	Fuel rail pressure (FRP) sensor – low input	**Wiring short to ground, FRP sensor**
16577	–	Fuel rail pressure (FRP) sensor – high input	**Wiring short to positive, FRP sensor**
16578	–	Fuel rail pressure (FRP) sensor – circuit intermittent	**Wiring open circuit, FRP sensor**
16581	–	Engine oil temperature (EOT) sensor – low input	**Wiring short to ground, EOT sensor**
16582	–	Engine oil temperature (EOT) sensor – high input	**Wiring short to positive, EOT sensor**
16585	–	Injector 1 – circuit malfunction	**Wiring, injector**
16586	–	Injector 2 – circuit malfunction	**Wiring, injector**
16587	–	Injector 3 – circuit malfunction	**Wiring, injector**
16588	–	Injector 4 – circuit malfunction	**Wiring, injector**
16589	–	Injector 5 – circuit malfunction	**Wiring, injector**
16590	–	Injector 6 – circuit malfunction	**Wiring, injector**
16603	–	Engine over speed condition	**Incorrect gear change**
16605	–	Throttle position (TP) sensor B – range/performance problem	**Wiring, TP sensor**
16606	–	Throttle position (TP) sensor B – low input	**Wiring short to ground, TP sensor**
16607	–	Throttle position (TP) sensor B – high input	**Wiring short to positive, TP sensor**
16610	–	Accelerator pedal position (APP) sensor A/B – range/performance problem	**Wiring, APP sensor**

Scan code 5-digit	OBD-II code	Fault location	Probable cause
16611	–	Accelerator pedal position (APP) sensor A – low input	Wiring short to ground, APP sensor
16612	–	Accelerator pedal position (APP) sensor A – high input	Wiring short to positive, APP sensor
16614	–	Fuel pump relay – circuit malfunction	Wiring, fuel pump relay
16618	–	Engine boost condition – limit exceeded	Hose connection(s), wiring, TC wastegate regulating valve, TC wastegate
16619	–	Engine boost condition – limit not reached	Hose connection(s), wiring, TC wastegate regulating valve, TC wastegate
16620	–	Manifold absolute pressure (MAP) sensor A, TC system – range/performance problem	Intake/exhaust leak, hose connection(s), MAP sensor
16621	–	Manifold absolute pressure (MAP) sensor A, TC system – low input	Wiring short to ground, MAP sensor
16622	–	Manifold absolute pressure (MAP) sensor A, TC system – high input	Wiring short to positive, MAP sensor
16627	–	Turbocharger (TC) wastegate regulating valve A – circuit malfunction	Wiring, TC wastegate regulating valve
16629	–	Turbocharger (TC) wastegate regulating valve A – circuit low	Wiring short to ground, TC wastegate regulating valve
16630	–	Turbocharger (TC) wastegate regulating valve A – circuit high	Wiring short to positive, TC wastegate regulating valve
16645	–	Injector 1 – circuit low	Wiring short to ground, injector
16646	–	Injector 1 – circuit high	Wiring short to positive, injector
16648	–	Injector 2 – circuit low	Wiring short to ground, injector
16649	–	Injector 2 – circuit high	Wiring short to positive, injector
16651	–	Injector 3 – circuit low	Wiring short to ground, injector
16652	–	Injector 3 – circuit high	Wiring short to positive, injector
16654	–	Injector 4 – circuit low	Wiring short to ground, injector
16655	–	Injector 4 – circuit high	Wiring short to positive, injector
16657	–	Injector 5 – circuit low	Wiring short to ground, injector
16658	–	Injector 5 – circuit high	Wiring short to positive, injector
16660	–	Injector 6 – circuit low	Wiring short to ground, injector
16661	–	Injector 6 – circuit high	Wiring short to positive, injector
16684	–	Random/multiple cylinder(s) – misfire detected	Spark plug(s), HT lead(s), injector(s), ignition coil(s), low compression, wiring
16685	–	Cylinder 1 – misfire detected	Spark plug, HT lead(s), injector, ignition coil, low compression, wiring
16686	–	Cylinder 2 – misfire detected	Spark plug, HT lead(s), injector, ignition coil, low compression, wiring
16687	–	Cylinder 3 – misfire detected	Spark plug, HT lead(s), injector, ignition coil, low compression, wiring
16688	–	Cylinder 4 – misfire detected	Spark plug, HT lead(s), injector, ignition coil, low compression, wiring

Scan code 5-digit	OBD-II code	Fault location	Probable cause
16689	–	Cylinder 5 – misfire detected	**Spark plug, HT lead(s), injector, ignition coil, low compression, wiring**
16690	–	Cylinder 6 – misfire detected	**Spark plug, HT lead(s), injector, ignition coil, low compression, wiring**
16705	–	Crankshaft position (CKP) sensor/engine speed (RPM) sensor – range/performance problem	**Air gap, metal particles, insecure sensor/rotor, wiring, CKP/RPM sensor**
16706	–	Crankshaft position (CKP) sensor/engine speed (RPM) sensor – no signal	**Wiring, CKP/RPM sensor**
16709	–	Knock sensor (KS) 1, bank 1 – circuit malfunction	**Wiring, poor connection, KS**
16710	–	Knock sensor (KS) 1, bank 1 – range/ performance problem	**Wiring, KS incorrectly tightened, KS**
16711	–	Knock sensor (KS) 1, bank 1 – low input	**Insecure KS, poor connection, wiring short to ground, incorrectly tightened, KS**
16712	–	Knock sensor (KS) 1, bank 1 – high input	**Wiring short to positive, KS incorrectly tightened, KS**
16716	–	Knock sensor (KS) 2, bank 2 – low input	**Insecure KS, poor connection, wiring short to ground, KS incorrectly tightened, KS**
16717	–	Knock sensor (KS) 2, bank 2 – high input	**Wiring short to positive, KS incorrectly tightened, KS**
16719	–	Crankshaft position (CKP) sensor – circuit malfunction	**Wiring, CKP/RPM sensor**
16724	–	Camshaft position (CMP) sensor A, bank 1 – circuit malfunction	**Wiring, CMP sensor**
16725	–	Camshaft position (CMP) sensor A, bank 1 – range/performance problem	**Insecure sensor/rotor, air gap, wiring, CMP sensor**
16726	–	Camshaft position (CMP) sensor A, bank 1 – low input	**Wiring short to ground, CMP sensor**
16727	–	Camshaft position (CMP) sensor A, bank 1 – high input	**Wiring short to positive, CMP sensor**
16730 2	–	Camshaft position (CMP) sensor A, bank 2 – range/performance problem	**Insecure sensor/rotor, air gap, wiring, CMP sensor**
16731 2	–	Camshaft position (CMP) sensor A, bank 2 – low input	**Wiring short to ground, CMP sensor**
16732 2	–	Camshaft position (CMP) sensor A, bank 2 – high input	**Wiring short to positive, CMP sensor**
16735	–	Ignition coil, cylinder 1, primary/ secondary – circuit malfunction	**Wiring, ignition amplifier, ignition coil**
16736	–	Ignition coil, cylinder 2, primary/ secondary – circuit malfunction	**Wiring, ignition amplifier, ignition coil**
16737	–	Ignition coil, cylinder 3, primary/ secondary – circuit malfunction	**Wiring, ignition amplifier, ignition coil**
16738	–	Ignition coil, cylinder 4, primary/ secondary – circuit malfunction	**Wiring, ignition amplifier, ignition coil**
16764	–	Glow plugs – circuit A malfunction	**Wiring, glow plug relay, fuse, glow plugs**

Scan code 5-digit	OBD-II code	Fault location	Probable cause
16784	–	Exhaust gas recirculation (EGR) system – flow malfunction	**Basic setting not carried out, EGR valve/ solenoid**
16785	–	Exhaust gas recirculation (EGR) system – insufficient flow detected	**Basic setting not carried out, EGR valve/ solenoid**
16786	–	Exhaust gas recirculation (EGR) system – excessive flow detected	**Basic setting not carried out, EGR valve/ solenoid**
16787	–	Exhaust gas recirculation (EGR) system – circuit malfunction	**Wiring, EGR valve/solenoid**
16788	–	Exhaust gas recirculation (EGR) system – range/performance problem	**Hose connection(s), wiring, EGR valve/ solenoid**
16791	–	Exhaust gas recirculation (EGR) valve position sensor – low input	**Wiring short to ground, EGR valve position sensor**
16792	–	Exhaust gas recirculation (EGR) valve position sensor – high input	**Wiring short to positive, EGR valve position sensor**
16795	–	Secondary air injection (AIR) system – incorrect flow detected	**AIR pump, AIR valve, AIR hose(s)**
16796	–	Secondary air injection (AIR) solenoid A – circuit malfunction	**Wiring, AIR solenoid**
16802	–	Secondary air injection (AIR) pump relay A – circuit malfunction	**Wiring, AIR pump relay**
16804	–	Catalytic converter system, bank 1 – efficiency below threshold	**Catalytic converter**
16806	–	Main catalytic converter, bank 1 – efficiency below threshold	**Catalytic converter**
16814	–	Catalytic converter system, bank 2 – efficiency below threshold	**Catalytic converter**
16824	–	Evaporative emission (EVAP) system – malfunction	**Hose connection(s), intake leak, EVAP canister purge valve**
16825	–	Evaporative emission (EVAP) system – incorrect flow detected	**Hose connection(s), intake leak, EVAP canister purge valve**
16826	–	Evaporative emission (EVAP) system – small leak detected	**Hose connection(s), intake leak, EVAP canister, EVAP canister purge valve**
16827	–	Evaporative emission (EVAP) canister purge valve – circuit malfunction	**Wiring, EVAP canister purge valve**
16828	–	Evaporative emission (EVAP) canister purge valve – open circuit	**Wiring open circuit, EVAP canister purge valve**
16829	–	Evaporative emission (EVAP) canister purge valve – short circuit	**Wiring short circuit, EVAP canister purge valve**
16839	–	Evaporative emission (EVAP) system – large leak detected	**Hose connection(s), intake leak, EVAP canister, EVAP canister purge valve**
16845	–	Fuel tank level sensor – range/ performance problem	**Wiring, fuel tank level sensor**
16864	–	Engine coolant blower motor 1 – circuit malfunction	**Wiring, engine coolant blower motor**
16865	–	Engine coolant blower motor 2 – circuit malfunction	**Wiring, engine coolant blower motor**
16885	–	Vehicle speed sensor (VSS) – range/ performance problem	**Wiring, speedometer, VSS, CAN data bus**

Scan code 5-digit	OBD-II code	Fault location	Probable cause
16887	–	Vehicle speed sensor (VSS) – intermittent/ erratic/high input	**Wiring, other connected system, instrument panel, VSS**
16890	–	Idle speed control (ISC) system – rpm lower than expected	**Throttle control unit**
16891	–	Idle speed control (ISC) system – rpm higher than expected	**Throttle control unit**
16894	–	Closed throttle position (CTP) switch – circuit malfunction	**Wiring, CTP switch**
16916	–	AC refrigerant pressure sensor – low input	**AC refrigerant pressure too low (incorrectly charged), wiring, AC refrigerant pressure sensor**
16917	–	AC refrigerant pressure sensor – high input	**AC refrigerant pressure too high (cooling fault/incorrectly charged), wiring, AC refrigerant pressure sensor**
16928	–	Exhaust gas recirculation temperature (EGRT) sensor, bank 1 – circuit malfunction	**Wiring, EGRT sensor**
16929	–	Exhaust gas recirculation temperature (EGRT) sensor, bank 1 – low input	**Wiring short to ground, EGRT sensor**
16930	–	Exhaust gas recirculation temperature (EGRT) sensor, bank 1 – high input	**Wiring short to positive, EGRT sensor**
16935	–	Power steering pressure (PSP) sensor/ switch – range/performance problem	**Wiring, PSP switch**
16944	–	System voltage – malfunction	**Fuse(s), battery, wiring, engine control relay**
16946	–	System voltage – low	**Fuse(s), battery, wiring, engine control relay**
16947	–	System voltage – high	**Alternator, wiring**
16952	–	Cruise control master/selector switch, SET signal – malfunction	**Wiring, cruise control master/selector switch**
16955	–	Stop lamp switch/brake pedal position (BPP) switch – circuit malfunction	**Wiring, stop lamp switch, BPP switch**
16983	–	CAN data bus – malfunction	**Wiring**
16985	–	Engine control module (ECM) – memory check sum error	**ECM**
16987	–	Engine control module (ECM) – KAM error	**ECM**
16988	–	Engine control module (ECM) – RAM error	**ECM**
16989	–	Engine control module (ECM) – ROM error	**ECM**
16990	–	Engine control module (ECM) – PCM processor fault	**ECM**
17022	–	Throttle actuator control, bank 1 – range/ performance problem	**Basic setting not carried out, throttle control unit, APP sensor**
17026	–	Engine control module (ECM), knock control – defective	**ECM**
17029	–	Air conditioning	**Wiring, AC system**

Scan code 5-digit	OBD-II code	Fault location	Probable cause
17034	–	Malfunction indicator lamp (MIL) – circuit malfunction	**Wiring, MIL**
17040	–	Instrument panel, fuel consumption signal – circuit malfunction	**Wiring**
17071	–	Engine control relay – short to ground	**Wiring short to ground, engine control relay**
17072	–	Engine control relay – short to positive	**Wiring short to positive, engine control relay**
17075	–	Engine coolant blower motor 1 – short to ground	**Wiring short to ground, engine coolant blower motor**
17076	–	Engine coolant blower motor 1 – short to positive	**Wiring short to positive, engine coolant blower motor**
17077	–	Engine coolant blower motor 2 – short to ground	**Wiring short to ground, engine coolant blower motor**
17078	–	Engine coolant blower motor 2 – short to positive	**Wiring short to positive, engine coolant blower motor**
17091	–	Transmission range (TR) sensor – low input	**Wiring short to ground, TR sensor**
17092	–	Transmission range (TR) sensor – high input	**Wiring short to positive, TR sensor**
–	P1009	Mass air flow (MAF) sensor 1/2 – implausible load detection signal	**Wiring, MAF sensor(s)**
17428	P1020	Fuel pressure – control limit exceeded	**Wiring, fuel pressure sensor, fuel pressure control valve, high pressure fuel pump**
17431	P1023	Fuel pressure control valve – short to ground	**Wiring short to ground, fuel pressure control valve**
17432	P1024	Fuel pressure control valve – open circuit	**Wiring open circuit, fuel pressure control valve**
17433	P1025	Fuel pressure control valve – mechanical fault	**Fuel pressure control valve**
17434	P1026	Intake manifold air control solenoid – short to positive	**Wiring short to positive, intake manifold air control solenoid**
17435	P1027	Intake manifold air control solenoid – short to ground	**Wiring short to ground, intake manifold air control solenoid**
17436	P1028	Intake manifold air control solenoid – open circuit	**Wiring open circuit, intake manifold air control solenoid**
17437	P1029	Intake manifold air control valve position sensor – upper limit not reached	**Air control flap tight/sticking, hose connection(s), intake manifold air control actuator**
17438	P1030	Intake manifold air control valve position sensor – lower limit not reached	**Air control flap tight/sticking, hose connection(s), intake manifold air control actuator**
17439	P1031	Intake manifold air control valve position sensor – specification not attained	**Air control flap tight/sticking, hose connection(s), intake manifold air control actuator, intake manifold air control solenoid**
17440 3	P1032	Nitrogen oxides (NOx) sensor – signal too high	**Catalytic converter, wiring short to positive, NOx sensor**

Scan code 5-digit	OBD-II code	Fault location	Probable cause
17441 3	P1033	Nitrogen oxides (NOx) sensor – signal too low	**Exhaust leak, wiring short to ground, NOx sensor**
17442 3	P1034	Nitrogen oxides (NOx) sensor – signal outside tolerance	**Catalytic converter, exhaust leak, wiring, NOx sensor**
17443 3	P1035	Nitrogen oxides (NOx) sensor – range/ performance problem	**Wiring, heating inoperative, NOx sensor**
17444 3	P1036	Nitrogen oxides (NOx) heater sensor – short to positive	**Wiring short to positive, NOx sensor**
17445 3	P1037	Nitrogen oxides (NOx) heater sensor – short to ground	**Wiring short to ground, NOx sensor**
17446 3	P1038	Nitrogen oxides (NOx) heater sensor – open circuit	**Wiring open circuit, NOx sensor**
17447	P1039	Injector 1, supply voltage – short circuit	**Wiring, injector**
17448	P1040	Injector 1, supply voltage – circuit malfunction	**Wiring, injector**
17449	P1041	Injector 2, supply voltage – short circuit	**Wiring, injector**
17450	P1042	Injector 2, supply voltage – circuit malfunction	**Wiring, injector**
17451	P1043	Injector 3, supply voltage – short circuit	**Wiring, injector**
17452	P1044	Injector 3, supply voltage – circuit malfunction	**Wiring, injector**
17453	P1045	Injector 4, supply voltage – short circuit	**Wiring, injector**
17454	P1046	Injector 4, supply voltage – circuit malfunction	**Wiring, injector**
17455	P1047	Camshaft position (CMP) actuator, exhaust/right/rear, bank 1 – circuit malfunction	**Wiring, camshaft position (CMP) actuator**
17456	P1048	Camshaft position (CMP) actuator, exhaust/right/rear, bank 1 – short to positive	**Wiring short to positive, camshaft position (CMP) actuator**
17457	P1049	Camshaft position (CMP) actuator, exhaust/right/rear, bank 1 – short to ground	**Wiring short to ground, camshaft position (CMP) actuator**
17458	P1050	Camshaft position (CMP) actuator, exhaust/right/rear, bank 1 – open circuit	**Wiring open circuit, camshaft position (CMP) actuator**
17471	P1063	Fuel pressure – control limit not reached	**Air in fuel system, fuel leak, fuel lift pump, fuel bypass valve, fuel pressure control valve, high pressure fuel pump**
17472	P1064	Fuel pressure – mechanical fault	**Air in fuel system, fuel leak, fuel lift pump, fuel bypass valve, fuel pressure control valve, high pressure fuel pump**
17473	P1065	Fuel pressure – system deviation	**Air in fuel system, fuel leak, fuel lift pump, fuel bypass valve, fuel pressure control valve, high pressure fuel pump**
17474	P1066	Intake manifold air control solenoid – short to positive	**Wiring short to positive, intake manifold air control solenoid**
17475	P1067	Intake manifold air control solenoid – short to ground	**Wiring short to ground, intake manifold air control solenoid**

Scan code 5-digit	OBD-II code	Fault location	Probable cause
17476	P1068	Intake manifold air control solenoid – open circuit	**Wiring open circuit, intake manifold air control solenoid**
17477 [3]	P1069	Nitrogen oxides (NOx) sensor – heater control – short to ground	**Wiring short to ground, NOx sensor**
17478 [3]	P1070	Nitrogen oxides (NOx) sensor – heater control – short to positive	**Wiring short to positive, NOx sensor**
17479 [3]	P1071	Nitrogen oxides (NOx) sensor – heater control – incorrect signal	**Wiring, NOx sensor**
17480 [3]	P1072	Nitrogen oxides (NOx) sensor – heater control – circuit malfunction	**Wiring, NOx sensor**
–	P1073	Mass air flow (MAF) sensor 2 – signal too low	**Wiring, MAF sensor**
–	P1074	Mass air flow (MAF) sensor 2 – signal too high	**Wiring, MAF sensor**
17501	P1093	Mixture control (MC), bank 1 – malfunction	**Fuel pressure/pump, injector(s) intake leak**
17509	P1101	Heated oxygen sensor (HO2S) 1, bank 1 – low voltage	**Intake/exhaust leak, fuel pressure/ pump, wiring short to ground, HO2S**
17510	P1102	Heated oxygen sensor (HO2S) 1, bank 1 – heater short to positive	**Wiring short to positive, HO2S**
17511	P1103	Heated oxygen sensor (HO2S) 1, bank 1 – heater output too low	**Wiring, HO2S**
17513	P1105	Heated oxygen sensor (HO2S) 2, bank 1 – heater short to positive	**Wiring short to positive, HO2S**
17514	P1106	Heated oxygen sensor (HO2S) 1, bank 2 – low voltage/air leak	**Intake/exhaust leak, fuel pressure/ pump, wiring short to ground, HO2S**
17515	P1107	Heated oxygen sensor (HO2S) 1, bank 2 – heater short to positive	**Wiring short to positive, HO2S**
17518	P1110	Heated oxygen sensor (HO2S) 2, bank 2 – heater short to positive	**Wiring short to positive, HO2S**
17519	P1111	Heated oxygen sensor (HO2S) control, bank 1 – system too lean	**Intake/exhaust leak, injector blocked, MAF sensor, fuel pressure/pump, HO2S**
17520	P1112	Heated oxygen sensor (HO2S) control, bank 1 – system too rich	**Excessive fuel in engine oil, injector leaking, fuel pressure, EVAP canister purge valve, MAF sensor, HO2S**
17521	P1113	Heated oxygen sensor (HO2S) 1, bank 1 – heater resistance too high	**Wiring, HO2S**
17522	P1114	Heated oxygen sensor (HO2S) 2, bank 1 – heater resistance too high	**Wiring, HO2S**
17523	P1115	Heated oxygen sensor (HO2S) 1, bank 1 – heater short to ground	**Wiring short to ground, HO2S**
17524	P1116	Heated oxygen sensor (HO2S) 1, bank 1 – heater open circuit	**Wiring open circuit, HO2S**
17525	P1117	Heated oxygen sensor (HO2S) 2, bank 1 – heater short to ground	**Wiring short to ground, HO2S**
17526	P1118	Heated oxygen sensor (HO2S) 2, bank 1 – heater open circuit	**Wiring open circuit, HO2S**

Scan code 5-digit	OBD-II code	Fault location	Probable cause
17527	P1119	Heated oxygen sensor (HO2S) 1, bank 2 – heater short to ground	Wiring short to ground, HO2S
17528	P1120	Heated oxygen sensor (HO2S) 1, bank 2 – heater open circuit	Wiring open circuit, HO2S
17529	P1121	Heated oxygen sensor (HO2S) 2, bank 2 – heater short to ground	Wiring short to ground, HO2S
17530	P1122	Heated oxygen sensor (HO2S) 2, bank 2 – heater open circuit	Wiring open circuit, HO2S
17535	P1127	Long term fuel trim, bank 1 – system too rich	Fuel pressure, EVAP canister purge valve, injector(s)
17536	P1128	Long term fuel trim, bank 1 – system too lean	Fuel pressure/pump, injector(s), intake/ exhaust leak, AIR system, hose leak
17537	P1129	Long term fuel trim, bank 2 – system too rich	Fuel pressure, EVAP canister purge valve, injector(s)
17538	P1130	Long term fuel trim, bank 2 – system too lean	Fuel pressure/pump, injector(s), intake/ exhaust leak, AIR system, hose leak
17540	P1132	Heated oxygen sensor (HO2S) 1, bank 1 & 2 – heater control – circuit high	Wiring short to positive, HO2S
17541	P1133	Heated oxygen sensor (HO2S) 1, bank 1 & 2 – heater control – circuit low	Wiring open circuit/short to ground, HO2S
17544	P1136	Long term fuel trim, idling, bank 1 – system too lean	Fuel pressure/pump, injector(s), intake/ exhaust leak, AIR system, hose leak
17545	P1137	Long term fuel trim, idling, bank 1 – system too rich	Fuel pressure, injector(s), EVAP canister purge valve
17546	P1138	Long term fuel trim, idling, bank 2 – system too lean	Fuel pressure/pump, injector(s), intake/ exhaust leak, AIR system, hose leak
17547	P1139	Long term fuel trim, idling, bank 2 – system too rich	Fuel pressure, injector(s), EVAP canister purge valve
17548	P1140	Heated oxygen sensor (HO2S) 2, bank 2 – heater resistance too high	Wiring, HO2S
17549	P1141	Load calculation – implausible value	Wiring, intake leak, MAP sensor, MAF sensor, throttle control unit
17550	P1142	Load calculation – too low	Wiring, MAP sensor, MAF sensor, APP sensor, throttle control unit
17551	P1143	Load calculation – too high	Wiring, intake leak, MAP sensor, MAF sensor, APP sensor, throttle control unit
17552	P1144	Mass air flow (MAF) sensor – open circuit/ short to ground	Wiring open circuit/short to ground, MAF sensor
17553	P1145	Mass air flow (MAF) sensor – short to positive	Wiring short to positive, MAF sensor
17554	P1146	Mass air flow (MAF) sensor – supply voltage	Operating voltage too high/low, wiring
17555	P1147	Heated oxygen sensor 1, bank 2 – lambda regulation, system too lean	Intake leak, wiring, HO2S
17556	P1148	Heated oxygen sensor 1, bank 2 – lambda regulation, system too rich	Exhaust leak, wiring, HO2S
17557	P1149	Heated oxygen sensor (HO2S) 1, bank 1 – implausible lambda control value	Exhaust leak, wiring, HO2S

Scan code 5-digit	OBD-II code	Fault location	Probable cause
17558	P1150	Heated oxygen sensor (HO2S) 1, bank 2 – implausible lambda control value	**Exhaust leak, wiring, HO2S**
17559	P1151	Long term fuel trim 1, bank 1 – below lean limit	**Fuel pressure/pump, injectors, intake/ exhaust leak, HO2S**
17560	P1152	Long term fuel trim 2, bank 1 – below lean limit	**Fuel pressure/pump, injectors, intake/ exhaust leak, AIR system, HO2S**
17561	P1153	Heated oxygen sensor (HO2S) 2, bank 1 & 2 – interchanged	**HO2S 2 on bank 1 & 2 incorrectly installed**
17563	P1155	Manifold absolute pressure (MAP) sensor – short to positive	**Wiring short to positive, MAP sensor**
17564	P1156	Manifold absolute pressure (MAP) sensor – open circuit/short to ground	**Wiring open circuit/short to ground, MAP sensor**
17565	P1157	Manifold absolute pressure (MAP) sensor – supply voltage	**Wiring, MAP sensor**
17566	P1158	Manifold absolute pressure (MAP) sensor – range/performance problem	**Wiring, hose connection(s), MAP sensor**
17567	P1159	Mass air flow (MAF) sensor, bank 1 & 2 – implausible ratio	**EGR system, intake leak, wiring, MAF sensor 1/2**
17568	P1160	Intake air temperature (IAT) sensor – short to ground	**Wiring short to ground, IAT sensor**
17569	P1161	Intake air temperature (IAT) sensor – open circuit/short to positive	**Wiring open circuit/short to positive, IAT sensor**
17570	P1162	Fuel temperature sensor – short to ground	**Wiring short to ground, fuel temperature sensor**
17571	P1163	Fuel temperature sensor – open circuit/ short to positive	**Wiring open circuit/short to positive, fuel temperature sensor**
17572	P1164	Fuel temperature sensor – range/ performance problem	**Wiring, fuel temperature sensor**
17573	P1165	Long term fuel trim 1, bank 1 – above rich limit	**Fuel pressure/pump, injectors, EVAP canister purge valve, EGR system, HO2S, intake/exhaust system**
17574	P1166	Long term fuel trim 2, bank 1 – above rich limit	**Fuel pressure/pump, injectors, EVAP canister purge valve, EGR system, HO2S, intake/exhaust system**
17579	P1171	Throttle motor position sensor 2 – range/ performance problem	**Wiring, throttle valve tight/sticking, throttle motor position sensor, throttle control unit**
17580	P1172	Throttle motor position sensor 2 – low input	**Wiring short to ground, throttle motor position sensor, throttle control unit**
17581	P1173	Throttle motor position sensor 2 – high input	**Wiring short to positive, throttle motor position sensor, throttle control unit**
17582	P1174	Fuel measurement system, bank 1 – injection timing incorrect	**Fuel pressure/pump, injector(s), intake/ exhaust leak, EGR system, EVAP canister purge valve, HO2S**
17584	P1176	Lambda correction after catalyst, bank 1 – control limit reached	**Wiring, intake leak, HO2S**
17585	P1177	Lambda correction after catalyst, bank 2 – control limit reached	**Wiring, intake leak, HO2S**

Scan code 5-digit	OBD-II code	Fault location	Probable cause
17586	P1178	Heated oxygen sensor (HO2S) 1, bank 1, pump current – open circuit	**Wiring open circuit, HO2S**
17587	P1179	Heated oxygen sensor (HO2S) 1, bank 1, pump current – short to ground	**Wiring short to ground, HO2S**
17588	P1180	Heated oxygen sensor (HO2S) 1, bank 1, pump current – short to positive	**Wiring short to positive, HO2S**
17589	P1181	Heated oxygen sensor (HO2S) 1, bank 1, reference voltage – open circuit	**Wiring open circuit, HO2S, HT leads, spark plugs, misfire detection**
17590	P1182	Heated oxygen sensor (HO2S) 1, bank 1, reference voltage – short to ground	**Wiring short to ground, HO2S, HT leads, spark plugs, misfire detection**
17591	P1183	Heated oxygen sensor (HO2S) 1, bank 1, reference voltage – short to positive	**Wiring short to positive, HO2S, HT leads, spark plugs, misfire detection**
–	P1184	Heated oxygen sensor (HO2S) 1, bank 1, ground wire – open circuit	**Wiring open circuit, HO2S**
–	P1185	Heated oxygen sensor (HO2S) 1, bank 1, ground wire – short to ground	**Wiring short to ground, HO2S**
–	P1186	Heated oxygen sensor (HO2S) 1, bank 1, ground wire – short to positive	**Wiring short to positive, HO2S**
17595	P1187	Heated oxygen sensor (HO2S) 1, bank 1 or 2 – circuit malfunction	**Wiring, HO2S**
17598	P1190	Heated oxygen sensor (HO2S) 1, bank 1, reference voltage – range/performance problem	**Wiring, HO2S, HT leads, spark plugs, misfire detection**
17599	P1191	Heated oxygen sensor (HO2S) 1, bank 1 & 2 – interchanged	**HO2S 1 on bank 1 & 2 incorrectly installed**
17600	P1192	Fuel pressure sensor – supply voltage	**Wiring, fuel pressure sensor**
17601	P1193	Fuel pressure sensor – open circuit/short to positive	**Wiring open circuit/short to positive**
17602	P1194	Fuel pressure control valve – short to positive	**Wiring short to positive, fuel pressure control valve**
17604	P1196	Heated oxygen sensor (HO2S) 1, bank 1, heater circuit malfunction	**Wiring, HO2S**
17605	P1197	Heated oxygen sensor (HO2S) 1, bank 2, heater circuit malfunction	**Wiring, HO2S**
17606	P1198	Heated oxygen sensor (HO2S) 2, bank 1, heater circuit malfunction	**Wiring, HO2S**
17607	P1199	Heated oxygen sensor (HO2S) 2, bank 2, heater circuit malfunction	**Wiring, HO2S**
–	P1200	Turbocharger (TC) bypass valve – mechanical malfunction	**Hose connection, TC bypass valve**
17609	P1201	Injector 1 – circuit malfunction	**Wiring, injector**
17610	P1202	Injector 2 – circuit malfunction	**Wiring, injector**
17611	P1203	Injector 3 – circuit malfunction	**Wiring, injector**
17612	P1204	Injector 4 – circuit malfunction	**Wiring, injector**
17613	P1205	Injector 5 – circuit malfunction	**Wiring, injector**
17614	P1206	Injector 6 – circuit malfunction	**Wiring, injector**

Scan code 5-digit	OBD-II code	Fault location	Probable cause
17621	P1213	Injector 1 – short to positive	Wiring short to positive, injector
17622	P1214	Injector 2 – short to positive	Wiring short to positive, injector
17623	P1215	Injector 3 – short to positive	Wiring short to positive, injector
17624	P1216	Injector 4 – short to positive	Wiring short to positive, injector
17625	P1217	Injector 5 – short to positive	Wiring short to positive, injector
17626	P1218	Injector 6 – short to positive	Wiring short to positive, injector
17633	P1225	Injector 1 – short to ground	Wiring short to ground, injector
17634	P1226	Injector 2 – short to ground	Wiring short to ground, injector
17635	P1227	Injector 3 – short to ground	Wiring short to ground, injector
17636	P1228	Injector 4 – short to ground	Wiring short to ground, injector
17637	P1229	Injector 5 – short to ground	Wiring short to ground, injector
17638	P1230	Injector 6 – short to ground	Wiring short to ground, injector
17645	P1237	Injector 1 – open circuit	Wiring open circuit, injector
17646	P1238	Injector 2 – open circuit	Wiring open circuit, injector
17647	P1239	Injector 3 – open circuit	Wiring open circuit, injector
17648	P1240	Injector 4 – open circuit	Wiring open circuit, injector
17649	P1241	Injector 5 – open circuit	Wiring open circuit, injector
17650	P1242	Injector 6 – open circuit	Wiring open circuit, injector
17653	P1245	Injector needle lift sensor – short to ground	Wiring short to ground, injector needle lift sensor
17654	P1246	Injector needle lift sensor – range/ performance problem	Injector needle lift sensor, injector pipe defective, fuel level low
17655	P1247	Injector needle lift sensor – open circuit/ short to positive	Wiring open circuit/short to positive, injector needle lift sensor
17656	P1248	Start of injection – control difference	Fuel injection timing solenoid, injector needle lift sensor, fuel level low, pump timing
17658	P1250	Fuel tank level sensor – low input	Fuel level too low
17659	P1251	Fuel injection timing solenoid – short to positive	Wiring short to positive
17660	P1252	Fuel injection timing solenoid – open circuit/short to ground	Wiring open circuit/short to ground, fuel injection timing solenoid
17661	P1253	Instrument panel, fuel consumption signal – short to ground	Wiring short to ground, instrument panel
17662	P1254	Instrument panel, fuel consumption signal – short to positive	Wiring short to positive, instrument panel
17663	P1255	Engine coolant temperature (ECT) sensor – short to ground	Wiring short to ground, ECT sensor
17664	P1256	Engine coolant temperature (ECT) sensor – open circuit/short to positive	Wiring open circuit/short to positive, ECT sensor
17668	P1260	Injector 1 – implausible signal	No control
17669	P1261	Injector 1 – control limit exceeded	Control period too long, wiring, injector

Scan code 5-digit	OBD-II code	Fault location	Probable cause
17670	P1262	Injector 1 – control limit not reached	**Control period too short, fuel level low, air in fuel system**
17671	P1263	Injector 2 – implausible signal	**No control**
17672	P1264	Injector 2 – control limit exceeded	**Control period too long, wiring, injector**
17673	P1265	Injector 2 – control limit not reached	**Control period too short, fuel level low, air in fuel system**
17674	P1266	Injector 3 – implausible signal	**No control**
17675	P1267	Injector 3 – control limit exceeded	**Control period too long, wiring, injector**
17676	P1268	Injector 3 – control limit not reached	**Control period too short, fuel level low, air in fuel system**
17677	P1269	Injector 4 – implausible signal	**No control**
17678	P1270	Injector 4 – control limit exceeded	**Control period too long, wiring, injector**
17679	P1271	Injector 4 – control limit not reached	**Control period too short, fuel level low, air in fuel system**
17686	P1278	Fuel metering solenoid – short to positive	**Wiring short to positive, fuel metering solenoid**
17687	P1279	Fuel metering solenoid – open circuit/ short to ground	**Wiring open circuit/short to ground, fuel metering solenoid**
17689	P1281	Fuel metering solenoid – short to ground	**Wiring short to ground, fuel metering solenoid**
17690	P1282	Fuel metering solenoid – open circuit	**Wiring open circuit, fuel metering solenoid**
17695	P1287	Turbocharger (TC) bypass valve – open circuit	**Wiring open circuit, TC bypass valve**
17696	P1288	Turbocharger (TC) bypass valve – short to positive	**Wiring short to positive, TC bypass valve**
17697	P1289	Turbocharger (TC) bypass valve – short to ground	**Wiring short to ground, TC bypass valve**
17698	P1290	Engine coolant temperature (ECT) sensor, ECM controlled cooling system – high input	**Wiring, ECT sensor**
17699	P1291	Engine coolant temperature (ECT) sensor, ECM controlled cooling system – high input	**Wiring, ECT sensor**
17700	P1292	Engine coolant thermostat – open circuit	**Wiring open circuit, engine coolant thermostat**
17701	P1293	Engine coolant thermostat – short to positive	**Wiring short to positive, engine coolant thermostat**
17702	P1294	Engine coolant thermostat – short to ground	**Wiring short to ground, engine coolant thermostat**
17703	P1295	Turbocharger (TC), bypass – flow malfunction	**TC wastegate regulating valve, hose connection(s), injector**
17704	P1296	Engine cooling system – malfunction	**ECT sensor, coolant thermostat**
17705	P1297	Turbocharger (TC)/throttle valve, hose connection – pressure loss	**Hose connection**

Scan code 5-digit	OBD-II code	Fault location	Probable cause
17707	P1299	Fuel metering solenoid – circuit malfunction	Wiring, fuel metering solenoid
17708	P1300	Random/multiple cylinder(s) – misfire detected	Fuel level low, fuel gauge tank sensor
–	P1321	Knock sensor (KS) 3 – signal too low	Wiring, KS incorrectly tightened/ defective
–	P1322	Knock sensor (KS) 3 – signal too high	Wiring, KS incorrectly tightened/ defective
–	P1323	Knock sensor (KS) 4 – signal too low	Wiring, KS incorrectly tightened/ defective
–	P1324	Knock sensor (KS) 4 – signal too high	Wiring, KS incorrectly tightened/ defective
17733	P1325	Knock control, cylinder 1 – control limit reached	Poor quality fuel, incorrect fuel, insecure engine component, KS incorrectly tightened/defective, shield wiring open circuit, poor connection
17734	P1326	Knock control, cylinder 2 – control limit reached	Poor quality fuel, incorrect fuel, insecure engine component, KS incorrectly tightened/defective, shield wiring open circuit, poor connection
17735	P1327	Knock control, cylinder 3 – control limit reached	Poor quality fuel, incorrect fuel, insecure engine component, KS incorrectly tightened/defective, shield wiring open circuit, poor connection
17736	P1328	Knock control, cylinder 4 – control limit reached	Poor quality fuel, incorrect fuel, insecure engine component, KS incorrectly tightened/defective, shield wiring open circuit, poor connection
17737	P1329	Knock control, cylinder 5 – control limit reached	Poor quality fuel, incorrect fuel, insecure engine component, KS incorrectly tightened/defective, shield wiring open circuit, poor connection
17738	P1330	Knock control, cylinder 6 – control limit reached	Poor quality fuel, incorrect fuel, insecure engine component, KS incorrectly tightened/defective, shield wiring open circuit, poor connection
17743	P1335	Engine torque control 1/2 – limit reached/exceeded	Throttle control unit, hose(s), TC system, MAP sensor, IAT sensor, MAF sensor, ECT sensor
17744	P1336	Engine torque monitoring – control limit exceeded	Hose(s), TC system, throttle control unit, IAT sensor, MAP sensor, MAF sensor, ECT sensor, APP sensor
17745	P1337	Camshaft position (CMP) sensor, bank 1 – short to ground	Wiring short to ground, CMP sensor
17746	P1338	Camshaft position (CMP) sensor, bank 1 – circuit malfunction	Wiring, CMP sensor
17747	P1339	Crankshaft position (CKP) sensor/engine speed (RPM) sensor – interchanged	Multi-plugs incorrectly connected

Scan code 5-digit	OBD-II code	Fault location	Probable cause
17748	P1340	Camshaft position (CMP) sensor 1/bank 1/ crankshaft position (CKP) sensor – out of sequence	**Wiring, valve timing, CKP/CMP sensor installation, CKP/CMP sensor defective, CKP sensor rotor**
17749	P1341	Ignition amplifier, primary circuit 1 – short to ground	**Wiring short to ground, ignition amplifier, CMP sensor, HT leads, spark plugs**
17750	P1342	Ignition amplifier, primary circuit 1 – short to positive	**Wiring short to positive, ignition amplifier, CMP sensor, HT leads, spark plugs**
17751	P1343	Ignition amplifier, primary circuit 2 – short to ground	**Wiring short to ground, ignition amplifier, CMP sensor, HT leads, spark plugs**
17752	P1344	Ignition amplifier, primary circuit 2 – short to positive	**Wiring short to positive, ignition amplifier, CMP sensor, HT leads, spark plugs**
17753	P1345	Ignition amplifier, primary circuit 3 – short to ground	**Wiring short to ground, ignition amplifier, CMP sensor, HT leads, spark plugs**
17754	P1346	Ignition amplifier, primary circuit 3 – short to positive	**Wiring short to positive, ignition amplifier, CMP sensor, HT leads, spark plugs**
17755	P1347	Camshaft position (CMP) sensor 2/bank 2/ crankshaft position (CKP) sensor – out of sequence	**Wiring, valve timing, CKP/CMP sensor installation, CKP/CMP sensor defective, CKP sensor rotor**
17756	P1348	Ignition amplifier, primary circuit 1 – open circuit	**Wiring short to ground, ignition amplifier, CMP sensor, HT leads, spark plugs**
17757	P1349	Ignition amplifier, primary circuit 2 – open circuit	**Wiring short to positive, ignition amplifier, CMP sensor, HT leads, spark plugs**
17758	P1350	Ignition amplifier, primary circuit 3 – open circuit	**Wiring short to positive, ignition amplifier, CMP sensor, HT leads, spark plugs**
17759	P1351	Camshaft position (CMP) sensor, bank 1 – range/performance problem	**Ignore trouble code, erase fault memory**
17762	P1354	Fuel quantity adjuster position sensor	**Wiring, fuel injection pump**
17763 4	P1355	Ignition coil/amplifier, cylinder 1 – open circuit	**Wiring open circuit, ignition coil/ amplifier**
17764 4	P1356	Ignition coil/amplifier, cylinder 1 – short to positive	**Wiring short to positive, ignition coil/ amplifier**
17765 4	P1357	Ignition coil/amplifier, cylinder 1 – short to ground	**Wiring short to ground, ignition coil/ amplifier**
17766 5	P1358	Ignition coil/amplifier, cylinder 2 – open circuit	**Wiring open circuit, ignition coil/ amplifier**
17767 5	P1359	Ignition coil/amplifier, cylinder 2 – short to positive	**Wiring short to positive, ignition coil/ amplifier**
17768 5	P1360	Ignition coil/amplifier, cylinder 2 – short to ground	**Wiring short to ground, ignition coil/ amplifier**

Scan code 5-digit	OBD-II code	Fault location	Probable cause
17769	P1361	Ignition coil/amplifier, cylinder 3 – open circuit	**Wiring open circuit, ignition coil/ amplifier**
17770	P1362	Ignition coil/amplifier, cylinder 3 – short to positive	**Wiring short to positive, ignition coil/ amplifier**
17771	P1363	Ignition coil/amplifier, cylinder 3 – short to ground	**Wiring short to ground, ignition coil/ amplifier**
17772	P1364	Ignition coil/amplifier, cylinder 4 – open circuit	**Wiring open circuit, ignition coil/ amplifier**
17773	P1365	Ignition coil/amplifier, cylinder 4 – short to positive	**Wiring short to positive, ignition coil/ amplifier**
17774	P1366	Ignition coil/amplifier, cylinder 4 – short to ground	**Wiring short to ground, ignition coil/ amplifier**
17775	P1367	Ignition coil/amplifier, cylinder 5 – open circuit	**Wiring open circuit, ignition coil/ amplifier**
17776	P1368	Ignition coil/amplifier, cylinder 5 – short to positive	**Wiring short to positive, ignition coil/ amplifier**
17777	P1369	Ignition coil/amplifier, cylinder 5 – short to ground	**Wiring short to ground, ignition coil/ amplifier**
17778	P1370	Ignition coil/amplifier, cylinder 6 – open circuit	**Wiring open circuit, ignition coil/ amplifier**
17779	P1371	Ignition coil/amplifier, cylinder 6 – short to positive	**Wiring short to positive, ignition coil/ amplifier**
17780	P1372	Ignition coil/amplifier, cylinder 6 – short to ground	**Wiring short to ground, ignition coil/ amplifier**
17793	P1385	Engine control module (ECM) – defective	**ECM**
17794	P1386	Engine control module (ECM), knock control – defective	**ECM**
17795	P1387	Engine control module (ECM), BARO sensor – defective	**ECM**
17796	P1388	Engine control module (ECM), ETS – defective	**ECM**
17799	P1391	Camshaft position (CMP) sensor 2/bank 2 – short to ground	**Wiring short to ground, CMP sensor**
17800	P1392	Camshaft position (CMP) sensor 2/bank 2 – open circuit/short to positive	**Wiring open circuit/short to positive, CMP sensor**
17801	P1393	Ignition amplifier, primary circuit 1 – circuit malfunction	**Wiring, ignition amplifier, HT leads, spark plugs**
17802	P1394	Ignition amplifier, primary circuit 2 – circuit malfunction	**Wiring, ignition amplifier, HT leads, spark plugs**
17803	P1395	Ignition amplifier, primary circuit 3 – circuit malfunction	**Wiring, ignition amplifier, HT leads, spark plugs**
17805	P1397	Crankshaft position (CKP) sensor/engine speed (RPM) sensor – control limit reached	**Insecure/damaged rotor, CKP/RPM sensor**
17806	P1398	Crankshaft position (CKP) sensor/engine speed (RPM) sensor – short to ground	**Wiring short to ground, CKP/RPM sensor**
17807	P1399	Crankshaft position (CKP) sensor/engine speed (RPM) sensor – short to positive	**Wiring short to positive, CKP/RPM sensor**

Scan code 5-digit	OBD-II code	Fault location	Probable cause
17808	P1400	Exhaust gas recirculation (EGR) valve/ solenoid, bank 1 – circuit malfunction	Wiring, EGR valve
17809	P1401	Exhaust gas recirculation (EGR) valve/ solenoid, bank 1 – short to ground	Wiring short to ground, EGR valve
17810	P1402	Exhaust gas recirculation (EGR) valve/ solenoid, bank 1 – short to positive	Wiring short to positive, EGR valve/ solenoid
17811	P1403	Exhaust gas recirculation (EGR) system – control difference	Basic setting not carried out, EGR system
17812	P1404	Exhaust gas recirculation (EGR) system – basic setting	Basic setting not carried out, EGR system
17815	P1407	Exhaust gas recirculation temperature (EGRT) sensor – low input	Wiring short to ground, EGRT sensor
17816	P1408	Exhaust gas recirculation temperature (EGRT) sensor – high input	Wiring short to positive, ground wire defective, EGRT sensor
17817	P1409	Evaporative emission (EVAP) canister purge valve – circuit malfunction	Wiring, EVAP canister purge valve
17818	P1410	Evaporative emission (EVAP) canister purge valve – short to positive	Wiring short to positive, EVAP canister purge valve
17819	P1411	Secondary air injection (AIR) system, bank 2 – insufficient flow detected	Intake leak, hose(s) blocked/leaking, AIR valve/solenoid
17822	P1414	Secondary air injection (AIR) system, bank 2 – leak detected	Intake leak, hose(s) leaking, AIR valve/ solenoid
17823	P1415	Exhaust gas recirculation (EGR) valve position sensor – lower limit exceeded	Basic setting not carried out
17824	P1416	Exhaust gas recirculation (EGR) valve position sensor – upper limit exceeded	Basic setting not carried out
17828	P1420	Secondary air injection (AIR) valve/ solenoid – circuit malfunction	Wiring, AIR solenoid
17829	P1421	Secondary air injection (AIR) valve/ solenoid – short to ground	Wiring short to ground, AIR valve/ solenoid
17830	P1422	Secondary air injection (AIR) valve/ solenoid – short to positive	Wiring short to positive, AIR valve/ solenoid
17831	P1423	Secondary air injection (AIR) system, bank 1 – insufficient flow detected	Hose connection(s), wiring, AIR pump relay, AIR valve/solenoid, AIR pump
17832	P1424	Secondary air injection (AIR) system, bank 1 – leak detected	Hose(s) leaking, AIR valve, exhaust leak
17833	P1425	Evaporative emission (EVAP) canister purge valve – short to ground	Wiring short to ground, EVAP canister purge valve
17834	P1426	Evaporative emission (EVAP) canister purge valve – open circuit	Wiring open circuit, EVAP canister purge valve
17835	P1427	Vacuum pump, brakes – short to positive	Wiring short to positive, vacuum pump
17836	P1428	Vacuum pump, brakes – short to ground	Wiring short to ground, vacuum pump
17837	P1429	Vacuum pump, brakes – open circuit	Wiring open circuit, vacuum pump
17838	P1430	Vacuum pump, brakes – open circuit/short to positive	Wiring open circuit/short to positive, vacuum pump
17839	P1431	Vacuum pump, brakes – open circuit/short to ground	Wiring open circuit/short to ground, vacuum pump

Scan code 5-digit	OBD-II code	Fault location	Probable cause
17840	P1432	Secondary air injection (AIR) valve/ solenoid – open circuit	Wiring open circuit, fuse, AIR valve/ solenoid
17841	P1433	Secondary air injection (AIR) pump relay – open circuit	Wiring open circuit, fuse, AIR pump relay
17842	P1434	Secondary air injection (AIR) pump relay – short to positive	Wiring short to positive, AIR pump relay
17843	P1435	Secondary air injection (AIR) pump relay – short to ground	Wiring short to ground, AIR pump relay
17844	P1436	Secondary air injection (AIR) pump relay – circuit malfunction	Wiring, AIR pump relay
17845	P1437	Exhaust gas recirculation (EGR) valve/ solenoid, bank 2 – short to positive	Wiring short to positive, EGR valve/ solenoid
17846	P1438	Exhaust gas recirculation (EGR) valve/ solenoid, bank 2 – open circuit/short to ground	Wiring open circuit/short to ground, EGR valve/solenoid
17847	P1439	Exhaust gas recirculation (EGR) valve position sensor – basic setting	Basic setting not carried out, EGR system
17848	P1440	Exhaust gas recirculation (EGR) valve – open circuit	Wiring open circuit, EGR valve
17849	P1441	Exhaust gas recirculation (EGR) valve/ solenoid, bank 1 – open circuit/short to ground	Wiring open circuit/short to ground, EGR solenoid
17850	P1442	Exhaust gas recirculation (EGR) valve position sensor – high input	Wiring short to positive, EGR valve position sensor
17851	P1443	Exhaust gas recirculation (EGR) valve position sensor – low input	Wiring short to ground, EGR valve position sensor
17852	P1444	Exhaust gas recirculation (EGR) valve position sensor – range/performance problem	Wiring, EGR valve position sensor
17858	P1450	Secondary air injection (AIR) system – short to positive	Wiring short to positive, AIR relay
17859	P1451	Secondary air injection (AIR) system – short to ground	Wiring short to ground, AIR relay
17860	P1452	Secondary air injection (AIR) system – open circuit	Wiring open circuit, AIR relay
17861	P1453	Exhaust gas recirculation temperature (EGRT) sensor 1 – open circuit/short to positive	Wiring open circuit/short to positive, EGRT sensor
17862	P1454	Exhaust gas recirculation temperature (EGRT) sensor 1 – short to ground	Wiring short to ground, EGRT sensor
17863	P1455	Exhaust gas recirculation temperature (EGRT) sensor 1 – range/performance problem	Exhaust leak, wiring, EGRT sensor
17864	P1456	Exhaust gas recirculation temperature (EGRT) control, bank 1 – control limit reached	EGRT sensor
17865	P1457	Exhaust gas recirculation temperature (EGRT) sensor 2 – open circuit/short to positive	Wiring open circuit/short to positive, EGRT sensor

Scan code 5-digit	OBD-II code	Fault location	Probable cause
17866	P1458	Exhaust gas recirculation temperature (EGRT) sensor 2 – short to ground	**Wiring short to ground, EGRT sensor**
17867	P1459	Exhaust gas recirculation temperature (EGRT) sensor 2 – range/performance problem	**Exhaust leak, wiring, EGRT sensor**
17868	P1460	Exhaust gas recirculation temperature (EGRT) control, bank 2 – control limit reached	**EGRT sensor**
17869	P1461	Exhaust gas recirculation temperature (EGRT) control, bank 1 – range/performance problem	**Exhaust leak/blockage, EGRT sensor**
17870	P1462	Exhaust gas recirculation temperature (EGRT) control, bank 2 – range/performance problem	**Exhaust leak/blockage, EGRT sensor**
17875	P1467	Evaporative emission (EVAP) canister purge valve – short to positive	**Wiring short to positive , EVAP canister purge valve**
17876	P1468	Evaporative emission (EVAP) canister purge valve – short to ground	**Wiring short to ground, EVAP canister purge valve**
17877	P1469	Evaporative emission (EVAP) canister purge valve – open circuit	**Wiring open circuit, EVAP canister purge valve**
17878	P1470	Evaporative emission (EVAP) leak detection pump/fuel tank vent system – circuit malfunction	**Wiring, EVAP leak detection pump**
17879	P1471	Evaporative emission (EVAP) leak detection pump – short to positive	**Wiring short to positive, EVAP leak detection pump**
17880	P1472	Evaporative emission (EVAP) leak detection pump – short to ground	**Wiring short to ground, EVAP leak detection pump**
17881	P1473	Evaporative emission (EVAP) leak detection pump/fuel tank vent system – Wiring open circuit	**Wiring open circuit, EVAP canister purge valve, EVAP leak detection pump**
17883	P1475	Evaporative emission (EVAP) leak detection pump/fuel tank vent system – no signal	**Wiring, EVAP canister purge valve, EVAP leak detection pump, ECM**
17884	P1476	Evaporative emission (EVAP) leak detection pump/fuel tank vent system – vacuum to low	**system leak, hose blockage, EVAP canister, EVAP leak detection pump**
17885	P1477	Evaporative emission (EVAP) leak detection pump/fuel tank vent system – malfunction	**Wiring, hose leak/blockage, EVAP canister purge valve, EVAP canister, EVAP leak detection pump**
17886	P1478	Evaporative emission (EVAP) leak detection pump/fuel tank vent system – hose blockage detected	**Hose blockage**
17887	P1479	Vacuum system, brakes – mechanical fault	**Vacuum pump**
–	P1490	Evaporative emission (EVAP) canister purge valve 2 – short to ground	**Wiring short to ground, EVAP canister purge valve**
–	P1491	Evaporative emission (EVAP) canister purge valve 2 – open circuit	**Wiring open circuit, EVAP canister purge valve**
17908	P1500	Fuel pump relay – circuit malfunction	**Wiring, fuel pump relay**

Scan code 5-digit	OBD-II code	Fault location	Probable cause
17909	P1501	Fuel pump relay – short to ground	Wiring short to ground, fuel pump relay
17910	P1502	Fuel pump relay – short to positive	Wiring short to positive, fuel pump relay
17911	P1503	Alternator load signal	Wiring, alternator
17912	P1504	Intake system – leak detected	Intake leak, EGR system, EVAP system, hose connection(s), throttle control unit
17913	P1505	Closed throttle position (CTP) switch – does not close	Throttle cable/valve, wiring open circuit/ short to positive, CTP switch adjustment/defective, ECM
17914	P1506	Closed throttle position (CTP) switch – does not open	Moisture ingress, wiring short to ground, CTP switch adjustment/ defective, ECM
17915	P1507	Idle speed control (ISC) – lower limit reached	Throttle control unit/basic setting, intake/exhaust leak, mechanical fault, AC signals
17916	P1508	Idle speed control (ISC) – upper limit reached	Throttle control unit/basic setting, intake/exhaust leak, mechanical fault, AC signals
17917	P1509	Idle air control (IAC) valve – circuit malfunction	Wiring, IAC valve
17918	P1510	Idle air control (IAC) valve – short to positive	Wiring open circuit/short to ground, IAC valve
17919	P1511	Intake manifold air control solenoid 1 – current circuit	Wiring, intake manifold air control solenoid
17920	P1512	Intake manifold air control solenoid 1 – short to positive	Wiring short to positive, intake manifold air control solenoid
17921	P1513	Intake manifold air control solenoid 2 – short to positive	Wiring short to positive, intake manifold air control solenoid
17922	P1514	Intake manifold air control solenoid 2 – short to ground	Wiring short to ground, intake manifold air control solenoid
17923	P1515	Intake manifold air control solenoid 1 – short to ground	Wiring short to ground, intake manifold air control solenoid
17924	P1516	Intake manifold air control solenoid 1 – open circuit	Wiring open circuit, intake manifold air control solenoid
17925	P1517	Engine control relay – circuit malfunction	Wiring, engine control relay
17926	P1518	Engine control relay – short to positive	Wiring short to positive, engine control relay
17927	P1519	Camshaft position (CMP) control, bank 1 – malfunction	Cylinder head oil pressure too low, CMP actuator sticking/defective
17928	P1520	Intake manifold air control solenoid 2 – open circuit	Wiring open circuit, intake manifold air control solenoid
17930	P1522	Camshaft position (CMP) control, bank 2 – malfunction	Cylinder head oil pressure too low, CMP actuator sticking/defective
17931	P1523	SRS crash signal received	Airbag triggered
17932	P1524	Fuel pump relay – open circuit/short to ground	Wiring open circuit/short to ground, fuel pump relay
17933	P1525	Camshaft position (CMP) actuator, bank 1 – circuit malfunction	Wiring, CMP actuator

Scan code 5-digit	OBD-II code	Fault location	Probable cause
17934	P1526	Camshaft position (CMP) actuator, bank 1 – short to positive	**Wiring short to positive, CMP actuator**
17935	P1527	Camshaft position (CMP) actuator, bank 1 – short to ground	**Wiring short to ground, CMP actuator**
17936	P1528	Camshaft position (CMP) actuator, bank 1 – open circuit	**Wiring open circuit, CMP actuator**
17937	P1529	Camshaft position (CMP) actuator – short to positive	**Wiring short to positive, CMP actuator**
17938	P1530	Camshaft position (CMP) actuator – short to ground	**Wiring short to ground, CMP actuator**
17939	P1531	Camshaft position (CMP) actuator – open circuit	**Wiring open circuit, CMP actuator**
17940	P1532	Idle control – lean running speed below specification	**Throttle control unit**
17941	P1533	Camshaft position (CMP) actuator, bank 2 – circuit malfunction	**Wiring, CMP actuator**
17942	P1534	Camshaft position (CMP) actuator, bank 2 – short to positive	**Wiring short to positive, CMP actuator**
17943	P1535	Camshaft position (CMP) actuator, bank 2 – short to ground	**Wiring short to ground, CMP actuator**
17944	P1536	Camshaft position (CMP) actuator, bank 2 – open circuit	**Wiring open circuit, CMP actuator**
17945	P1537	Fuel shut-off solenoid – malfunction	**Fuel shut-off solenoid (leaking/sticking)**
17946	P1538	Fuel shut-off solenoid – open circuit/short to ground	**Wiring open circuit/short to ground, fuel shut-off solenoid**
17947	P1539	Clutch pedal position (CPP) switch – range/performance problem	**Wiring, CPP switch**
17948	P1540	Vehicle speed signal – high input	**Excessive vehicle speed, instrument panel defective**
17949	P1541	Fuel pump relay – open circuit	**Wiring open circuit, fuel pump relay**
17950	P1542	Throttle motor position sensor 1 – range/performance problem	**Throttle valve requires cleaning, wiring, throttle motor position sensor, throttle control unit**
17951	P1543	Throttle motor position sensor 1 – low input	**Wiring short to ground, throttle motor position sensor, throttle control unit**
17952	P1544	Throttle motor position sensor 1 – high input	**Wiring short to positive, throttle motor position sensor, throttle control unit**
17953	P1545	Throttle valve control – malfunction	**Throttle valve tight/sticking, wiring, throttle control unit**
17954	P1546	Turbocharger (TC) wastegate regulating valve – short to positive	**Wiring short to positive, TC wastegate regulating valve**
17955	P1547	Turbocharger (TC) wastegate regulating valve – short to ground	**Wiring short to ground, TC wastegate regulating valve**
17956	P1548	Turbocharger (TC) wastegate regulating valve – open circuit	**Wiring open circuit, TC wastegate regulating valve**
17957	P1549	Turbocharger (TC) wastegate regulating valve – open circuit/short to ground	**Wiring open circuit/short to ground, TC wastegate regulating valve**

Scan code 5-digit	OBD-II code	Fault location	Probable cause
17958	P1550	Turbocharger (TC) pressure – control difference	Intake/exhaust leak, hoses interchanged/not connected, MAP sensor, TC wastegate regulating valve, turbocharger (TC) wastegate actuator, TC
17961	P1553	Manifold absolute pressure (MAP) sensor/ barometric pressure (BARO) sensor – range/performance problem	Intake/exhaust leak, EGR system, EVAP canister purge valve, throttle control unit, wiring, MAP sensor, BARO sensor
17962	P1554	Throttle control unit – basic setting conditions	Basic setting conditions not met
17963	P1555	Turbocharger (TC) pressure – upper limit exceeded	Hoses interchanged/not connected, TC wastegate regulating valve, turbocharger (TC) wastegate actuator, TC
17964	P1556	Turbocharger (TC) pressure – control limit not reached	TC wastegate regulating valve, intake leak, TC defective
17965	P1557	Turbocharger (TC) pressure – control limit exceeded	Hose connection interchanged/not connected
17966	P1558	Idle speed control (ISC) actuator/throttle motor – circuit malfunction	Wiring, ISC actuator/throttle motor
17967	P1559	Throttle control unit – basic setting malfunction	Accelerator pedal or starter motor operated during basic setting
17968	P1560	Maximum engine RPM exceeded	Incorrect gear shift, wiring open circuit, CKP/RPM sensor
17969	P1561	Fuel quantity adjuster – control difference	Wiring, fuel injection pump
17970	P1562	Fuel quantity adjuster – upper stop value	Fuel quantity adjuster blocked/ defective, stop value reached
17971	P1563	Fuel quantity adjuster – lower stop value	Fuel quantity adjuster blocked/ defective, stop value reached
17972	P1564	Throttle control unit – voltage low during basic setting	Battery, wiring
17973	P1565	Throttle control unit – lower stop not reached	Throttle valve tight/sticking, ISC actuator, throttle control unit
17974	P1566	AC compressor, load signal – implausible signal	Wiring, AC system
17976	P1568	Throttle control unit – mechanical fault	Throttle valve tight/sticking, throttle control unit
17977	P1569	Cruise control master switch	Wiring, cruise control master switch
17978	P1570	Engine control module (ECM) – immobilizer active	Incorrect/damaged key, incorrectly coded, ECM/immobilizer replacement without coding, wiring, immobilizer defective
17979	P1571	Engine mounting control solenoid, bank 2 – short to positive	Wiring short to positive, engine mounting control solenoid
17980	P1572	Engine mounting control solenoid, bank 2 – short to ground	Wiring short to ground, engine mounting control solenoid
17981	P1573	Engine mounting control solenoid, bank 2 – open circuit	Wiring open circuit, engine mounting control solenoid

Scan code 5-digit	OBD-II code	Fault location	Probable cause
17983	P1575	Engine mounting control solenoid, bank 1 – short to positive	Wiring short to positive, engine mounting control solenoid
17984	P1576	Engine mounting control solenoid, bank 1 – short to ground	Wiring short to ground, engine mounting control solenoid
17985	P1577	Engine mounting control solenoid, bank 1 – open circuit	Wiring open circuit, engine mounting control solenoid
17987	P1579	Throttle control unit – basic setting	Basic setting not carried out
17988	P1580	Throttle motor, bank 1 – circuit malfunction	Wiring, throttle motor
17989	P1581	Throttle control unit – basic setting	Basic setting not carried out
17990	P1582	Idle speed adaptation – limit reached	Intake/exhaust leak, AIR system, fuel pressure/pump, injector(s), EVAP canister purge valve, throttle control unit
17994	P1586	Engine mounting control solenoid, bank 1 & 2 – short to positive	Wiring short to positive, engine mounting control solenoid
–	P1587	Engine mounting control solenoid, bank 1 & 2 – short to ground	Wiring short to ground, engine mounting control solenoid
–	P1588	Engine mounting control solenoid, bank 1 & 2 – open circuit	Wiring open circuit, engine mounting control solenoid
17997	P1589	AC/heater air temperature control switch – short to ground	Wiring short to ground, AC/heater air temperature control switch
17998	P1590	AC/heater air temperature control switch – open circuit	Wiring open circuit, AC/heater air temperature control switch
18000	P1592	Barometric pressure (BARO) sensor/ manifold absolute pressure (MAP) sensor – implausible ratio	TC system, MAP sensor
18001	P1593	Altitude adaption – signal outside tolerance	Intake leak, MAF sensor, throttle control unit
18007	P1599	Idle control – lean running speed above specification	IAC valve
18008	P1600	Engine control module (ECM) – supply voltage low from ignition switch	Battery, alternator, wiring open circuit
18009	P1601	Engine control module (ECM) – supply voltage	Wiring, engine control relay
18010	P1602	Engine control module (ECM) – supply voltage low from battery	Battery was disconnected, battery discharged, alternator, wiring open circuit, fuse
18011	P1603	Engine control module (ECM) – defective	ECM
18012	P1604	Engine control module (ECM) – defective	ECM
18014	P1606	Rough road signal – circuit malfunction	ABS control module trouble code(s) stored, CAN data bus
18016	P1608	Power steering pressure (PSP) switch – circuit malfunction	Wiring, PSP switch
18017	P1609	Engine control module (ECM) – crash switch-off triggered	Airbag triggered
18018	P1610	Engine control module (ECM) – defective	ECM

Scan code 5-digit	OBD-II code	Fault location	Probable cause
18019	P1611	Malfunction indicator lamp (MIL) – short to ground	Wiring short to ground
18020	P1612	Engine control module (ECM) – coding	Incorrectly coded
18021	P1613	Malfunction indicator lamp (MIL) – open circuit/short to positive	Wiring open circuit/short to positive
18023	P1615	Engine oil temperature (EOT) sensor – range/performance problem	Engine oil level, wiring, EOT sensor
18024	P1616	Glow plug warning lamp – short to positive	Wiring short to positive
18025	P1617	Glow plug warning lamp – open circuit/ short to ground	Bulb, wiring open circuit/short to ground
18026	P1618	Glow plug relay – short to positive	Wiring short to positive, glow plug relay
18027	P1619	Glow plug relay – open circuit/short to ground	Wiring open circuit/short to ground, glow plug relay
18028	P1620	Instrument panel, ECT signal – open circuit/short to positive	Wiring open circuit/short to positive, instrument panel
18029	P1621	Instrument panel, ECT signal – short to ground	Wiring short to ground, instrument panel
18030	P1622	Instrument panel, ECT signal – implausible signal	Wiring, instrument panel, ECT sensor
18031	P1623	CAN data bus – no signal	Trouble code(s) stored in other system(s), wiring, matching resistor in ECM
18032	P1624	Malfunction indicator lamp (MIL) – request signal active	Trouble code(s) stored in other system(s)
18033	P1625	CAN data bus, TCM – incorrect signal	TCM trouble code(s) stored, TCM incorrectly coded, wiring, matching resistor in ECM
18034	P1626	CAN data bus, TCM – no signal	TCM trouble code(s) stored, TCM incorrectly coded, wiring, matching resistor in ECM
18037	P1629	CAN data bus, cruise control – no signal	Cruise control trouble code(s) stored, wiring, matching resistor in ECM
18038	P1630	Throttle/accelerator pedal position (APP) sensor 1/2 – low input	Wiring short to ground, APP sensor, TP sensor
18039	P1631	Throttle/accelerator pedal position (APP) sensor 1/2 – high input	Wiring short to positive, APP sensor, TP sensor
18040	P1632	Accelerator pedal position (APP) sensor – supply voltage	Operating voltage too high/low, wiring
18041	P1633	Accelerator pedal position (APP) sensor 2 – low input	Wiring short to ground, APP sensor
18042	P1634	Accelerator pedal position (APP) sensor 2 – high input	Wiring short to positive, APP sensor
18043	P1635	CAN data bus, AC – no signal	AC control module trouble code(s) stored, wiring, matching resistor in ECM
18044	P1636	CAN data bus, SRS – no signal	SRS control module trouble code(s) stored, wiring, matching resistor in ECM

Scan code 5-digit	OBD-II code	Fault location	Probable cause
18045	P1637	CAN data bus, electronic CE – no signal	Trouble code(s) stored, wiring, matching resistor in ECM
18047	P1639	Throttle/accelerator pedal position (APP) sensor 1/2 – range/performance problem	Wiring, APP sensor, TP sensor
18048	P1640	Engine control module (ECM) – defective	ECM
18050	P1642	SRS control module – system malfunction	Trouble code(s) stored
18053	P1645	CAN data bus, 4WD – no signal	4WD trouble code(s) stored, wiring, matching resistor in ECM
–	P1647	CAN data bus, ECM coding	Incorrectly coded ECM
18056	P1648	CAN data bus – defective	Wiring, matching resistor in ECM
18057	P1649	CAN data bus, ABS – no signal	ABS control module trouble code(s) stored, wiring, matching resistor in ECM
18058	P1650	CAN data bus, instrumentation – no signal	Instrumentation control module trouble code(s) stored, wiring, matching resistor in ECM
18060	P1652	Transmission control module (TCM) – system malfunction	Trouble code(s) stored
18061	P1653	ABS control module – system malfunction	Trouble code(s) stored
18062	P1654	Instrumentation control module – system malfunction	Trouble code stored for engine oil level/ temperature sensor
18064	P1656	AC signal – short to ground	Wiring short to ground
18065	P1657	AC signal – short to positive	Wiring
18066	P1658	CAN data bus, cruise control – incorrect signal	Cruise control trouble code(s) stored, wiring, matching resistor in ECM
18067	P1659	Engine coolant blower motor, speed 1 – short to positive	Wiring short to positive, engine coolant blower motor control module, engine coolant blower motor
18068	P1660	Engine coolant blower motor, speed 1 – short to ground	Wiring short to ground, engine coolant blower motor control module, engine coolant blower motor
18069	P1661	Engine coolant blower motor, speed 2 – short to positive	Wiring short to positive, engine coolant blower motor control module, engine coolant blower motor
18070	P1662	Engine coolant blower motor, speed 2 – short to ground	Wiring short to ground, engine coolant blower motor control module, engine coolant blower motor
18071	P1663	Injector, activation – short to positive	Wiring short to positive, ECM
18072	P1664	Injector, activation – current circuit	Wiring open circuit/short to ground
18073	P1665	Injector – mechanical fault	Injector
18074	P1666	Injector 1 – current circuit	Wiring open circuit/short to ground
18075	P1667	Injector 2 – current circuit	Wiring open circuit/short to ground
18076	P1668	Injector 3 – current circuit	Wiring open circuit/short to ground
18077	P1669	Injector 4 – current circuit	Wiring open circuit/short to ground

Scan code 5-digit	OBD-II code	Fault location	Probable cause
18080	P1672	Engine coolant blower motor, speed 1 – open circuit/short to ground	**Wiring open circuit/short to ground, engine coolant blower motor control module, engine coolant blower motor**
18082	P1674	CAN data bus, instrumentation – incorrect signal	**Wiring, instrumentation control module trouble code(s) stored, matching resistor in ECM**
18084	P1676	ETS warning lamp – circuit malfunction	**Instrumentation control module trouble code(s) stored, wiring, matching resistor in ECM**
18085	P1677	ETS warning lamp – short to positive	**Instrumentation control module trouble code(s) stored, wiring short to positive, matching resistor in ECM**
18086	P1678	ETS warning lamp – short to ground	**Instrumentation control module trouble code(s) stored, wiring short to ground, matching resistor in ECM**
18087	P1679	ETS warning lamp – open circuit	**Instrumentation control module trouble code(s) stored, wiring open circuit, matching resistor in ECM**
18088	P1680	Limp-home mode – active	**Throttle control unit, APP sensor**
18089	P1681	Engine control module (ECM) – programming incomplete	**ECM**
18090	P1682	CAN data bus, ABS – implausible signal	**ABS control module trouble code(s) stored, wiring, matching resistor in ECM**
18091	P1683	CAN data bus, SRS – implausible signal	**SRS control module trouble code(s) stored, wiring, matching resistor in ECM**
18098	P1690	Malfunction indicator lamp (MIL) – circuit malfunction	**Instrumentation control module trouble code(s) stored, wiring, matching resistor in ECM**
18099	P1691	Malfunction indicator lamp (MIL) – open circuit	**Instrumentation control module trouble code(s) stored, wiring open circuit, matching resistor in ECM**
18100	P1692	Malfunction indicator lamp (MIL) – short to ground	**Instrumentation control module trouble code(s) stored, wiring short to ground, matching resistor in ECM**
18101	P1693	Malfunction indicator lamp (MIL) – short to positive	**Instrumentation control module trouble code(s) stored, wiring short to positive, matching resistor in ECM**
18104	P1696	CAN data bus, steering column electronics – incorrect signal	**Wiring, matching resistor in ECM**
–	P1702	Transmission control module (TCM) – malfunction	**TCM**
–	P1778	Transmission (AT), solenoid valve 7 – circuit malfunction	**Wiring, solenoid valve**
–	P1780	Torque reduction – malfunction	**ATF level , mechanical fault**
–	P1847	ABS control module	**trouble code(s) stored**
–	P1850	CAN data bus, ECM – missing information	**Wiring, matching resistor in ECM**
18259	P1851	CAN data bus, ABS – incorrect signal	**ABS trouble code(s) stored, wiring, matching resistor in ECM**

Scan code 5-digit	OBD-II code	Fault location	Probable cause
18261	P1853	CAN data bus, ABS – incorrect signal	**ABS trouble code(s) stored, wiring, matching resistor in ECM**
18262	P1854	CAN data bus, ABS – defective	**ABS trouble code(s) stored, wiring, matching resistor in ECM**
–	P1855	CAN data bus, software monitoring	**wiring, matching resistor in ECM**
–	P1857	Engine control module (ECM) – load signal	**Wiring, intake leak , MAF sensor**
–	P1861	Engine control module (ECM) – APP/TP sensor signal malfunction	**Wiring, APP sensor, TP sensor**
–	P1866	CAN data bus, missing information	**Wiring, matching resistor in ECM**
18308	P1900	Engine coolant blower motor, speed 2 – open circuit/short to ground	**Wiring open circuit/short to ground, engine coolant blower motor control module, engine coolant blower motor**
18309	P1901	Engine coolant blower motor run-on relay – short to positive	**Wiring short to positive, engine coolant blower motor run-on relay**
18310	P1902	Engine coolant blower motor run-on relay – open circuit/short to ground	**Wiring open circuit/short to ground, engine coolant blower motor run-on relay**
18320	P1912	Brake servo pressure sensor – open circuit/short to positive	**Wiring open circuit/short to positive, brake servo pressure sensor**
18321	P1913	Brake servo pressure sensor – short to ground	**Wiring short to ground, brake servo pressure sensor**
18322	P1914	Brake servo pressure sensor – range/performance problem	**Vacuum leak, wiring, brake servo pressure sensor**
18328	P1920	Engine mounting control solenoid, bank 1 & 2 – open circuit/short to ground	**Wiring open circuit/short to ground, engine mounting control solenoid**
–	P1925	Engine coolant pump relay – short to positive	**Wiring short to positive , engine coolant pump relay**
–	P1926	Engine coolant pump relay – short to ground	**Wiring short to ground , engine coolant pump relay**
–	P1927	Engine coolant pump relay – open circuit	**Wiring open circuit , engine coolant pump relay**
–	P1953	Turbocharger (TC) control module 2 – malfunction	**Turbocharger**
–	P1954	Glow plug control module 2 – circuit malfunction	**Wiring, glow plug control module**
–	P1955	Glow plug control module 2, ECM communication – range/performance problem	**Wiring, glow plug control module, ECM**
19456	P3000	CAN data bus, instrumentation – glow plug warning lamp	**Wiring, matching resistor in ECM**
19458	P3002	Accelerator pedal position (APP) sensor – transmission kick-down switch	**APP sensor**
19459	P3003	Engine coolant heater relay 1, low output	**Wiring, engine coolant heater relay**
19461	P3005	Engine coolant heater relay 2, high output	**Wiring, engine coolant heater relay**
19463	P3007	Camshaft position (CMP) sensor – no signal	**Air gap, insecure sensor/rotor, wiring, CMP sensor**

Scan code 5-digit	OBD-II code	Fault location	Probable cause
19464	P3008	Camshaft position (CMP) sensor – signal limit exceeded	**Insecure rotor, camshaft alignment**
19465	P3009	Fuel cooling pump relay – short to positive	**Wiring short to positive, fuel cooling pump relay**
19466	P3010	Fuel cooling pump relay – open circuit/ short to ground	**Wiring open circuit/short to ground, fuel cooling pump relay**
19467	P3011	Fuel pump relay 1/2 – short to positive	**Wiring short to positive, fuel pump relay**
19468	P3012	Fuel pump relay 1/2 – open circuit/short to ground	**Wiring open circuit/short to ground, fuel pump relay**
19469	P3013	Turbocharger (TC) wastegate regulating valve B – short to positive	**Wiring short to positive, TC wastegate regulating valve**
19470	P3014	Turbocharger (TC) wastegate regulating valve B – open circuit/short to ground	**Wiring open circuit/short to ground, TC wastegate regulating valve**
19471	P3015	Fuel bypass valve – short to positive	**Wiring short to positive, fuel bypass valve**
19472	P3016	Fuel bypass valve – open circuit/short to ground	**Wiring open circuit/short to ground, fuel bypass valve**
–	P3018	Exhaust gas control flap solenoid 1 – short to ground	**Wiring short to ground, flap solenoid**
–	P3019	Exhaust gas control flap solenoid 1 – open circuit	**Wiring open circuit, flap solenoid**
19496	P3040	Gear ratio – implausible	**Transmission fault**
19497	P3041	CAN data bus, instrumentation – implausible ECT signal	**Wiring, matching resistor in ECM**
–	P3047	Starter motor relay 2 – short to positive	**Wiring short to positive, relay**
–	P3048	Starter motor relay 2 – short to ground	**Wiring short to ground, relay**
–	P3049	Starter motor relay 2 – open circuit	**Wiring open circuit, relay**
–	P3050	Starter motor relay 2 – circuit malfunction	**Wiring , relay**
–	P3058	Exhaust gas recirculation (EGR) system – insufficient flow detected	**Hose leak/blockage, wiring, EGR valve/ solenoid**
–	P3059	Exhaust gas recirculation (EGR) system – excessive flow detected	**Wiring, EGR valve/solenoid**
–	P3062	Turbocharger (TC) – boost limit exceeded	**Hoses interchanged/not connected, hoses blocked/leaking, intake leak, MAP sensor, TC wastegate control motor , TC**
–	P3063	Turbocharger (TC) – boost limit not reached	**Hoses interchanged/not connected, hoses blocked/leaking, intake leak, MAP sensor, TC wastegate control motor , TC**
–	P3081	Engine temperature too low	**Allow engine to warm up, cooling system fault**
–	P3092	Engine control module (ECM) – internal fault	**ECM**
–	P3093	Engine control module (ECM) – internal fault	**ECM**
–	P3096	Engine control module (ECM) – internal fault	**ECM**

Scan code 5-digit	OBD-II code	Fault location	Probable cause
–	P3097	Engine control module (ECM) – internal fault	**ECM**
19560	P3104	Intake manifold air control solenoid – short to positive	**Wiring short to positive, intake manifold air control solenoid**
19561	P3105	Intake manifold air control solenoid – open circuit/short to ground	**Wiring open circuit/short to ground, intake manifold air control solenoid**
–	P3211	Heated oxygen sensor (HO2S) 1, bank 1 – heater circuit malfunction	**Wiring, HO2S, ECM**
–	P3212	Heated oxygen sensor (HO2S) 1, bank 2 – heater circuit malfunction	**Wiring, HO2S, ECM**
–	P3255	Heated oxygen sensor (HO2S) 1, bank 1, heater circuit – upper limit reached	**Wiring, HO2S**
–	P3256	Heated oxygen sensor (HO2S) 1, bank 1, heater circuit – lower limit reached	**Wiring, HO2S**
–	P3257	Heated oxygen sensor (HO2S) 1, bank 2, heater circuit – upper limit reached	**Wiring, HO2S**
–	P3258	Heated oxygen sensor (HO2S) 1, bank 2, heater circuit – lower limit reached	**Wiring, HO2S**
19717	P3262	Heated oxygen sensor (HO2S) 2, bank 1 & 2 – interchanged	**HO2S 2 on bank 1 & 2 incorrectly installed**
–	P3266	Heated oxygen sensor (HO2S) 1, bank 1 – internal resistance too high	**Wiring, HO2S**
–	P3267	Heated oxygen sensor (HO2S) 1, bank 2 – implausible internal resistance	**Wiring, HO2S**
–	P3328	Intake manifold flap motor 2 – short to positive	**Wiring short to positive, intake flap motor**
–	P3329	Intake manifold flap motor 2 – open/short to ground	**Wiring open/short to ground, intake flap motor**
–	P3330	Intake manifold flap motor 2 – no signal	**Wiring, intake flap motor**
–	P3331	Intake manifold flap motor 2 – malfunction	**Wiring, flap tight/sticking, intake flap motor**
–	P3339	Glow plug timer relay 2 – circuit malfunction	**Wiring, relay**
65280	–	CAN data bus, ABS – defective	**ABS trouble code(s) stored, wiring, matching resistor in ECM**
65535	–	Engine control module (ECM) – defective	**ECM**

1 Trouble code may be displayed if engine is not idling during self-diagnosis due to missing CKP sensor signal. Ignore trouble code if engine starts.
2 Located at rear of exhaust camshaft.
3 Incorporates heated oxygen sensor (HO2S) 2.
4 May also produce HT voltage for cylinder 4.
5 May also produce HT voltage for cylinder 3.

VOLKSWAGEN

Model:	**Jetta 2.0L • Passat 2.0L**
Year:	**1992-93**
Engine identification:	**9A**
System:	**Bosch KE-Motronic 1.2/1.2.1/1.2.2**

Data link connector (DLC) locations

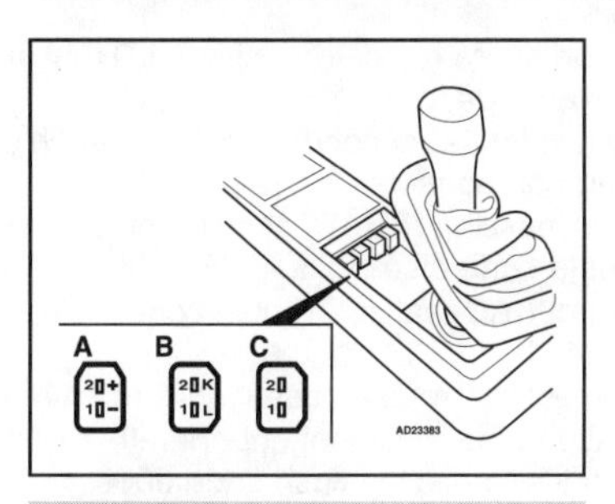

Near gear lever

Self-diagnosis

General information

- Refer to the front of this manual for general test conditions, terminology, detailed descriptions of wiring faults and a general trouble shooter for electrical and mechanical faults.
- Trouble codes are displayed by using an LED connected to the data link connector (DLC).
- The ECM fault memory can also be checked and erased using diagnostic equipment connected to the data link connector (DLC).
- Self-diagnosis using flash type trouble codes may not display all available diagnostic information.
- Carry out road test for at least 5 minutes.
- Briefly fully depress throttle pedal to increase engine speed over 3000 rpm.
- Allow to idle for 2 minutes.
- If engine does not start: Crank engine for 6 seconds.

Accessing

- Ensure ignition switched OFF.
- Connect LED test lamp between black data link connector (DLC) terminal 2 **1** [**A**] and blue data link connector (DLC) terminal 2 **1** [**C**].
- Jump black data link connector (DLC) terminal 1 **1** [**A**] and brown/white/yellow data link connector (DLC) terminal 1 **1** [**B**] with a switched lead – contacts normally open.
- Switch ignition ON.
- Operate switch for 4-6 seconds.
- Release switch. Check that LED flashes.
- Count LED flashes. Note trouble code.
- Each trouble code consists of four groups of one to four flashes.
- A 2.5 second pause separates each trouble code group **2** [**A**].
- Repeat switch operation. Note trouble codes. Compare with trouble code table.
- End of test sequence indicated by trouble code 0000 (long flashes).
- Switch ignition OFF. Rectify faults as necessary.

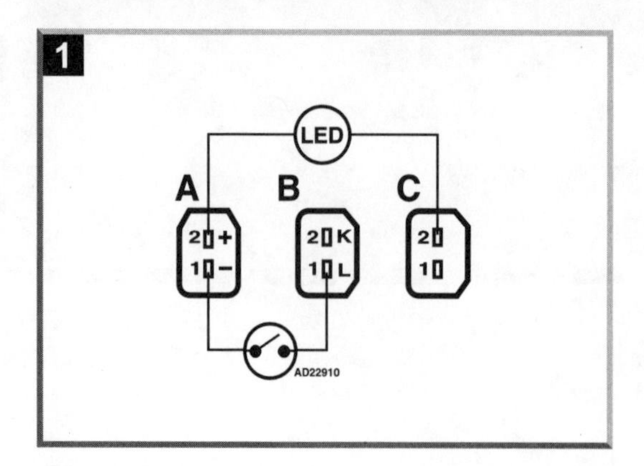

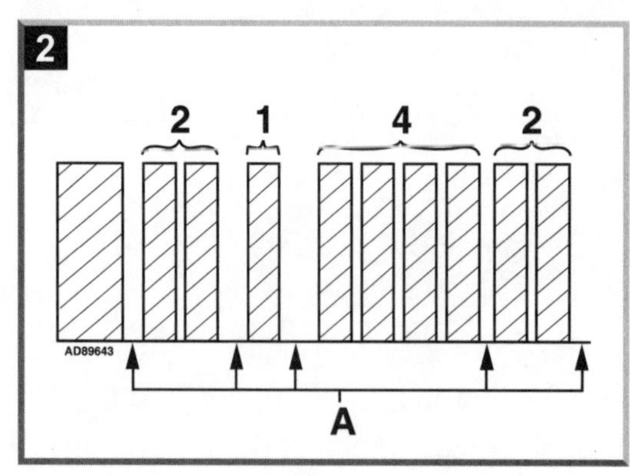

Erasing

- Ensure ignition switched OFF.
- Connect LED test lamp between black data link connector (DLC) terminal 2 **1** [**A**] and blue data link connector (DLC) terminal 2 **1** [**C**].
- Jump black data link connector (DLC) terminal 1 **1** [**A**] and brown/white/yellow data link connector (DLC) terminal 1 **1** [**B**] with a switched lead – contacts normally open.
- Access trouble codes. Switch ignition OFF.
- Operate switch. Switch ignition ON after 4-6 seconds.
- Release switch after 4-6 seconds. Check that LED flashes.
- Differential pressure regulator activated. Trouble code displayed: 4341.
- Operate switch for 4-6 seconds to activate each of the following components:
- Evaporative emission (EVAP) canister purge valve. Trouble code displayed: 4343.
- Idle air control (IAC) valve. Trouble code displayed: 4431.
- Cold start injector. Trouble code displayed: 4443.
- After activation of last component operate switch for 4-6 seconds to display flash code 0000 (long flashes).
- Operate switch for 4-6 seconds to erase fault memory.
- Repeat checking procedure to ensure no data remains in ECM fault memory.

Trouble code identification

Flash code 4-digit	Fault location	Probable cause
0000	End of test sequence	–
1111	Engine control module (ECM)	**ECM**
1231	Vehicle speed sensor (VSS)	**Wiring, speedometer, VSS**
2112	Camshaft position (CMP) sensor, cylinder 4 HT lead	**CMP sensor tabs not facing towards distributor, spark plug(s), HT lead(s), distributor cap/rotor, wiring, CMP sensor**
2113	Crankshaft position (CKP) sensor	**Insecure rotor, wiring, CKP sensor, volume air flow (VAF) sensor plate tight/sticking**
2121	Closed throttle position (CTP) switch	**Accelerator cable adjustment, CTP switch adjustment, throttle valve tight/sticking, wiring, CTP switch**
2123	Wide open throttle (WOT) switch	**Accelerator cable adjustment, WOT switch adjustment, wiring, WOT switch**

Flash code 4-digit	Fault location	Probable cause
2141	Knock control – control limit exceeded	Incorrect fuel/compression, ignition timing, screening of KS wiring, insecure engine component
2142	Knock sensor (KS) 1	Wiring, KS, ECM
2144	Knock sensor (KS) 2	Wiring, KS, ECM
2231	Idle speed control	Ignition timing, EVAP canister purge valve 1/2, intake leak, VAF sensor, throttle valve adjustment, AC signal missing, wiring
2232	Volume air flow (VAF) sensor	Wiring, VAF sensor
2312	Engine coolant temperature (ECT) sensor	Wiring, poor connection, ECT sensor
2341	Heated oxygen sensor (HO2S) – lambda control	Mixture adjustment, HO2S heater/wiring, cold start injector, EVAP canister purge valve, intake/exhaust leak, ECM ground wire to intake manifold
2342	Heated oxygen sensor (HO2S)	Wiring, HO2S, heating inoperative
4431	Idle air control (IAC) valve	Wiring, IAC valve, ECM
4444	No fault found	–

VAG code 5-digit	Fault location	Probable cause
00000	No fault found	–
00281	Vehicle speed sensor (VSS)	Wiring, speedometer, VSS
00514	Camshaft position (CMP) sensor, cylinder 4 HT lead	CMP sensor tabs not facing towards distributor, spark plug(s), HT lead(s), distributor cap/rotor, wiring, CMP sensor
00515	Crankshaft position (CKP) sensor	Insecure rotor, wiring, CKP sensor, volume air flow (VAF) sensor plate tight/sticking
00516	Closed throttle position (CTP) switch	Accelerator cable adjustment, CTP switch adjustment, throttle valve tight/sticking, wiring, CTP switch
00517	Wide open throttle (WOT) switch	Accelerator cable adjustment, wiring, WOT switch
00520	Volume air flow (VAF) sensor	Wiring, VAF sensor
00522	Engine coolant temperature (ECT) sensor	Wiring, poor connection, ECT sensor
00524	Knock sensor (KS) 1	Wiring, KS, ECM

VAG code 5-digit	Fault location	Probable cause
00525	Heated oxygen sensor (HO2S)	Wiring, HO2S, heating inoperative
00533	Idle speed control	Ignition timing, EVAP canister purge valve 1/2, intake leak, VAF sensor, throttle valve adjustment, AC signal missing, wiring
00535	Knock control – control limit exceeded	Incorrect fuel/compression, ignition timing, screening of KS wiring, insecure engine component
00537	Heated oxygen sensor (HO2S) – lambda control	Mixture adjustment, HO2S heater/wiring, cold start injector, EVAP canister purge valve, intake/exhaust leak, ECM ground wire to intake manifold
00540	Knock sensor (KS) 2	Wiring, KS, ECM
00558	Mixture control (MC) – mixture lean	ECM ground wire to intake manifold, intake leak, idle speed adjustment
00559	Mixture control (MC) – mixture rich	Injector(s), cold start injector, idle speed adjustment
00587	Mixture control (MC)	ECM ground wire to intake manifold, intake leak, idle speed adjustment, injector(s), cold start injector
01257	Idle air control (IAC) valve	Wiring, IAC valve, ECM
17978	Engine control module (ECM) – immobilizer active	Incorrect/damaged key, incorrectly coded, ECM/immobilizer replacement without coding, wiring, immobilizer defective
65535	Engine control module (ECM)	ECM

VOLKSWAGEN

Model:	Engine identification:	Year:
Phaeton 4.2/6.0L	**BAP, BGH, BGJ,**	**2004**

System: **Bosch Motronic ME7.1.1**

Self-diagnosis

General information

- Refer to the front of this manual for general test conditions, terminology, detailed descriptions of wiring faults and a general trouble shooter for electrical and mechanical faults.

Accessing and erasing

- The engine control module (ECM) fault memory can only be accessed and erased using diagnostic equipment connected to the data link connector (DLC) **1**.

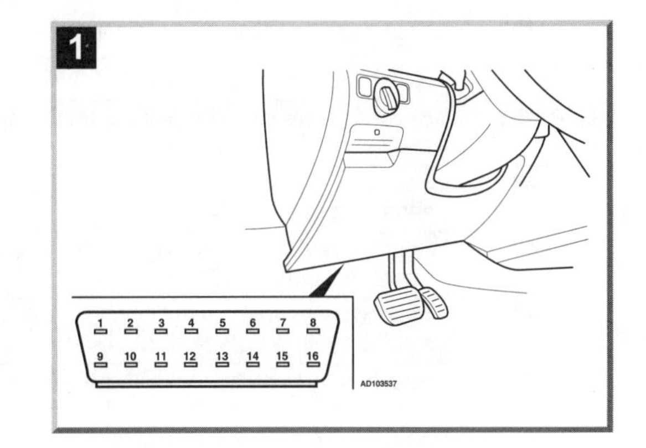

Trouble code identification

OBD-II code	Fault location	Probable cause
All P0, P2 and U0 codes	Refer to OBD-II trouble code tables at the front of this manual	–
P1009	Mass air flow (MAF) sensor 1/2 – implausible load detection signal	**Wiring, MAF sensor(s)**
P1073	Mass air flow (MAF) sensor 2 – signal too low	**Wiring, MAF sensor**
P1074	Mass air flow (MAF) sensor 2 – signal too high	**Wiring, MAF sensor**
P1081	Long term fuel trim 1, bank 3 – above lean limit	**Wiring, intake leak, HO2S**
P1082	Long term fuel trim 2, bank 3 – above lean limit	**Wiring, intake leak, HO2S**
P1083	Long term fuel trim 1, bank 3 – above rich limit	**Wiring, intake leak, HO2S**
P1084	Long term fuel trim 2, bank 3 – above rich limit	**Wiring, intake leak, HO2S**
P1085	Long term fuel trim 1, bank 4 – above lean limit	**Wiring, intake leak, HO2S**
P1086	Long term fuel trim 2, bank 4 – above lean limit	**Wiring, intake leak, HO2S**
P1087	Long term fuel trim 1, bank 4 – above rich limit	**Wiring, intake leak, HO2S**
P1088	Long term fuel trim 2, bank 4 – above rich limit	**Wiring, intake leak, HO2S**
P1167	Mass air flow (MAF) sensor 2 – range/ performance problem	**Intake leak, wiring, MAF sensor**
P1176	Lambda correction after catalyst, bank 1/2 – control limit reached	**Intake leak, wiring, HO2S**

OBD-II code	Fault location	Probable cause
P1177	Lambda correction after catalyst, bank 3/4 – control limit reached	**Intake leak, wiring, HO2S**
P1207	Injector 7 – circuit malfunction	**Wiring, injector**
P1321	Knock sensor (KS) 3 – signal too low	**Wiring, KS incorrectly tightened/defective**
P1322	Knock sensor (KS) 3 – signal too high	**Wiring, KS incorrectly tightened/defective**
P1323	Knock sensor (KS) 4 – signal too low	**Wiring, KS incorrectly tightened/defective**
P1324	Knock sensor (KS) 4 – signal too high	**Wiring, KS incorrectly tightened/defective**
P1482	Secondary air injection (AIR) pump relay 2 – short to positive	**Wiring short to positive, AIR pump relay**
P1483	Secondary air injection (AIR) pump relay 2 – short to ground	**Wiring short to ground, AIR pump relay**
P1484	Secondary air injection (AIR) pump relay 2 – circuit malfunction	**Wiring, AIR pump relay**
P1497	Secondary air injection (AIR) system, bank 3 – insufficient flow detected	**Hose(s) blocked/leaking, wiring, AIR pump relay, AIR valve/solenoid, AIR pump**
P1498	Secondary air injection (AIR) system, bank 4 – insufficient flow detected	**Hose(s) blocked/leaking, wiring, AIR pump relay, AIR valve/solenoid, AIR pump**
P1647	CAN data bus, ECM coding	**Incorrectly coded ECM**
P1702	Transmission control module (TCM) – malfunction	**TCM**
P1857	Engine control module (ECM) – load signal malfunction	**Wiring, intake leak, MAF sensor**
P1924	Engine control module (ECM), pin coding – implausible signal	**ECM**
P3025	Throttle motor position sensor 1, throttle control unit 2 – implausible signal	**Wiring, throttle motor position sensor, throttle control unit**
P3026	Throttle motor position sensor 1, throttle control unit 2 – signal too low	**Wiring, throttle motor position sensor, throttle control unit**
P3027	Throttle motor position sensor 1, throttle control unit 2 – signal too high	**Wiring, throttle motor position sensor, throttle control unit**
P3028	Throttle motor position sensor 2, throttle control unit 2 – implausible signal	**Wiring, throttle motor position sensor, throttle control unit**
P3029	Throttle motor position sensor 2, throttle control unit 2 – signal too low	**Wiring, throttle motor position sensor, throttle control unit**
P3030	Throttle motor position sensor 2, throttle control unit 2 – signal too high	**Wiring, throttle motor position sensor, throttle control unit**
P3031	Throttle motor 2 – circuit malfunction	**Wiring, throttle motor, throttle control unit**
P3032	Throttle control unit 2 – basic setting	**Basic setting not carried out**
P3033	Intake air temperature (IAT) sensor 2 – signal low	**Wiring, IAT sensor**
P3034	Intake air temperature (IAT) sensor 2 – signal high	**Wiring, IAT sensor**

OBD-II code	Fault location	Probable cause
P3035	Throttle control unit 2 – mechanical fault	**Throttle valve tight/sticking, throttle motor, throttle control unit**
P3038	Throttle control unit 2 – lower stop not reached	**Throttle valve tight/sticking, throttle valve requires cleaning, throttle control unit**
P3081	Engine temperature too low	**Allow engine to warm up, cooling system fault**
P3096	Engine control module (ECM) – internal fault	**ECM**
P3097	Engine control module (ECM) – internal fault	**ECM**
P3148	Heated oxygen sensor (HO2S) 1, bank 3, heater circuit – upper limit reached	**Wiring, HO2S, ECM**
P3149	Heated oxygen sensor (HO2S) 1, bank 3, heater circuit – lower limit reached	**Wiring, HO2S, ECM**
P3150	Heated oxygen sensor (HO2S) 1, bank 4, heater circuit – upper limit reached	**Wiring, HO2S, ECM**
P3151	Heated oxygen sensor (HO2S) 1, bank 4, heater circuit – lower limit reached	**Wiring, HO2S, ECM**
P3152	Heated oxygen sensor (HO2S) 1, bank 3 – implausible internal resistance	**Wiring, HO2S**
P3153	Heated oxygen sensor (HO2S) 1, bank 4 – implausible internal resistance	**Wiring, HO2S**
P3200	Heated oxygen sensor (HO2S) 1, bank 3 – heater short to ground	**Wiring short to ground , HO2S**
P3201	Heated oxygen sensor (HO2S) 1, bank 3 – heater short to positive	**Wiring short to positive , HO2S**
P3202	Heated oxygen sensor (HO2S) 1, bank 3 – heater open circuit	**Wiring open circuit , HO2S**
P3207	Heated oxygen sensor (HO2S) 1, bank 3 – circuit malfunction	**Wiring , HO2S**
P3208	Heated oxygen sensor (HO2S) 1, bank 3 – no activity detected	**Wiring, HO2S**
P3209	Heated oxygen sensor (HO2S) 1, bank 3 – signal too slow	**Wiring, HO2S**
P3211	Heated oxygen sensor (HO2S) 1, bank 1 – heater circuit malfunction	**Wiring, HO2S**
P3212	Heated oxygen sensor (HO2S) 1, bank 2 – heater circuit malfunction	**Wiring, HO2S**
P3213	Heated oxygen sensor (HO2S) 1, bank 3 – heater circuit malfunction	**Wiring, HO2S**
P3214	Heated oxygen sensor (HO2S) 1, bank 4 – heater circuit malfunction	**Wiring, HO2S**
P3215	Heated oxygen sensor (HO2S) 2, bank 3 – heater short to ground	**Wiring short to ground , HO2S**
P3216	Heated oxygen sensor (HO2S) 2, bank 3 – heater short to positive	**Wiring short to positive , HO2S**

OBD-II code	Fault location	Probable cause
P3217	Heated oxygen sensor (HO2S) 2, bank 3 – heater open circuit	**Wiring open circuit , HO2S**
P3218	Heated oxygen sensor (HO2S) 2, bank 3 – heater circuit malfunction	**Wiring, HO2S**
P3220	Heated oxygen sensor (HO2S) 2, bank 3 – voltage too low	**Wiring, HO2S**
P3221	Heated oxygen sensor (HO2S) 2, bank 3 – voltage too high	**Wiring, HO2S**
P3222	Heated oxygen sensor (HO2S) 2, bank 3 – circuit malfunction	**Wiring, HO2S**
P3223	Heated oxygen sensor (HO2S) 2, bank 3 – no activity detected	**Wiring, HO2S**
P3224	Heated oxygen sensor (HO2S) 2, bank 3 – signal too slow	**Wiring, HO2S**
P3224	Heated oxygen sensor (HO2S) 2, bank 3 – signal too slow	**Wiring, HO2S**
P3225	Lambda correction after catalyst, bank 3 – control limit reached	**Wiring, HO2S 1, catalytic converter**
P3226	Lambda correction after catalyst, bank 4 – control limit reached	**Wiring, HO2S 1, catalytic converter**
P3230	Heated oxygen sensor (HO2S) 1, bank 4 – heater short to ground	**Wiring short to ground , HO2S**
P3231	Heated oxygen sensor (HO2S) 1, bank 4 – heater short to positive	**Wiring short to positive , HO2S**
P3232	Heated oxygen sensor (HO2S) 1, bank 4 – heater open circuit	**Wiring open circuit , HO2S**
P3237	Heated oxygen sensor (HO2S) 1, bank 4 – circuit malfunction	**Wiring, HO2S**
P3238	Heated oxygen sensor (HO2S) 1, bank 4 – no activity detected	**Wiring, HO2S**
P3239	Heated oxygen sensor (HO2S) 1, bank 4 – signal too slow	**Wiring, HO2S**
P3245	Heated oxygen sensor (HO2S) 2, bank 4 – heater short to ground	**Wiring short to ground , HO2S**
P3246	Heated oxygen sensor (HO2S) 2, bank 4 – heater short to positive	**Wiring short to positive , HO2S**
P3247	Heated oxygen sensor (HO2S) 2, bank 4 – heater open circuit	**Wiring open circuit , HO2S**
P3248	Heated oxygen sensor (HO2S) 2, bank 4 – heater circuit malfunction	**Wiring, HO2S**
P3250	Heated oxygen sensor (HO2S) 2, bank 4 – voltage too low	**Wiring, HO2S**
P3251	Heated oxygen sensor (HO2S) 2, bank 4 – voltage too high	**Wiring, HO2S**
P3252	Heated oxygen sensor (HO2S) 2, bank 4 – circuit malfunction	**Wiring, HO2S**

OBD-II code	Fault location	Probable cause
P3253	Heated oxygen sensor (HO2S) 2, bank 4 – no activity detected	Wiring, HO2S
P3254	Heated oxygen sensor (HO2S) 2, bank 4 – signal too slow	Wiring, HO2S
P3261	Heated oxygen sensor (HO2S) 1, bank 3 & 4 – interchanged	Wiring, HO2S 1 on bank 3 & 4 incorrectly installed
P3263	Heated oxygen sensor (HO2S) 2, bank 3 & 4 – interchanged	Wiring, HO2S 2 on bank 3 & 4 incorrectly installed
P3278	Heated oxygen sensor (HO2S) 1, bank 3, pump current – open circuit	Wiring, HO2S
P3281	Heated oxygen sensor (HO2S) 1, bank 3, reference voltage – open circuit	Wiring open circuit, HO2S
P3285	Heated oxygen sensor (HO2S) 1, bank 3, ground wire – open circuit	Wiring open circuit, HO2S
P3288	Heated oxygen sensor (HO2S) 1, bank 4, pump current – open circuit	Wiring, HO2S
P3291	Heated oxygen sensor (HO2S) 1, bank 4, reference voltage – open circuit	Wiring open circuit, HO2S
P3295	Heated oxygen sensor (HO2S) 1, bank 4, ground wire – open circuit	Wiring open circuit, HO2S
P3298	Catalytic converter, bank 3 – efficiency below threshold	Wiring, HO2S, catalytic converter
P3299	Catalytic converter, bank 4 – efficiency below threshold	Wiring, HO2S, catalytic converter

VOLVO

Model:	Engine identification:	Year:
C70 2.3L Turbo	B5234T3	1998
V70 2.3L Turbo	B5234T3/T6	1998
S90/V90 2.9L Turbo	B6304FS2	1997-98

System: **Bosch Motronic 4.4**

Self-diagnosis

General information

- Refer to the front of this manual for general test conditions, terminology, detailed descriptions of wiring faults and a general trouble shooter for electrical and mechanical faults.
- Malfunction indicator lamp (MIL) will illuminate if certain faults are recorded.
- ECM operates in backup mode if sensors fail, to enable vehicle to be driven to repair shop.

Accessing

- Trouble codes can only be displayed by using a Scan Tool connected to the data link connector (DLC) :
  - C70/V70 **1**
  - S/V90 – **2**

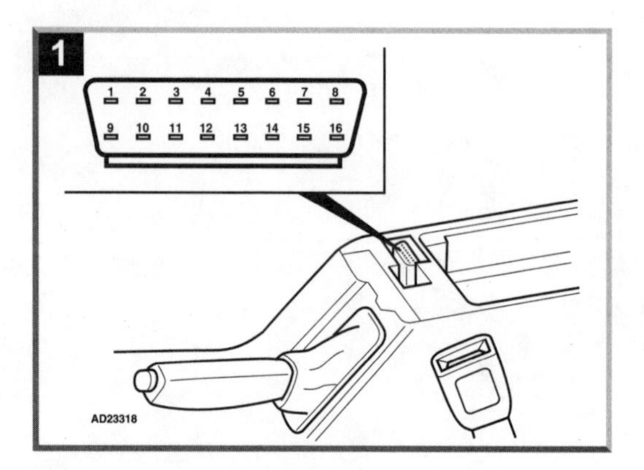

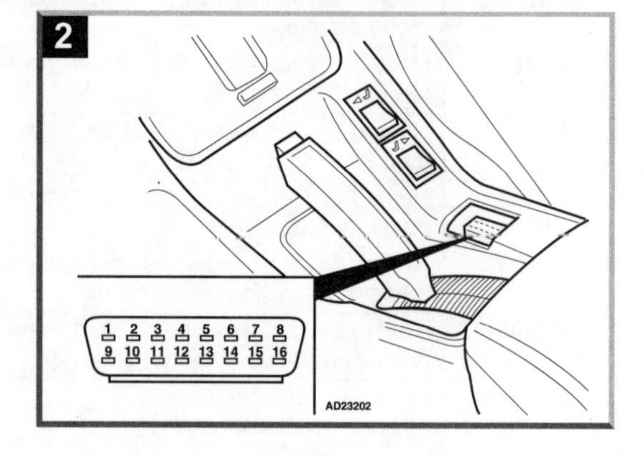

Erasing

- Trouble codes can only be erased by using a Scan Tool connected to the data link connector (DLC).

Trouble code identification

Scan code Volvo type	Scan code P type	Fault location	Probable cause
–	*All P0, P2 and U0 codes*	Refer to OBD-II trouble code tables at the front of this manual	–
EFI-112	**P1326**	Engine control module (ECM) – malfunction	**ECM**
EFI-112	**P1327**	Engine control module (ECM) – malfunction	**ECM**
EFI-112	**P1328**	Engine control module (ECM) – malfunction	**ECM**
EFI-112	**P1329**	Engine control module (ECM) – malfunction	**ECM**
EFI-112	**P1401**	Engine control module (ECM) – malfunction	**ECM**
EFI-112	**P1403**	Engine control module (ECM) – malfunction	**ECM**
EFI-112	**P1404**	Engine control module (ECM) – malfunction	**ECM**
EFI-115	–	Injector 1	**Wiring, injector**
EFI-121	–	Mass air flow (MAF) sensor	**Intake leak/blockage, hose leak, wiring, MAF sensor**
EFI-123	–	Engine coolant temperature (ECT) sensor	**Wiring, ECT sensor**
EFI-125	–	Injector 2	**Wiring, injector**
EFI-131	–	Crankshaft position (CKP) sensor/ engine speed (RPM) sensor – no signal	**Wiring, CKP/RPM sensor**
EFI-132	–	Engine control module (ECM) – supply voltage	**Battery, alternator, wiring**
EFI-135	–	Injector 3	**Wiring, injector**
EFI-143	–	Knock sensor (KS) 1	**Wiring, KS**
EFI-145	–	Injector 4	**Wiring, injector**
EFI-153	–	Heated oxygen sensor (HO2S) 2	**Exhaust leak, wiring, HO2S 1/2**
EFI-155	–	Injector 5	**Wiring, injector**
EFI-165	–	Injector 6	**Wiring, injector**
EFI-212	–	Heated oxygen sensor (HO2S) 1	**Wiring, HO2S 1/2**
EFI-214	–	Crankshaft position (CKP) sensor/ engine speed (RPM) sensor – incorrect/ intermittent signal	**Poor connection, shield wiring, flywheel/drive plate damaged, CKP/ RPM sensor**
EFI-223	**P1505**	Idle air control (IAC) valve – opening signal	**Wiring, IAC valve**
EFI-223	**P1506**	Idle air control (IAC) valve – opening signal	**Wiring, IAC valve**
EFI-225	–	AC refrigerant pressure sensor	**Wiring, AC refrigerant pressure sensor**

Scan code Volvo type	Scan code P type	Fault location	Probable cause
EFI-231	–	Long term fuel trim, part load	Intake/exhaust leak, fuel pressure, excessive fuel in engine oil, engine oil level too high, injector(s), EVAP canister purge valve, wiring, MAF sensor, ECT sensor, HO2S
EFI-232	–	Long term fuel trim, idling	Intake/exhaust leak, fuel pressure, excessive fuel in engine oil, engine oil level too high, injector(s), EVAP canister purge valve, wiring, MAF sensor, ECT sensor, HO2S
EFI-233	–	Long term fuel trim, idle air adjustment	Intake leak/blockage, throttle valve adjustment, accelerator cable adjustment, wiring, IAC valve
EFI-242	–	Turbocharger (TC) wastegate regulating valve	Wiring, TC wastegate regulating valve
EFI-245	P1507	Idle air control (IAC) valve – closing signal	Wiring, IAC valve
EFI-245	P1508	Idle air control (IAC) valve – closing signal	Wiring, IAC valve
EFI-311	–	Vehicle speed signal	Wiring, speedometer
EFI-314	–	Camshaft position (CMP) sensor	Wiring, poor connection, CMP sensor
EFI-315	–	Evaporative emission (EVAP) canister purge valve – leak detected	EVAP canister purge valve
EFI-335	P1617	Transmission control module (TCM) – MIL ON request	Wiring
EFI-335	P1618	Transmission control module (TCM) – MIL ON request	Wiring
EFI-411	–	Throttle position (TP) sensor	Wiring, TP sensor
EFI-414	–	Turbocharger (TC) pressure – control –	Hose blocked, TCM wire (TC pressure reduction signal), wiring, MAF sensor, TC wastegate actuator/regulating valve
EFI-422	–	Barometric pressure (BARO) sensor	Wiring, sensor supply, BARO sensor, AC pressure sensor, fuel tank pressure sensor, G-force sensor
EFI-425	–	Heated oxygen sensor (HO2S) 2	Intake/exhaust leak, uneven compression, fuel pressure, HO2S
EFI-432	P1406	Engine control module (ECM) – overheating	Engine coolant blower motor, air supply blocked
EFI-433	–	Knock sensor (KS) 2	Wiring, KS
EFI-435	–	Heated oxygen sensor (HO2S) 1 – slow response	Intake/exhaust leak, uneven compression, fuel pressure, MAF sensor, heating inoperative, HO2S
EFI-436	–	Heated oxygen sensor (HO2S) 2 – compensation	Intake/exhaust leak, uneven compression, fuel pressure, HO2S 1/2
EFI-443	–	Catalytic converter – efficiency below threshold	Intake/exhaust leak, uneven compression, fuel pressure, catalytic converter

Scan code Volvo type	Scan code P type	Fault location	Probable cause
EFI-444	P1307	G-force sensor – signal low	Wiring short to ground, G-force sensor
EFI-444	P1308	G-force sensor – signal high	Wiring, sensor supply, AC pressure sensor, fuel tank pressure sensor, G-force sensor
EFI-445	–	Pulsed secondary air injection (PAIR) pump relay	Wiring, PAIR pump relay
EFI-446	–	Pulsed secondary air injection (PAIR) valve – leak detected	PAIR valve/solenoid, exhaust leak
EFI-447	–	Pulsed secondary air injection (PAIR) solenoid	Wiring, PAIR solenoid
EFI-448	–	Pulsed secondary air injection (PAIR) system – incorrect flow detected	PAIR valve/solenoid, PAIR pump/relay, hoses/pipes blocked, exhaust leak, PAIR pump wiring
EFI-451	–	Cylinder 1 – misfire detected	Spark plug, HT lead, distributor rotor/ cap, moisture ingress, injector/wiring, air leak, compression, head gasket, low-tension wiring, repeated cold starts (flooding)
EFI-452	–	Cylinder 2 – misfire detected	Spark plug, HT lead, distributor rotor/ cap, moisture ingress, injector/wiring, air leak, compression, head gasket, low-tension wiring, repeated cold starts (flooding)
EFI-453	–	Cylinder 3 – misfire detected	Spark plug, HT lead, distributor rotor/ cap, moisture ingress, injector/wiring, air leak, compression, head gasket, low-tension wiring, repeated cold starts (flooding)
EFI-454	–	Cylinder 4 – misfire detected	Spark plug, HT lead, distributor rotor/ cap, moisture ingress, injector/wiring, air leak, compression, head gasket, low-tension wiring, repeated cold starts (flooding)
EFI-455	–	Cylinder 5 – misfire detected	Spark plug, HT lead, distributor rotor/ cap, moisture ingress, injector/wiring, air leak, compression, head gasket, low-tension wiring, repeated cold starts (flooding)
EFI-456	–	Cylinder 6 – misfire detected	Spark plug, HT lead, distributor rotor/ cap, moisture ingress, injector/wiring, air leak, compression, head gasket, low-tension wiring, repeated cold starts (flooding)
EFI-514	P1619	Engine coolant blower motor relay, low speed	Wiring, engine coolant blower motor relay
EFI-514	P1620	Engine coolant blower motor relay, low speed	Wiring, engine coolant blower motor relay

Scan code Volvo type	Scan code P type	Fault location	Probable cause
EFI-521	–	Heated oxygen sensor (HO2S) 1 – heater control	**Wiring, HO2S**
EFI-522	–	Heated oxygen sensor (HO2S) 2 – heater control	**Wiring, HO2S**
EFI-532	P1602	Engine control module (ECM) – power stage – group B	**Wiring, IAC valve, TC wastegate regulating valve, engine coolant blower motor relay, ECM**
EFI-533	P1603	Engine control module (ECM) – power stage – group C	**Wiring short circuit at terminal B19, ECM**
EFI-534	P1604	Engine control module (ECM) – power stage – group D	**Wiring short to positive, MIL, engine coolant blower motor relay, fuel pump, ignition coil/amplifier, TCM (load signal), ECM**
EFI-535	P1605	Engine control module (ECM) – power stage – group E	**Wiring short to positive, AC relay/ PAIR pump relay, ECM**
EFI-541	–	Evaporative emission (EVAP) canister purge valve – signal	**Wiring, EVAP canister purge valve**
EFI-543	–	Misfire detected, at least 1 cylinder	**Spark plug, HT lead, distributor rotor/ cap, moisture ingress, injector/wiring, air leak, compression, head gasket, low-tension wiring, repeated cold starts (flooding)**
EFI-545	–	Misfire detected – at least 1 cylinder – TWC damage	**Spark plug, HT lead, distributor rotor/ cap, moisture ingress, injector/wiring, air leak, compression, head gasket, low-tension wiring, repeated cold starts (flooding)**
EFI-551	–	Cylinder 1, misfire detected – TWC damage	**Spark plug, HT lead, distributor rotor/ cap, moisture ingress, injector/wiring, air leak, compression, head gasket, low-tension wiring, repeated cold starts (flooding)**
EFI-552	–	Cylinder 2, misfire detected – TWC damage	**Spark plug, HT lead, distributor rotor/ cap, moisture ingress, injector/wiring, air leak, compression, head gasket, low-tension wiring, repeated cold starts (flooding)**
EFI-553	–	Cylinder 3, misfire detected – TWC damage	**Spark plug, HT lead, distributor rotor/ cap, moisture ingress, injector/wiring, air leak, compression, head gasket, low-tension wiring, repeated cold starts (flooding)**
EFI-554	–	Cylinder 4, misfire detected – TWC damage	**Spark plug, HT lead, distributor rotor/ cap, moisture ingress, injector/wiring, air leak, compression, head gasket, low-tension wiring, repeated cold starts (flooding)**

Scan code Volvo type	Scan code P type	Fault location	Probable cause
EFI-555	–	Cylinder 5, misfire detected – TWC damage	**Spark plug, HT lead, distributor rotor/ cap, moisture ingress, injector/wiring, air leak, compression, head gasket, low-tension wiring, repeated cold starts (flooding)**
EFI-556	–	Cylinder 6, misfire detected – TWC damage	**Spark plug, HT lead, distributor rotor/ cap, moisture ingress, injector/wiring, air leak, compression, head gasket, low-tension wiring, repeated cold starts (flooding)**
EFI-621	–	Fuel tank pressure sensor	**Wiring, sensor supply, G-force sensor, AC pressure sensor, fuel tank pressure sensor**
EFI-666	P1621	Transmission control module (TCM) – access fault memory	**Trouble code(s) stored**

VOLVO

Model:	Engine identification:	Year:
C70 2.3L/2.4L Turbo	B5234T3/T9/B5244T/T7	1999-04
S/V40 2.4L	B5244S4	2004
S/V40 2.5L Turbo	B5254T3	2004
S60 2.3L/2.4L/2.5L Turbo	B5234T3/B5244T3/B5254T2/T4	2001-04
S60 2.4L	B5244S	2001-04
S80 2.5L/2.8L/2.9L Turbo	B5254T2/B6284T/B6294T	1999-04
S80 2.9L	B6294S/S2	1999-04
V70 2.3L/2.4L/2.5L Turbo	B5234T3/T8/B5244T/T2/T3/B5254T2/T4	1998-04
V70 2.4L	B5244S/S6	1998-04
XC70 2.5L Turbo	B5254T2	2003-04
XC90 2.5L/2.9L Turbo	B5254T2/B6294T	2003-04

System: **Denso ○ Bosch Motronic ME 7.0**

Self-diagnosis

General information

- Refer to the front of this manual for general test conditions, terminology, detailed descriptions of wiring faults and a general trouble shooter for electrical and mechanical faults.
- Malfunction indicator lamp (MIL) will illuminate if certain faults are recorded.
- ECM operates in backup mode if sensors fail, to enable vehicle to be driven to repair shop.

Accessing

- Trouble codes can only be displayed by using a Scan Tool connected to the data link connector (DLC):
  - S40/V40/S60/V70 00→/XC70/S80 – **1**
  - C70/V70 →00 – **2**
  - XC90 – **3**

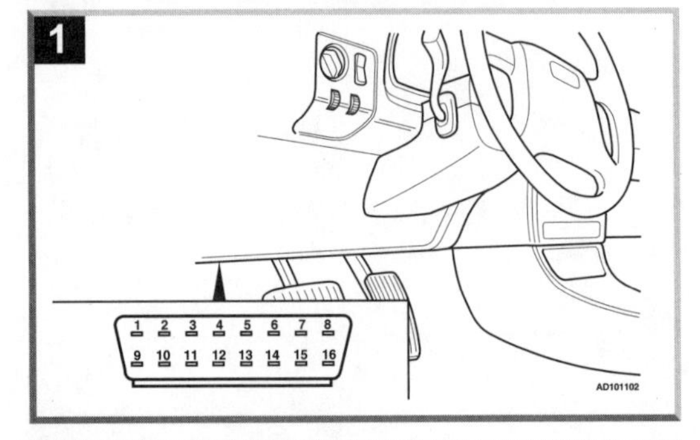

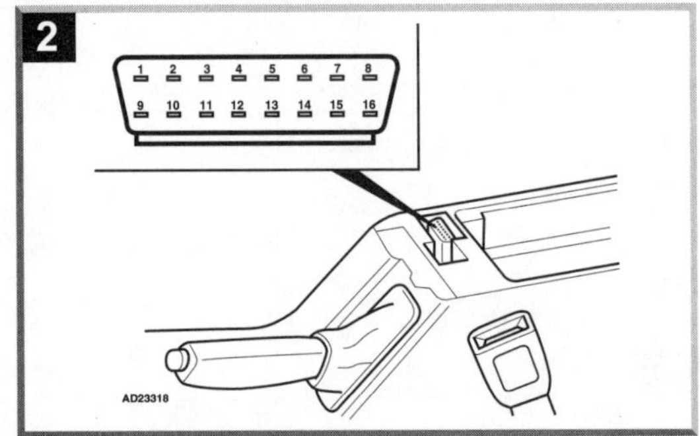

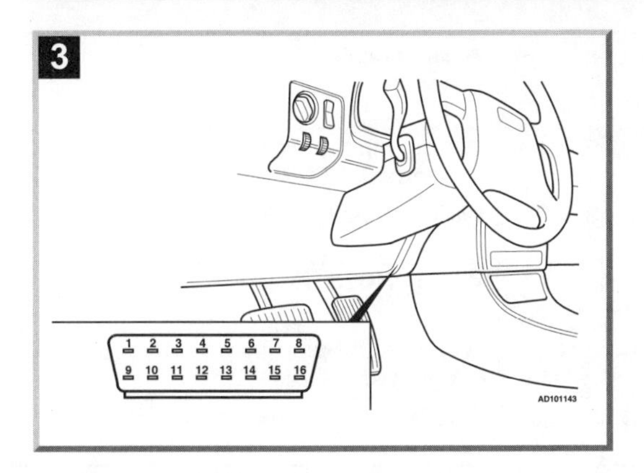

Erasing

- Trouble codes can only be erased by using a Scan Tool connected to the data link connector (DLC).

Trouble code identification

Scan code P type	Scan code Volvo type	Fault location	Probable cause
All P0, P2 and U0 codes	-	Refer to OBD-II trouble code tables at the front of this manual	–
P1108	9818	Manifold absolute pressure (MAP) sensor/ mass air flow (MAF) sensor – signal high	**Mechanical fault, intake leak, wiring, MAF sensor, MAP sensor, ECM**
P1111	110A	Outside air temperature sensor – range/performance problem	**Wiring, poor connection, outside air temperature sensor**
P1111	1100	Outside air temperature sensor – range/performance problem	**Wiring, poor connection, outside air temperature sensor**
P1112	1100	Outside air temperature sensor – signal malfunction	**Wiring, poor connection, outside air temperature sensor**
P1113	1100	Outside air temperature sensor – signal malfunction	**Wiring, poor connection, outside air temperature sensor**
P1115	100A	Barometric pressure (BARO) sensor – range/performance problem	**Wiring, poor connection, MAP sensor, BARO sensor**
P1116	1000	Barometric pressure (BARO) sensor – signal low	**Wiring short to ground, poor connection, BARO sensor**
P1117	1000	Barometric pressure (BARO) sensor – signal high	**Wiring short to positive, BARO sensor**
P1118	150A	Manifold absolute pressure (MAP) sensor/ mass air flow (MAF) sensor, adaptation – range/performance problem, lower limit	**Wiring, intake leak, MAF sensor, MAP sensor**
P1119	150A	Manifold absolute pressure (MAP) sensor/ mass air flow (MAF) sensor, adaptation – range/performance problem, higher limit	**Wiring, intake leak, MAF sensor, MAP sensor**
P1131	280A	Heated oxygen sensor (HO2S) 1, bank 1 – signal malfunction	**Wiring, HO2S, ECM**
P1132	280A	Heated oxygen sensor (HO2S) 1, bank 1 – signal high	**Wiring, HO2S, ECM**

Scan code P type	Scan code Volvo type	Fault location	Probable cause
P1133	280A	Heated oxygen sensor (HO2S) 1, bank 1 – no signal	Wiring, multi-plug incorrectly connected, HO2S
P1151	290A	Heated oxygen sensor (HO2S) 1, bank 2 – signal malfunction	Wiring, HO2S, ECM
P1152	290A	Heated oxygen sensor (HO2S) 1, bank 2 – signal high	Wiring, HO2S, ECM
P1153	290A	Heated oxygen sensor (HO2S) 1, bank 2 – no signal	Wiring, multi-plug incorrectly connected, HO2S
P1169	D00E	Heated oxygen sensor (HO2S) 1, bank 1 – no signal	Wiring, heating inoperative, HO2S
P1171	261A	System too lean, bank 1 – part load	Intake/exhaust leak, MAF sensor, MAP sensor, fuel pressure low, fuel pump (FP), injector(s), HO2S
P1172	261A	System too rich, bank 1 – part load	Intake blocked, EVAP canister purge valve, fuel pressure high, EGR system, injector(s), HO2S
P1174	260A	System too lean, bank 1 – full load	Intake/exhaust leak, MAF sensor, MAP sensor, fuel pressure low, fuel pump (FP), injector(s), HO2S
P1175	260A	System too rich, bank 1 – full load	Intake blocked, EVAP canister purge valve, fuel pressure high, EGR system, injector(s), HO2S
P1177	268A	System too lean, bank 1 – full load	Intake/exhaust leak, MAF sensor, MAP sensor, fuel pressure low, fuel pump (FP), injector(s), HO2S
P1178	268A	System too rich, bank 1 – full load	Intake blocked, EVAP canister purge valve, fuel pressure high, EGR system, injector(s), HO2S
P1179	268A	System too rich, bank 1 – signal malfunction	Wiring, HO2S, ECM
P1181	271A	System too lean, bank 2 – part load	Intake/exhaust leak, MAF sensor, MAP sensor, fuel pressure low, fuel pump (FP), injector(s), HO2S
P1182	271A	System too rich, bank 2 – part load	Intake blocked, EVAP canister purge valve, fuel pressure high, EGR system, injector(s), HO2S
P1184	270A	System too lean, bank 2 – full load	Intake/exhaust leak, MAF sensor, MAP sensor, fuel pressure low, fuel pump (FP), injector(s), HO2S
P1185	270A	System too rich, bank 2 – full load	Intake blocked, EVAP canister purge valve, fuel pressure high, EGR system, injector(s), HO2S
P1187	278A	System too lean, bank 2 – full load	Intake/exhaust leak, MAF sensor, MAP sensor, fuel pressure low, fuel pump (FP), injector(s), HO2S

Scan code P type	Scan code Volvo type	Fault location	Probable cause
P1188	278A	System too rich, bank 2 – full load	Intake blocked, EVAP canister purge valve, fuel pressure high, EGR system, injector(s), HO2S
P1189	278A	System too rich, bank 2 – signal malfunction	Wiring, HO2S, ECM
P1234	6814	Turbocharger (TC) boost pressure sensor – boost control malfunction	Hose connection(s), intake leak/ blockage, wiring, TC boost pressure sensor
P1237	6806	Turbocharger (TC) boost control deviation	Turbocharger (TC) overboost
P1238	6806	Turbocharger (TC) boost pressure sensor – boost control malfunction	Hose connection(s), intake leak/ blockage, wiring, TC wastegate regulating valve, TC boost pressure actuator, TC control solenoid, TC boost pressure sensor, TC wastegate, turbocharger
P1273	9117	Electronic throttle system (ETS) – malfunction	Wiring, throttle valve tight/sticking
P1281	917F	Electronic throttle system (ETS) – range/performance problem	Wiring, throttle valve tight/sticking, ETS
P1285	913F	Electronic throttle system (ETS) – range/performance problem	Wiring, throttle valve tight/sticking, ETS
P1286	913C	Electronic throttle system (ETS) – range/performance problem	Wiring, throttle valve tight/sticking, ETS
P1287	913C	Electronic throttle system (ETS) – range/performance problem	Wiring, throttle valve tight/sticking, ETS
P1293	914F	Electronic throttle system (ETS) control module – malfunction	Wiring, ETS control module
P1325	34F0	Engine control module (ECM), knock control detection – circuit test pulse	ECM
P1326	34E0	Engine control module (ECM), knock control detection – circuit offset	ECM
P1329	3400	Engine control module (ECM), knock control detection – circuit zero-test	ECM
P1330	640F	Camshaft position (CMP) actuator, inlet camshaft – valve timing advanced for too long	Wiring, timing belt/chain, mechanical fault, CMP actuator
P1336	5200	Starter request circuit – malfunction	Wiring, immobilizer system, relay
P1350	3300	Ignition coil, cylinder 1, primary – signal low	Wiring short to ground, ignition coil
P1351	3300	Ignition coil, cylinder 1, primary – signal high	Wiring short to positive, ignition coil
P1352	3310	Ignition coil, cylinder 2, primary – signal low	Wiring short to ground, ignition coil
P1353	3310	Ignition coil, cylinder 2, primary – signal high	Wiring short to positive, ignition coil
P1354	3320	Ignition coil, cylinder 3, primary – signal low	Wiring short to ground, ignition coil
P1355	3320	Ignition coil, cylinder 3, primary – signal high	Wiring short to positive, ignition coil

Scan code P type	Scan code Volvo type	Fault location	Probable cause
P1356	3330	Ignition coil, cylinder 4, primary – signal low	Wiring short to ground, ignition coil
P1357	3330	Ignition coil, cylinder 4, primary – signal high	Wiring short to positive, ignition coil
P1358	3340	Ignition coil, cylinder 5, primary – signal low	Wiring short to ground, ignition coil
P1359	3340	Ignition coil, cylinder 5, primary – signal high	Wiring short to positive, ignition coil
P1360	3350	Ignition coil, cylinder 6, primary – signal low	Wiring short to ground, ignition coil
P1361	3350	Ignition coil, cylinder 6, primary – signal high	Wiring short to positive, ignition coil
P1440	4338	Evaporative emission system – leak detected (fuel cap loose/off)	Mechanical fault, hose connection(s), EVAP pressure sensor, EVAP canister
P1457	4038	Evaporative emission system – leak detected (fuel cap loose/off)	Mechanical fault, hose connection(s), EVAP pressure sensor, EVAP canister
P1480	6100	Engine coolant blower motor control module – signal low	mechanical fault, engine coolant blower motor control module
P1481	6110	Engine coolant blower motor relay – signal low	Wiring short to ground, engine coolant blower motor relay
P1482	6110	Engine coolant blower motor relay – signal high	Wiring short to positive, engine coolant blower motor relay
P1486	6120	Engine coolant blower motor relay – signal low	Wiring short to ground, engine coolant blower motor relay
P1487	6120	Engine coolant blower motor relay – signal high	Wiring short to positive, engine coolant blower motor relay
P1513	9400	Brake pedal position (BPP) sensor – range/performance problem	Wiring, BPP sensor adjustment, BPP sensor
P1520	917F	Accelerator pedal position (APP) sensor – implausible signal	Wiring, APP sensor, ECM
P1520	918F	Accelerator pedal position (APP) sensor – implausible signal	Wiring, APP sensor, ECM
P1520	9180	Accelerator pedal position (APP) sensor – implausible signal	Wiring, APP sensor, ECM
P1525	91DF	Brake pedal position (BPP) switch – range/performance problem	Wiring, BPP switch adjustment, BPP switch
P1526	980A	Engine control module (ECM) – malfunction	ECM
P1528	91CF	Cruise control, plausibility – range/performance problem	Wiring, BPP switch, cruise control master switch, ECM
P1550	510F	Vehicle speed, serial communication – implausible signal	ABS trouble code(s) stored, CAN data bus, ABS control module, ECM
P1551	510F	Vehicle speed, serial communication – implausible signal	ABS trouble code(s) stored, CAN data bus, ABS control module, ECM
P1554	801A	Engine control relay – signal malfunction	Charging system, wiring, engine control relay, ECM
P1555	801A	Engine control relay – signal malfunction	Charging system, wiring, engine control relay, ECM

Scan code P type	Scan code Volvo type	Fault location	Probable cause
P1560	928C	Cruise control master switch – signal high	**Wiring, cruise control master switch, cruise control module/multifunction control module**
P1560	9200	Cruise control master switch – circuit malfunction	**Wiring, cruise control master switch, cruise control module/multifunction control module**
P1561	928C	Cruise control master switch – signal low	**Wiring, cruise control master switch, cruise control module/multifunction control module**
P1562	928C	Cruise control master switch – no signal	**Wiring, cruise control master switch, cruise control module/multifunction control module**
P1562	9200	Cruise control master switch – circuit malfunction	**Wiring, cruise control master switch, cruise control module/multifunction control module**
P1600	91A7	Electronic throttle system (ETS) – malfunction	**Wiring, throttle valve tight/sticking**
P1600	9190	Electronic throttle system (ETS) – malfunction	**Wiring, throttle valve tight/sticking**
P1601	91F0	Electronic throttle system (ETS), throttle motor – circuit malfunction	**Wiring, throttle motor**
P1602	91B7	Electronic throttle system (ETS), throttle return spring – malfunction	**ETS**
P1603	91A7	Electronic throttle system (ETS) – range/performance problem	**Wiring, throttle valve tight/sticking, ETS**
P1604	910E	Electronic throttle system (ETS) control module – malfunction	**Wiring, ETS control module**
P1605	911A	CAN data bus, ETS – CAN data bus OFF	**CAN data bus**
P1606	912A	CAN data bus, ETS – range/performance problem	**Wiring, ETS, ECM**
P1607	91E0	Electronic throttle system (ETS) – supply voltage	**Wiring, alternator, battery**
P1608	903F	Electronic throttle system (ETS) – throttle valve command/actual throttle valve position, correlation	**Wiring, throttle valve tight/sticking, ETS**
P1609	903F	Electronic throttle system (ETS) – throttle valve command/actual throttle valve position, correlation	**Wiring, throttle valve tight/sticking, ETS**
P1610	903F	Electronic throttle system (ETS) – throttle valve command/actual throttle valve position, correlation	**Wiring, throttle valve tight/sticking, ETS**
P1611	902A	CAN data bus, ETS – no update signals	**Wiring, ETS, ECM**
P1612	902A/B	Electronic throttle system (ETS) – no signal	**Wiring, ETS**

Scan code P type	Scan code Volvo type	Fault location	Probable cause
P1613	902A/B	Electronic throttle system (ETS) – malfunction	**Wiring, ETS**
P1614	913F	Engine control module (ECM) – malfunction	**ECM**
P1615	900E	Engine control module (ECM) – malfunction	**ECM**
P1615	901E	Engine control module (ECM) – malfunction	**ECM**
P1618	530D	Transmission control module (TCM) – MIL request	**Automatic transmission malfunction**
P1619	531D	CAN data bus, digital multifunction display – MIL request	**CAN data bus, digital multifunction display**
P1620 1	532D	Maximum torque exceeded – fuel cut-off activated/MIL activation requested by ETS	**Wiring, MAF sensor, MAP sensor, TC regulating valve, ETS, ECM**
P1620 2	6300	ETS warning lamp output – intermittent signal	**Wiring, ETS, ECM**
P1621	710A	Instrumentation control module, real time clock – communication malfunction	**CAN data bus, instrumentation control module**
P1622	7110	Knock control – malfunction	**ECM**
P1623	7120	Engine control module (ECM), HO2S 1 – circuit malfunction	**ECM**
P1625	6B00	Malfunction indicator lamp (MIL) – signal low	**Wiring short to ground, ECM**
P1626	6B00	Malfunction indicator lamp (MIL) – signal high	**Wiring short to positive, ECM**
P1627	901A	CAN data bus – communication error	**CAN data bus**
P1628	901A	CAN data bus – communication	**CAN data bus**
P1629	901A	CAN data bus – malfunction	**CAN data bus**
P1630	E003	Control module – programming error	**CAN data bus, serial communication, ECM**
P1631	530B	CAN data bus, TCM – communication error	**Wiring, CAN data bus, TCM, ECM**
P1632	710A	Multifunction control module, real time clock – malfunction	**Multifunction control module**
P1633	510F	CAN data bus, ABS (VSS signal) – not received in time	**Wiring, CAN data bus, ABS control module**
P1635	7140	Engine control module (ECM), internal temperature sensor – signal low	**ECM, temperature sensor voltage**
P1635	7180	Engine control module (ECM), internal temperature sensor – signal low	**ECM, temperature sensor voltage**
P1636	7140	Engine control module (ECM), internal temperature sensor – signal high	**ECM, temperature sensor voltage**
P1636	7180	Engine control module (ECM), internal temperature sensor – signal high	**ECM, temperature sensor voltage**
P1637	715D	Engine control module (ECM), internal temperature high - stage 1	**Wiring, module cooling fan**

Scan code P type	Scan code Volvo type	Fault location	Probable cause
P1638	715D	Engine control module (ECM), internal temperature high - stage 2	**Wiring, module cooling fan**
P1638	716D	Engine control module (ECM), internal temperature high - stage 2	**Wiring, module cooling fan**
P1646	2800	Heated oxygen sensor (HO2S) 1, bank 1, heater control – defective	**Wiring, HO2S, ECM**
P1646	6000	A/C relay – signal low	**Wiring short to ground, A/C relay**
P1647	6000	A/C relay – signal high	**Wiring short to positive, A/C relay**
P1649	6440	Camshaft position (CMP) actuator, inlet camshaft – signal low	**Wiring short to ground, poor connection, CMP actuator**
P1650	6440	Camshaft position (CMP) actuator, inlet camshaft – signal high	**Wiring short to positive, CMP actuator**
P1651 1	980F	Mass air flow (MAF) sensor/manifold absolute pressure (MAP) sensor – implausible signal with throttle position	**Wiring, MAF sensor, MAP sensor, TP sensor**
P1651 3	981A	Maximum torque exceeded – fuel cut-off activated	**Wiring, MAF sensor, MAP sensor, TC regulating valve, ETS, ECM**
P1652	982A	Engine control module (ECM), primary fuel cut-off – implausible signals between sensors	**Wiring, MAF sensor, MAP sensor, ETS, ECM**
P1655	6440	Camshaft position (CMP) actuator, exhaust – signal low	**Wiring short to ground, poor connection, CMP actuator**
P1656	6440	Camshaft position (CMP) actuator, exhaust – signal high	**Wiring short to positive, CMP actuator**
P1657	6300	Electronic throttle system (ETS), MIL request OFF – signal low (MIL ON)	**Wiring, ETS**
P1658	6310	Electronic throttle system (ETS), MIL request ON – signal high (MIL OFF)	**Wiring, ETS**
P1671	6400	Camshaft position (CMP) actuator – range/ performance problem	**Wiring, CMP actuator**
P1660	922A	Cruise control master switch – signal high	**Wiring, cruise control master switch, cruise control module/multifunction control module**
P1660	928C	Cruise control master switch – signal high	**Wiring, cruise control master switch, cruise control module/multifunction control module**
P1661	922A	Cruise control master switch – signal low	**Wiring, cruise control master switch, cruise control module/multifunction control module**
P1661	928C	Cruise control master switch – signal low	**Wiring, cruise control master switch, cruise control module/multifunction control module**
P1662	922A	Cruise control master switch – no signal	**Wiring, cruise control master switch, cruise control module/multifunction control module**

Scan code P type	Scan code Volvo type	Fault location	Probable cause
P1662	928C	Cruise control master switch – no signal	**Wiring, cruise control master switch, cruise control module/multifunction control module**
P1663	922A	Cruise control master switch – signal malfunction	**Wiring, cruise control master switch, cruise control module/multifunction control module**
P1670	720A	Immobilizer control module – communication error/tampering/code error	**Wiring, immobilizer control module/ multifunction control module, ECM**
P1672	640A	Camshaft position (CMP) actuator – range/ performance problem	**Wiring, CMP actuator, incorrect valve timing**
P1672	8410	Engine control module (ECM) – 5 volt output low	**Wiring, MAF sensor, MAP sensor, TC boost pressure sensor, A/C pressure sensor**
P1673	8410	Engine control module (ECM) – 5 volt output high	**Wiring, MAF sensor, MAP sensor, TC boost pressure sensor, A/C pressure sensor**
P1680	720A	CAN data bus, Immobilizer control module – no signal	**Wiring, immobilizer control module, ECM**
P1721	6220	Module cooling fan – signal high	**Wiring, module cooling fan**
P1722	6220	Module cooling fan – signal low	**Wiring, module cooling fan**
P1723	6220	Module cooling fan – no signal	**Wiring, module cooling fan**
P1729	9170	Throttle position (TP) sensor – implausible signal between circuit 1 & 2	**Wiring, ETS**
P1740	4030	Evaporative emission (EVAP) leak detection pump – no signal	**Wiring, poor connection, EVAP leak detection pump**
P1741	4030	Evaporative emission (EVAP) leak detection pump – signal low	**Wiring short to ground, EVAP leak detection pump**
P1742	4030	Evaporative emission (EVAP) leak detection pump – signal high	**Wiring short to positive, EVAP leak detection pump**
P1770	3020	Camshaft position (CMP) sensor, intake/ exhaust – signal malfunction	**Wiring, CMP sensor(s), ECM**

1 B5244S/S6.
2 B5244T7.
3 Except B5244S.

VOLVO

Model:	Engine identification:	Year:
S40/V40 2.0L Turbo	**B4204T2/T3/T4**	**2000-04**

System: **Siemens EMS 2000**

Self-diagnosis

General information

- Refer to the front of this manual for general test conditions, terminology, detailed descriptions of wiring faults and a general trouble shooter for electrical and mechanical faults.
- Malfunction indicator lamp (MIL) will illuminate if certain faults are recorded.
- ECM operates in backup mode if sensors fail, to enable vehicle to be driven to repair shop.

Accessing

- Trouble codes can only be displayed by using a Scan Tool connected to the data link connector (DLC) **1**.

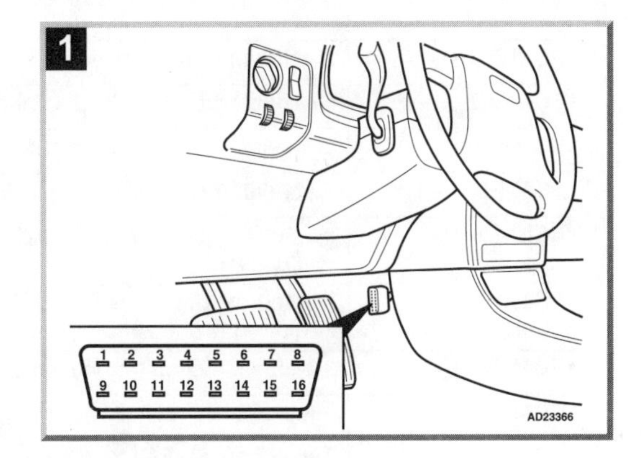

Erasing

- Trouble codes can only be erased by using a Scan Tool connected to the data link connector (DLC).

Trouble code identification

Scan code Volvo type	Scan code P type	Fault location	Probable cause
–	*All P0, P2 and U0 codes*	Refer to OBD-II trouble code tables at the front of this manual	–
ECM-AC	–	Turbocharger (TC) boost pressure sensor 1 – boost control malfunction	**Hose connection(s), intake leak/ blockage, wiring, TC wastegate regulating valve, TC boost pressure actuator, TC control solenoid, TC boost pressure sensor 1, TC wastegate, turbocharger, MAF sensor, ECM**
ECM-AD	–	Turbocharger (TC) boost pressure sensor 1 – boost control malfunction	**Hose connection(s), intake leak/ blockage, wiring, TC wastegate regulating valve, TC boost pressure actuator, TC control solenoid, TC boost pressure sensor 1, TC wastegate, turbocharger, MAF sensor, ECM**
ECM-AF	P1446	Evaporative emission (EVAP) leak detection pump – signal malfunction	**Wiring, poor connection, EVAP leak detection pump**

Scan code Volvo type	Scan code P type	Fault location	Probable cause
ECM-A0	–	Engine coolant temperature (ECT) sensor – circuit intermittent	**Wiring, poor connection, ECT sensor, ECM**
ECM-A2	–	Insufficient coolant temperature for closed loop fuel control	**Wiring, cooling system, coolant thermostat, ECT sensor**
ECM-A3	–	System voltage – malfunction	**Wiring, poor connection, battery, alternator, engine control relay, ECM**
ECM-DC	–	CAN data bus, TCM – signal malfunction	**Wiring, TCM, ECM**
ECM-DD	–	CAN data bus, TCM – signal malfunction	**Wiring, TCM, ECM**
ECM-DE	–	CAN data bus, TCM – signal malfunction	**Wiring, TCM, ECM**
ECM-E0	–	CAN data bus, TCM – signal malfunction	**Wiring, TCM, ECM**
ECM-1A	–	Heated oxygen sensor (HO2S) 2, bank 1 – malfunction	**Wiring, heating inoperative, HO2S, ECM**
ECM-1C	–	Heated oxygen sensor (HO2S) 2, bank 1, heater control – circuit malfunction	**Wiring, HO2S, ECM**
ECM-1E	–	Injector 1 – circuit malfunction	**Wiring, injector, ECM**
ECM-1F	–	Injector 2 – circuit malfunction	**Wiring, injector, ECM**
ECM-2D	–	Turbocharger (TC) wastegate regulating valve – circuit malfunction	**Wiring, TC wastegate regulating valve, ECM**
ECM-2E	– 1	Engine boost condition – signal low	**Hose connection(s), wiring, TC wastegate regulating valve, TC wastegate, TC, MAF sensor**
ECM-2E	P1236 2	Engine boost condition – signal high	**Hose connection(s), wiring, TC wastegate regulating valve, TC wastegate, MAF sensor**
ECM-2E	P1241 1	Engine boost condition – signal high	**Hose connection(s), wiring, TC wastegate regulating valve, TC wastegate, MAF sensor**
ECM-3A	–	Ignition coil B, primary/secondary – circuit malfunction	**Wiring, ignition coil, ECM**
ECM-4D	–	Random/multiple cylinder(s) – misfire detected	**Spark plug(s), HT lead(s), injector(s), ignition coil(s), low compression, wiring**
ECM-5A	–	Catalytic converter system, bank 1 – efficiency below threshold	**Catalytic converter, wiring, HO2S 2**
ECM-5C	–	Evaporative emission (EVAP) canister purge valve – circuit malfunction	**Wiring, EVAP canister purge valve, ECM**
ECM-5D	–	Camshaft position (CMP) actuator, intake, bank 1 – circuit malfunction	**Wiring, CMP actuator, ECM**
ECM-5E	–	Camshaft position (CMP) actuator, exhaust, bank 1 – timing over-advanced/ system performance	**Valve timing, engine mechanical fault, CMP actuator**
ECM-5F	–	Camshaft position (CMP) actuator, exhaust, bank 1 – timing over-retarded	**Valve timing, engine mechanical fault, CMP actuator**
ECM-6A	–	Evaporative emission (EVAP) system – incorrect flow detected	**Hose connection(s), intake leak, EVAP canister purge valve**

Scan code Volvo type	Scan code P type	Fault location	Probable cause
ECM-6B	–	A/C refrigerant pressure sensor – malfunction	**A/C refrigerant pressure too low/high (cooling fault/incorrectly charged), wiring, A/C refrigerant pressure sensor, ECM**
ECM-6F	–	Vehicle speed signal (VSS) – circuit malfunction	**Wiring, ABS trouble code(s) stored, ABS control module, ECM**
ECM-7D	–	A/C relay – circuit malfunction	**Wiring, A/C relay**
ECM-7F	–	Malfunction indicator lamp (MIL) – circuit malfunction	**Wiring, MIL, ECM**
ECM-8A	–	Idle speed control (ISC) system – malfunction	**Wiring, IAC valve, throttle valve tight/ sticking, ECM**
ECM-8D	–	Idle speed control (ISC) system – malfunction than expected	**Wiring, IAC valve, throttle valve tight/ sticking, ECM**
ECM-9A	P1630 1	Immobilizer control module – communication error/tampering/code error	**Wiring, poor connection, immobilizer control module, ECM**
ECM-9A	P1513 2	Immobilizer control module – communication error/tampering/code error	**Wiring, poor connection, immobilizer control module, ECM**
ECM-9B	–	VIN not programmed or mismatch – ECM/immobilizer control module	**Immobilizer control module, ECM**
ECM-10	–	Intake air temperature (IAT) sensor – circuit intermittent	**Wiring, poor connection, IAT sensor, ECM**
ECM-13	–	Mass air flow (MAF) sensor – circuit malfunction	**Wiring, MAF sensor, ECM**
ECM-14	–	Mass air flow (MAF) sensor – range/performance problem	**Intake leak/blockage, MAF sensor**
ECM-15	–	Engine coolant temperature (ECT) sensor – circuit malfunction	**Coolant thermostat, wiring, ECT sensor**
ECM-16	–	Heated oxygen sensor (HO2S) 1, bank 1 – circuit malfunction	**Heating inoperative, poor connection, wiring, HO2S, ECM**
ECM-18	–	Heated oxygen sensor (HO2S) 1, bank 1, heater control – circuit malfunction	**Wiring, HO2S, ECM**
ECM-20	–	Injector 3 – circuit malfunction	**Wiring, injector, ECM**
ECM-21	–	Injector 4 – circuit malfunction	**Wiring, injector, ECM**
ECM-24	–	Engine control relay – circuit malfunction	**Wiring, engine control relay, ECM**
ECM-25	–	System too lean/rich, bank 1	**Intake/exhaust leak, AIR system, Intake blocked, EVAP canister purge valve, MAF sensor, fuel pressure/pump, injector(s), HO2S**
ECM-28	–	Heated oxygen sensor (HO2S) 1, bank 1 – slow response	**Heating inoperative, wiring, HO2S**
ECM-29	–	Heated oxygen sensor (HO2S) 2, bank 1 – slow response	**Heating inoperative, wiring, HO2S**
ECM-32	–	Camshaft position (CMP) sensor A, bank 1 – malfunction	**Insecure sensor/rotor, air gap, wiring, CMP sensor, ECM**

Scan code Volvo type	Scan code P type	Fault location	Probable cause
ECM-36	–	Heated oxygen sensor (HO2S) 2, bank 1, heater control – circuit malfunction	Wiring, HO2S, ECM
ECM-38	–	Ignition coil A, primary/secondary – circuit malfunction	Wiring, ignition coil, ECM
ECM-41	–	Knock sensor (KS) 1, bank 1 – circuit malfunction	Wiring, poor connection, KS
ECM-43	–	Knock control system error	Wiring, poor connection, KS, ECM
ECM-44	–	Random/multiple cylinder(s) – misfire detected	Spark plug(s), HT lead(s), injector(s), ignition coil(s), low compression, wiring
ECM-45	–	Random/multiple cylinder(s) – misfire detected	Spark plug(s), HT lead(s), injector(s), ignition coil(s), low compression, wiring
ECM-53	P1336	Crankshaft position (CKP) sensor/engine speed (RPM) sensor – range/performance problem	Air gap, metal particle contamination, insecure sensor/rotor, wiring, CKP/RPM sensor
ECM-60	–	Camshaft position (CMP) actuator, exhaust, bank 1 – timing over-advanced/ system performance	Valve timing, engine mechanical fault, CMP actuator
ECM-61	–	Camshaft position (CMP) actuator, exhaust, bank 1 – timing over-retarded	Valve timing, engine mechanical fault, CMP actuator
ECM-62	– 1	Camshaft position (CMP) actuator, exhaust, bank 1 – timing over-advanced/ system performance	Valve timing, engine mechanical fault, CMP actuator
ECM-62	P1014 2	Camshaft position (CMP) actuator, exhaust, bank 1 – timing over-advanced/ system performance	Valve timing, engine mechanical fault, CMP actuator
ECM-65	P1449	Evaporative emission (EVAP) leak detection pump – circuit malfunction	Mechanical fault, hose connection(s), wiring, EVAP leak detection pump
ECM-66	P1440	Evaporative emission (EVAP) leak detection pump – signal low	Mechanical fault, hose connection(s), EVAP canister, EVAP canister purge valve, wiring, EVAP leak detection pump
ECM-68	–	Evaporative emission (EVAP) system – leak detected	Mechanical fault, hose connection(s), intake leak, EVAP canister, EVAP canister purge valve, EVAP pressure sensor
ECM-70	P1600	Throttle position (TP) sensor – communication error	Wiring, poor connection, TP sensor, ESP control module, ECM
ECM-73	–	Throttle position (TP) sensor – circuit malfunction	Wiring, TP sensor, ECM
ECM-74	–	Throttle position (TP) sensor – range/performance problem	Accelerator cable adjustment, TP sensor
ECM-76	P1618	Transmission control module (TCM) – MIL request	AT trouble code(s) stored, automatic transmission malfunction
ECM-78	–	Crankshaft position (CKP) sensor/engine speed (RPM) sensor – malfunction	Insecure sensor/rotor, air gap, wiring, CKP/RPM sensor, ECM

Scan code Volvo type	Scan code P type	Fault location	Probable cause
ECM-82	P1483 [1]	A/C condenser blower motor relay – circuit malfunction	**Wiring, A/C condenser blower motor relay, ECM**
ECM-82	P1482 [2]	A/C condenser blower motor relay – circuit malfunction	**Wiring, A/C condenser blower motor relay, ECM**
ECM-85	–	Engine coolant blower motor relay, low speed – circuit malfunction	**Wiring, engine coolant blower motor relay, ECM**
ECM-87	–	Engine coolant blower motor relay, high speed – circuit malfunction	**Wiring, engine coolant blower motor relay, ECM**
ECM-91	P1600	Torque reduction communication error	**ESP trouble code(s) stored, wiring open circuit/short circuit, poor connection, electronic stability program (ESP) control module, ECM**
ECM-0A	P1107	Barometric pressure (BARO) sensor – low input	**Wiring short to ground, BARO sensor, ECM**
ECM-0A	P1108	Barometric pressure (BARO) sensor – high input	**Wiring short to positive, BARO sensor, ECM**
ECM-0B	P1106	Barometric pressure (BARO) sensor – range/performance problem	**Intake/exhaust leak, wiring, BARO sensor**
ECM-0C	–	Turbocharger (TC) boost pressure sensor 2 – boost control malfunction	**Hose connection(s), intake leak/ blockage, wiring, TC wastegate regulating valve, TC boost pressure actuator, TC control solenoid, TC boost pressure sensor 2, TC wastegate, turbocharger, ECM**
ECM-0D	–	Turbocharger (TC) boost pressure sensor 2 – boost control malfunction	**Hose connection(s), intake leak/ blockage, wiring, TC wastegate regulating valve, TC boost pressure actuator, TC control solenoid, TC boost pressure sensor 2, TC wastegate, turbocharger, ECM**
ECM-0F	–	Intake air temperature (IAT) sensor – circuit malfunction	**Wiring, IAT sensor, ECM**

[1] 2000.
[2] 2001→.

VOLVO

Model:	**240 2.3 • 740 2.3 • 740 2.3 Turbo • 940 2.3/Turbo**
Year:	**1992-95**
Engine identification:	**B200F, B200FT, B200G, B204E, B204FT, B230F, B230FB, B230FD, B230FK, B230FT, B230FX, B234F, B5204S, B5254S**
System:	**Bosch LH-Jetronic 2.4/3.2**

Data link connector (DLC) locations

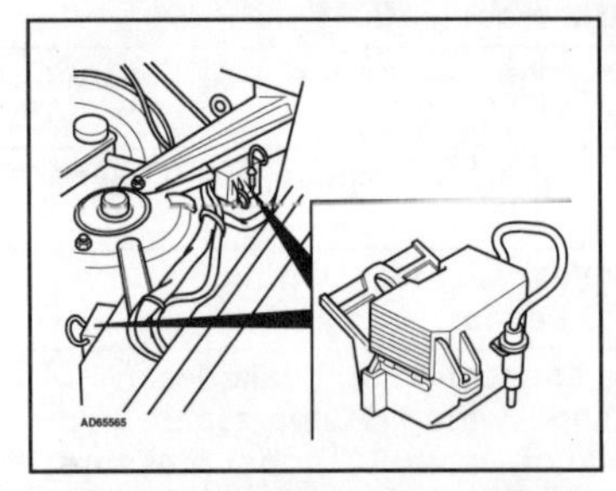

240, 740/760, 940/960 – engine bay, LH rear

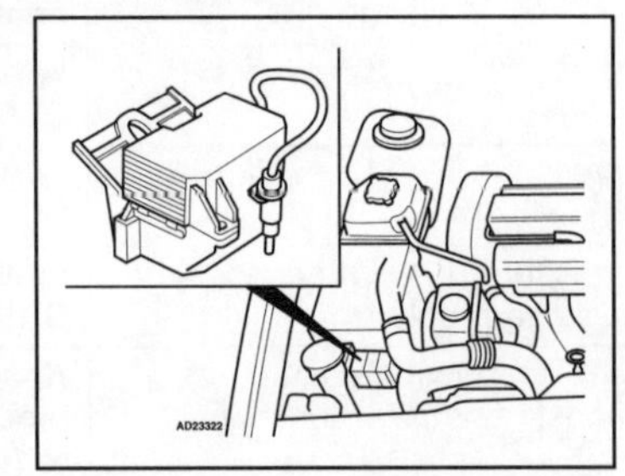

850 – engine bay, RH front

Self-diagnosis

General information

- Refer to the front of this manual for general test conditions, terminology, detailed descriptions of wiring faults and a general trouble shooter for electrical and mechanical faults.
- Trouble codes are displayed by the LED on the data link connector (DLC).
- The fuel/ignition/turbocharger control module fault memory can also be checked and erased using diagnostic equipment connected to the data link connector (DLC).

Accessing

Fuel system

- Ensure ignition switched OFF.
- Open cover on black data link connector (DLC).
- Insert selector cable in socket 2 **1** [**1**].
- Switch ignition ON.
- Depress test button for approximately 1 second **1** [**2**].
- Count LED flashes **1** [**3**]. Note trouble code.
- Each trouble code consists of three groups of flashes **2**.
- A 3 second pause separates each group **2** [**A**].
- For example: Trouble code 223 displayed **2** – idle air control (IAC) valve.
- Repeat switch operation **1** [**2**]. Note trouble codes. Compare with trouble code table.
- Switch ignition OFF. Rectify faults as necessary.

NOTE: ***Maximum of 3 trouble codes can be stored by memory at one time.***

1

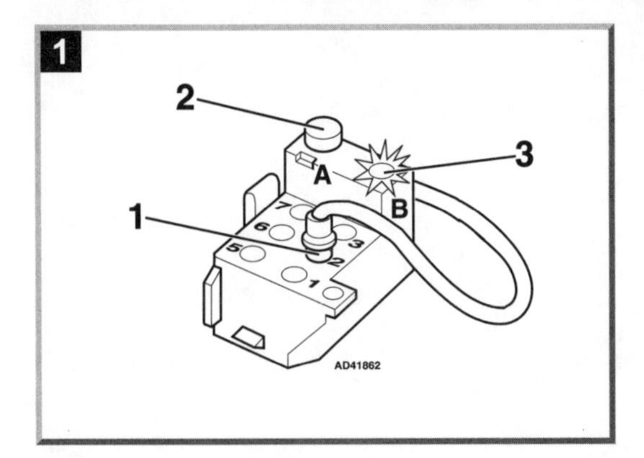

2

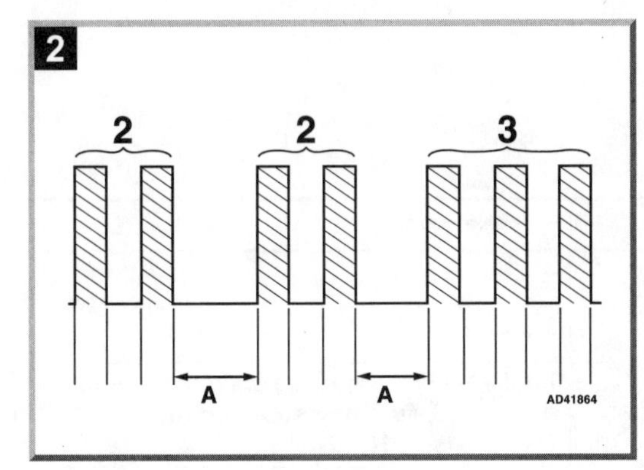

Ignition system

- Ensure ignition switched OFF.
- Open cover on black data link connector (DLC).
- Insert selector cable in socket 6 3 [**1**].
- Switch ignition ON.
- Depress test button for approximately 1 second 3 [**2**].
- Count LED flashes 3 [**3**]. Note trouble code.
- Trouble codes are displayed in the same way as those from the fuel system:
  - Each trouble code consists of three groups of flashes 2.
  - A 3 second pause separates each group 2 [**A**].
  - For example: Trouble code 223 displayed 2 – idle air control (IAC) valve.
- Repeat switch operation 3 [**2**]. Note trouble codes. Compare with trouble code table.
- Switch ignition OFF. Rectify faults as necessary.

NOTE: ***Maximum of 3 trouble codes can be stored by memory at one time.***

3

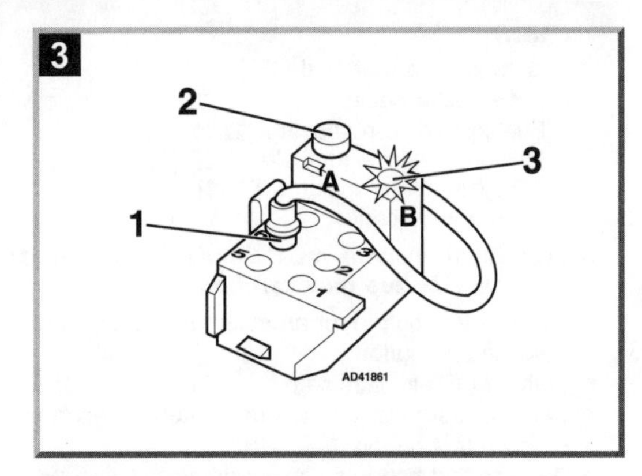

Turbocharger (TC) system - B204FT

- Ensure ignition switched OFF.
- Open cover on black data link connector (DLC).
- Insert selector cable in socket 5 4 [**1**].
- Switch ignition ON.
- Depress test button for approximately 1 second 4 [**2**].
- Count LED flashes 4 [**3**]. Note trouble code.
- Trouble codes are displayed in the same way as those from the fuel system:
  - Each trouble code consists of three groups of flashes 2.
  - A 3 second pause separates each group 2 [**A**].
  - For example: Trouble code 223 displayed 2 – idle air control (IAC) valve.
- Repeat switch operation 4 [**2**]. Note trouble codes. Compare with trouble code table.
- Switch ignition OFF. Rectify faults as necessary.

NOTE: ***Maximum of 7 trouble codes can be stored by memory at one time.***

4

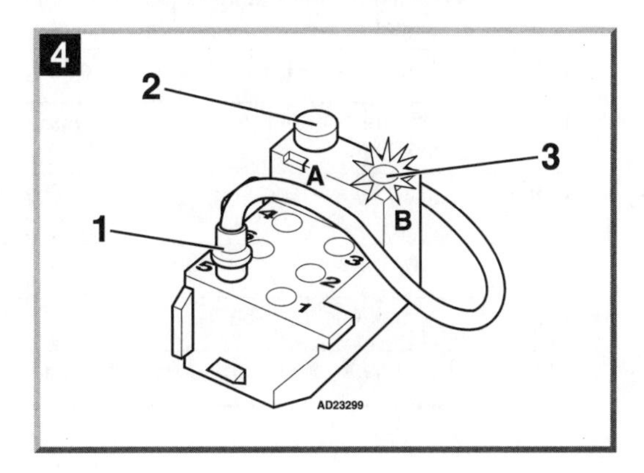

Erasing

- Ensure ignition switched OFF.
- Insert selector cable:
  - Fuel system – in socket 2 **1** [**1**].
  - Ignition system – in socket 6 **2** [**1**].
  - TC system – in socket 5 **4** [**1**].
- Access trouble codes.

NOTE: ***Trouble codes must be displayed at least once before they can be erased.***

- Depress test button for approximately 5 seconds.
- Release test button.
- Wait for LED to illuminate.
- Depress test button for approximately 5 seconds.
- Release test button.
- Depress test button for approximately 1 second.
- Ensure trouble code 111 displayed.

Trouble code identification

Flash code	Fault location	Probable cause
111	No fault found	–
112	Engine control module (ECM)	ECM
113 **1**	Fuel trim (FT) – too weak/rich	Intake leak/blockage, excessive fuel in engine oil, fuel pressure, wiring, MAF sensor, HO2S
113 **2**	Fuel trim (FT) – too weak	Intake leak, fuel pressure low, wiring, MAF sensor, HO2S
121	Mass air flow (MAF) sensor	Wiring, MAF sensor
123	Engine coolant temperature (ECT) sensor	Wiring, ECT sensor
131 **3**	Ignition control module (ICM) – engine RPM signal	Wiring
131 **4**	Engine speed (RPM) sensor – no signal	Wiring, air gap, RPM sensor
132	Engine control module (ECM) – supply voltage low/high	Battery, alternator
133	Throttle position (TP) switch – CTP contacts	TP switch adjustment, accelerator cable adjustment, wiring, TP switch
142	Ignition control module (ICM)	ICM
143	Knock sensor (KS) 1	Wiring, KS
144	Engine control module (ECM) – load signal	Wiring, ECM

Flash code	Fault location	Probable cause
154	Exhaust gas recirculation (EGR) system – excessive flow detected	**Wiring, EGR valve, EGR control module, engine coolant thermostat**
212	Heated oxygen sensor (HO2S)	**Wiring, heating inoperative, MAF sensor, HO2S**
213	Throttle position (TP) switch – WOT contacts	**TP switch adjustment, accelerator cable adjustment, wiring, TP switch**
214	Engine speed (RPM) sensor – intermittent signal	**Wiring, flywheel/drive plate damaged, air gap, RPM sensor**
221	Fuel trim (FT), part load – too weak	**Intake/exhaust leak, fuel pressure, wiring, MAF sensor, HO2S**
223	Idle air control (IAC) valve	**Wiring, IAC valve**
224	Engine coolant temperature (ECT) sensor	**Wiring, ECT sensor**
231 [1]	Fuel trim (FT), part load – too weak/rich	**Intake/exhaust leak, fuel pressure, excessive fuel in engine oil, injector(s), EVAP canister purge valve, wiring, MAF sensor, ECT sensor, HO2S**
231 [2]	Fuel trim (FT), part load – too rich	**Fuel pressure high, excessive fuel in engine oil, injector(s), EVAP canister purge valve, wiring, MAF sensor, ECT sensor, HO2S**
232	Fuel trim (FT), idling – weak/rich	**Intake/exhaust leak, fuel pressure, excessive fuel in engine oil, injector(s), EVAP canister purge valve, wiring, MAF sensor, ECT sensor, HO2S**
234	Throttle position (TP) switch – CTP contacts	**TP switch adjustment, accelerator cable adjustment, wiring, TP switch**
241	Exhaust gas recirculation (EGR) system – insufficient flow detected	**Hose leak/blockage, wiring, EGR valve, EGR control module, EGRT sensor, engine coolant thermostat**
242	Turbocharger (TC) wastegate regulating valve	**Wiring, TC wastegate regulating valve**
311	Vehicle speed sensor (VSS)	**Wiring, speedometer, VSS**
312	Ignition control module (ICM) – KS signal missing	**Wiring, ICM**
314	Camshaft position (CMP) sensor – no/incorrect signal	**Wiring, CMP sensor**
322	Mass air flow (MAF) sensor – no filament burn-off signal	**Wiring, ECM**
324	Camshaft position (CMP) sensor – intermittent signal	**Wiring, CMP sensor**
342	AC compressor clutch cut-off relay	**Wiring, AC compressor clutch cut-off relay**
344	Exhaust gas temperature sensor	**Wiring, EGRT sensor**

Flash code	Fault location	Probable cause
411	Throttle position (TP) sensor	**Wiring, TP sensor**
413	Exhaust gas recirculation temperature (EGRT) sensor	**Wiring, EGRT sensor**
421	Manifold absolute pressure (MAP) sensor	**Hose leak/blockage, TC, TC control module**
422 5	Exhaust gas temperature sensor	**Wiring, exhaust gas temperature sensor**
423	Throttle position (TP) sensor	**Wiring, TP sensor**
424	Engine control module (ECM) – load signal	**Wiring, ECM if trouble code also stored in ICM**
431	Engine coolant blower motor temperature sensor	**Wiring, engine coolant blower motor temperature sensor**
432	Engine control module (ECM) – overheating	**Engine coolant blower motor, air supply blocked, module cooling fan**
433	Knock sensor (KS) 2	**Wiring, KS**
512	Fuel trim (FT) – too rich	**Intake blockage, excessive fuel in engine oil, fuel pressure high, wiring, MAF sensor, HO2S**
513	Engine control module (ECM) – overheating	**Engine coolant blower motor, air supply blocked, module cooling fan**

1 Bosch LH-Jetronic 2.4
2 Bosch LH-Jetronic 3.2
3 Engine control module (ECM).
4 Ignition control module (ICM).
5 Has no function on early models. Erase trouble code and check fuel system for trouble code 344.

Model: **850 2.3L Turbo**

Engine identification: **B5234T/T2/T3/T4/T6**

Year: **1994-97**

System: **Bosch Motronic 4.4 ○ Siemens fenix 5.1**

Self-diagnosis

General information

- Refer to the front of this manual for general test conditions, terminology, detailed descriptions of wiring faults and a general trouble shooter for electrical and mechanical faults.

Accessing and erasing

- The engine control module (ECM) fault memory can only be accessed and erased using diagnostic equipment connected to the data link connector (DLC) **1**.

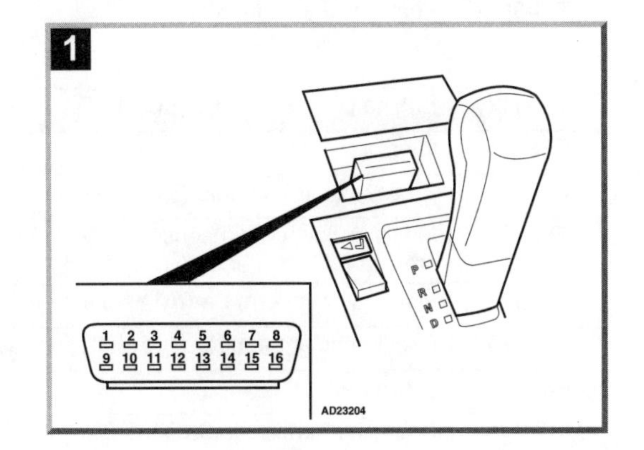

Trouble code identification

Scan code	Fault location	Probable cause
EFI-0011	Heated oxygen sensor (HO2S)	Wiring, HO2S
EFI-0012	Mass air flow (MAF) sensor	Wiring, MAF sensor
EFI-0013	Intake air temperature (IAT) sensor	Wiring, IAT sensor
EFI-0014	Throttle position (TP) sensor	Wiring, TP sensor, IAC valve signal exceeded 2 V for 4 seconds
EFI-0021	Engine coolant temperature (ECT) sensor	Wiring, ECT sensor
EFI-0022	Crankshaft position (CKP) sensor/ engine speed (RPM) sensor	Wiring, poor connection, CKP/RPM sensor
EFI-0023	Camshaft position (CMP) sensor	Wiring, poor connection, CMP sensor
EFI-0024	Speedometer	Wiring, speedometer
EFI-0025	Barometric pressure (BARO) sensor	Wiring, BARO sensor
EFI-0031	Knock sensor (KS)	Wiring, KS
EFI-0036	Engine control module (ECM) – ignition adjustment signal	Activation of ignition adjustment incomplete
EFI-0041	Injectors	Wiring, injector(s), injector control module
EFI-0044	Ignition coil	Spark plug(s), wiring, ignition coil
EFI-0054	Immobilizer control module – communication error	Wiring

Scan code	Fault location	Probable cause
EFI-0056	Fuel pressure sensor	**Wiring, fuel pressure sensor**
EFI-0058	Fuel trim (FT) – abnormal intake air volume	**Wiring, IAC valve, intake air by-pass valve**
EFI-0064	Alternator	**Wiring, alternator**
EFI-0066	Brake vacuum sensor	**Wiring, brake vacuum sensor**
EFI-0089	Fuel trim (FT) – high/low	**Intake leak/blockage, fuel pressure/pump, wiring, IAT sensor, BARO sensor, MAF sensor**
EFI-112	Engine control module (ECM)	**ECM**
EFI-115	Injector 1	**Wiring, injector**
EFI-116	Engine control module (ECM) – internal fault	**ECM**
EFI-121	Mass air flow (MAF) sensor/manifold absolute pressure (MAP) sensor	**Intake leak/blockage, hose leak, wiring, MAF/MAP sensor**
EFI-122	Intake air temperature (IAT) sensor	**Wiring, IAT sensor**
EFI-123	Engine coolant temperature (ECT) sensor	**Wiring, ECT sensor**
EFI-125	Injector 2	**Wiring, injector**
EFI-131	Crankshaft position (CKP) sensor/ engine speed (RPM) sensor – no signal	**Wiring, CKP/RPM sensor**
EFI-132	Engine control module (ECM) – supply voltage	**Battery, alternator, wiring**
EFI-137	Crankshaft position (CKP) sensor/ engine speed (RPM) sensor	**Wiring, flywheel/drive plate damaged, CKP/RPM sensor insecure/defective**
EFI-135	Injector 3	**Wiring, injector**
EFI-143	Knock sensor (KS) 1	**Wiring, KS**
EFI-145	Injector 4	**Wiring, injector**
EFI-153	Heated oxygen sensor (HO2S) 2	**Exhaust leak, wiring, HO2S 1/2**
EFI-155	Injector 5	**Wiring, injector**
EFI-165	Injector 6	**Wiring, injector**
EFI-211	Mixture adjustment resistor	**Wiring, mixture adjustment resistor incorrectly adjusted/defective**
EFI-212	Heated oxygen sensor (HO2S) 1	**Wiring, HO2S 1/2**
EFI-214	Crankshaft position (CKP) sensor/ engine speed (RPM) sensor – incorrect/ intermittent signal	**Poor connection, shield wiring, flywheel/ drive plate damaged, CKP/RPM sensor**
EFI-216	Heated oxygen sensor (HO2S)	**Wiring, HO2S**
EFI-222	Engine control relay	**Wiring, engine control relay**
EFI-223	Idle air control (IAC) valve – opening signal	**Wiring, IAC valve**
EFI-225	AC refrigerant pressure sensor	**Wiring, AC refrigerant pressure sensor**
EFI-226	Idle air control (IAC) valve	**Wiring, IAC valve**
EFI-227	Idle speed control (ISC) valve	**Wiring, ISC valve**

Scan code	Fault location	Probable cause
EFI-231	Long term fuel trim, part load	Intake/exhaust leak, fuel pressure, excessive fuel in engine oil, engine oil level too high, injector(s), EVAP canister purge valve, wiring, MAF sensor, ECT sensor, HO2S
EFI-232	Long term fuel trim, idling	Intake/exhaust leak, fuel pressure, excessive fuel in engine oil, engine oil level too high, injector(s), EVAP canister purge valve, wiring, MAF sensor, ECT sensor, HO2S
EFI-233	Long term fuel trim, idle air adjustment	Intake leak/blockage, throttle valve adjustment, accelerator cable adjustment, wiring, IAC valve
EFI-235	Exhaust gas recirculation (EGR) solenoid	Wiring, EGR solenoid
EFI-242	Turbocharger (TC) wastegate regulating valve	Wiring, TC wastegate regulating valve
EFI-245	Idle air control (IAC) valve – closing signal	Wiring, IAC valve
EFI-251	Outside air temperature sensor	Wiring, outside air temperature sensor
EFI-254	Engine control relay	Wiring, engine control relay
EFI-311	Vehicle speed signal	Wiring, speedometer
EFI-314	Camshaft position (CMP) sensor	Wiring, poor connection, CMP sensor
EFI-315	Evaporative emission (EVAP) canister purge valve – leak detected	EVAP canister purge valve
EFI-323	Malfunction indicator lamp (MIL)	Wiring, MIL
EFI-325	Engine control module (ECM) – memory fault	ECM disconnected, battery discharged/ disconnected, wiring
EFI-335	Transmission control module (TCM) – MIL ON request	Wiring
EFI-336	Transmission control module (TCM) – gear recognition	Wiring
EFI-337	Transmission control module (TCM)/traction control module – torque reduction signal	Wiring, TCM, traction control module
EFI-338	Engine control module (ECM) – immobilizer authorisation signal missing	Immobilizer trouble code IMM-322 stored, wiring
EFI-342	AC relay	Wiring, AC relay
EFI-343	Fuel pump relay	Wiring, fuel pump relay
EFI-353	Immobilizer control module – communication error	Wiring, poor connection, immobilizer control module
EFI-355	Mass air flow (MAF) sensor – incorrect signal	MAF sensor, TP sensor
EFI-411	Throttle position (TP) sensor	Wiring, TP sensor
EFI-414	Turbocharger (TC) pressure – control	Hose blocked, TCM wire (TC pressure reduction signal), wiring, MAF sensor, TC wastegate actuator/regulating valve
EFI-415	Manifold absolute pressure (MAP) sensor, TC system	Wiring, MAP sensor, BARO sensor

Scan code	Fault location	Probable cause
EFI-422	Barometric pressure (BARO) sensor	**Wiring, sensor supply, BARO sensor, AC pressure sensor, fuel tank pressure sensor, G-force sensor**
EFI-425	Heated oxygen sensor (HO2S) 2	**Intake/exhaust leak, uneven compression, fuel pressure, HO2S**
EFI-432	Engine control module (ECM) – overheating	**Engine coolant blower motor, air supply blocked**
EFI-433	Knock sensor (KS) 2	**Wiring, KS**
EFI-435	Heated oxygen sensor (HO2S) 1 – slow response	**Intake/exhaust leak, uneven compression, fuel pressure, MAF sensor, heating inoperative, HO2S**
EFI-436	Heated oxygen sensor (HO2S) 2 – compensation	**Intake/exhaust leak, uneven compression, fuel pressure, HO2S 1/2**
EFI-442	Pulsed secondary air injection (PAIR) system – excessive flow detected	**PAIR valve/solenoid, exhaust leak**
EFI-443	Catalytic converter – efficiency below threshold	**Intake/exhaust leak, uneven compression, fuel pressure, catalytic converter**
EFI-444	G-force sensor	**Wiring, sensor supply, AC pressure sensor, fuel tank pressure sensor, G-force sensor**
EFI-445	Pulsed secondary air injection (PAIR) pump relay	**Wiring, PAIR pump relay**
EFI-446	Pulsed secondary air injection (PAIR) valve – leak detected	**PAIR valve/solenoid, exhaust leak**
EFI-447	Pulsed secondary air injection (PAIR) solenoid	**Wiring, PAIR solenoid**
EFI-448	Pulsed secondary air injection (PAIR) system – incorrect flow detected	**PAIR valve/solenoid, PAIR pump/relay, hoses/pipes blocked, exhaust leak, PAIR pump wiring**
EFI-451	Cylinder 1 – misfire detected	**Spark plug, HT lead, distributor rotor/cap, moisture ingress, injector/wiring, air leak, compression, head gasket, low-tension wiring, repeated cold starts (flooding)**
EFI-452	Cylinder 2 – misfire detected	**Spark plug, HT lead, distributor rotor/cap, moisture ingress, injector/wiring, air leak, compression, head gasket, low-tension wiring, repeated cold starts (flooding)**
EFI-453	Cylinder 3 – misfire detected	**Spark plug, HT lead, distributor rotor/cap, moisture ingress, injector/wiring, air leak, compression, head gasket, low-tension wiring, repeated cold starts (flooding)**
EFI-454	Cylinder 4 – misfire detected	**Spark plug, HT lead, distributor rotor/cap, moisture ingress, injector/wiring, air leak, compression, head gasket, low-tension wiring, repeated cold starts (flooding)**
EFI-455	Cylinder 5 – misfire detected	**Spark plug, HT lead, distributor rotor/cap, moisture ingress, injector/wiring, air leak, compression, head gasket, low-tension wiring, repeated cold starts (flooding)**

Scan code	Fault location	Probable cause
EFI-456	Cylinder 6 – misfire detected	Spark plug, HT lead, distributor rotor/cap, moisture ingress, injector/wiring, air leak, compression, head gasket, low-tension wiring, repeated cold starts (flooding)
EFI-512	Short term fuel trim	Intake/exhaust leak/blockage, MAF sensor, fuel pressure
EFI-513	Engine control module (ECM) – overheating	Engine coolant blower motor, air supply blocked
EFI-514	Engine coolant blower motor relay, low speed	Wiring, engine coolant blower motor relay
EFI-515	Engine coolant blower motor relay, high speed	Wiring, engine coolant blower motor relay
EFI-516	AC condenser blower motor relay	Wiring, AC condenser blower motor relay
EFI-521	Heated oxygen sensor (HO2S) 1 – heater control	Wiring, HO2S
EFI-522	Heated oxygen sensor (HO2S) 2 – heater control	Wiring, HO2S
EFI-525	Ignition coil 1, cylinders 1 & 4	Wiring, ignition coil
EFI-526	Ignition coil 2, cylinders 2 & 3	Wiring, ignition coil
EFI-532	Engine control module (ECM) – power stage – group B	Wiring, IAC valve, TC wastegate regulating valve, engine coolant blower motor relay, ECM
EFI-533	Engine control module (ECM) – power stage – group C	Wiring short circuit at terminal B19, ECM
EFI-534	Engine control module (ECM) – power stage – group D	Wiring short to positive, MIL, engine coolant blower motor relay, fuel pump, ignition coil/ amplifier, TCM (load signal), ECM
EFI-535	Engine control module (ECM) – power stage – group E	Wiring short to positive, AC relay/PAIR pump relay, ECM
EFI-541	Evaporative emission (EVAP) canister purge valve – signal	Wiring, EVAP canister purge valve
EFI-543	Misfire detected, at least 1 cylinder	Spark plug, HT lead, distributor rotor/cap, moisture ingress, injector/wiring, air leak, compression, head gasket, low-tension wiring, repeated cold starts (flooding)
EFI-545	Misfire detected – at least 1 cylinder – TWC damage	Spark plug, HT lead, distributor rotor/cap, moisture ingress, injector/wiring, air leak, compression, head gasket, low-tension wiring, repeated cold starts (flooding)
EFI-551	Cylinder 1, misfire detected – TWC damage	Spark plug, HT lead, distributor rotor/cap, moisture ingress, injector/wiring, air leak, compression, head gasket, low-tension wiring, repeated cold starts (flooding)
EFI-552	Cylinder 2, misfire detected – TWC damage	Spark plug, HT lead, distributor rotor/cap, moisture ingress, injector/wiring, air leak, compression, head gasket, low-tension wiring, repeated cold starts (flooding)

Scan code	Fault location	Probable cause
EFI-553	Cylinder 3, misfire detected – TWC damage	Spark plug, HT lead, distributor rotor/cap, moisture ingress, injector/wiring, air leak, compression, head gasket, low-tension wiring, repeated cold starts (flooding)
EFI-554	Cylinder 4, misfire detected – TWC damage	Spark plug, HT lead, distributor rotor/cap, moisture ingress, injector/wiring, air leak, compression, head gasket, low-tension wiring, repeated cold starts (flooding)
EFI-555	Cylinder 5, misfire detected – TWC damage	Spark plug, HT lead, distributor rotor/cap, moisture ingress, injector/wiring, air leak, compression, head gasket, low-tension wiring, repeated cold starts (flooding)
EFI-556	Cylinder 6, misfire detected – TWC damage	Spark plug, HT lead, distributor rotor/cap, moisture ingress, injector/wiring, air leak, compression, head gasket, low-tension wiring, repeated cold starts (flooding)
EFI-611	Fuel tank system – large leak detected	Fuel tank, filler cap/pipe, EVAP system, fuel tank pressure sensor
EFI-612	Fuel tank system – small leak detected	Fuel tank, filler cap/pipe, EVAP system, fuel tank pressure sensor
EFI-614	Evaporative emission (EVAP) canister shut-off valve – incorrect flow detected	Hose blocked, EVAP canister shut-off valve/ filter, EVAP canister purge valve, fuel tank pressure sensor
EFI-616	Evaporative emission (EVAP) canister shut-off valve – signal	Wiring, EVAP canister shut-off valve
EFI-621	Fuel tank pressure sensor	Wiring, sensor supply, G-force sensor, AC pressure sensor, fuel tank pressure sensor
EFI-666	Transmission control module (TCM) – access fault memory	Trouble code(s) stored
EFI-724	Engine coolant heater relay	Wiring, engine coolant heater relay, engine coolant heater wiring
EFI-725	Engine control relay	Wiring, engine control relay
EFI-726	Engine control module (ECM) – supply voltage	Wiring, engine control relay
EFI-730	Brake pedal position (BPP) switch	Brake pedal depressed for more than 3 minutes, wiring, BPP switch
EFI-732	Accelerator pedal position (APP) sensor – signal	Wiring, APP sensor
EFI-733	Accelerator pedal position (APP) sensor – supply voltage	Wiring short to ground, excessive load, ECM
EFI-742	Transmission control module (TCM) – communication error	Wiring, TCM, ECM
EFI-743	Cruise control master/selector switch	Wiring, cruise control master/selector switch

VOLVO

Model:	960 3.0L
Year:	1992
Engine identification:	B6304F
System:	Bosch Motronic 1.8 • Bosch Motronic 4.3 • Siemens Fenix 5.2

Data link connector (DLC) locations

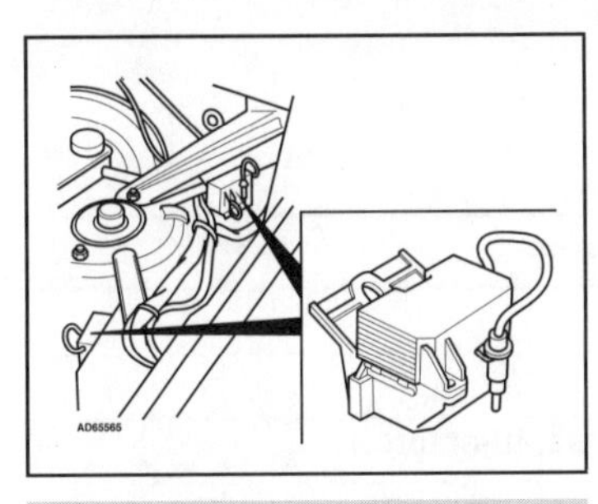

960 – engine bay, LH rear

Self-diagnosis

General information

- Refer to the front of this manual for general test conditions, terminology, detailed descriptions of wiring faults and a general trouble shooter for electrical and mechanical faults.
- Trouble codes are displayed by the LED on the data link connector (DLC).
- The ECM fault memory can also be checked and erased using diagnostic equipment connected to the data link connector (DLC).

Accessing

- Ensure ignition switched OFF.
- Open cover on black data link connector (DLC).
- Insert selector cable in socket 2 **1** [**1**].
- Switch ignition ON.
- Depress test button for approximately 1 second **1** [**2**].
- Count LED flashes **1** [**3**]. Note trouble code.
- Each trouble code consists of three groups of flashes **2**.
- A 3 second pause separates each group **2** [**A**].
- For example: Trouble code 223 displayed **2** – idle air control (IAC) valve.
- Repeat switch operation **1** [**2**]. Note trouble codes. Compare with trouble code table.
- Switch ignition OFF. Rectify faults as necessary.

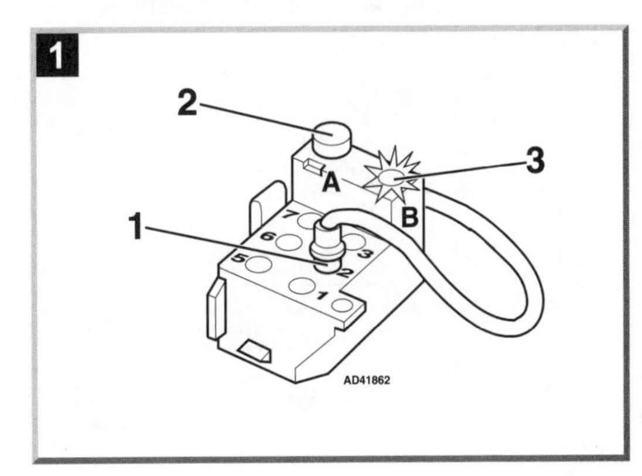

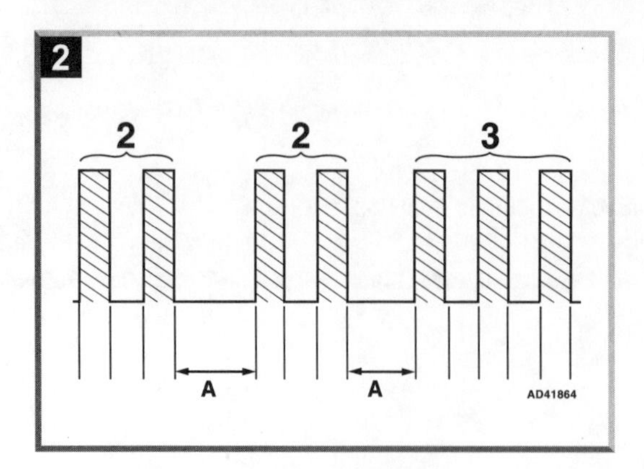

Erasing

- Ensure ignition switched OFF.
- Insert selector cable in socket 2 **1** [**1**].
- Access trouble codes.

NOTE: ***Trouble codes must be displayed at least once before they can be erased.***

- Depress test button for approximately 5 seconds.
- Release test button.
- Wait for LED to illuminate.
- Depress test button for approximately 5 seconds.
- Release test button.
- Depress test button for approximately 1 second.
- Ensure trouble code 111 displayed.

Trouble code identification

Flash code	Fault location	Probable cause
111	No fault found	–
112	Engine control module (ECM)	**ECM**
113 **1**	Early models: Injector(s)	**Wiring, fuse, injector(s)**
	Late models: Injectors 1, 2 & 4	**Wiring, fuse, injector(s)**
113 **2**	1993: Short term fuel trim	**No vacuum to MAP sensor, HO2S/signal short to ground, MAP sensor/ground wire**
	1994 →: Short term fuel trim – upper limit reached	**Wiring, HO2S**
115 **1**	Injectors 3, 5 & 6	**Wiring, fuse, injector(s)**
115 **3**	Injector 1	**Wiring, injector**
121	Mass air flow (MAF) sensor/manifold absolute pressure (MAP) sensor	**Wiring, hose connection, MAF/MAP sensor**
122	Intake air temperature (IAT) sensor	**Wiring, IAT sensor**
123	Engine coolant temperature (ECT) sensor	**Wiring, ECT sensor**
125	Injector 2	**Wiring, injector**
131	Engine speed (RPM) sensor – no signal	**Wiring, RPM sensor**
132	Engine control module (ECM) – supply voltage low/high	**Battery, alternator**
135	Injector 3	**Wiring, injector**
143	Knock sensor (KS) 1	**Wiring, KS**
145	Injector 4	**Wiring, injector**
152	Secondary air injection (AIR) solenoid	**Wiring, AIR solenoid**
153	Heated oxygen sensor (HO2S) 2	**Wiring, heating inoperative, HO2S**

Flash code	Fault location	Probable cause
154	Exhaust gas recirculation (EGR) system – excessive flow detected	**Wiring, EGR valve, EGR control module, engine coolant thermostat**
155	Injector 5	**Wiring, injector**
212	Heated oxygen sensor (HO2S) 1	**Wiring, heating inoperative, MAF sensor, HO2S**
214	Engine speed (RPM) sensor – intermittent signal	**Wiring, flywheel/drive plate damaged, air gap, RPM sensor**
221	Long term fuel trim, part load – upper limit reached	**Intake/exhaust leak, fuel pressure low, injector(s) blocked, hose connection, wiring, MAF/MAP sensor, ECT sensor, HO2S**
222	Engine control relay	**Wiring, engine control relay**
223 [1]	Early models: Idle air control (IAC) valve	**Wiring, IAC valve**
	Late models: Idle air control (IAC) valve – opening signal	**Wiring, IAC valve**
225	AC refrigerant pressure sensor	**Wiring, AC refrigerant pressure sensor**
231	Long term fuel trim, part load – lower limit reached	**Fuel pressure high, excessive fuel in engine oil, engine oil level high, injector(s) leaking, EVAP canister purge valve, hose connection, wiring, MAF/MAP sensor, ECT sensor, HO2S**
232	Long term fuel trim, idling	**Intake/exhaust leak, fuel pressure, excessive fuel in engine oil, injector(s), EVAP canister purge valve, wiring, MAF sensor, ECT sensor, HO2S**
233	Long term fuel trim, idle air adjustment	**Intake leak/blockage, throttle valve adjustment, wiring, IAC valve**
235	Exhaust gas recirculation (EGR) solenoid	**Wiring, EGR solenoid**
241	Exhaust gas recirculation (EGR) system – incorrect flow detected	**Hose leak/blockage, wiring, EGR valve, EGR control module, engine coolant thermostat**
243	Throttle position (TP) sensor	**Wiring, TP sensor**
245	Idle air control (IAC) valve – closing signal	**Wiring, IAC valve**
311	Vehicle speed sensor (VSS)	**Wiring, speedometer, VSS**
314	Camshaft position (CMP) sensor	**Wiring, CMP sensor**
315	Evaporative emission (EVAP) system	**EVAP canister, hose leak/blockage, EVAP canister purge valve**
322	Mass air flow (MAF) sensor – no filament burn-off signal	**Wiring, ECM**
323	Malfunction indicator lamp (MIL)	**Wiring, MIL**
325	Engine control module (ECM) – supply voltage, battery	**Battery discharged/disconnected, wiring, ECM disconnected**
335	Transmission control module (TCM) – MIL ON request	**Wiring**
342	AC relay	**Wiring, AC relay**

Flash code	Fault location	Probable cause
343	Fuel pump relay	**Wiring, fuel pump relay**
411	Throttle position (TP) sensor	**Wiring, TP sensor**
413	Exhaust gas recirculation temperature (EGRT) sensor	**Wiring, EGRT sensor**
414	Turbocharger (TC) pressure – control	**TC wastegate regulating valve hose leak/ blockage, wiring, MAF sensor, TCM wire (TC boost reduction), TC wastegate regulating valve**
416	Transmission control module (TCM) – TC boost reduction signal	**Wiring short to ground**
432	Engine control module (ECM) – overheating	**Engine coolant blower motor, air supply blocked, module cooling fan**
433	Knock sensor (KS) 2	**Wiring, KS**
435	Heated oxygen sensor (HO2S) 1 – slow response	**Intake/exhaust leak, fuel pressure, compression, HO2S**
436	Heated oxygen sensor (HO2S) 2 – compensation	**Intake/exhaust leak, fuel pressure, compression, HO2S 1/2**
442	Secondary air injection (AIR) pump relay	**Wiring, AIR pump relay**
443	Catalytic converter – efficiency	**Intake/exhaust leak, fuel pressure, compression, catalytic converter**
444	Accelerator pedal position (APP) sensor	**Wiring, APP sensor**
451	Cylinder 1 – misfire detected	**Spark plug, HT lead, distributor rotor/cap, moisture ingress, injector/wiring, air leak, compression, head gasket, low-tension wiring, repeated cold starts (flooding), puncture in front tire**
452	Cylinder 2 – misfire detected	**Spark plug, HT lead, distributor rotor/cap, moisture ingress, injector/wiring, air leak, compression, head gasket, low-tension wiring, repeated cold starts (flooding), puncture in front tire**
453	Cylinder 3 – misfire detected	**Spark plug, HT lead, distributor rotor/cap, moisture ingress, injector/wiring, air leak, compression, head gasket, low-tension wiring, repeated cold starts (flooding), puncture in front tire**
454	Cylinder 4 – misfire detected	**Spark plug, HT lead, distributor rotor/cap, moisture ingress, injector/wiring, air leak, compression, head gasket, low-tension wiring, repeated cold starts (flooding), puncture in front tire**
455	Cylinder 5 – misfire detected	**Spark plug, HT lead, distributor rotor/cap, moisture ingress, injector/wiring, air leak, compression, head gasket, low-tension wiring, repeated cold starts (flooding), puncture in front tire**

Flash code	Fault location	Probable cause
511	Long term fuel trim, idling	**Intake/exhaust leak, fuel pressure, excessive fuel in engine oil, injector(s), EVAP canister purge valve, wiring, MAF sensor, ECT sensor, HO2S**
512 2	Short term fuel trim – lower limit reached	**HO2S signal wire short to ground, EGR system leak, MAP sensor, HO2S, MAP sensor ground wire open circuit/poor connection**
512 1	Short term fuel trim	**Fuel pressure, wiring, MAF sensor**
513	Engine control module (ECM) – overheating	**Engine coolant blower motor, air supply blocked, module cooling fan**
514	Engine coolant blower motor, low speed	**Wiring, engine coolant blower motor relay**
515	Engine coolant blower motor, high speed	**Wiring, engine coolant blower motor relay**
521	Heated oxygen sensor (HO2S) 1 – heater control	**Wiring, HO2S**
522	Heated oxygen sensor (HO2S) 2 – heater control	**Wiring, HO2S**
523	Module cooling fan	**Wiring, module cooling fan**
524	Transmission control module (TCM) – torque reduction signal	**Wiring**
531	Engine control module (ECM) – power stage – group A	**Wiring, injector(s), EVAP canister purge valve, ECM**
532	Engine control module (ECM) – power stage – group B	**Wiring, IAC valve, TC wastegate regulating valve, engine coolant blower motor relay, ECM**
533	Engine control module (ECM) – power stage – group C	**Wiring short to positive, EGR control module, engine coolant blower motor relay, TCM (load signal), instrument panel (ECT/RPM/fuel consumption signals), ECM**
534	Engine control module (ECM) – power stage – group D	**Wiring short to positive, MIL, AC relay, fuel pump, ignition amplifier, ECM**
535	Turbocharger (TC) wastegate regulating valve	**Wiring, TC wastegate regulating valve**
541	Evaporative emission (EVAP) canister purge valve	**Wiring, EVAP canister purge valve**
542	Misfire detected, more than 1 cylinder	**Spark plug, HT lead, distributor rotor/cap, moisture ingress, injector/wiring, air leak, compression, head gasket, low-tension wiring, fuel pressure, fuel pump/wiring, fuel level low, incorrect/contaminated fuel, engine oil level high, repeated cold starts (flooding), puncture in front tire**
543	Misfire detected, at least 1 cylinder	**Spark plug, HT lead, distributor rotor/cap, moisture ingress, injector/wiring, air leak, compression, head gasket, low-tension wiring, repeated cold starts (flooding), puncture in front tire**

Flash code	Fault location	Probable cause
544	Misfire detected, more than 1 cylinder – TWC damage	**Spark plug, HT lead, distributor rotor/cap, moisture ingress, injector/wiring, air leak, compression, head gasket, low-tension wiring, fuel pressure, fuel pump/wiring, fuel level low, incorrect/contaminated fuel, engine oil level high, repeated cold starts (flooding), puncture in front tire**
545	Misfire detected – at least 1 cylinder – TWC damage	**Spark plug, HT lead, distributor rotor/cap, moisture ingress, injector/wiring, air leak, compression, head gasket, low-tension wiring, repeated cold starts (flooding), puncture in front tire**
551	Cylinder 1, misfire detected – TWC damage	**Spark plug, HT lead, distributor rotor/cap, moisture ingress, injector/wiring, air leak, compression, head gasket, low-tension wiring, repeated cold starts (flooding), puncture in front tire**
552	Cylinder 2, misfire detected – TWC damage	**Spark plug, HT lead, distributor rotor/cap, moisture ingress, injector/wiring, air leak, compression, head gasket, low-tension wiring, repeated cold starts (flooding), puncture in front tire**
553	Cylinder 3, misfire detected – TWC damage	**Spark plug, HT lead, distributor rotor/cap, moisture ingress, injector/wiring, air leak, compression, head gasket, low-tension wiring, repeated cold starts (flooding), puncture in front tire**
554	Cylinder 4, misfire detected – TWC damage	**Spark plug, HT lead, distributor rotor/cap, moisture ingress, injector/wiring, air leak, compression, head gasket, low-tension wiring, repeated cold starts (flooding), puncture in front tire**
555	Cylinder 5, misfire detected – TWC damage	**Spark plug, HT lead, distributor rotor/cap, moisture ingress, injector/wiring, air leak, compression, head gasket, low-tension wiring, repeated cold starts (flooding), puncture in front tire**

1 Bosch Motronic 1.8
2 Siemens Fenix 5.2
3 Bosch Motronic 4.3